Handbook of Research on 5G Networks and Advancements in Computing, Electronics, and Electrical Engineering

Augustine O. Nwajana
University of Greenwich, UK

Isibor Kennedy Ihianle
Nottingham Trent University, UK

IGI Global
PUBLISHER of TIMELY KNOWLEDGE

A volume in the Advances in Computer and Electrical Engineering (ACEE) Book Series

Published in the United States of America by
IGI Global
Engineering Science Reference (an imprint of IGI Global)
701 E. Chocolate Avenue
Hershey PA, USA 17033
Tel: 717-533-8845
Fax: 717-533-8661
E-mail: cust@igi-global.com
Web site: http://www.igi-global.com

Library of Congress Cataloging-in-Publication Data

Names: Nwajana, Augustine Onyenwe, 1979- editor. | Ihianle, Isibor, 1976- editor.
Title: Handbook of research on 5G networks and advancements in computing, electronics, and electrical engineering / Augustine Nwajana, and Isibor Ihianle, editors.
Description: Hershey PA : Engineering Science Reference, [2021] | Includes bibliographical references and index. | Summary: "This book investigates the advent of the emerging fifth generation (5G) networks and how they are changing the paradigm of how computing, electronics and electrical (CEE) systems are interconnected. and ways to seamlessly link together to turn the globe into a digital world"-- Provided by publisher.
Identifiers: LCCN 2020058411 (print) | LCCN 2020058412 (ebook) | ISBN 9781799869924 (h/c) | ISBN 9781799869948 (ebook)
Subjects: LCSH: 5G mobile communication systems.
Classification: LCC TK5103.25 .H36 2021 (print) | LCC TK5103.25 (ebook) | DDC 621.39/81--dc23
LC record available at https://lccn.loc.gov/2020058411
LC ebook record available at https://lccn.loc.gov/2020058412

This book is published in the IGI Global book series Advances in Computer and Electrical Engineering (ACEE) (ISSN: 2327-039X; eISSN: 2327-0403)

British Cataloguing in Publication Data
A Cataloguing in Publication record for this book is available from the British Library.

All work contributed to this book is new, previously-unpublished material. The views expressed in this book are those of the authors, but not necessarily of the publisher.

For electronic access to this publication, please contact: eresources@igi-global.com.

Advances in Computer and Electrical Engineering (ACEE) Book Series

Srikanta Patnaik
SOA University, India

ISSN:2327-039X
EISSN:2327-0403

Mission

The fields of computer engineering and electrical engineering encompass a broad range of interdisciplinary topics allowing for expansive research developments across multiple fields. Research in these areas continues to develop and become increasingly important as computer and electrical systems have become an integral part of everyday life.

The **Advances in Computer and Electrical Engineering (ACEE) Book Series** aims to publish research on diverse topics pertaining to computer engineering and electrical engineering. **ACEE** encourages scholarly discourse on the latest applications, tools, and methodologies being implemented in the field for the design and development of computer and electrical systems.

Coverage

- Power Electronics
- VLSI Design
- Computer Architecture
- Chip Design
- Algorithms
- Sensor Technologies
- Optical Electronics
- Programming
- Computer Science
- VLSI Fabrication

IGI Global is currently accepting manuscripts for publication within this series. To submit a proposal for a volume in this series, please contact our Acquisition Editors at Acquisitions@igi-global.com or visit: http://www.igi-global.com/publish/.

The Advances in Computer and Electrical Engineering (ACEE) Book Series (ISSN 2327-039X) is published by IGI Global, 701 E. Chocolate Avenue, Hershey, PA 17033-1240, USA, www.igi-global.com. This series is composed of titles available for purchase individually; each title is edited to be contextually exclusive from any other title within the series. For pricing and ordering information please visit http://www.igi-global.com/book-series/advances-computer-electrical-engineering/73675. Postmaster: Send all address changes to above address.

Titles in this Series

For a list of additional titles in this series, please visit: http://www.igi-global.com/book-series/advances-computer-electrical-engineering/73675

Hydrogen Fuel Cell Technology for Stationary Applications
Gheorghe Badea (Technical University of Cluj-Napoca, Romania) Raluca-Andreea Felseghi (Ştefan cel Mare University of Suceava, Romania) and Ioan Aşchilean (Technical University of Cluj-Napoca, Romania)
Engineering Science Reference • © 2021 • 332pp • H/C (ISBN: 9781799849452) • US $215.00

Computational Methodologies for Electrical and Electronics Engineers
Rajiv Singh (G.B. Pant University of Agriculture and Technology, Pantnagar, India) Ashutosh Kumar Singh (Indian Institute of Information Technology, Allahabad, India) Ajay Kumar Dwivedi (SIET, India & Dr. A.P.J. Abdul Kalam Technical University, Lucknow, India) and P. Nagabhushan (Indian Institute of Information Technology, Allahabad, India)
Engineering Science Reference • © 2021 • 281pp • H/C (ISBN: 9781799833277) • US $225.00

Innovations in the Industrial Internet of Things (IIoT) and Smart Factory
Sam Goundar (British University Vietnam, Vietnam) J. Avanija (Sree Vidyanikethan Engineering College, India) Gurram Sunitha (Sree Vidyanikethan Engineering College, India) K. Reddy Madhavi (Sree Vidyanikethan Engineering College, India) and S. Bharath Bhushan (Sree Vidyanikethan Engineering College, Tirupati, India)
Engineering Science Reference • © 2021 • 311pp • H/C (ISBN: 9781799833758) • US $225.00

Innovative Applications of Nanowires for Circuit Design
Balwinder Raj (National Institute of Technical Teachers Training and Research, Chandigarh, India)
Engineering Science Reference • © 2021 • 263pp • H/C (ISBN: 9781799864677) • US $225.00

Optimizing and Measuring Smart Grid Operation and Control
Abdelmadjid Recioui (Université M'hamed Bougara de Boumerdes, Algeria) and Hamid Bentarzi (Université M'hamed Bougara de Boumerdes, Algeria)
Engineering Science Reference • © 2021 • 400pp • H/C (ISBN: 9781799840275) • US $225.00

New Methods and Paradigms for Modeling Dynamic Processes Based on Cellular Automata
Stepan Mykolayovych Bilan (State University of Infrastructure and Technology, Ukraine) Mykola Mykolayovych Bilan (Mayakskaya Secondary School, Moldova) and Ruslan Leonidovich Motornyuk (Main Information and Computing Center, Ukraine)
Engineering Science Reference • © 2021 • 326pp • H/C (ISBN: 9781799826491) • US $195.00

IGI Global
PUBLISHER of TIMELY KNOWLEDGE

701 East Chocolate Avenue, Hershey, PA 17033, USA
Tel: 717-533-8845 x100 • Fax: 717-533-8661
E-Mail: cust@igi-global.com • www.igi-global.com

To Our Lovely Families

Editorial Advisory Board

List of Contributors

Table of Contents

Detailed Table of Contents

replaced by nanoelectronics in the future. The concept of nanoscale network on chip system is expected to play an important role in the field of communication systems for designing new devices of ultra-high speed for long and short-range communication links, power efficient computing devices, high-density memory and logic, and ultrafast interconnects. This chapter focuses on the mechanism of tailoring, patterning, and manipulating optical signals using nanometer-scale structures that may play the role of lumped nanocircuit elements at optical domain when selected properly with tremendous promise for application for fifth-generation communication systems.

In the past few years, UAVs have evolved considerably towards real-world applications, going beyond entertaining activities. The unique features that UAVs can provide, such as flexible and easy deployment, high mobility, and mostly autonomous operation, tend to be appealing solutions in a wide variety of applications. With the number of UAVs expected to increase significantly in the forthcoming years, it is deemed mandatory to integrate innovative features of communication technologies in order to tackle several limitations and challenges that exist to date. To this end, the amalgamation of UAVs and 5G cellular networks is an auspicious solution in order to realise the full potential that the UAVs can provide as well as to effectively open up new fields of applications. The unique features that 5G cellular networks offer are utilised for supporting and providing reliable and secure UAV command and control. This chapter provides an overview of the 5G technology to support communication services for the UAV ecosystem.

This chapter presents the development of a user-friendly MATLAB-based electronic design automation (EDA) tool for RF/microwave antenna design. The development steps of EDA using MATLAB 'guide' and MATLAB 'appdesigner' design environments have been explained in detail. In this chapter, seven EDAs of different antennas are developed and described. The EDA is capable of predicting the dimensions of the antenna by inputting the desired operating specifications of the antenna, such as operating frequency. The rigorous analytical formulations that are implemented in this EDA are briefly described. The developed EDAs have simple and standalone guide user interface (GUI) display and fast program running time compared to iterative solution or numerical-based EDAs. The developed EDAs can be packaged and accessed either as standalone desktop app or web app. The accuracy of the EDA output results is validated by comparing to measurement and simulation results obtained from a commercial simulator.

This chapter presents the basic approach of microwave bandpass filter design for 5G network applications. The chapter serves as a reference source to microwave stakeholders with little or no filter design experience. It should help them to design and implement their first filter device using microstrip technology. A three-pole Chebyshev bandpass filter with centre frequency of 2.6 GHz, fractional bandwidth of 3%, passband ripple of 0.04321 dB, and return loss of 20 dB has been designed. The designed filter implementation is based on the Rogers RT/Duroid 6010LM substrate with a 10.7 dielectric constant and 1.27 mm thickness. The circuit model and microstrip layout results of the BPF are presented and show good agreement. The microstrip layout simulation results show that a less than 1.8 dB minimum insertion loss and a greater than 25 dB in-band return loss were achieved. The overall device size of the BPF is 18.0 mm by 10.7 mm, which is equivalent to 0.16λg x 0.09λg, where λg is the guided wavelength of the 50 Ohm microstrip line at the filter centre frequency.

Two-dimensional materials have already demonstrated their potential in electronic applications. In this chapter, the authors propose and investigate a non-reciprocal series-fed microstrip patch antenna array based on graphene-black phosphorus for THz applications. The modeling and simulation of this structure are made with the wave concept iterative process method, in which the 2D materials are incorporated as conductive surface boundaries. Combining the advantages of graphene and BP, the antenna exhibits both strong anisotropic and plasmon responses that are not available in either individual graphene or black phosphorus layer. The authors envision that this strategy of hybridization of graphene and BP may pave the way towards non-reciprocal plasmonic components with enhanced functionalities at THz band with important applications in imaging and communications.

A solar-powered omnidirectional wheelchair is implemented for physically challenged persons. The framework was mounted on the wheels that were connected with two direct current (DC) motors. The ratings of the battery and solar module were determined using system voltage (12V). A 7,805-voltage regulator was used to supply 5VDC to the AT89352 microcontroller. The microcontroller was programmed to provide a reference signal to the motor. The motor provides the needed torque to drive the wheels through interconnected relays. The relays are energized by the microcontroller and omnidirectional movement achieved through relays connected with microprocessor and micro switches, eliminating the need for joysticks and complex control mechanisms. System performance test result showed that the auxiliary solar power supply of the wheelchair increased the travel range by approximately 86% compared with that of a wheelchair powered by battery alone.

Digital multimedia broadcasting is a technology for transmitting multimedia signals to devices like mobile phones, PC, laptops, etc. This allows everyone to watch TV on mobile devices. With the fifth-generation network, one can download a movie in a few seconds. Mobile TV offers round-the-clock participation in communication and broadcasting around the globe. This has become possible by watching a live TV program on handheld devices, as it facilitates real-time updates. Digital terrestrial television broadcasting aimed at transmitting two types of information: global and local content. The global content aims at broadcasting over the entire network, and the local content is location specific. Broadcasting the local content using multiple transmitters at different frequencies in multifrequency network requires additional spectrum. Global content can be transmitted through single frequency network. Therefore, if both global and local information are transmitted in a single frequency network (SFN), the system will become highly spectrally efficient and cost-effective.

This chapter aims to serve as an introduction to the integration process of filters and antennas into a single component, a filtering antenna, or filtenna, which is a topic of rising scientific interest and importance in recent years. As most receiving and transmitting devices nowadays call for a smaller footprint, better power management, and simplification of fabrication, this chapter showcases all the steps the reader must take to have a holistic approach to the integration of the aforementioned two components. Therefore, the fundamental principles of antenna, microstrip line, and filter theories are explored, and they are applied to a proof-of-concept patch filtering antenna operating in the S-Band, and at a frequency suitable for allocation to the 5G communications applications, while notable examples from other authors also demonstrate different ways of tackling this domain.

On a global scale, the telecommunication industry is experiencing tremendous growth in mobile phones. Mobile phones communicate with base stations that are erected by the telecommunication industry. The base station produces radio frequency and exposes the people near the base stations to radiation. The effect of electromagnetic radiation from four base stations located at the residential area was studied by measuring their electric field strength and calculating their magnetic field strength and power density at different distances covering a frequency range of 900MHz to 2100MHz. The obtained values showed that the four cellular base stations are operating above the standard values of the International Commission on Non-Ionizing Radiation Protection Electromagnetic Field Radiation. The specific absorption rate was measured to determine the amount of radio frequency electromagnetic radiation absorbed by the human body. The result which is in the range of 3.22-3.70 W/kg is higher than the acceptable 2 W/kg for localized specific absorption rate.

This chapter proposes the use of asynchronously coupled-resonator junctions (ACJ) in the design of a multi-input multi-output (MIMO) filtering network and a masthead combiner (MHC). By employing the resonator junctions, miniaturised circuits are achieved without using any transmission-line-based impedance matching circuits. The main challenge in the designs is the control and implementation of the external couplings at the common ports of this all-resonator-based MIMO filtering network and MHC. Both devices are four ports-based with the MIMO filtering network operating at 1.8, 2.1, and 2.6 GHz, while the MHC operates at the two channels of 1.8 and 2.1 GHz. The demonstrated designs achieved fractional bandwidths of 1.764 GHz to 1.836 GHz, 2.058 GHz to 2.142 GHz, and 2.548 GHz to 2.652 GHz, respectively. Good agreements have been achieved between the measurements of the prototype devices and the simulations.

Due to pressure to meet consumer requirements, miniaturization, reliability, and increased functionality have become more important. To achieve good assembly of electronic devices, there must be smooth deposition of tiny and consistent paste deposits on the board which must be up to an acceptable height and volume based on desirable process parameters in the output. The printing performance of the solder paste is dependent on conditions such as paste deformation behavior, good paste roll, complete aperture filling, and paste release onto the substrate pads. Eighteen tests were performed using two types of lead-free solder paste. Three temperatures were investigated under three different time intervals. Results analysis showed that variability of the paste distribution was lowest and more uniform for paste stored at 25°C for 48 hours. From the analysis, it is obvious that temperature has tremendous effects on the printing performance of solder paste. This study recommends 25°C as a good temperature for solder paste printing.

that enables recognition of simple activities of daily living in the home environment. The proposed approach is evaluated and validated using the Kasteren and Ordonez datasets.

Patrice Seuwou, Northampton University, UK
Vincent F. Adegoke, London South Bank University, UK

The opportunities offered by digital technology are enormous. The global social and economic system is being reconfigured at an incredible rate. Connectivity is increasingly reshaping our world and redefining the way we interact with our environment. The rise of digital technologies is transforming almost every aspect of modern life. More and more of our interactions are mediated by machines. Along with the rapid evolution comes the risks, threats, and vulnerabilities in the system for those who plan to exploit it. In this chapter, firstly, the authors explore the role of 5G, big data, the internet of things (IoT), artificial intelligence (AI), autonomous vehicles (AV), and cloud computing play in the context of smart societies; secondly, they analyse how the synergy between these technologies will be used by governments and other stakeholders around the world to improve the safety of citizens albeit increasingly relinquishing privacy rights and encouraging mass surveillance at the expense of liberty.

Kehinde Oluwafemi Olusuyi, Federal University Oye Ekiti, Nigeria
Paul Kehinde Olulope, Ekiti State University, Ado Ekiti, Nigeria
Abiodun Ernest Amoran, Federal University Oye Ekiti, Nigeria
Eno Edet Peter, Federal University Oye Ekiti, Nigeria

The present-day electric power system is an evolving cyber-physical system. Researchers and industry players in the energy world continue to deploy new technologies towards making the electric power system a smarter grid. This involves the integration of information, communication, and control technologies into the existing power grid in order to improve its stability, security, and operational efficiency. Reliance of the modern power system's applications such as state estimation, sequential control and data acquisition (SCADA) systems, phasor measurement units (PMUs), etc. on open communication technologies including the internet has exposed the smart grid to various vulnerabilities, threats, and cyber-physical attacks. This chapter seeks to exploit the robust synergy which exists between artificial intelligence (AI) and fifth-generation (5G) technology to mitigate these challenges. A comprehensive review of techniques which have hitherto proven efficient and/or effective in mitigating identified challenges was carried out with a view to availing researchers of future directions.

Richard Otuka, University of East London, UK

Presently, SMEs are finding it difficult to adopt cloud services for their businesses due to various service providers offering similar services. In addition, little work has been carried out in regards to the cloud services adoption process by SMEs. In this chapter, the authors propose CLOUDSME, a novel

Preface

The advent of the emerging fifth generation (5G) networks has changed the paradigm of how computing, electronics, and electrical (CEE) systems are interconnected. CEE devices and systems, with the help of the 5G technology, can now be seamlessly linked in a way that is rapidly turning the globe into a digital world. Smart cities and internet of things have come to stay but not without some challenges, which must be discussed.

The Handbook of Research on 5G Networks and Advancements in Computing, Electronics, and Electrical Engineering focuses on current technological innovations as the world rapidly heads towards becoming a global smart city. It covers important topics such as power systems, electrical engineering, mobile communications, network, security, and more. This book examines vast types of technologies and their roles in society with a focus on how each works, the impacts it has, and the future for developing a global smart city. This book is ideal for both industrial and academic researchers, scientists, engineers, educators, practitioners, developers, policymakers, scholars, and students interested in 5G technology and the future of engineering, computing, and technology in human society.

The book is organized in two sections and 19 chapters. The first section (Section 1) contains 14 chapters covering selected topics in electronics and electrical engineering. The second section (Section 2) holds five chapters covering selected topics in computing and computer science. A brief description of each chapter is as follows:

Chapter 1 covers recent advancements on the Internet of Vehicles (IoV). Progress in technological research have seen the use of Mobile Data Collectors (MDCs) or data MULEs for Wireless Sensor Network (WSN) applications. In the context of Smart City (SC) waste management scenarios, vehicular networks, or IoV can be exploited as MDCs or data Mules for data collection and transmission purposes from the sparsely distributed smart sensors that are attached to the smart bins to an access point or sink node and further deployed for waste management operations. A major challenge with the traditional methods of data collection using static sink nodes is the high energy consumption of the sensor-nodes. The use of MDCs has been well studied and shown to be energy efficient. To the best of our knowledge, this scheme has not been exploited for waste management operations in a SC. Compared to the centralized schemes, the data Mule scheme presents several advantages for data collection in WSN applications. This paper proposes an energy-efficient model for opportunistic data collection in IoV-enabled SC waste management operations.

Chapter 2 focuses on the relevance of passive circuit components in 5G networks. The passive circuit elements resistor, inductor, and capacitor, which are the basic building blocks of an electronic circuit are needed to be scaled down for applications in the fifth-generation wireless communication network. Due to the growing demands in memory and computational capacities of integrated circuits along with high processing and transmission data speed for next-generation, microelectronics will be replaced by nanoelectronics in the future. The concept of nanoscale Network on Chip system is expected to play an important role in the field of communication systems for designing new devices of ultra-highspeed for long- and short-range communication links, power efficient computing devices, high-density memory and logic, ultrafast interconnects. This chapter focuses on the mechanism of tailoring, patterning, and manipulating optical signals using nanometer scale structures which may play the role of lumped nano-circuit elements at optical domain when selected properly with tremendous promise for application for fifth-generation communication systems.

Chapter 3 discusses Unmanned Aerial Vehicles (UAVs). In the past few years, UAVs have evolved considerably towards real world applications, going beyond entertaining activities. The unique features that UAVs can provide, such as flexible and ease deployment, high mobility and mostly the autonomous operation, tend to find appealing solutions in a wide variety of applications. With the number of UAVs expected to increase significantly in the forth-coming years, it is deemed mandatory to integrate innovative features of communication technologies in order to tackle several limitations and challenges that exist to date. To this end, the amalgamation of UAVs and 5G cellular networks is an auspicious solution to realise the full potentials that the UAVs can provide as well as to open effectively, new fields of applications. The unique features that 5G cellular networks offer are utilised not only for supporting and providing reliable and secure UAVs' command and control. This chapter provides an overview of the 5G technology to support communication services for the UAV ecosystem.

Chapter 4 presents the development of user-friendly MATLAB-based electronic design automation (EDA) tool for RF/microwave antenna design. The development steps of EDA using MATLAB 'guide' and MATLAB 'appdesigner' design environments have been explained in detail. In this chapter, seven EDAs of different antennas are developed and described. The EDA can predict the dimensions of the antenna by inputting the desired operating specifications of the antenna, such as operating frequency. The rigorous analytical formulations which are implemented in this EDA are briefly described. The developed EDAs have simple and standalone guide user interface (GUI) display and fast program running time compared to iterative solution or numerical-based EDAs. Besides, the developed EDAs can be packaged and accessed either as standalone desktop app or web app. The accuracy of the EDA output results is validated by comparing to measurement and simulation results obtained from commercial simulator.

Chapter 5 presents the basics of microwave bandpass filter design for 5G network applications. The chapter serves as a reference source to microwave stakeholders with little or no filter design experience. It should help them to design and implement their first filter device using the microstrip technology. A 3-pole Chebyshev bandpass filter (BPF) with centre frequency of 2.6 GHz, fractional bandwidth of 3%, passband ripple of 0.04321 dB and return loss of 20 dB has been designed, implemented, and simulated. The designed filter implementation is based on the Rogers RT/Duroid 6010LM substrate with a 10.7 dielectric constant and 1.27 mm thickness. The BPF was also fabricated using the same substrate material used for the design simulation. The circuit model and microstrip layout results of the BPF are presented and show good agreement. The microstrip layout simulation results show that a less than 1.8 dB minimum insertion loss and a greater than 25 dB in-band return loss were achieved. The overall device size of the BPF is 18.0 mm by 10.7 mm, which is equivalent to $0.16\lambda g$ x $0.09\lambda g$, where λg is the guided wavelength of the 50 Ohm microstrip line at the filter centre frequency.

Chapter 6 proposes and investigates a non-reciprocal series-fed microstrip patch antenna array based on graphene-black phosphorus for THz applications. The modelling and simulation of this structure are made with the wave concept iterative process method, which the 2D materials are incorporated as conductive surface boundaries. Combining the advantages of graphene and BP, the antenna exhibits both strong anisotropic and plasmon responses that are not available in either individual graphene or black phosphorus layer. We envision that this strategy of hybridization of graphene and BP may pave the way towards non-reciprocal plasmonic components with enhanced functionalities at THz band, with important applications in imaging and communications.

Chapter 7 discusses a solar-powered omnidirectional wheelchair implemented for physically challenged persons. The framework was mounted on the wheels connected with two Direct Current (DC) motors. The ratings of the battery and solar module were determined using system voltage (12V). A 7805-voltage regulator was used to supply 5VDC to the AT89352 microcontroller. The microcontroller was programmed to provide a reference signal to the motor. The motor provides the needed torque to drive the wheels through interconnected relays. The relays are energized by the microcontroller and omnidirectional movement achieved through relays connected with microprocessor and micro switches, eliminating the need for joysticks and complex control mechanisms. System performance test result showed that the auxiliary solar power supply of the wheelchair increased the travel range by approximately 86% compared with that of a wheelchair powered by battery alone.

Chapter 8 focuses on digital multimedia broadcasting, a technology for transmitting multimedia signals to devices like mobile phones, PC, Laptops, etc. This enhances everyone to watch TV on mobile devices. With the Fifth-generation network, one can download a movie in a few seconds. Mobile TV offers round the clock participation in communication, and broadcasting around the globe. This has become possible by watching a live TV program on handheld devices, as it facilitates real-time updates. Digital terrestrial television broadcasting aimed at transmitting two types of information that is global and local content. The global content aims at broadcasting over the entire network and the local content is location specific. Broadcasting the local content using multiple transmitters at different frequencies in multifrequency network requires additional spectrum. Global Content can be transmitted through single frequency network. Therefore, if both global and local information are transmitted in a Single Frequency Network (SFN), the system will become highly spectrally efficient and cost-effective.

Chapter 9 serves as an introduction to the integration process of filters and antennas into a single component, a filtering antenna (or filtenna), which is a topic of rising scientific interest and importance in recent years. As most receiving and transmitting devices nowadays call for a smaller footprint, better power management, and simplification of fabrication, this chapter showcases all the steps the reader must take to have a holistic approach to the integration of the two components. Therefore, the fundamental principles of antenna, microstrip line, and filter theories are explored, and they are applied to a proof-of-concept patch filtering antenna operating in the S-Band, and at a frequency suitable for allocation to the 5G communications applications, while notable examples from other authors also demonstrate different ways of tackling this domain.

Chapter 10 investigates the adverse effect of electromagnetic radiation from cellular base station in Nigeria. On a global scale, the telecommunication industry is experiencing tremendous growth in mobile phones. Mobile phones communicate with base stations that are erected by the telecommunication industry. The base station produces radio frequency and exposes the people near the base stations to radiation. The effect of electromagnetic radiation from four base stations located at the residential area was studied by measuring their electric field strength and calculating their magnetic field strength and power density at different distances covering a frequency range of 900MHz to 2100MHz. The obtained values showed that the four cellular base stations are operating above the standard values of the International Commission on Non-ionizing Radiation Protection Electromagnetic Field Radiation.

The specific absorption rate was measured to determine the amount of radio frequency electromagnetic radiation absorbed by the human body. The result which is in the range of 3.22-3.70 W/kg is higher than the acceptable 2 W/kg for localized Specific absorption rate.

Chapter 11 proposes the use of asynchronously coupled-resonator junctions (ACJ) in the design of a multi-input multi-output (MIMO) filtering network, and a masthead combiner (MHC). By employing the resonator junctions, miniaturised circuits are achieved without using any transmission-line based impedance matching circuits. The main challenge in the designs is the control and implementation of the external couplings at the common ports of this all-resonator-based MIMO filtering network and MHC. Both devices are four ports based with the MIMO filtering network operating at 1.8, 2.1, and 2.6 GHz, while the MHC operates at the two channels of 1.8 and 2.1 GHz. The demonstrated designs achieved fractional bandwidths of 1.764 GHz to 1.836 GHz, 2.058 GHz to 2.142 GHz, and 2.548 GHz to 2.652 GHz, respectively. Good agreements have been achieved between the measurements of the prototype devices and the simulations.

Chapter 12 investigates the effects of temperature, paste height, and time on solder paste to its printing performance. Due to the pressure to meet consumer requirements; miniaturization, reliability and increased functionality have become more important. To achieve good assembly of electronic devices, there must be smooth deposition of tiny and consistent paste deposits on the board which must be up to an acceptable height and volume based on desirable process parameters in the output. The printing performance of the solder paste is dependent on conditions such as paste deformation behaviour, good paste roll, complete aperture filling, and paste release onto the substrate pads. Eighteen tests were performed using two types of lead-free solder paste. Three temperatures were investigated under three different time intervals. Results analysis showed that variability of the paste distribution was lowest and more uniform for paste stored at 25°C for 48 hours. From the analysis, it is obvious that temperature has tremendous effects on the printing performance of solder paste. This study recommends 25°C as a good temperature for solder paste printing.

Chapter 13 discusses the comparative study of transmission line junction versus asynchronously coupled junction diplexers. A conventional diplexer with a T-junction transmission line is designed in this work and compared with a new approach of designing a diplexer using the asynchronously coupled resonant junction (ACJ). The use of ACJ is to explore its features associated with enhancing the output response of the diplexer as well as its ease in control and implementation of the external couplings at the common port of an all-resonator-base diplexer. The measured and simulated responses of the diplexers were compared and they both agree with each other.

Chapter 14 explores the enhanced simulation tools for analyzing signal processing techniques in the optical transmission path. With emerging applications and needs of ever-increasing bandwidth, it is anticipated that 5G networks with much higher bandwidth present a natural path forward to satisfy these demands. For 5G network operators, it is interesting and profitable to develop valuable broadband optical networks at the same time. Broadband optical networks naturally utilize the optical transmission medium and for increasing transmission rates of broadband applications and services in 5G networks, it is necessary to characterize and investigate some specific features of advanced signal processing techniques utilized in the optical transmission path. For this case, advanced techniques must be also analyzed and evaluated for real long-distance, metropolitan and access optical networks as an integral part of 5G networks. In this chapter, enhanced simulation tools for analyzing advanced signal processing techniques in the optical transmission path are presented together with possibilities for their practical implementations in real single-channel and multi-channel optical transmission systems.

Chapter 15 discusses the patterns of object use for human activity recognition. The accurate recognition of Activities of Daily Living (ADL) is fundamental in the support and provision of assistance for the elderly and cognitively and the cognitively impaired. Current ontology-based techniques and knowledge-driven model object concepts from assumptions and everyday knowledge of object used for activities. Activities modelled from assumptions and everyday knowledge can lead to incorrect recognition results of routine activities and possible failure to detect abnormal activity trends. A significant step to the accurate recognition of activities of daily living is the discovery of the object usage for specific routine activities. This paper presents an approach which discovers object usage for routine activities using Latent Dirichlet Allocation (LDA) topic modelling. The object usage discovery augments an activity ontology which enables recognition of simple activities of daily living in the home environment. The proposed approach is evaluated and validated using the Kasteren and Ordonez datasets.

Chapter 16 investigates the numerous opportunities offered by digital technology in relation to 5G networks. The global social and economic system is being reconfigured at an incredible rate. Connectivity is increasingly reshaping our world and redefining the way we interact with our environment. The rise of digital technologies is transforming almost every aspect of modern life. More and more of our interactions are being mediated by machines. Along with the rapid evolution comes the risks, threats, and vulnerabilities in the system for those who plan to exploit it. In this paper, firstly, we will be exploring the role of 5G, Big data, the Internet of Things (IoT), Artificial Intelligence (AI), Autonomous vehicles (AV) and cloud computing play in the context of smart societies, secondly, analysing how the synergy between these technologies will be used by governments and other stakeholders around the world to improve the safety of citizens albeit increasingly relinquishing our privacy rights and encouraging mass surveillance at the expense of our liberty.

Chapter 17 looks at the present-day electric power system as an evolving cyber-physical system. Researchers and industry players in the energy world continue to deploy new technologies towards making the electric power system a smarter grid. This involves the integration of information, communication, and control technologies into the existing power grid to improve its stability, security, and operational efficiency. Reliance of the modern power system's applications such as state estimation, Sequential Control and Data Acquisition (SCADA) systems, Phasor Measurement Units (PMUs), etc, on open communication technologies including the Internet has exposed the smart grid to various vulnerabilities, threats, and cyber-physical attacks. This paper seeks to exploit the robust synergy which exists between Artificial Intelligence (AI) and fifth generation (5G) technology to mitigate these challenges. A comprehensive review of techniques which have hitherto proven efficient and/or effective in mitigating identified challenges was carried out with a view to availing researchers of future directions.

Chapter 18 presents a framework for evaluating SMEs adoption of SAAS in the cloud. Presently SMEs are finding it difficult to adopt cloud services for their businesses due to various service providers offering similar services. In addition, little work has been carried out regarding cloud services adoption process by SMEs. This chapter propose CLOUDSME as a novel framework that aids in the adoption process of SAAS cloud services. Accordingly, the chapter implements a decision support system, which includes an ontology of cloud services knowledge within the proposed framework. Analytical hierarchical process (AHP) is used to determine the weight of each cloud service attribute and a benchmark is set to determine the acceptability of each cloud service based on its ability to meet the acceptable benchmark for each criterion. It can also help in a healthy competition to improve the quality of service among cloud service providers. The CLOUDSME semantic model will guide SME owners answering user requirement towards decision-making in cloud service adoption process.

Chapter 19 discusses ontology-based travel recommender system. Recommender systems are designed to suggest information to users according to their preferences, items could be movies, books, and various kinds of products. Most of the existing recommender systems are based on a database with limited advantages. However, in this paper, we propose a knowledge-driven travel recommender system to integrate semantic data built based using Web Ontology Language (OWL) ontology to allow users to find suitable destinations that fulfil users' travel preferences. This work aims to develop a travel recommendation tool and to examine the reliability, the usability of the system and satisfaction rate responded from users. We are also able to demonstrate that users can obtain desired results through queries on the ontology-based system. The overall evaluation of the system shows that users are happy and satisfied with the recommendation results.

Augustine O. Nwajana
University of Greenwich, UK

Acknowledgement

The editors would like to acknowledge and appreciate the contributions of all the stakeholders who helped and ensured that all the chapter manuscripts submitted to this book were professionally peer-reviewed in a timely manner.

The authors would also like to demonstrate their sincere gratitude to all the staff at IGI Global for their support throughout the stages of this book project.

Augustine O. Nwajana
University of Greenwich, UK

Isibor Kennedy Ihianle
Nottingham Trent University, UK

Section 1
Electronics and Electrical Engineering

Chapter 1
An Energy-Efficient Model for Opportunistic Data Collection in IoV-Enabled SC Waste Management

Gerald K. Ijemaru
https://orcid.org/0000-0001-5146-6829
University of the Sunshine Coast, Australia

Ericmoore T. Ngharamike
Federal University Oye-Ekiti, Nigeria

Emmanuel U. Oleka
North Carolina A&T State University, USA

Augustine O. Nwajana
https://orcid.org/0000-0001-6591-5269
University of Greenwich, UK

ABSTRACT

Recent advancements in technological research have seen the use of mobile data collectors (MDCs) or data MULEs for wireless sensor network (WSN) applications. In the context of smart city (SC) waste management scenarios, vehicular networks or the internet of Vehicles (IoV) can be exploited as MDCs or data MULEs for data collection and transmission purposes from the sparsely distributed smart sensors that are attached to the smart bins to an access point or sink node and further deployed for waste management operations. A major challenge with the traditional methods of data collection using static sink nodes is the high energy consumption of the sensor-nodes. The use of MDCs has been well studied and shown to be energy efficient. To the best of the authors' knowledge, this scheme has not been exploited for waste management operations in a SC. Compared to the centralized schemes, the data MULE scheme presents several advantages for data collection in WSN applications. This chapter proposes an energy-efficient model for opportunistic data collection in IoV-enabled SC waste management operations.

DOI: 10.4018/978-1-7998-6992-4.ch001

1. INTRODUCTION

The increasing connection of vehicles to the Internet of Things (IoT) has ostensibly given rise to the recent technological advancements which have paved way for the emergence of a new era of vehicular networks known as the Internet of Vehicles (IoV). In the context of Smart City (SC) scenarios, these vehicular networks can be exploited as Mobile Data Collectors (MDCs) or data MULEs for waste management strategies. The era of IoT technology has given birth to the development of several innovative applications for smart cities. Essentially, IoT is expected to be the bedrock for novel applications and technological platforms for the emergence of the IoV from the existing vehicular ad hoc networks (VANETs), characterized by a seamless integration of sensor-driven mobile vehicles, and mobile devices within the network environment. Hence, IoT is expected to permit the evolution of IoV with new technological features and transformations in SCs. One such category of applications is the use of smart vehicles (IoV) as data MULEs for data collection in a SC waste management, which is the essence of this study. Other applications of the use of vehicular networks for data collection in smart cities may be found in transportation (i.e. intelligent transport), traffic light, environmental data sensing, monitoring, surveillance etc. The electronic toll which involves a communication between vehicles and roadside (V2R) in which vehicles send signals to the toll to make payments to the highway authorities is also another form of IoV applications. As mentioned earlier, to the best of our knowledge, this scheme is yet to be exploited in the context of SC waste management scenarios. Essentially, vehicular networks can be exploited for data collection meant for SC applications. Hence, Berrilio (2014) proposed the use of a network infrastructure for Data Collection for Low Energy Devices (DC4LED) integrated with sensors for data transmission to the internet and capable of providing connectivity to vehicular-sensors and access point so that data generated from the sensors could be transmitted to the internet via the access point. Other applications of IoV may include traffic & routing information (also known as a GPS application) that informs individuals of the route to a destination, security and collision avoidance where the driver is warned about a possible danger ahead such as an oil spill, a car accident, and so on (Berrilio, 2014).

Silva et al. (2017) proposed the use of ITS-based Decision Maker (DM) architecture for opportunistic networking based on heterogeneous and dynamic access network. The proposed scheme is expected to monitor a variety of information and will enable information exchange between vehicles and their environments for accuracy and better decision strategies. A feature of this scheme is its ability to provide the best route for a given process. Researchers (Er et al., 2017; Er et al., 2019) conducted the performance evaluation of four basic Vehicular Delay-Tolerant Networks (VDTN) routing protocols using the Opportunistic Networking Environment (ONE) simulator in terms of V2X communications for data delivery in SCs. DC4LED is a simple hierarchical forwarding routing scheme in the context of VDTN network. The focus of their work is the use of VDTN as an opportunistic network that can provide a data collection service by exploiting its V2X communication abilities. The results of their work showed that vehicles with predetermined routes presented high data delivery performance. Further to this was the use of DC4LED forwarding algorithm to determine the mobility features of vehicles by considering three kinds of vehicles (e.g. buses, taxis and cars), which can be considered advantageous for VDTN routing as well as sensor-driven mobile collectors.

Based on perceived dissimilarities in the mobility patterns of the selected vehicles, it is observed that buses follow almost fixed routes especially during their service hours, thus making them the best fit for VDTN routing applications, if the internet Point-of-Presence (PoP) is planted along their paths. On the other hand, cars and taxis traverse the streets which are not bus-routes, and thus collect data from the sensors planted along the streets in the city. However, compared to cars, taxis appear to traverse the cities much more. Cars make few trips with longer stationary times. Hence, based on this comparison, buses are found to be the best bet for sensor-driven mobile data collectors followed by taxis while cars are the least reliable in the hierarchical routing scheme.

A transformation from Vehicular ad hoc networks (VANETs) to the IoV became necessary (Contreras-Castillo et al., 2017) owing to the mobility constraints associated with VANETs. These constraints include lack of processing capability of all the data collected by the sensors and mobile devices within the network as a result of vehicles' prone to disconnection as they move outside the network coverage distance. Chaqfeh et al. (2016) envisioned the application of Vehicular Cloud Computing (VCC) in data collection for ITS. The proposed model is targeted at bringing traffic solutions and ensuring driver's convenience. Some of the solutions include congestion detection/avoidance, efficient route planning, etc. Accordingly, VCC would leverage the advantages of cloud computing to provide efficient service to VANETs with focus on providing solutions by minimizing the day-to-day road accidents, traffic time and congestion.

Owing to the need for real-time information processing and network expansion for massive data collection, the traditional VANETs are rapidly transforming to the IoVs, which hold better prospects for the future transportation system. Big data and IoV are inseparable within the context of vehicular network environment, and the underlying problems relating to big data enabled-IoV need to be investigated. To this end, researchers (Xu et al., 2017) proposed the design of efficient schemes to minimize data redundancy, transmission overhead, storage and processing. A feature of the work is the mutualism existing between big data and IoV development in which data acquisition, transmission, storage and computing are supported by IoV. Teng et al. (2019) proposed the use of mobile vehicles as code MULEs to opportunistically collect and distribute data codes at low cost within the city.

Fig.1 shows a typical Data Mule architecture which can be exploited for data collection for waste management operations in SCs. Waste management scenario in a smart city is characterized by the deployment of a large number of sensor-nodes in a sparsely distributed sensor network across spatially separated geographic locations of a SC. The proposed data Mule architecture combines the concepts of embedded intelligence, cyber physical systems and data communication systems. The layout of the architecture is designed to enable autonomous, simultaneous and synergetic operations among the various concepts and features.

This is discussed in detail in the new architecture of data Mule (IoV) for SC waste management section.

In the traditional and centralized scheme (as shown in Fig. 2), a sensor with a data to transmit will send it to a neighboring sensor-node, which performs the task of routing the data to another neighboring node using the single-hop communication until the data finally gets to an Access Point (AP) or sink node. One of the major challenges of the traditional scheme is that too much energy will be expended in data transmission between neighboring nodes as the distance between them increases more and more (Anastasi et al., 2008). Consequently, this may result in an unprecedented gap in communication between sensor-nodes. On the other hand, the Mule approach which is associated with MDCs is a characteristic representation of the multi-hop communication (Ang et al., 2018). A sensor-node wishing to transmit its data to the collector will send the data to a neighboring node which then performs data routing using the multi-hop communication path until the data gets to the collector and is finally transmitted to the access point (AP) or sink node. For the current scenario, data collection is achieved more efficiently using the data Mules (MDCs) which pick up data from the static sensors and deposit same to an AP or sink node (Anastasi, et al., 2008) as shown in Fig. 1. As mentioned earlier, Mules are characterized with low energy-consumption requirements compared to their traditional counterparts and can effectively handle data collection from sensor-nodes across spatially separated regional areas. This is because of the opportunistic style or mode of data collection. Compared with the senor-nodes which are typically energy-constrained occasioned by the nature of their deployments, Mules are power renewable. This ultimately makes them the best-fit as data collectors for our proposed scheme.

Figure 1. Data Mule Architecture

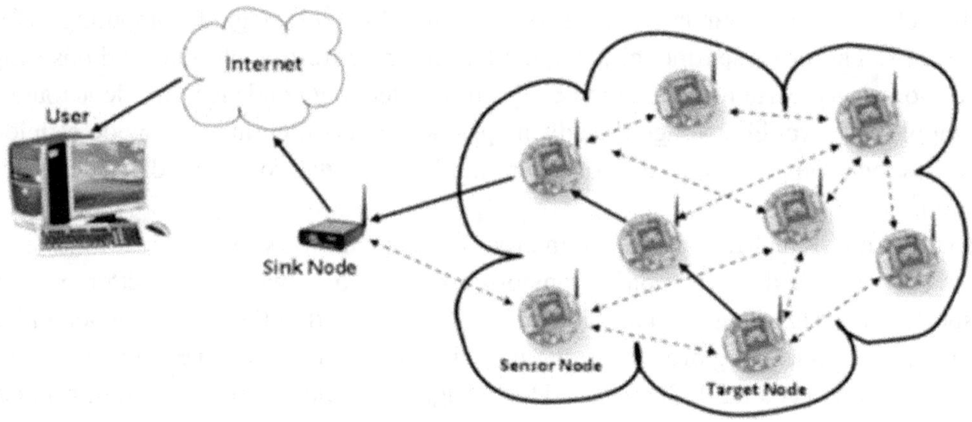

Figure 2. Centralized Architectural Scheme

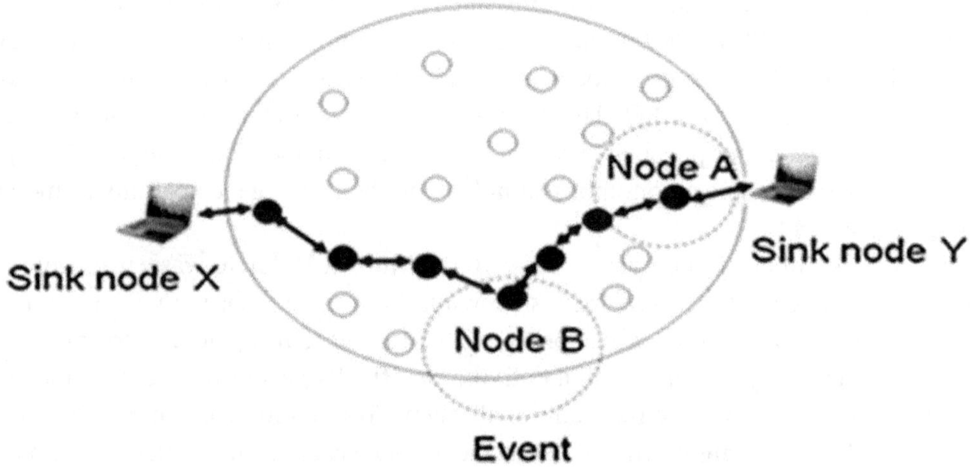

Several IoV-based techniques for SC solutions have been discussed. In Ang et al. (2018), Ang et al. (2017) and Teng et al. (2019), we see that vehicles can be exploited to serve as data MULEs in an opportunistic manner for data gathering and distribution for SC solutions. Other application-techniques of IoV may be found in intelligent transportation system (ITS), environmental monitoring, urban surveillance, traffic management and infotainment (Berrilio, 2014; Silva et al., 2017; and Ang et al., 2018). However, none of these techniques has been exploited for SC waste management strategies. Hence, the potentials of IoV (used as data MULEs) for data collection and distribution are yet to be comprehensively explored in the context of SC waste management.

This paper describes and analyzes a simulation-based comparison of both the traditional and swarm-based (Mule) scenarios by evaluating their performance metrics for a reliable and energy-efficient data collection scheme for waste management strategies in a SC. The focus is on the methods of data collection: the traditional method and the use of data Mule (MDC) approach. The data Mule technique involves the use of vehicles embedded with sensors for data collection and distribution in an opportunistic manner (Ang et al., 2018; Ang et al., 2017) and which can be further exploited for waste management operations in a SC.

In order to evaluate the performance of our proposed model, we have employed a simulation-based technique using both classical-based and swarm-based algorithms. The motivation for the simulation-based approach compared to the analytical-based approaches is drawn from the angle that our major focus is to propose an energy-efficient and cost-effective data collection methods for smart city waste management operations. Hence, a simulation-based approach will enable a proper comparative analysis of the different performance metrics such as energy consumption, success rate, throughput and latency. In addition, simulations are required to give credence to the performance of a proposed model, where analytical-based approaches cannot be employed. Where both simulation-based and analytical-based approaches are adopted, the latter only serves as a baseline for comparison with the former (Ang et al., 2018). Zhao and Tong (2005) opine that the results obtained from the analytical-based approach can be viewed as a baseline-benchmark for the simulation-based approach. Besides, simulation-based approaches are mostly adopted by many researchers to evaluate and validate algorithms, and for network systems. Despite these advantages, simulation-based approaches are not without some drawbacks, which include running long simulation times, difficulty in handling a large number of nodes for LS-WSNs and having inconsistent results (Ang et al., 2017).

2. RELATED WORK

IoT-based technologies have found several applications in SC waste management operations. Waste collection and transportation is the costliest phase in a waste management program. Waste management operation has become a critical issue due to the rapid increase in population especially in urban areas. To address the issues arising from this, it has become very necessary and expedient to deploy appropriate waste management techniques in order to improve the environmental conditions and the wellbeing of residents. This technique involves the use of sensors such as IR sensors, Ultrasonic sensors, RFID, Wi-Fi and GSM/GPRS, for real-time data generation and gathering from the waste bins placed at different distant locations. The major targets are to provide a smart and efficient waste collection management solutions, economic feasibility, collecting real-time information about the status of waste bins, cost reduction in transportation and dynamic route optimization. For instance, IR sensors have been recommended for real-time data gathering to help municipal authorities in routing & scheduling of waste collection operations. IR level sensor can also be used for garbage-level detection to minimize indiscriminate garbage disposal along the roadside which often results in air pollution and environmental hazards. Other techniques employed include the use of water sensors for wet-garbage detection, measuring sensors for waste-volume measurement (Shyam et al., 2017; Bharadwaj et al., 2017) and the adoption of ICT and internet technologies for an efficient garbage monitoring system (Anagnostopoulos et al., 2017; Fallavi et al., 2017).

In (Kumar et al., 2017; Jain and Bagherwal, 2017; Haribabu et al., 2017), sensor systems and GSM/GPRS technologies were adopted as efficient IoT-based tools for garbage monitoring, collection and disposal in SCs whereas Karthikeyan et al. (2017) proposed the use of ZigBee networks. Both techniques have the underlying objective of detecting the filled status of dustbins with the help of sensors and then transmitting information to the control room through the ZigBee or GSM/GPRS networks. Both technologies proved effective and recorded improvement in waste-collection efficiency and cost reduction in

Table 1. Summary of IoT-Based Solutions for Waste Management in a Smart City

Ref.	IoT Sensors	IoT Technologies	Sensor Applications
(Chaudhari et al., 2019; Rao et al., 2020; Catarinucci et al., 2019)	Ultrasonic & Weight Sensors	Raspberry Pi, GSM, Wi-Fi (Node MCU), Cloud system, & RFID tag	Used for waste level monitoring & weight measurement/calculation.
(Al-Masri et al., 2018a; Chang et al., 2019)	Ultrasonic sensor; Screw Cap sensor	Edge computing device (Recycle.io), Raspberry Pi, LPWAN, Cloud	Used to detect disposal violations; detects illegal disposal of liquid waste.
(Rohit et al., 2018; Durrani et al., 2019; Xenya et al., 2020)	Ultrasonic & Infrared Sensors	GPS, GSM/GPRS, ATmega328/Arduino,	Used for waste level monitoring & detection.
(Chen et al., 2018; Aravindaraman and Ranjana, 2019)	Gas sensor, IR & Ultrasonic sensors	Node MCU, Wi-Fi, LoRa, LED, Raspberry Pi	Smell detection & waste level monitoring.
(Memon et al., 2020; Tan et al., 2019; Soh et al., 2019; Chaudhari et al., 2018; Kang et al., 2020)	Ultrasonic sensor Ethanol gas senor	Raspberry Pi, Wi-Fi, GPS, Android device, Arduino/ATmega, RFID, Cloud system	Waste level/capacity monitoring & detection; Decomposition level sensing.
(Arinta et al., 2020; Jagtap et al., 2020; Jaikumar et al., 2020; Nehete et al., 2018)	Moisture sensors, Infrared and Ultrasonic Sensors	GPS, Wi-Fi, Android device, Arduino, GSM	Detection of moisture content of garbage, garbage capacity/volume & disposal violations
(Elhassan et al., 2019; Lopes and Machado, 2019; Ajay et al., 2020; Paul et al., 2018)	Metal sensor, IR & Ultrasonic sensors, Proximity sensor, Image sensor	GSM/GPRS, GPS, Wi-Fi, Arduino IDE, Arduino Uno (ATmega)	Used for waste separation/segregation; level sensing; metallic & non-metallic detection
(Awawdeh et al., 2019)	Temperature, Smoke & Gas sensors	GPS/GSM, Wi-Fi, Cloud server, Atmega 2560	Measures temp. values; detects smoke level to trigger off an alarm
(Sreejith et al., 2019)	Gas sensor, Rain sensor, IR sensor,	Wi-Fi, Arduino,	Detects dangerous smell created by wastes; Detects the presence of rain & the threshold level of waste inside bins and automatically closes the top door of the bin.
(Raj et al., 2020)	IR sensor, Moisture & Metal sensors	Cloud, Wi-Fi, Arduino	Waste segregation; detection of wet, metallic and non-biodegradable wastes
(Aggarwal et al., 2019; Chellaswamy et al., 2019; Rekha et al., 2020; Kodali et al., 2019)	PH, Temperature, Turbidity, Colour, Flow, Level & Oxygen, sensors	Arduino UNO, Wi-Fi, Cloud server, Arduino Mega 2560	Keeps track of quality of water supply; for data collection w.r.t water bodies; monitors water quality parameters.
(Zahur et al., 2019; Sha et al., 2018)	RGB LED, Ultrasonic sensor, Temperature & Humidity sensor	GSM/GPS, Arduino Uno, Wi-Fi, RFID, Cloud system, Android device, GPRS	Used to show the level of waste inside waste bin; used for waste level detection/monitoring; used to measure the temperature & humidity.
(Devi et al., 2018)	Ultrasonic sensor DHT22 sensor MQ135 sensor	Wi-Fi (Node MCU), Cloud server, Bluetooth, ZigBee, Android app.	Used for garbage level monitoring and measurement; measures the temperature, humidity and the air quality.

Table 2. Summary of Previous Works on Smart Waste Management (SWM) Techniques

Reference	Technique/Approach	IoT Application Domain and Devices	Feature/Focus of Work
(Shah et al., 2018)	A stochastic optimization scheme to improve the waste recovery & collection processes	Health, Education, Entertainment, Urban living, Workplace	Focused on reducing operational costs, energy dissipation, transportation & pollution emissions
(Anagnostopoulos et al., 2015)	A novel algorithm with the aim of providing effective and scalable solutions to the dynamic waste collection	Health (Hospitals), Education, Urban living, Gas stations, Factories	Sensor-attached (Smart) bins located in different areas of a SC, Routing functionalities for several trucks
(Ting et al., 2017)	Application of Data Envelopment Analysis (DEA) to quantify & benchmark waste generation intensity	Machine, Energy, Internet network, Logistics, Mobile devices, Monitoring	The focus is on real-time benchmarking analysis of waste monitoring to realize waste reduction potentials in the floor of a manufacturing shop
(Gutierrez et al., 2015; Marques et al., 2019; Aleyadeh and Taha, 2018; Chen et al., 2018; Adam et al., 2018)	IoT-Based Technologies (android application, Cloud, GSM/GPRS, Sensors, etc.)	Workplace, GSM, Logistics, Sensors, Surveillance, GPS, RFID, Urban living,	Focused on providing efficient waste collection/management solutions & economic feasibility; getting real-time information about the status of dustbins
(Medvedev et al., 2015; Wen et al., 2018; Soh et al., 2019; Lokuliyana et al., 2018)	A Cloud-Based advanced Decision Support System (DSS) for dynamic scheduling & routing models	Automation, RFID, Monitoring, Sensors, Surveillance cameras.	Focused on efficient waste collection & dynamic route optimization in SCs
(Siregar et al., 2017)	Simulation-Based approach using load cell sensors	Load cell sensors, GPS, GSM, Web	Focus is to determine the maximum waste load capacity inside waste bins.
(Warrier, 2018)	IoT-Based technology using Ultrasonic level sensor to determine the load capacity of garbage in bins	Ultrasonic sensor	Focused on tracking the amount of garbage in a trash can & to reduce transportation costs with higher optimized routes.
(Hartono et al., 2018)	A Robust Counter Model of Open Capacitated Vehicle Routing Problem	Workplace	Focus is to determine the optimal routes for garbage transportation
(Bakhshi and Ahmed, 2018; Ray et al., 2018; Rutqvist et al., 2019)	IoT-Based sensor and machine learning analytics (Automated Machine Learning approach)	Raspberry Pi, GPRS, Ultrasonic sensors, Urban environment	Focused on real-time waste monitoring & collection; optimizing time costs & operational efficiency
(Steyn and Willemse, 2018)	Application of Big Data Analysis on vehicle routing using GPS data	GPS, Big data	Focused on vehicle routing & optimization for waste collection
(Rohit et al., 2018)	A smart dual dustbin model using ultrasonic & IR sensors, GSM, Servo Uno, Arduino	Ultrasonic sensor, IR obstacle sensor, GSM, Servo, Arduino Uno,	Focused on a smart & efficient waste mgt. with advanced dynamic routing & higher optimized routes
(Nehete et al., 2018)	A Smart (IoT-Based) dustbin model using IR sensor & LCD	IR sensor, GSM, Monitoring system,	Focused on garbage monitoring & control
(Saha et al., 2018; Aravindaraman and Ranjana, 2019)	IoT-Based smart garbage monitoring & clearance alert system using RGB LED, Ultrasonic Sensor, GSM, Arduino UNO, GPS, Wi-Fi,	LED light, Sensors, GSM, Arduino, GPS, Wi-Fi	Focused on garbage level monitoring for disposal, clearance & routing optimization. Detecting the max. capacity level of garbage & alerting waste managers
(Venkatesan et al., 2020)	Application of Lab-scale Landfill model	Sensors, LM35 temp. sensor, Soil moisture sensor	Focused on monitoring & processing of food waste
(Al-Masri et al., 2018a; Al-Masri et al., 2018b)	Application of Recycle.io IoT-Based waste management technique based on a serverless architecture	Camera, Sensor, Microsoft Azure IoT Hub, Urban, IR, Raspberry Pi, Cloud	Focused on real-time monitoring of recycling & disposal violations & behaviours prior to waste collection.
(Al-Jubori and Gazder, 2018)	A framework that uses ArcGIS network analysis for optimal routing	GIS, Urban areas, Education, Household	Focused on MSW routing within urban context; Static & Dynamic routing models
(Gu et al., 2017)	Survey,Implementation of Big data technologies for WEEE Mgt: Big data collecting, Big data processing, Big data utilization & Big data analytics	Logistics, Electric road vehicles & roadside charging, Electric cars, Monitoring, RFID Tags, Waste recycling & Mgt	The focus is on potential application of IoT & Big data technologies in WEEE Mgt.
(Elia et al., 2016)	Application of Productive-Service System (PSS) approach for WEEE collection services	Electric road vehicles and roadside charging, Logistics, Weight monitoring,	Evaluating the cost efficiency & capacity of a PSS-based model for e-waste collection, enabling dynamic scheduling & routing schemes
(Kang et al., 2020)	Design of a smart household e-waste collection box fitted with e-waste level measurement sensors to record the disposal data	Household, Sensors, GSM, RFID, Microcontroller,	To employ smart household e-waste management systems pertaining to the Malaysia e-waste management
(Jayaranjan et al., 2018)	A smart phone monitoring system for e-wastes using image processing technique (MATLAB program)	Electronic gadgets, Monitoring, Sensor,	The focus is on management of e-wastes disposal & collection, monitoring of e-wastes through a dedicated android application
Masuduzzaman et al., 2018)	Review approach. A review of the potential use of e-waste in concrete & its environmental impacts	Environment, Water, Air pollution, Soil pollution,	The focus is on the potential use of e-waste in concrete and its environmental impacts

transportation. Sensor-based technologies can be used in path planning, scheduling and routing optimization, since sensors can read the level of wastes to determine when the bins are low, medium or high. And information about the status of waste bins can be remotely accessed by the municipal authorities via their server to be properly guided to make informed decisions on the deployment of just the sufficient number of waste collectors in order to minimize unnecessary collection trips as well as transportation costs, while increasing collection efficiency and dynamic route optimization (Medvedev et al., 2015; Anagnostopoulos et al., 2015a; Soh et al., 2019; Warrier, 2018; Lokuliyana et al., 2018; Wen et al., 2018).

An IoT-enabled smart waste monitoring & collection architecture that incorporated real-time dynamic scheduling model driven by sensors was proposed in (Likotiko et al., 2017; Anagnostopoulos et al., 2015b; Malapur and Pattanshetti, 2017) to address the challenges of waste and trip management in cities and drastically minimize operational cost/time and achieve path/route optimization for waste collection. The use of IoT systems to provide intelligence to trashcans for transmitting trash volume of data over the internet was assessed in (Gutierrez et al., 2015), to monitor trash, vehicles drivers and customers thus optimizing many elements of the essential services offered and providing real-time information to drivers & managers for routing scheduling, data gathering and processing (Padilla et al., 2018; Khan et al., 2017) have been well discussed. Table 1 shows a summary of the applications of IoT-based technologies for efficient waste management strategies in a SC while Table 2 shows previous works on Smart Waste Management (SWM) techniques.

3. COMPARING TRADITIONAL VS DATA MULE SCHEMES

The use of data Mules for data collection is gaining considerable momentum for large-scale wireless sensor network (LS-WSN) applications. This may be because of the need to optimize sensor-node energy consumption. Researchers are now focusing on the use of data Mules as MDCs for smart city applications. The data Mule scheme presents significant advantages over the traditional methods using static sinks. Some of the benefits of using MDCs include: time reduction, higher coverage data collection, cost reduction, low-energy consumption of the sensor-nodes, improved operational efficiency, error reduction at the point of data collection, improved transparency and data visibility, reduction of risks of data leaks, cheaper and quicker access to a more efficient analyzable data than the traditional methods and increased big data collection, storage, processing and real-time data analysis and transmission over the internet. Unlike the traditional methods, mobile data collection involves the use of mobile devices with IoT-enabled technologies for data collection aimed at minimizing time, cost, energy consumption and optimizing collection-efficiency. The use of data Mules ensures that data is directly exported into a centralized database or the platform server where it can be accessed in real-time for analysis and further processing/evaluation.

According to Ang et al. (2018) and Ang et al. (2017), the use of data Mules would effectively address two major concerns of data collection for LS-WSN, which included monitoring of spatially separated regional areas (unconnected regions) and optimizing the energy efficiency of the battery-powered sensor nodes. Accordingly, the first challenge has to do with the issue of divided network which prevents sensor nodes across different regions to have physical communications. This has made it very difficult for a single collector or sink from one region to be used for data collection for another region. The other challenge is the problem of optimization of the lifespan of the battery-powered sensor-nodes which are non-rechargeable with a limited energy supply. Because many sensor-nodes used in IoT-IoV applications are battery-powered with limited lifecycle of operations and cannot be recharged after being deployed, it becomes necessary to adopt new methods of data collection that are not only cost-effective but also energy-efficient. Compared to the conventional methods, the data Mule scheme is more advantageous because the Mules can be used to roam over the different regions (Ang et al., 2018), collect data from the sensors, buffer it and deliver it to the big data collection center/access points (Shah et al., 2003). Essentially, the opportunistic style of the MDCs has been shown to be cost-effective with low node-energy consumption value, because the control/collection protocol overheads can be effectively shifted from the sensor-nodes to the MDCs. MULEs in close range can collect data from the sensors and deliver to the access points thereby saving substantial power at the sensors and cutting down their transmission overheads. Our target is to determine the node-energy consumption and present an energy-efficient model for opportunistic data collection for SC waste management strategies.

In the context of waste management, especially for SC scenarios, a data Mule can be exploited to collect data from the largely distributed sensor-nodes deployed across spatially separated regional areas. A crucial design consideration would be focused on the methods deployed for data collection from the sensor-nodes. Compared to the centralized (traditional) methods using static sinks, the data mule approach presents several advantages including:

- Mules can effectively handle data collection from sensor-nodes across spatially separated regional areas (Ang et al., 2018)
- Mules have lower requirements for node-energy consumption. This is because of reduced collection protocol overhead occasioned by of short-range data communication.
- Mules are cost effective due to the opportunistic style in which data is collected and can traverse the city within a short time duration.
- Contrary to the typical energy-constrained nature of sensor-nodes, data Mules are power renewable (Anastasi et al., 2008). The communication process therefore between the Mule and the sensor must be energy efficient in order to prolong the lifespan of the battery-powered sensor-nodes.
- Mule system is robust, scalable, and flexible; and the opportunistic approach used in data collection from the smart bins allows the same vehicle to be used simultaneously for other applications.
- Data transmission between the mules and the smart sensors (smart bins) occurs at short ranges resulting in huge power savings at the sensors.

However, some of the drawbacks of the Mule system include:

- High latency: This includes the waiting time during which sensors await the arrival of the Mule for data exchange (Shah et al., 2003). Delays can also occur between the time of data collection by the Mule from the smart sensors and the time of data delivery/transmission to the access points. However, the above drawback can be limited by choosing Mules that traverse designated routes. Studies have shown that vehicles with predetermined routes present high data delivery performance with low latency (Er et al., 2017; Er et al., 2019).
- Unexpected network failures can affect efficient data delivery at the AP and unexpected change in weather conditions can affect mobility patterns of the Mules.
- Loss of a Mule to an unexpected occurrence (e.g. accidents) can lead to data loss.

4. A UNIVERSAL ARCHITECTURE OF DATA MULE (IOV) FOR SC WASTE MANAGEMENT

This section proposes a Universal data Mule (U-Mule) architecture for waste management in a SC ecosystem. **Figure 3** shows the proposed data Mule system architecture for SC waste management operations, which combines the concepts of embedded intelligence, cyber physical systems and data communications systems. The layout of the architecture is designed to enable autonomous, simultaneous and synergetic operations among the various concepts and features. The model is defined by three concepts: the embedded intelligence, the cyber physical systems, and the communication systems. The layout of the architecture is designed to enable the concepts to converge. Depending on the application scenario, data Mules may be either part of the cyber physical system (e.g. buses, taxis, etc.) or part of the embedded intelligence (e.g. mobile robots). The cyber physical system consists of the houses, laptops, data processing centers, garbage trucks, waste materials (smart bins), waste management authorities, truck drivers, and so on. The embedded intelligence consists of the smart sensors embedded in the smart bins which communicate at short-range with the data Mules and internet of vehicles known as the data MULEs used for data collection and transmission, while the communication network includes the satellite, internet or GPRS systems through which collected data can be deposited to the access points or sink nodes.

Figure 3. U-Mule Architecture for SC Waste Management

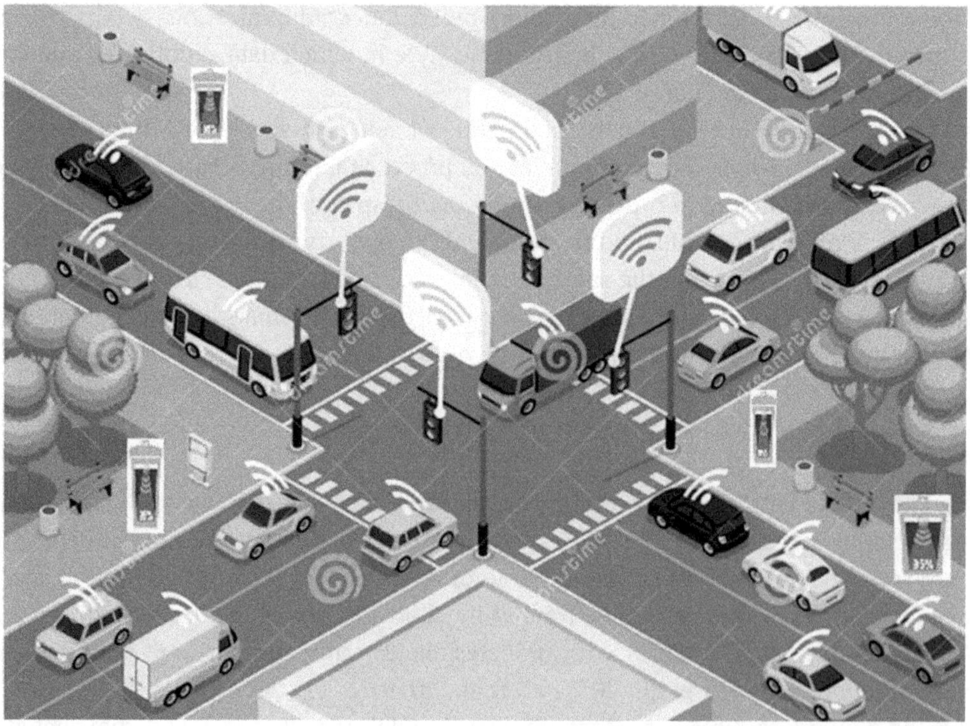

As shown in **Fig. 3** above, the smart waste bins are planted along the routes where the vehicles traverse opportunistically within the smart cities. The sensor-driven mobile data collectors known as the data Mules are used in an opportunistic way for data collection from the smart waste bins. Three kinds of vehicles (data Mules) can be selected for this purpose – buses, taxis and cars. Finding reveal that vehicles with predetermined routes present high data delivery performance. Based on the perceived disparities in the mobility patterns of the selected vehicles (mobile data collectors), buses present the best fit for use as MDCs as their mobility follows almost a fixed route and they are continuously on the move especially during their service hours This analysis is detailed in (Er et al., 2017; Er et al., 2019).

Real-time monitoring, routing and scheduling operations will be efficient and optimized, if accurate data is collected and delivered in real-time with respect to determining the real status of the smart waste bins. From Fig. 3, we see the vehicles (data Mules) or MDCs communicating with the smart bins stationed along their paths. The data collected from the smart bins are transmitted to the sink or access points via the internet or satellite and can be accessed by the waste authorities at the big data center. The data will help waste managers in decision making with respect to real-time routing and scheduling operations. Waste collectors (truck drivers) can be promptly scheduled and deployed for waste collection operations.

The three concepts are seen in the data Mule operations. The MULEs collect data from the smart bins and deposit the data at the big data collection center or access points from where waste authorities can access the data via the internet and be able to deploy waste collectors in a timely and efficient manner. The motion of the Mule is opportunistic or unpredicted and is expected to cut down the cost often incurred in frequent deployments of waste collectors (as seen in the traditional and centralized methods) which often result in high costs. If the smart bins are planted along the routes traversed by the Mules (MDCs), data exchange can occur at a short-range thereby minimizing the energy consumption of the sensors which is occasioned by high transmission overheads. When the Mules approach the vicinity of the smart sensors at close range, they collect data from the sensors and deliver the data to the access points at the next available opportunistic movement. Compared to the sensors, Mules have big storage capabilities, are power-renewable and can exchange communication with both sensors and access points. Importantly, Mules can be advanced to exchange data with each other to allow the formation of a multi-hop Mule network, which can be exploited to further minimize the latency times associated with Mule system as well as reduce the probabilities of data replication. Compared to the traditional and centralized solutions, Mule architecture is robust and scalable.

5. SIMULATIONS, RESULTS AND ANALYSIS

This section presents some graphical analysis obtained from the results of several simulations performed using different routing protocols. The following routing protocols were used in the simulation (1) Ad hoc On Demand Distance Vector (AODV) Routing Protocol (Perkins and Royer, 1999), (2) Flooded Forward (FF) Ant Routing Protocol (Zhang et al., 2004), (3) Improved Energy Efficient Ant-based Routing (IEEABR) Protocol and (4) BeeSensor Routing Protocol (Ketshabetswe et al., 2019). Simulations were performed for the different routing protocols and comparisons were drawn using the different metrics for ease of reporting. 200 mobile nodes of varying speeds were used in the simulations to represent mobile data collectors (data Mules) which were randomly distributed in a network topology size of $200m \times 200m$. We considered a case of static sink nodes (source & destination) and mobile sensor-nodes in the network.

The results are shown in **figures 4-6.** The result is an average of 10 simulations performed for each of the routing protocols. The results obtained show that in general, swarm-based routing protocols such as EEABR, and BeeSensor performed optimally in terms energy efficiency, throughput and packet delivery ratio (success rate) compared to their classical-based counterparts such as AODV. Below are the analyses of the performance metrics of the routing protocols used in our experiment.

5.1 Energy Efficiency

This is simply the ratio of the total data packets transmitted to the destination to that consumed by sensor nodes in the network. That is

$$EE = \left(\frac{Success\ rate \times Total\ packets\ sent\ to\ the\ sink}{Total\ energy\ consumed} \right) \text{.Kbits/J)}$$

This is denoted by the ratio of the total size of received non-overhead data packet by the destination node to the total energy consumed by the network (Li et al., 2019). It is measured in kilobits per joules (Kbits/J). A highly efficient routing algorithm would tend to transmit more data packets per consumed energy unit. **Figure 4** shows the graphical performance of each of the routing protocols in terms of energy efficiency. From the simulations, it is evident that the improved Energy Efficient Ant-based Routing (EEABR) outperformed all other routing protocols in terms of energy efficiency while main-

Figure 4. Effects of increasing the number of sensor-nodes in Swarm and Classical-based routing protocols

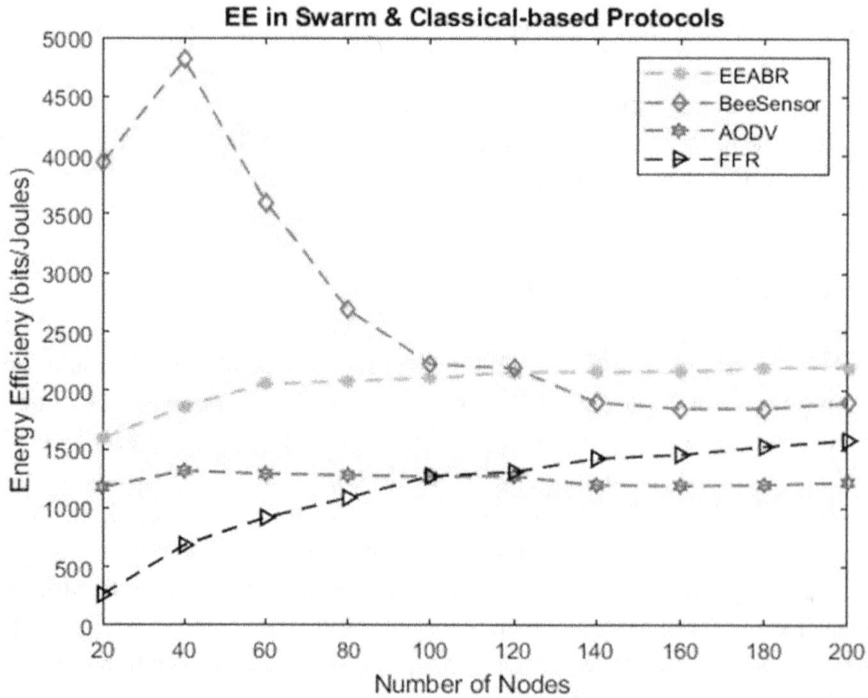

taining the least in terms of energy consumption. This is followed by the BeeSensor and then FFR. The reason for the optimal performance of EEABR may be due to its proactive nature. In proactive routing, all connections between sensor-nodes and including the routes between source and destination nodes are maintained and always up to date notwithstanding the data traffic. This enables data to travel along a predetermined route to its destination. Our proposed scheme presupposes that vehicles with predetermined routes have high data delivery performance. AODV being a classical-based routing protocol is seen in the least performance cadre. This may be ostensibly because of its reactive (on-demand) nature, in which it adopts a randomly proportional route discovery. As a result, greater energy is dissipated in its route exploratory process. The results are therefore evident that swarm-based routing algorithms outperformed their classical-based counterparts in terms of energy efficiency.

5.2 Throughput

This metric shows the measurement of the number of data packets successfully transmitted through the network in each time. It refers to the average rate of successfully delivered packets over the network. Throughput is improved by increasing the number of sensor nodes in the network. Figures 5 shows the performance of the each of the protocols in terms of throughput. Again, nature-inspired routing protocols (e.g. EEABR and BeeSensor,) performed better than their classical-based counterparts in terms of throughput.

5.3 Latency

Latency is the time required to send and receive a request between the sender and the receiver. It can be used to describe a delay arising from the transmission of a packet from the source node to the destination node. A low latency describes a network that is optimized to process a huge volume of data with minimal latency (delay). Latency is a function of time. This time is expected to remain as small as possible. However, several factors at play may prevent the network latency times to remain as low as possible. Latency may be affected by distance, data aggregation and multi-hop relays, and is significantly impacted by propagation delay, routing & switching, and queuing & buffering. Figure 6 shows the performance of four routing protocols (EEABR, BeeSensor, AODV & FFR) in terms of end-to-end delay. The latency in this scenario describes the time required for the data MULEs to opportunistically pick up data from the smart bins (smart sensors) and deposit them at the access point (AP) or sink node. This time is expected to be as small as possible. However, high latency may be expected in our case scenario owing to several factors including the opportunistic style of data collection by the Mules. This probably justifies the reason for the high latency observed in BeeSensor. However, our scheme presupposes to predominantly use vehicles with predetermined routes to minimize the latency times since such vehicles are presumed to have high data delivery performance including low latencies. Hence, this probably justifies why EEABR recorded the least latency as shown in the diagram of Fig.6. Latency (delay) is affected by an increase in the number of packet transmissions. A network experiences high latency as a result of network congestions occasioned by an increase in the number of transmitted packets. Hence, throughput is greatly impacted by an increase in latency. As a result, packets experience high delays in the queues and many RREQs fail to reach their destinations. This obviously explains why swarm-based routing protocols still outperformed their conventional-based routing counterparts

Figure 5. Throughput vs Speed in Swarm-based & Conventional Routing Protocols

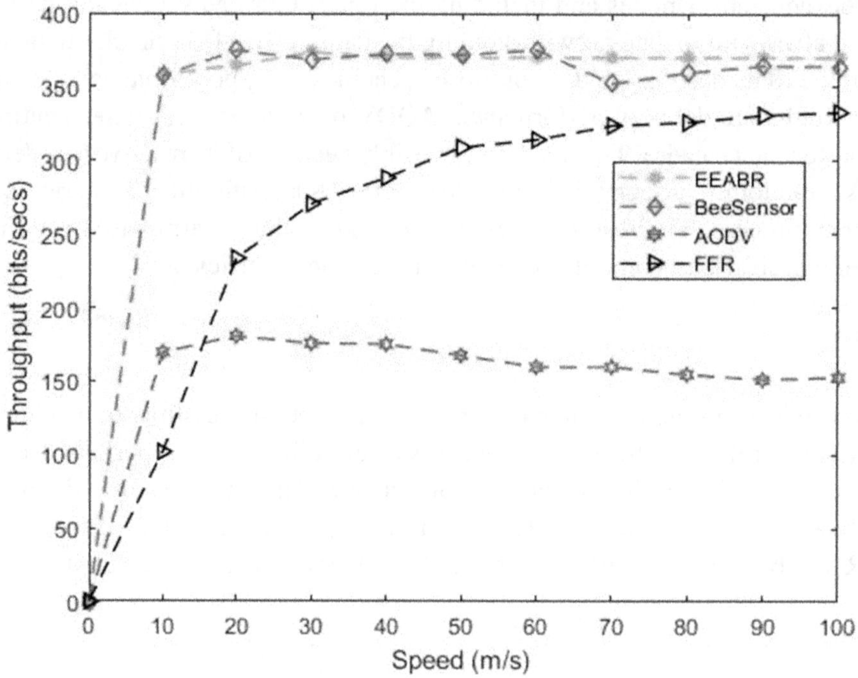

Figure 6. Latency vs Time in Swarm-based & Conventional Routing Protocols

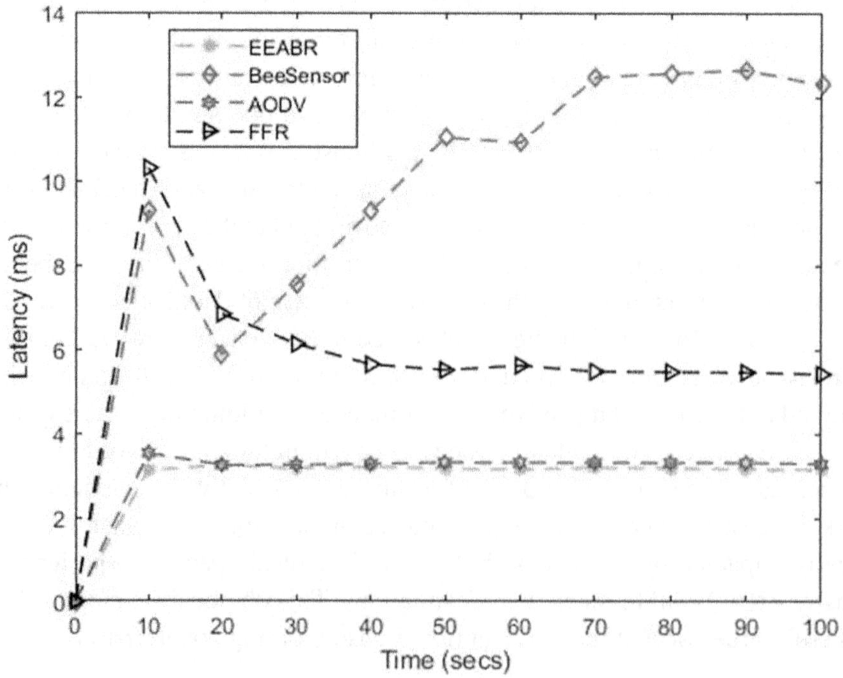

6. DISCUSSIONS

From the results of the experiments, we can make the following deductions:

- Nature-inspired routing algorithms can indeed transmit more valid sensing data packets with minimal node-energy requirements. This further justifies the claim that energy efficiency can be optimized in a dynamic routing scenario. This clearly underscores our proposed model. The general perspectives that the traditional or centralized approach offers significant improvement in a large network of sensor-nodes, since the distance between neighboring nodes will be reduced and lesser energy will be required for data transmission to the AP or sink, can be justified in Fig. 4 where the BeeSensor first reached an optimal point before falling due to probably an increase in the distance between neighboring nodes in the network. However, this position is also contradicted as exemplified in Fig.4 where EEABR outperformed the others, thus justifying the essence of our proposed scheme.
- When the nodes are a few meters away from each other and sparsely distributed, it is found that the Mule approach performs better than the traditional methods, because Mules will be able to traverse the routes along which the smart bins are planted at short-range to enable data exchange with the smart bins. This is not possible with the traditional methods where sensor-nodes would have to expend a substantial amount of energy to transmit their data directly to the access point (AP) or sink node thus leading to high energy consumption occasioned by high transmission and control overheads.

7. CONCLUSION

This paper has presented an energy-efficient model for opportunistic data collection using data Mules for SC waste management operations. A universal architecture of data Mule scheme is hereby proposed. The proposed architecture exploits a combination of three major concepts in data collection scheme, namely an embedded intelligence (e.g. smart sensors embedded in the waste bins), a cyber physical system (e.g. buses, taxis) and data communication system (e.g. internet, satellite, GPRS). These systems are designed to enable autonomous and simultaneous operations of the different features. The vehicles (data Mules) will be used opportunistically for data collection to minimize costs. Buses and taxis are known to have predetermined routes and often operate at peak hours. This positions them well for this purpose, if the smart bins could be planted along their routes or paths.

The performance of our proposed scheme was tested by running several simulations using both classical-based and nature-inspired routing protocols, and comparisons with the traditional methods of data collection were drawn. The results show high performance with our proposed scheme using energy-efficient routing protocols. A notable challenge is the high latency times. However, this can be minimized by using predominantly vehicles with predetermined routes as MDCs. In future, we will consider exploiting data Mules (MDCs) as sources of energy maintenance to the sensor-nodes to further optimize energy efficiency.

REFERENCES

Anagnostopoulos, T., Kolomvatsos, K., Anagnostopoulos, C., Zaslavsky, A., & Hadjiefthymiades, S. (2015a). Assessing dynamic models for high priority waste collection in smart cities. *Journal of Systems and Software*, *110*, 178–192. doi:10.1016/j.jss.2015.08.049

Anagnostopoulos, T., Zaslavsky, A., Kolomvatsos, K., Medvedev, A., Amirian, P., Morley, J., & Hadjieftymiades, S. (2017). Challenges and opportunities of waste management in IoT-enabled smart cities: A survey. *IEEE Transactions on Sustainable Computing*, *2*(3), 275–289. doi:10.1109/TSUSC.2017.2691049

Anagnostopoulos, T., Zaslavsy, A., Medvedev, A., & Khoruzhnicov, S. (2015b). Top--k query based dynamic scheduling for IoT-enabled smart city waste collection. In *2015 16th IEEE International Conference on Mobile Data Management* (Vol. 2, pp. 50-55). IEEE.

Anastasi, G., Conti, M., & Di Francesco, M. (2008). Data collection in sensor networks with data mules: An integrated simulation analysis. In *2008 IEEE Symposium on Computers and Communications* (pp. 1096-1102). IEEE. 10.1109/ISCC.2008.4625629

Ang, K. L. M., Seng, J. K. P., & Zungeru, A. M. (2017). Optimizing energy consumption for big data collection in large-scale wireless sensor networks with mobile collectors. *IEEE Systems Journal*, *12*(1), 616–626. doi:10.1109/JSYST.2016.2630691

Ang, L. M., Seng, K. P., Ijemaru, G. K., & Zungeru, A. M. (2018). Deployment of IoV for smart cities: Applications, architecture, and challenges. *IEEE Access: Practical Innovations, Open Solutions*, *7*, 6473–6492. doi:10.1109/ACCESS.2018.2887076

Ang, L. M., Seng, K. P., Zungeru, A. M., & Ijemaru, G. K. (2017). Big sensor data systems for smart cities. *IEEE Internet of Things Journal*, *4*(5), 1259–1271. doi:10.1109/JIOT.2017.2695535

Berrilio, P. D. (2014). *Exploiting Vehicular Networks for Data Collection in Smart Cities* (Doctoral dissertation). University of Rennes, France.

Bharadwaj, B., Kumudha, M., & Chaithra, G. (2017, February). Automation of Smart waste management using IoT to support "Swachh Bharat Abhiyan"-a practical approach. In *2017 2nd International Conference on Computing and Communications Technologies (ICCCT)* (pp. 318-320). IEEE.

Chaqfeh, M., Mohamed, N., Jawhar, I., & Wu, J. (2016, December). Vehicular cloud data collection for intelligent transportation systems. In 2016 3rd Smart Cloud Networks & Systems (SCNS) (pp. 1-6). IEEE. doi:10.1109/SCNS.2016.7870555

Contreras-Castillo, J., Zeadally, S., & Guerrero-Ibañez, J. A. (2017). Internet of vehicles: Architecture, protocols, and security. *IEEE Internet of Things Journal*, *5*(5), 3701–3709. doi:10.1109/JIOT.2017.2690902

Er, N. I., Singh, K., & Bonnin, J. M. (2017). On the performance of VDTN routing protocols with V2X communications for data delivery in smart cities. Academic Press.

Er, N. I., Singh, K. D., & Bonnin, J. M. (2019). CityView: A Vehicle Based Image and Data Collection for Smart Cities. In *2019 International Conference on Wireless and Mobile Computing, Networking and Communications (WiMob)* (pp. 1-7). IEEE. 10.1109/WiMOB.2019.8923547

Fallavi, K. N., Kumar, V. R., & Chaithra, B. M. (2017, February). Smart waste management using Internet of Things: A survey. In *2017 International Conference on I-SMAC (IoT in Social, Mobile, Analytics and Cloud)(I-SMAC)* (pp. 60-64). IEEE. 10.1109/I-SMAC.2017.8058247

Gutierrez, J. M., Jensen, M., Henius, M., & Riaz, T. (2015). Smart waste collection system based on location intelligence. *Procedia Computer Science*, *61*, 120–127. doi:10.1016/j.procs.2015.09.170

Haribabu, P., Kassa, S. R., Nagaraju, J., Karthik, R., Shirisha, N., & Anila, M. (2017, December). Implementation of a smart waste management system using IoT. In *2017 International Conference on Intelligent Sustainable Systems (ICISS)* (pp. 1155-1156). IEEE. 10.1109/ISS1.2017.8389367

Jain, A., & Bagherwal, R. (2017, July). Design and implementation of a smart solid waste monitoring and collection system based on Internet of Things. In *2017 8th International Conference on Computing, Communication and Networking Technologies (ICCCNT)* (pp. 1-5). IEEE. 10.1109/ICCCNT.2017.8204165

Karthikeyan, S., Rani, G. S., Sridevi, M., & Bhuvaneswari, P. T. V. (2017, May). IoT enabled waste management system using ZigBee network. In *2017 2nd IEEE International Conference on Recent Trends in Electronics, Information & Communication Technology (RTEICT)* (pp. 2182-2187). IEEE. 10.1109/RTEICT.2017.8256987

Ketshabetswe, L. K., Zungeru, A. M., Mangwala, M., Chuma, J. M., & Sigweni, B. (2019). Communication protocols for wireless sensor networks: A survey and comparison. *Heliyon*, *5*(5), e01591. doi:10.1016/j.heliyon.2019.e01591 PMID:31193432

Khan, M., Babar, M., Ahmed, S. H., Shah, S. C., & Han, K. (2017). Smart city designing and planning based on big data analytics. *Sustainable Cities and Society*, *35*, 271–279. doi:10.1016/j.scs.2017.07.012

Kumar, B. S., Varalakshmi, N., Lokeshwari, S. S., Rohit, K., & Sahana, D. N. (2017, December). Eco-friendly IOT based Waste segregation and management. In *2017 International Conference on Electrical, Electronics, Communication, Computer, and Optimization Techniques (ICEECCOT)* (pp. 297-299). IEEE. 10.1109/ICEECCOT.2017.8284686

Li, X., Keegan, B., Mtenzi, F., Weise, T., & Tan, M. (2019). Energy-efficient load balancing ant-based routing algorithm for wireless sensor networks. *IEEE Access: Practical Innovations, Open Solutions*, *7*, 113182–113196. doi:10.1109/ACCESS.2019.2934889

Likotiko, E. D., Nyambo, D., & Mwangoka, J. (2017). *Multi-agent based IoT smart waste monitoring and collection architecture.* arXiv preprint arXiv:1711.03966.

Lokuliyana, S., Jayakody, A., Dabarera, G. S. B., Ranaweera, R. K. R., Perera, P. G. D. M., & Panangala, P. A. D. V. R. (2018, August). Location based garbage management system with iot for smart city. In *2018 13th International Conference on Computer Science & Education (ICCSE)* (pp. 1-5). IEEE. 10.1109/ICCSE.2018.8468682

Lotfinezhad, M., & Liang, B. (2005). Energy efficient clustering in sensor networks with mobile agents. In *IEEE Wireless Communications and Networking Conference*, 2005 (Vol. 3, pp. 1872-1877). IEEE. 10.1109/WCNC.2005.1424797

Luo, C., Wang, Y., Hong, Y., Chen, W., Ding, X., Zhu, Y., & Li, D. (2019). Minimizing data collection latency with unmanned aerial vehicle in wireless sensor networks. *Journal of Combinatorial Optimization, 38*(4), 1019–1042. doi:10.100710878-019-00434-w

Malapur, B. S., & Pattanshetti, V. R. (2017, August). IoT based waste management: An application to smart city. In *2017 International Conference on Energy, Communication, Data Analytics and Soft Computing (ICECDS)* (pp. 2476-2486). IEEE. 10.1109/ICECDS.2017.8389897

Massaguer, D., Fok, C. L., Venkatasubramanian, N., Roman, G. C., & Lu, C. (2006). Exploring sensor networks using mobile agents. In *Proceedings of the fifth international joint conference on Autonomous agents and multiagent systems* (pp. 323-325). 10.1145/1160633.1160688

Medvedev, A., Fedchenkov, P., Zaslavsky, A., Anagnostopoulos, T., & Khoruzhnikov, S. (2015). Waste management as an IoT-enabled service in smart cities. In *Internet of Things, Smart Spaces, and Next Generation Networks and Systems* (pp. 104–115). Springer. doi:10.1007/978-3-319-23126-6_10

Mergen, G., Zhao, Q., & Tong, L. (2006). Sensor networks with mobile access: Energy and capacity considerations. *IEEE Transactions on Communications, 54*(11), 2033–2044. doi:10.1109/TCOMM.2006.884845

Padilla, D. D. (2020). The Internet of Things (IoT) Making Waste Collection Part of a Smarter Future. *Looking Ahead – IoT, Smart Cities and Refuse Collection.* https://txswana.org/images/meeting/040917/Presentations/don_diego_padilla_the_internet_of_things__lot_.pdf

Perkins, C. E., & Royer, E. M. (1999). Ad-hoc on-demand distance vector routing. In *Proceedings WMCSA'99. Second IEEE Workshop on Mobile Computing Systems and Applications* (pp. 90-100). IEEE.

Shah, R. C., Roy, S., Jain, S., & Brunette, W. (2003). Data mules: Modeling and analysis of a three-tier architecture for sparse sensor networks. *Ad Hoc Networks, 1*(2-3), 215–233. doi:10.1016/S1570-8705(03)00003-9

Shyam, G. K., Manvi, S. S., & Bharti, P. (2017, February). Smart waste management using Internet-of-Things (IoT). In *2017 2nd international conference on computing and communications technologies (ICCCT)* (pp. 199-203). IEEE.

Silva, R., Bonnin, J. M., & Ernst, T. (2017). *An ITS-based Architecture for Opportunistic Networking.* Academic Press.

Soh, Z. H. C., Husa, M. A. A. H., Abdullah, S. A. C., & Shafie, M. A. (2019, April). Smart Waste Collection Monitoring and Alert System via IoT. In *2019 IEEE 9th Symposium on Computer Applications & Industrial Electronics (ISCAIE)* (pp. 50-54). IEEE.

Teng, H., Liu, Y., Liu, A., Xiong, N. N., Cai, Z., Wang, T., & Liu, X. (2019). A novel code data dissemination scheme for Internet of Things through mobile vehicle of smart cities. *Future Generation Computer Systems, 94*, 351–367. doi:10.1016/j.future.2018.11.039

Tong, L., Zhao, Q., & Adireddy, S. (2003). Sensor networks with mobile agents. In *IEEE Military Communications Conference, 2003. MILCOM 2003.* (Vol. 1, pp. 688-693). IEEE. 10.1109/MILCOM.2003.1290187

Warrier, A. (2018). *Smart Transportation & Smart Waste Management.* https://medium.com/bolt-iot/smart-transportation-smart-waste-management-d35f1e141629

Wen, Z., Hu, S., De Clercq, D., Beck, M. B., Zhang, H., Zhang, H., Fei, F., & Liu, J. (2018). Design, implementation, and evaluation of an Internet of Things (IoT) network system for restaurant food waste management. *Waste Management (New York, N.Y.)*, *73*, 26–38. doi:10.1016/j.wasman.2017.11.054 PMID:29242117

Xu, W., Zhou, H., Cheng, N., Lyu, F., Shi, W., Chen, J., & Shen, X. (2017). Internet of vehicles in big data era. *IEEE/CAA Journal of Automatica Sinica, 5*(1), 19-35.

Zhang, Y., & Fromherz, M. (2004). Message-initiated constraint-based routing for wireless ad-hoc sensor networks. In *First IEEE Consumer Communications and Networking Conference, 2004. CCNC 2004* pp. 648-650 IEEE. 10.1109/CCNC.2004.1286943

Zhang, Y., Kuhn, L. D., & Fromherz, M. P. (2004). Improvements on ant routing for sensor networks. In *International Workshop on Ant Colony Optimization and Swarm Intelligence* (pp. 154-165). Springer. 10.1007/978-3-540-28646-2_14

Zhao, Q., & Tong, L. (2004). Distributed opportunistic transmission for wireless sensor networks. In *2004 IEEE International Conference on Acoustics, Speech, and Signal Processing* (vol. 3, pp. iii-833). IEEE.

Zhao, Q., & Tong, L. (2005). Energy efficiency of large-scale wireless networks: Proactive versus reactive networking. *IEEE Journal on Selected Areas in Communications*, *23*(5), 1100–1112. doi:10.1109/JSAC.2005.845411

Chapter 2
Nanoelectronics:
A Tool for Next-Generation 5G Wireless Communication Systems

Jutika Devi
https://orcid.org/0000-0003-0110-5886
Gauhati University, India

Pranayee Datta
Gauhati University, India

ABSTRACT

The passive circuit elements resistor, inductor, and capacitor, which are the basic building blocks of an electronic circuit, need to be scaled down for application in fifth-generation wireless communication networks. Due to the growing demands in memory and computational capacities of integrated circuits along with high processing and transmission data speed for next-generation, microelectronics will be replaced by nanoelectronics in the future. The concept of nanoscale network on chip system is expected to play an important role in the field of communication systems for designing new devices of ultra-high speed for long and short-range communication links, power efficient computing devices, high-density memory and logic, and ultrafast interconnects. This chapter focuses on the mechanism of tailoring, patterning, and manipulating optical signals using nanometer-scale structures that may play the role of lumped nanocircuit elements at optical domain when selected properly with tremendous promise for application for fifth-generation communication systems.

INTRODUCTION

Microelectronics has pervaded our lives for the past fifty years, with tremendous impact on health, materials & manufacturing, computers, agricultures, environment, energy storage, information technologies, mobility, security, communications, education, entertainment, and virtually on every aspect of human lives. The 1947 invention of the transfer resistor, or transistor (Schaller, 1997) with feature size about 100 micrometre, by William Shockley, John Bardeen and Walter Brattain at Bell Laboratories, Murray

DOI: 10.4018/978-1-7998-6992-4.ch002

Hill, N. J. on December 16, made the starting of solid-state era of electronics. The concept was based on the discovery that the flow of electricity through a solid such as silicon can be controlled by adding impurities with the appropriate electronic configurations. The vacuum tube (which was also known as the thermionic valve) was the dominant technology (which may be termed as macroelectronics) for this task at the time; but the transistor proved to be significantly more reliable, required much less power, and above all, could be made incredibly smaller. And in fact, the miniaturization of the device was to become the hallmark of the semiconductor industry and the basis for Moore's Law (Moore, 1965).

Moore's law has been a reliable predictor of the pace of electronic technology advancement since 1965 (Wu et al., 2013). However, quantum physics may provide the ultimate limit to this mathematical model. Gordon Moore's general prediction is that the density of transistors and computing power double every one to two years. Moore noted that the number of components in integrated circuits had doubled every year from the invention of the integrated circuit in 1958 until 1965 and predicted that the trend would continue "for at least ten years". His prediction has proven to be uncannily accurate in part, because the law is now used in the semiconductor industry to guide long-term planning and to set targets for research and development (Powell, 2008). This amazing predictive history has emboldened to state that technological evolution often follows a fairly predictable path (Moore, 1965). Moore's law continues to be an accurate means of predicting the future development of technology till date (Wu et al., 2013). Determining an ultimate physics limit to Moore's law would mark out a future boundary to electronics miniaturization. According to Stephen Hawking, the fundamental limits to microelectronics are the speed of light and the atomic nature of matter (Wu et al., 2013). The capabilities of many digital electronic devices are strongly linked to Moore's law in terms of processing speed, memory capacity, sensors and even the number and size of pixels in digital cameras. The world's biggest chipmaker Intel is now using an element called hafnium and metal gates in its chip making processes, which will let Moore's Law to continue further. Heisenberg's uncertainty would be the fundamental limit to Moore's law, and if the electron is the smallest possible transistor component, then the year 2036 is a reasonable prediction of convergence of Moore's law and quantum physics.

Semiconductor manufacturers will be able to produce chips on the 16 nm manufacturing process, expected by conservative estimates to arrive in 2018. However, semiconductor makers won't be able to shrink transistors much at all, beyond 2021, according to a recent paper from Intel (Zhirnov et al., 2003). Transistors are essentially microscopic on/off switches that consist of a source (S), where electrons come from, a drain (D), where the electrons target to, and a gate (G) that mainly controls the flow of electrons through a channel that connects the source and the drain. When the length of the gate gets smaller than 5 nanometers, tunneling effect will begin to kick into play. Electrons will simply pass through the channel on their own without any driver voltage, because the source and the drain will be extremely close. Therefore, a transistor becomes unreliable as a source of basic data, because the probability of spontaneous transmission is about 50 percent. In other words, Heisenberg's uncertainty principle is in action, because the location of the electrons can't be accurately predicted based on Heisenberg's uncertainty principle.

Although there has been an impressive advancement at the device and technology level during the last decade, the progress in the development of new architectures dealing with billions of transistors is scarce. It is expected that around 1 terabyte of data will be downloaded by the mobile user annually. Due to the growing demands in memory & computational capacities of integrated circuits, high processing & transmission data speed along with portability from industries like communications, information, military, space etc. microelectronics will be replaced by nanoelectronics in the near future. The limitation to continuous miniaturization of electronic devices further is the basis for the transition from microelectronics domain to nanoelectronics domain. Nanotechnology is the only alternative to take Moore's law for next 50 years and hence, nanoelectronics is the driving force behind the future next-generation electronics industry.

As discussed by Hao et al. (2020), antennas are the essential element in 5G network system which must support the adaptation to 5G-oriented network transition and have the flexible coordination with other equipments of the intelligent heterogeneous network system. The 5G wireless can work on the range of millimetre waves 30–100 GHz (Panwar et al., 2016). As the bandwidth is inversely proportional to the antenna size, the size of the antennas in 5G communication network is reduced to nanoscale. Patil (2012), Jamthe and Bhande (2017) and Rojas et al.(2017) discussed the application of nanotechnology for developing different set of nanoscale components which are widely used in wireless communication network.

NANOSTRUCTURES AND SCALING OF ELECTRONIC DEVICE SIZE

Dimensions below 100 nm are called nanoscale dimensions (Hanson, 2008). Materials consisting of particles having nanoscale dimensions are called nanomaterials. Nanomaterials show novel physical, chemical, electrical and biological properties not observed at the macro & microscopic levels. A nanodevice is a functional structure having nanoscale dimensions with ability to perform some useful operation. As nanostructure have dimensions comparable to the electron de Broglie wavelength, the behaviour of the electrons in such structures should be described quantum mechanically.

Quantum effect nanoelectronic devices are subdivided in to two broad areas- solid state nanoelectronic devices and molecular electronic devices. The taxonomy of nanoelectronic devices (Dongale et al., 2013) is shown in figure 1.

Development of Single Electron Transistor (SET) partially completed the search as SET having smaller size provides excellent speed with low power dissipation and is considered as very attractive devices for future large scale integration. SET is three terminal (source, drain and gate) devices having a tiny conductive island coupled to a gate electrode and source & drain electrodes are connected to the island through a tunnel barrier. SETs have been used as ultrasensitive electrometer device for precise measurement of electric charges and also it can be used as read-out devices for quantum bits (Schoelkopf et al.,1998). Cavin et al. (2012) discussed the promising application of SET as nano-flash memory.

Resonant tunneling diodes (RTDs) are fabricated from layers of two different III/V semiconductor alloys termed as binary semiconductor (such as the pair of GaAs and AlAs). Resonant tunneling diode (RTD) and resonant tunneling transistor (RTT) are the simplest type of resonant tunneling devices. Successful applications of RT devices are reported in multi-valued logic and in various logic circuits, and in memory circuits (Mazumder et al., 1998). Compañò (2001) emphasized on the potential markets of RTDs which include digital to analogue converters (DACs), clock quantizers, shift registers and ultra-low-power SRAM.

Rapid single-flux-quantum (RSFQ) logic is a digital superconductor circuit technology offering high speed in the GHz regime at feature sizes in the micrometre range. RSFQ has market potential for those applications where Si CMOS could not achieve the same frequency due to the fabrication limitation. Possible application may be in high- speed analogue to digital and digital to analogue conversion (Compañò, 2001).

Figure 1. Taxonomy of quantum effect nanoelectronics devices (Dongale et al., 2013).

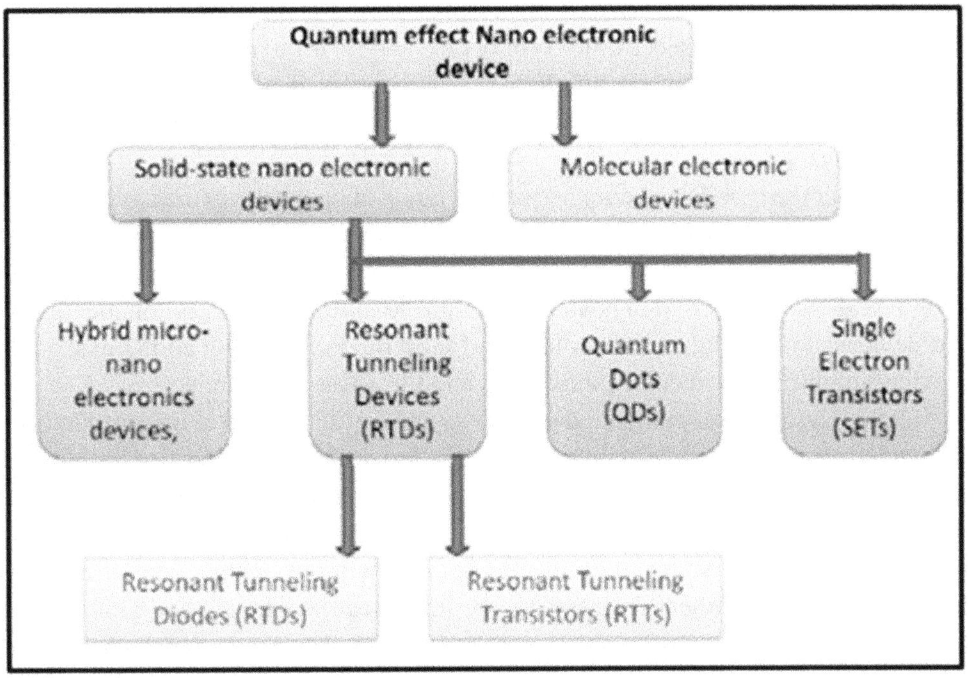

One nanostructure paradigm is Quantum Dots having 3D quantum confinement (QDs). Due to the very small size of QD, by using arrays of QD to implement Boolean logic functions, extremely high packing densities can be achieved with simplified interconnection and extremely low-power delay product.

Spintronics (spin transport electronics or spin-based electronics) is a new technology where electron spin carries the information and it has considerable potential to design new devices that combine standard microelectronics with spin - dependent effects. As discussed by Ohno et al. (2016) spintronics may offer a way for the development of hybrid devices that efficiently combine charged – based information processing with spin-based storage elements.

If the source and drain are ferromagnetic, then spin-polarized electrons might be injected into a semiconductor. Devices of this type have not yet been demonstrated, but current research is examining how to combine ferromagnetic metals and semiconductors, how to inject spin-polarized electrons into the semiconductor, and how to maintain the spin polarization once the electrons are in the semiconductor (Graziosi and Neophytou, 2018). If this type of device could be realized, faster switching with lower switching energy and manipulation of spins of individual electrons (single electron spin transistor) will be possible, which could lead to the realization of quantum computers.

Molecular electronics is the most futuristic area of research with large potentials for reduction of electronic device sizes. Molecular electronics utilizes single molecule or small groups of molecules as components in electronic devices with feature size as small as 1 nm. These new devices can be integrated into complex information processing systems with billions and eventually trillions of parts, all at low cost (Ohno et al., 2016).

NEED OF NANOPARTICLES FOR NEXT-GENERATION TECHNOLOGY

The term ''nanonetworks" refer to electronic components and their interconnection within a single chip at nano-scale (Bjerregaard and Mahadevan, 2006). This concept is also known as Network on Chip (NoC). This term is also referred to as the network-like interconnection of nanomaterials as well, e.g., carbon nanotubes arrays.

Capsulating together the mobile and core network in mode of operation of the nanotechnology as well as perfection in security and its better impact on the sensors make nanotechnology the most significant technology for next generation mobile communication systems. Production of novel materials on the nanometre length scale is expected to play an important role in the field of communication systems for designing new devices of ultra-high-speed for long and short range communication links, power efficient computing devices, high-density memory and logic, and ultra-fast interconnects.

In communication systems, an embedded mobile system with high level of computation and communication when interacting with other human environments such as home, office, public place require an intelligent way to sensing, computing and communicating. Intelligence of this kind of embedded mobile devices strictly requires that the devices must be of limited size with less power consumption, autonomous, robust and can be deployed easily, and survive without explicit care or management. For intelligent mobile systems, the other requirements are intelligent interfacing and interaction with other devices and environments, sensing, context, awareness and increased data rates which require more memory and computing power. All these requirements for an intelligent mobile device cannot be resolved with current technologies and hence, there is an urgent need of next-generation smart technology. The only technology to give a solution for these requirements is nanotechnology as it could provide solutions for sensing, actuation, radio, embedding intelligence into the environment, power efficient computing and memory (Islam and Logeeswaran, 2010). The advent of intelligence and nanotechnology concepts in the mobile devices will assist in embedding the devices inside the human environments that can create a brand new platform to achieve capabilities like self-powering, sensible to the environment or smart interaction with other system. According to Ali (2012), these mobiles will be referred as Nano Devices in 5th generation telecommunications as they are geared up with nanotechnology. Mobile company Nokia together with the University of Cambridge, UK developed a concept called Morph (Ferrari et al., 2015) that demonstrates how future mobile devices might be stretchable and flexible, allowing the user to transform their mobile device into radically different shapes. As stated by Akyildiz and Nie (2018), terahertz band communications are envisioned as a key wireless technology to satisfy the future demands within 5G and beyond as the total consecutive available bandwidth for millimeter waves is not enough to support the terabit - per second data rate. Nanotechnology might be capable of delivering transparent electronics, flexible materials, self-cleaning surfaces and self-powered mobile devices with capabilities of sensing the environment in the terahertz band. Carbon nanotube will be added soon in the cellular phone to enhance its performance. From technology perspective, it is high time to focus on the electronic behaviour of different metal, dielectric single component and double component nanoparticles.

Advancing to the nanoscale electronics is not just a step toward miniaturization, but requires the introduction and consideration of many additional phenomena as compared to microelectronics devices. At the nanoscale, most phenomena and processes are dominated by quantum physics and they exhibit unique behaviour. Advances in design of new nanodevices and progresses in manufacturing of next-generation nanodevices will depend on the ability to model, design & measure the properties and performances of nanodevices over a wide frequency range.

Concept of Optical Signal Manipulation for Wireless Communication

A great deal of work has been focused on improving conventional interconnect technology for on-chip, chip-to-chip, and broad-to-broad communications and novel interconnection techniques using nanomaterials can resolve the interconnect issues. Plasmonics offer the ability to combine the superior technical advantages of photonics and electronics on the same chip. Nanophotonics interconnects can potentially revolutionize future computing and communication by bringing parallelism, immense interconnection bandwidths and wavelength routing capabilities without electrical impedance or crosstalk limitations. Gold nanoparticles have been used as conductor to interconnect the resistors, capacitors and other elements inside an electronic chip.

Electromagnetic antennas, a key enabling technology for devices such as cellular phones, laptops and televisions, are mostly used in radio-wave or micro-wave regime of the electromagnetic spectrum. At optical frequencies, on the contrary, electromagnetic fields are controlled by re-directing the wave fronts of propagating radiation by means of lenses, mirrors, and diffractive elements. Over the past few years, an exciting development in nano-optical research has set in with the recognition that metallic nanoparticles with their ability to tailor surface plasmon resonance wavelengths can be used in a fashion very similar to the radio frequency (RF) antennas which introduces the concept of optical nanoantennas.

Experts in telecommunication and computer engineering have already envisioned the interconnection of nanodevices in nanonetworks and their wide applications in different fields. It is likely that optical nanoantennas will have a broad impact on future technology as because nanoscale devices need to interface with optical radiation for optimized energy transfer and also for communicating the information signals. Optical nanoantennas provide a promising pathway toward advanced manipulation of light waves, such as directional scattering, polarization conversion, emission of light and plasmonic-assisted fluorescence enhancement. Due to tailorable resonating optical response of nanostructured optical antennas, they can enable extreme control of their scattering strength and directionality depending on the properties of the exciting optical wave, such as frequency, polarization and direction. The capabilities of light manipulation offered by nanoantennas has important applications in areas including microscopy, spectroscopy, photodetection, sensing, quantum computing, nano-imaging, on-chip optical communication, ultra-compact polarized demultiplexing etc. Optical nanoantennas make processes more efficient by increasing the specificity of gathered information. New ideas and developments are emerging at a rapid pace and it is clear that optical nanoantenna concept will provide new opportunities for optoelectronic architectures and devices. In a 5G network identification of best materials for nanoantennas is a challenge since each nanoantenna element acts as both a transmitter and a receiver (Liu et al., 2012). As discussed by Greffet (2005) and Devi et al. (2018) metallic nanoparticles can be used as nanoantennas in 5G network.

The passive lumped circuit elements e.g. resistors (R), inductors(L) and capacitors (C), which are the basic building blocks of an electronic circuit, need to be scaled down to the optical frequencies to obtain high density, high speed analogues of the traditional circuits.

We present here the nano scale resistor, capacitor and inductor concept in the optical frequency and is given below.

Let us consider a nanosphere of radius R made of a homogeneous nonmagnetic material with dielectric constant $\varepsilon(\omega)$. The sphere is assumed to be much smaller than the wavelength of operation (λ) in a particular medium. If the medium is vacuum, the radius R $\ll \lambda_0, \lambda_0$ .eing the wavelength in vacuum, and if the medium is any other material R $\ll \lambda_0 / \sqrt{|R_e(\varepsilon)| / \varepsilon_0}$ .where $|R_e(\varepsilon)|$ .is the magnitude of real part of permittivity in that medium & ε_0 .is the permittivity in vacuum. Illuminating the sphere with an optical signal / by an incident electromagnetic wave E_0 .under a monochromatic optical excitation $e^{-i\omega t}$. as shown in figure2,

Figure 2. Illumination of a nanosphere of permittivity $\varepsilon(\omega)$.with an optical signal.

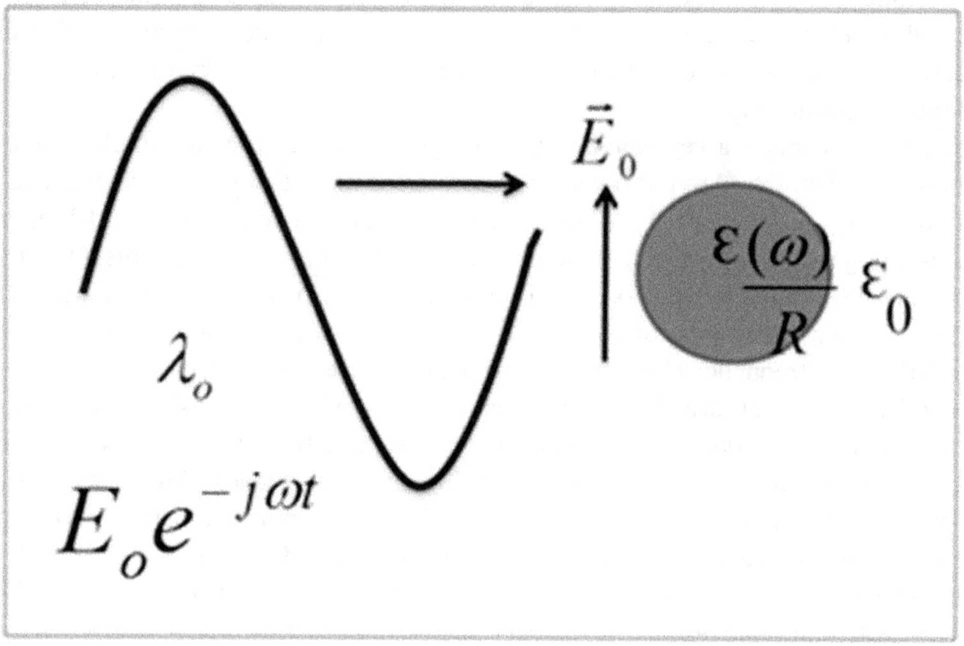

the total fields and scattered electromagnetic far field inside and outside the sphere respectively, is estimated by solving the Maxwell equations using the time-harmonic, quasi static approach (Engheta et al., 2005) and is given below.

$$\overrightarrow{E_{int}} \cdot = 3\varepsilon_0 \overrightarrow{E_0} / \left(\varepsilon + 2\varepsilon_0\right) \cdot \quad (1)$$

$$\overrightarrow{E_{ext}} \cdot = \overrightarrow{E_0} \cdot + \left[3\vec{u}\left(\vec{p}.\vec{u}\right) - \vec{p}\right] / 4\pi\varepsilon_0 r^3 . \quad (2)$$

with $\vec{p} .= 4\pi\varepsilon_0 R^3 \left(\varepsilon - \varepsilon_0\right) \vec{E}_0 / \left(\varepsilon + 2\varepsilon_0\right)$. $\vec{u} .\vec{r} / r$. $\vec{r}$ .being the position vector from the sphere's centre to the point of observation and $r = \left|\vec{r}\right|$. The second term of equation (2) represents displacement of electric field and is denoted by $\vec{E}_{dip}$ ..On the surface of the sphere, at every point, the normal component of the displacement $-i\omega D_n$ .is continuous, implying that

$$-i\omega\left(\varepsilon - \varepsilon_0\right) \vec{E}_0 . \hat{n} .= -i\omega\varepsilon_0 \vec{E}_{dip} . \hat{n} . + i\omega\varepsilon \vec{E}_{res} . \hat{n} . \quad (3)$$

Here, $\vec{E}_{res}$. $\vec{E}_{dip} - \vec{E}_0$ .epresents the residual field internal to the nanosphere and $\hat{n}$. is the local outward unit vector normal to the surface of the sphere.

If $\vec{E}_0$ .s oriented as shown in figure1, we get the total displacement current, integrating equation (3) over the hemispherical surface as

$$-i\omega(\varepsilon-\varepsilon_0)\pi R^2\left|\overrightarrow{E_0}\right| . \frac{-i\omega\varepsilon\pi R^2(\varepsilon-\varepsilon_0)\left|\overrightarrow{E_0}\right|}{(\varepsilon+2\varepsilon_0)} - \frac{i\omega\varepsilon_0 2\pi R^2(\varepsilon-\varepsilon_0)\left|\overrightarrow{E_0}\right|}{(\varepsilon+2\varepsilon_0)} . \tag{4}$$

In equation (4) the left hand term is the impressed displacement current source (I_{imp} ., the first term of right hand side is the displacement current circulating in the nanosphere (I_{sph} ., and the second term of right hand side is the fringe dipolar displacement current (I_{fringe}), .espectively and all are related to the surface polarization charges of the nanosphere induced by the excitation (Bharadwaj, 2009). Equation (4) can be interpreted as branch currents at a node in a parallel circuit as shown in figure3. These currents of equation (4) obey Kirchoff's current law and Kirchoff's voltage law is also satisfied by these currents since in quasistatic approximation, $\nabla \times E$ .is locally near zero.

Figure 3. A basic nanoscale circuit in the optical regime. Left: a nonplasmonic sphere with ε .0, which provides a nanocapacitor and a nanoresistor; right: a plasmonic sphere with ε .0, which gives a nanoinductor and a nanoresistor. Solid black arrows show the incident electric field, and the thinner field lines together with the gray arrows represent the fringe dipolar electric field from the nanosphere (Engheta et al., 2005).

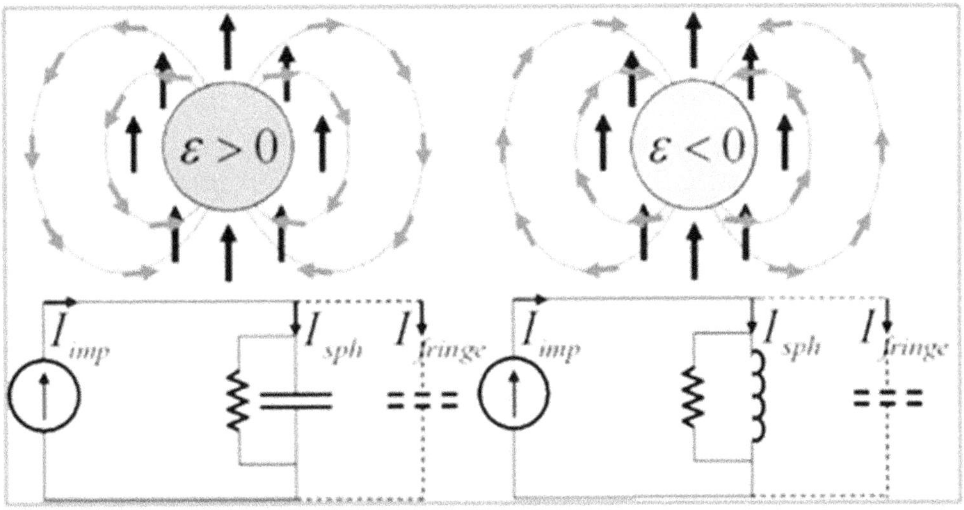

Now, the average potential difference (due to $\overrightarrow{E}_{res}$. between the upper and lower hemispherical surfaces of the sphere is

$$V_{sph} \cdot = V_{fringe} \cdot = \frac{R(\varepsilon - \varepsilon_0)\left|\overrightarrow{E_0}\right|}{\varepsilon + 2\varepsilon_0} .5)$$

Therefore, from (4) & (5) equivalent impedance of the nanosphere is

$$Z_{sph} = \frac{R(\varepsilon - \varepsilon_0)\left|\overrightarrow{E_0}\right| \Big/ (\varepsilon + 2\varepsilon_0)}{-i\omega\varepsilon(\varepsilon - \varepsilon_0)\pi R^2 \left|\overrightarrow{E_0}\right| \Big/ (\varepsilon + 2\varepsilon_0)} .= (i\omega\varepsilon\pi R)^{-1} . \qquad (6)$$

and, equivalent impedance of the fringe branch is

$$Z_{fringe} = \frac{R(\varepsilon - \varepsilon_0)\left|\overrightarrow{E_0}\right| \Big/ (\varepsilon + 2\varepsilon_0)}{-i\omega\varepsilon_0(\varepsilon - \varepsilon_0)2\pi R^2 \left|\overrightarrow{E_0}\right| \Big/ (\varepsilon + 2\varepsilon_0)} = (-i\omega 2\pi R\varepsilon_0)^{-1} . \qquad (7)$$

Since relative permittivity of the nanosphere ε .is a complex quantity, it can be written as $\varepsilon = \varepsilon_1 + i * \varepsilon_2$ .where ε_1 .is the real part and ε_2 .s the imaginary part of the relative permittivity, respectively. Thus, equation (6) can be written as

$$Z_{sph} = -\frac{1}{i\omega\pi R(\varepsilon_1 + i * \varepsilon_2)} . = \frac{1}{\omega\pi R\varepsilon_2 - i\omega\pi R\varepsilon_1} . = \frac{1}{\omega\pi R\left(\varepsilon_2 + \frac{\varepsilon_1^2}{\varepsilon_2}\right)} + i\frac{1}{\omega\pi R\left(\varepsilon_1 + \frac{\varepsilon_2^2}{\varepsilon_1}\right)} . \qquad (8)$$

For plasmonic nanosphere (such as Au, Ag) , ε_1 .is $-ve$. $\varepsilon_2 \ll |\varepsilon_1|$. Thus, in equation (8) $\left(\varepsilon_1 + \frac{\varepsilon_2^2}{\varepsilon_1}\right) \approx \varepsilon_1$ .and the reactive part becomes

$Z_{sph} = \frac{i}{\omega\pi R\varepsilon_1}$., which is *–ve.*

Again from the typical expression for impedance (Shi et al, 2014), $Z = R - i\left(L\omega - \frac{1}{C\omega}\right)$ .where R, L & C is the resistance, inductance & capacitance and hence, the reactive part for inductance is $Z = -iL\omega$.

Therefore, $L_{sph} = \frac{-1}{\omega^2 \pi R \varepsilon_1}$. (9)

& $Z_{fringe} = \frac{-1}{i\omega 2\pi R\varepsilon_0}$. .s $+ve$ since ε_0 .is $+ve$ .and Z_{fringe} .s capacitive.

Thus, $Z_{fringe} = \frac{i}{C\omega}$ .Therefore, $C_{fringe} = 2\pi R\varepsilon_0$. (10)

For non-plasmonic (dielectric/semiconductor) nanosphere ε_1 .is $+ve$. $\varepsilon_2 \ll \varepsilon_1$. Thus, in equation (8), $\left(\varepsilon_1 + \frac{\varepsilon_2^2}{\varepsilon_1}\right) \approx \varepsilon_1$ .and the reactive part becomes

$Z_{sph} = \frac{i}{\omega \pi R \varepsilon_1}$. , is $+ve$ and hence capacitive.

$= \frac{i}{\omega C_{sph}}$ .Therefore, $C_{sph} = \pi R \varepsilon_1$. (11)

and, $C_{fringe} = 2\pi R\varepsilon_0$. (12)

It is seen from (11) & (12) that there are two capacitive elements for non-plasmonic nanosphere and hence, no resonance is present in this case which is consistent with observed behaviour of non-plasmonic nanosphere interacting with optical wave. The first term in equation (8) gives resistive part of impedance for both plasmonic & non-plasmonic spheres.

In case of the metal nanosphere, since there is an inductor in parallel to the fringe capacitor, the circuit may exhibit resonance which corresponds to surface plasmon resonance.

Using (9) & (10), we get

$$L_{sph} \times C_{fringe}. = \frac{-1}{\omega^2 \pi R \varepsilon_1} \times 2\pi R\varepsilon_0. = -\frac{2\varepsilon_0}{\omega^2 \varepsilon_1}. \quad (13)$$

As for an LCR circuit, condition for resonance is

$$L \times C = \frac{1}{\omega^2}. \quad (14)$$

To achieve resonance, from (13) using (14)

$$\varepsilon_1 = -\frac{2\varepsilon_0}{\omega^2 \times \frac{1}{\omega^2}}. = -2\varepsilon_0. \tag{15}$$

which is the well- known condition of plasmonic resonance for a nanosphere.

Thus, a small nanosphere excited by an optical signal may effectively behave as a nanocapacitor or a nanoinductor at optical frequencies if the sphere is made of non-metallic (dielectric or semiconductor) or metallic nanparticles, respectively. The imaginary part of relative complex permittivity provides an equivalent nanoresistor.

It is well known that at lower frequencies, an inductor requires the form of "wound wires". But at optical frequencies (i.e. higher) plasmonic characteristics of natural noble metals provide us with an effective intrinsic inductance which can be controlled by size, shape and material of the nanostructures.

These concepts i.e. theoretical findings leads to new possibilities for miniaturization of electrical circuits at optical frequencies using nanoparticles as lumped circuit elements. Small colloidal metallic and dielectric nanoparticles synthesized by conventional chemical process are the ideal platforms to translate these concepts of lumped element to optical frequencies.

Thus, we get three basic circuit elements of nanodimensions – nanoinductors, nanocapacitors & nanoresistors (shown in figure 4) in the optical frequencies and these are the building blocks for designing more complex circuits at optical wavelengths.

Figure 4. Sub wavelength nanoparticles as lumped nanocircuit elements at optical frequencies (N. Engheta, 2007).

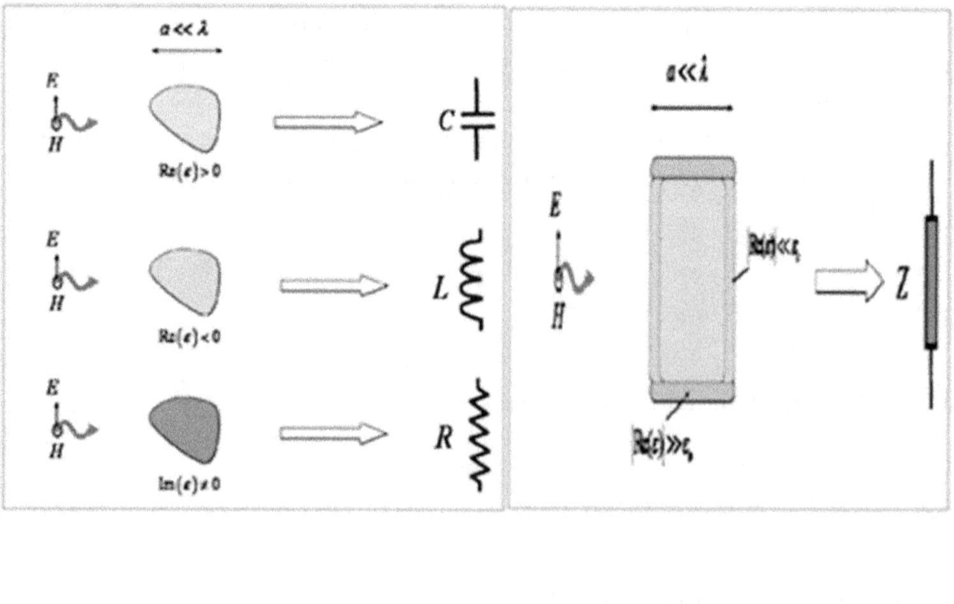

Expanding these concepts, we consider a configuration consisting of more than one nanoparticle. Considering two nanospheres with different radius say, R_1 & R_2 .ade of different materials having permittivities $\varepsilon_1 . \varepsilon_2$ .respectively and placing it a certain distance d .part, coupled nanocircuits has been configured as shown in figure 4. To form series and parallel circuit elements with these nanoparticles, two or more numbers of these nanoparticles are placed closely together in specific orientations and are excited by an electric field. The left circuit in figure 4 can be treated as a parallel resonant L-C circuit and the right circuit can be treated as series resonant L-C circuit. The resonant behavior is achieved due to the use of oppositely signed permittivity with equal magnitudes for both halves.

Figure 5. Parallel and series nanoelements. Top: two fused lossless semi cylinders with positive and negative permittivities, illuminated by an optical field; bottom: equivalent circuits showing parallel and series elements representing the fused structure as seen from the outside.

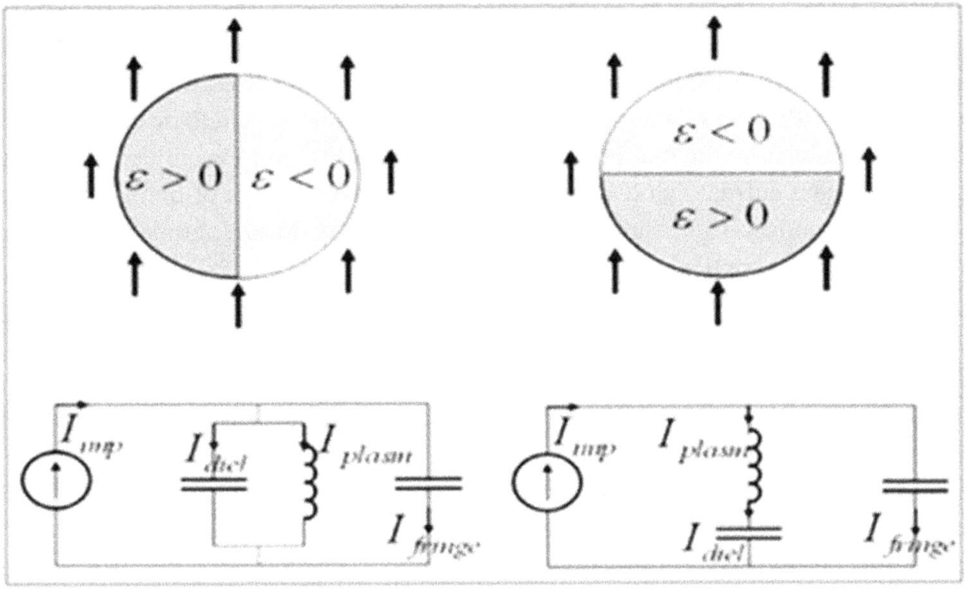

Figure 6. Top: conceptual nanoscale circuit formed by rectangular blocks of plasmonic and nonplasmonic segments; bottom: its equivalent circuit.

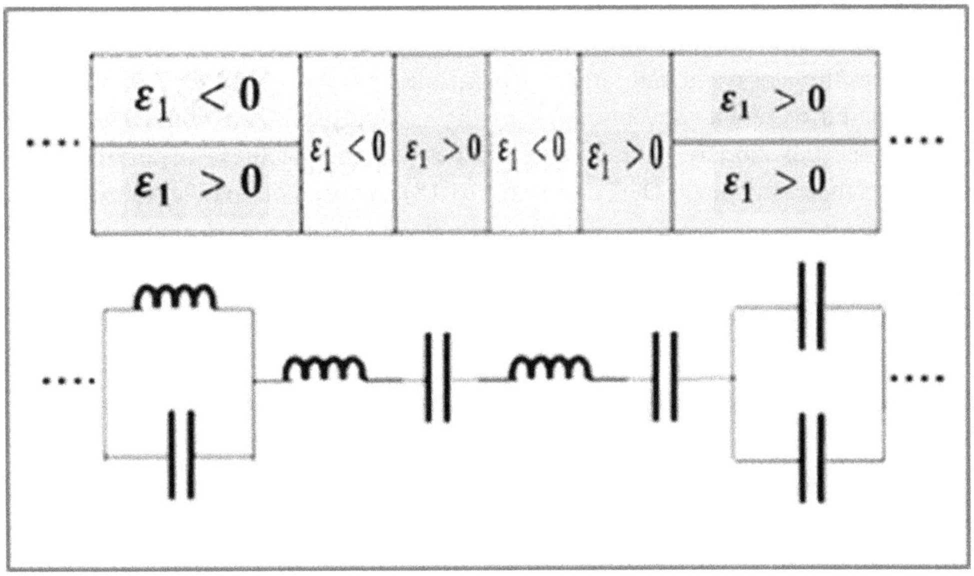

Depending on the pairing configuration and orientation with the external excitation different pairs would behave as nonresonant series or parallel elements. Apart from nanosphere, other nanostructures such as rectangular, triangular etc. can also be paired up to act as series and parallel nanocircuits and a conceptually formed rectangular nanostructure based circuit is shown in figure 6. Such nanoscale circuits will be applicable in nanobarcodes and in plasmonic data storage systems.

By properly arranging nanoscale circuit elements, optical nanotransmission line can also be designed.

DISCUSSION

In standard RF circuits, nonconducting insulating materials such as dielectric or air are used to insulate the individual lumped elements from each other and conducting wires are used at the terminal of lumped elements to connect it with other elements. Thus, usually there is no reason to be concerned about leakage of conduction current from the middle of the lumped elements. But for lumped elements at optical domain, the displacement current can leak into and out of different parts of the nanostructures which can lead to coupling among various tightly packed nanoelements. To strengthen the analogy between conventional RF circuits and optical nanocircuits, use of additional material with proper values of permittivity around the nanoparticle is required. To obtain the negative permittivity values, the electrically negative metamaterials use the thin metallic wires and the magnetically negative metamaterials can have a structure of split ring resonator, which is composed of two concentric metallic rings and separated by a gap (Zheludev and Kivshar, 2012). By altering the shape, size, and orientation of individual metamaterial resonators or by manipulating the near-field interactions between them, tunability of electromagnetic characteristics of metamaterials is achieved which may be applied as nanoelements. To prevent leakage and allow negligibly small displacement current, the layer must act as insulators and thus, the layer should be of relative permittivity with real part very small i.e. epsilon-near-zero (ENZ) materials. To allow very easy flow of displacement current without introducing a strong electric field, layer materials must be of high real part of relative permittivity i.e. epsilon-very-large ((EVL) materials so as to act as nanoscale connector for the optical displacement current analogous to a conducting metallic wire in the RF domains. Thus the ENZ and EVL materials play the insulator and conductor role at optical domain analogous to insulator and conductor of RF domain.

Graphene, Carbon NanoTubes, metallic nanomaterials, and metamaterials are the most promising materials used for millimeter wave and terahertz band frequency which is expected to utilize smaller chip area as compared to other conventional metallic counterparts (Dash and Patnaik, 2018). The electromagnetic properties namely, permittivity and permeability of these materials are tunable and highly dependent on their structures (Devi and Datta, 2018). A speed of up to terabits per second can be achieved in the terahertz frequency range by using graphene – based nanostructures (Dash & Patnaik, 2020) which is envisioned as a key wireless technology to satisfy the future demands within 5G wireless network and even beyond.

FUTURE RESEARCH DIRECTIONS

The need for development of nanostructures based fifth-generation wireless communication network is the highly discussed area of research in engineering due to rapidly increasing number of mobile devices, voluminous data, and requirement of higher data rate. The 5G Wireless Communication Technology has the bright future because of the high bit rate, multiple data transfer path, no traffic at the time of accessing the data. The 5G wireless can work on very high frequencies and incorporating nanostructures as different lumped circuit elements in the 5G technology make it smart and intelligent offering tremendous data capabilities, unrestricted call volumes, and infinite data broadcast. While storing the information in a wireless communication network, one important parameter to be considered is frequency allocation. 5G is considered as next-generation mobile communication technologies beyond 2020(). This chapter emphasis on the application of graphene, single component (metal) and core/shell metal/dielectric nanoparticles as 5G circuit components which need more research. Due to the advance switch and router technology, high connectivity is expected to be provided by 5G Network. Experiment on global utilization of 5G network is going on and is expected to be a new revolution in mobile tele-phone market. The present discussion opens up the prospects of nanobased fifth-generation technology for application in mobile communication reducing the data traffic problem.

CONCLUSION

A composite nanostructure i.e. a nanoparticle with a thin layer of epsilon-near-zero material or with a thin layer of epsilon-very-large material can act approximately as a modularized lumped nanoelement at optical frequencies depending on the permittivity of the core nanoparticle.

Nano shell particle with dielectric core and plasmonic shell resembles RLC resonant circuit at optical frequencies. These lumped elements concept of nanoparticles based on their permittivity at optical frequencies indeed form a new paradigm of nanoelectronic devices that is analogous to RF circuits and systems.

Plasmonic nanoparticle and core/shell nanoparticle constitutes of dielectric core and metallic shell result in a plasmon derived optical resonance which, make it suitable for efficient coupling of electromagnetic radiation into nanodevices for a board range of applications such as communication, quantum optics, information processing and in several linear and non-linear next-generation wireless communication processes. Graphene based nanostructures can be used to achieve a speed of up to terabits per second in the terahertz band which is one of the key feature of fifth-generation wireless communication network.

ACKNOWLEDGMENT

This research was supported by the Department of Science and Technology, Government of India [grant number SR/WOS-A/ET-1102/2015].

REFERENCES

Akyildiz, I. F., Han, C., & Nie, S. (2018). Combating the distance problem in the millimeter wave and Terahertz frequency band. *IEEE Communications Magazine*, *56*, 102–108.

Ali, E., Ismail, M., Nordin, R., & Abdulah, N. F. (2017). Beamforming techniques for massive MIMO systems in 5G: Overview, classification, and trends for future research. *Front Inform Tech EL, 18*, 753–772.

Bharadwaj, P., Deutsch, B., & Novotny, L. (2009). Optical antennas. *Advances in Optics and Photonics, 1*, 438–483.

Bjerregaard, T., & Mahadevan, S. (2006). A survey of research and practices of network-on-chip. *ACM Computing Surveys, 38*, 1–50.

Cavin, R. K., Lugli, P., & Zhirnov, V. V. (2012). Science and Engineering Beyond Moore's Law. *Proceedings of the IEEE, 100*, 1720–1749.

Compañó, R. (2001). Trends in Nanoelectronics. *Nanotechnology, 12*, 85.

Dash, S., & Patnaik, A. (2018). Material selection for THz antennas. *Microwave and Optical Technology Letters, 60*, 1183–1187.

Dash, S., & Patnaik, A. (2020). Behavior of Graphene Based Planar Antenna at Microwave and Terahertz Frequency, Photonics and Nanostructures - Fundamentals and Applications. *Photonics*, 100800.

Devi, J., & Datta, P. (2018). Yagi–Udananoantenna For NIR domain. *Journal of Computational Electronics, 17*, 406–418.

Devi, J., Datta, P. (n.d.). A Study on Dielectric Properties of Cadmium Sulfide–Zinc Sulfide Core–Shell Nanocomposites for Application as Nanoelectronic Filter Component in the Microwave Domain. *Journal of Elec Materi, 47*, 3529–3542.

Dongale, T. D., Magdum, S. K., Goilkar, K. Y., Chougule, N. A., & Ghatage, S. R. (2013). Shifting towards Nanoelectronics: A device level overview. *Journal of Academia and Industrial Research, 1*, 467–471.

Engheta, N., Salandrino, A., & Alu, A. (2005). Circuit elements at optical frequencies: Nano-inductors, nanocapacitors and nano-resistors. *Physical Review Letters, 95*, 095504.

Ferrari, A. C., & … . (2015). Science and technology roadmap for graphene, related two –dimensional crystals, and hybrid systems. *Nanoscale, 7*, 4598–4810.

Graziosi, P., & Neophytou, N. (2018). Simulation study of ballistic spin-MOSFET devices with ferromagnetic channels based on some Heusler and oxide compounds. *Journal of Applied Physics, 123*, 084503.

Greffet, J. J. (2005). Nanoantennas for light emission. *Science, 308*, 1561–153.

Hanson, G. W. (2008). *Fundamentals of Nanoelectronics* (1st ed.). Prentice Hall.

Hao, H., Hui, D., & Lau, D. (2020). Material advancement in technological development for the 5G wireless communications. *Nanotechnology Reviews, 9*, 683–699.

Islam, M. S., & Logeeswaran, V. J. (2010). Nanoscale materials and devices for future communication networks. *Communications Magazine, IEEE, 48*, 112–120.

Jamthe, D. V., & Bhande, S. A. (2017). Nanotechnology in 5G wireless communication network – an approach. *Int Res J Eng Tech., 4*, 58–61.

Liu, D. Q., Zhang, M., Luo, H. J., Wen, H. L., & Wang, J. (2018). Dual – band platform - free PIFA for 5G MIMO application of mobile devices. *IEEE Transactions on Antennas and Propagation*, *66*, 6328–6333.

Mazumder, P., Kulkarni, S., Bhattacharya, M., Sun, J. P., & Haddad, G. I. (1998). Digital Circuit Applications of Resonant Tunneling Devices. *Proceedings of the IEEE*, *86*, 664–686.

Moore, G. E. (1965). Cramming more components onto integrated circuits. *Electronics Magazine*, *38*, 4.

Ohno, H., Stiles, M. D., & Dieny, B. (2016). Spintronics. *Proc IEEE Inst Electr Electron Eng.*, *104*, 1782–1786.

Panwar, N., Sharma, S., & Singh, A. K. (2016). A survey on 5G: The next generation of mobile communication. *Physical Communication*, *18*, 64–84.

Patil, G. R., & Wankhade, P. S. (2014). 5G wireless technology. *Int J Comput Sci Mobile Comput*, *3*, 203–207.

Powell, J. R. (2008). The quantum limit to Moore's law. *Proceedings of the IEEE*, *96*, 1247–1248.

Rojas, J. P., Singh, D., Inayat, S. B., Sevilla, G. T., Fahad, H. M., & Hussain, M. M. (2017). Review – micro and nano –engineering enabled new generation of thermoelectric generator devices and applications. *ECS Journal of Solid State Science and Technology: JSS*, *6*, 3036–3044.

Schaller, R. R. (1997). Moore's Law: Past, Present, and Future. *IEEE Spectrum*, *34*(6), 52–59. doi:10.1109/6.591665

Schoelkopf, R. J., Wahlgren, P., Kozhevnikov, A. A., Delsing, P. D., & Prober, E. (1998). The Radio-Frequency Single-Electron Transistor (RF-SET): A Fast and Ultrasensitive Electrometer. *Science*, *280*, 1238–1242.

Wu, J., Shen, Y., Reinhardt, K., Szu, H., & Dong, B. (2013). A Nanotechnology Enhancement to Moore's Law. *Applied Computational Intelligence and Soft Computing*, 1–14.

Zheludev, N. I., & Kivshar, Y. S. (2012). From metamaterials to metadevices. *Nature Materials*, *11*, 917–924.

Zhirnov, V. V., Cavin, R. K., Hutchby, J. A., & Bourianoff, G. I. (2003). Limits to Binary Logic Switch Scaling—A Gedanken Model. *Proceedings of the IEEE*, *91*, 1934–1939.

Chapter 3
5G and Unmanned Aerial Vehicles (UAVs) Use Cases:
Analysis of the Ecosystem, Architecture, and Applications

Georgios Makropoulos
NCSR "Demokritos", Greece

Harilaos Koumaras
https://orcid.org/0000-0003-0486-2700
NCSR "Demokritos", Greece

Fotini Setaki
Cosmote S.A., Greece

Konstantinos Filis
Cosmote S.A., Greece

Thomas Lutz
Frequentis, Austria

Pawel Montowtt
Involi, Switzerland

Lechoslaw Tomaszewski
Orange Polska S.A., Poland

Piotr Dybiec
DroneRadar, Poland

Tanel Järvet
CafaTech, Estonia

ABSTRACT

In the past few years, UAVs have evolved considerably towards real-world applications, going beyond entertaining activities. The unique features that UAVs can provide, such as flexible and easy deployment, high mobility, and mostly autonomous operation, tend to be appealing solutions in a wide variety of applications. With the number of UAVs expected to increase significantly in the forthcoming years, it is deemed mandatory to integrate innovative features of communication technologies in order to tackle several limitations and challenges that exist to date. To this end, the amalgamation of UAVs and 5G cellular networks is an auspicious solution in order to realise the full potential that the UAVs can provide as well as to effectively open up new fields of applications. The unique features that 5G cellular networks offer are utilised for supporting and providing reliable and secure UAV command and control. This chapter provides an overview of the 5G technology to support communication services for the UAV ecosystem.

DOI: 10.4018/978-1-7998-6992-4.ch003

INTRODUCTION

Unmanned Aerial Vehicles (UAVs) also commercially known as drones, are aircrafts which can either be remotely operated by a person or autonomously navigated by pre-programmed operation systems. The past few years there is a rapid increase on the use of UAVs that have become invaluable assets for end-users, enterprises, and governments in a wide range of applications. The fact that the global market for unmanned aerial vehicles (UAVs) is estimated at USD 19.3 billion in 2019 and is projected to reach USD 45.8 billion by 2026, reflects the significant investments into the UAVs' space (Godage, 2019).

The recent technological advances in the UA ecosystem have paved the way for an ever-increasing number of use cases in various vertical industries concerning emergency response, healthcare, agriculture and more. However, the technology has yet to be fully exploited. One of the limiting factors is the inability to have direct control of a UAV, beyond the range of the inherent communication channel. This factor restricts the services and capabilities available for missions that require beyond visual line of sight (BVLOS) operation.

In addition, it is off the essence to develop and integrate features and technologies, not only towards communication and control, but also in respect to high-capacity information transmission, that will enable UAVs to execute more complex and sophisticated missions.

The deployment of fifth generation wireless technology envisions to achieve a wider variety of objectives in terms of higher multi Gbps data speeds, ultra-low latency, advanced reliability, increased network capacity and availability, as well as greater bandwidth and throughput. In turn these innovative characteristics are being considered as ideal enablers for enhanced UAVs' capabilities towards cellular connected UAVs' communications and UAVs' assisted communications.

In the light of the above the goal of this chapter is to advocate the optimal exploitation of the 5G technology by the UAV industry, for the provision of enriched and niche UAV services and applications. Towards this direction, section 2 introduces the current 5G and UAV ecosystem, section 3 provides a background analysis on the key points of 5G and UAV technology as well as the specifications towards the amalgamation of the two domains. Section 4 details a proposed system architecture for an end-to-end integrated platform to automate UAV trials on 5G infrastructures, and section 5 presents and analyses the overall regulatory framework, including UAV flights and 5G spectrum policies. Key observations and future directions as well as concluding remarks are presented in section 6 and 7 accordingly.

BACKGROUND

5G Ecosystem

5G, the cornerstone of future communications, offers ubiquitous, super-fast, reduced-latency connectivity and seamless service delivery in all circumstances, paving the way for revolutionary applications, such as autonomous and remote driving, drone-operated logistics, remote real-time surgeries and 4D holographic virtual reality. These do not come with a simple upgrade from the previous LTE generation, and mandate radical transformation in all levels, in the access, transport and core networks. New Radio specifications and convergence of telecom & IT technologies inaugurate elastic, programmable network deployments and on-demand expansions that are fully adaptable to customer demands. In brief impressive advancements are introduced:

- **Flexible and modular architecture** with technologies such as Network Functions Virtualization (NFV), Software Defined Network (SDN), Network Slicing, and Cloud RAN.
- **5G Service Based Architecture (SBA)**, using HTTP-based APIs, incorporating principles such as modularity, reusability, and self-containment of network functions, while simplifying integration with third-party software.
- **MEC (Multi Access Edge Computing)** deployments distributing core functions at the edge and close to the end-user that together with the New Radio capabilities, achieve low latency performance.
- **Support for many frequency bands,** licensed and unlicensed, including existing cellular bands and frequencies above 6 GHz (i.e. mmWave). Low bands are used for coverage, mid-band frequencies for a blend of coverage and capacity, and mmWave bands for extremely high capacity but with restricted coverage.
- **Massive Multiple-Input and Multiple-Output (mMIMO)** antennas and beamforming techniques offering increased capacity and data throughput. mMIMO utilizes large antenna arrays to send and receive more data simultaneously. The benefit to users is that more people can simultaneously connect to the network and maintain high throughput. Beamforming technique can control the direction of the antenna signal towards a specific user providing a stronger signal and reducing energy consumption and interference. It can be regarded as a subset of mMIMO.
- **Mass deployment of small cells** towards network densification for the provision of high data rates both in indoor & outdoor environments.
- **Device-to-Device (D2D)** communication allowing devices to communicate with each other directly, with minimal involvement of the network, supporting the development of edge-centric applications.

With such capabilities, 5G does not only offer remarkable throughput and latency performance indicators, but promises to provide a set of premium service classes in a dedicated, guaranteed and extremely reliable manner, such as:

- **Enhanced Mobile Broadband (eMBB)** addressing bandwidth intensive services and applications, including Augmented Reality (AR)/Virtual Reality (VR), Ultra-High Definition (UHD) Video sharing in heavily crowded hotspots, TV programs broadcasting, etc. The bandwidth requirements of this type of services are expected to be on average about 100 Mbps per user and can reach up to 10 Gbps (for broadcast services).
- **Massive Machine Type Communications (mMTC)** referring to massive IoT services extending the Long-Term Evolution (LTE) IoT (Narrow Band–IoT) to support huge numbers of devices with lower costs, enhanced coverage, and long battery life. With regard to ITU objectives, 5G will support ten times as many devices per area as LTE. This category includes eHealth, wearables, industrial control and factory automation, and sensor networks.
- **Ultra-Reliable and Low-Latency Communications (URLLC)** focusing to latency sensitive, wireless applications and services, some of which are impossible to be supported by existing network deployments. These services are referred also as "mission-critical" communications or critical Machine-Type Communications (cMTC), and include public safety lifeline and situational awareness, industrial automation, drone control, new medical applications, autonomous vehicles, etc. The latency requirements for this type of services are expected to range between 1ms-2ms for the radio interface and less than 10ms for the end-to-end data plane.

At the end of Q3 2020, 5G was commercially available from 107 operators in 47 markets worldwide, and the forecast for 2025 considers 1.7 billion connections, and an adoption rate of 19.5% (Global 5G Landscape, 2020). While 5G has appeared on the mass market with consumer facing offers, such as eMBB for Cloud Gaming not possible with 4G, it is the industrial applications that target to reveal the true potentials of the technology by exploring URLLC and mMTC capabilities. The 5G Infrastructure Public Private Partnership (5G PPP), the European initiative that is actively promoting 5G innovative solutions per industry, or also called "'Vertical", has launched, through 5GPP projects funded by the European Union, more than 18- trials across 28 member states (5GPPP, 2020), in order to validate, evaluate and promote feasible business potentials for different vertical industries. In the process 5G PPP has revealed, as early as 2016, the extended map of involved stakeholder categories, graphically depicted in Figure 1, and is currently working on attaching priorities in each group.

Figure 1. 5G PPP projects target stakeholders – created by the Euro-5G project with contributions from the 5G PPP Phase 1 projects Source: 5GPPP

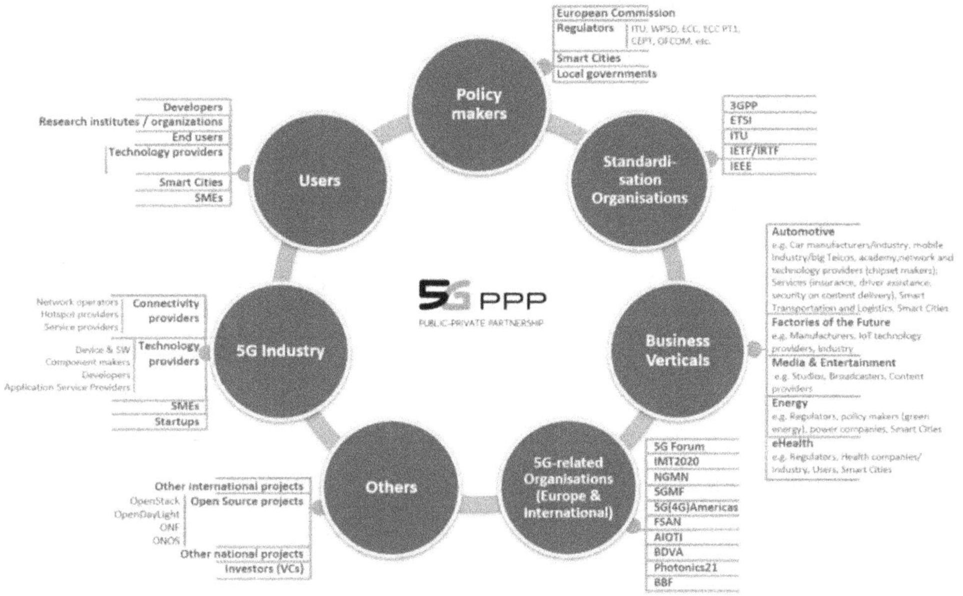

As shown in the Figure above, the 5G adopters include the mobile subscribers, who are mainly interested in voice and data communications, but primarily consider the business verticals that benefit significantly from the global digital transformation that is accelerated by 5G. In vertical sectors such as automotive, transportation, media and entertainment, smart cities, factories of the future, energy, public safety, (air)ports, tourism, agri-food, and e-Health innovative use cases are identified and realized through the advanced features of 5G and are expected to become the revenue source to justify the significant investments necessary. In such an ecosystem, various business models are expected to arise, and 5G PPP assesses through its trials the viability of 5G infrastructure investments, either in global or localized opportunities (ex. Highways), and the foreseen role of private, public stakeholders, third-parties and SMEs.

Moving from the existing network-oriented provisioning towards the 5G application-driven, dedicated service instantiation, the technical activities of 5G become interrelated with the analysis of the vertical stakeholders and their service requirements. A number of important 5G enablers are identified as part of the 5G PPP Verticals stakeholders' analysis (5GPPP, 2020), to be included for relevant verification and demonstration as part of the 5GPPP projects' trials, as depicted in Table 1. It is noteworthy that the 5G New Radio (NR) interface and the network slicing functionality, are the prevailing enablers across the majority of vertical categories.

Table 1. Demonstrated and Planned 5G Functionalities in Verticals

5G Features	Automotive	Transportation	Media	Smart City	Healthcare	Smart Factories	Public Safety	(Air) Ports	Tourism	Agri-food
Network Slicing	x	x	x	x	x	x	x	x	x	x
Mobile Edge Computing	x	x	x	x	x			x	x	x
Functional Split in RAN		x	x							
Advanced Security	x			x				x		
Smart network management			x	x	x		x	x		
Location Services & Context Awareness	x	x	x		x	x		x		
5G NR Capabilities	x	x	x	x	x	x	x	x	x	x
Softwarization	x		x		x			x		
Service chaining		x	x		x					
Traffic steering		x								
Spectrum coverage	x	x	x							
Guaranteed QoS	x	x		x	x	x	x			

From the verticals' perspective, automotive and transportation (including cars, trucks, buses, trains, and drones) are the two verticals benefitting the most from the new functionalities that 5G brings. In the very near future vehicles will have to communicate with other vehicles (V2V), with the surrounding infrastructure (V2I), with pedestrians (V2P), with networks (V2N), and the list will keep growing. Even though today LTE-based C-V2X (cellular vehicle to anything) communication is available, the C-V2X evolution to 5G will offer the superior performance required to support connected and automated vehicle communications, resulting in less congestion, reduced emissions, and a smoother and safer driving experience (Mondal et al., 2020). Among the new functionalities, the high-bandwidth and low-latency connectivity offered by 5G are key to the Cooperative, Connected and Automated Mobility (CCAM) of future vehicles (Hetzer et al., 2019). Several use cases are envisioned and tested by ongoing research projects including tele-operating driving, cooperative collision avoidance, cooperative manoeuvring, high-definition mapping, platooning, situation awareness, video streaming and infotainment, green driving, to mention a few. The new market for automated and connected vehicles is expected to grow exponentially and large economic benefits are expected, with for instance revenues exceeding EUR 620 billion by 2025 for the EU automotive industry and EUR 180 billion for the EU electronic sector (European Commission, 2018). At the same time, business and market research reports (PwC, 2016) (Goldman Sachs, n.d.) clearly state that the UAV technology, that can be seen as part of the automotive & transportation vertical, drives innovation to various other vertical industries, apart from the obvious impact on military, civil and governmental applications. Considering the Unmanned Aerial Systems (UAS) interest in using cellular connectivity is strong, as the 5G ecosystem offers excellent benefits for UAS operation (3GPP, 2018) it becomes obvious that these vertical industries and opportunities are also relevant in the 5G perspective. More specifically, and as noted in (PwC, 2016) Infrastructure, Agriculture, Transport, Insurance, Mining industries shall cumulatively create a market value of $127bn due to the drone's disruption in their business.

UAV and U-space Ecosystem

UAVs (Unmanned Aerial Vehicles), or drones, as they are commonly referred to, have evolved significantly over the last decade. They are used in many different domains, by many different users - ranging from the general public to highly specialized UAS and their population in the air is expected to highly increase.

Main key actors identified in UAS operations framework are:

- **UAS Operators:** focused on flying UAS.
- **U-space service providers (USSPs) including local administration units (LAU):** focused on providing safety relevant services to UAS Operators; each USSP should provide at least 4 basic services: Network Identification Service (NIS), Geo-awareness service, UAS Flight Authorization Service and Traffic Information Service (TIS)
- **Common Information Service (CIS) providers**: focused on establishing a "single source of truth", providing fair access for USSPs, integrating relevant services & components (e.g., Air Traffic Management - ATM, Authorities, etc.).

Apart from the main key actors, there are also complement actors:

- **End users:** consumers of the services provided by UAS operators
- **Supplemental U-space Service Providers (S USSPs):** service providers, who do not need to provide mentioned above basic services but instead provides complementary services like weather information, altitude/height conversion services, etc.

As a consequence, the need to provide a regulation framework was identified and several initiatives have started in this respect. Work was done in the U.S. by NASA / FAA and industry partners, establishing the so called LAANC system (Wallace et al., 2020). China established first regulations in 2017 (CAAC, 2017). In Europe, with the so-called Warsaw Declaration on drones in November 2016, the notion of U-space was introduced and described in the U-space Blueprint. These are examples of lots of international, national, and local programmes to progress in UTM (Unmanned Traffic Management). This section has its focus on the European programme, U-space.

Even though the name U-space might suggest a strong geographical focus, indicating a segregated airspace to be used by UAVs exclusively, it is meant to be more. "U-space is a set of new services and specific procedures designed to support safe, efficient and secure access to airspace for large numbers of drones" (SESAR Joint Undertaking, 2017). Looking at this large scope, an approach on a roll-out has been sketched in the U-space blueprint. The sketch highlights the U-space evolution with new services and procedures, built upon continuous progression of two enablers, automation and connectivity.

Figure 2. U-space blueprint rollout Source: SESAR blueprint

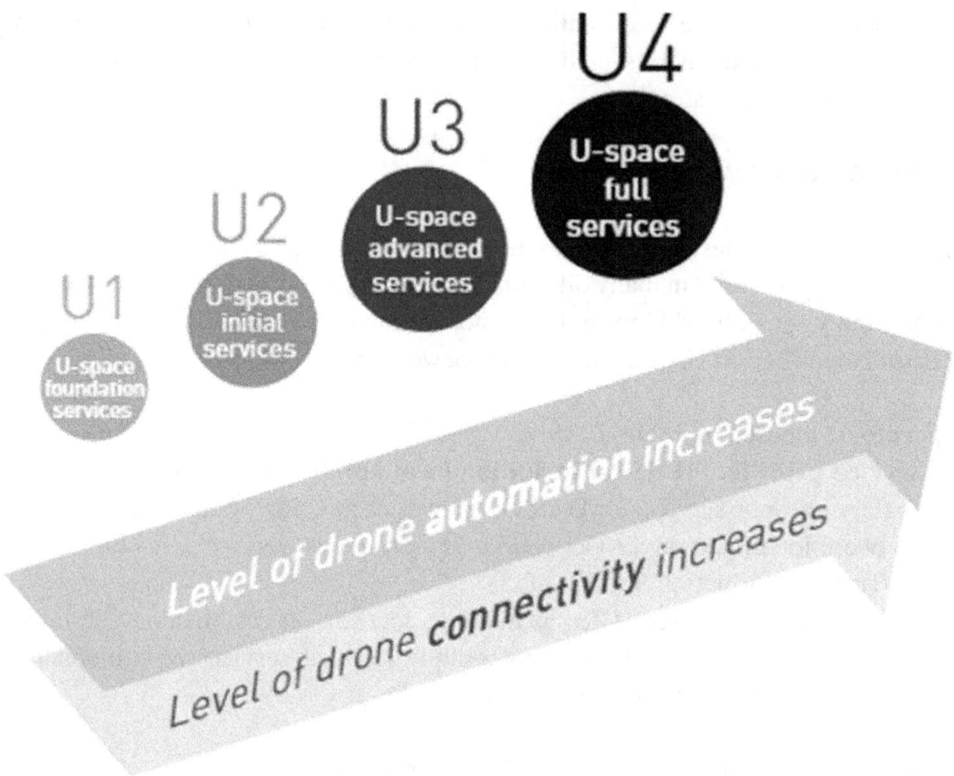

Are drones the next big thing? Alone, they are likely not. But, increasingly automated or even autonomous UAVs, connected by networks which are continuously improved in performance and reliability, allow for operational procedures and services so far not possible in aviation. New deployment options, e.g. cloud deployment or edge computing, allow us to scale up, following the increasing number of drones that will require access. Applied in combination with U-space, these technologies together, could be the foundation for the next big thing. New stakeholders and roles are and will be part of this ecosystem. Some of them are briefly explained in the subsequent paragraphs and an overview on U-space participants is given in CORUS (SESAR, 2019a). At the heart of U-space, drone operators are an important part of the value chain. They apply the combined technologies described above to offer commercial services to end users, such as to name just few, airborne video streams, efficient land survey, and delivery services.

Drone operators are supported by U-space service providers. Entities providing “new services” such as drone registration, flight plan preparation assistance and management, and supporting operational approval processes. These enable the identification of drones and provide geo-awareness to drone operators. At the time of writing, these services, and other U1 and U2 level services, are put in operations already, and work is ongoing on progressing U3 and U4 level services towards higher levels of technical maturity. CORUS lists more types of service providers supporting U-space services and drone operators, e.g. supplemental data service provider (SDSP), providing weather and ground risk information. Especially relevant for the C2 Link (command and control link) are communication service providers, “responsible for the provision of a reliable and safe communication link between systems”.

Drone operators and service providers need to share a large amount of data among them as well as with other stakeholders. This requires focal points to facilitate information exchange and service discovery. These focal points are called CIS (Common Information Service). They not only connect services; they also integrate with existing systems in ATM (Air Traffic Management). Often CIS is referred to as “single source of truth”, ensuring consistent and high-quality information is available in U-space.

Stakeholders, roles, and responsibilities have been subject to research and validation in Europe, as reported in the SESAR U-space programme (SESAR, 2020) and a draft U-space regulation is already put in review (EASA, 2020). More information on U-space regulation is provided in a following section.

Primarily regulation is focused on safety aspects and its initial version is looking at U1 and U2 level services. Still, sufficient connectivity is essential for those services that are expected to be mandated in the initial U-space. Network identification service, geo-awareness service, UAS flight authorisation service, and traffic information service require safe and reliable infrastructure to share information.

Slicing, QoS, Priority, Mobility Management are key enablers to support these services in operations. Complemented with high performance mobile networks, low latency, and high bandwidth will facilitate market growth of commercial services. Still, even the most performant mobile network cannot ensure coverage everywhere and anytime, technical and commercial constraints apply. This is a risk for aviation, which needs to be mitigated. Knowing upfront which service level is available where, using information on network capabilities in flight planning, will enable drone operators “safe, efficient and secure access to airspace for large numbers of drones.” (SESAR Joint Undertaking, 2017).

U-Space Services

U-space is divided into four different levels of services ranging from Initial U1 services (registration, electronic identification and geo-awareness (here to fly and where not to fly) to U4 full integration with manned aviation. U2-level services are mainly targeting the drone pilot, where U3-level services have a higher degree of automation and interface directly with the drone system. SESAR JU executive director presented U-space roll-out timeline in 2019. Current U-space services with initial timeline being worked on include:

1. U1 Level (2019+):
    a. Registration
    b. E-identification
    c. Geo-awareness

2. U2 Level (2022+):
    a. Drone Aeronautical Information Publication
    b. Geo-fencing provision
    c. Incident / accident reporting
    d. Weather information
    e. Position report submission sub-service
    f. Tracking
    g. Drone operation plan processing
    h. Emergency management
    i. Monitoring
    j. Procedural interface with ATC
    k. Strategic conflict resolution
    l. Legal recording
    m. Digital logbook
    n. Traffic information
    o. Geospatial information service
    p. Population density map
    q. Electromagnetic interference information
    r. Navigation coverage information
    s. Communication coverage information
3. U3 Level (2027+):
    a. Collaborative interface with ATC
    b. Dynamic capacity management
    c. Tactical conflict resolution
4. U4 Level (2030+):
    a. Full integration with manned aviation

UAV TECHNOLOGY AND CELLULAR NETWORKS STANDARDISATION

UAV Technology

As of December 2020, there are no drone industry standards specific to radio link for drones. From a bandwidth perspective, the different drone controllers use portions of the radio spectrum reserved internationally for purposes other than telecommunications (the ISM bands), primarily the ISM 2.4 GHz or 5.8 GHz bands defined for global use, or region-specific ISM bands, like 433 MHz for the EU or 915 MHz for the USA. The Wi-Fi standard is used in some of the commercial and toy drones, but very often the ISM bands 2.4 GHz or 5.8 GHz are used by proprietary solutions.

Despite the fact that the standards for radio link for drones are still not finalized, one can nonetheless cite other relevant standards:

- Communication over 2.4 GHz band, following the IEEE 802.15.4 standard (Low Rate Wireless Personal Area Network - LR WPAN)

- Communication over 902-928 MHz ISM band, following the following standards: RTCA/DO-160G, RTCA/DO-178C Level C, RTCA/DO-254 Level C and FCC 47 CFR Part 15.247.

Figure 3. Remote command and control of drones flying BVLOS

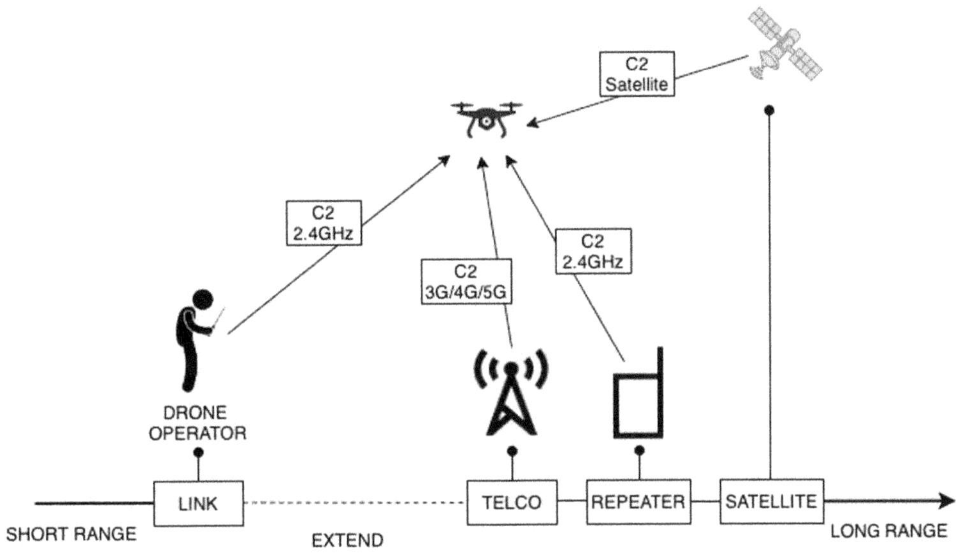

Unmanned Aircraft Systems are using several methods to communicate between the drone while air bound and the pilot on the ground:

- Radio Command (RC) link, which works only one way – from ground to the vehicle in the air and is used to send radio signals from a transmitter to a receiver to control motors and servos. These signals require little bandwidth, have very low latency and packet loss and usually use the 2.4 GHz band. RC is used to download the flight plan and as backup in case the telemetry link should be broken.
- Telemetry, contrary to the RC link is a two-way communication between drone and pilot, which allows to better understand what is happening with the airborne vehicle. It is possible to receive status updates like speed, altitude, position etc., and send the commands in uplink, like take-off, landing, change the flight mode and upload the mission plan for autonomous flights.
- Real time video feed from the drone, done as analogue transmission from the drone to the ground. This signal, when received on the ground, can be converted to digital or connected to the analogue monitor and used to steer the drone or perform observations. The advantage of the analogue transmission is very low latency.

The C2 link range can be extended using the technologies described below:

· LTE/5G connectivity (requires a modification of the drone with an LTE/5G modem). Commands from the drone controller to the drone are relayed using the mobile communication network.

· 2.4 GHz or 902-928 MHz repeater: hardware equipped with 2.4 GHz or 902-928 MHz transmitters/receivers, in order to extend the range of the drone controller. Commands and controls travel over the internet to reach the drone, which is not modified in any way.

· Satellite communication (requires a modification of the drone with a satellite modem). Commands from the drone controller are delivered to the drone via satellite link. Such a link is effective but expensive.

Limitations of Technology to Date

- Jamming
- Interference
- Saturation
- Security
- Range, reliability of the link, high interferences caused by very extensive use of ISM bands in populated areas, lack of reliable monitoring

Limitations and problems mentioned above, were solved by different companies: one can cite for instance DJI's Lightbridge and OcuSync proprietary link solutions. The second is used for the low latency, high resolution (1080P/720P) long range video transmission up to 8km. The table below gives the overview of solutions used in DJI's products – Wi-Fi, Lightbridge and OcuSync.

Table 2. Overview of solutions used in DJI's products

Solution	Frequency band	Models	Range [m]	Video res. for Live View	Encryption
Wi-Fi	2.4 or 5.8 GHz	Spark	500	720p HD Ready	WPA2-PSK
Enhanced Wi-Fi		Mavic Air	2000		
Lightbridge	2.4 or 5.8 GHz	Phantom 4 Pro	3500	720p @30 fps	
Lightbridge 2	2.4 or 5.8 GHz	Inspire 2	3500	1080p Full HD	
OcuSync	2.4 GHz	Mavic Pro	4000	720p or 1080p @30 fps	WPA2-PSK AES-256
OcuSync 2.0	2.4 or 5.8 GHz	Phantom 4 Pro V2.0	8000	1080p	WPA2-PSK AES-256

Many of these problems can be overcome by using omnipresent mobile networks, especially the new 5G technology. The table below summarizes the identified benefits of the 5G network compared to already used solutions.

Table 3. Identified benefits of 5G network compared to actually used solutions

Subject	Legacy Solutions	5G
Pilot/GCS presence close to the flight area	Required – the Ground Control Station (GCS) must be close enough to the UAV to ensure direct radio link connectivity. In practice, it can be up to 7-8 km	Not required – GCS can be remote or in MEC to assure low latency
First Person View support	Existing proprietary solutions.	Standard broadband data transmission over dedicated data link, having its own QoS.
Impact of radio interferences - dedicated, licensed band usage	The legacy solutions usually work in ISM bands, used simultaneously by many different use cases, where the risk of interference or jamming is high. Especially, in areas with high drone traffic density such operation will be impossible	5G networks will use dedicated licensed bands and features enhancing interference management
Omnipresence of the service	Not feasible – local service only	Service area is dependent on 5G coverage of the Mobile Network Operator (MNO), which ultimately will be countrywide, or even beyond – with seamless roaming. Ad-hoc 5G coverage deployment on demand can be also imaginable. In case the constant QoS cannot be provided everywhere, the lower level QoS (minimum) can be agreed in regions of lower network performance
Multiple UAV handling	Not designed for this purpose. No mechanisms for coordination of simultaneous radio links for each UAV.	Possible to handle huge number of UAVs in the same time and the same space. Additionally, if the 5G network-based connectivity is treated as one of U-space resources (similarly to e.g. airspace), the efficient U-space resources management can be targeted to optimize the flight planning process due to active rearrangement of demands to accommodate as many flights as possible in order to maximize the approval ratio and balance the average utilization of U-space resources
Quality of Service (QoS)	Not ensured	Standardized, can be guaranteed with SLA. QoS may be differentiated depending on the location. 3D QoS-indicating coverage estimation data may be shared between the MNO and U-space Service Provider (USS)
Ability to move over a large area	Radio link between UAV and GCS is limiting factor – up to 7-8 km	Handovers and mobility are part of the 3GPP standard. UAV can move freely in the mobile network coverage area
Service recovery mechanisms	Basic – return to the point where signal was better	Highly sophisticated – on the different protocol layers, with dedicated features (i.e. RRC Connection Reestablishment), using intra- and inter-technology mobility, data retransmission, buffering etc.
Immediate access to the payload	Yes, with wideband analogue or digital link. Usually, the payload is accessible after finishing the mission.	Yes, with different QoS, specified on the slice level and in accordance to SLA.
Tracking of UAV	Based on info received by telemetry link	5G System can support and use Location Services (LCS) for locating UAV independently of its GNSS system. However, the non-UE-based, Network-Assisted Location mode needs implementation of LCS support in gNBs.
Modulation and Coding Scheme (MCS) adaptation and scheduling of the data performed in Packet Scheduler (PS)	Basic or proprietary solutions.	Sophisticated - data stream type-dependent. Defined by 3GPP.
Support of distributed data processing architectures	No	With selective User Plane traffic routing and implementation of Edge Computing/MEC both C2 and payload stream can be processed (enriched, augmented, transcoded, aggregated etc.) in transit to optimize data transmission, volume, latency etc.
Increased UAV autonomy and operational time	Not possible through cellular networks	Possible thanks to a shift of computing power, to Edge or MEC, transcoding or other data processing offloading is possible.
Connectivity to U-space	No, unless using cellular network to connect GCS to U-space.	Yes, the 5GS can expose Control Plane services to other system. In case of the User Plane traffic, the additional security level can be offered (5G operator can be Trusted 3rd Party, testifying integrity of information sent from/to the U-space).
Open market – little or no barriers to new entrants	Single standard not existing. Proprietary solutions are dominating.	Worldwide 3GPP standard, accessible for everybody.
Identity handling and privacy	UAV specific identities can be locally used, if UAV has them. Remote ID is a standard recently introduced.	Telecom identities: IMEI and IMSI can be used in parallel with other UAV-specific IDs. They all can be communicated to centralised e-registers, the 5G network can certify them, privacy is kept to the highest degree.

All these advantages related to utilisation of 5G mobile networks can make the BVLOS drone flights and related use cases easier to deploy, safer and more attractive from business point of view. Potentially they can boost benefits of drone operators and mobile networks.

Fundamental 5G Technology and UAV Support Requirements

The 5th Generation mobile network has been considered as a communication platform for UAS since its early visions. Naturally, one of the fundamental advantages of the mobile network over other communication platforms (especially those using a P2P transmission between UAC and UAV in ISM bands) is its omnipresence and common availability as a universal platform for mobile access to the Internet. The fundamental service performance and quality demands for 5G (ITU, 2015) are also applicable to the UAS use case, especially the peak and user experienced data rate, the supported mobility (motion velocity) or latency. These two issues have led to an increasing effort to integrate the requirements of the unmanned aviation into support through the 3GPP 5G network. A recent publication highlights how 5G network can interact with UAV ecosystem by exploiting the U-space architecture (Si-Mohammed et al., 2021). In fact, standardization endeavours are progressively incorporating UAVs with 5G networks as the work of Marojevic et al. (2020) states. Analyzing the 3GPP 5GS ability and readiness to support communication services for UAS, two aspects should be taken into account – the generic service requirements, mechanisms, functionalities and architectural solutions that will be applicable in the context of supporting communication services for UAS and those that result from specific problematics in the field of unmanned aviation.

Generic Aspects

The general 5G service requirements are standardized (3GPP, 2020b) and in the context of UAS services, the following requirements are of importance:

- **Network Slicing:** the paradigm shifts from one multi-purpose and multi-service communication network ensuring unsatisfactory compromise between service requirements to a federation of specialized networks tailored to the specificity of services and processing their fractions of the overall service traffic in optimal way – a UAS needs several communication links both for non-payload (aviation domain-related, e.g. C2, telemetry, UTM or FPV) and payload (use case-related, e.g. real-time video, still images, sensoric information, remote control of cargo etc.). These links will have different requirements.
- **Mobility Management:** optimization of network behavior according to mobility patterns, maintaining the IP address during the session if required, minimization of interruption time and packet loss – a UAV with a 5G UE will be characterized by high mobility across individual base stations coverage areas and high demand of service continuity.
- **Efficient user plane:** optimization of user plane path for communication between UEs or from a UE to service hosting environment (e.g. edge cloud) for low latency or low bandwidth pressure – both non-payload and payload links may need implementation of their service enablers near to the location of a UAV and its further following.

- **Priority, QoS and policy control:** different services may share the same QoS characteristics, but have different priority requirements – different UAS communication links or UAS links and other services.
- **Connectivity models:** the 5G network connectivity may be achieved either directly, or via other UE – in case of UAVs such feature may help in maintaining transmission during a momentary loss of network coverage.
- **Network capability exposure:** ability to expose APIs to trusted 3rd parties for creation, modification, deletion or monitoring of network slices as well as defining or updating services or capabilities and UEs supported by these slices – the functionality very important both for U-space Service Provider and UAV Service Provider.
- **Context aware network:** network behavior optimization based on UE behavior – adaptability of 5G network may increase UAS services reliability.
- **Extreme long-range coverage in low density areas:** extreme long-range coverage (up to 100 km) at remote rural areas or at sea, with minimum DL throughput of 1 Mbps – important for providing omnipresence of the 5G services for UAS.
- **Multi-network connectivity and service delivery across operators:** ability to maintain the continuity of home network services in case of its coverage gaps through visited networks in a "national roaming".
- **QoS monitoring:** exposure of a mechanism for supporting real time E2E QoS monitoring – dependability of network and services for UAS.
- **Positioning services:** by different single and hybrid positioning methods (3GPP and non-3GPP positioning technologies) – to provide validation of position reported by UAV.

Unmanned Aviation-Specific Aspects

The specificity of unmanned aviation requirements, has been separately standardized and includes the following:

- **Remote Identification of UAS:** enabling the UAS-UTM communication with the network user and UE identity (IMEI, IMSI, MSISDN, IP address), UE capabilities and other aviation/flight-related information, "certification" or augmenting of data sent by UAS provided by the network operator, supplementing them with location information, and involving the UTM in the authentication.
- **UAS traffic management:** communication between UAS and UTM with latency less than 500 ms, short-range (up to 600 m) broadcasting of UAV identification and position for collision avoidance (also cross-network) with latency less than 100 ms between UAVs with relative speed up to 320 km/h.
- **Security:** protection against spoofing attacks of the UAS identities, non-repudiation, integrity and privacy of UAS-UTM exchange.
- **UAV service status and restriction:** exposure of real-time monitoring of UAV UE status information, providing 4D (3D space + time) information about UAV services availability, network-based 3D space positioning in a given range of altitudes and early warning of possible UAV connectivity loss.

Additionally, the service performance requirements both for non-payload and payload transmission specific to individual UAV use cases are defined. In case of payload transmission services, other than those defined in the initial report for the specific aspects (3GPP, 2020b), it may be useful to refer to service requirements for the 5G system or other standards devoted to video, imaging and audio for professional applications (3GPP, 2020p) or cyber-physical control applications in vertical domains (3GPP, 2020q). It is worth to mention that 3GPP, despite the fact that UAVs are "vehicles", considers V2X in the context of automotive domain only.

Standardization of 5G Networks for UAVs Support

ITU-R has defined the main usage scenarios for IMT 2020 and beyond in their Recommendation (ITU, 2015). All key capabilities may to some extent be important for most use cases, the relevance of certain key capabilities may be significantly different, depending on the use cases/scenario. The importance of each key capability for the usage scenarios eMBB, URLLC and mMTC is illustrated in Figure 3. This is done using an indicative scaling in three steps as "high", "medium" and "low".

Figure 4. The importance of key capabilities in different usage scenarios
Source Recommendation ITU-R M.2083-0

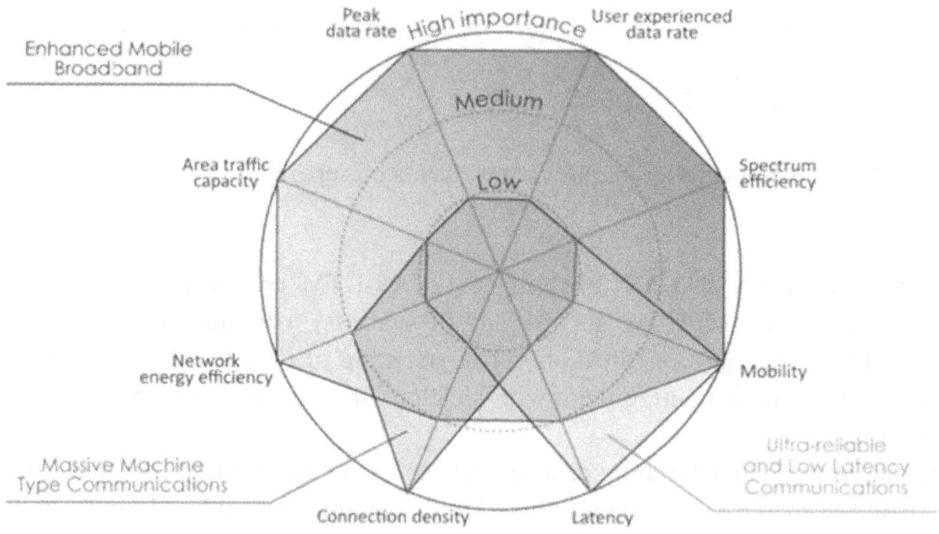

In Europe, there is a fast-growing demand to operate UAS under beyond- visual-line-of-sight (BVLOS) conditions mainly for professional purposes. To enable these intended range of use cases, there is the need for communication links between the UAS, its operator and an intended UAV Traffic Management (UTM) system.

The European Conference of Postal and Telecommunication Administrations (CEPT) has identified existing Mobile Fixed Communications Network (MFCN) bands as a potential means to provide connectivity to UAS, via existing LTE mobile networks for command and control links. It is also evaluating the current regulatory framework of MFCN bands for this use case. One possibility for professional UAS applications is to use existing mobile MFCN networks to provide connectivity to UAS by usual (unmodified) mobile networks with LTE technology provided, so that the command and control link(s), where appropriate, meet the relevant aviation safety requirements prevalent in the country of concern. This can be realised either by an external LTE device attached to UAS, or in future by implementing SIM-cards installed within UAS. Such a connectivity could be used both for serving the payloads such as video or other collected data via sensors and for the command and control function of UAS (Electronic Communication Committee, 2018).

The ECC Report 309 (Electronic Communication Committee, 2020) has the aim to evaluate the use of MFCN for the communication links of Unmanned Aircrafts (UAs) within the current MFCN spectrum harmonised regulatory framework for the different MFCN bands, including potential impact of such use on MFCN networks and other systems and services and possible regulatory considerations. The intention is to use already existing MFCN Base Stations (BSs)which are typically deployed to provide effective coverage at ground level. At this stage, mobile operators do not intend to develop specific network planning to respond to these new usages. Due to this, coexistence studies are mostly required for uplink due to the elevated position of aerial UEs. No specific studies are required in the downlink for non-AAS (Active Antenna Systems) base station, since the emissions characteristics are not modified. Studies in the downlink would only be required for the case of AAS-Base Station, where beam steering may lead to beam pointing above the horizon and may modify the emission characteristics.

3GPP is actively leading technical studies and standards related to 5G usage by UAVs. To address the needs of a new and quickly maturing sector, there has been a lot of activity in the 3GPP Working Groups to ensure that the 5G system will meet the connectivity needs of Unmanned Aerial Systems (UAS). The foundations of UAS communication services support by the 3GPP 5G system (5GS) are primarily in definition of the 5GS architecture (3GPP, 2020) followed by definition of 5GS procedures, which provided various concepts and functions important for UAS. The latest version of (3GPP, 2020a) has been released on Dec 2020.

On-going 5G Standardization, Open Issues and Gaps

Apart from general 5GS architecture framework mechanisms, which are of importance for UAS support, the 3GPP also works on issues specific to UAS support by the 3GPP system:

- **TS 22.125:** this technical standard defines a variety of UAS use cases and their requirements (aviation domain-related, including C2, UTM and remote identification, as well as specific to drone usage), especially functional and in terms of performance targets: data rates, E2E delays, reliability rates, allowed altitudes above ground levels and ground speeds and positioning latencies/accuracies (3GPP, 2020b).
- **TS 22.261:** this technical standard provides supplementary information, as it is dedicated to general 5G service use cases and requirements. Despite the fact that the UAS support area has been excluded from it and moved to the separate aforementioned document TS 22.125, here the overall picture is complemented (e.g. tampering detection and prevention) (3GPP, 2020c).

- **TR 23.754:** this architectural study follows the standardized requirements for UAS connectivity, identification and tracking, and analyses necessary enhancements for their support. It provides the architectural approach of the aviation ecosystem containing UAS components, UTM, authorized 3rd parties (e.g. police, aviation authorities) and the 5GS. The vision outlines interfaces for these actors' interaction with 5GS steering mechanisms in the 5GS CP (3GPP, 2020d).
- **TR 23.755** this report identifies the application architecture aspects to support UAS, and corresponding architectural solutions (3GPP, 2020e). The study includes identifying architecture requirements that are necessary to ensure efficient use and deployment of UAS services and applications over 3GPP systems and deals with:
    - applicability and possible needs for enhancements of service enabler architecture layer (SEAL – set of functions for management of location, group, configuration, identity, key and network resources) – common to all vertical industry applications – for UAS services;
    - broadcast communication amongst UAVs both in off-network and on-network scenarios;
    - UAV location information – both verification of location reported by UAV, which cannot be fully-trusted, by the 3GPP network and possible supplementing of 3GPP system- based location information to the location reporting towards UTM;
    - UAS services capabilities exposure – real-time monitoring of the UAV status information (e.g. location of UAV, communication link status) and exposure to a 3rd party of the information about UAV service status in a certain geographical area and/or at a certain time.

On the subject of key points for UAV 5G standardisation support in relation to the aforementioned support requirements, it must be noted that network slicing (NS) is an inherent part of the 5GS. The support of network slicing in 5G system is constantly evolving and generic features are still developed within the standardization documents. The CP support of network slicing will be enhanced within the Release 17 and the study work has been started on slicing mechanisms (3GPP, 2020f) and its management (3GPP, 2020g) to feed further standardization. In the RAN area, the network slicing is currently recognized at the level of general principles and requirements (3GPP, 2020h), but the mechanisms of network slicing support by RAN (especially at the level of radio resources management) are also a part of the Release 17 scope.

The mechanisms of Proximity-based Services (ProSe) and Sidelink relay are fundamental for providing direct, cross-network communication between UAVs for presence and location broadcasting and use of a neighbouring UAV as a network connectivity relay, in case of coverage loss. They are, however, currently missing in the 5GS standardization and belong to the scope of the 3GPP Release 17, currently at the stage of study (3GPP, 2020i; cf. 3GPP, 2020j).

Sharing of 3D information about network coverage and performance, is a very important aspect of the integration of the 5G network with the U-space ecosystem, especially in terms of flight plans validation and environmental awareness, needed for real-time UAV traffic management. 5G RAN physical layer measurements at UE, have been defined in (3GPP, 2020k), but no mechanisms similar to 3G/4G measurements collection (Minimization of Drive Tests – MDT, a part of 4G Self-Organizing Network – SON, (3GPP, 2020l) have been specified in the 5G SON framework (3GPP, 2020m), yet. It is a part of the 3GPP Release 17. Therefore, the only way to provide such information is interfacing network operator's RAN planning tool (outside of 5GS) with the U-space domain. Alternatively, the U-space provider may collect such data from UAVs using a special system and telemetry agents embedded in

UAVs, based on a U-space service agreement terms. The data in this system would have only a historical and illustrative meaning.

One of the fundamental enablers of UAV use cases is integration of the 5G network with the MEC framework, upgraded with network slicing support. It is commonly accepted that the Multi-access Edge Platform (MEP) will be integrated with the 5G CP as a specific case of AF (Mp2 interface exposed as Naf interface). The currently proposed Standards Developing Organisations (SDOs) visions assume loose integration of ETSI NFV, ETSI MEC and 5GS frameworks.

Hence there are many architectural inconsistencies, duplicate functionalities and competing mechanisms. The integrated framework needs to be thoroughly rethought and redesigned with reuse of valuable mechanisms for simplification and synergy. There are already some proposed approaches (Ksentini, 2020) (Tomaszewski et al., 2020). Additionally, the MEC has been initially defined for LTE and MEP APIs for MEC applications support LTE network. Either 5G APIs or radio access technology-agnostic APIs should be defined. 3GPP also works on integration of MEC within the Release 17. Currently, there are only non-normative studies reports available (3GPP, 2019), (3GPP, 2020n), (3GPP, 2020o), (3GPP, 2019b) -no documents, provided yet.

The deep integration of 5GS and unmanned aviation domains into one vertical ecosystem is very important also from the business processes architecture and their cross-domain span (Si-Mohammed et al., 2020). Involvement of the 5G network operator in the U-space processes, as well as view of the 5G network resources as a part of the U-space ecosystem resources, can be considered as one of the opportunities of the network operator. Especially the flight planning, validation and acceptance process, fed by 5G network coverage data exposure, will have a positive impact on the temporal 5G network resources planning for optimization of utilization and overloading/degradation risk minimization (Tomaszewski et al., 2020a). In the future, the common multi-operator and multi-UAV services provider ecosystem around the U-space service provider, exposing a portal for selection of the best offer by the UAV services customers, will be possible (Kuklinski et al., 2020).

A special concern should be taken of the problem of the UAV operations outside of the country of registration, especially cross-border operations (Tomaszewski et al., 2020b). In the European Union, according to the principle of free movement of goods and free competition, as well as the European Commission initiative of the "Single European sky", the final ecosystem will be composed of multiple U-space service providers operating in a specific area and serving both home and visiting drones. It will therefore deal with two-tier roaming; the U-space level and 5G network level. The architecture of the ecosystem will contain numerous entities and interfaces for interactions between them, both in the aviation layer processes and the mobile network control procedures.

Although GSMA is working on the standardization of Generic/Network Slice Templates (NG.116) (GSMA, 2020) despite the generic equivalents of network slice instances at different operators, there will also be a need to provide mobility of specific proprietary slice templates, in particular, ensuring their confidentiality. During the UAV operation outside the country of registration, both Local Break-Out (LBO) and Home-Routed roaming scenarios will be potentially used at the same time in User Planes of different NSIs. Another very important issue is providing a seamless network change at the country border for transmission continuity in all involved NSIs. This may involve redefinition of network (re) registration procedures in roaming. The currently available work-around for this problem is dual/multi-SIM terminal and gradual reregistration, which is disputable due to limited UAV battery capacity.

Support of UAV use Case Requirements by the 5G System

The generic 5GS architectural framework (3GPP, 2020) provides fundamental mechanisms and functionalities, which answer the majority of UAV use cases requirements:

- Network slicing is supported within the Core Network (CN) by the various Control Plane (CP) functions: AMF (Access and Mobility Management Function), which coordinates the entire UE-related signalling and involves Network Slice Selection Function (NSSF) during the attach to specific Network Slice Instance (NSI) requested by UE, with optional support of Network Slice-Specific Authentication and Authorization Function (NSSAAF), based on user profile stored in Unified Data Management (UDM). The signalling between UE and CN-gateway point (AMF) is enriched with S-NSSAI parameter (Single-Network Slice Selection Assistance Information), which is composed of Slice Service Type (SST) and optional Slice Differentiator (SD) to provide unique identifier of an NSI within the network. Currently 3GPP defines 4 types of SST: eMBB, URLLC, MIoT, and V2X, but the list may be extended in the future. It is assumed that a UE may be connected to up to 8 different NSIs at the same time. The NSI specific User Plane Function is controlled by its individual counterpart (CP), Session Management Function.
- Priority, QoS, and policy control (including also network slicing, roaming, and mobility management policies) are supported by Policy Control Function (PCF). The 5GS QoS framework defines standardized 5G QoS Indicators (5QI) of service QoS classes, which are characterized by resource type (the issue of flow Guaranteed Bit Rate: non-GBR, GBR, delay critical GBR), default priority level, packet delay budget, packet error, default maximum data burst volume, and default averaging window. The 5QI, received from PCF, is enforced by SMF for its UPF. The detailed specification of the policy and charging control framework for the 5GS, including interactions of PCF with other Network Functions (NFs), has been defined in (Tomaszewski et al, 2020).
- Network capability exposure and integration of 5G CP with an external environment of e.g. U-space Service Provider or UAV Service Provider is enabled through Network Exposure Function (NEF), which exposes CP services based on catalogue stored in Network Repository Function (NRF) and Service-Based Architecture of CP communication bus. The specific functional UAV enablers may be also embedded within the CP as special cases of generic Application Function (AF), either still using NEF as a proxy, due to security reasons, or interacting with CP functions directly, if considered as trusted (subject to network operator's policy).
- UTM Network context awareness, QoS monitoring, and /UAV Service Provider early warning will be based on Network Data Analytics Function (NWDAF) and Management Data Analytics Function (MDAF), a part of the 5G Management and Orchestration Architecture Framework. Specific monitoring of managed objects of interest (related events, performance indicators) may be subscribed and e.g. threshold-crossing notifications may be exposed through NEF or trigger special algorithms implemented as AFs in CP or Management Functions.
- Augmenting/supplementing UAS data and security requirements can be met by the UPF, which is defined, in opposition to LTE UP – a mobility-supporting data pipe anchored at the IP network gateway point, as a complete chain of functions for versatile UP data processing, including those (firewall, Deep Packet Inspection, packet selection/marking/enrichment for data content billing, anti-viral protection, parental control, images down-sampling, selective traffic redirection etc.), which are placed in SGi LAN of LTE.

The 3GPP Location Services (LCS) framework (3GPP, 2020r) is a 5GS CP extension to support UE positioning. The architecture has been defined both for home network and roaming scenarios. The procedures of UE-based/assisted (with UE involvement as a position data source) and Network-assisted (UE-independent) positioning are supported. The positioning requests can be network-induced (by AMF for some regulatory service), mobile-terminated (by LCS client or other AF), mobile-originated (by UE itself) bot in immediate or deferred (in case of occurrence of a specific event). The LCS architecture contains a Gateway Mobile Location Centre (GMLC) answering the positioning requests, which is supported by a Location Retrieval Function (LRF) responsible for retrieval and validation of location data.

The extreme long-range coverage in low density areas may be achieved by the use of Sub-GHz bands, but at the expense of the transmission speed, as the channel widths are low in these bands. However, in specific use cases, where there are no excessive requirements for transmission in payload channels, this solution may be sufficient. It should be mentioned, however, that in the low bands the dimensions and thus the mass of the antennas (especially MIMO) will increase, which may be important in terms of the drone's load-bearing capacity and its battery capacity.

5G AND UAV SYSTEM COMPONENTS AND TRIAL ARCHITECTURE

It is obvious that apart from the theoretical analysis of the aforementioned approach of UAVs and cellular technology, field trials are also necessary in order to understand the performance of the 5G cellular networks in real world operation conditions. A recent publication (Koumaras et al., 2021) describes the amalgamation of UAVs and 5G cellular networks as an auspicious solution for realising energy efficiency of UAVs and presents a proof-of-concept implementation of such a 5G-enabled UAV with softwarised Flight Control System (FCS) component at the edge of the 5G network. Moreover, the work of Muzaffar et al. (2020) analyses 5G radio links for use with drones via experiments performed between a drone flying at different heights and a commercial 5G base station.

On the other hand, to design and deploy a system, that would also support UAS operations in 5G networks, it is a very challenging and complex task. It would need to bring together completely different worlds of aviation and telecommunication. It must be emphasized that 5G based infrastructure would need to support and comply with very strict procedures and policies specific to UAS operations, as defined by the regulatory bodies. As for the time of writing this document (December 2020) there were no publicly available complete regulatory frameworks for the U-space operations. Further considerations are based on the available documents, opinions and recommendations (EASA, 2020) (SESAR, 2019).

Current trial implementations of 5G services supporting UAS operations, should meet the requirements as described in the previous section. As they are mostly related to the UAS mission execution (drone flight, called also as tactical phase) and are related to the provision of communication services, it must be underlined, that support will also be needed from the 5G network operator's side during the flight/mission preparation stage (mission planning, pre-tactical phase). One of the most obvious examples is the ability to obtain or validate a flight plan from the perspective of the network's coverage quality along with the route of planned flight and expected SLA. This kind of information might be required to perform risk analysis of the flight by the designated USSP (U-space Service Provider). The other example of the 5G operator's involvement might be the situation, where specific, dedicated services for UAS operations, are hosted on operator's edge servers (e.g., MEC). This case can include ground control station (GCS) application for controlling drones as well as specific applications like system for calculating common reference altitude (recalculation of height measurements performed in different reference systems) or system broadcasting information about local obstacles in order for the drones to avoid collisions with them. Taking the above into consideration, from the U-space perspective, the UAS ecosystem can be modelled as depicted in the Figure 5.

Figure 5. UAS Trials Platform Ecosystem

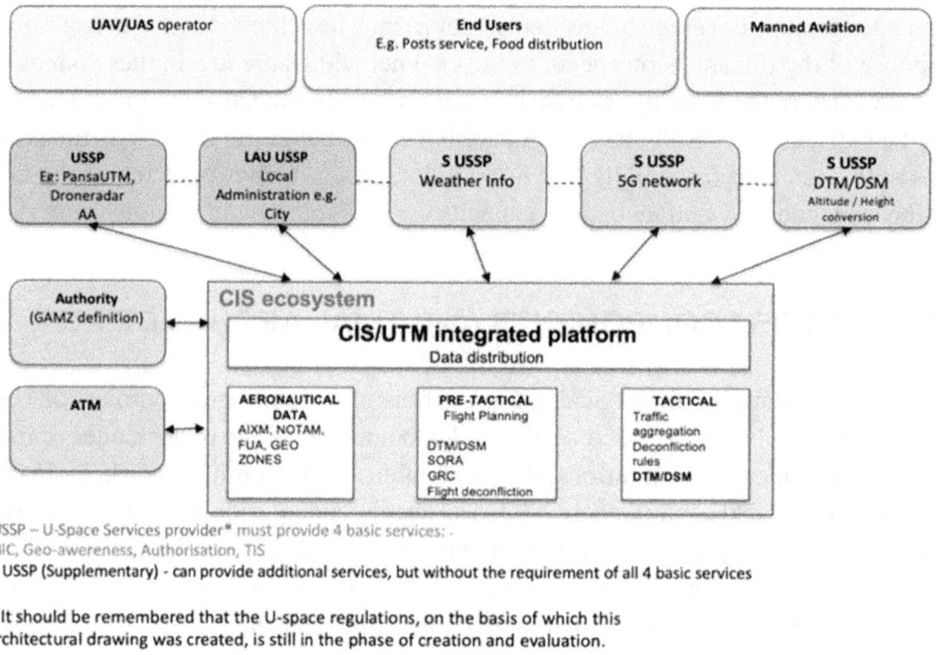

Figure 6. Reference Architecture of the integration of 5G with U-space
Source:(Si-Mohammed et. al, 2020)

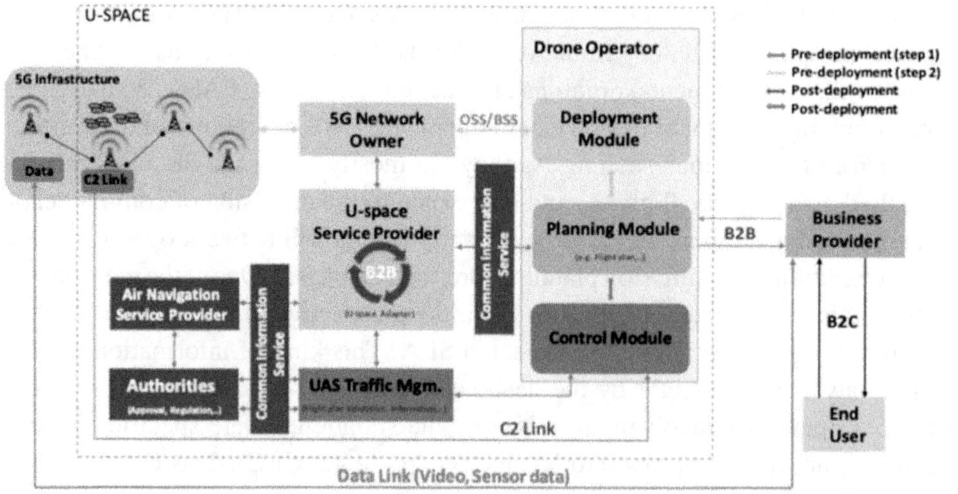

In this diagram, CIS/UTM is represented as an integrated platform. It gathers and processes key aeronautical data and pre-tactical and tactical information. It is connected to various entities: ATM, authorities, SUSSPs and USSP. In our example, we also introduce a few SUSSPs like weather information service and altitude/height conversion system for CARS (Common Altitude Reference System) support. It is also worth to notice, that the latter mentioned SUSSP, might be implemented as a service deployed on 5G network's edge servers. The main conclusion from the diagram above, is that 5G network from U-space perspective would be treated as another SUSSP. Hence, this is one of the components among others, which is used to successfully complete the UAS mission. In our example, it will provide communication service for C2 link and payload, coverage information for flight planning with QoS levels, telemetry, computing power on edge servers for SUSSPs and UAV applications, KPI data and network monitoring information, external application deployment and management (for platforms hosted on 5G infrastructure's edge servers), billing and accounting information, etc. Many researches have been performed on how to design and deploy an architecture integrating 5G infrastructure with U-space, to fulfil so complex interactions. One of the proposed solutions is described in "Supporting UAV Services in 5G Networks: New High-Level Architecture integrating 5G with U-space" (Si-Mohammed et al., 2020) and will be briefly summarized here.

The reference architecture of such an integration is presented in Figure 6. This diagram represents the view from "outside" of the U-space and concentrates on entities that would participate in the information flow between 5G network, UAS and U-space. The term "U-space" represents here the set of federated services which are depicted in more details in the Figure 5.

As the UAS mission from planning perspective is a multi-stage process, in which different entities and their functionalities are used on each stage, different colours are used to illustrate the involvement of the different components on each stage:

Step 1: Pre-deployment corresponds to the mission planning phase (pre-tactical phase), in which flight plan is defined and validated from the perspective of network's resources and capabilities as well as risk analysis on UTM side; the main modules involved at this stage is Planning Module and U-space Service Provider.

Step 2: Pre-deployment corresponds to the network deployment stage, in which all associated network slices and applications necessary to serve the UAS mission are deployed; main activities at this stage are under the responsibility of the Deployment Module.

Step 3: Post-deployment corresponds to the mission execution (tactical phase), in which the actual UAS flight is performed; during the flight, applications acting as Control Module are used and UAS Traffic Management functionality to control the traffic in U-space.

One of the implementations of the proposed reference integration architecture is the Trial Controller system designed and currently being developed by the 5G!Drones project consortium members (https://5gdrones.eu/). The project's aim is to trial several UAV use-cases covering eMBB, URLLC, and mMTC 5G services, and to validate 5G KPIs for supporting such challenging use-cases. The feature of Network Slicing will be used as the key component to simultaneously run the three types of UAV services on the same 5G infrastructure (including the RAN, back/fronthaul, Core), demonstrating that each UAV application runs independently and does not affect the performance of other UAV applications, while covering different 5G services.

The high-level architecture of the solution is presented in Figure 7. The trial controller consists of few components, which support an end to end UAS mission preparation and execution: starting from the pre-tactical mission planning, going through mission deployment, mission execution (flight) and ending with the post-mission KPI analysis. The system is designed and built based on the micro service architecture. The trial controller exposes APIs to client's applications (Northbound APIs) for the purpose of mission's preparation, execution, control and for collecting results of KPI measurements as well as APIs to 5G facilities (Southbound APIs) used to provision network slices and applications to 5G network facilities and also for the purpose of collecting KPI related measurements. There is also a specialized U-space adapter module, which implements interfaces to different U-space components (different USSPs/SUSSPs).

Figure 7. High level architecture of the trial controller
Source: (5G!Drones, 2019)

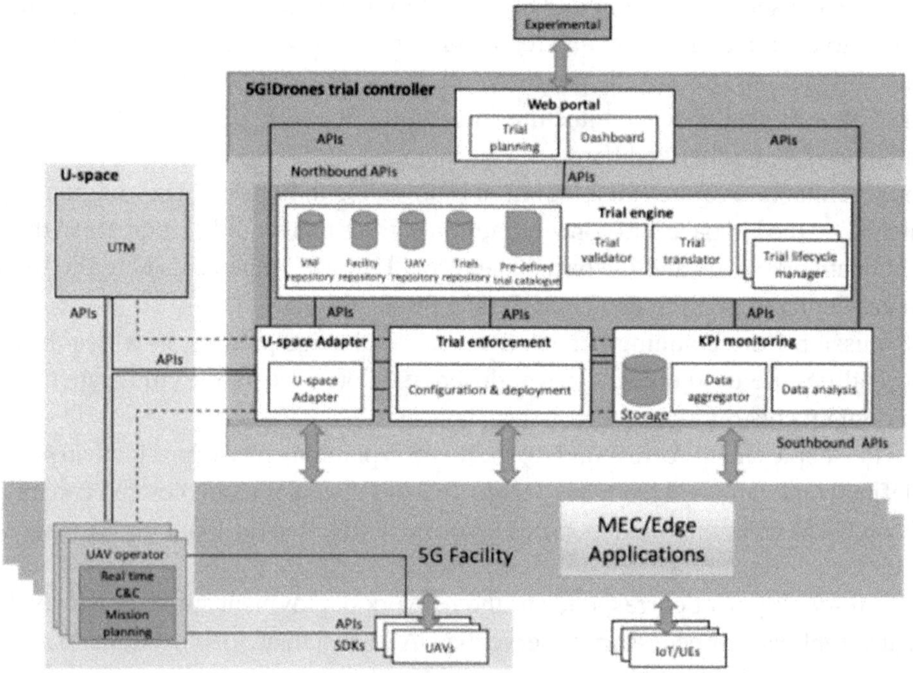

The trial controller is accessed by the user through the dedicated web portal. Through this portal, the user is able to define mission parameters and requirements, which are mapped and "translated" into the drone flight plan and the network slice template. Based on this information, the mission is "validated" by the U-space and the selected facilities. After the successful validation, the trial controller handles the resource provisioning. When mission is supposed to be commenced, the portal through the specialized dashboard allows the users to monitor the progress of the mission.

The trial engine is the main part, the "heart" of the system, which is responsible for maintaining the overall end to end process of the trial. It consists of the following:

- Various registries, which act as persistent repositories for the platform (VNF, Facility, UAV and trial repository)

- Trial validator module, responsible for handling the initial validation process through the U-space (risk analysis of the prepared flight plan) and the facility (checking the availability of the required resources/capabilities)
- Trial translator, which supports the preparation process of the network's slice template
- Life Cycle Manager (LCM), which is a module responsible for maintaining the overall end to end mission execution

The other components of the Trial Controller are:

- Trial Enforcement module, responsible for provisioning the facilities with prepared slice templates
- KPI monitoring module, responsible for gathering, collecting and processing measurements from facilities and other systems; based on the collected data, defined KPIs are calculated and the results are presented to the user via the web portal.

5G AND UAV REGULATORY FRAMEWORK

Spectrum Allocation

In 5G networks, radio spectrum will be required to support services ranging from narrow-band IoT, through enhanced mobile broadband (eMBB) and fixed-wireless access (FWA), to ultra-high-speed and extremely low latency connectivity, thus it needs to be divided across low, mid and high spectrum ranges. Figure 8 depicts 5G spectrum band classes and relevant use cases.

Figure 8. 5G spectrum band classes & use cases

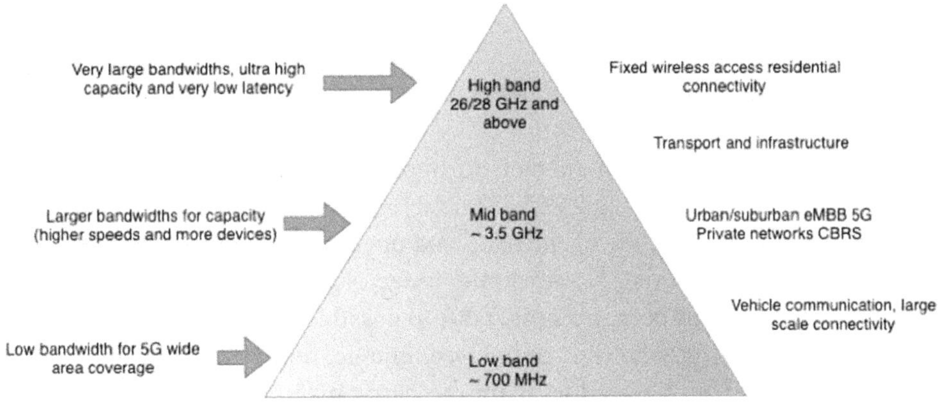

In fact, 5G radio spectrum is categorized into the following two different frequency ranges (3GPP, 2020s) (3GPP, 2020t) which have capacity and performance characteristics that suit some service models better than others. More specifically:

- Frequency Range 1 (FR1) covers spectrum bands from 410 MHz to 7125 MHz and can be further divided into the following sub-band categories:
    - Sub-1 GHz that supports widespread coverage across urban, suburban and rural areas
    - 1-6 GHz that offers a combination of coverage and capacity benefits.
        - This includes spectrum within the 3.3-3.8 GHz range which is expected to form the basis of many initial 5G services. The 3.4-3.6 GHz range is almost globally harmonized, which can drive the economies of scale needed for low-cost devices.
        - It also includes others which may be assigned to, or reframed by, operators for 5G including 1800 MHz, 2.3 GHz and 2.6 GHz, etc. In the long term, more spectrum is needed to maintain 5G quality of service and growing demand, in bands between 3 and 24 GHz.
- Frequency Range 2 (FR2) covers frequency range above 6 GHz and thus includes frequency bands from 24.25 GHz to 52.6 GHz and is needed to meet the ultra-high broadband speeds envisioned for 5G. Currently, the 26 GHz and/or 28 GHz bands have the most international support in this range for facilitating development of the fifth generation (5G) mobile networks.

At WRC-19, countries supported a globally harmonised identification of 26 GHz, 40 GHz and 66 GHz for IMT. In some counties, there is also the possibility to start using 50 GHz. However, it should be noted that radio access in FR2 has an effective coverage radius of a few hundred meters maximum. Hence, the high capacity of 5G network in FR2 is achieved at the expense of high density of base stations. Despite of this cost, the attractiveness of the FR2 bands comes from the much higher channel bandwidths in these bands (standardized values between 50 MHz and 400 MHz, while in FR1 it is 5-100 MHz).

It the light of the above, the European Union has taken a number of decisions with the aim to promote availability of new spectrum bands and ensure harmonization of suitable spectrum for 5G. In January 2019, the European Commission adopted an Amending Implementation Decision to harmonize radio spectrum at 3400–3800 MHz (3.6 GHz band) and updated the technical conditions to support 5G future deployment.

5G Spectrum Towards Services of UAVs

In May 2019, the European Commission adopted an Implementing Decision (European Commission, 2019) to harmonize radio spectrum in the 24.25–27.5 GHz (26 GHz) band, enabling Member States to set common technical conditions for use of the band and open it up for use (with 5G in mind, but on a technology- and service- neutral basis). However, the usage of 24.25-27.5 GHz band for ground-UAV (i.e. downlink) communication has been prohibited due to possible impact on existing satellite receivers in the FSS and ISS; while UAV-ground (i.e., uplink) communication is still allowed.

In addition, there will also be the need to adapt the harmonized regulatory framework of Mobile/Fixed Communications Networks (MFCN) in the frequency bands ranging from 694 MHz to 3.8 GHz to account for 5G (e.g. to accommodate the needs of networks based on small cells and Active Antenna Systems). CEPT ECC conducted technical analysis for the 900 & 1800 MHz bands (Electronic Communication Committee, 2019) and the 2.1 GHz band (Electronic Communication Committee, 2019a). Similar analysis for the 2.6 GHz band is provided in a report of the ECC (Electronic Communication Committee, 2020a), and an update to the harmonised utilization of spectrum for Mobile/Fixed Communications Networks (Electronic Communication Committee 2019b) was agreed in July 2019.

The ECC Report 309 (Electronic Communication Committee, 2020) approved by ECC plenary meeting in July 2020, provides the analysis on use of MFCN for the command & control and payload links of UAVs within the current MFCN harmonised regulatory framework. The conclusions of ECC Report 309 studies on the conditions of use of drones in the frequency bands 700 MHz, 800 MHz, 900 MHz, 1800 MHz, 2 GHz, 2.6 GHz and 3.5 GHz open promising prospects for several types of applications. The constraints remain moderate and linked to the protection of certain uses in the adjacent strip:

- In the 700 MHz band, drones will have to fly more than 30 meters from the ground in order to avoid interference to TV broadcasting receivers.
- National conditions will have to be established to protect earth stations in the 3.4-4.2 GHz bands (based on minimum radius of UAV or antenna ERP).

Studies have therefore shown that enough frequencies would be usable for communications with long-range drones, including when roaming. This work will make it possible to define harmonized conditions for the use of these drones in Europe. Each frequency manager will then set up national rules and then authorization frameworks (France Radio Communication Agency, 2020).

It should be noted that interference mitigation is complex problem and solutions come via 3GPP specifications in 2020-2021, where network operators and device vendors (Nokia, Ericsson etc.) participating actively. 3GPP Release 16 Stage 3 ended in July 2020 and it describes related to UAV services:

- Interference mitigation,
- Location and positioning

Release 17 that is planned for delivery in September 2021, includes some prioritised areas and topics related to UAV and UTM integration into 3GPP systems, that are particularly relevant for 5G!Drones. More specifically these topics are:

- Aerial UE-specific reporting, such as height, location and flight path reporting.
- Remote identification, including UAV-UAV, UAV-ground identification and tracking;
- UAV connectivity requirements and enhancements;
- UAV restrictions

5G Spectrum Auctions and Commercial Launch

Regulators around the world are actively developing their 5G spectrum plans and some have completed the first assignments. The key focus is on new mobile bands including spectrum in the 3.5 GHz band (i.e. 3.3-3.8 GHz) that has been assigned in numerous countries. However, other bands are also being considered. For example:

- Several countries plan to use spectrum in the 4.5-5 GHz range for 5G, including China and Japan;
- A growing number of countries are considering the 3.8-4.2 GHz range, and 5925/6425-7125 MHz;
- There is also interest in assigning the 2.3 GHz and 2.5/2.6 GHz bands for 5G.

On 18 September 2020, the Commission published a Recommendation calling Member States to boost investment in very high-capacity broadband connectivity infrastructure, including 5G. The spectrum-related actions covered by the Recommendation may support the preparation of the future Commission's updated action plan for Europe on 5G and 6G, that is announced in the Commission's Communication 'Shaping Europe's digital future. In its Conclusions from 1 and 2 October 2020, the European Council has asked the Commission to support the acceleration for the deployment of very high capacity and secure network infrastructures such as 5G all over the EU. To ensure this, it also urges all Member States to submit their national plans on the roll-out of 5G to the Commission by the end of 2020, as set out in the 5G Action Plan and to consider the 5G cybersecurity toolbox adopted on 29 January 2020 while doing this.

Even though the COVID-19 pandemic disrupted the preset 5G auction timelines and several spectrum auctions worldwide were postponed, at EU level the 5G spectrum auctions that have already taken place in the three pioneer bands have assigned spectrum for 5G in one or two bands for most of the countries. More specifically:

- The 700 MHz band has been assigned in 7 EU Member States so far including Denmark, France, Finland, Germany, Hungary, Italy and Sweden.
- The 3.4-3.8 GHz band has been assigned in 9 EU Member States: Austria, Czech Republic, Finland, Germany, Hungary, Ireland, Italy, Latvia, Spain.
- The 26 GHz band has been assigned in Italy.

The latest findings under the 5G observatory show overall good progress towards the 5G Action Plan (European Commission, 2016) targets with the commercial launch of 5G services in more than half of EU countries and further launches expected by the end of 2020.

More specifically:

- Telecom Italia (TIM) launched commercial 5G services in Italy after the activation of its 5G network in parts of Rome and Turin on June 2020. In Spain, all major operators have switched on their 5G networks.
- Hrvatski Telekom (HT) on Croatia launched the first commercial 5G network in late October 2020 in six major cities. The launch of the 5G network covered 18% of the Croatian population, and is planned to increase coverage up to 25% of the population.
- In Sweden Telenor also launched commercial 5G services in Stockholm in October 2020.
- Bulgarian operator Vivacom launched commercially 5G network in September 2020. The operator uses dynamic spectrum sharing techniques on the existing 4G network, in order to reach 5G coverage in the 27 biggest cities in the country.
- Vodafone in Czech Republic announced in October 2020 the launch of its NSA 5G network in the 7 major cities (Prague, Brno, Usti and Labem, Jesenik, and Karlovy Vary), using also Dynamic Spectrum Sharing technology.

Europe's 5G Action Plan identifies a number of concrete areas where further actions can contribute to a swift 5G roll-out and adoption in the EU. The Action Plan sets ambitious objectives for pan-European deployment of 5G connectivity infrastructure along major transport paths by 2025. These "5G corridors" are expected to be a key enabler for connected and automated mobility and for the development of innovative ecosystems around cars and other means of transport. Subsequent versions of the 5G Strategic Deployment Agenda will be coordinated under the European Partnership on Smart and Network Services, which is planned to be launched by the first half of 2021.

UAVs and U-Space Regulation

The regulation affects UAV flights in three parts:

- Requirements of UAVs and operators
- Spectrum regulation of drone communication module and payload communication
- Requirements of flight routes and locations (incl. No-Fly-Zones) and U-space regulation

Since 2015 most of the European countries have established UAV related regulation or have amended a pre-existing regulation. From December 31, 2020, the overall European Union UAV regulation- Commission Delegated Regulation (European Union, 2019a) and Commission Implementing Regulation (European Union, 2019b), gradually enter into force for all EU member states. The national rules will remain gradually in force until the 2nd of December 2022 and a transitional arrangement is considered after this date. The rules and procedures for the operation of unmanned aircrafts divide UAVs´ operations into three classes based on their weight and potential effects:

- **Open Category:** Open category operations are limited in visual line of sight (VLOS) and below 120 m altitude of AGL and has a maximum take-off mass of less than 25 kg and during flight, the unmanned aircraft does not carry dangerous goods and does not drop any material.
- **Specific Category:** Where one of the requirements laid down in Open category is not met, a UAS operator shall be required to obtain an operational authorisation from the competent authority in the Member State where is registered. The operator shall perform a risk assessment and submit it along with the application, including adequate mitigating measures.
- **Certified Category:** Where one of the requirements laid down in Open or Specific category is not met, then the operations shall be classified as UAS operations in the "certified" category if they are conducted under one of the following conditions:
    - Over assemblies of people
    - Involves the transport of people
    - Involves the carriage of dangerous goods, that may result in high risk for third parties in case of accident

In summary, BVLOS flights for UAVs supported by 5G communications, fall into the Specific category, with the exception of the case of air-taxi services that fall into the Certified category. On 13 May 2020 EC Commission Implementing Regulation (European Commission, 2020) amended the EC Regulation 2019/947 (European Commission, 2019b). The amendments include definitions, implementing provisions, annex and standard scenarios for operations executed in or beyond the visual line of sight. The objective of the amending regulation is to increase the cost-effectiveness for drone operators, manufacturers and competent authorities, and to improve the harmonisation of drone operations throughout Europe. The amendments cover two standard scenarios to facilitate some UAS operations posing a low risk in the specific category. For those, drone operators will be allowed to just send a declaration to the respective authority instead of applying and waiting for an authorisation. The standard scenarios include Urban VLOS (Visual Line of Sight) and Rural BVLOS (Beyond Visual Line of Sight) above control ground area operations (EASA, 2019). These European standard scenarios (STS) intend to replace the national scenarios S1, S2 and S3. They are based on a declarative regime, and require the use of class CE marked drones (not yet available on the market). It should be noted that the development of standard scenarios for BVLOS flights in populated areas, is the next important step to support the wider deployment of cellular UAV services.

U-space Regulation

In addition to regulation on drones, EU is working on a regulatory framework for unmanned traffic management (UTM) system solutions, that in Europe are called U-space. EASA (European Union Aviation Safety Agency) published an Opinion on March 13, 2020 "High-level regulatory framework for the U-space" as well as an Appendix (EASA,2020a). The objective of this opinion is to create and harmonise the necessary conditions for manned and unmanned aircraft to operate safely in the U-space airspace, to prevent collisions between aircraft and to mitigate the air and ground risks. Therefore, the U-space regulatory framework, supported by clear and simple rules, should permit safe aircraft operations in all areas and for all types of unmanned operations.

This Opinion contains a draft regulation and is submitted to the European Commission, which will use it as a technical basis in order to prepare an EU regulation and Appendices 1, 2 and 3 to draft Commission Implementing regulation. For information, EASA published the draft text for the related EASA decision that contains acceptable means of compliance (AMC) and guidance material (GM) (EASA, 2020b). The final decision that issues the AMC & GM will be published by EASA once the European Commission has adopted the regulation and once the necessary consultation with the affected stakeholders has been performed. The Opinion is therefore, a first regulatory step to allow immediate implementation of the U-space after the entry into force of the Regulation and to let the unmanned aircraft systems and U-space technologies evolve. The Opinion with appendices is published on EASA website and also Terms of reference for rulemaking task (EASA, 2018). EC regulations regarding U-space are expected to enter into force in 2022 based on past regulatory development experience. This regulation provides the basis for the development of U-space Initial services to achieve the functionality of U2, which is planned for the period 2022-2027, as described in the section of UAV and U-space Ecosystem.

FUTURE RESEARCH DIRECTIONS

Applications of unmanned aerial vehicles (UAVs) have been fast growing during the past few years. However, traditional UAV systems mainly rely on simple direct communication between the UAV and the ground pilot over unlicensed spectrum (e.g., ISM 2.4 GHz), which is typically of low data rate, unreliable, insecure, vulnerable to interference, difficult to monitor and manage, and has very short operation range. To overcome the above limitations, there has been significant interest in integrating UAVs into cellular networks, e.g., 5G networks., Future research directions include the use of advanced machine learning techniques utilizing the edge computing nodes of the 5G network, in order to support the UAV missions in sophisticated use-cases, such as how to determine their optimal positioning or trajectory, how to perform resource management or manage interference to name a few. One way of solving these issues is by analysing data coming from the network to learn patterns and react accordingly, resulting in an autonomous AI-driven UAV, capable of taking decisions. Furthermore, another area of applicability is the realization of IoT, since integrating UAVs in IoT networks can bring several benefits, not only in terms of the aerial network, but also regarding the lifetime and networking of the ground sensors.

CONCLUSION

In this chapter, we provided a comprehensive overview of the latest research towards the integration of the UAVs into cellular networks, by covering aspects in both UAV and 5G ecosystem. Moreover we described in details an implementation (Trial Controller) related to the reference integration architecture (UAS-5G) that is being designed and developed by the European Research project 5G!Drones. However, it is worth mentioning that the majority of 5G implementations worldwide (and probably all European ones) are NSA, and the 5G mechanisms that can provide the UAV-specific support within the 5GS are under development. Studies on architecture and application layer support are ready, but the TS phase is ahead and the actual implementation by vendors will need some additional time. Thus, taking into account the advancements in cellular technologies and the deployment of 5G cellular infrastructure to date, integrating unmanned aerial vehicles (UAVs) into the fifth-generation (5G) cellular networks is a promising solution in order to enhance and support UAV operation as well as enabling diversified applications with mission-specific characteristics.

ACKNOWLEDGMENT

This work was partially supported by the European Union's Horizon 2020 Research and Innovation Program under the 5G! Drones project (Grant No. 857031).

REFERENCES

GSM Association. (2020). *Generic Network Slice Template.* https://www.gsma.com/newsroom/wp-content/uploads//NG.116-v3.0.pdf

CAAC. (2017, May). *Regulations on Real-name Registration of Civil Unmanned Aerial Vehicles.* http://www.caac.gov.cn/en/XWZX/201705/t20170524_44222.html

Drones Project. (2019). *D2.1 Initial definition of the trial controller architecture mechanisms and APIs.* https://5gdrones.eu/wp-content/uploads/2020/06/D2.1-Initial-definition-of-the-trial-controller-architecture-mechanisms-and-APIs_v1.1.pdf

Electronic Communication Committee. (2018). *ECC Report 268: Technical and Regulatory Aspects and the Needs for Spectrum Regulation for Unmanned Aircraft Systems (UAS).* https://docdb.cept.org/download/efbd2f57-f227/ECCRep268.pdf

Electronic Communication Committee. (2019). *ECC Report 298: Analysis of the suitability and update of the regulatory technical conditions for 5G MFCN and AAS operation in the 1920 -1980 MHz and 2110-2170 MHz band.* https://docdb.cept.org/download/5b21b308-d1e7/ECC%20Report%20298.pdf

Electronic Communication Committee. (2019a). *ECC Report 297: Analysis of the suitability and update of the regulatory technical conditions for 5G MFCN and AAS operation in the 900 MHz and 1800 MHz bands.* https://docdb.cept.org/download/5949a9c0-66bd/ECC%20Report%20297.pdf

Electronic Communication Committee. (2019b). *ECC Decision (05) 05. Harmonised utilization of spectrum for Mobile/Fixed Communications Networks (MFCN) operating within the band 2500-2690 MHz.* https://docdb.cept.org/download/217614c9-9edd/ECCDec0505.pdf

Electronic Communication Committee. (2020). *ECC Report 309: Analysis of the usage of aerial UE for communication in current MFCN harmonised bands.* https://docdb.cept.org/download/9f4305fb-aa77/ECC%20Report%20309.pdf

Electronic Communication Committee. (2020a). *ECC Report 308: Analysis of the suitability and update of the regulatory technical conditions for 5G MFCN and AAS operation in the 2500-2690 MHz band.* https://docdb.cept.org/download/cd436696-88f5/ECC%20Report%20308.pdf

European Commission. (2016). *Communication from the Commission to the European Parliament, the Council, the European Economic and Social Committee and the Committee of the Regions: "5G for Europe: An Action Plan".*

European Commission. (2018). *On the road to automated mobility: An EU strategy for mobility of the future.* https://ec.europa.eu/transport/sites/transport/files/3rd-mobility pack/com20180283_en.pdf

European Commission. (2019). *Commission Implementing Decision (EU) 2019/784 on harmonisation of the 24,25-27,5 GHz frequency band for terrestrial systems capable of providing wireless broadband.* https://eur-lex.europa.eu/legal-content/EN/TXT/?uri=CELEX%3A32019D0784

European Commission. (2019a). *Commission Delegated Regulation (EU) 2019/945 on unmanned aircraft systems and on third-country operators of unmanned aircraft systems.* https://eur-lex.europa.eu/legal-content/EN/TXT/?uri=CELEX:32019R0945

European Commission. (2019b). *Commission Implementing Regulation (EU) 2019/947 on the rules and procedures for the operation of unmanned aircraft electronic communications services in the Union.* https://eur-lex.europa.eu/eli/reg_impl/2019/2072/oj

European Commission. (2020). *Commission Implementing Regulation (EU) 2020/639 amending Implementing Regulation (EU) 2019/947 as regards standard scenarios for operations executed in or beyond the visual line of sight.* https://www.easa.europa.eu/document-library/regulations/commission-implementing-regulation-eu-2020639

European Union Aviation Safety Agency. (2018). *Terms of reference for rulemaking task RMT.0230. Regulatory framework to accommodate unmanned aircraft systems in the European aviation system.* https://www.easa.europa.eu/sites/default/files/dfu/ToR%20RMT.0230%20%E2%80%93%20Issue%202.pdf

European Union Aviation Safety Agency. (2019). *Opinion 05/2019 Standard scenarios for UAS operations in the 'specific' category*. https://www.easa.europa.eu/document-library/opinions/opinion-052019

European Union Aviation Safety Agency. (2020). *Opinion No 01/2020 High-level regulatory framework for the U-space*. https://www.easa.europa.eu/sites/default/files/dfu/Opinio n%20No%2001-2020.pdf

European Union Aviation Safety Agency. (2020a). *Appendix to Opinion No 01/2020. Comment-response document (CRD) to draft Opinion*. https://www.easa.europa.eu/sites/default/files/dfu/Appendix to Opinion No 01-2020.pdf

European Union Aviation Safety Agency. (2020b). *Draft acceptable means of compliance (AMC) and Guidance Material (GM) to Opinion No 01/2020 on a high-level regulatory framework for the U-space*. https://www.easa.europa.eu/sites/default/files/dfu/Draft%20AMC%20%26%20GM%20to%20the%20 U-space%20Regulation%20%E2%80%94%20for%20info%20only.pdf

Godage, L. (2019). *Global Unmanned Aerial Vehicle Market (UAV) Industry Analysis and Forecast (2018-2026)*. https://www.inkwoodresearch.com/reports/unmanned-aerial-vehicles-market/

Goldman Sachs. (n.d.) *Drones reporting for work*. https://www.goldmansachs.com/insights/technology-driving-innovation/drones/

GPP TR 23.558 (2020o) "*Architecture for enabling Edge Applications (EA)*" Rel. 17 v2.0.0

GPP TR 23.700-40 (2020f) "*Study on enhancement of network slicing; Phase 2*" Rel. 17 v2.0.0

GPP TR 23.748 (2020n) "*Study on enhancement of support for Edge Computing in 5G Core network (5GC)*" Rel. 17 v17.0.0

GPP TR 23.752 (2020i) "*Study on system enhancement for Proximity based Services (ProSe) in the 5G System (5GS)*" Rel. 17 v2.0.0

GPP TR 23.754 (2020d) "*Study on supporting Unmanned Aerial Systems (UAS) connectivity, Identification and tracking*" Rel. 17 v17.0.0

GPP TR 23.755 (2020e) "*Study on application layer support for Unmanned Aerial Systems (UAS)*" Rel. 16 v2.0.0

GPP TR 23.758 (2019) "*Study on application architecture for enabling Edge Applications*" Rel. 17 v17.0.0

GPP TR 23.948 (2019a) "*Deployment guidelines for typical edge computing use cases in 5G Core network (5GC)*" Rel. 17, no published document, yet

GPP TR 28.811 (2020g) "*Study on network slice management enhancement*" Rel. 17 v0.4.0

GPP TR 36.777 (2018) "*Enhanced LTE support for aerial vehicles*" Rel. 15, v15.0.0

GPP TR 38.836 (2020j) "*Study on NR sidelink relay*" Rel. 17 v2.0.0

GPP TS 22.104 (2020q) "*Service requirements for cyber-physical control applications in vertical domains*" Rel. 17 v17.4.0

GPP TS 22.125 (2020b) "*Unmanned Aerial System (UAS) support in 3GPP*" Rel. 16 v17.2.0

GPP TS 22.261 (2020c) "*Service requirements for the 5G system*" Rel. 15 v18.1.1

GPP TS 22.263 (2020p) "*Service requirements for Video, Imaging and Audio for Professional Applications (VIAPA)*" Rel. 17 v17.3.0

GPP TS 23.273 (2020r) "*5G System (5GS) Location Services (LCS); Stage 2*" Rel. 16 v16.5.0

GPP TS 23.501 (2020) "*System architecture for the 5G System (5GS)*" Rel. 15 v16.7.0

GPP TS 23.501 (2020a) "*System architecture for the 5G System (5GS)*" Rel. 15 v16.7.0

GPP TS 28.313 (2020m) "*Self-Organizing Networks (SON) for 5G networks*" Rel. 16 v17.0.0

GPP TS 37.320 (2020l) "*Universal Terrestrial Radio Access (UTRA) and Evolved Universal Terrestrial Radio Access (E-UTRA); Radio measurement collection for Minimization of Drive Tests (MDT); Overall description; Stage 2*" Rel. 10, v16.3.0

GPP TS 38.101-1, (2020s) "*NR; User Equipment (UE) radio transmission and reception; Part 1: Range 1 Standalone*", Rel. 15 v17.0.0

GPP TS 38.101-2, (2020t) "*NR; User Equipment (UE) radio transmission and reception; Part 2: Range 2 Standalone*", Rel. 15 v17.0.0

GPP TS 38.215 (2020k) "*NR; Physical layer measurements*" Rel. 15 v16.4.0

GPP TS 38.300 (2020h) "*NR; NR and NG-RAN Overall description; Stage-2*" Rel. 15 v16.4.0

GPPP. (2020) *5GPPP Empowering Vertical Industries through 5G Networks - Current Status and Future Trends.* https://5g-ppp.eu/wp-content/uploads/2020/09/5GPPP-VerticalsWhitePaper-2020-Final.pdf

GSMA. (2020). *Global 5G Landscape, Q3 2020.* https://data.gsmaintelligence.com/research/research/research-2020/global-5g-landscape-q3-2020

Hetzer, D., Muehleisen, M., Kousaridas, A., & Alonso-Zarate, J. (2019, June). *5G Connected and Automated Driving: Use Cases and Technologies in Cross-border Environments.* In *2019 European Conference on Networks and Communications (EuCNC)* (pp. 78-82). IEEE.

ITU –R. (2015). *IMT Vision – Framework and overall objectives of the future development of IMT for 2020 and beyond.* https://www.itu.int/dms_pubrec/itu-r/rec/m/R-REC-M.2083-0-201509-I!!PDF-E.pdf

Koumaras, H., Makropoulos, G., Batistatos, M., Kolometsos, S., Gogos, A., Xilouris, G., Sarlas, A., & Kourtis, M. A. (2021). 5G-Enabled UAVs with Command and Control Software Component at the Edge for Supporting Energy Efficient Opportunistic Networks. *Energies, 14*(5), 1480.

Ksentini, A., & Frangoudis, P. A. (2020). Towards Slicing-Enabled Multi-access Edge Computing in 5G. *IEEE Network, 34*(2), 99–105.

Kuklinski, S., Tomaszewski, L., Korzec, P., & Kołakowski, R. (2020). 5G-UASP: 5G-based multi-provider UAV platform architecture. *6th IEEE International Conference on Network Softwarization (NetSoft 2020).*

Marojevic, V., Guvenc, I., Dutta, R., Sichitiu, M. L., & Floyd, B. A. (2020). Advanced Wireless for Unmanned Aerial Systems: 5G Standardization, Research Challenges, and AERPAW Architecture. *IEEE Vehicular Technology Magazine*, *15*(2), 22–30.

Mondal, S., Nandi, D., & Bera, R. (2020, February). V2X Communication Test Bed for Smart Electrical Vehicle with 5G IOV Technology. In *2020 URSI Regional Conference on Radio Science (URSI-RCRS)* (pp. 1-4). IEEE.

Muzaffar, R., Raffelsberger, C., Fakhreddine, A., Luque, J. L., Emini, D., & Bettstetter, C. (2020, June). First experiments with a 5G-Connected drone. In *Proceedings of the 6th ACM Workshop on Micro Aerial Vehicle Networks, Systems, and Applications* (pp. 1-5). ACM.

PwC. (2016). *Global report on the commercial applications of drone technology.* https://www.pwc.pl/pl/pdf/clarity-from-above-pwc.pdf

SESAR Joint Undertaking. (2017). *U-space: Blueprint.* https://op.europa.eu/en/publication-detail/-/publication/f8613e25-cf38-11e7-a7df-01aa75ed71a1/language-en

SESAR Joint Undertaking. (2019). *Concept of Operations for U-space.* https://www.sesarju.eu/node/3411

SESAR Joint Undertaking. (2020). *U-space.* https://www.sesarju.eu/U-space

Si-Mohammed, S., Bouaziz, M., Hellaoui, H., Bekkouche, O., Ksentini, A., Taleb, T., Tomaszewski, L., Lutz, T., Srinivasan, G., Jarvet, T., & Montowtt, P. (2020). Supporting UAV Services in 5G Networks: New High-Level Architecture integrating 5G with U-space. *IEEE Vehicular Technology Magazine.*

Tomaszewski, L., Kołakowski, R., & Korzec, P. (2020b, June). On 5G support of cross-border UAV operations. In *2020 IEEE International Conference on Communications Workshops (ICC Workshops)* (pp. 1-6). IEEE.

Tomaszewski, L., Kołakowski, R., & Kukliński, S. (2020a, June). Integration of U-space and 5GS for UAV services. In *2020 IFIP Networking Conference (Networking)* (pp. 767-772). IEEE.

Tomaszewski, L., Kukliński, S., & Kołakowski, R. (2020, June). A new approach to 5G and MEC integration. In *IFIP International Conference on Artificial Intelligence Applications and Innovations* (pp. 15-24). Springer.

Wallace, R. J., Robbins, J. M., Loffi, J. M., Holliman, J. K., Metscher, D. S., & Rogers, T. R. (2020). Evaluating LAANC Utilization & Compliance for Small Unmanned Aircraft Systems in Controlled Airspace. *International Journal of Aviation, Aeronautics, and Aerospace*, *7*(2), 4.

Chapter 4
Development Electronic Design Automation for RF/Microwave Antenna Using MATLAB GUI

Kok Yeow You
https://orcid.org/0000-0001-5214-7571
Universiti Teknologi Malaysia, Malaysia

ABSTRACT

This chapter presents the development of a user-friendly MATLAB-based electronic design automation (EDA) tool for RF/microwave antenna design. The development steps of EDA using MATLAB 'guide' and MATLAB 'appdesigner' design environments have been explained in detail. In this chapter, seven EDAs of different antennas are developed and described. The EDA is capable of predicting the dimensions of the antenna by inputting the desired operating specifications of the antenna, such as operating frequency. The rigorous analytical formulations that are implemented in this EDA are briefly described. The developed EDAs have simple and standalone guide user interface (GUI) display and fast program running time compared to iterative solution or numerical-based EDAs. The developed EDAs can be packaged and accessed either as standalone desktop app or web app. The accuracy of the EDA output results is validated by comparing to measurement and simulation results obtained from a commercial simulator.

INTRODUCTION

Nowadays, radio (RF) and microwave devices have been widely used and have become indispensable necessities in our daily work, such as 5G communication networks, Wi-Fi, Bluetooth, mobile phones, the Internet, and RFID-based security system in work offices. The rapid development of communication technology is mainly attributed to the existing electronic design automation (EDA). Electronic design automation can design and modify communication components and devices in a short period of time, such as making 5G improvements to 4G technologies. The manufacturer's production process time and cost (no need to build trial prototypes) can be significantly reduced and obtain product prices that the general public can afford.

DOI: 10.4018/978-1-7998-6992-4.ch004

Now, the competition for EDA microwave business is accelerating, EDA manufacturers are trying to promote their products to the university to populate the use of EDA among university students. This is because future graduates' students will indirectly determine the brand and type of EDA used in the manufacturer in which they work. At the same time, the university is also trying to study the most popular types of EDA used by manufacturers before purchasing the EDA software for student learning and research. Many types of microwave EDA have been sold on the market, some specially for solving planar shape (printed circuit board) problems, and some provide a wide range of solutions for all situations as listed in Table 1. Due to the increase in PC technology, the complexity and functions of EDA have also increased from 2D virtual view to 3D virtual view, and now has animation simulation capabilities. In addition, the dimensions of the microwave device to be designed can be automatically optimized and adjusted through the latest version of EDA.

Nowadays, emerging 4G/5G communication circuit tends to assembled with nanoscale monolithic-microwave-integrated circuits (MMICs), hence designers can no longer design their circuit parts independently, since designers need to consider specifications and performance of the MMICs on the circuit. Therefore, most EDAs have built-in MMIC packages/modules, allowing designers to simulate microwave circuits containing certain brands and models of MMICs. However, if the designer only focuses on the antenna part, the designer can still design the antenna independently and ignore the circuit assembly effect.

In this chapter, the development of simple EDA (so-called antenna calculator) using MATLAB GUI for several antennas has been presented, such as monopole antenna, Yagi-Uda array antenna, log-periodic antenna, rectangular waveguide antenna, helical antenna, pyramidal horn antenna, and conical monopole antenna. Although, there are free antenna calculators that microwave manufacturers provide online or download, but those calculators are usually simple and only shows one or two calculated outputs. In fact, the latest version of MATLAB software has been attached with "antenna designer" app. The complexity and accuracy for the "antenna designer" app is quite high since the numerical-based solutions have been used in the app.

There may be a question to be asked why it is necessary to review analytical formulas since now advanced numerical-based EDAs are available? Indeed, the main factor is the cost of the simulators and computer workstations. In addition, the analytical analysis is particularly useful when designer would like to use for initial design an antenna that operates at very low frequencies ($f < 100$ MHz) or at very high frequencies ($f > 20$ GHz). Some low-frequency antenna has a size up to few meters. On the other hand, the size for high-frequency antenna may only a few millimeters. Those situations will provide difficulties in numerical-based simulation (require very high density and large amount of mesh in simulation as well as time consuming) using a regular personal computer (PC) or high-medium end PC.

Since analytical solutions are implemented in this study EDA, hence the program running time is significantly reduced (less than 1 second using a PC with an i5 processor and 4 GB RAM). Besides, the EDA's GUI is created in a simple and compact form, which is less input parameters and at the same time provides high accuracy of simulation results. The calculated results from this MATLAB-based EDA are compared with the results obtained from commercial EDA for accuracy verification. The analytical formulas used in the study EDA will be described in detail. In this chapter, open-source and commercial microwave EDA are comprehensively reviewed.

In fact, the GUI is an appropriate element that can be incorporated into some engineering curriculum nowadays. For instance, teaching content for antenna engineering subject, mostly involves formulas for antenna design. If the GUI is applied in these formulas, this will increase students' interest in learning the subject. Besides, the teaching element of the subject will be closer to the current era of automation.

The chapter is expected can be used as a brief guideline/reference to revise some telecommunication engineering curriculum in the future.

COMPUTATIONAL METHODS IN MICROWAVE ANTENNAS AND PASSIVE DEVICES

Brief History of Computational Methods in Antennas

In the beginning, many antenna inventions were modified from regular waveguides and had simple structures, such as linear monopole antennas (Marconi, 1896), dipole antennas (Hertz, 1888), Yagi-Uda antennas (Yagi and Uda, 1926), slotted rectangular waveguide antennas (Alan, 1938), and wire loop antenna. The operating frequency and performance of these simple antennas mainly depends on their physical size and structure. The antenna size design can be achieved by referring to specific simple design guidelines.

During the second world war, microwave was widely applied in the development of telecommunication equipment and facilities for military use. Hence, in the 1940s, a lot of rigorous analytical formulas were intensively developed for the microwave antennas and other passive devices design purpose (Atta, 1981), and such work began to be published in the early 1950s. At that time the analytical approach was still very useful for antenna analysis. A lot of these analytical formulas are collected in the 'Waveguide Handbook' (Marcuvitz, 1964). The analytical formulas for various typical forms of antennas/waveguides can also be found in some textbooks (Stratton, 1941; Silver, 1949; Collin, 1985; Wadell, 1991; King, 2013; Balanis, 2015).

In the 1960s, more and more antennas/microwave devices were built in planar-circuit form to reduce the size and cost of the devices. Thus, research and derivation of analytical formulations for various types of strip transmission lines is carried out rapidly for computer-aided design (CAD) purpose. Until now, the analytical formulations are still very useful in the initial design of antenna in planar-circuit structure. Analytical theories and formulas of the planar transmission lines have been comprehensively collected in one of the research book entitled 'Analysis Methods for RF, Microwave, and Millimeter-Wave Planar Transmission Line Structures' (Nguyen, 2001).

In the late 1960s, various types of modern numerical methods, such as method of moments (MoM) (Harrington, 1967), finite difference time domain (FDTD) (Yee, 1966), and finite element method (FEM) (Silvester, 1969), began to be proposed for antennas and waveguides analysis. The numerical-based computation methods intensively studied and developed to replace the existing analytical methods, which are not adequate for use in the invention of the latest broadband and high-frequency (up to few GHz) antennas, such as phased-array antennas for radar applications. Later, other emerging numerical methods, such as uniform theory of diffraction (UTD/GTD) (Kouyoumjian and Pathak, 1974), finite integration technique (FIT) (Weiland, 1977), transmission line matrix (TLM), and physical optics (PO), were studied depending on the operating frequency, geometrical complexity of the antennas, and applications, which had to be resolved. Until the end of 1990s, a lot of EDAs from the results of the study have been commercialized. For instance, the FEKO is commercialized in 1997, while the CST Microwave Studio and the COMSOL Multiphysics (Formerly FEMLAB) are respectively founded in 1998. Most of commercial EDA simulators involves a combination of several numerical methods so-called hybrid technique, in order to solve various cases of antenna problems as listed in Table 1.

Table 1. Commercial RF/microwave electronic design automation (EDA) list

Manufacturer	EDA Software	Place
Ansoft Corporation	**High-Frequency Structure Simulator (HFSS)** 3D full-wave finite element code	Pittsburgh, PA, USA
COMSOL Multiphysics	**COMSOL RF Module (Formerly FEMLAB)** 3D static and full-wave finite element/MoM code	Stockholm, Sweden
Sonnet Software Inc.	**Sonnet Suites** 3D method of moments (MoM) code for planar structures	Liverpool, NY, USA
Altair Engineering Inc. Formerly EM Software and Systems	**FEKO** 3D hybrid moment-method / UTD-physical optics code Finite element code, multi-level fast multipole code, geometrical optics (GO) and physical optics. (Recently)	Troy, Michigan, USA Stellenbosch, South Africa
Mician	**Microwave Wizard (μWave Wizard)** 2D/3D finite element method and mode matching techniques.	Bremen, Germany
Dassault Systèmes Simulia Corp. Formerly Computer Simulation Technology (CST)	**Computer Simulation Technology (CST) Microwave Studio** 3D full electromagnetic analysis (FEM, MoM, MLFMM), Perfect boundary approximation/ Finite integration method	Vélizy-Villacoublay, France Darmstadt, Germany
Keysight Technologies. Formerly Hewlett-Packard (HP) and Agilent Technologies	**Momentum** Method of moments (MoM) for 3D planar structures	Santa Rosa, California, USA
Keysight Technologies	**Advanced Design System (ADS)** 2D and 3D RF planar circuit	Santa Rosa, California, USA
Remcom, Inc.	**XFDTD** 3D Full wave finite difference time domain (FDTD)	315 S. Allen St. Suite 222 State College, PA 16801
Poynting Innovations Ltd.	**SuperNEC** Hybrid method of moment (MoM)/Unified theory of diffraction (UTD)	Wendywood, 2144 South Africa
Cadence Design Systems. Formerly Applied Wave Research, Inc. (AWR) and National Instruments.	**Microwave Office** Method of moments for planar structures	Los Angeles, California, USA; San Jose, CA, USA; Austin, Texas, USA
EM Scientific, Inc.	**MiniNEC Professional** Method of moments for thin wire structures	Carson City, NV
Zeland Software, Inc.	**Integral Equation Three-Dimensional (IE3D)** Method of moments for planar circuit structures	Zeland
Dassault Systèmes Simulia Corp	**Antenna Magus** Up to few hundred antenna structures design	Vélizy-Villacoublay, France
FILPAL(S)Pte. Ltd.	**FILPAL** Topological matrix for 2D planar RF circuits	Malaysia

Analytical and Numerical Methods for Antenna Analysis

Most analytical formulations are derived from Maxwell's equation with the aid of other mathematical techniques, such as method of images, Gauss's theorem, Stokes's theorem, Green's theorem, Poynting' theorem, cavity theory, and conformal mapping. For instance, the equation (1) of microstrip line was derived using conformal mapping technique (Wheeler, 1965; Hammerstad, 1975). In addition, the integral equation (4) of monopole antenna impedance was derived based on method of images (Brillouin, 1922; Weiner, 2003).

Since the wave propagating through the antennas are the scattered wave, most analytical formulas of the antennas are expressed in integral form, which contains the Green function term, G (the so-called kernel function of the integral equation), which is satisfied by the unbounded scattered wave as (You, 2016):

$$G = \frac{e^{-jk|r_2 - r_1|}}{4\pi |r_2 - r_1|}$$

Hence, the exponential properties of above expression term can almost be found in the electromagnetic integral equations, such as equations (4) and (33).

Sometimes, the analytical formulas are alternatively expressed in Bessel's integral form in order to facilitate mathematically solving such problems. The derived integral formulations are normally solved by some classical numerical methods, such as series expansion and numerical integral methods (trapezium's rules, Simpson's rules or Gaussian quadrature rules). Normally, in analytical analysis, it is assumed that all variables in the analytical formula are ideal and known. Sometimes, the variables are expressed using closed-form functions.

With the advent of computer technology in 1960, attention to the research and usage of numerical methods (instead of analytical methods) in antennas and other microwave passive devices design has raised increasingly. In numerical analysis approach, at least one of the variables in the theoretical formula is assumed to be an unknown value. Usually, the variable is a set of unknown values which is solved using modern numerical methods, such as matrix methods. For instance, Harrington (1967) was first proposed the matrix method to solve the Fredholm integral equation of the first kind for dipole antenna problems. Recently, the numerical matrices method has been named as method of moments (MoM) and has been extended to solve various types of antenna problems. Sometimes, for other applications, the method is called the boundary element method (BEM).

Let the monopole antenna case be used as an example to understand the difference between analytical and numerical MoM. The integral equation of input impedance, Z_{in}, formula of the z-component electrical field, E_z, and distributed current, I_z on the antenna surface, which are used in analytical and numerical analysis, are listed in Table 1. In the analytical method, the sinusoidal distributed current, I_z is assumed to be a real value and its value become zero at the end of the monopole. In addition, the amplitude of the distributed current, I_o is assumed to be constant. Hence, the closed-form distributed current, I_z is only agreeable in cases where the monopole antenna has a small diameter (thin monopole) and operates at low frequencies.

On the other hand, in the numerical MoM approach, the amplitude of the current, I_o is an unknown variable to be determined. The I_z is not guarantee to be zero at the end of the monopole antenna. The I_o value maybe in complex depends on the antenna size and operating frequencies. In addition, the solved I_o value took into account the effects of self- and mutual-coupling at each pointed position on the antenna surface. Thus, the numerical approach provides calculated results that are closer to the actual situation (practical measurement) than the analytical approach. However, there is a close relationship between the analytical and numerical MoM approaches, due to the fact that MoM cannot be used if these is no established analytical integral equation (impedance integral equation) serves as a governing function in monopole antenna problem. In MoM, the set of unknown I_o value can only be found by discretizing the analytical integral equation of Z_{in} as shown in Table 2.

In fact, the main difference between the analytical and numerical MoM is the complexity of the solution and the accuracy of the solution. For example, the input impedance, Z_{in} of the monopole antenna can be analytically determined by solving the impedance integral equation using Simpson's rules (see Table 2). Besides, the input impedance, Z_{in} of the monopole antenna can also be obtained indirectly using MoM, which the Z_{in} is required to be determined based on the solved I_n value as:

$$Z_{in} = \frac{V_o}{I_{(N+1)/2}}$$

where V_o (= 1 V) is the electric potential at the excitation position of the monopole antenna. On the other hand, $I_{(N+1)/2}$ is the solved I_n at the excitation position of the monopole antenna. It should be noted that the $I_{(N+1)/2}$ is a complex value. The analytical method to determine Z_{in} is simpler, but its value is not as accurate as the Z_{in} value obtained from MoM. For a detail description of the relationship between analytical method and MoM method, can refer to 'RF Coaxial Slot Radiators: Modeling, Measurements and Applications' (You, 2016).

FEM and FDTD are numerical methods that use partial differential equations (wave equation or Helmholtz equation) as the governing equations. Normally, iterative techniques are used to solve the governing partial differential equations. However, initially, compared to MoM, the FEM and FDTD are rarely used for the unbounded scattered wave problems, such as antenna radiation problems. The reason is that FEM and FDTD need to discretize the entire computation domain including the boundary on the antenna surface and also the scattered wave or radiation region. On the other hand, MoM only needs to discretize the boundary on the antenna surface. Hence, the stepwise iterative process in FEM/FDTD requires to cover a large unbounded scattered wave region (must far from the excitation source). The computational load is much higher for FEM/FDTD compared with MoM under the same antenna problem. In addition, in FEM/FDTD computation, the scattered wave region is limited into finite space, thus the virtual outer boundary conditions in the scattered wave region need to be precisely defined.

Previously, in FEM/FDTD computation, the scattered wave region is sealed with the Neumann boundary. However, the Neumann boundary is not very reliable for accurately simulating unbounded radiation characteristics. In 1981, Mur proposed an effective virtual outer boundary model for scattered wave in finite computation space/domain, namely absorbing boundary condition (ABC) (Mur, 1981). The ABC reduces the effect of the scattering field reflection at the outer boundary of the scattered wave region. At the same time, ABC truncates the computation space. Later, the most efficient ABC was developed by Berenger, who assumed that the scattered wave region is surrounded by a thin artificial absorbing layer, called perfectly matched layer (PML) (Berenger, 1994). Subsequently, a lot of researchers modified Mur and Berenger's work to improve the simplicity and the efficiency of absorption at the outer boundary of the scattered wave region. Therefore, nowadays, many FDTD- and FEM-based EDAs have been used to solve the antenna problems.

Table 2. Comparison formulations between analytical method and MoM-based numerical method (You, 2016)

Analytical method	Method of moments (MoM)
Impedance integral equation is given as: $$Z_{in} = \frac{j\eta_o}{4\pi\sin^2(kh)} \times \left\{ \int_0^h \left\{ \sin\left[k(h-z)\right] \times \left[\frac{e^{-jkr_1}}{r_1} + \frac{e^{-jkr_2}}{r_2} - 2\cos(kh)\frac{e^{-jkr_3}}{r_3} \right] \right\} dz \right\}$$ where $r_1 = \sqrt{\rho^2+(z-h)^2}$; $r_2 = \sqrt{\rho^2+(z+h)^2}$; $r_3 = \sqrt{\rho^2+z^2}$ The impedance integral equation can be solved by numerical integration, such as Simpson's rules.	Discretization of impedance integral equation Z_{mn} is given as: $$Z_{mn} = \frac{j\eta_o}{4\pi\sin^2(k\Delta z)} \times \left\{ \int_{z_{m-1}}^{z_{m+1}} \sin\left[k\left(\Delta z - \lvert z' - z_m \rvert\right)\right] K_{mn} dz \right\}$$ where K_{mn} is expressed as: $$K_{mn} = \frac{e^{-jkr_{n+1}}}{r_{n+1}} + \frac{e^{-jkr_{n-1}}}{r_{n-1}} - 2\cos(k\Delta z)\frac{e^{-jkr_n}}{r_n}$$ and $r_{n-1} = \sqrt{a^2+(z-z_{n-1})^2}$; $r_n = \sqrt{a^2+(z-z_n)^2}$; $r_{n+1} = \sqrt{a^2+(z-z_{n+1})^2}$
The electric potential, V_z can be analytically solved as: $$V_z = -\int_0^h E_z dz$$ where the E_z is given as: $$E_z = \frac{j\eta_o I_o \sin\left[k(h-z)\right]}{4\pi\sin(kh)} \times \left[\frac{e^{-jkr_1}}{r_1} + \frac{e^{-jkr_2}}{r_2} - 2\cos(kh)\frac{e^{-jkr_3}}{r_3} \right]$$ where $r_1 = \sqrt{\rho^2+(z-h)^2}$; $r_2 = \sqrt{\rho^2+(z+h)^2}$; $r_3 = \sqrt{\rho^2+z^2}$	Discretization electric potential, V_m along the monopole antenna can be rewritten as: $$V_m = \int_{z_{m-1}}^{z_{m+1}} E_{zm} \sin\left[k\left(\Delta z - \lvert z - z_m \rvert\right)\right] dz$$ Discretization electric field, E_{zm} is expressed as: $$E_{zm} = \frac{-j\eta_o}{4\pi\sin(k\Delta z)} \times \left[\frac{e^{-jkr_{m+1}}}{r_{m+1}} + \frac{e^{-jkr_{m-1}}}{r_{m-1}} - 2\cos(k\Delta z)\frac{e^{-jkr_m}}{r_m} \right]$$ and $r_{m-1} = \sqrt{a^2+(z-z_{m-1})^2}$; $r_m = \sqrt{a^2+(z-z_m)^2}$; $r_{m+1} = \sqrt{a^2+(z-z_{m+1})^2}$
The distributed current, I_z along the linear monopole antenna is expressed in closed-form equation as: $$I_z = I_o \frac{\sin\left[k(h-z)\right]}{\sin(kh)}, \text{ for } 0 \le z \le h$$ where z is the observation point along the monopole antenna. On the other hand, I_o is the amplitude of maximum current.	Using Ohm's law and the Kirchhoff's network, the relationship between Z_{mn}, V_m, and I_n: $$[Z_{mn}][I_n] = [V_m]$$ The unknown current, I_n at each position (n = 1, 2, 3,.., N) along the monopole antenna can be solved using matrix solution: $$[I_n] = [Z_{mn}]^{-1}[V_m]$$

MATLAB GUI AND APP

In this section, the steps of GUI development using MATLAB 'guide' and 'appdesigner' antenna design and measurement have been described in detail. In fact, the MATLAB 'appdesigner' is the new feature in version R2016a as alternative for 'guide'. The comparison user access and sharing features for both GUI environment is tabulated in Table 3.

Table 3. Comparison user access and sharing features between the MATLAB 'guide' and MATLAB 'appdesigner'

App Sharing and Access	Design Environment	
	guide	appdesigner
Share as MATLAB App Install File	/	/
Share as MATLAB Standalone Desktop App	/	/
Share as MATLAB Web App		/
Create, Edit, and Run Apps in MATLAB Online		/
Export as Standalone MATLAB Class File		/

Since the microstrip transmission line calculator is an important, simple, and common program tool in planar antenna design as well as high-frequency circuit design, hence the GUI for microstrip line calculator has been taken as an example in this demonstration.

Microstrip Line Calculator GUI Using MATLAB 'guide'

Microstrip Line Formulations

The electrical characteristics of the microstrip transmission line is strongly dependent on the physical width, *W* and length, *L* of the microstrip line and the printed circuit board (PCB) substrate used. Therefore, almost all high-frequency or microwave planar form circuit designs involve the calculation of *W* and *L*. In this calculator, the values of *W* and *L* of the microstrip line are calculated using explicit Equations (1) and (2) (Hammerstad, 1975; Pozar, 2012).

$$W = \begin{cases} \dfrac{8h\exp(A)}{\exp(2A)-2} & \text{for } \dfrac{W}{h} < 2 \\ \dfrac{2h}{\pi}\left\{B-1-\ln(2B-1)+\dfrac{\varepsilon_r-1}{2\varepsilon_r}\left[\ln(B-1)+0.39-\dfrac{0.61}{\varepsilon_r}\right]\right\} & \text{for } \dfrac{W}{h} > 2 \end{cases} \quad (1)$$

where

$$A = \frac{Z_o}{60}\sqrt{\frac{\varepsilon_r + 1}{2}} + \frac{\varepsilon_r - 1}{\varepsilon_r + 1}\left(0.23 + \frac{0.11}{\varepsilon_r}\right) \text{ and } B = \frac{377\pi}{2Z_o\sqrt{\varepsilon_r}}$$

where Z_o is the characteristic impedance of the microstrip line. Symbols ε_r and h are the dielectric constant and thickness of used PCB substrate. Once the values of W are obtained from Equation (1), the length, L of the microstrip transmission line can be calculated as (Hammerstad, 1975; Pozar, 2012):

$$L = \frac{c}{f_o\sqrt{\varepsilon_{eff}}} - 0.412h\left[\frac{(\varepsilon_{eff} + 0.3)(W/h + 0.262)}{(\varepsilon_{eff} - 0.258)(W/h + 0.813)}\right] \quad (2)$$

The velocity of light in vacuum, $c = 299792458$ ms^{-1} and the effective relative permittivity, ε_{eff} of the microstrip line is given as (Pozar, 2012):

$$\varepsilon_{eff} = \begin{cases} \frac{\varepsilon_r + 1}{2} + \frac{\varepsilon_r - 1}{2}\left[\frac{1}{\sqrt{1 + 12h/W}} + 0.04\left(1 - \frac{W}{h}\right)^2\right] & \text{for } \frac{W}{h} \le 1 \\ \frac{\varepsilon_r + 1}{2} + \frac{\varepsilon_r - 1}{2}\left(\frac{1}{\sqrt{1 + 12h/W}}\right) & \text{for } \frac{W}{h} > 1 \end{cases} \quad (3)$$

Getting Started with MATLAB 'guide'

The MATLAB-based calculator allows users to predict the physical width, W and length, L of the microstrip line once the PCB substrate parameters (ε_r and h) and desired transmission line parameters (f, Z_o, and θ) are provided. Therefore, in this GUI calculator, it must have five inputs and two outputs as:

Five inputs: Two outputs:

1 Dielectric constant substance: ε_r 1 Microstrip line width: W
2 Thickness substance: h 2 Microstrip line length: L
3 Operating frequency: f
4 Microstrip line impedance: Z_o
5 Microstrip line electrical length: θ

Next, the steps to build the GUI of the microstrip line calculator using MATLAB 'guide' are described in detail.

Step 1: Open MATLAB command window and MATLAB 'guide'

Open MATLAB. Go to the command window and type in 'guide' (as shown in Figure 1) and click enter.

Figure 1. MATLAB command window

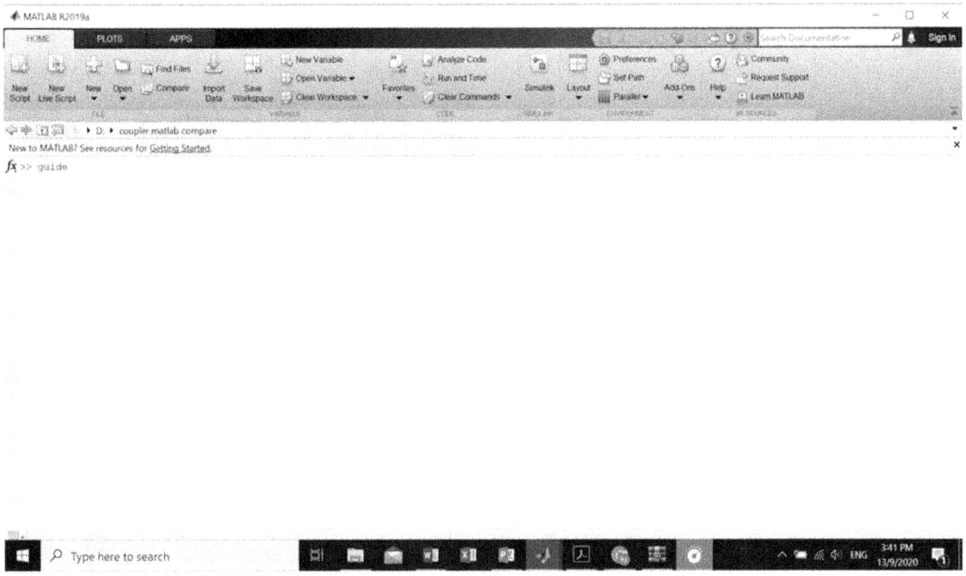

Step 2: Open MATLAB 'guide' empty canvas

Choose the first option "Blank GUI (Default)" and click 'OK' button as shown in Figure 2.

Figure 2. GUIDE quick start

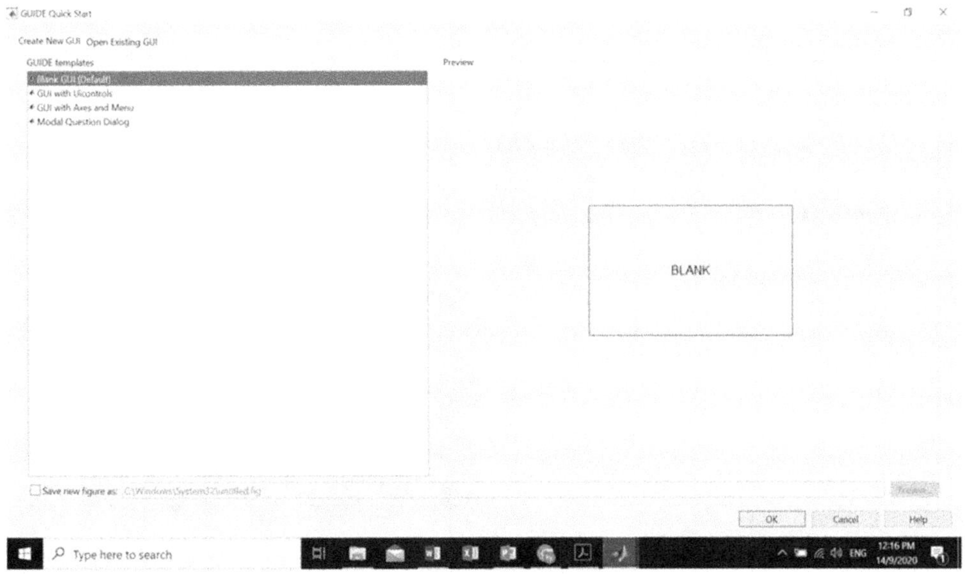

Then, the following screen is shown (in Figure 3).

Figure 3. MATLAB GUI canvas

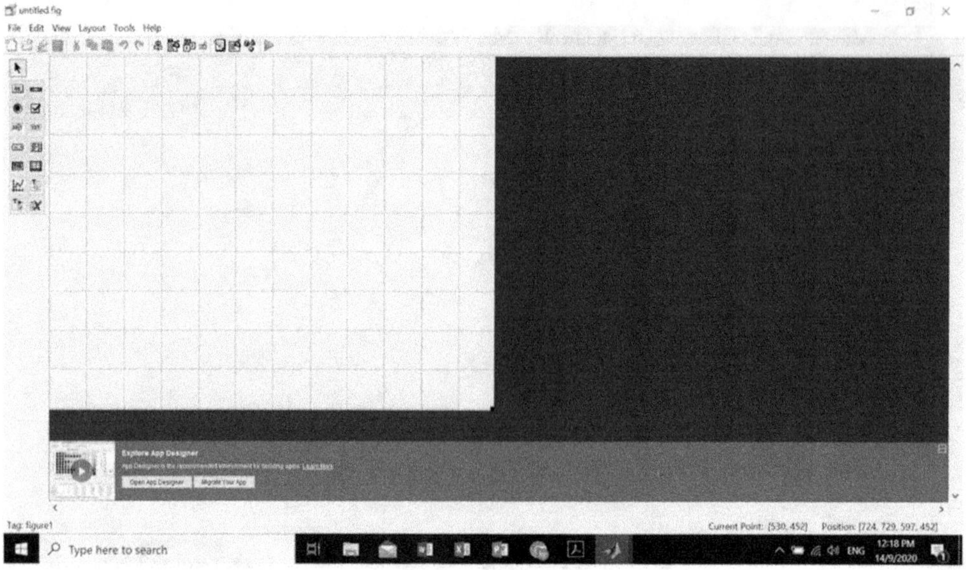

Step 3: Select user interface (UI) components in GUI

The microstrip line calculator GUI has five input and two output parameters. Thus, need the following components (in Figure 4):

i) Five 'Edit Text' components
ii) Two 'Static Text' components
iii) One 'Pushbutton' component

Figure 4. The user interface components of the MATLAB GUI

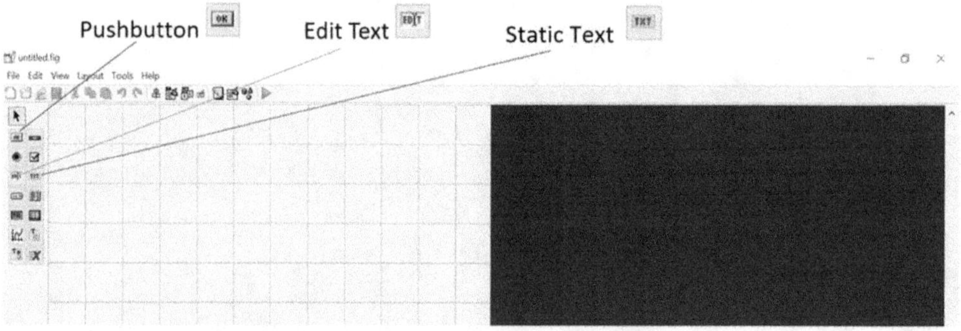

Drag the eight components onto the canvas as shown in Figure 5.

Step 4: Edit the properties of the two 'Static Text' components

Edit the properties of the 'Static Text' components by double click one of the 'Static Text' components. Then, the property 'Inspector' will pop out as shown in Figure 6.

Figure 5. Drag and drop components from the palette to the canvas

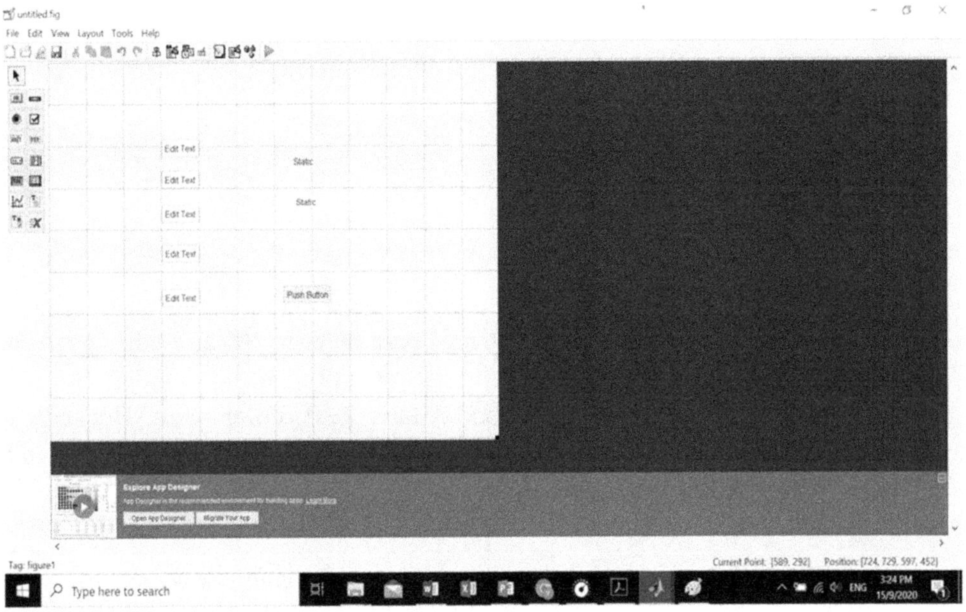

Figure 6. The property 'Inspector' of the 'Static Text' components

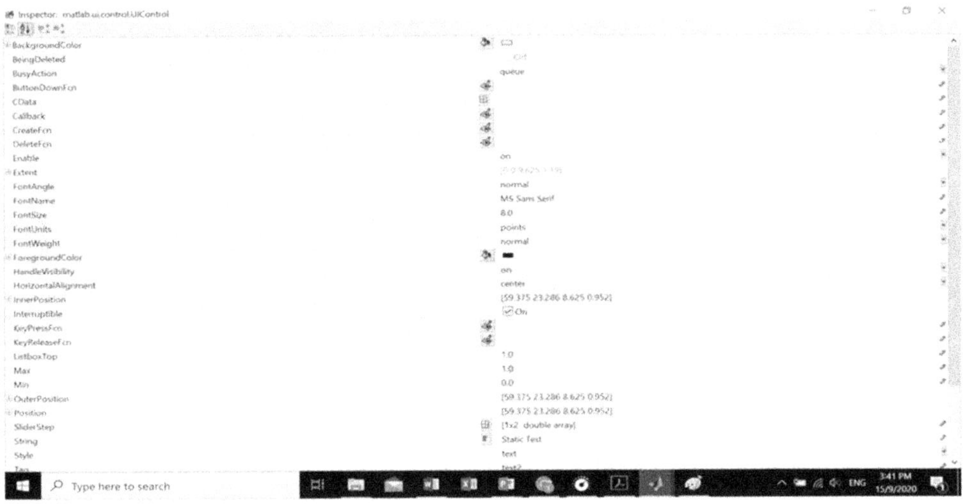

Figure 7. Edit 'String' and 'Tag' properties of the first 'Static Text' component

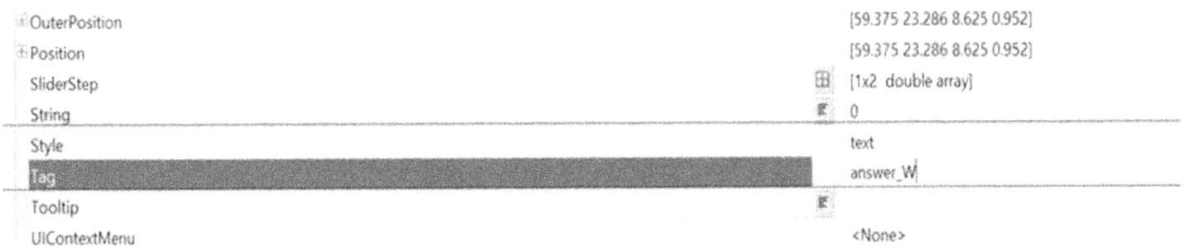

Figure 8. Edit 'String' and 'Tag' properties of the second 'Static Text' component

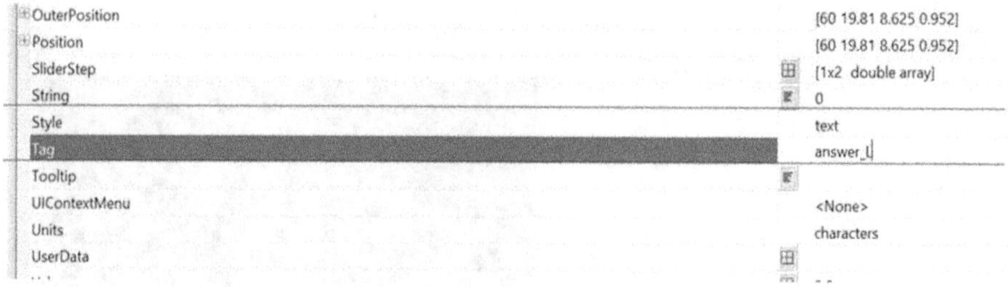

Just edit 'String' and 'Tag' properties. Edit 'String' parameter to '0' and edit 'Tag' parameter to 'answer_W' (in Figure 7).

Similar, the same procedure is applied to second 'Static Text' component. But, edit its 'String' parameter to '0' and 'Tag' parameter as 'answer_L' for second 'Static Text' component (in Figure 8).

Then, the platform will look like (in Figure 9):

Figure 9. The default initial output value of both 'Static Text' components is set to zero

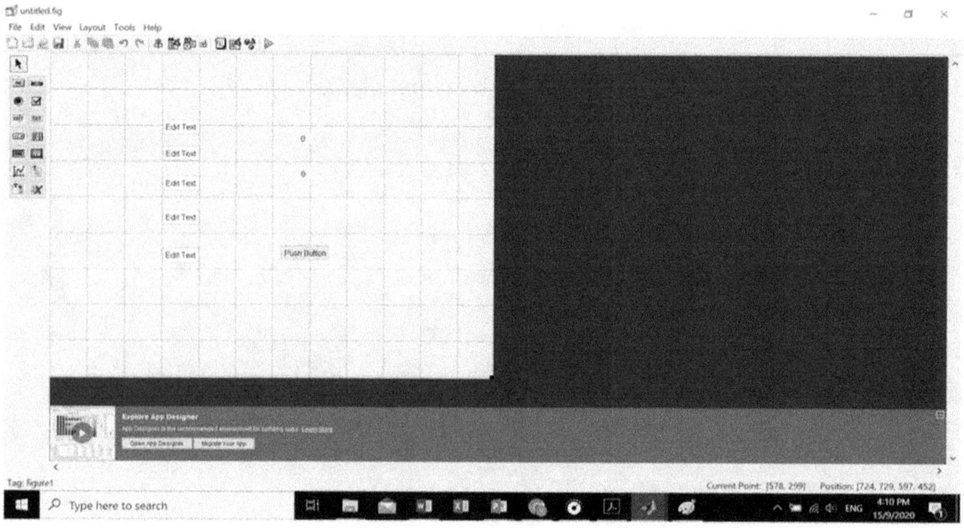

Step 5: Edit the properties of the five 'Edit Text' components

Edit the properties of the first 'Edit Text' component by double click one of the 'Edit Text' components. Then, the property 'Inspector' will pop out. Edit 'String' parameter to '0' and 'Tag' parameter to 'input_er'. Similarly, edit the properties of the other 'Edit Text' components by double click the 'Edit Text' components:

i) For second 'Edit Text' components, edit 'String' parameter to '0' and 'Tag' parameter to 'Input_h', respectively.
ii) For third 'Edit Text' components, edit 'String' parameter to '0' and 'Tag' parameter to 'Input_f'.

iii) For forth 'Edit Text' components, edit 'String' parameter to '0' and 'Tag' parameter to 'Input_Zo'.
iv) For firth 'Edit Text' components, edit 'String' parameter to '0' and 'Tag' parameter to 'Input_theta'.

Click to save the name of the canvas as 'Microstrip_Line_Calculator'.

Step 6: Edit the MATLAB code of the GUI

After saving the canvas or GUI as the 'Microstrip_Line_Calculator.m' file, the MATLAB window with the generated MATLAB code will automatically open as (in Figure 10):

Figure 10. The automatically generated MATLAB codes of the GUI

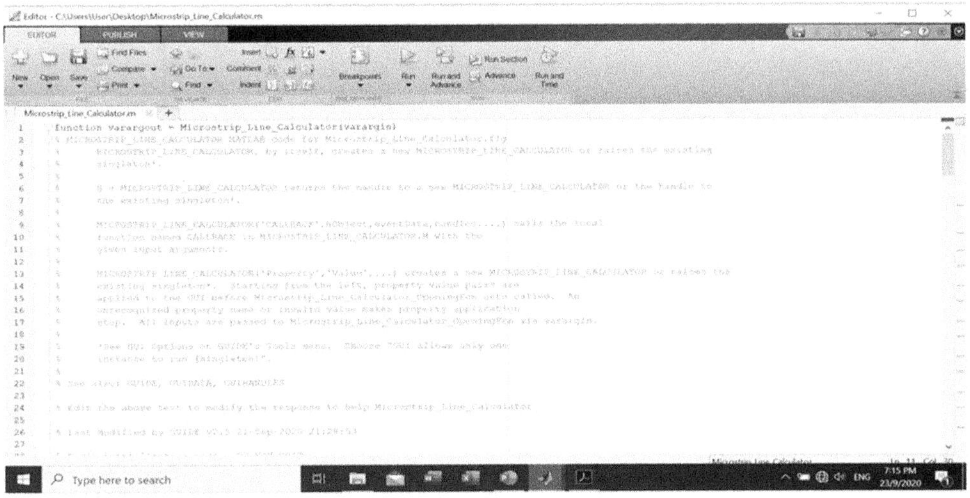

Go to 'function Input_er_Callback(hObject, eventdata, handles)' line. Add the several line of MATLAB code to the bottom of that code block as (in Figure 11):

Figure 11. MATLAB code to accept only numbers in the 'Edit Text' box

```
function Input_er_Callback(hObject, eventdata, handles)
% hObject handle to Input_er (see GCBO)
% eventdata reserved - to be defined in a future version of MATLAB
% handles structure with handles and user data (see GUIDATA)

% Hints: get(hObject,'String') returns contents of Input_er as text
%        str2double(get(hObject,'String')) returns contents of Input_er as
a double

input=str2num(get(hObject,'String'));
if(isempty(input))
set(hObject,'String','0')
end
```

Next, add the same code to the bottom of all call-back functions, namely 'function Input_h_Callback(hObject, eventdata, handles)', 'function Input_f_Callback(hObject, eventdata, handles), 'function Input_Zo_Callback(hObject, eventdata, handles)', and 'function Input_theta_Callback(hObject, eventdata, handles)' lines.

Finally, go to 'function pushbutton1_Callback(hObject, eventdata, handles)' line as shown in Figure 12.

Figure 12. Include commands and instructions under 'function pushbutton1_Callback(hObject, eventdata, handles)'

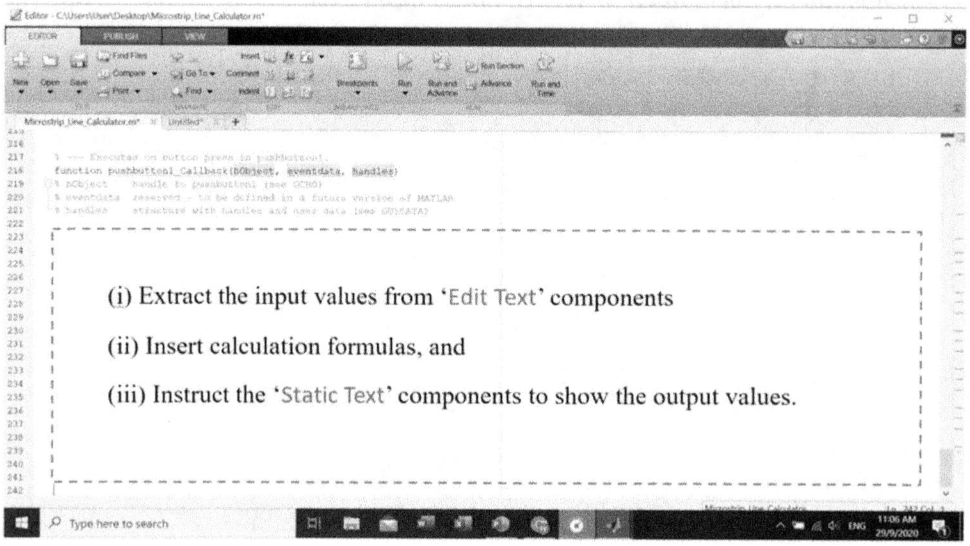

Several major operations under the 'function pushbutton1_Callback(hObject, eventdata, handles)' section are required to be programmed, namely

1. Extract the input values from 'Edit Text' components.
2. Insert calculation formulas, and
3. Instruct the 'Static Text' components to show the output values.

The MATLAB code to extract the input values from 'Edit Text' components is given as:

```
er=str2num(get(handles.Input_er,'String'));
h=str2num(get(handles.Input_h,'String')).*1e-3;
f=str2num(get(handles.Input_f,'String')).*1e9;
Zo=str2num(get(handles.Input_Zo,'String'));
Theta=str2num(get(handles.Input_theta,'String')).*pi./180;
```

The MATLAB code of formulations of (1), (2), and (3) are explained below. Based on (1), the value of W is involved in the calculation condition and at the same time is also the calculated output value. The equivalent MATLAB code should be written as:

```
if W./h > 2
A=(Zo./60).*sqrt((er+1)./2)+((er-1)./(er+1)).*(0.23+0.11./er);
W=8.*h.*exp(A)./(exp(2.*A)-2);
else
B=377.*pi./(2.*Zo.*sqrt(er));
W=(2.*h./pi).*(B-1-log(2.*B-1)+((er-1)./2.*er).*(log(B-1)+0.39-(0.61./er)));
End
```

However, the above MATLAB code is illogical in *W* calculation. Consequently, the MATLAB code for the condition of *W/h*>2 is required to be modified. The modification can be done by inserting the value of *W/h* = 2 into Equations (2) and (3), yields

$$\varepsilon_{eff\_o} = \frac{\varepsilon_r + 1}{2} + \frac{\varepsilon_r - 1}{2}\left(\frac{1}{\sqrt{7}}\right) \text{ and } Z_{o1} = \frac{120\pi}{\sqrt{\varepsilon_{eff\_o}}\left[3.393 + \frac{2}{3}\ln(3.444)\right]}$$

And the condition that input impedance value, Z_o is less than Z_{o1} implies that *W/h*>2:

$$Z_{o1} - Z_o \geq 0 \, (Z_o \text{ less than } Z_{o1})$$

Hence, the MATLAB code of *W* calculation using (1) can be modified as:

```
effo=((er+1)./2)+((er-1)./2).*(1./sqrt(7));
Zo1=(120.*pi)./(sqrt(effo).*(3.393+(2./3).*log(3.444)));
if Zo1-Zo >= 0
B=376.9911.*pi./(2.*Zo.*sqrt(er));
W=(2.*h./pi).*(B-1-log(2.*B-1)+((er-1)./(2.*er)).*(log(B-1)+0.39-(0.61./er)))
else
A=(Zo./60).*sqrt((er+1)./2)+((er-1)./(er+1)).*(0.23+(0.11./er));
W=8.*h.*exp(A)./(exp(2.*A)-2)
End
```

Once the value of *W* is obtained, the MATLAB code of *L* calculation using (2) is written as:

```
if W =< h
eff=((er+1)./2)+((er-1)./2).*(1./sqrt(1+(12.*h./W))+0.04.*(1-W./h).^2);
else
eff=((er+1)./2)+((er-1)./2).*(1./sqrt(1+(12.*h./W)));
end
L=((theta.*3e8)./(2.*pi.*f.*sqrt(eff)))...
    -0.412.*h.*(((eff+0.3).*((W./h)+0.264))./((eff-0.258).*((W./h)+0.8)));
```

Figure 13. MATLAB codes and operations under 'function pushbutton1_Callback(hObject, eventdata, handles)'

```
% --- Executes on button press in pushbutton1.
function pushbutton1_Callback(hObject, eventdata, handles)
% hObject    handle to pushbutton1 (see GCBO)
% eventdata  reserved - to be defined in a future version of MATLAB
% handles    structure with handles and user data (see GUIDATA)

er=str2num(get(handles.Input_er,'String'));
h=str2num(get(handles.Input_h,'String'));
f=str2num(get(handles.Input_f,'String'));
Zo=str2num(get(handles.Input_Zo,'String'));
Theta=str2num(get(handles.Input_theta,'String'));

effo=((er+1)./2)+((er-1)./2).*(1./sqrt(7));
Zo1=(120.*pi)./(sqrt(effo).*(3.393+(2./3).*log(3.444)));

if Zo1-Zo >= 0
B=376.9911.*pi./(2.*Zo.*sqrt(er));
W=(2.*h./pi).*(B-1-log(2.*B-1)+((er-1)./(2.*er)).*(log(B-1)+0.39-(0.61./er)))
else
A=(Zo./60).*sqrt((er+1)./2)+((er-1)./(er+1)).*(0.23+(0.11./er));
W=8.*h.*exp(A)./(exp(2.*A)-2)
end

if W <= h
eff=((er+1)./2)+((er-1)./2).*(1./sqrt(1+(12.*h./W))+0.04.*(1-W./h).^2);
else
eff=((er+1)./2)+((er-1)./2).*(1./sqrt(1+(12.*h./W)));
end

L=((Theta.*3e8)./(2.*pi.*f.*sqrt(eff)))...
   -0.412.*h.*(((eff+0.3).*((W./h)+0.264))./((eff-0.258).*((W./h)+0.8)));

set(handles.answer_W,'String',sprintf('%.2f',W.*1000));
set(handles.answer_L,'String',sprintf('%.2f',L.*1000));
```

Extract the input values from 'Edit Text' components.

MATLAB code of *W* calculation.

MATLAB code of *L* calculation.

Instruct the 'Static Text' components to show the output values in unit millimeter.

Finally, all above MATLAB codes are inserted below 'function pushbutton1_Callback(hObject, eventdata, handles)' line as shown in Figure 13.

Figure 14. Primary GUI display

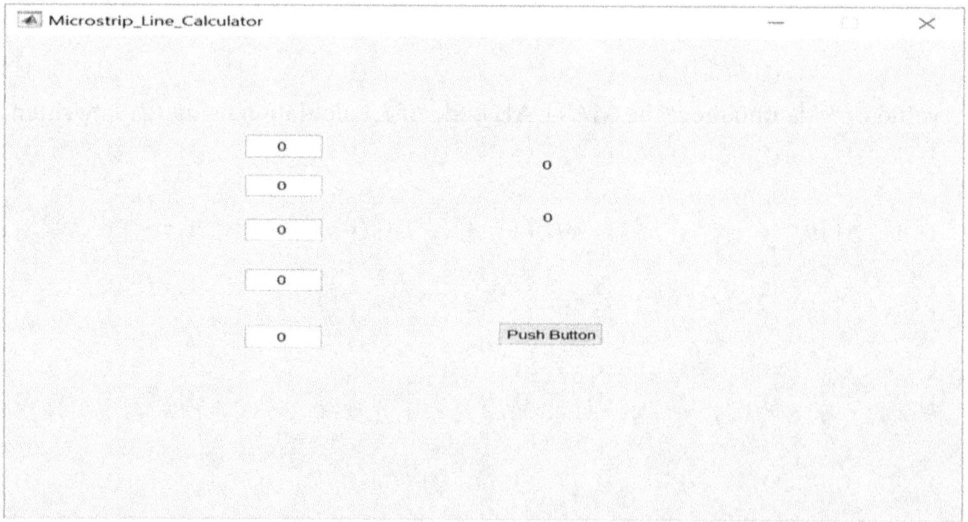

Figure 15. GUI label preparation for parameters and units

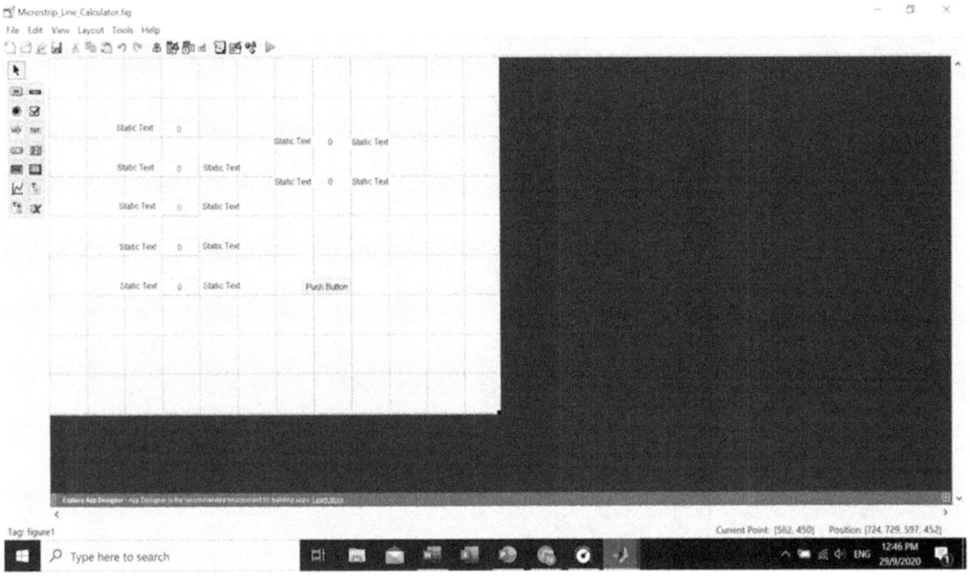

Then, click the run button, the following GUI will pop-out as (in Figure 14):

Step 7: Labelling and Beautification of the GUI

Opening the GUI canvas 'Microstrip_Line_Calculator.fig' again. Drag thirteen 'Static Text' components onto GUI canvas and arrange as shown in Figure 15.

Edit the labeling of the 'Static Text' components by double click one of the 'Static Text' components. Then, the 'property Inspector' will pop out. Edit 'String' parameter from 'Static Text' word to 'er'.

Figure 16. The default initial input value of the all labelled 'Edit Text' components is set to zero

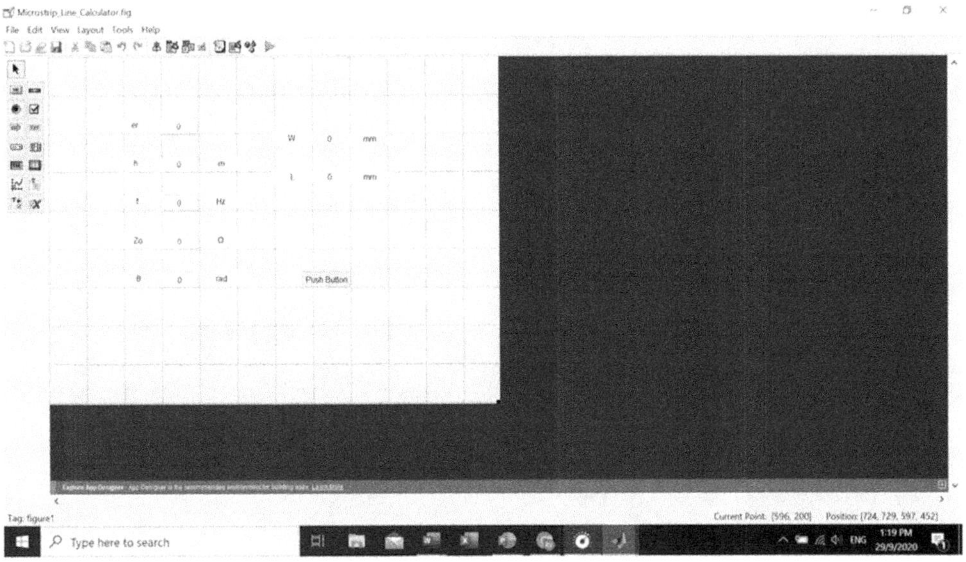

Figure 17. Labelled GUI display

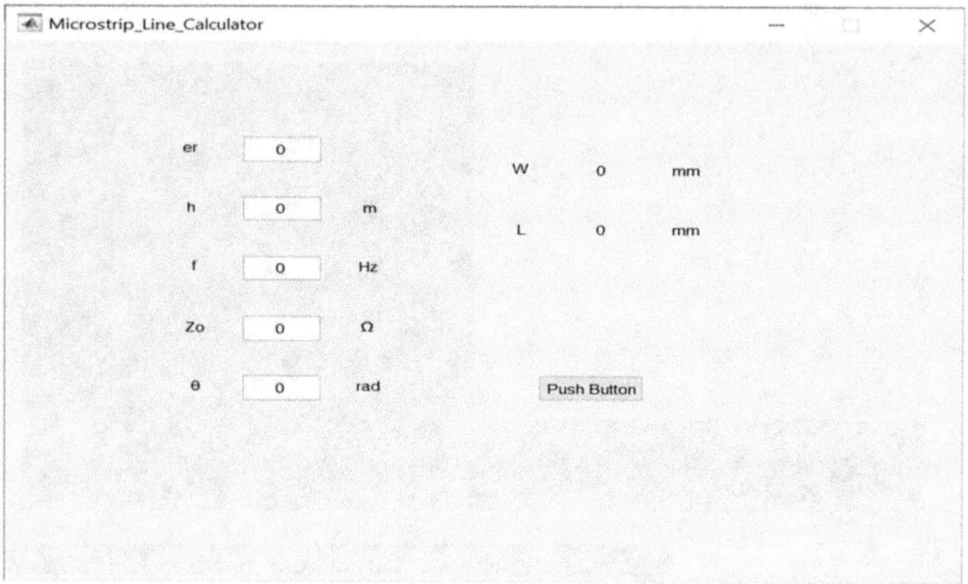

Similarly, other 'Static Text' components onto the GUI canvas can also be edited at 'String' properties as shown in Figure 16.

Then, click again the run button, the following GUI will pop-out as (in Figure 17):

Figure 18. Trial calculation using the designed GUI

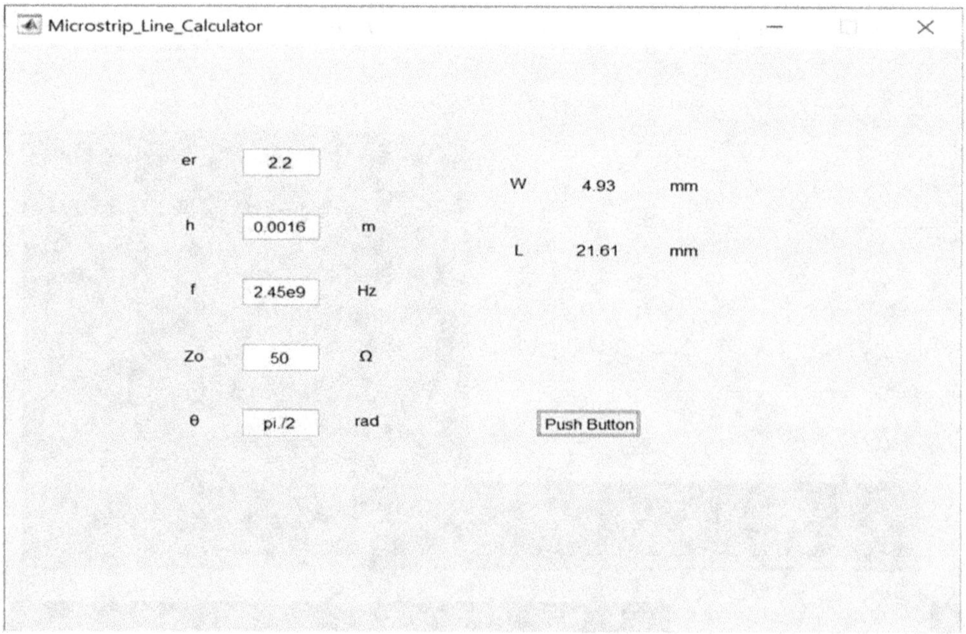

Figure 19. Simple microstrip line

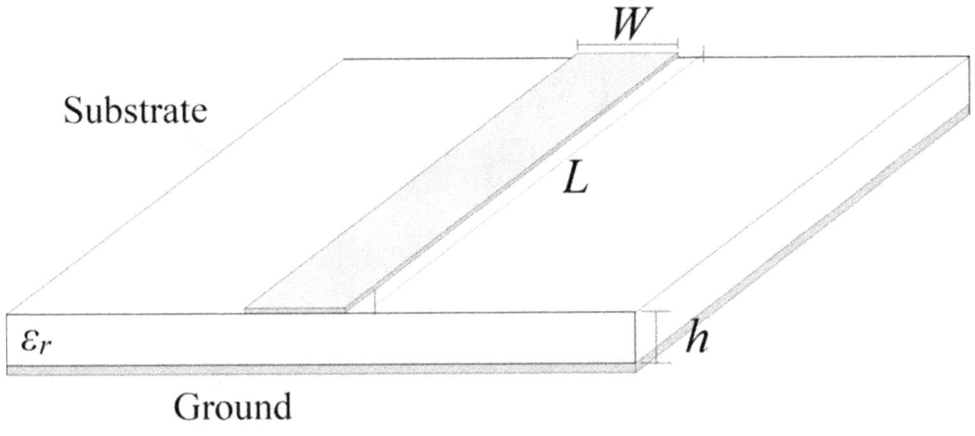

Let input $\varepsilon_r = 2.2$, $h = 1.6$ mm, $f = 2.45$ GHz, $Zo = 50\ \Omega$, and $\theta = 90$ deg. Then, click the 'Push Button' button to calculate the dimension of the microstrip line and the output yield: width, $W = 4.93$ mm and length, $L = 21.61$ mm as shown in Figure 18.

Finally, we would like to include image (like Figure 19) or annotations in the GUI. In this work, the image of Figure 19 is saved as JPG format and named the image file as 'Microstrip_Line'.

Open back the GUI canvas and drag 'Axes' component into GUI canvas. The default 'Tag' in the 'property Inspector' of the 'Axes' component is labelled as 'axes1'. In MATLAB window, at the below line of 'function Microstrip_Line_Calculator_OpeningFcn(hObject,eventdata,handles,varargin)', two lines of MATLAB code, namely

Figure 20. The complete GUI of the microstrip line calculator

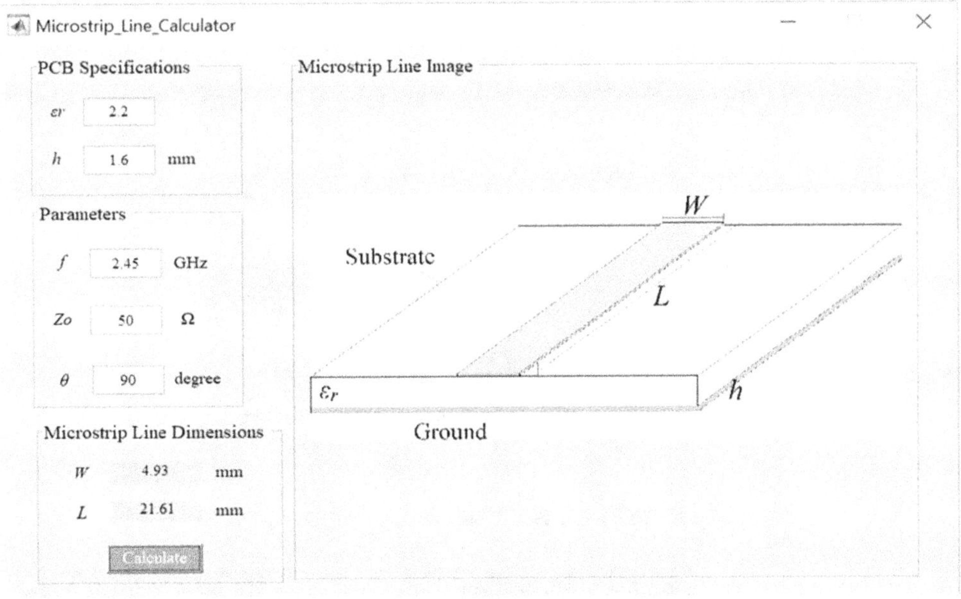

Figure 21. Built-in MATLAB application (App) categories and lists

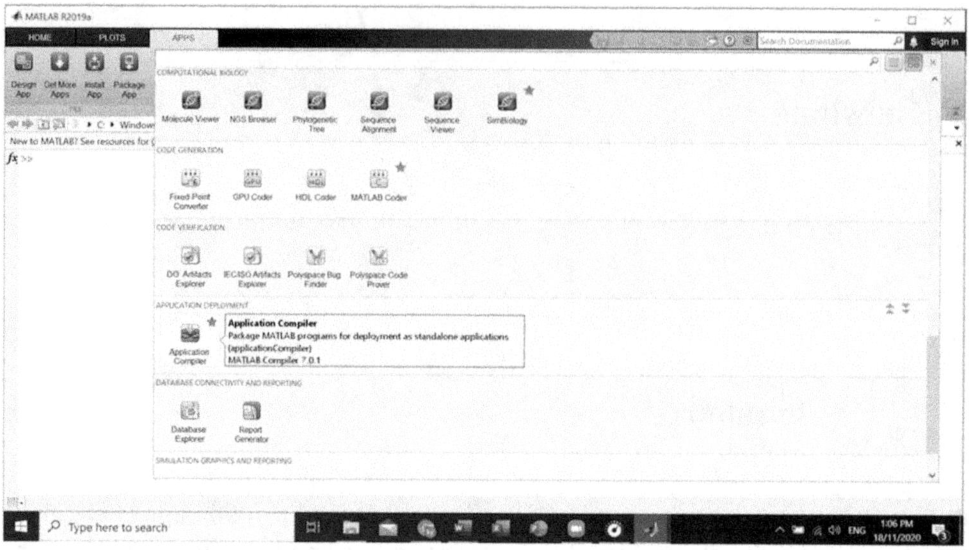


```
axes(handles.axes1);
imshow('Microstrip_Line.jpg');
```

are added in order to insert the image of Figure 19 in GUI. The GUI can be further enhanced by dragging the four 'Panel' components in the canvas and then dragging all existing components to the 'Panel'. Revise the tittle of the four 'Panel' components as 'PCB Specifications', 'Parameters', 'Microstrip Line Dimensions', and 'Microstrip Line Image', respectively. The final version of the simple microstrip line GUI is shown as in Figure 20.

Figure 22. 'Application Compiler' application (App) display screen

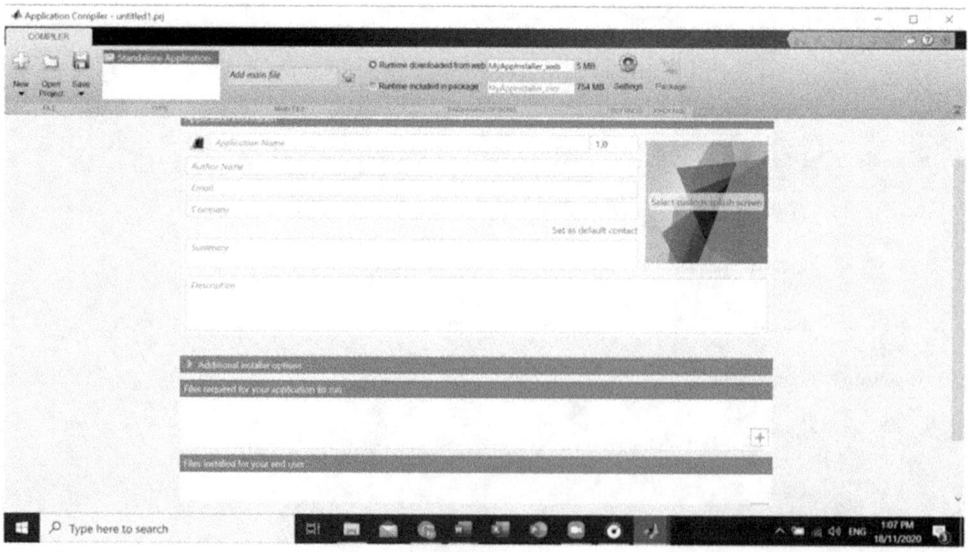

Figure 23. 'MAIN FILE' section of the 'Application Compiler'

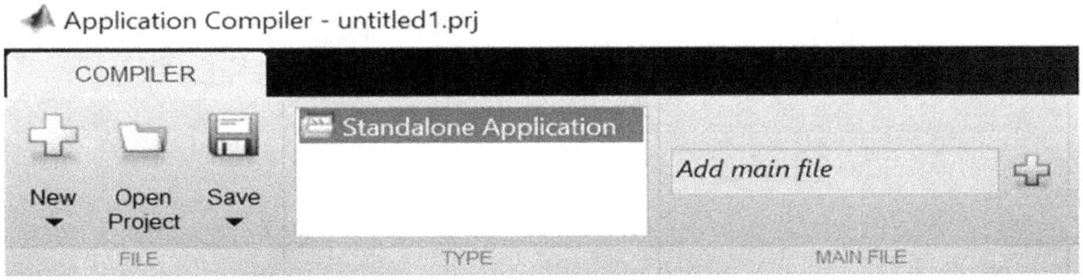

Figure 24. Select 'Microstrip_Line_Calculator.m' file and browse the file in 'MAIN FILE' section

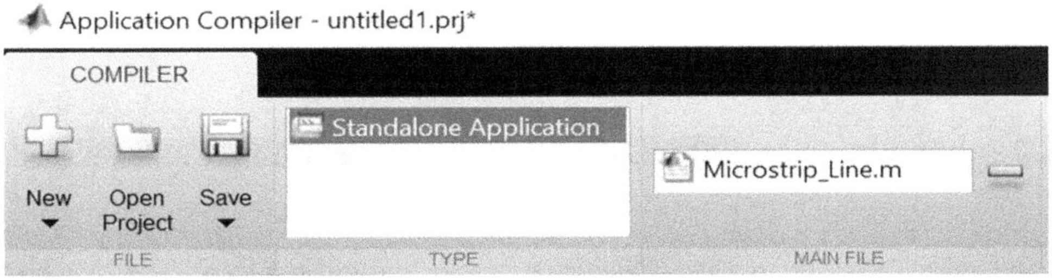

Step 8: Create the GUI calculator as standalone desktop app

On the MATLAB 'APPS' tab, on the far right of the 'APPS' section, click the arrow. In 'APPLICATION DEPLOYMENT', click 'Application Compiler' button, as shown in Figure 21.

After clicking on the 'Application Compiler', the following interface appears as (in Figure 22):

In the 'MAIN FILE' section of the toolstrip (in Figure 23), click.

Figure 25. MATLAB command window

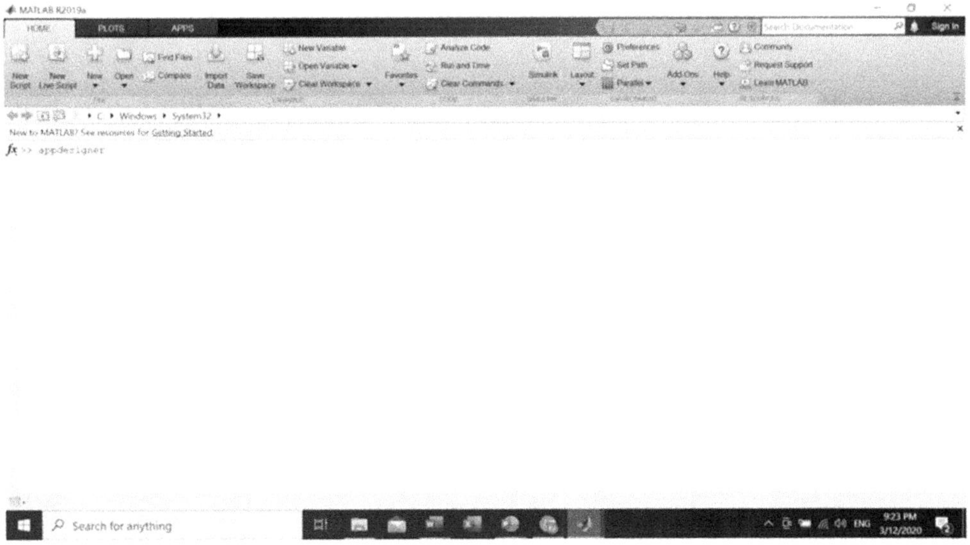

Figure 26. MATLAB App Designer

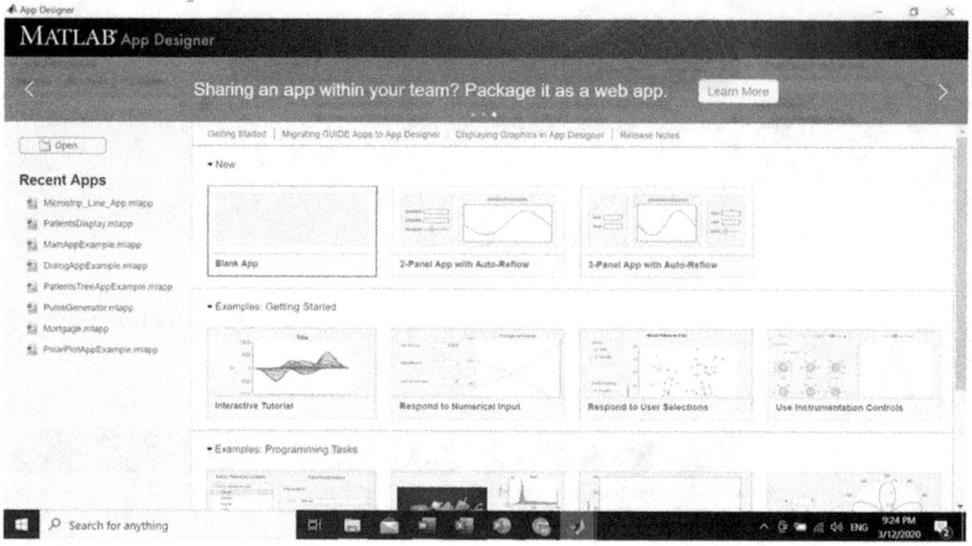

Browse and select 'Microstrip_Line_Calculator.m' file and click 'Open'. The 'Microstrip_Line_ Calculator.m' file is added to the list of main files as shown in Figure 24.

For the MATLAB Runtime installer part, there are two options in the 'PACKAGING OPTIONS' section:

Figure 27. App Designer empty template (canvas)

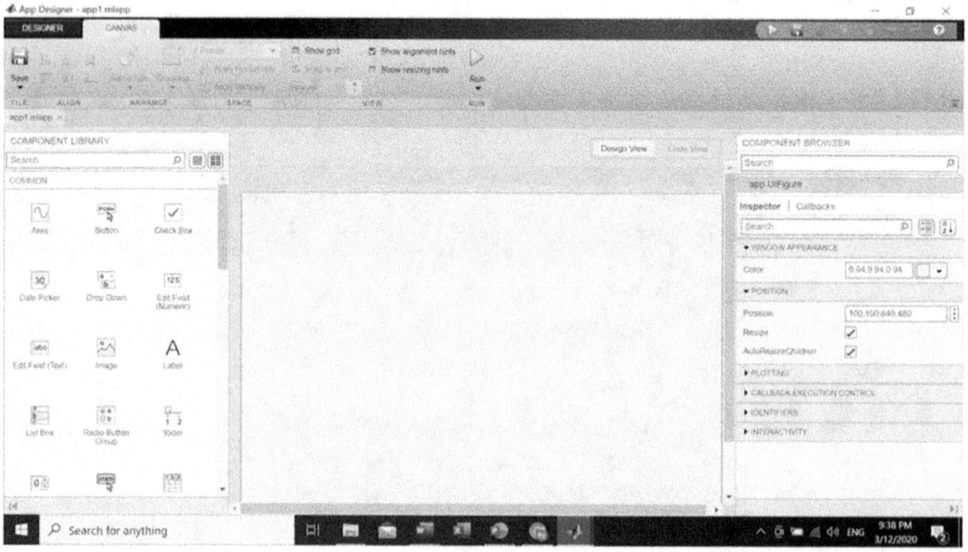

Figure 28. Drag and drop night components from the 'COMPONENT LIBRARY' to the canvas

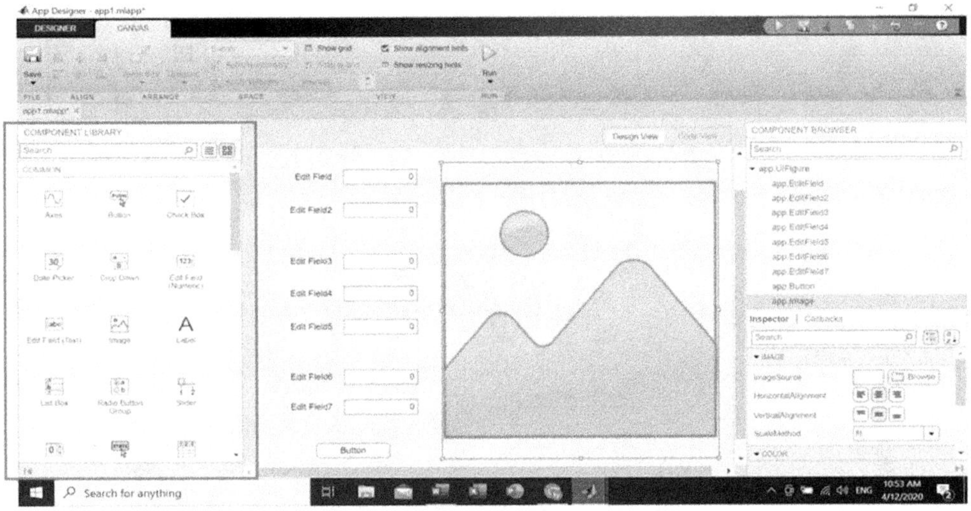

i) 'Runtime downloaded from web' - Generates an installer that downloads the MATLAB Runtime and installs it along with the deployed MATLAB app (size: ~ 5 MB).

ii) 'Runtime included in package' - Generates an installer that includes the MATLAB Runtime installer (size: ~ 900 MB).

App designer can select one of the above two options. Besides, app designer can also customize the packaged app and its appearance (as shown in Figure 22). Finally, app designer can click 'Package' button, to generate the packaged application.

Figure 29. Defined label text of 'Edit Field (Numeric)' components

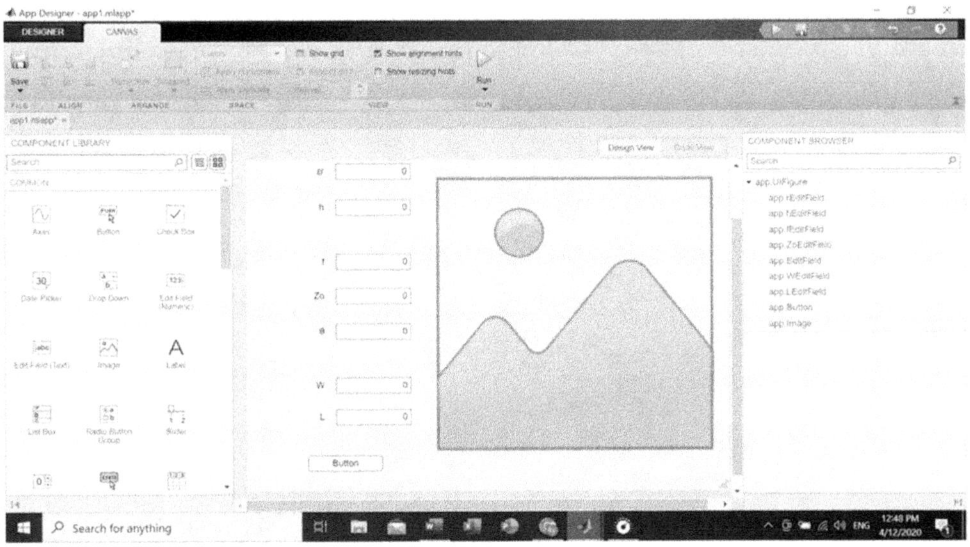

Figure 30. View the automatically generated MATLAB code by clicking on the 'Code View' tab

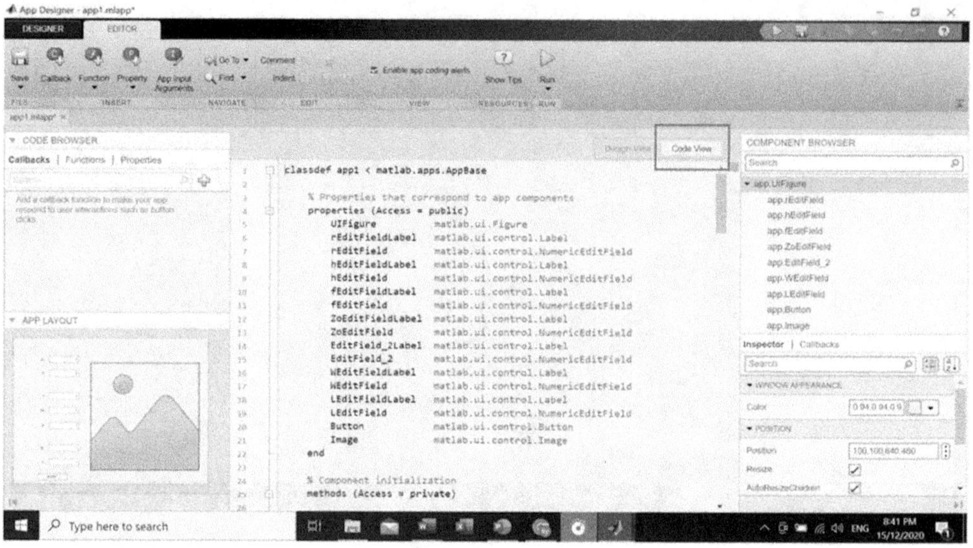

Microstrip Line Calculator GUI Using MATLAB 'appdesigner'

Besides MATLAB 'guide', the microstrip line calculator can also be created using MATLAB 'appdesigner'. In this section, the calculator app construction using 'appdesigner' is demonstrated in detail.

Step 1: Open MATLAB command window and MATLAB 'appdesigner'

Open MATLAB. Go to the command window and type in 'appdesigner' and click enter.

Step 2: Open MATLAB 'appdesigner' empty canvas

Figure 31. Add a callback function that executes MATLAB commands whenever the user presses the 'Button' component

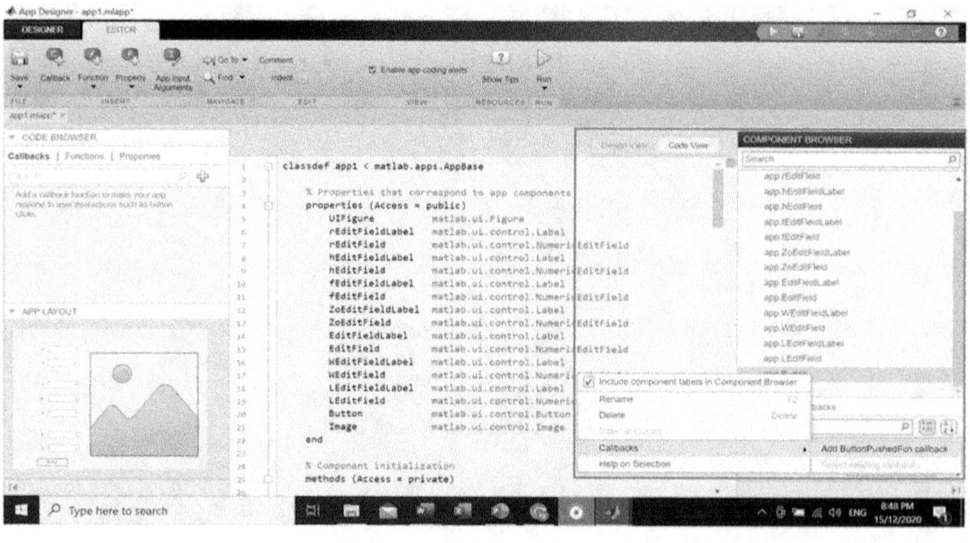

Figure 32. There is cursor under the 'function ButtonPushed(app, event)' function

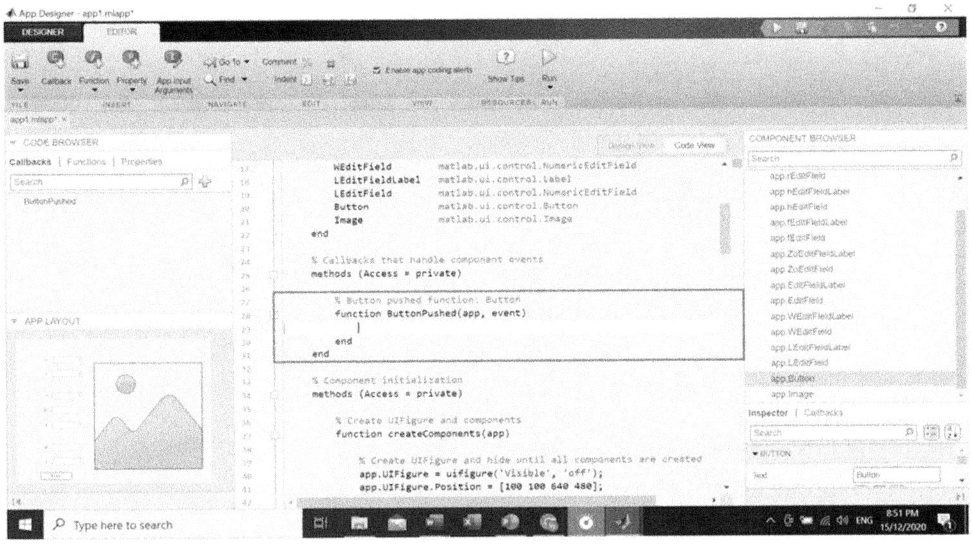

Choose the new option "Blank App" and double click on the icon.
Then, the following screen (empty canvas) is shown as Figure 27.
Step 3: Select user interface (UI) components in GUI
Drag night components from 'COMPONENT LIBRARY' onto the canvas (in Figure 28):

Figure 33. Insert MATLAB code under 'function ButtonPushed(app, event)'

```
    er=app.rEditField.Value;
    Zo = app.ZoEditField.Value;
    h = app.hEditField.Value.*1e-3;
    f = app.fEditField.Value.*1e9;
    Theta = app.EditField.Value.*pi./180;

    effo=((er+1)./2)+((er-1)./2).*(1./sqrt(7));
    Zo1=(120.*pi)./(sqrt(effo).*(3.393+(2./3).*log(3.444)));

if Zo1-Zo >= 0
   B=376.9911.*pi./(2.*Zo.*sqrt(er));
   W=(2.*h./pi).*(B-1-log(2.*B-1)+((er-1)./(2.*er)).*(log(B-1)+0.39-(0.61./er)))
else
   A=(Zo./60).*sqrt((er+1)./2)+((er-1)./(er+1)).*(0.23+(0.11./er));
   W=8.*h.*exp(A)./(exp(2.*A)-2)
end

if W <= h
   eff=((er+1)./2)+((er-1)./2).*(1./sqrt(1+(12.*h./W))+0.04.*(1-W./h).^2);
else
   eff=((er+1)./2)+((er-1)./2).*(1./sqrt(1+(12.*h./W)));
end

   L=((Theta.*3e8)./(2.*pi.*f.*sqrt(eff)))...
    -0.412.*h.*(((eff+0.3).*((W./h)+0.264))./((eff-0.258).*((W./h)+0.8)));

 app.WEditField.Value=W.*1e3;
 app.LEditField.Value=L.*1e3;
```

Extract the input values from 'Edit Field (Numeric)' components.

MATLAB code of *W* calculation.

MATLAB code of *L* calculation.

Instruct the 'Edit Field (Numeric)' components to show the output values in unit millimeter.

Figure 34. (a) Browse and insert picture/image into the 'Image' component. (b) Microstrip line image is inserted into the 'Image' component

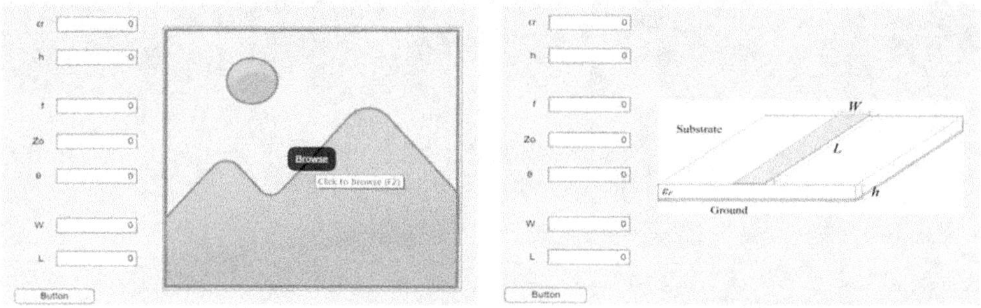

(a) (b)

i) Seven 'Edit Field (Numeric)' components
ii) One 'Button' component
iii) One 'Image' component

Revise the 'Edit Field (Numeric)' component's label text. Double click the label and replace the word 'Edit Field' with '*εr*'. Similarly, other 'Edit Field (Numeric)' component's label text are replaced by ' *h* ', ' *f* ', ' *Zo* ', ' *θ* ', ' *W* ', and ' *L* ', respectively as shown in Figure 29.

Step 4: Edit the MATLAB code of the GUI

Click 'Code View' tab (located above the canvas), the generated MATLAB code will automatically open as shown in Figure 30.

The MATLAB code can be further added and edited by creates a callback function. Right click on 'app.Button' in the 'COMPONENT BROWSER'. Then select 'Callbacks' and 'Add ValueChangedF-cncallback' in the context menu as shown in Figure 31.

After clicking 'Add ValueChangedFcncallback', the MATLAB code will move to the 'function ButtonPushed(app, event)' line, and the cursor will be existed and located below the line as shown in Figure 32.

The MATLAB code of formulations of (1), (2), and (3) as shown Figure 33 are added into the cursor location.

Figure 35. (a) Provision of GUI labels. (b) The units of the parameters are labelled

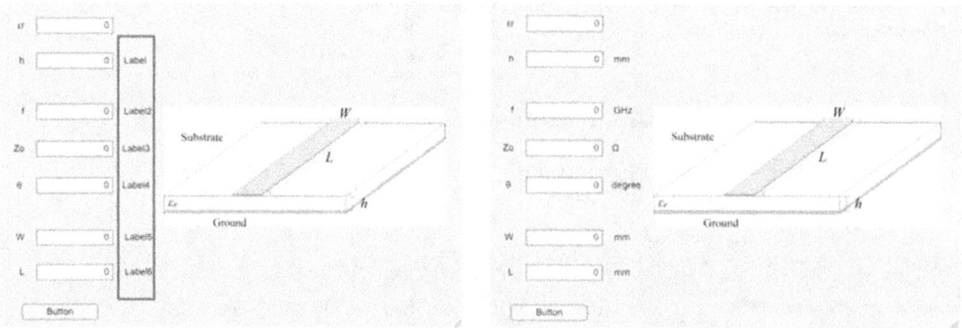

(a) (b)

Figure 36. (a) The initial stage design of the microstrip line calculator. (b) Trial calculation using the calculator

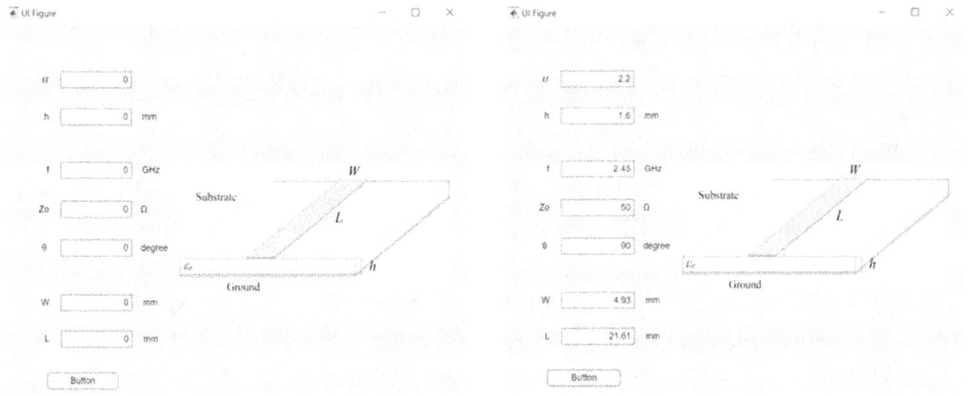

(a) (b)

Step 5: Insert the microstrip line image into GUI

The microstrip line image (JPG format) in Figure 19 can be inserted into the GUI by pointing the cursor to the centre of the 'Image' component and clicking 'Browse' as shown in Figure 34 (a). Select the microstrip line image file and insert it into the 'Image' component as shown in Figure 34 (b).

Drag six 'Label' components onto the previous canvas as shown in Figure 35 (a). Edit the text of each 'Label' component as 'mm', 'GHz', 'Ω', and 'degree', respectively as shown in Figure 35 (b).

Figure 37. The complete MATLAB-based GUI of the microstrip line calculator

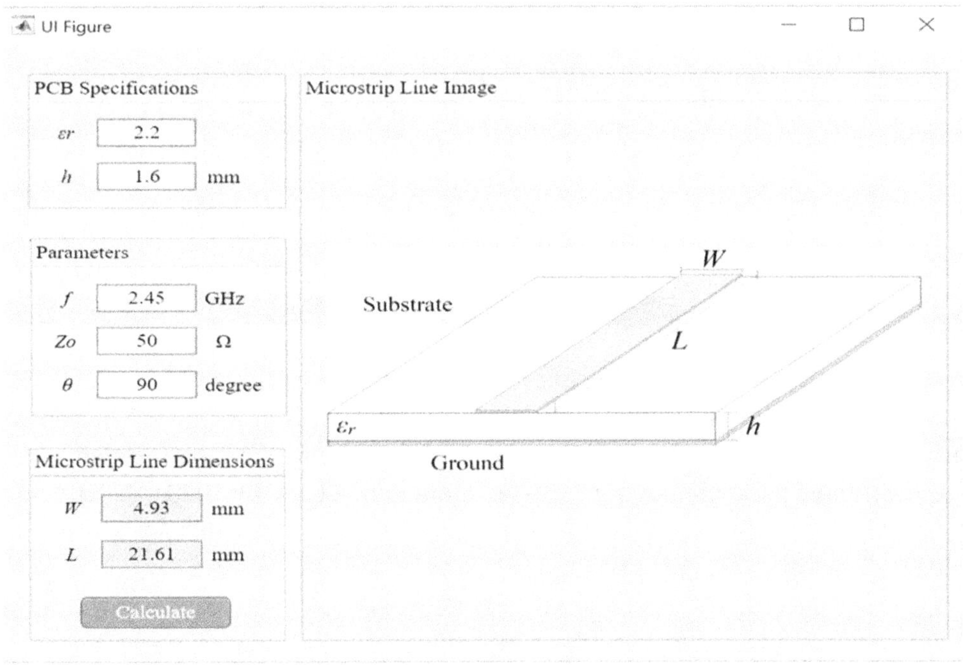

Click ‘Run’ button to save and run the GUI. Save the GUI’s name as ‘Microstrip_Line_App’. Finally, the GUI of the microstrip line calculator is generated as shown in Figure 36 (a). Let input $\varepsilon_r = 2.2$, $h =$ 1.6 mm, $f = 2.45$ GHz, $Z_o = 50\ \Omega$, and $\theta = 90$ deg. Then, click the ‘Button’ button and the outputs are generated: width, $W = 4.93$ mm and length, $L = 21.61$ mm as shown in Figure 36 (b).

The final version of the simple microstrip line GUI using App Designer environment is shown as in Figure 37.

Step 6: Package microstrip line GUI as single app installer file

After creating an microstrip line calculator GUI in MATLAB ‘appdesigner’ environment, the GUI can be packaged into a single app installer file for easy sharing with other users in three ways (in Figure 38):

1. MATLAB App – Create an app installation file to share your app with MATLAB users.
2. Web App – Create a deployed web app using MATLAB Compiler.
3. Standalone Desktop App – Create a standalone desktop app using MATLAB Compiler.

The packaged installation file allows users to install the app and access it from the app gallery without concern for installation details or the MATLAB path. The single app installer file can be created by clicking ‘Share’ in the ‘DESIGNER’ tab as shown in Figure 38.

Figure 38. Package the GUI as an app through the three access options in the package application dialog

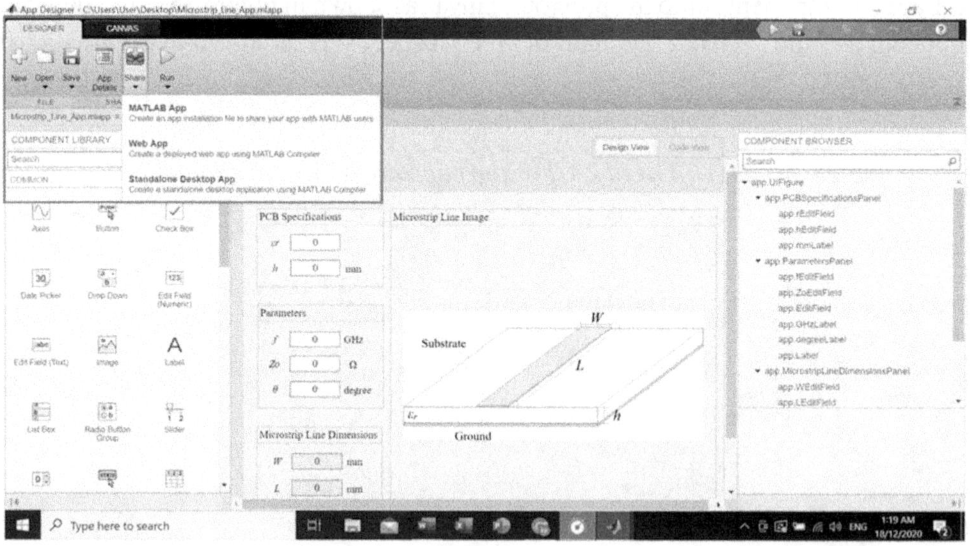

Then, a package app dialog box with three packaging options will drop down. The app designer can use either one of the three packaging options to create the single application installer file. A detailed description of these three packaging options can be found in MATLAB R2020a, (2020).

ANTENNA DESIGN

Monopole and Dipole Antennas

Dimension of Monopole Driven from Coaxial Line

Monopole and dipole antennas are the simplest antenna structure compared to other small antennas. In practice, normally, the bare monopole is driven from coaxial line as shown in Figure 39.

Input Impedance

For dipole or monopole structures, the energy of current is released through the extended conductor from an excitation point, thus, the inductive properties are normally dominant at the point ($z = 0$). Hence, the analysis and prediction of the performance of the dipole or monopole are often referred as the input impedance $Z_{in} = \Re e(Z_{in}) + j\,\Im m(Z_{in})$ instead of input admittance, Y_{in}.

Figure 39. (a) Dimensions of monopole antenna driven from coaxial line. (b) Distribution of azimuth magnetic fields, H_ϕ of the monopole antenna

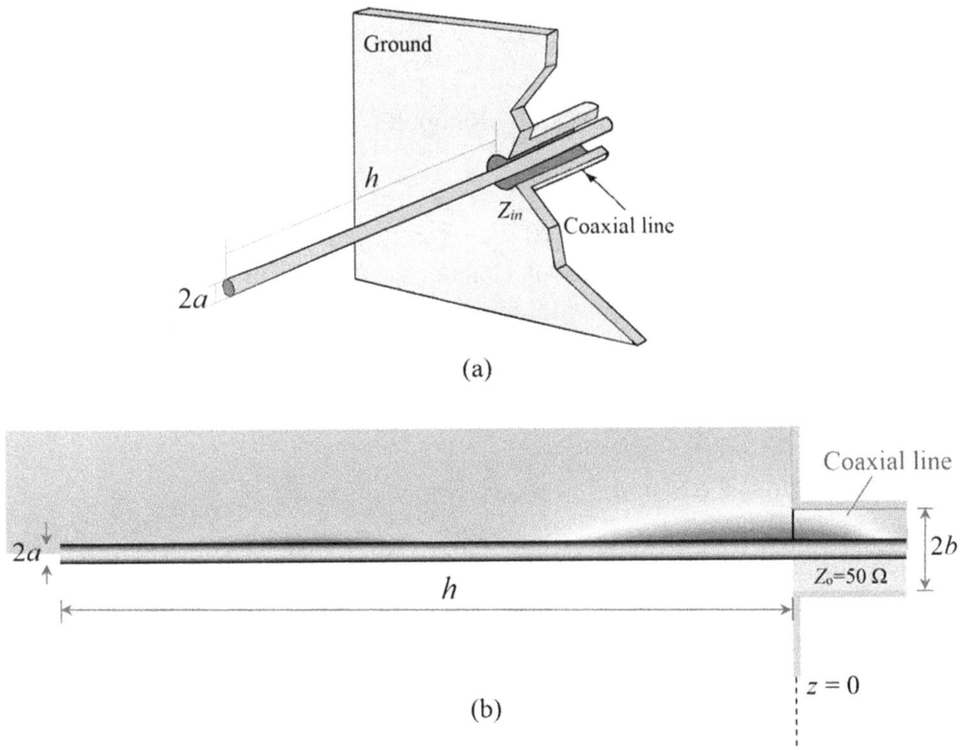

The induced electromagnetic field (EMF) method was introduced by Brillouin (1922) to determine the input impedance, Z_{in} at the driving point of linear monopole:

$$Z_{in} = \frac{j\eta_o}{4\pi\sqrt{\varepsilon_r}\sin^2(kh)}\int_0^h \left\{\sin\left[k(h-z)\right]\times\left[\frac{e^{-jkr_1}}{r_1}+\frac{e^{-jkr_2}}{r_2}-2\cos(kh)\frac{e^{-jkr_3}}{r_3}\right]\right\}dz \text{ (Monopole)} \quad (4)$$

The formulation of Z_{in} is also given as (Weiner, 2003):

$$Z_{in} = \frac{j\eta_o}{4\pi\sqrt{\varepsilon_r}\sin^2(kh)}\int_0^h \left\{\sin\left[k(h-z)\right]\times\left[\begin{array}{c}\frac{e^{-jkr_1}}{r_1}-\cos(kh)\frac{e^{-jkr_3}}{r_3}\\ -z\sin(kh)e^{-jkr_3}\left(\frac{jk}{r_3^2}+\frac{1}{r_3^3}\right)\end{array}\right]\right\}dz \text{ (Monopole)} \quad (5)$$

The single integrations of (4) and (5) can be solved analytically using a Simpson's rule ($n = 32$). A semi-empirical formulation of input impedance, Z_{in} for dipole with $2h$ of length, has been proposed by Tai's book chapter in 2007 (Tai and Long, 2007), as:

$$Z_{in} = R + jX - j120\left[\ln\left(\frac{2h}{a}\right)-1\right]\cot(kh) \text{ (Dipole)} \quad (6a)$$

$$Z_{in} = \frac{1}{2}(R + jX) - j60\left[\ln\left(\frac{2h}{a}\right)-1\right]\cot(kh) \text{ (Monopole)} \quad (6b)$$

The accuracy of (6) is limited in the range ($0 < kh \leq \pi/2$). The values of R and X are obtained by fitting with numerical simulation data. In his book chapter, the R and X in (6) are expressed as a function of (kh) in third-order polynomials as (Tai and Long, 2007):

$$R = -0.4787 + 7.3246(kh) + 0.3963(kh)^2 + 15.6131(kh)^3 \quad (7a)$$

$$X = -0.4456 + 17.0082(kh) + 8.6793(kh)^2 + 9.6031(kh)^3 \quad (7b)$$

The calculated normalized resistance, $\Re e(Z_{in})/Z_o$, and the normalized reactance, $\Im m(Z_{in})/Z_o$, of monopole for free space at room temperature using (4), (5), and (6). Z_o is the characteristic impedance in coaxial line which can be calculated as:

$$Z_o = \frac{1}{Y_o} = \frac{60\ln(b/a)}{\sqrt{\varepsilon_c}} \quad (8)$$

where ε_c is a relative permittivity of a material filled in the coaxial line ($\varepsilon_c = 2.06$ for Teflon).

Table 3. Coefficients, β for weight parameter function, α (You et al., 2010)

Coefficients, β					
β_1	-18.19648039	β_{14}	116.668095	β_{27}	84.06831414
β_2	12.76199981	β_{15}	-52.9378065	β_{28}	2.755085456×10^3
β_3	-3.312103266	β_{16}	6.270991827	β_{29}	33.82544182
β_4	0.3003225802	β_{17}	-0.118979601	β_{30}	72.12824746
β_5	$3.590117582 \times 10^{-3}$	β_{18}	$4.855052808 \times 10^{-2}$	β_{31}	-17.93696706
β_6	$3.085732072 \times 10^{-4}$	β_{19}	$-5.013624408 \times 10^{-3}$	β_{32}	$5.533645369 \times 10^{-2}$
β_7	$-1.162266887 \times 10^{-4}$	β_{20}	$4.370988471 \times 10^{-5}$	β_{33}	-0.1312955126
β_8	$1.261236962 \times 10^{-6}$	β_{21}	$-1.468965157 \times 10^{-5}$	β_{34}	$2.684299667 \times 10^{-2}$
β_9	$-1.627446234 \times 10^{-6}$	β_{22}	$1.078813248 \times 10^{-6}$	β_{35}	$-3.389494003 \times 10^{-5}$
β_{10}	$2.334588069 \times 10^{-7}$	β_{23}	7.893108958×10^5	β_{36}	$5.840124572 \times 10^{-5}$
β_{11}	3.056186734×10^3	β_{24}	6.417179527×10^5	β_{37}	$-1.138928243 \times 10^{-5}$
β_{12}	281.1896375	β_{25}	-2.656170879×10^5		-
β_{13}	-89.38714019	β_{26}	-5.541042383×10^4		-

Normally, small monopole antenna is driven from coaxial line. Hence, the input impedance, Z_{in} of the monopole antennas is strongly affected by the fringing field at driving point (at the end of coaxial line), especially for short monopole antenna which is operated at frequencies more than 6 GHz. The fringing field is caused by coupling effect between the outer conductor of the coaxial line and the monopole near the end of coaxial line. Thus, the ideal input impedance, Z_{in} needs an end correction term to correlate the analytical solution of (4) to the actual situation.

Recently, You *et al.* (2009) proposed a semi-empirical capacitance correction for the analytical admittance formula over a wide range of length, *h*, radius, *a*, aspect ratio, *b/a* and frequency, *f* as well as the relative permittivity, ε_r of the medium which is surrounding the monopole. The apparent admittance, Y_{in} is written as:

$$Y_{in} = \frac{1}{Z_{in}} - j\alpha\omega\varepsilon_r C \tag{9}$$

The Z_{in} in (9) is calculated from (4). Symbol α is the weight parameter of the end correction, and C is the aperture capacitance between the inner cylindrical conductor and the ground plane at the open end of the coaxial line which is expressed as (Gajda and Stuchly, 1983):

$$C = 2.48\varepsilon_o\left(b-a\right) \tag{10}$$

The weight parameter, α in (9) is expressed in second order polynomial form, which includes three variables (a, b and h) as:

$$\alpha = A_3 a^2 + A_2 a + A_1 \tag{11}$$

where

$$A_3 = \sum_{n=0}^{4} B_n''' h^{-n},\ A_2 = \sum_{n=0}^{3} B_n'' h^{-n},\ A_1 = \sum_{n=0}^{2} B_n' h^{-n}$$

and

$$\begin{aligned}
&B_0' = \beta_4 R_t^3 + \beta_3 R_t^2 + \beta_2 R_t + \beta_1 \ ; \quad B_1' = \beta_7 R_t^2 + \beta_6 R_t + \beta_5 \ ; \quad B_2' = \beta_{10} R_t^2 + \beta_9 R_t + \beta_8 \ ; \\
&B_0'' = \beta_{13} R_t^2 + \beta_{12} R_t + \beta_{11} \ ; \quad B_1'' = \beta_{16} R_t^2 + \beta_{15} R_t + \beta_{14} \ ; \quad B_2'' = \beta_{19} R_t^2 + \beta_{18} R_t + \beta_{17} \ ; \\
&B_3'' = \beta_{22} R_t^2 + \beta_{21} R_t + \beta_{20} \ ; \quad B_0''' = \beta_{25} R_t^2 + \beta_{24} R_t + \beta_{23} \ ; \quad B_1''' = \beta_{28} R_t^2 + \beta_{27} R_t + \beta_{26} \ ; \\
&B_2''' = \beta_{31} R_t^2 + \beta_{30} R_t + \beta_{29} ;\ B_3''' = \beta_{34} R_t^2 + \beta_{33} R_t + \beta_{32} ;\ B_4''' = \beta_{37} R_t^2 + \beta_{36} R_t + \beta_{35}
\end{aligned}$$

where $R_t = b/a$ and the coefficients of β are the fitting values. Up to 37 fitting coefficients (β_1 to β_{37}) are chosen to provide a greater degree of freedom for the regression fitting. The values of β_1 to β_{37} are listed in Table 3.

The values of weight parameter, α is found by optimizing with up to 1700 calculated admittance values obtained from Finite Element Method (FEM). The (9) is applicable for low conducting materials at low frequencies satisfying the conditions: ($2.3 \le b/a \le 4$), ($0.22 \le a \le 0.65$) mm and ($1 \le h \le 10$) mm and $kh \le 3\pi/4$. For arbitrary values of size ratio, b/a, the weight parameter, α is close to unity when $h = (10\times a)$ mm; whereas, when $h \ge (10\times b)$ mm, α becomes zero.

Radiation Pattern

Study of the pattern of far field radiation is important for the dipole which is implemented as an antenna in communication applications. The radiation pattern describes the relative strength of the radiated field through various directions, at a fixed distance, r from the antenna. Thus, the transmission/reception of the RF signal by the dipole antenna from a particular direction can be identified. The radiation field can be considered as a far field when the observation distance, r is far away from the dipole antenna and a condition of $kr >> 1$ is achieved. The radiation performance is often described by E_θ and H_ϕ fields plane-pattern in spherical coordinates. The E_θ field of the monopole antenna can be computed as:

$$E_\theta \approx jk\eta_o \frac{e^{-jkr}}{4\pi r} \underbrace{\sin\theta \frac{\Delta z \sin\left(\frac{k\Delta z\cos\theta}{2}\right)}{\left(\frac{k\Delta z\cos\theta}{2}\right)} \sum_{n=1}^{N} I_n e^{jkz_n\cos\theta}}_{\textit{Radiation pattern}} \tag{12}$$

The θ component electric field, E_θ can be alternatively expressed as:

$$E_\theta \approx j60\frac{e^{-jkr}}{r}\underbrace{\frac{\cos\left(k\Delta z\cos\theta\right)-\cos\left(k\Delta z\right)}{\sin\theta\sin\left(k\Delta z\right)}\sum_{n=1}^{N} I_n e^{jkz_n\cos\theta}}_{Radiation\ pattern} \quad (13)$$

The far-field for E_θ decay as $1/r$, but the radiation pattern is dependent on the distance, r ($r \approx$ radius of spherical boundary, ρ). Based on (13), the far-field of the monopole antenna with length h and infinite ground plane can be further reduced as:

$$E_\theta \approx \begin{cases} \dfrac{j\eta_o I_o e^{-jkr}}{2\pi r}\left[\dfrac{\cos\left(\dfrac{kh}{2}\cos\theta\right)-\cos\left(\dfrac{kh}{2}\right)}{\sin\theta}\right] & 0 \le \theta \le \dfrac{\pi}{2} \\ 0 & \dfrac{\pi}{2} \le \theta \le \pi \end{cases} \quad (14a)$$

and

$$H_\phi = \frac{E_\theta}{\eta_o} \quad (14b)$$

where η_o ($=120\pi\ \Omega$) is the intrinsic impedance and θ is the polar angle from axis of antenna (z-axis).

Directivity and Gain of Monopole Antenna

From equation (14a), the general power density, P of the antenna can be written as:

$$P = \frac{\left|E_\theta\right|^2}{2\eta_o} = \frac{15\left|I_o\right|^2}{\pi r^2}\left[\frac{\cos\left(\dfrac{kh}{2}\cos\theta\right)-\cos\left(\dfrac{kh}{2}\right)}{\sin\theta}\right]^2 \quad (15)$$

From (15), the directivity, D pattern of the antenna is given as:

$$D = \frac{P_{maximum}}{P_{average}} = \frac{2\left[\dfrac{\cos\left(\dfrac{kh}{2}\cos\theta\right) - \cos\left(\dfrac{kh}{2}\right)}{\sin\theta}\right]^2}{\displaystyle\int_0^{\pi} \frac{\left[\cos\left(\dfrac{kh}{2}\cos\theta\right) - \cos\left(\dfrac{kh}{2}\right)\right]^2}{\sin\theta} d\theta} \tag{16}$$

The relationship between power gain, G and directivity, D is expressed as:

$$G = \chi D$$

where χ is the antenna efficiency factor ($0 \leq \chi \leq 1$) which is independent on *{\displaystyle \theta }* θ. Hence, the gain, G in unit decibel (dB) can be written as:

$$G(\text{dB}) = 10\log_{10}(\chi D) \tag{17}$$

For lossless antenna, the $\chi = 1$. In practice, the gain, G is always less than the directivity, D. Due to the singularity issue, the single integration of (16) cannot be solved using the Newton-Cotes quadrature, such as Simpson's rule. Hence, Gaussian quadrature rule (n = 16) is implemented to solve the integral equation (16). The complete GUI of monopole antenna EDA or calculator is shown in Figure 40.

Figure 40. MATLAB-based GUI of the monopole antenna calculator

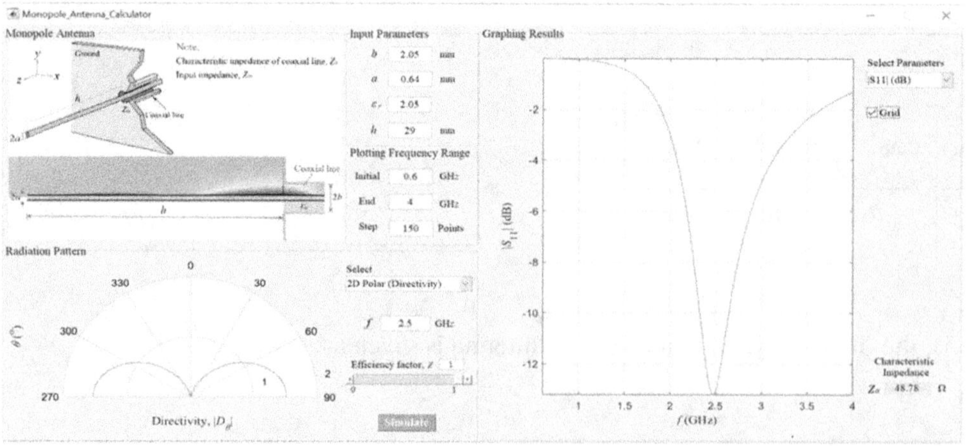

Rectangular Waveguide Antenna

Dimension of Rectangular Waveguide Antenna

The open-ended rectangular slot antenna is a rectangular metal pipe with height of a and width of b, respectively as shown in Figure 41. This antenna guides high frequency TE_{10} mode wave from one place to slot aperture without significant loss in intensity.

Figure 41. (a) Dimensions of rectangular waveguide. (b) Open-ended rectangular antennas (You, 2019)

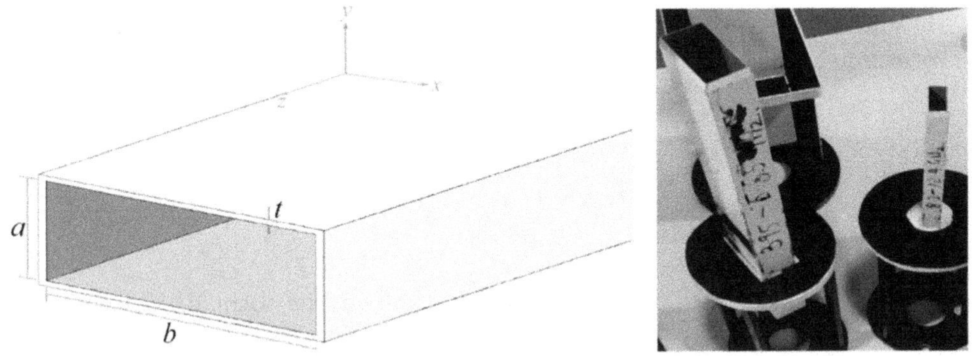

(a) (b)

Cutoff Frequency of Rectangular Waveguide Antenna

The lower cutoff frequency, f_c and the lower cutoff wavelength, λ_c of dominant TE_{10} mode of the open-ended rectangular slot antenna can be determined as:

$$f_c = \frac{c}{2b} \tag{18a}$$

and

$$\lambda_c = \frac{c}{f_c} \tag{18b}$$

where c (= 299792458 ms^{-1}) is the velocity of light in the free space and b (in unit meter) is the width of the slot aperture. Normally, the operating bandwidth of the rectangular waveguide antenna can be simply determined as:

$$1.23 f_c \leq f \leq 1.83 f_c \tag{19}$$

The effective wavelength, λ_g of propagated wave in the antenna can be calculated as:

$$\lambda_g = \frac{\lambda_o}{\sqrt{1-(\lambda_o/\lambda_c)^2}} = \frac{c}{f\sqrt{1-\{c/(2bf)\}^2}}$$

where f and λ_o (= c/f) are the operating frequency and the wavelength in free space, respectively.

Gain of Rectangular Waveguide Antenna

Stratton, (1941) was initially expressed the on-axis gain, G formulation of the rectangular slot antenna based on the reflection coefficient, S_{11} at slot aperture as:

$$G = \frac{\pi k^2 ab}{8\tilde{\beta}\left(1-|S_{11}|\right)^2}\left|\left[1+\tilde{\beta}+S_{11}\left(1-\tilde{\beta}\right)\right]\left(\frac{2}{\pi}\right)^2\right|^2 \tag{20}$$

where k (= $2\pi/\lambda_o$) is propagation constant in free space and $\tilde{\beta}$ is the normalized propagation constant of the TE_{10} mode which is given as:

$$\tilde{\beta} = \sqrt{1-\left(\frac{\pi}{kb}\right)^2} \tag{21}$$

Normally, the calculated gain using (20) has lower than the actual value of 1 dB to 2 dB. Yaghjian, (1984) has been improved the (20) by adding a correction term, C_o as:

$$G = \frac{\pi k^2 ab}{8\tilde{\beta}\left(1-|S_{11}|\right)^2}\left|\left[1+\tilde{\beta}+S_{11}\left(1-\tilde{\beta}\right)\right]\left(\frac{2}{\pi}\right)^2 + C_o\right|^2 \tag{22}$$

where C_o is the positive real constant due to edge current and mutual coupling effects at the slot aperture, which value is depended on the operating wavelength, λ_o and the width of slot aperture, b. For far-fields gain estimation, the C_o in (22) is simply expressed as (Selvan, 1998):

$$C_o \simeq \frac{8}{\pi^2\sqrt{2}}\left(\frac{1}{\tilde{\beta}}-1\right) \tag{23}$$

Equation (23) can also be interpreted as:

$$C_o \simeq \frac{3}{4\pi}\sqrt{\frac{\lambda}{2b}}\left(\frac{\sin(\theta/2)}{\cos\theta}\right) \tag{24}$$

where $\theta = \cos^{-1}\left(\tilde{\beta}\right)$. Besides formulations of (20) and (22), the empirical expression of gain in unit dB is published by Enkhbayar *et al.* (2009) and given as:

$$G(\text{dB}) = \begin{Bmatrix} -10.02819 + 41.30528\left(\frac{f}{f_c}\right) - 51.80386\left(\frac{f}{f_c}\right)^2 + 36.23459\left(\frac{f}{f_c}\right)^3 \\ -12.70621\left(\frac{f}{f_c}\right)^4 + 1.751408\left(\frac{f}{f_c}\right)^5 + 5.33333\left(\frac{t}{b}\right)\left(\frac{f_c}{f}\right)^2 \end{Bmatrix} \tag{25}$$

where t is the wall thickness of the waveguide and valid range of t/b is given as $0.008 \le t/b \le 0.1$. On the other hand, f is the operating frequency which is obeyed the range of (19).

Directivity of Rectangular Waveguide Antenna

In Collin, (1985), the directivity, D_o expression of the open-ended rectangular slot antennas is given as:

$$D_o \simeq \frac{64ab}{\lambda_o^3\sqrt{k^2 - (\pi/b)^2}} \tag{26}$$

On the other hand, in Balanis (2015), the maximum directivity, D_o is expressed as:

$$D_o = \frac{8ab}{\pi^3}k^2 \tag{27}$$

The directivity, D as a function of elevation, θ and azimuth, ϕ angles is given as (Collin, 1985):

$$D(\theta,\phi) = \frac{4\pi}{P_r}U(\theta,\phi) \tag{28}$$

where $U(\theta, \phi)$ and P_r are the radiation intensity and radiated power, respectively.

$$U(\theta,\phi) = \frac{r^2}{2\eta}\left(|E_\theta|^2 + |E_\phi|^2\right) \tag{29}$$

$$P_r = \frac{a \times b}{4\eta}\left|E_o\right|^2 \tag{30}$$

where far-field E_θ and E_ϕ in (29) are expressed as (Balanis, 2015):

$$E_\theta = jkabE_o\frac{e^{-jkr}}{4r}\left\{\sin\phi\frac{\sin\left(\frac{ka}{2}\sin\theta\sin\phi\right)}{\frac{ka}{2}\sin\theta\sin\phi}\bullet\frac{\cos\left(\frac{kb}{2}\sin\theta\cos\phi\right)}{\left(\frac{\pi}{2}\right)^2-\left(\frac{kb}{2}\sin\theta\cos\phi\right)^2}\right\} \tag{31a}$$

$$E_\phi = jkabE_o\frac{e^{-jkr}}{4r}\left\{\cos\phi\cos\theta\frac{\sin\left(\frac{ka}{2}\sin\theta\sin\phi\right)}{\frac{ka}{2}\sin\theta\sin\phi}\bullet\frac{\cos\left(\frac{kb}{2}\sin\theta\cos\phi\right)}{\left(\frac{\pi}{2}\right)^2-\left(\frac{kb}{2}\sin\theta\cos\phi\right)^2}\right\} \tag{31b}$$

From (31a) to (31b), the E-plane and H-plane directivity patterns ($0 \le \theta \le \pi/2$) for the radiating rectangular aperture can be written as:

$$D\left(\theta,\frac{\pi}{2}\right) = \frac{8abk^2}{\pi^3}\left|\frac{\sin\left(\frac{ka}{2}\sin\theta\right)}{\frac{ka}{2}\sin\theta}\right|^2 \quad (E\text{-plane}) \tag{32a}$$

$$D(\theta,0) = \frac{\pi abk^2}{2}\left|\frac{\cos\theta\cos\left(\frac{kb}{2}\sin\theta\right)}{\left(\frac{\pi}{2}\right)^2-\left(\frac{kb}{2}\sin\theta\right)^2}\right|^2 \quad (H\text{-plane}) \tag{32b}$$

It should be noted that (32a) and (32b) does not include the effects of diffraction at the waveguide edges nor the far-field contribution from field distribution on the outer portions of a radiating waveguide.

Aperture Admittance and Return Loss

The normalized input admittance of the rectangular aperture waveguide can be calculated as (Compton, 1963):

Figure 42. Variation in normalized aperture conductance, $Re(Y_{in})/Y_o$, and normalized susceptance, $Im(Y_{in})/Y_o$, of air using X-band rectangular waveguide antenna (You, 2017)

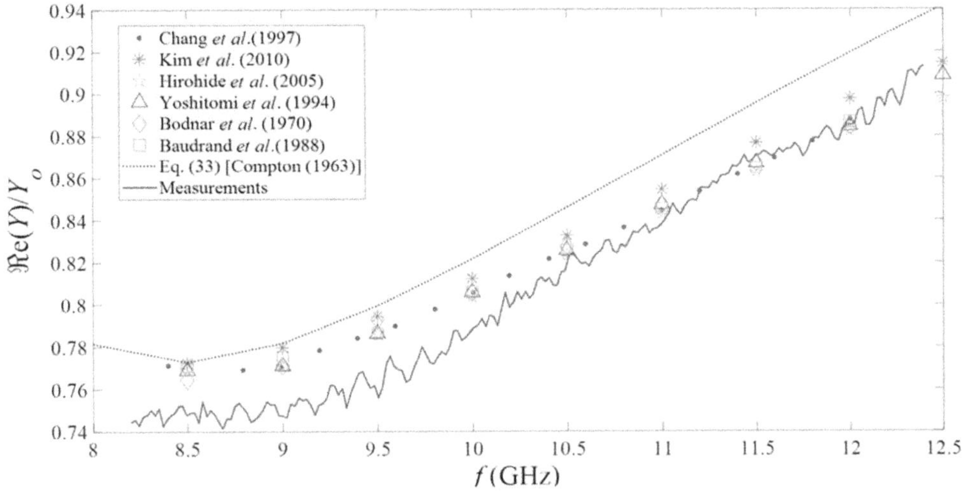

(a)

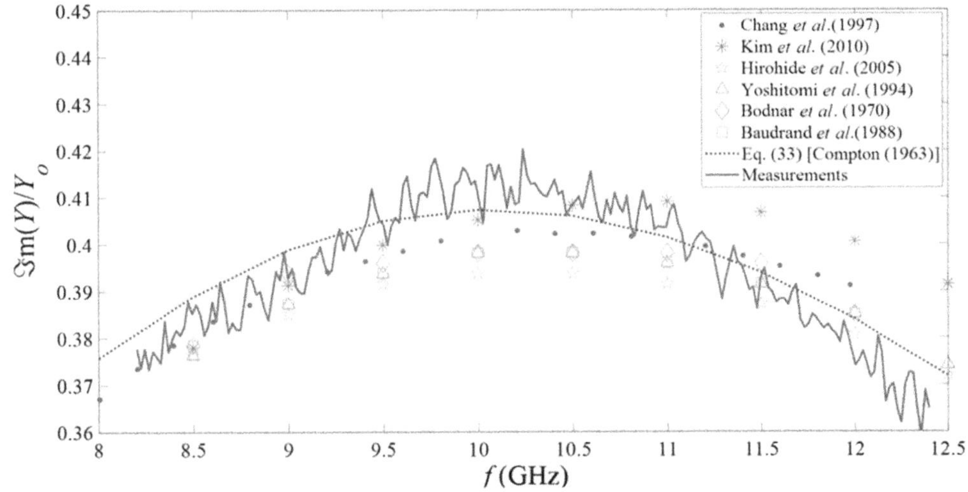

(b)

$$\tilde{Y}_{in} = \Re e\left(Y/Y_o\right) + j\Im m\left(Y/Y_o\right)$$
$$= \frac{j8b}{k_1 a}\int_0^a \int_0^b (a-x)\left[D_1(b-y)\cos\left(\frac{\pi y}{b}\right) + D_2 \sin\left(\frac{\pi y}{b}\right)\right]\frac{e^{-jk_2\sqrt{x^2+y^2}}}{\sqrt{x^2+y^2}}dxdy \quad (33)$$

The D_1 and D_2 parameters in (33) are given as:

Figure 43. (a) Simulated radiation field from rectangular waveguide aperture using COMSOL simulator. (b) Reflection coefficient, $|S_{11}|$ versus operating frequency of X-band rectangular waveguide.

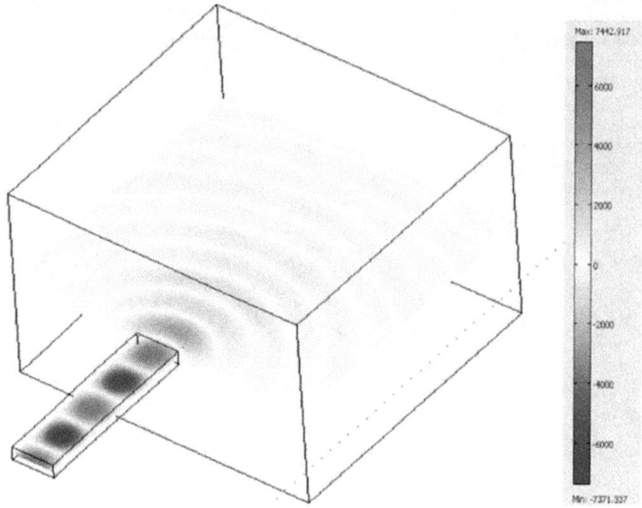

(a)

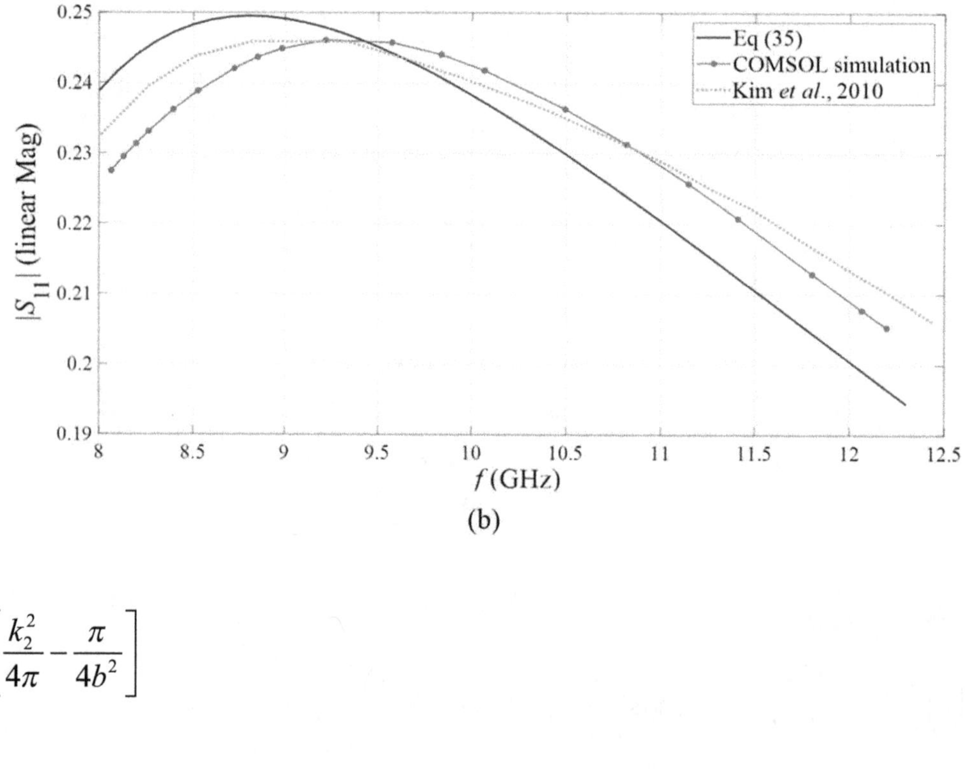

(b)

$$D_1 = \frac{1}{b^2}\left[\frac{k_2^2}{4\pi} - \frac{\pi}{4b^2}\right] \tag{34a}$$

$$D_2 = \frac{1}{\pi b}\left[\frac{k_2^2}{4\pi} + \frac{\pi}{4b^2}\right] \tag{34b}$$

Figure 44. MATLAB-based GUI of the rectangular waveguide antenna calculator

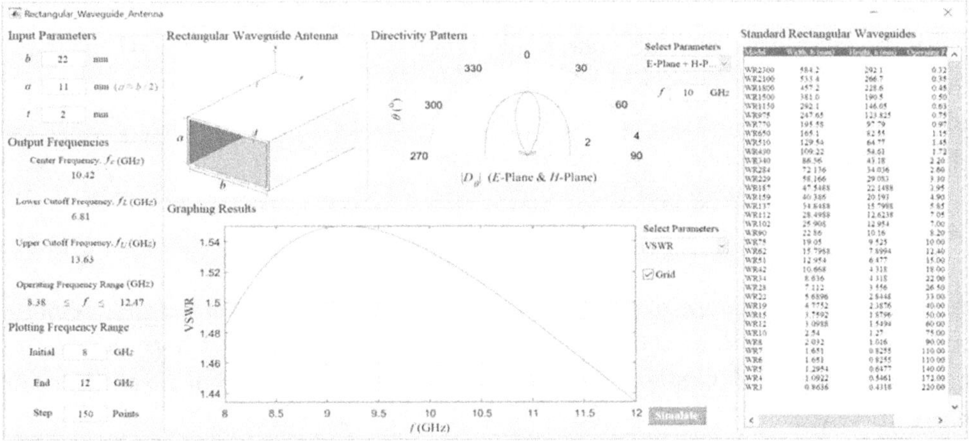

where $k_1 = k$ and $k_2 = k\sqrt{\varepsilon_r}$ are the propagation constants for rectangular waveguide line, respectively. Symbol ε_r is the relative permittivity for the external medium (outer region). The reflection coefficient, S_{11} at the aperture waveguide can be calculated from (33) as:

$$S_{11} = \frac{1-\tilde{Y}_{in}}{1+\tilde{Y}_{in}} \tag{35}$$

Figure 45. (a) Dimension of pyramidal horn antenna. (b) E-plane and H-plane views of horn antenna.

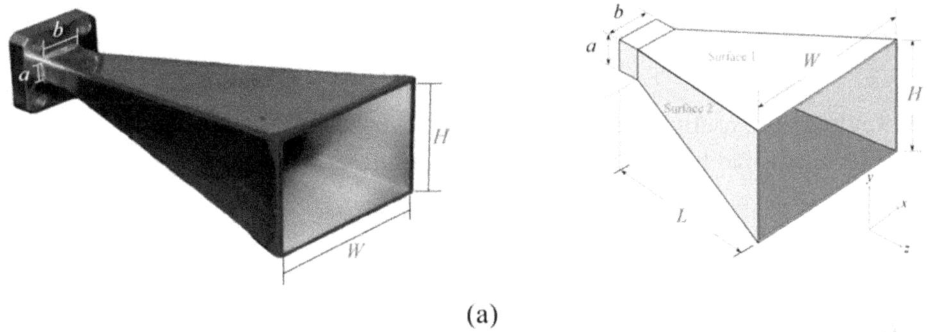

(a)

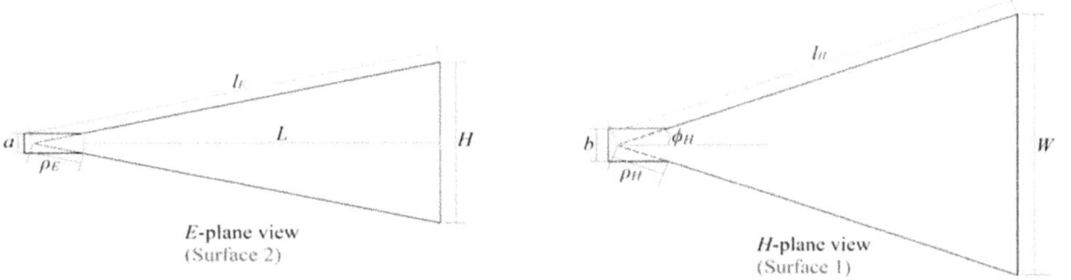

(b)

The double integral of equation (33) is solved by the 8-term Gaussian integration method. The calculated normalized conductance, $\Re e(Y_{in})/Y_o$ and the $\Im m(Y_{in})/Y_o$ using (33) are found to be in good agreement with literature data over the operational range of frequencies. The GUI of the rectangular waveguide antenna EDA created using MATLAB 'guide' is shown in Figure 44.

Pyramidal Horn Antenna

Horn Dimensions

Pyramidal horn antenna is one type of horn antenna. The shape of the horn is a quadrangular pyramid with a rectangular cross section as shown in Figure 45 (a). The pyramidal horn consists of a pair of flared sides and another pair of parallel sides, so-called *E*-field and *H*-field plane surfaces, respectively as shown in Figure 45 (b).

The pyramidal horn antenna is capable of providing wide bandwidth, high directivity, and high gain, G (10 dBi $\leq G \leq$ 25 dBi). The gain, G of the pyramidal horn depends on the size of the antenna.

The first step in designing the horn antenna is defining the necessary far-field gain, G_o at desired operating frequency. Then, the width, W, height, H, and length, L dimensions of the horn can be determined so that can be reached the desired gain, G. Apply the inverse solution to objective equation (36) to obtain the optimum value of χ (Balanis, 2015).

$$\left(\sqrt{2\chi}-a\frac{f}{c}\right)^2(2\chi-1)-\left(\frac{\sqrt{3}G_o}{2\pi\sqrt{2\pi\chi}}-b\frac{f}{c}\right)^2\left(\frac{G_o^2}{6\pi^3\chi}-1\right)=0 \tag{36}$$

The initial/trial value can be calculated from:

$$\chi_{Initial}=\frac{G_o}{2\pi\sqrt{2\pi}}$$

In this work, finding zero routine is realized using the MATLAB fzero command. Once the value of χ is obtained, the width, W and height, H of the horn aperture can be determined as:

$$W=\frac{\sqrt{3}cG_o}{2\pi f\sqrt{2\pi\chi}} \tag{37a}$$

$$H=\frac{c\sqrt{2\chi}}{f} \tag{37b}$$

The l_E and l_H are the slant lengths of the side of the *E*-plane and *H*-plane surfaces (in Figure 45(b)) which can be calculated as:

$$l_E = \frac{H}{2}\sqrt{1+\left(\frac{H}{\lambda}\right)^2}\ ;\ l_H = W\sqrt{\frac{1}{4}+\left(\frac{W}{3\lambda}\right)^2}$$

and

$$L = (H-a)\sqrt{\left(\frac{l_E}{H}\right)^2 - \frac{1}{4}} \text{ or } L = (W-b)\sqrt{\left(\frac{l_H}{W}\right)^2 - \frac{1}{4}}$$

Table 4 lists the comparison predictions of *W* and *H* values between these three methods. These three methods are consistent with each other and give approximate values of *W* and *H*.

Table 4. Three different design methods are used to predict W and H values.

Waveguide model	*f* (GHz)	G_o (dB)	Güney, 2001		Selvan, 1999		Eq (37a), (37b)	
			W (cm)	*H* (cm)	*W* (cm)	*H* (cm)	*W* (cm)	*H* (cm)
WR650	1.431	15.93	58.95	45.18	58.28	44.62	59.58	44.88
WR430	2.163	16.53	41.69	32.10	41.27	31.76	42.10	31.92
WR340	2.736	17.45	36.53	28.31	36.22	28.06	36.83	28.18
WR284	3.275	18.00	32.52	25.19	32.22	24.94	32.77	25.09
WR229	4.061	18.75	28.47	22.24	28.28	22.09	28.67	22.17
WR187	4.968	19.81	26.27	20.55	26.08	20.39	26.43	20.51
WR159	5.849	20.75	24.76	19.53	24.64	19.43	24.89	19.51
WR137	6.779	21.50	23.31	18.36	23.15	18.23	23.42	18.35
WR112	8.290	22.10	20.41	16.10	20.26	15.98	20.50	16.09
WR90	10.34	22.60	17.31	13.69	17.19	13.59	17.38	13.68
WR75	12.40	23.12	15.28	12.16	15.21	12.10	15.34	12.16
WR62	14.95	23.45	13.16	10.48	13.09	10.43	13.21	10.48
WR51	18.24	23.82	11.25	8.97	11.19	8.92	11.29	8.97
WR34	27.36	24.25	7.87	6.29	7.84	6.25	7.90	6.29
WR28	33.22	24.58	6.73	5.38	6.70	5.35	6.76	5.38

Return Loss of Horn Antenna

The total reflection coefficient, S_{11} of the horn antenna is contributed by the reflection coefficient, Γ_1 at waveguide junction and the reflection coefficient, Γ_2 at horn aperture as (Jull, 1972):

$$S_{11} = \Gamma_1 + \Gamma_2 \tag{38}$$

where

$$\Gamma_1 = \frac{H_1^{(2)}(\beta\rho_H) - jH_o^{(2)}(\beta\rho_H)}{H_1^{(2)}(\beta\rho_H) + jH_o^{(2)}(\beta\rho_H)} \tag{39a}$$

The reflection coefficient, Γ_2 due to single and double diffraction at the aperture horn, is given as (Jull, 1972):

$$\Gamma_2 = \frac{-\exp\left[-j2(\beta l_H - \phi + \delta)\right]}{4\beta\phi_H l_H}\left\{1 + \sqrt{\frac{2}{\pi}}\sum_{n=0}^{m}\frac{\exp\left[-j(kd_n - 3\pi/4)\right]}{(kd_n)^{1/2}\left\{1 + \sin\left[(n+1)\phi_H\right]\right\}}\right\} \tag{39b}$$

and $d_n = 2l_H \sin\left[(n+1)\phi_H\right]$. However, the largest series term, m in (39b) is required to be less than $\frac{\pi}{2\phi_H} - 1$. Generally, for the 10 dB, 15 dB, and 20 dB gains of the pyramidal horn, the maximum integer of m is equal to 1, 2, and 3, respectively.

In fact, the (39b) can be simplified as:

$$\Gamma_2 = -\exp\left[-j2(\beta l_H - \phi + \delta)\right]\left\{\begin{array}{l}\dfrac{1}{4\beta\phi_H l_H} + \dfrac{1}{2\beta b}\left(\dfrac{k}{\beta} - 1\right) \\ \cdot\left\{1 + 2\sqrt{\dfrac{bf}{c}}\dfrac{\exp\left[-j(kb - 3\pi/4)\right]}{\beta b}\right\}\end{array}\right\} \tag{39c}$$

The alternative expression of (39c) can be written as (Selvan *et al.*, 1999):

$$\Gamma_2 = -\exp\left[-j2(\beta l_H - \phi + \delta)\right]\left\{\frac{1}{4\beta\phi_H l_H} + \frac{1}{2\beta b}\left(\frac{k}{\beta} - 1\right)\exp\left\{\frac{1}{\beta}\sqrt{\frac{2k}{\pi b}}\exp\left[j(kb + \pi/4)\right]\right\}\right\} \tag{39d}$$

where the propagation constant, β in waveguide, horn flare angle, ϕ_H of H-plane, and phase shifts ϕ and δ at horn-waveguide junction are given as:

$$\phi_H = \sin^{-1}\left(\frac{W}{2l_H}\right)$$

$$\beta = \sqrt{k^2 - \left(\frac{\pi}{b}\right)^2}$$

Table 5. Comparison of computed value of gain, G at 4.1 GHz between CST simulation and equation (40) for b= 5.8166 cm and a=2.912 cm.

Dimensions (cm)	Expected Gain, G_o (dB)	Gain, G (dB) at 4.1 GHz	
		CST simulation	Equation (42)
$W = 9.26$ $H = 6.29$ $L = 1.45$	10	10.1	9.1
$W = 17.53$ $H = 13.29$ $L = 9.36$	15	15.1	14.7
$W = 31.86$ $H = 25.02$ $L = 37.79$	20	20.1	19.9

$$\phi = \text{phase of } H_1^{(1)}(\beta\rho_o)$$

$$\delta = \tan^{-1}\frac{A\cos[\psi - \phi]}{1 + A\sin[\psi - \phi]}$$

where A and ψ are given as:

$$A = \left|\frac{H_o^{(1)}(\beta\rho_o)}{H_1^{(1)}(\beta\rho_o)}\right| \text{ and } \psi = \text{phase of } H_o^{(1)}(\beta\rho_o)$$

On-Axis Gain of Horn Antenna

The near-field on-axis gain of the pyramidal horn is given as:

$$\begin{aligned} G &= G_o R_E R_H \\ &= \frac{32WH}{\pi}\left(\frac{f}{c}\right)^2 R_E R_H \end{aligned} \tag{40}$$

where G_o is the far-field gain of a rectangular aperture with a uniform distribution in the y direction, a cosinusoidal variation in the x direction and uniform phase across the aperture (Rao *et al.*, 1997). The R_E and R_H are the E- and H-planes gain reduction factors, which is represented the effect of the aperture phase taper due to the quadratic phase errors in the principal planes (Maybell *et al.*, 1993).

Figure 46. Comparison of computed value of gain, G (in dB) versus operating frequency, f using CST simulation and equation (40) for b= 5.8166 cm and a=2.912 cm

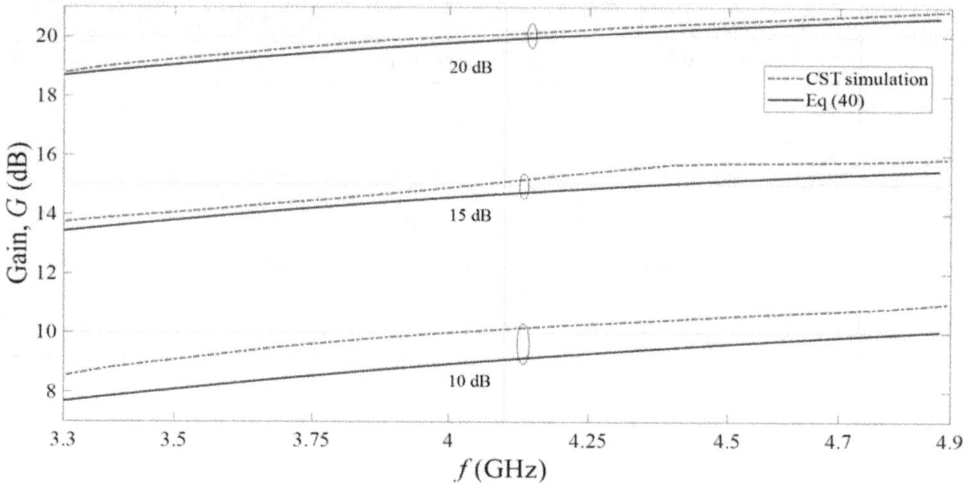

$$R_E = \left| \frac{2}{H} \int_0^{H/2} \exp\left\{ -jkL_E \left(\sqrt{1 + \left(\frac{y}{L_E} \right)^2} - 1 \right) \right\} \partial y \right|^2 \tag{41a}$$

$$R_H = \left| \frac{\pi}{W} \int_0^{W/2} \cos\left(\frac{\pi x}{W} \right) \exp\left\{ -jkL_H \left(\sqrt{1 + \left(\frac{x}{L_H} \right)^2} - 1 \right) \right\} \partial x \right|^2 \tag{41b}$$

where

$$L_H = \sqrt{l_H^2 - \left(\frac{W}{2} \right)^2} \text{ and } L_E = \sqrt{l_E^2 - \left(\frac{H}{2} \right)^2}$$

The integral of equations (41a) and (41b) are solved by the 16-term Gaussian integration method. Dimensions W and H in the Table 5 are trial values so that the simulated gain, G value from CST simulator approaches the expected G_o value. When the same values of W and H are inserted in equation (40), the calculated gain, G is lower than the expected G_o value at the centre frequency of 4.1 GHz. For a 10-dB pyramid horn design, there is a ~1 dB deviation between the expected G_o value and the calculated G value using (40). However, the accuracy of (40) is increased for high gain (> 15 dB) pyramidal horn designs. Comparison between calculated and simulated values of gain, G for frequency range from 3.3 GHz to 4.9 GHz is shown in Figure 46.

Radiation Pattern of Horn Antenna

The far-field pattern of the pyramidal horn in E- and H-planes can be computed as (Balanis, 2015):

$$E_{\theta} = j\frac{kE_{o}e^{-jkr}}{4\pi r}\left(\frac{\pi}{2k}\sqrt{L_{E}L_{H}}\right)\left[\sin\phi\left(1+\cos\theta\right)I_{1}I_{2}\right] \tag{42a}$$

$$E_{\phi} = j\frac{kE_{o}e^{-jkr}}{4\pi r}\left(\frac{\pi}{2k}\sqrt{L_{E}L_{H}}\right)\left[\cos\phi\left(1+\cos\theta\right)I_{1}I_{2}\right] \tag{42b}$$

where I_1 and I_2 functions in (42) are given as:

$$I_{1} = \left\{\begin{array}{l} \exp\left(j\frac{k_{x}^{2}L_{H}}{2k}\right)\left\{\left[C(t_{2})-C(t_{1})\right]-j\left[S(t_{2})-S(t_{1})\right]\right\} \\ +\exp\left(j\frac{k_{x1}^{2}L_{H}}{2k}\right)\left\{\left[C(t_{4})-C(t_{3})\right]-j\left[S(t_{4})-S(t_{3})\right]\right\} \end{array}\right\} \tag{43a}$$

$$I_{2} = \exp\left(j\frac{k_{y}^{2}L_{E}}{2k}\right)\left\{\left[C(t_{6})-C(t_{5})\right]-j\left[S(t_{6})-S(t_{5})\right]\right\} \tag{43b}$$

where $C(t)$ and $S(t)$ are the cosine and sine Fresnel integral functions as:

$$C(t) = \int_{0}^{t}\cos\left(\frac{\pi}{2}x^{2}\right)\partial x \text{ and } S(t) = \int_{0}^{t}\sin\left(\frac{\pi}{2}x^{2}\right)\partial x$$

where

$$k_{x} = k\sin\theta\cos\phi + \frac{\pi}{W};\ k_{x1} = k\sin\theta\cos\phi - \frac{\pi}{W}$$

$$t_{1} = \sqrt{\frac{1}{\pi kL_{H}}}\left(-\frac{kW}{2} - k_{x}L_{H}\right);\ t_{2} = \sqrt{\frac{1}{\pi kL_{H}}}\left(+\frac{kW}{2} - k_{x}L_{H}\right)$$

Figure 47(a) Simulated radiation field from pyramidal horn antenna using CST simulator. Calculated E-plane and H-plane radiation patterns using equations (43a) and (43b) for pyramidal horn with (b) G_o=10 dB, (c) G_o=15 dB, and (d) G_o= 20 dB, respectively.

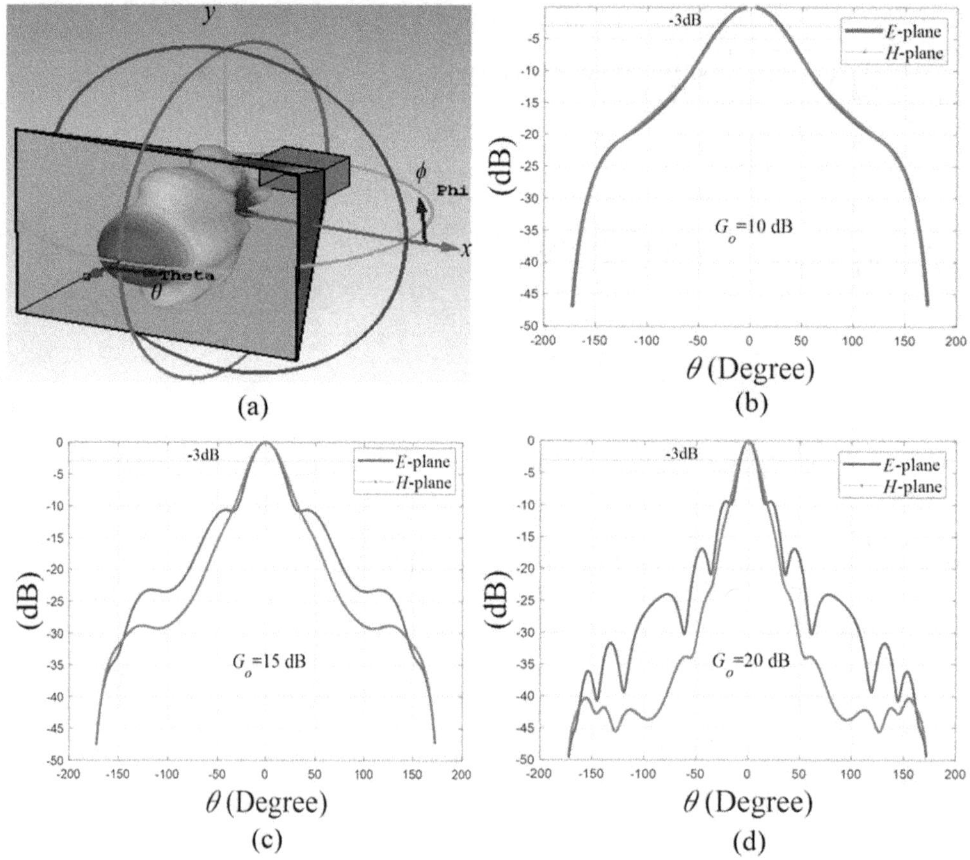

$$t_3 = \sqrt{\frac{1}{\pi k L_H}}\left(-\frac{kW}{2} - k_{x1} L_H\right) \quad ; \quad t_4 = \sqrt{\frac{1}{\pi k L_H}}\left(+\frac{kW}{2} - k_{x1} L_H\right) \quad k_y = k\sin\theta \quad ;$$

$$t_5 = \sqrt{\frac{1}{\pi k L_E}}\left(-\frac{kH}{2} - k_y L_E\right);\ t_6 = \sqrt{\frac{1}{\pi k L_E}}\left(+\frac{kH}{2} - k_y L_E\right)$$

From (43a) and (43b), the principal *E*-plane ($\phi = \pi/2$) and *H*-plane ($\phi = 0$) radiation patterns of the pyramidal horn can be simplified as:

$$F_E(\theta) = \frac{|1+\cos\theta|}{2}\frac{|I_2(\theta)|}{|I_2(\theta = 0^o)|} \text{ E-plane } (\phi = \pi/2) \tag{43a}$$

and

Figure 48. MATLAB-based GUI of the pyramidal horn antenna calculator

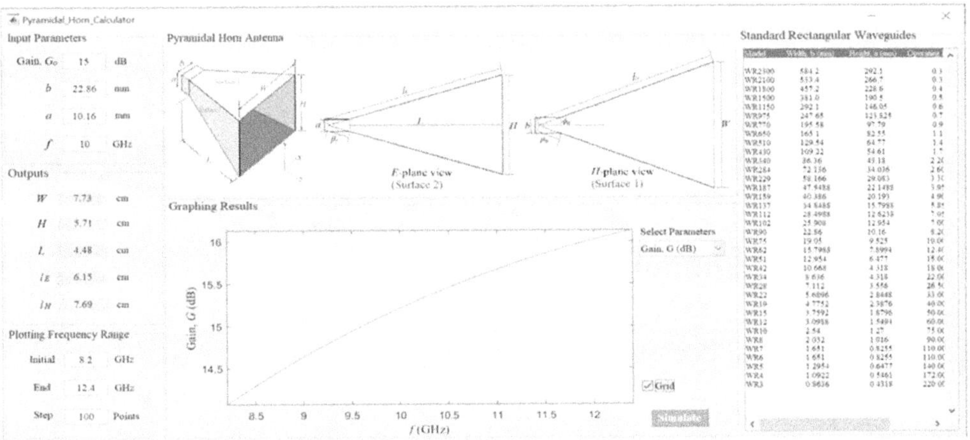

$$F_H = \frac{|1+\cos\theta|}{2}\frac{|I_1(\theta)|}{|I_1(\theta = 0^o)|}\ H\text{-plane } (\phi = 0) \tag{43b}$$

Figure 49. (a) Standard log periodic antenna (Balmain and Nkeng, 1976). (b) Four array of log periodic antenna. (c) Planar form of log periodic antenna

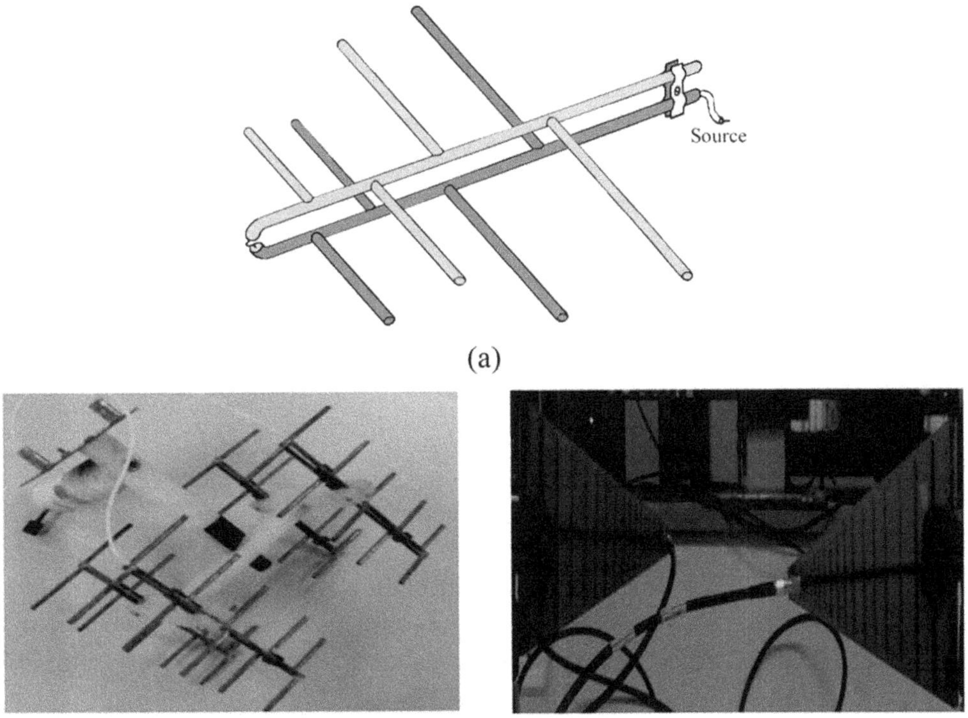

(a)

(b)

(c)

Figure 50. The relationship between the scale factor τ, space factor δ and gain of the log periodic antenna (Cebik, 2007). The limitation range of τ and δ are 0.8 < τ < 0.98 and 0.03 ≤ δ ≤ (0.243τ - 0.051)

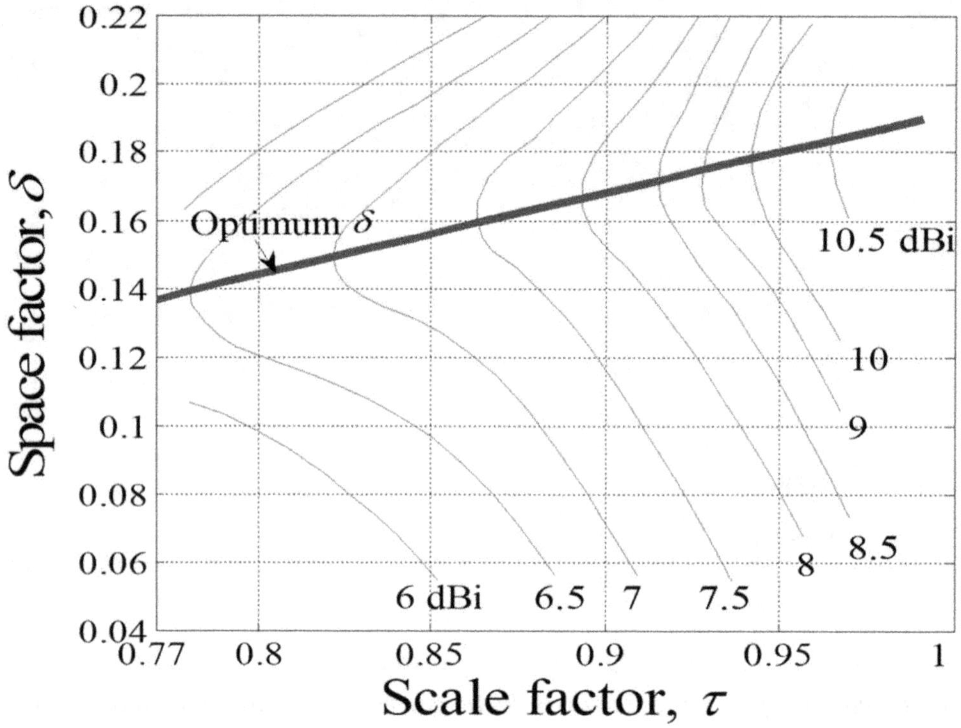

where $I_1(\theta)$ is the formula of (43a) with only $\phi = 0^o$ and $I_1(\theta = 0^o)$ is the formula of (43a) with $\theta = 0^o$ and $\phi = 0^o$. In fact, the I_1 and I_2 can be directly calculated as:

$$I_1 = \int_{-W/2}^{W/2} \cos\left(\frac{\pi x'}{W}\right) e^{-jk\sqrt{L_H^2 + x'^2}} e^{jk\sin\theta x'} dx' \tag{44a}$$

and

$$I_2 = \int_{-H/2}^{H/2} e^{-jk\sqrt{L_E^2 + y'^2}} e^{jk\sin\theta y'} dy' \tag{44b}$$

Figure 47 shows the simulated and calculated radiation patterns at 4.1 GHz using CST simulator and equations (43a)-(43b), respectively. From Figures 47 (b), (c), and (d), the half-power beamwidth, HPBW (3dB beamwidth) of the pyramidal horn with near-field gain, G_0 of 10 dB, 15 dB, and 20 dB are given as 57.6°, 29.6°, and 17.6°, respectively. On the other hand, when the same dimension of the pyramidal horn antenna is simulated using CST simulator, the simulated NPBW are given as 57.7°, 29.3°, and 15.4°, respectively. Figure 48 shows the GUI of the pyramidal horn antenna EDA which is created using MATLAB 'guide'.

Figure 51. The schematic structure and dimensions of the log periodic antenna.

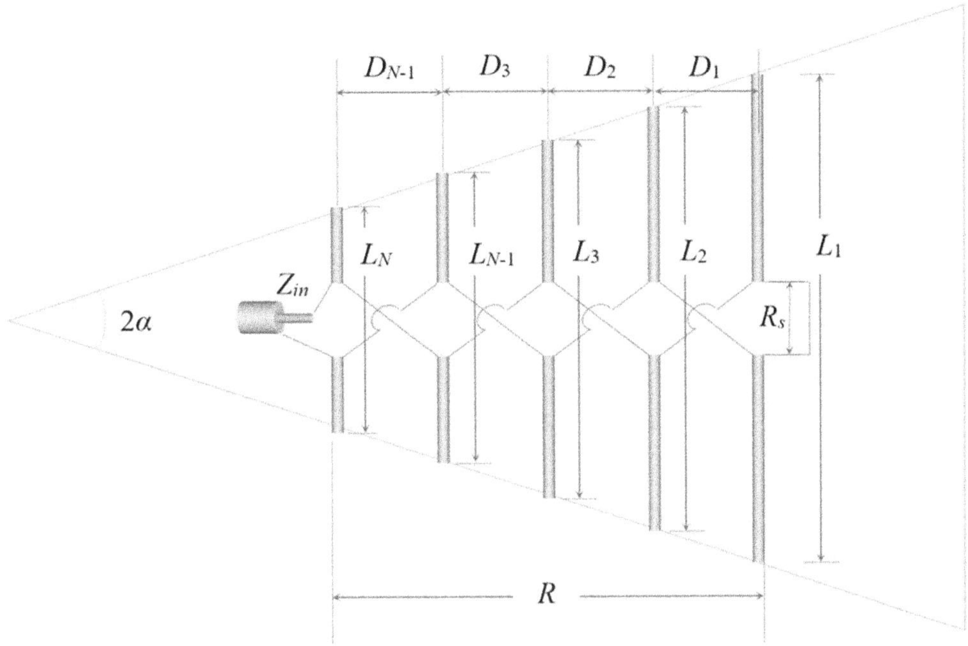

Log Periodic Antenna

The log periodic antenna consists of a set of dipoles with different lengths in the array as shown in Figure 49. The log periodic antenna is directional antenna designed to operate over a wide band of frequencies. The smaller scale length of the dipole element in the array contributes to higher frequencies. On the other hand, the larger scale length of the dipole element contributes to lower frequencies. However, this kind of antenna requires many of the dipole elements in the array to increase the gain value of the antenna.

For log periodic antenna design, the first step is to identify the gain, G value of the designed log periodic antenna. Figure 50 shows the relationship between the scale factor τ, space factor δ, and gain, G of the antenna. The values of τ and δ for the particular selected gain can be determined by referring to Figure 50. Based on Figure 50, the optimum space factor, δ_{opt} can be represented by linear equation as:

$$\delta_{opt} = 0.243\tau - 0.051 \text{ for } 0.8 \le \tau \le 0.98 \quad (46)$$

Besides, based on Figure 50, the relationship between the gain, G (in unit dB) and δ_{opt} can be expressed as polynomial equation as:

$$G = \left\{ \begin{matrix} 9.62050137\times10^5\delta_{opt}^4 - 5.62371751\times10^5\delta_{opt}^3 + 1.23335126\times10^5\delta_{opt}^2 \\ -1.19808109\times10^4\delta_{opt} + 440.09176778 \end{matrix} \right\} 6.5\,\text{dB} \le G \le 11\,\text{dB} \quad (47)$$

Identify the lowest and highest operating frequencies (f_{min} and f_{max}) of the designed log periodic antenna. From the values of τ, δ, f_{min} and f_{max}, the number of dipole elements, N in the log periodic array can be found as:

$$N = 1 + \frac{\ln\left[\left(\frac{f_{\max}}{f_{\min}}\right)\left(1.1 + 7.7\left(1-\tau\right)^2 \frac{4\delta}{1-\tau}\right)\right]}{\ln\left(1/\tau\right)} \tag{48}$$

Determine the length L_1 of the longest dipole element (in Figure 51) and the spacing $(R_1 - R_2)$ between the two longest dipoles as:

$$L_1 = \frac{c}{2f_{\min}} \text{ and } R_1 - R_2 = \frac{L_1 - L_2}{2}\left(\frac{4\delta}{1-\tau}\right)$$

where c (= 299792458 m/s) is the velocity of light in free space and $L_2 = \tau L_1$. The lengths and spacing of the other dipole elements can be calculated as:

$$L_{n+1} = \tau L_n \tag{49a}$$

$$D_n = 2\delta L_n \tag{49b}$$

Determine the short position at a distance R_S (in Figure 51) behind the longest dipole element. The short circuit acts as a reflector and its distance from the longest element should be $R_S = L_1/4$.

The apex half-angle, α in Figure 51 is given as:

$$\alpha = \cot^{-1}\left(\frac{4\delta}{1-\tau}\right) \tag{50}$$

Calculate the characteristic impedance, Z_o of feeder connecting to the shortest dipole element:

$$Z_o = \frac{R_{rad}^2}{8\delta' Z_{mean}} + R_{rad}\sqrt{\left(\frac{R_{rad}}{8\delta' Z_{mean}}\right)^2 + 1} \tag{51}$$

where R_{rad} is the mean radiation resistance level, this value can find in Figure 52 or equation (53). On the other hand, Z_{mean} is the mean characteristic impedance of a shorter dipole element as:

Figure 52. Mean radiation resistance level, R_{rad} versus pitch angle, α for τ = 0.81, 0.89 and 0.95, respectively (Isbell, 1960).

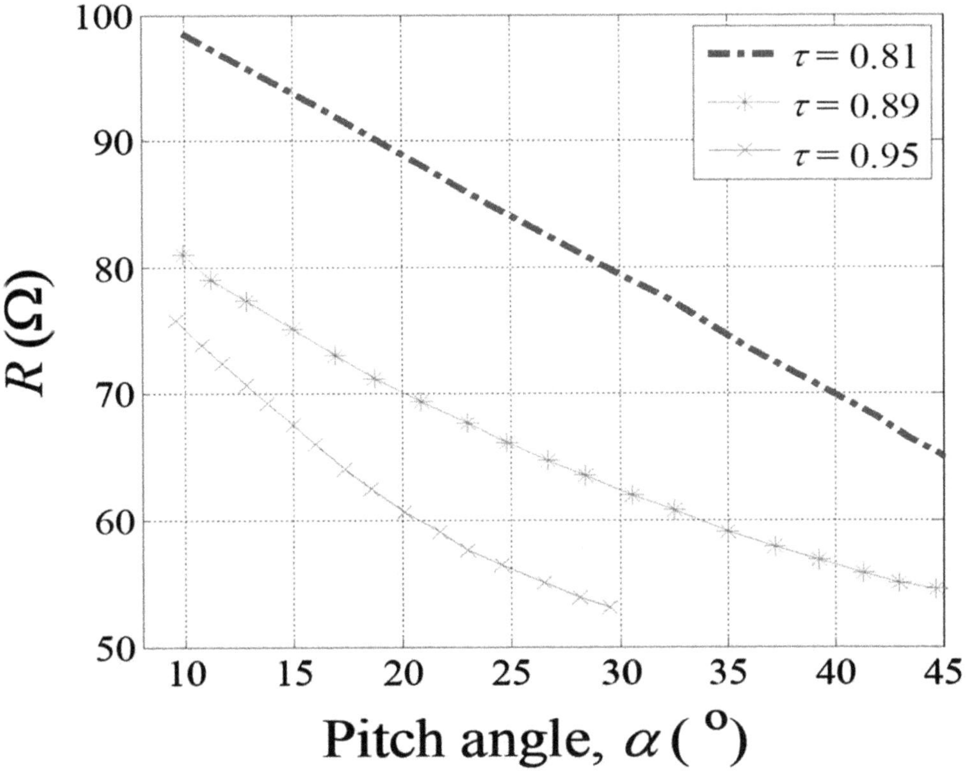

$$Z_{mean} = 120\left[\ln\left(\frac{L_N}{2a}\right) - 2.55\right] \tag{52}$$

Figure 53. MATLAB-based GUI of the log periodic antenna calculator

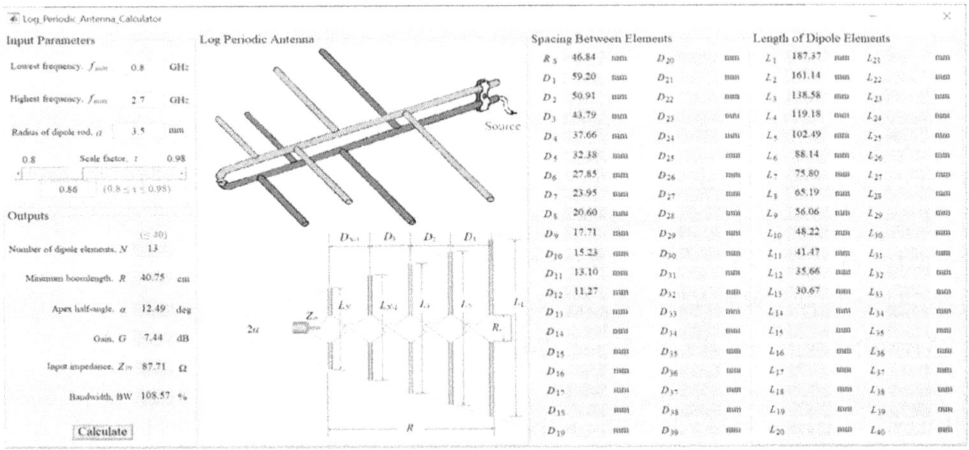

Figure 54. (a) The basic three elements (N = 3) Yagi-Uda antenna. (b) Commercial eighteen elements (N = 18) Yagi-Uda antenna with operating frequencies of 2.4 – 2.483 GHz and gain, G of 18 dBi. (c) Planar form of Yagi-Uda antennas.

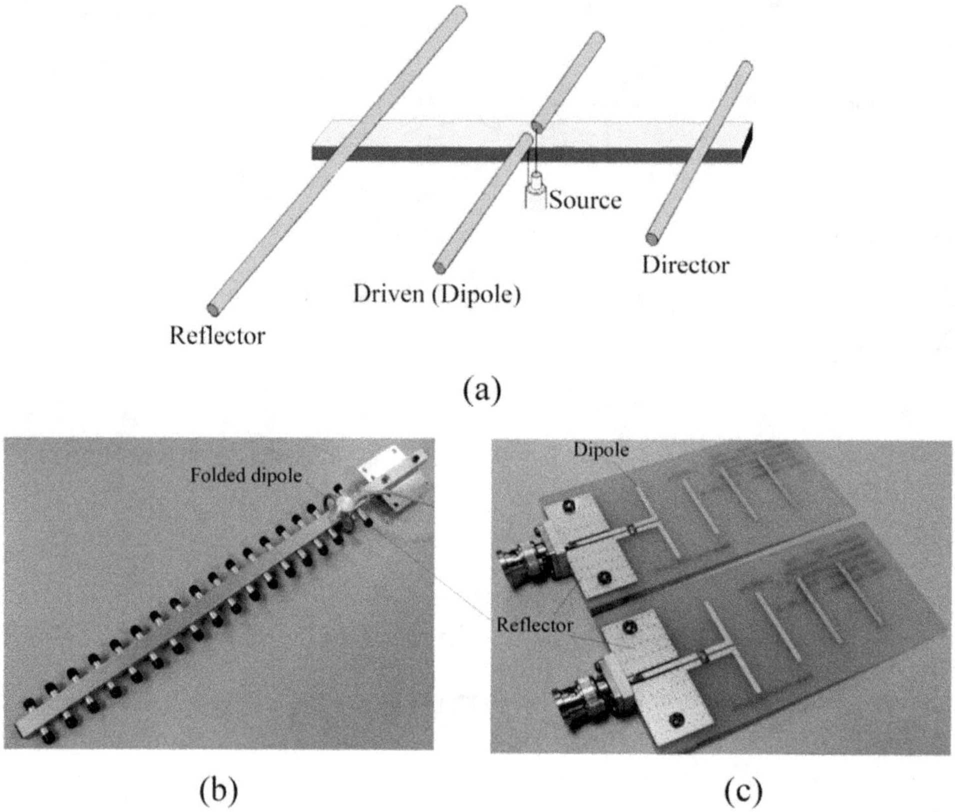

where a is the radius of the shortest dipole element and δ' is the mean spacing factor as:

$$\delta' = \frac{\delta}{\sqrt{\tau}}$$

$$R_{rad} = A_3\alpha^2 + A_2\alpha + A_1 \tag{53}$$

where the polynomial coefficients A_1, A_2, and A_3 are expressed as function of scale factor, τ as:

$$A_1 = 0.9558561\tau^2\text{-}1.4653282\tau\text{+}0.55953898$$
$$A_2 = \text{-}61.9115357\tau^2\text{+}99.0961646\tau\text{-}40.59162667$$
$$A_3 = 1421.0316405\tau^2\text{-}2590.6092854\tau\text{+}1274.0652285$$

Finally, the MATLAB-based GUI of the log periodic antenna EDA is illustrated in Figure 53.

Yagi-Uda Antenna

The Yagi-Uda antenna consists of one driven element, one reflector element (behind of the driven element) and at least one director element (in front of the driven element) as shown in Figure 54. The number of director elements in the array is used as parasitic elements which is determined the gain, G of antenna (up to 20 dBi). However, conventional Yagi-Uda antenna has narrow bandwidth.

In order to understand each symbol and parameter, an example of the $N = 5$ Yagi-Uda antenna is used as a reference schematic of the antenna, as shown in the Figure 55.

Figure 55. The schematic structure and dimensions of the Yagi-Uda antenna (Example: five elements (N = 5), namely one reflector, one driven element and three directors).

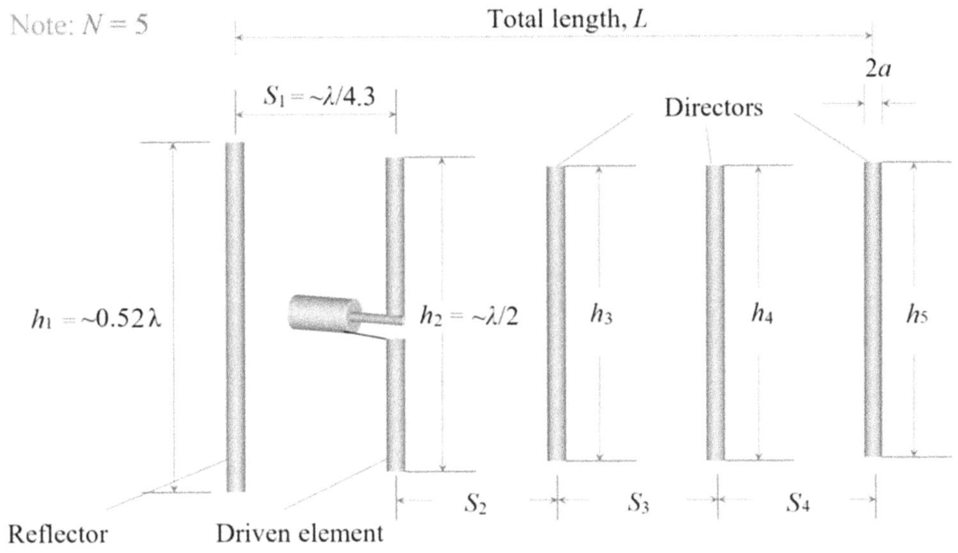

The mutual coupling effect between the parasitic elements is based on the length, h of each director and spacing distance, S between the director elements. There are several design guidelines for the Yagi-Uda antenna:

a) The length, h_1 ($> 0.5\lambda$) of the reflector is normally varied between 0.51 λ to 0.55 λ.
b) b) The length, h_2 of driven element is determined as:

$$h_2 \approx \frac{\lambda}{2} \approx \frac{c}{2f} \tag{54}$$

c) The lengths, h_n ($n = 3, 4, 5, \ldots, N$) of each director must less than 0.5λ and normally only varied between 0.3λ to 0.45λ which is depended on the size of $2a$. To facilitate antenna design, all director lengths, h_n are normally considered to be the same ($h_n = h_3 = h_4 = h_5, \ldots, h_N$).
d) d) The separation distance, S_1 between the reflector and driven element is given as:

Figure 56. (a) Optimum total length, L and (b) gain, G of the Yagi-Uda array versus number of dipole elements, N in the array (Greenblum, 1956a, 1956b).

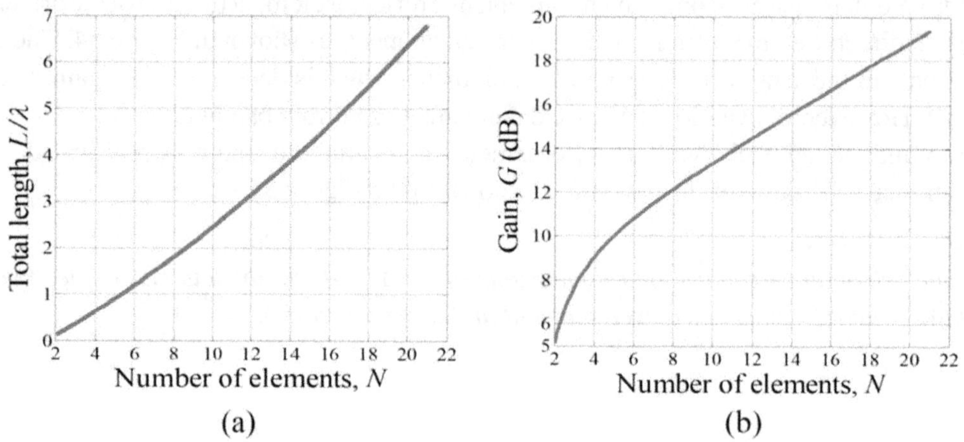

$$S_1 \approx \frac{\lambda}{4.3} \approx \frac{c}{4.3f} \tag{55}$$

e) The separation distance between the directors, S_n (n = 2, 3, 4, …, N) normally only varied from 0.1λ to 0.4λ. The specific values of S_n are tabulated in Table 6.

Figure 57. Optimum length of director, h versus number of director elements, (N-2) for various size of diameter, 2a for the corresponding director element.

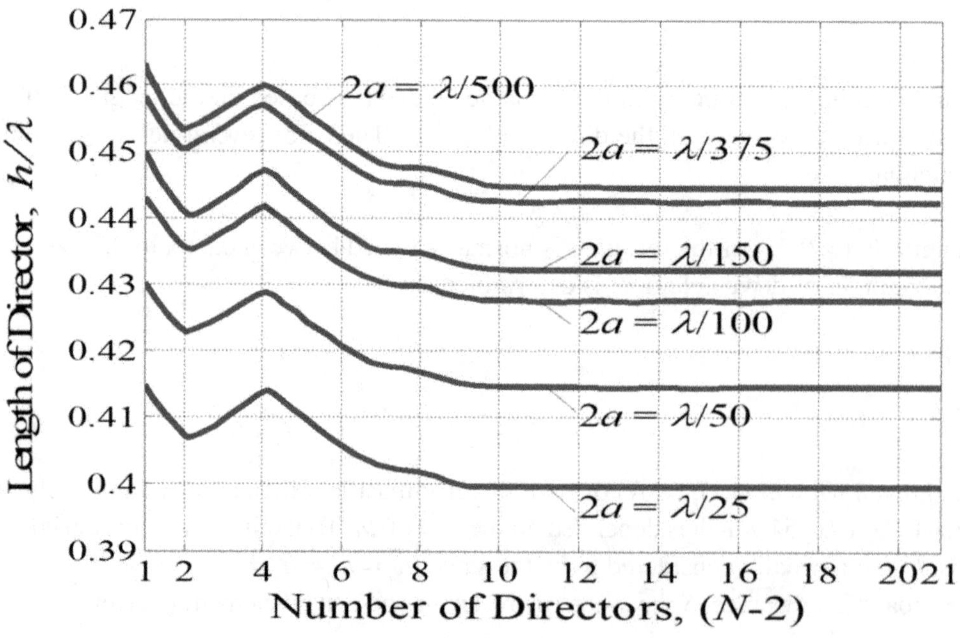

The total length, L (in unit meter) of the Yagi-Uda array can be predicted using (56) with a given number of dipole elements, N in the array (You, 2016).

$$L = \lambda \left\{ \begin{matrix} -4.7909454 \times 10^{-6} N^3 + 5.6329477 \times 10^{-3} N^2 \\ +0.22573937N - 0.3666209 \end{matrix} \right\} \text{for } 2 \leq N \leq 21 \quad (56)$$

In addition, the gain, G (in unit dB) of the Yagi-Uda array can be estimated using (57) with a given number of dipole elements, N in the array (You, 2016).

$$G = \left\{ \begin{matrix} 13.4802351\left(\log_{10}(N)\right)^3 - 28.606668\left(\log_{10}(N)\right)^2 \\ +30.2363326\log_{10}(N) - 1.7759598 \end{matrix} \right\} \text{for } 2 \leq N \leq 21 \quad (57)$$

Equations (56) and (57) are derived based on the data in Figure 56 (a) and (b), respectively. The relationship between the element number, N and its typical gain, G (dB) of the commercial Yagi-Uda array antenna are listed in Table 7.

For Yagi-Uda antenna with element, N less than 12, the lengths, h_n ($n = 3, 4, 5,\ldots, N$) of the directors depends on the number of elements, N and the diameter, $2a$ of the director rod as shown in Figure 57. Hence, the h_n (in unit meter) is expressed as a function of N, $2a$, and λ as:

$$h_n = \lambda\left(\beta_7 N^6 + \beta_6 N^5 + \beta_5 N^4 + \beta_4 N^3 + \beta_3 N^2 + \beta_2 N + \beta_1\right) \text{for } 3 \leq N \leq 12 \quad (58a)$$

where $h_n = h_3 = h_4 = h_5,\ldots, h_{12}$ and the polynomial coefficients, β_n are given as:

$$\beta_1 = 176.0060392\left(\frac{2a}{\lambda}\right)^2 - 11.1108389\left(\frac{2a}{\lambda}\right) + 1.2154460;$$

$$\beta_2 = -130.8517312\left(\frac{2a}{\lambda}\right)^2 + 7.7807503\left(\frac{2a}{\lambda}\right) - 0.6856983;$$

$$\beta_3 = 44.5236461\left(\frac{2a}{\lambda}\right)^2 - 2.6805942\left(\frac{2a}{\lambda}\right) + 0.2440262;$$

$$\beta_4 = -7.5703117\left(\frac{2a}{\lambda}\right)^2 + 0.4632869\left(\frac{2a}{\lambda}\right) - 0.0438291;$$

$$\beta_5 = 0.68388256\left(\frac{2a}{\lambda}\right)^2 - 0.04261015\left(\frac{2a}{\lambda}\right) + 0.00422037;$$

$$\beta_6 = -0.03131249\left(\frac{2a}{\lambda}\right)^2 + 0.00198723\left(\frac{2a}{\lambda}\right) - 0.000208295;$$

$$\beta_7 = 0.0005696647\left(\frac{2a}{\lambda}\right)^2 - 3.6835 \times 10^{-5}\left(\frac{2a}{\lambda}\right) + 4.14439 \times 10^{-6};$$

On the other hand, for $N \geq 13$, the lengths, h_n ($n = 13, 14,\ldots, 21$) of the directors are independent on the number of elements, N of the antenna, thus the empirical formulation of lengths, h_n (in unit meter) can be simplified as:

$$h_n = \lambda \left\{ \begin{array}{l} -982.4800824\left(\frac{2a}{\lambda}\right)^3 + 84.6223055\left(\frac{2a}{\lambda}\right)^2 - 3.0778998\left(\frac{2a}{\lambda}\right) \\ +0.45040134 \end{array} \right\} \text{ for } 12 < N \leq 21 \quad (58b)$$

Radiation Pattern of Yagi-Uda Antenna

The E-field radiation pattern of the single element Yagi-Uda antenna can be computed as (Thiele, 1969):

$$E_\theta = \frac{j\omega\mu}{4\pi r_o} e^{-jkr_o} \sin\theta \int_{-h/2}^{h/2} I(z')\exp(jkz\cos\theta)\,dz' \quad (59)$$

Consider the distribution current, $I(z')$ represents by Fourier series with M series term, the solution of (59) can be written as (Thiele, 1969):

$$E_\theta = -j\sqrt{\frac{\mu}{\varepsilon}}\frac{h}{\pi r_o} e^{-jkr_o} \sin\theta \sum_{m=1}^{M} (-1)^m \frac{I_m (2m-1)\cos\left(\frac{\pi h}{\lambda}\cos\theta\right)}{(2m-1)^2 - \left(\frac{2h}{\lambda}\cos\theta\right)^2} \quad (60)$$

where $r_o = r - (x\sin\theta\cos\phi + y\sin\theta\sin\phi + z\cos\theta)$. Each current amplitude, I_m term in (60) is normalized with maximum current, . Hence, the radiation pattern factor, $F(\theta, \phi)$ of single element (N=1) Yagi-Uda antenna with dipole length of h can be simplified as:

$$F(\theta,\phi) = h \left\{ \begin{array}{l} \exp\left[jk(x\sin\theta\cos\phi + y\sin\theta\sin\phi + z\cos\theta)\right] \\ \sum_{m=1}^{M} (-1)^m \frac{I_o (2m-1)\cos\left(\frac{\pi h}{\lambda}\cos\theta\right)}{(2m-1)^2 - \left(\frac{2h}{\lambda}\cos\theta\right)^2} \end{array} \right\}$$

Where I_o is relative amplitude current (constant value). Hence, the pattern factor, $F(\theta, \phi)$ of N elements Yagi-Uda antenna can be written as:

Figure 58. Relative current amplitude of each element of Yagi-Uda array antenna (Thiele, 1969)

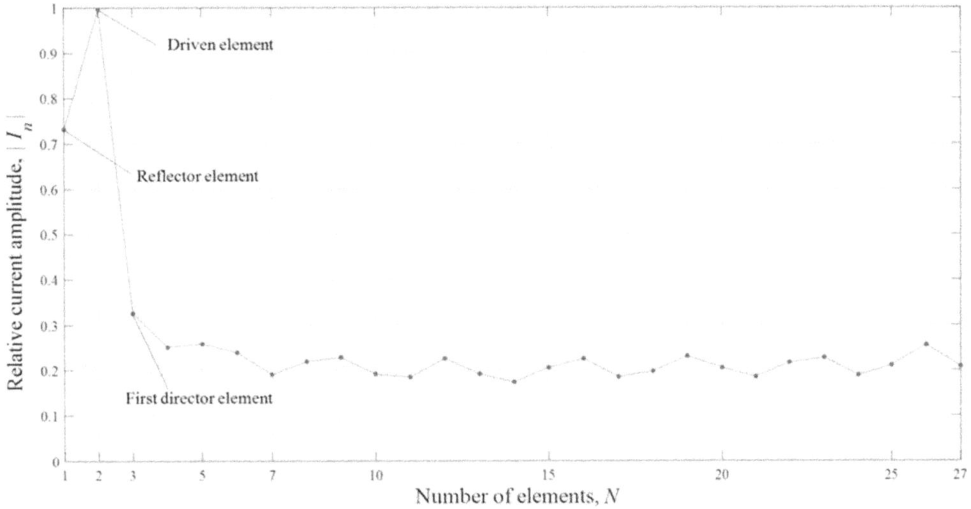

$$F(\theta,\phi)=\left\{\begin{array}{l}\sum_{n=1}^{N} h_n \exp\left[jk\left(x_n \sin\theta\cos\phi + y_n \sin\theta\sin\phi + z_n\cos\theta\right)\right] \\ \times\sum_{m=1}^{M}(-1)^m \dfrac{I_{on}(2m-1)\cos\left(\dfrac{\pi h_n}{\lambda}\cos\theta\right)}{(2m-1)^2-\left(\dfrac{2h_n}{\lambda}\cos\theta\right)^2}\end{array}\right\} \quad (61a)$$

where N is the total elements (reflector + driven + director) number of the Yagi-Uda antenna. The (61a) can be simplified as:

$$F(\theta,\phi)\approx\left\{\begin{array}{l}\sum_{n=1}^{N} h_n \exp\left[jk\left(\sum_{\kappa=1}^{n} S_\kappa \sin\theta\cos\phi\right)\right] \\ \times\sum_{m=1}^{M}(-1)^m \dfrac{I_{on}(2m-1)\cos\left(\dfrac{\pi h_n}{\lambda}\cos\theta\right)}{(2m-1)^2-\left(\dfrac{2h_n}{\lambda}\cos\theta\right)^2}\end{array}\right\} \quad (61b)$$

Assume coordinates, $y_n = 0$ and $z_n = 0$ along the axial axis (x-axis). On the other hand, the coordinate, $x_n = \sum_{\kappa=1}^{n} S_\kappa$. For instance, if the element with h_3 length is at the location of x_3, then the coordinate location of the element can be calculated as $x_3 = S_1 + S_2 + S_3$. Based on Figure 58, the distribution of relative amplitude currents, $I_{o1} = 0.73$, $I_{o2} = 1$, $I_{o3} = 0.33$, $I_{o4} = 0.25$, $I_{o5} = 0.25$, $I_{o6} = 0.24$, and $I_{o7} = I_{o8} = \ldots = I_{oN} = 0.2$. The MATLAB-based GUI of the Yagi-Uda antenna EDA is shown in Figure 59.

Table 6. Optimum element spacing for Yagi-Uda arrays (Greenblum, 1956).

Spacing Distance	Number of elements, N (reflector + driven element + directors)							
	2	2	3	4	5	6	8	8 to N
Reflector to driven element	0.175λ (0.15λ -0.2λ)	-	0.195λ (0.16λ -0.23λ)	0.2λ (0.18λ -0.22λ)	0.2λ (0.18λ -0.22λ)	0.18λ (0.16λ -0.20λ)	0.18λ (0.16λ -0.20λ)	0.18λ (0.16λ -0.20λ)
Driven element to first director, S_1	-	0.09λ (0.07λ -0.11λ)	0.175λ (0.16λ -0.19λ)	0.15λ (0.13λ -0.17λ)	0.155λ (0.14λ -0.17λ)	0.155λ (0.14λ -0.17λ)	0.15λ (0.14λ -0.16λ)	0.15λ (0.14λ -0.16λ)
S_2	-	-	-	0.16λ (0.14λ -0.18λ)	0.175λ (0.15λ -0.20λ)	0.205λ (0.16λ -0.25λ)	0.215λ (0.18λ -0.25λ)	0.215λ (0.18λ -0.25λ)
S_3	-	-	-	-	0.2λ (0.17λ -0.23λ)	0.26λ (0.22λ -0.30λ)	0.3λ (0.25λ -0.35λ)	0.3λ (0.25λ -0.35λ)
S_4	-	-	-	-	-	0.285λ (0.25λ -0.32λ)	0.295λ (0.27λ -0.32λ)	0.295λ (0.27λ -0.32λ)
S_5	-	-	-	-	-	-	0.3λ (0.27λ -0.33λ)	0.3λ (0.27λ -0.33λ)
S_6	-	-	-	-	-	-	0.35λ (0.30λ -0.40λ)	0.385λ (0.35λ -0.42λ)
S_7 to S_N	-	-	-	-	-	-	-	*All* 0.385λ (0.35λ -0.42λ)

Table 7. Gain values of commercial Yagi-Uda antennas

Number of elements, N	Gain (dBi)	
	Typical	Tolerance
3	3	3 to 4
5	8	7 to 9
6	10	10 to 11
7	11	8 to 11
8	12	11 to 16
9	13	7 to 16
12	16	12 to 16
13	16	-
15	-	13 to 16
16	-	14 to 20
18	18	-

Figure 59. MATLAB-based GUI of the Yagi-Uda array antenna calculator

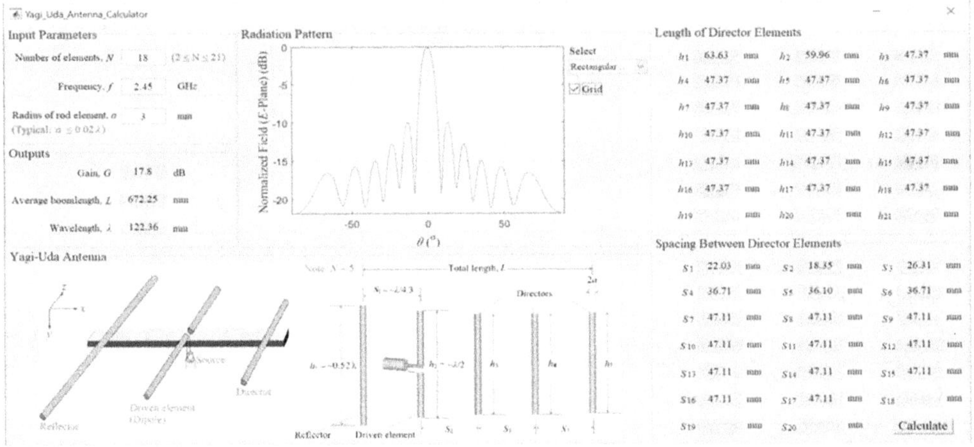

Figure 60. (a) Helical monopole antenna. (b) Critical dimensions of helical monopole antenna

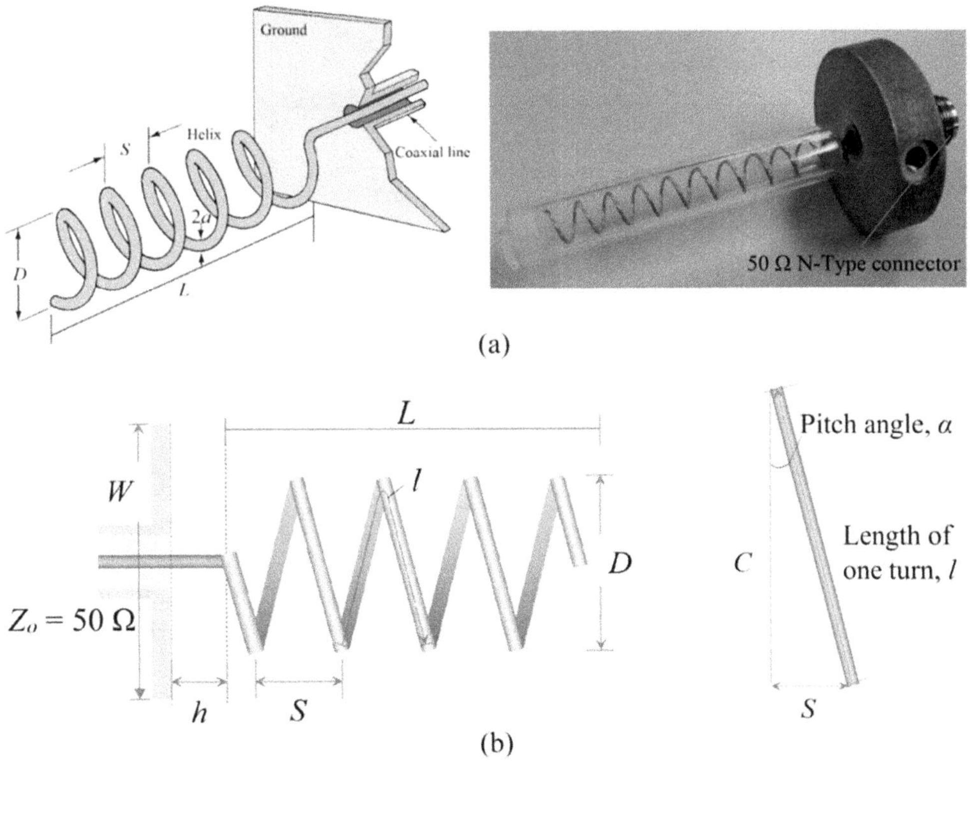

Table 8. Design parameters for axis mode and normal mode of helical antenna

Design Parameters	Axis Mode $0.75 \le C/\lambda \le 1.33$ $12° \le \alpha \le 15°$	Normal Mode $L << \lambda$ $D << \lambda$
Operating wavelength, λ (m)	$\lambda = \frac{c}{f}$ $c = 299792458$ m/s	$\lambda = \frac{c}{f}$ $c = 299792458$ m/s
Diameter of the helix, D (m)	$D = \frac{\lambda}{\pi}$	$D = K\frac{\lambda}{\pi}$ for $K < 0.5$. Typical $K = 0.2$
Circumference of the turn, C (m)	$C = \pi D$	$C = \pi D$
Spacing distance between the turn, S (m)	$S = \frac{C}{4}$	$S = \frac{C^2}{2\lambda}$
Total axis length of helical antenna, L (m)	$L = N \times S$	$L = N \times S$
Pitch angle, α (rad)	$\alpha = \tan^{-1}\frac{S}{\pi D}$	$\alpha = \tan^{-1}\left(\frac{\pi D}{2\lambda}\right)$
Axial ratio, AR (polarization)	$\text{AR} \approx 1 + \frac{1}{2N}$	$\text{AR} \approx \frac{2S\lambda}{(\pi D)^2}$
Radiation resistance, R_{rad} (Ω)	$R_{rad} \approx \frac{140C}{\lambda}$	$R_{rad} \approx \left(\frac{25L}{\lambda}\right)^2$

Helical Antenna

Dimensions of Helical Antenna

Helical monopole antenna is an antenna consisting of one conducting wire wound in the helix form as shown in Figure 60. The helical monopole antenna can operate in one of two principal modes, namely normal mode or axial mode, based on the designed dimensions of the antenna.

In the initial helical antenna design, two parameters need to be determined first, namely the operating frequency, f and the number of helical turns, N. Once the values of f and N are determined, the other dimensions as listed in Table 8 for the helical antenna can be further determined.

Radiation Pattern of Helical Antenna

The effective wavelength, λ_{eff} of normal mode helical antenna is assumed to be (Zacharia, 1998):

$$\lambda_{eff} = \frac{\lambda}{K} \tag{62}$$

where λ and K are the free space wavelength and axial compression factor of the normal mode helical antenna, respectively. The value of K is determined by the axial length, L (or radiation resistance, R_{rad}) of the helical antenna. The relationship between the K and R_{rad} is plotted in Figure 61 (Zacharia, 1998) and represents in Equation (63) which is only valid in the range of $0\ \Omega \le R_{rad} \le 37\ \Omega$. When $R_{rad} > 37\ \Omega$, the value of K is assumed to be unity.

$$\frac{1}{K} = \left\{ \begin{array}{l} 9.3450648\times10^{-8} R_{rad}^{5} - 9.712378\times10^{-6} R_{rad}^{4} + 3.7860958\times10^{-4} R_{rad}^{3} \\ -7.0592717\times10^{-3} R_{rad}^{2} + 8.4852228\times10^{-2} R_{rad} + 7.4327756\times10^{-2} \end{array} \right\} \text{for } R_{rad} \le 37\ \Omega \tag{63}$$

Figure 61. Radiation resistance, R_{rad} of resonant quarter-wave normal mode helical monopole antenna (Zacharia, 1998)

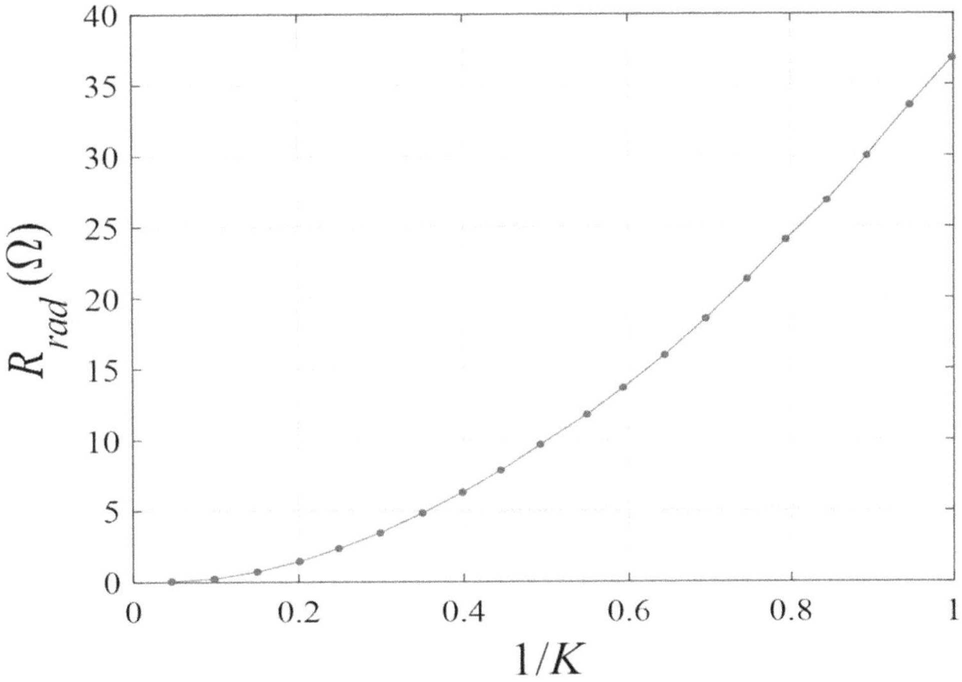

The radiation field components of the normal mode helical antenna are expressed as:

$$H_{\phi} = \frac{jI_o K}{2\pi r} e^{-jkr} \left[\frac{\cos\left(\frac{\pi}{2K}\cos\theta\right)\sin\theta}{K^2 - \cos^2\theta} \right] \tag{64a}$$

$$E_\theta = \eta H_\phi \tag{64b}$$

From (64b), the radiation distribution of normalized electric field, E_{norm} can be written as:

$$\begin{aligned} E_{norm} &= \left| \frac{E_\theta(r,\theta)}{E_\theta(r,\pi/2)} \right| \\ &= \left| \frac{K^2 \cos\left(\dfrac{\pi}{2K}\cos\theta\right)\sin\theta}{K^2 - \cos^2\theta} \right| \end{aligned} \tag{65}$$

On the other hand, the normalized far-field, E_{norm} pattern of the axial mode helical antenna is written as:

$$E_{norm} = \cos\theta \sin\left(\frac{\pi}{2N}\right)\left\{\frac{\sin\left(\dfrac{N\psi}{2}\right)}{\sin\left(\dfrac{\psi}{2}\right)}\right\} \tag{66}$$

where $\psi = kS\cos\theta - \dfrac{kL}{p}$. The p is the endfire radiation. The ordinary endfire radiation and Hansen –Woodyard endfire radiation are given as:

$$p = \frac{L}{S+\lambda} \text{ (Ordinary endfire radiation)}$$

$$p = \frac{L}{S+\lambda\left(1+\dfrac{1}{2N}\right)} \text{ (Hansen –Woodyard endfire radiation)}$$

Besides, the half-power beamwidth (3dB beamwidth), HPBW (in unit degrees) of the axial mode helical antenna can be calculated as:

$$\text{HPBW} \approx \frac{52}{C}\sqrt{\frac{\lambda^3}{NS}} \tag{67}$$

On the other hand, the beamwidth between first nulls, FNBW (in unit degrees) of the axial mode helical antenna is given as:

Figure 62. Comparison of antenna gain, G as a function of the normalized antenna length, L/λ

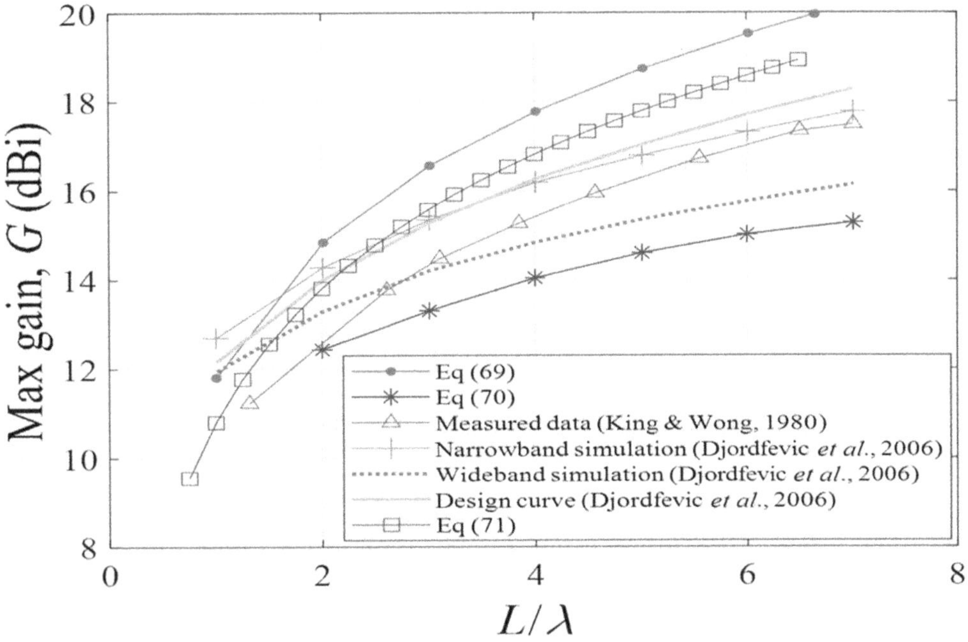

$$\text{FNBW} \approx \frac{115}{C}\sqrt{\frac{\lambda^3}{NS}} \tag{68}$$

On-Axis Gain of Axial Mode Helical Antenna

Kraus, (1941) was first proposed an empirical relationship between the helical antenna gain, G and its axial length, L, operating wavelength, λ, and circumference of the turn, C as:

$$G(\text{dBi}) = 10\log_{10}\left[15\left(\frac{C}{\lambda}\right)^2\left(\frac{L}{\lambda}\right)\right] \tag{69}$$

Later, Emerson, (1995) expresses the antenna gain, G as an empirical quadratic equation:

$$G(\text{dBi}) = 10.25 + 1.22\left(\frac{L}{\lambda}\right) - 0.0726\left(\frac{L}{\lambda}\right)^2 \text{ for } 2 < L/\lambda < 7 \tag{70}$$

Equation (70) is created by fitting with numerical simulation results. However, as the antenna axial length, L increases, the uncertainty in the gain, G prediction using Equations (69) and (70) will increase as shown in Figure 62. Equations (69) and (70) have opposite differences in the antenna gain predictions with respect to actual measurements. Equation (69) overestimates the antenna gain, G by nearly 4 dBi. In contrast, equation (70) underestimates the antenna gain, G up to 2 dBi (Djordfević *et al.*, 2006).

Hence, recently the numerical factor in Equation (69) is reduced from 15 to 12 and it is often used into various helical antenna-design calculators as:

$$G(\text{dBi}) = 10\log_{10}\left[12\left(\frac{C}{\lambda}\right)^2\left(\frac{L}{\lambda}\right)\right] = 10.8 + 10\log_{10}\left[\left(\frac{C}{\lambda}\right)^2\left(\frac{L}{\lambda}\right)\right] \tag{71}$$

Figure 63 shows the GUI of the helical monopole antenna calculator which is created using MATLAB 'guide'.

Figure 63. MATLAB-based GUI of the helical monopole antenna calculator

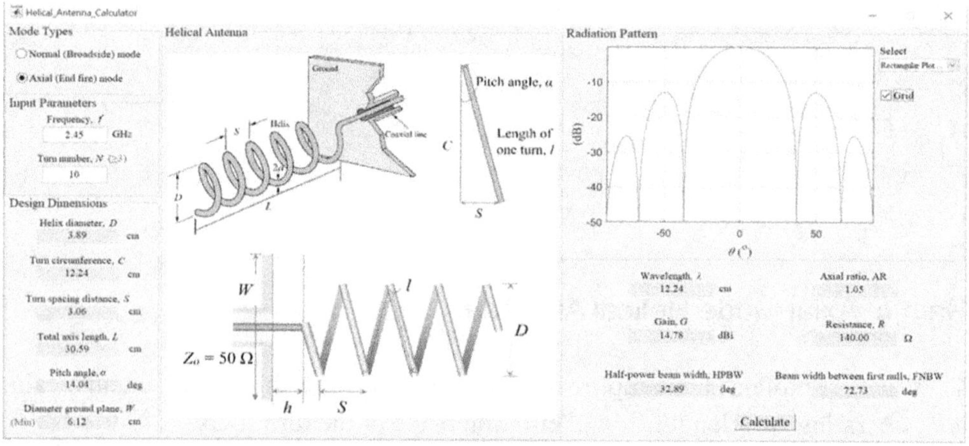

Conical Monopole Antenna

Dimensions of Conical Antenna

A broadband monopole antenna so-called conical monopole antenna had been firstly invented by Carter since 1939. Conical monopole has two primary design parameters, namely cone angle, θ_o and cone length, l. The conical monopole shows broadband characteristic as the cone angle is increased as shown in Figure 64. The width, D and length, H of the conical part of the antenna can be calculated as:

$$H = l\cos\theta_o \tag{72}$$

$$D = 2H\tan\theta_o + 2a \tag{73}$$

Figure 64. (a) Conical monopole antenna. (b) Dimensions of the conical antenna

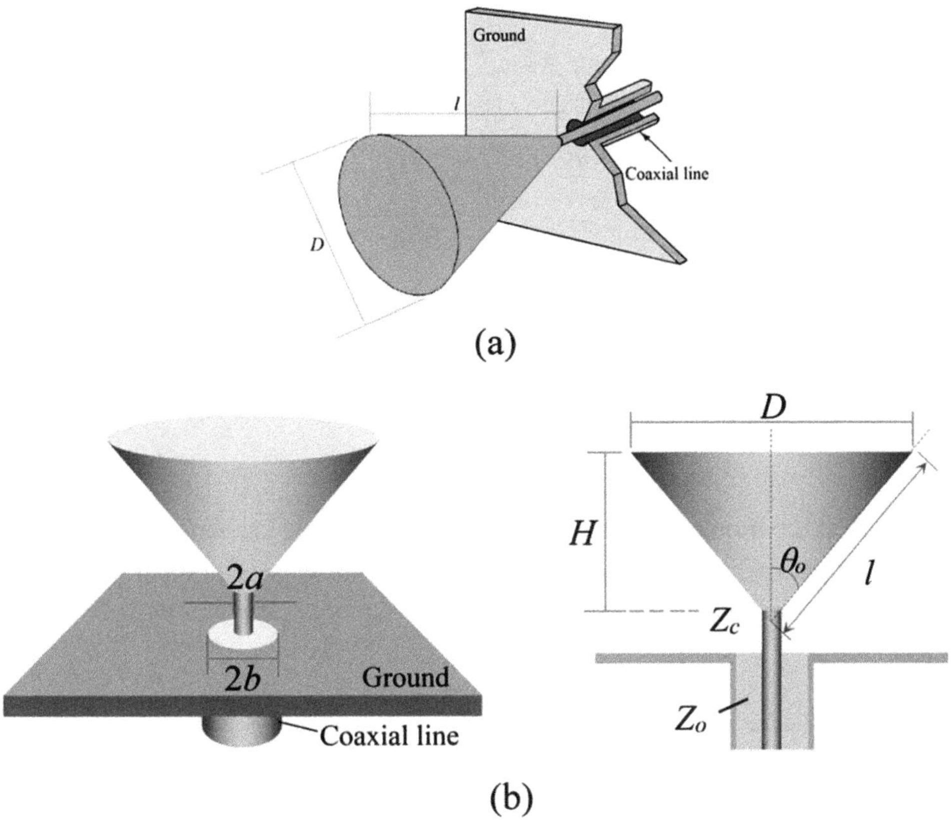

Input Impedance and Return Loss of Conical Antenna

The theoretical formulation of the input impedance, Z_{in} is expressed as (Papas and King, 1949):

$$Z_{in} = Z_c \frac{1-\dfrac{\beta}{\alpha}}{1+\dfrac{\beta}{\alpha}} \tag{74}$$

where Z_c is the characteristic impedance of the cone above ground plane which is given as:

$$Z_c = \frac{\eta_o}{2\pi}\ln\left[\cot\left(\frac{\theta_o}{2}\right)\right] \tag{75}$$

where $\eta_o = \sqrt{\mu_o/\varepsilon_o}$ ($\approx$ 376.73 Ω) is the intrinsic impedance of free space. The ratio β/α in (74) represents the reflection coefficient, S_{11} is expressed as:

$$S_{11} = \frac{\beta}{\alpha}$$

$$= e^{-2jkl} \frac{1 + j\frac{60}{Z_c}\sum_{n=1}^{\infty}\frac{2n+1}{n(n+1)}\left[P_n(\cos\theta_o)\right]^2 \zeta_n}{-1 + j\frac{60}{Z_c}\sum_{n=1}^{\infty}\frac{2n+1}{n(n+1)}\left[P_n(\cos\theta_o)\right]^2 \zeta_n} \quad \text{for } n = 1, 3, 5, \ldots, \infty \tag{76}$$

and

$$\zeta_n = \frac{H_n^{(2)}(kl)}{H_{n-1}^{(2)}(kl) - \frac{n}{kl}H_n^{(2)}(kl)} \quad \text{for } n = 1, 3, 5, \ldots, \infty$$

where $H_n^{(2)}(kl)$ is the n-order spherical Hankel function of the second kind and $P_n(\cos\theta_o)$ is the Legendre polynomial of order n. The (74) accurately predicts the equivalent input impedance of conical monopole (impedance matching) if the characteristic impedance, Z_c is equal to Z_o ($\approx$ 50 Ω) of coaxial feed line as:

$$Z_c = Z_o = 50\,\Omega \text{ (Impedance matching)}$$

Thus, the high accuracy of input impedance, Z_{in} calculation can be achieved when θ_o = 47°. The optimum cone angle, θ is normally varied between 30° to 50°. When the value of θ_o is too large (θ_o > 50°), the impedance mismatch between coaxial feed line and the cone shape of monopole will become significant. Comparison of calculated input impedance, Z_{in} using (74) and COMSOL simulation results with a = 1.5 mm, b = 4.916 mm, $\theta_o \approx$ 45°, D = 40 cm, and H = 20 cm is plotted in Figure 65.

Radiation Pattern of Conical Antenna

The radiation distribution of normalized electric field, $R(\theta)$ is given as (Papas and King, 1950; Sandler and King, 1994):

Figure 65. (a) Distribution of simulated H_ϕ field of conical monopole antenna using COMSOL simulator.(b) Input resistance, $Re(Z_{in})$ and input reactance, $Im(Z_{in})$ of conical monopole antenna with $a = 1.5$ mm, $b = 4.916$ mm, $\theta_o \approx 45^o$, $D = 40$ cm, and $H = 20$ cm

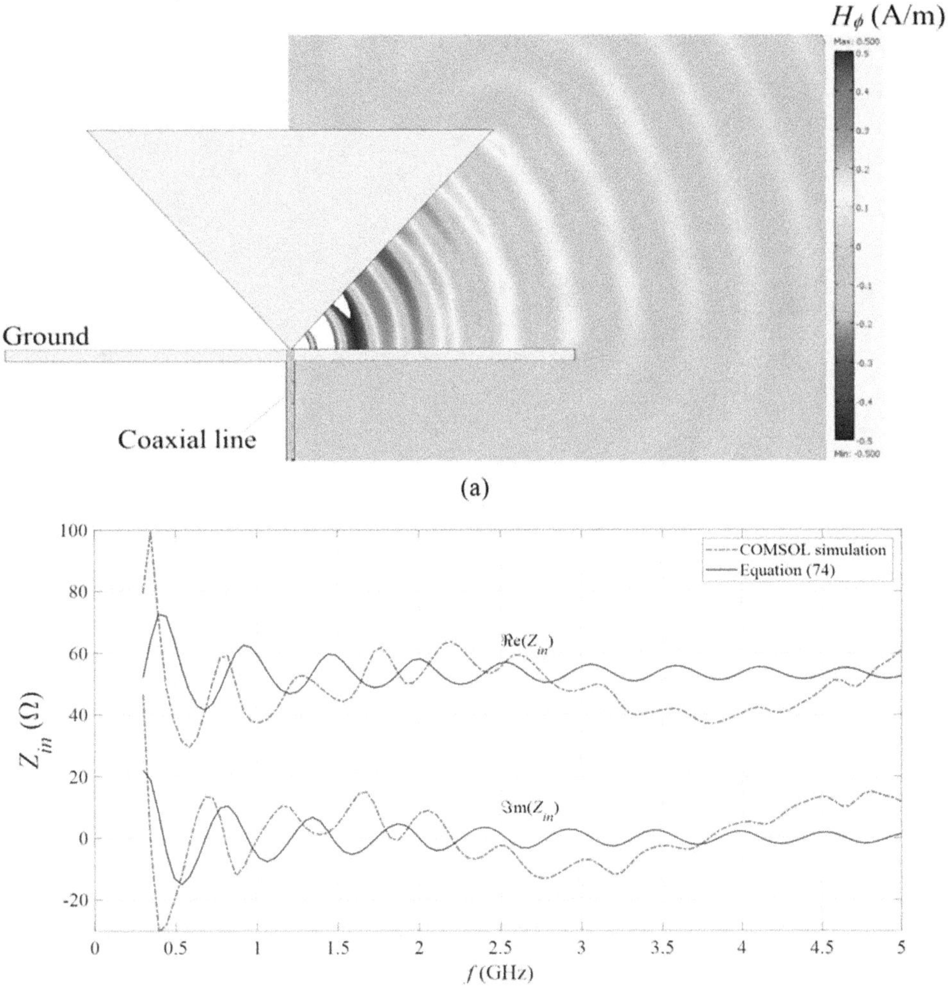

(b)

$$R(\theta) = \left| \frac{E_\theta(r,\theta)}{E_\theta(r,\pi/2)} \right| = \left| \frac{\sum_{n=1}^{\infty} P_n(\cos\theta_o) P_n'(\cos\theta) \xi_n}{\sum_{n=1}^{\infty} P_n(\cos\theta_o) P_n'(0) \xi_n} \right| \text{ for } n = 1, 3, 5, \ldots, \infty \tag{77}$$

where $P_n'(\cos\theta)$ is the differentiate of P_n $(\cos\theta)$ with respect to angle, θ as:

Figure 66. MATLAB-based GUI of the conical monopole antenna calculator. (a) Calculated input resistance, Re(Z_{in}) and input reactance, Im(Z_{in}) and (b) calculated normalized electric field, R(θ) radiation pattern.

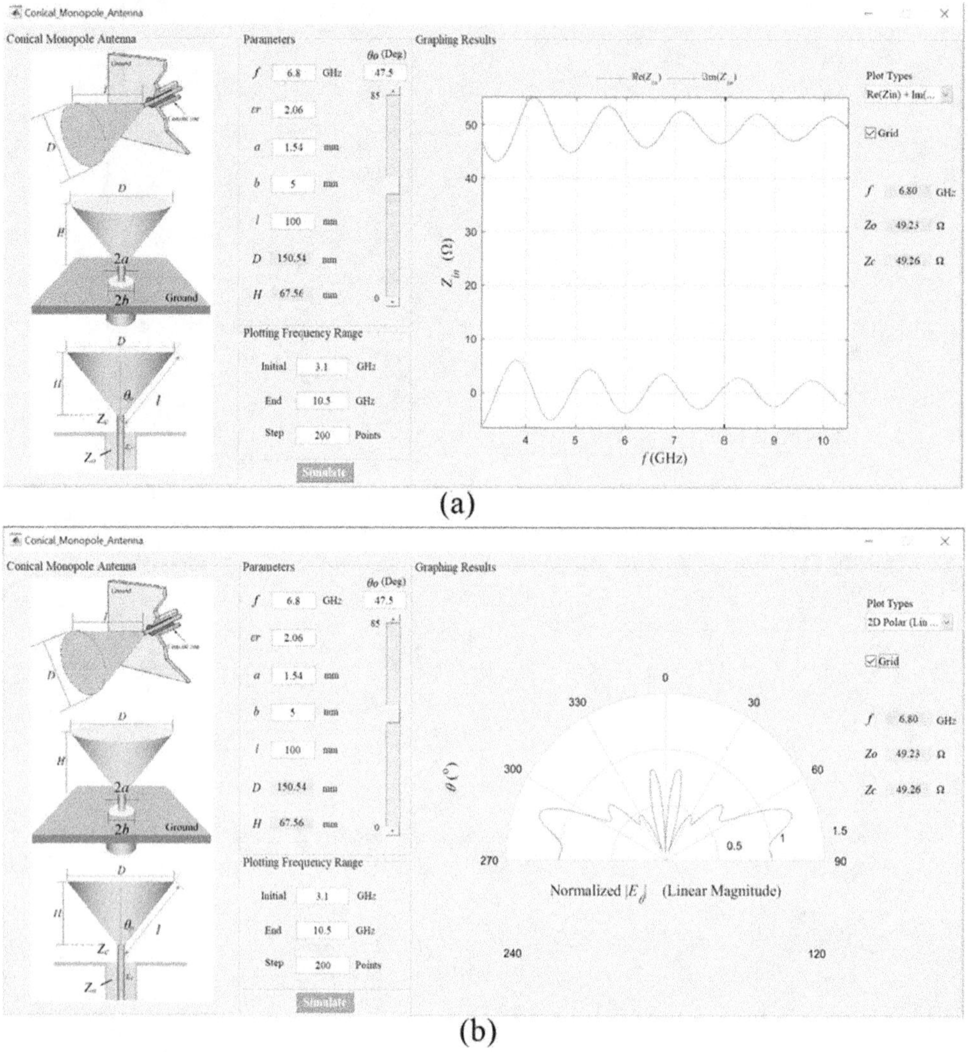

$$P_n'(\cos\theta) = -\frac{\partial P_n(\cos\theta)}{\partial\theta} \text{ for } n = 1, 3, 5, \ldots, \infty$$

and the function of ξ_n in (77) is expressed as:

$$\xi_n = j^n \frac{2n+1}{n(n+1)} \left(\frac{1}{H_{n-1}^{(2)}(kl) - \frac{n}{kl} H_n^{(2)}(kl)} \right) \text{ for } n = 1, 3, 5, \ldots, \infty$$

Figure 66 shows the GUI of the conical monopole antenna EDA which is created using MATLAB 'guide'.

LATEST AND FUTURE POTENTIAL OF RF/MICROWAVE EDA TOOLS

Since 2019, the wireless communication technology is increasingly pointing towards fifth generation (5G). The fifth-generation (5G) wireless technologies specifies new frequency bands to enable higher data throughput for applications like streaming HD video and allow for more network capacity. Today and in the near future, more wireless devices for daily use, such as mobile internet, Internet of Things (IoT), smart home, smart healthcare, and the fourth industrial revolution (Industry 4.0) are operated based on 5G wireless regulations/platforms. The recent 5G operating frequencies implemented by several countries are tabulated in Table 9. The 5G operating frequency of general public users is usually below 6 GHz (called sub-6 GHz). However, there are currently attempts to extend operating frequencies under 6 GHz to 8 GHz. In addition, in the next 2030, it is inferred that the sixth-generation (6G) wireless technology will be released. In fact, at a time when 5G technology is being used, preliminary research on 6G has already begun, with the goal of operating at frequencies above 100 GHz.

Since the antenna is one of the important components in 5G wireless devices, indirectly, the field of antenna design has become more and more important. Therefore, the development of wireless devices urgently needs EDA tools to provide assistance in the design of their antennas. Recently, we can observe that about 50% of IoT wireless devices are manufactured by companies less than 3 years old. Most of the new companies are small, low-profit manufacturers, and unable to purchase existing commercial EDAs on the market (too expensive). Hence, simple open-source EDAs are their main choice.

Table 9. 5G operating frequencies

Country	Operating Frequency
USA	600 MHz, 700 MHz, 2.5 GHz, 3.3-3.8 GHz, 4-5 GHz, 24-28 GHz, 37-40 GHz, 47-50 GHz
Mexico	2.5 GHz
South Korea	3.3-3.8 GHz, 28 GHz
Japan	3.6-5GHz, 28 GHz
China	2.5 GHz, 3.3-3.8 GHz, 4.9 GHz, 24-28 GHz
Taiwan	3.3-3.8 GHz
India	3.3-3.8 GHz
Australia	3.3-3.8 GHz
EMEA	600 MHz, 700 MHz, 3.3-3.8 GHz, 24-27 GHz
Russia	3.3-3.8 GHz

Open-source EDAs are usually developed to solve common RF/microwave problems, not just particular for antenna design. Normally, the development of open-source EDAs is supported by pledged supporters, funds from universities and governments, or donations from RF/microwave enthusiasts. Many numerical-based open-source EDAs has been released recently. For instance, open-source FDTD-based RF/microwave EDA software package, namely MEEP (Oskooi *et al.*, 2010), openEMS (Liebig *et al.*, 2013) and gprMax (Warren *et al.*, 2016) from Massachusetts Institute of Technology (MIT), Duissenberg University and Edinburgh University, respectively. Furthermore, the other examples of the FEM-, MoM- and UTD/GTD-based open-source RF/Microwave EDAs are MaxFEM, FEMM, NEC, GLMoM, MMANA-GAL, EM3DS, newFASANT, EMCoS, WedgeGUI (Hacıvelioğlu *et al.*, 2011). In fact, those numerical-based EDAs (iterative solution) has relatively complex operations and require long simulation runtimes compared to analytical-based EDA. Nevertheless, numerical-based EDA can give more accurate simulation results compared to analytical-based EDA. Apart from that, other open-source EDAs, such as GNU Radio and KidCad are very beneficial to the development of current RF/Microwave devices. However, the RF/microwave problems that can be solved by the open-source EDAs are limited compared to the commercial one.

In addition to the industry sector, microwave/antenna EDA also plays an important role in the education of engineering design skills in universities. Currently, a large number of DIY IoT devices or modules are sold online, such as Bluetooth camera, remote sensing devices, SX/JC1278A wireless LoRa modules, and NodeMCU Lua WiFi modules, to facilitate secondary development. And by 2020, the number of IoT wireless devices is expected to reach 2 to 35 billion. By 2025, IoT wireless devices will exceed 50 billion, and global IoT revenue will quadruple to reach 1.1 trillion US dollars. Therefore, more and more graduate engineers will directly or indirectly participate in work related to wireless device development. Here, universities play an important role in training future engineers to utilize EDA as well as develop their own EDA (create app). Hence, in the future, microwave/antenna EDA may become one of the elective courses for undergraduates in electrical engineering.

CONCLUSION

Recently, antennas have been widely used in daily life around the world, such as antennas used on televisions, mobile phones, and internet routers. In the future, Internet of Things (IoT) and Fifth Generation Wireless Technology (5G) applications will lead to increased use of antennas. Furthermore, antenna design has become an elective course of study in some universities. Design guideline and developed simple analytical formulations have been used as teaching materials for the course.

In fact, the analytical analysis of some antennas is more complicated compared with other microwave passive devices (such as branch-line couplers, Wilkinson power splitters, and filters) due to the antenna deals with scattered wave propagation. Therefore, several analytical techniques, such as transmission line theory, *ABCD*-matrix, and even-odd mode symmetric circuit decomposition, are not suitable implement for antenna analysis and design. Fortunately, for over eighty years, a lot of analytical formulations have been developed and proposed for various types of antennas. In this chapter, some analytical formulas have been successfully revisited and compared with commercial numerical-based simulators in terms of ease and accuracy.

In this chapter, the MATLAB-based apps for several types of antennas has been developed. Since the main purpose of this chapter is to demonstrate the app development of the antenna, thus only simple analytical methods rather than complex numerical methods are used in the development of the GUI. Recently, MATLAB has provided two types of apps design environment, namely MATLAB 'appdesigner' and MATLAB 'guide'.

Building an app with MATLAB 'appdesigner' environment is easier and user-friendly than MATLAB 'guide'. In fact, MATLAB 'appdesigner' (2016) is a new app development environment to replace MATLAB 'guide' (1996). However, due to the fact that MATLAB 'appdesigner' is still new, there is only a few information or examples of apps developed using 'appdesigner' that can be found in literature or on the internet compared to 'guide'. In the near future, more and more examples of app development with MATLAB 'appdesigner' environment will be shared on the internet, especially in 'YouTube' platform.

ACKNOWLEDGMENT

This research received no specific grant from any funding agency in the public, commercial, or not-for-profit sectors.

REFERENCES

Alan, B. (1938). *Improvements in or relating to high frequency electrical conductors or radiators*. British Patent no. 515684.

Atta, L. C. V. (1981). A history of early microwave antenna development. *IEEE Antennas and Propagation Society Newsletter*, *23*(5), 10–14. doi:10.1109/MAP.1981.27564

Balanis, C. A. (2015). *Antenna Theory: Analysis and Design* (4th ed.). Wiley.

Balmain, K. G., & Nkeng, J. N. (1976). Asymmetry phenomenon of log-periodic dipole antennas. *IEEE Transactions on Antennas and Propagation*, *24*(4), 402–410. doi:10.1109/TAP.1976.1141380

Berenger, J. P. (1994). A perfectly matched layer for the absorption of electromagnetic waves. *Journal of Computational Physics*, *114*(2), 185–200. doi:10.1006/jcph.1994.1159

Brillouin, L. (1922). Origin of radiation resistance. *Radioėlectricitė*, *3*, 147–152.

Carter, P. S. (1939). *Wide band, short wave antenna and transmission line system*. U.S. Patent 2,181,870.

Cebik, L. B. (2007). Log periodic arrays. In *W4RNL, The ARRL Antenna Book* (21st ed., pp. 10-1–10-31). The American Radio Relay League, Inc.

Collin, R. E. (1985). *Antennas and Radiowave Propagation*. McGraw-Hill.

Compton, R. T., Jr. (1963). The aperture admittance of a rectangular waveguide radiating into a lossy half-space. Technical Report, 1691-1. Ohio State University.

Dinh, N. Q., Michishita, N., Yamada, Y., & Nakatani, K. (2011). Deterministic equation for self-resonant structures of very small normal-mode helical antennas. *IEICE Transactions on Communications*, *E94-B*(5), 1276–1279. doi:10.1587/transcom.E94.B.1276

Djordfević, A. R., Zajić, A. G., Ilic, M. M., & Stuber, G. L. (2006). Optimization of helical antennas. *IEEE Antennas & Propagation Magazine*, *48*(6), 107–115. doi:10.1109/MAP.2006.323359

Emerson, D. T. (1995). The gain of the axial-mode helix antenna. *Antenna Compendium*, *4*, 64–68.

Enkhbayar, B., Bang, J. H., Cha, E. J., & Ahn, B. C. (2009). Formula for gain of open-ended rectangular waveguide. *Electronics Letters*, *45*(24), 1203–1204. doi:10.1049/el.2009.2237

Gajda, G., & Stuchly, S. S. (1983). Numerical analysis of open-ended coaxial lines. *IEEE Transactions on Instrumentation and Measurement*, *31*(5), 380–384.

Greenblum, C. (1956a). Notes on the development of Yagi arrays – Part I. *QST*, 11–17, 114, 116.

Greenblum, C. (1956b). Notes on the development of Yagi arrays – Part II. *QST*, *23–26*, 122.

Güney, K. (2001). Simple design method for optimum gain pyramidal horns. *International Journal of Electronics and Communications*, *55*(3), 205–208. doi:10.1078/1434-8411-00031

Hacıvelioğlu, F., Alper Uslu, M., & Sevgi, L. (2011). A MATLAB based virtual tool for the electromagnetic wave scattering from a perfectly reflecting wedge. *IEEE Antennas & Propagation Magazine*, *53*(6), 234–243. doi:10.1109/MAP.2011.6157766

Hammerstad, E. O. (1975). Equations for microstrip circuit design. *Proceedings European Microwave Conference*, 268–272.

Harrington, R. F. (1967). Matrix methods for field problems. *Proceedings of the IEEE*, *55*(2), 136–149. doi:10.1109/PROC.1967.5433

Hertz, H. (1888). On the finite velocity of propagation of electromagnetic actions. *Electric Waves*, 110.

Isbell, D. E. (1960). Log periodic dipole arrays. *IRE Transactions on Antennas and Propagation*, *8*(3), 260–267. doi:10.1109/TAP.1960.1144848

Jull, E. V. (1972). Reflection from the aperture of a long E-plane sectoral horn. *IEEE Transactions on Antennas and Propagation*, *20*(1), 62–68. doi:10.1109/TAP.1972.1140137

King, H. E., & Wong, J. L. (1980). Characteristics of 1 to 8 wavelength uniform helical antennas. *IEEE Transactions on Antennas and Propagation*, *28*(2), 291–296. doi:10.1109/TAP.1980.1142322

King, R. W. P. (2013). *The Theory of Linear Antennas: With Charts and Tables for Practical Applications*. Harvard University Press. doi:10.4159/harvard.9780674182189

Kouyoumjian, R. G., & Pathak, P. H. (1974). A uniform geometrical theory of diffraction for an edge in a perfectly conducting surface. *Proceedings of the IEEE*, *62*(11), 1448–1461. doi:10.1109/PROC.1974.9651

Kraus, J. D. (1947). Helical beam antennas. *Electronics (Basel)*, *20*, 109–111.

Liebig, T., Rennings, A., Held, S., & Erni, D. (2013). openEMS-a free and open source equivalent-circuit (EC) FDTD simulation platform supporting cylindrical coordinates suitable for the analysis of traveling wave MRI applications. *International Journal of Numerical Modelling: Electronic Networks, Devices and Fields*, *26*(6), 680–696.

Marconi, G. (1896). *Transmitting electrical signals*. U.S. Patent 586193.

Marcuvitz, N. (1964). *Waveguide Handbook*. Boston Technical Publishers.

MATLAB R2015b. (2015). *MATLAB Creating Graphical User Interfaces*. MathWorks.

MATLAB R2020a. (2020). *MATLAB App Building*. MathWorks.

Maybell, M. J., & Simon, P. S. (1993). Pyramidal horn gain calculation with improved accuracy. *IEEE Transactions on Antennas and Propagation*, *41*(7), 884–889. doi:10.1109/8.237618

Mur, G. (1981). Absorbing boundary conditions for the finite-difference approximation of the time-domain electromagnetic field equations. *IEEE Transactions on Electromagnetic Compatibility*, *23*(4), 377–382. doi:10.1109/TEMC.1981.303970

Nguyen, C. (2001). *Analysis Methods for RF, Microwave, and Millimeter-Wave Planar Transmission Line Structures*. John Wiley & Sons, Inc. doi:10.1088/0957-0233/12/10/705

Oskooi, A. F., Roundy, D., Ibanescu, M., Bermel, P., Joannopoulos, J. D., & Johnson, S. G. (2010). MEEP: A flexible free-software package for electromagnetic simulations by the FDTD method. *Computer Physics Communications*, *181*(3), 687–702. doi:10.1016/j.cpc.2009.11.008

Papas, C. H., & King, R. (1949). Input impedance of wide-angle conical antennas fed by a coaxial line. *Proceedings of the IRE*, *37*(11), 1269–1271. doi:10.1109/JRPROC.1949.234607

Papas, C. H., & King, R. (1951). Radiation from wide-angle conical antennas fed by a coaxial line. *Proceedings of the IRE*, *39*(1), 49–51. doi:10.1109/JRPROC.1951.230420

Pozar, D. M. (2012). *Microwave Engineering* (4th ed.). Wiley.

Rao, P. H., Selvan, K. T., & Das, S. K. (1997). Finite-range gain prediction of pyramidal horn with improved accuracy. *IEEE Transactions on Electromagnetic Compatibility*, *39*(4), 374–376. doi:10.1109/15.649840

Robertson, I. D., Karacaoglu, U., & Sánchez-Hernández, D. (1996). CAD techniques for microwave circuits. *IEEE Electronics & Communication Engineering Journal*, *8*(6), 245–256. doi:10.1049/ecej:19960601

Sánchez-Hernández, D., Karacaoglu, U., Gokdemir, T., Khalid, H. A., Rezazadeh, A. A., Ferrando, M., Reyes, E., & Robertson, I. D. (1997). CAD techniques for the electromagnetic design of monolithic millimetre-wave integrated circuits. In G. Pujolle (Ed.), *Annals of Telecommunications* (pp. 164–173). Springer Nature.

Sandler, S. S., & King, R. W. P. (1994). Compact conical antennas for wide-band coverage. *IEEE Transactions on Antennas and Propagation*, *42*(3), 436–439. doi:10.1109/8.280735

Selvan, K. T. (1998). Simple formulas for the gain and far-field of open-ended rectangular waveguides. *IEE Proceedings. Microwaves, Antennas and Propagation*, *145*(1), 80–84. doi:10.1049/ip-map:19981552

Selvan, K. T. (1999). Accurate design method for optimum gain pyramidal horns. *Electronics Letters*, *35*(4), 249–250. doi:10.1049/el:19990240

Selvan, K. T., Rao, P. H., & Geetha, S. (1999). High-frequency reflection from the aperture of a long E-plane sectoral horn. *IEEE Antennas & Propagation Magazine*, *41*(3), 58–59. doi:10.1109/74.775247

Silver, S. (1949). *Microwave Antenna Theory and Design*. McGraw-Hill.

Silvester, P. P. (1969). Finite element solution of homogeneous waveguide problems. *Alta Frequenza*, *38*, 313–317.

Steer, M. B., Bandler, J. W., & Snowden, C. M. (2002). Computer-aided design of RF and microwave circuits and systems. *IEEE Transactions on Microwave Theory and Techniques*, *50*(3), 996–1005. doi:10.1109/22.989983

Stratton, J. A. (1941). *Electromagnetic Theory*. McGraw-Hill.

Tai, C. T., & Long, S. A. (2007). Dipoles and monopoles. In J. L. Volakis (Ed.), *Antenna Engineering Handbook* (pp. 4-3–4-32). McGraw-Hill Inc.

Thiele, G. A. (1969). Analysis of Yagi-Uda-type antennas. *IEEE Transactions on Antennas and Propagation*, *17*(1), 24–31. doi:10.1109/TAP.1969.1139356

Wadell, B. C. (1991). *Transmission Line Design Handbook*. Artech House.

Warren, C., Giannopoulos, A., & Giannakis, I. (2016). gprMax: Open source software to simulate electromagnetic wave propagation for Ground Penetrating Radar. *Computer Physics Communications*, *209*(12), 163–170. doi:10.1016/j.cpc.2016.08.020

Weiland, T. (1977). A discretization method for the solution of Maxwell's equations for six-component fields. *Electronics and Communications, AEU.*, *31*(3), 116–120.

Weiner, M. M. (2003). *Monopole Antennas*. Marcel Dekker. doi:10.1201/9780203912676

Wheeler, H. A. (1965). Transmission-line properties of parallel strips separated by a dielectric sheet. *IEEE Transactions on Microwave Theory and Techniques*, *13*(2), 172–185. doi:10.1109/TMTT.1965.1125962

Yaghjian, A. D. (1984). Approximate formulas for the far fields and gain of open ended rectangular waveguide. *IEEE Transactions on Antennas and Propagation*, *32*(4), 378–384. doi:10.1109/TAP.1984.1143332

Yagi, H., & Uda, S. (1926). Projector of the sharpest beam of electric waves. *Proceedings of the Imperial Academy (Japan)*, *2*(2), 49–52. doi:10.2183/pjab1912.2.49

Yamada, Y. (2011). Design of a very small antenna for metal-proximity applications. In S. Preradovic (Ed.), *Advanced Radio Frequency Identification Design and Applications* (pp. 77–114). InTech Open Access. doi:10.5772/14734

Yee, K. S. (1966). Numerical solution of initial boundary value problems involving Maxwell's equations in isotropic media. *IEEE Transactions on Antennas and Propagation*, *14*(3), 302–307. doi:10.1109/TAP.1966.1138693

You, K. Y. (2016). *RF Coaxial Slot Radiators: Modeling, Measurements and Applications*. Artech House.

You, K. Y. (2017). Materials characterization using microwave waveguide system. In S. Goudos (Ed.), *Microwave Systems and Applications* (pp. 341–358). InTech Open Access. doi:10.5772/66230

You, K. Y. (2019). Analytical and calibration models for slot antennas. In J. Zbitou & A. Errkik (Eds.), *Emerging Innovations in Microwave and Antenna Engineering* (pp. 57–100). IGI Global. doi:10.4018/978-1-5225-7539-9.ch003

You, K. Y., Abbas, Z., Khalid, K., & Kong, N. F. (2009). Improved formulation for admittance of thin and short monopole driving from coaxial line into dissipative media. *IEEE Antennas and Wireless Propagation Letters*, *8*, 1246–1249. doi:10.1109/LAWP.2009.2035645

You, K. Y., Abbas, Z., Malek, M. F. A., & Cheng, E. M. (2014). Non-destructive dielectric measurements and calibration for thin materials using waveguide-coaxial adaptors. *Measurement Science Review*, *14*(1), 16–24. doi:10.2478/msr-2014-0003

Zacharia, K. P. (1998). Radiation resistance of resonant normal mode helices. *IETE Journal of Education*, *39*(1), 5–7. doi:10.1080/09747338.1998.11689746

ADDITIONAL READING

Baudrand, H., Tao, J., & Atechian, J. (1988). Study of radiation properties of open-ended rectangular waveguides. *IEEE Transactions on Antennas and Propagation*, *36*(8), 1071–1077. doi:10.1109/8.7219

Bodnar, D. G., & Paris, D. T. (1970). New variational principle in electromagnetics. *IEEE Transactions on Antennas and Propagation*, *18*(2), 216–223. doi:10.1109/TAP.1970.1139642

Chang, C. W., Chen, K. M., & Qian, J. (1997). Nondestructive determination of electromagnetic parameters of dielectric materials at X-band frequencies using a waveguide probe system. *IEEE Transactions on Instrumentation and Measurement*, *46*(5), 1084–1092. doi:10.1109/19.676717

Kim, J. H., Enkhbayar, B., Bang, J. H., Ahn, B. C., & Cha, E. J. (2010). New formulas for the reflection coefficient of an open-ended rectangular waveguide radiating into air including the effect of wall thickness or flange. *Progress in Electromagnetics Research M. Pier M*, *12*, 143–153. doi:10.2528/PIERM10033104

Kraus, J. D. (1949). The helical antenna. *Proceedings of the IRE*, *37*(3), 263–272. doi:10.1109/JRPROC.1949.231279

Serizawa, H., & Hongo, K. (2005). Radiation for a flanged rectangular waveguide. *IEEE Transactions on Antennas and Propagation*, *53*(12), 3953–3962. doi:10.1109/TAP.2005.859748

Venkatarayalu, N. V., & Ray, T. (2004). Optimum design of Yagi-Uda antennas using computational intelligence. *IEEE Transactions on Antennas and Propagation*, *52*(7), 1811–1818. doi:10.1109/TAP.2004.831338

Yoshitomi, K., & Sharobim, H. R. (1994). Radiation from a rectangular waveguide with a lossy flange. *IEEE Transactions on Antennas and Propagation*, *42*(10), 1398–1403. doi:10.1109/8.320746

You, K. Y., Lee, C. W., & Lee, C. Y. (2017). Commercial and open-source electromagnetic simulators for education, research and industrial design. *International Journal of Advances in Microwave Technology*, *2*(3), 117–121.

You, K. Y., Salleh, J., Abdul Malek, M. F., Abbas, Z., Cheng, E. M., & Lee, K. Y. (2012). Modeling of coaxial slot waveguides using analytical and numerical approaches: Revisite. *International Journal of Antennas and Propagation*, *2012*, 1–12. doi:10.1155/2012/701532

KEY TERMS AND DEFINITIONS

Analytical Solution: A closed-form mathematical solution in the form of mathematical expression can be solved analytically.

Antenna: A transducer designed to transmit or receive electromagnetic waves.

Computer Simulation: A process of mathematical modelling, performed on a computer, which is designed to predict the behaviour of or the outcome of a real-world or physical system.

Computer-Aided Design (CAD): The use of computer software to aid in the creation, modification, analysis, or optimization of a design.

Electronic Design Automation (EDA): A computer-aided design tools particular for designing electronic systems.

Fifth Generation Wireless Technology (5G): Digital cellular mobile communication networks that began wide deployment in 2019.

Graphical User Interface (GUI): A form of user interface that allows users to interact with electronic devices through graphical icons and visual indicators such as secondary notation, instead of text-based user interfaces, typed command labels or text navigation.

Internet of Things (IoT): A system of interrelated computing devices, mechanical, and digital machines provided with unique identifiers (UIDs) and the ability to transfer data over a network without requiring human-to-human or human-to-computer interaction.

Matrix Laboratory (MATLAB): A proprietary multi-paradigm programming language and numerical computing environment developed by MathWorks.

Microwave (MW): A form of electromagnetic radiation with wavelengths ranging from 1 m to 1 mm, which is corresponding to operating frequencies ranging from 300 MHz to 300 GHz.

Mobile Application (App): A computer program or software application designed to run on mobile devices, such as laptop, smartphone, tablet, or watch.

Numerical Method: A matrix or iterative technique for solving mathematical problems.

Radio Frequency (RF): A part of the electromagnetic spectrum from about 3 kHz to 300 MHz, which is corresponding to wavelengths ranging from 100 km to 1 m.

Chapter 5
Fundamentals of RF/Microwave Bandpass Filter Design

Augustine O. Nwajana
https://orcid.org/0000-0001-6591-5269
University of Greenwich, UK

Emenike Raymond Obi
RaySoft AssetAnalytics, Canada

Gerald Kelechi Ijemaru
University of the Sunshine Coast, Australia

Emmanuel U. Oleka
North Carolina A&T State University, USA

Destiny Chidi Anthony
Torrens University, Australia

ABSTRACT

This chapter presents the basic approach of microwave bandpass filter design for 5G network applications. The chapter serves as a reference source to microwave stakeholders with little or no filter design experience. It should help them to design and implement their first filter device using microstrip technology. A three-pole Chebyshev bandpass filter with centre frequency of 2.6 GHz, fractional bandwidth of 3%, passband ripple of 0.04321 dB, and return loss of 20 dB has been designed. The designed filter implementation is based on the Rogers RT/Duroid 6010LM substrate with a 10.7 dielectric constant and 1.27 mm thickness. The circuit model and microstrip layout results of the BPF are presented and show good agreement. The microstrip layout simulation results show that a less than 1.8 dB minimum insertion loss and a greater than 25 dB in-band return loss were achieved. The overall device size of the BPF is 18.0 mm by 10.7 mm, which is equivalent to 0.16λg x 0.09λg, where λg is the guided wavelength of the 50 Ohm microstrip line at the filter centre frequency.

DOI: 10.4018/978-1-7998-6992-4.ch005

1. INTRODUCTION

A channel filter, also known as bandpass filter (BPF) passes frequencies within a single band and rejects all other frequencies outside the band (Nwajana, Dainkeh, & Yeo, 2017). This type of filter is widely used as the building block in the design of complex and multi-port circuits and systems. Some of the more complex devices that can be formed from BPFs include filtering antennas (Liu, Leung, & Yang, 2020; Wei, Zhao, & Shi, 2019), multi-band filters (Nwajana, 2020; Yeo & Nwajana, 2013; Hou, Liu, B. Zhang, Song, Wu, J. Zhang, & He, 2020), filtering power dividers (Nwajana, Ijemaru, Ang, Seng, & Yeo, 2021; Dainkeh, Nwajana, & Yeo, 2016), diplexers (Nwajana & Yeo, 2016; Nwajana, Okafor, & Arinze, 2020), etc. Figure 1 shows the response from a bandpass filter that passes all signal components between a lower frequency limit, f_L and an upper frequency limit, f_H, while attenuating and rejecting all other signal components that fall outside the f_L and f_H band. A bandpass filter can be formed by combining a lowpass filter with a highpass filter. Bandpass filters are widely used in radio frequency (RF) front end of cellular radio base station transceivers. Its main function in the transmitter is to limit the bandwidth of the output signal to the band assigned for the transmission. By this, the transmitter is prevented from interfering with other stations. In the receiver, a bandpass filter permits signals within a certain band of frequencies to be received and decoded, while stopping signals at undesirable frequencies from getting through.

Many authors have reported BPFs designed and implemented using various transmission line technologies including waveguides (Shi, Zhang, Zhou, Feng, Cao, & Che, 2019; Dahle, Laforge, & Kuhling, 2017; AbuHussain & Hasar, 2020), microstrip (Huang, Wang, & Zhu, 2016; Nwajana & Yeo, 2016; Cai, Wang, Zhu, & Wu, 2016; Zhang, Liu, Chen, Weng, & Yang, 2020), and substrate integrated waveguide (Qui, Wu, Xie, Yin, & Mao, 2018; Azad & Mohan, 2018; S. Hu, Y. Hu, Zheng, Zhu, Gao, & Zhang, 2020). The BPF presented in this paper is based on the microstrip technology. The filter relies on the microstrip hairpin resonator to achieve compact size. It is also of high selectivity and sharp roll-off. Some filter design characteristics such as selectivity, cost, size, sensitivity to environmental effects, power handling capacity, in-band and out-of-band performance metrics, are critical specifications in the development of RF and microwave communication front end devices. Filter developers are often required to make compromise between several conflicting requirements as it is rather difficult or even physically and/or electrically impossible to simultaneously achieve all design criteria or specifications. For instance, achieving higher channel selectivity usually requires the use of more resonators, which will result in higher insertion loss along the transmission path since insertion loss is approximately proportional to the number of resonators used in the construction of a filter (Nwajana, 2020). Hence, care must be taken when selecting design specifications to meet the most critical design targets.

Some popular manufacturing techniques that have been employed in fabricating filters include printed circuit board (PCB) (Nwajana, Yeo, & Dainkeh, 2016), low temperature co-fired ceramic (LTCC) (Wong, Wang, Chen, & Chu, 2014) and liquid crystal polymer (LCP) (Dalmia, White, Sundaram, & Swaminathan, 2004). In terms of low cost and commercial availability, the PCB wins and hence, has been utilized in the fabrication of the BPF reported in this paper.

Figure 1. Channel/bandpass filter characteristics.

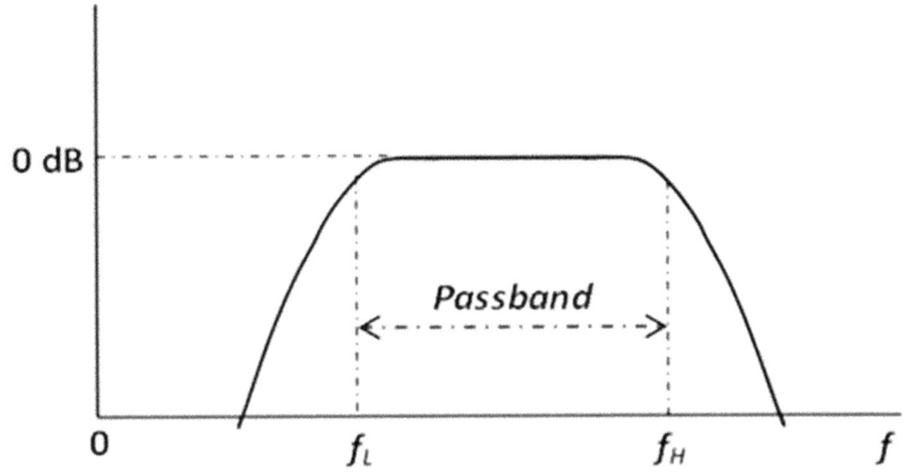

2. FILTER CIRCUIT MODEL

The BPF circuit model was established from the standard normalized 3-pole Chebyshev lowpass prototype filter shown in Figure 2 (Nwajana, 2017), where g is the element value for the filter. Some typical element values for Chebyshev lowpass prototype filters are given in Table 1 (Hong, 2011), for different passband ripples L_{Ar}, and filter order of 1 to 9. The proposed BPF is designed to have a center frequency, f_0 of 2.6 GHz, a fractional bandwidth of 3%, a passband ripple of 0.04321 dB, and a passband return loss of 20 dB.

Figure 2. Standard normalized 3-pole/3rd order lowpass prototype filter ($g_0 = g_4 = 1.0$, $g_1 = g_3 = 0.8516$, $g_2 = 1.1032$).

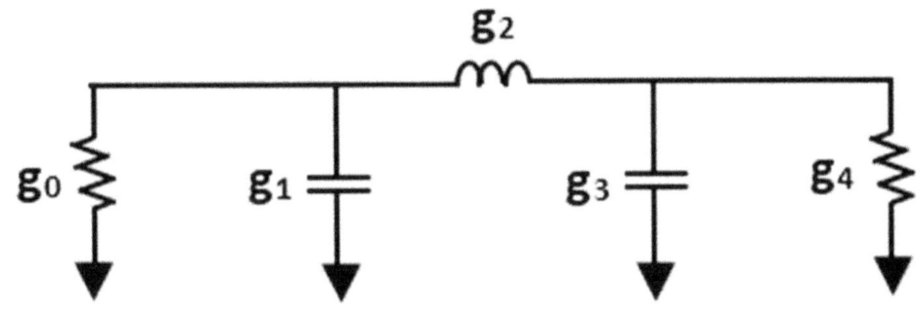

The first step to designing the BPF circuit model is to transform the 3rd order lowpass prototype filter of Figure 2 into the BPF circuit shown in Figure 3. The transformation is based on Equations (1) and (2) (Hong, 2011), where Z_0 is the 50 Ohms characteristics impedance at the input and output terminations, *FBW* is the fractional bandwidth, ω_o is the angular center frequency of the BPF, C_s and L_s are the series capacitance and inductance respectively, while C_p and L_p are the parallel capacitance and inductance respectively.

The second step of the BPF circuit model design is to use J-inverters to convert the series elements (i.e. C_s and L_s) in Figure 3 into shunt/parallel elements as shown in Figure 4. Hence, the BPF circuit model shown in Figure 4 contains only parallel *LC* elements, where $L = L_p$ and $C = C_p$. The numerical values for the J-inverters in Figure 4 were calculated using Equation (3) (Hong, 2011).

$$C_p = g_1 / \omega_0 Z_0 FBW \,;\, L_p = Z_0 FBW / g_1 \omega_0 \,. \tag{1}$$

$$C_s = FBW / g_2 \omega_0 Z_0 \,;\, L_s = g_2 Z_0 / \omega_0 FBW \,. \tag{2}$$

$$J_{01} = g_0 / Z_0 \,;\, J_{12} = \sqrt{g_1^2 / g_1 g_2} \Big/ Z_0 \,. \tag{3}$$

$$C_J = J / \omega_0 \,. \tag{4}$$

The third step of the BPF circuit model design is to replace each J-inverter in Figure 4 with a pi-network of capacitors or a pi-network of inductors as shown in Figure 5 (Hong, 2011). Working with Figure 5 (b), the BPF circuit in Figure 4 is then transformed into the final BPF circuit model shown in Figure 6 using Equation (4) (Hong, 2011). This final circuit model is suitable for direct simulation in any commercially available computer-aided design (CAD) tool. Some examples of CAD tools that can be used for the simulation include the electronic design automation (EDA) circuit simulators listed in Table 2.

The fourth and final step is to simulate the BPF circuit model. The Keysight ADS circuit simulator was used in the circuit model simulation. The simulation results are shown in Figure 7. The results clearly show that the design targets were met as the center frequency is at 2.6 GHz as designed. The return loss is also better than 20 dB as expected. The insertion loss is at 0 dB as expected for an ideal (i.e. circuit model) filter.

Table 1. Element values for Chebyshev Lowpass Prototype Filters (g_0 = 1.0). Hong, 2011

For passband ripple L_{Ar} = 0.01 dB

n	g_1	g_2	g_3	g_4	g_5	g_6	g_7	g_8	g_9	g_{10}
1	0.0960	1.0								
2	0.4489	0.4078	1.1008							
3	0.6292	0.9703	0.6292	1.0						
4	0.7129	1.2004	1.3213	0.6476	1.1008					
5	0.7563	1.3049	1.5773	1.3049	0.7563	1.0				
6	0.7814	1.3600	1.6897	1.5350	1.4970	0.7098	1.1008			
7	0.7970	1.3924	1.7481	1.6331	1.7481	1.3924	0.7970	1.0		
8	0.8073	1.4131	1.7825	1.6833	1.8529	1.6193	1.5555	0.7334	1.1008	
9	0.8145	1.4271	1.8044	1.7125	1.9058	1.7125	1.8044	1.4271	0.8145	1.0

For passband ripple L_{Ar} = 0.04321 dB

n	g_1	g_2	g_3	g_4	g_5	g_6	g_7	g_8	g_9	g_{10}
1	0.2000	1.0								
2	0.6648	0.5445	1.2210							
3	0.8516	1.1032	0.8516	1.0						
4	0.9314	1.2920	1.5775	0.7628	1.2210					
5	0.9714	1.3721	1.8014	1.3721	0.9714	1.0				
6	0.9940	1.4131	1.8933	1.5506	1.7253	0.8141	1.2210			
7	1.0080	1.4368	1.9398	1.6220	1.9398	1.4368	1.0080	1.0		
8	1.0171	1.4518	1.9667	1.6574	2.0237	1.6107	1.7726	0.8330	1.2210	
9	1.0235	1.4619	1.9837	1.6778	2.0649	1.6778	1.9837	1.4619	1.0235	1.0

For passband ripple L_{Ar} = 0.1 dB

n	g_1	g_2	g_3	g_4	g_5	g_6	g_7	g_8	g_9	g_{10}
1	0.3052	1.0								
2	0.8431	0.6220	1.3554							
3	1.0316	1.1474	1.0316	1.0						
4	1.1088	1.3062	1.7704	0.8181	1.3554					
5	1.1468	1.3712	1.9750	1.3712	1.1468	1.0				
6	1.1681	1.4040	2.0562	1.5171	1.9029	0.8618	1.3554			
7	1.1812	1.4228	2.0967	1.5734	2.0967	1.4228	1.1812	1.0		
8	1.1898	1.4346	2.1199	1.6010	2.1700	1.5641	1.9445	0.8778	1.3554	
9	1.1957	1.4426	2.1346	1.6167	2.2054	1.6167	2.1346	1.4426	1.1957	1.0

3. MICROSTRIP LAYOUT IMPLEMENTATION

This section covers the microstrip implementation of the proposed channel filter using the microstrip hairpin resonator. This type of resonator is well-known for its simple structure, flexibility, and compact size. The resonator is achieved by folding the conventional half-wavelength resonator into a "U" shape. Hence, hairpin resonator filters are designed based on the same set of equations employed in the design of parallel-coupled half-wavelength resonator filters (Hong, 2011). When folding the half-wavelength resonator to achieve a hairpin resonator, it is important to consider the reduction of the coupled-line lengths, which reduces the coupling between resonators. Should the two arms of a hairpin be closely spaced, they will function as a pair of coupled lines, and this can also influence the coupling (Hong, 2011). This section will be divided into subsections, with each subsection covering a step in the microstrip implementation of the BPF circuit model achieved in section 2.

Figure 3. Bandpass filter circuit both series and parallel LC resonators (C_p = 34.7529 pF, L_p = 0.1078 nH, C_s = 0.0333 pF, L_s = 112.551 nH).

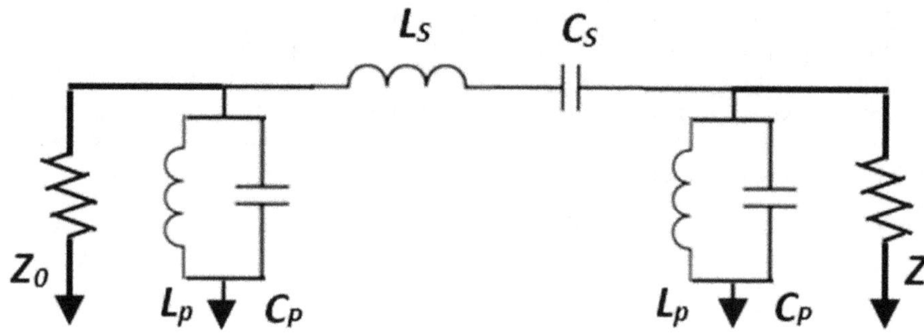

3.1. Microstrip Hairpin Resonator Dimensions

The microstrip hairpin resonator (HPR) is designed to resonate at the proposed BPF center frequency, i.e. f_0. The design is made on Rogers RT/Duroid 6010LM substrate with a dielectric constant, ε_r, of 10.7, a thickness, h, of 1.27 mm and a loss tangent of 0.0023. The width (w) and length (l) of the microstrip were estimated from (Hong, 2011) using Equations (5) and (6), respectively. Note that ε_{eff} is the effective dielectric constant of the design material, c_0 (i.e. 3 x 10^8 m/s) is the speed of light in free space and $l = \lambda_g/2$, where λ_g is the guided wavelength. Using Equations (5) and (6), the design values for w and l were estimated to be 1.14 mm and 21.61 mm, respectively. The actual dimensions of the microstrip was achieved by using the estimated values on various rounds of simulation and optimization in ADS momentum. The actual dimensions were reached at the point where the microstrip HPR started resonating at f_o as shown in Figure 8.

Figure 4. Bandpass filter circuit model with shunt/parallel-only identical LC resonators and J-inverters ($C = C_p$ and $L = L_p$).

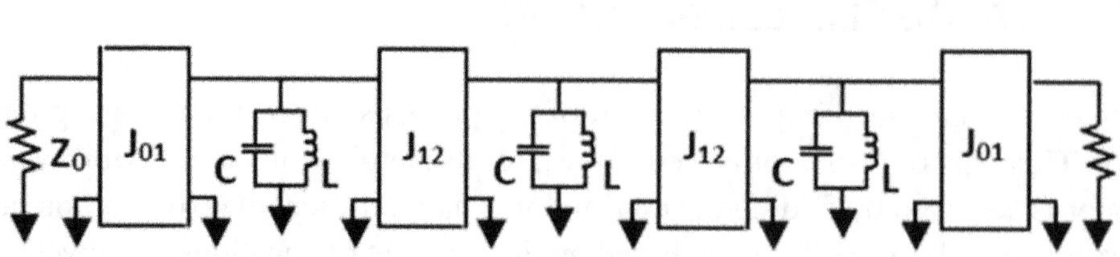

Figure 5. Reactive circuit elements modelling of admittance inverter: (a) J-inverter to pi-network of inductors; (b) J-inverter to pi-network of capacitors.

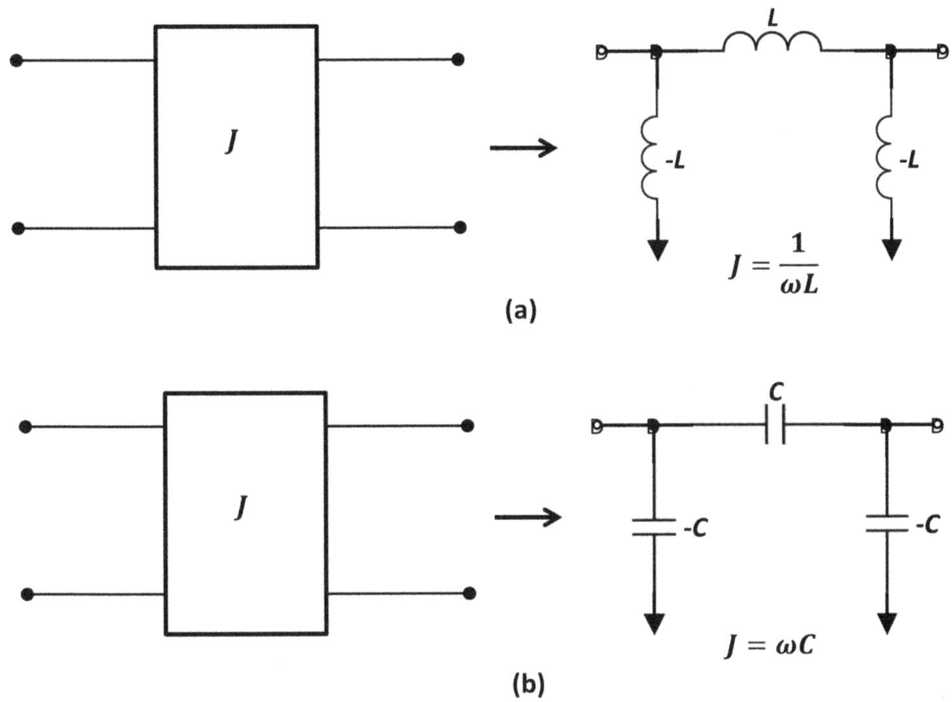

$$w = \frac{8he^{A}}{e^{2A}-2};\, A = \frac{Z_0}{60}\sqrt{\frac{\varepsilon_r+1}{2}} + \frac{\varepsilon_r-1}{\varepsilon_r+1}\left[0.23+\frac{0.11}{\varepsilon_r}\right]. \quad (5)$$

$$l = \frac{c_0}{2f_0\sqrt{\varepsilon_{eff}}};\, \varepsilon_{eff} = \frac{\varepsilon_r+1}{2} + \frac{\varepsilon_r-1}{2}\left(1+12\frac{h}{w}\right)^{-0.5}. \quad (6)$$

3.2. Coupling Coefficient Extraction

The coupling arrangement for the 3-pole channel filter is shown in Figure 9, where Q_{ext} is the external quality factor (to be discussed in the next subsection). The theoretical coupling between adjacent microstrip hairpin resonators is calculated to be 0.031 using Equation (7) (Hong, 2011), while the practical coupling value is achieved using Equation (8) (Hong, 2011) and the simulation technique shown in Figure 10. The separation distance, *s*, between adjacent resonators as shown in Figure 10, is the distance in mm that corresponds to the theoretical coupling value of 0.031, f_1 and f_2 are the eigen-modes from simulating the pair of coupled microstrip resonators. The matrix of k for the proposed bandpass filter is given in Equation (9).

Table 2. Some commercially available electronic design automation tools

Company	Typical product	Type
Ansoft www.ansoft.com	HFSS	3D EM simulator
Applied Wave Research (AWR) www.awr.com	Microwave Office	Integrated package including linear and nonlinear circuit simulators, optimizers, and EM analysis tools
CST www.cst.com	CST Microwave Studio	3D EM simulator
EM Works www.emworks.com	HFWorks	3D EM simulator
Keysight Technologies www.keysight.com	Advanced Design System (ADS)	Integrated package including 3D EM simulators
	Electromagnetic Professional (EMPro)	3D EM simulator
	Genesys	Integrated EM simulation package
Sonnet Software www.sonnetsoftware.com	Sonnet Suites	3D planar EM software
	Sonnet Lite	Free 3D planar EM simulator
QWED www.qwed.com.pl	Quick Wave-3D	3D EM simulator
Zeland Software www.zeland.com	IE3D	Planar and 3D EM simulation, optimization, and synthesis package

$$Theoretical\,coupling : k_{1,2} = k_{2,3} = {FBW}\Big/{\sqrt{g_1 g_2}}. \tag{7}$$

$$Practical\,coupling : k = {\left(f_2^2 - f_1^2\right)}\Big/{\left(f_2^2 + f_1^2\right)}. \tag{8}$$

Figure 6. Bandpass filter circuit model in ADS circuit simulator, with shunt/parallel-only identical LC resonators and pi-network of capacitors replacing J-inverters.

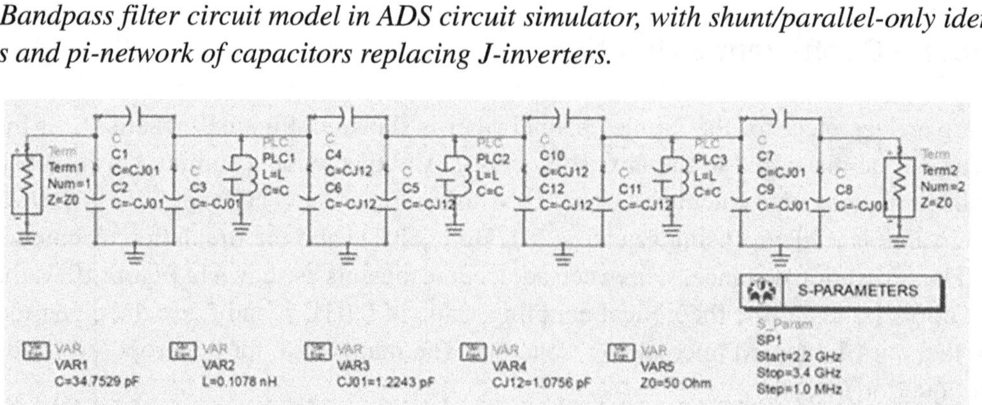

Figure 7. Simulation results of the bandpass filter circuit model.

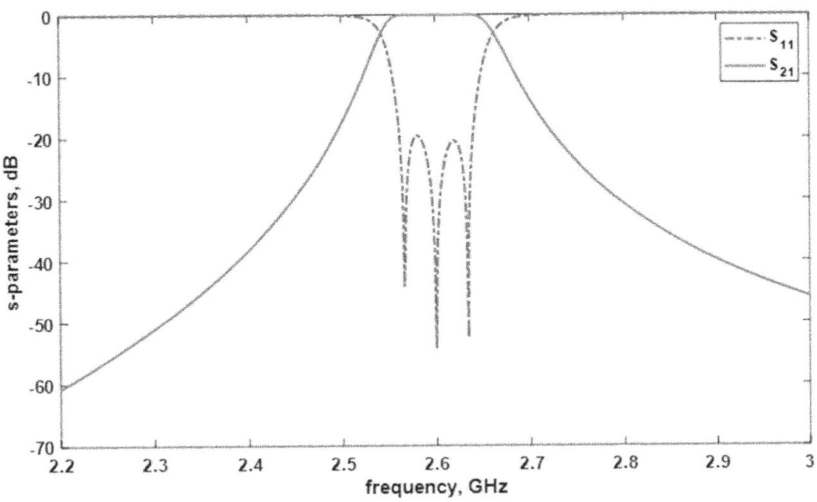

$$k = \begin{bmatrix} k_{1,1} & 0.031 & k_{1,3} \\ 0.031 & k_{2,2} & 0.031 \\ k_{3,1} & 0.031 & k_{3,3} \end{bmatrix}. \tag{9}$$

Figure 8. Hairpin resonator layout and response at 2.6 GHz (all dimensions in mm).

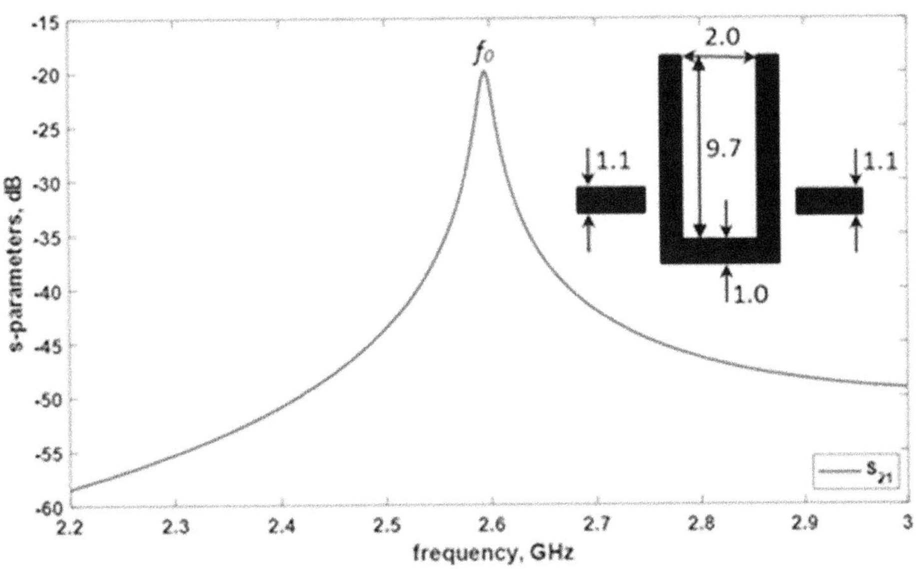

Figure 9. Coupling arrangement for the proposed 3-pole bandpass filter.

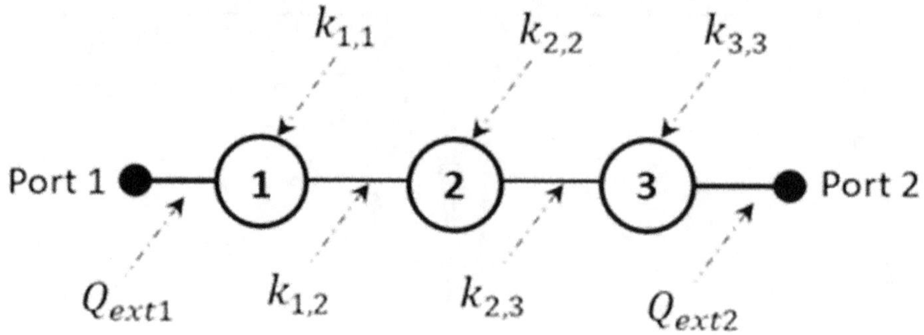

Figure 10. Coupling coefficient extraction technique for adjacent hairpin resonators (s = 1.9 mm).

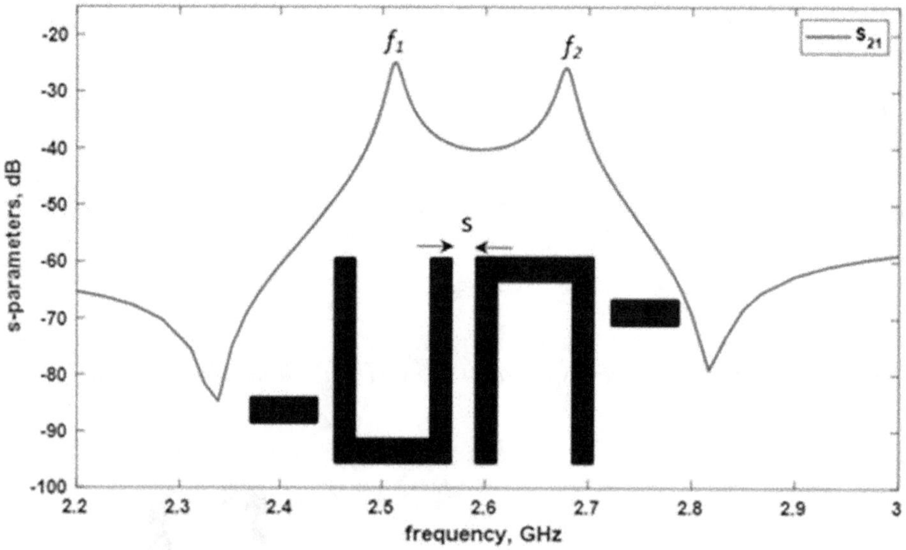

3.3. External Quality Factor Extraction

The theoretical external quality factor value is calculated to be 28.387 using Equation (10). The practical value of the Q_{ext} is determined using Equation (11) and the technique shown in Figure 11. The separation distance, t, between the first (or the last) resonator and the input (or the output) port as shown in Figure 11, is the distance in mm that corresponds to the theoretical Q_{ext} value of 28.387. Hence, both Q_{ext1} and Q_{ext2} shown in Figure 9 are equal to 28.387.

$$\text{Theoretical}\, Q_{ext} = {g_0 g_1}/{FBW}. \tag{10}$$

$$Practical\,Q_{ext} = {f_0}/{(f_2 - f_1)}. \tag{11}$$

Figure 11. External quality factor extraction technique (t = 0.7, all dimensions in mm).

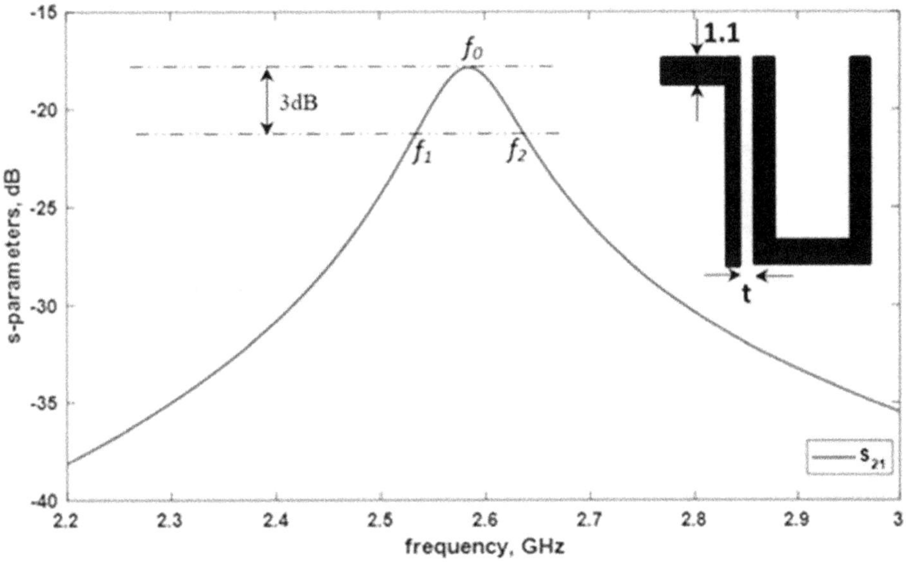

3.4. Filter Layout and Simulation

The proposed channel filter was put together based on the coupling arrangement of Figure 9. The filter layout was simulated using the Keysight ADS momentum full-wave electromagnetic simulator, with the layout and simulation results shown in Figure 12. The simulation results are in good agreement with the circuit model results previously presented in Figure 7. However, unlike the circuit model design that is in ideal condition (i.e. lossless), the microstrip layout implementation of the BPF included both conductor and dielectric losses. The loss tangent of the dielectric materials was kept at 0.0023, while the copper conductivity used for the simulation was maintained at 5.8 x 10^7 S/m, with a thickness of 35 micron for both the top and bottom metals of the microstrip line. Surface roughness and thickness variation of the substrate material were not considered. As a result of the lossy simulation, a simulated minimum insertion loss of better than 1.8 dB was recorded, with a better than 25 dB minimum return loss. The simulation center frequency is at 2.6 GHz as designed.

Figure 12. Layout and simulation results of the proposed 3-pole bandpass/channel filter.

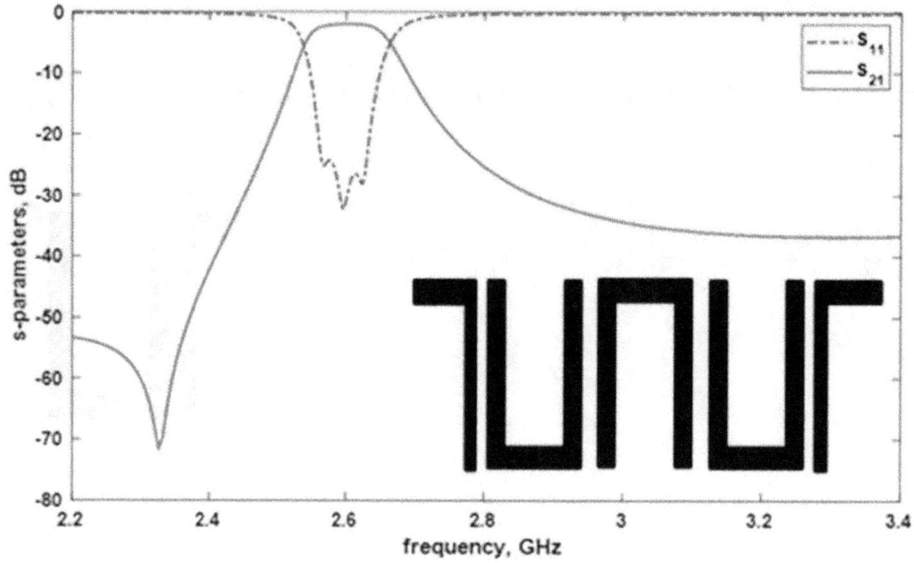

4. FABRICATION

The proposed BPF was fabricated using the same material employed in the electromagnetic simulation of the microstrip layout. Radio frequency printed circuit board (PCB) milling process technique was utilised for the circuit fabrication. A photograph of the fabricated microstrip BPF is shown in Figure 13. The overall size of the fabricated filter is 18.0 mm by 10.7 mm, i.e. 0.16λg x 0.09λg, where λg is the guided wavelength of the 50 Ω microstrip line at 2.6 GHz. A 50 Ohms SMA (sub-miniature version A) connector was attached to each port of the fabricated filter to facilitate measurement using Keysight Vector Network Analyzer.

5. DISCUSSION

The circuit model and microstrip layout results of the proposed BPF are co-presented in Figure 14 for easy comparison. The proposed filter performance was compared with some related published works using Table 3. The comparison with recently published works (Zhang, Liu, Chen, Weng, & Yang, 2020; S. Hu, Y. Hu, Zheng, Zhu, Gao, & Zhang, 2020; Jung & Hwang, 2007), clearly shows that the proposed BPF is of compact size, with a good insertion loss. The minimum return loss is excellent at 25 dB.

Figure 13. Photograph of the fabricated bandpass/channel filter.

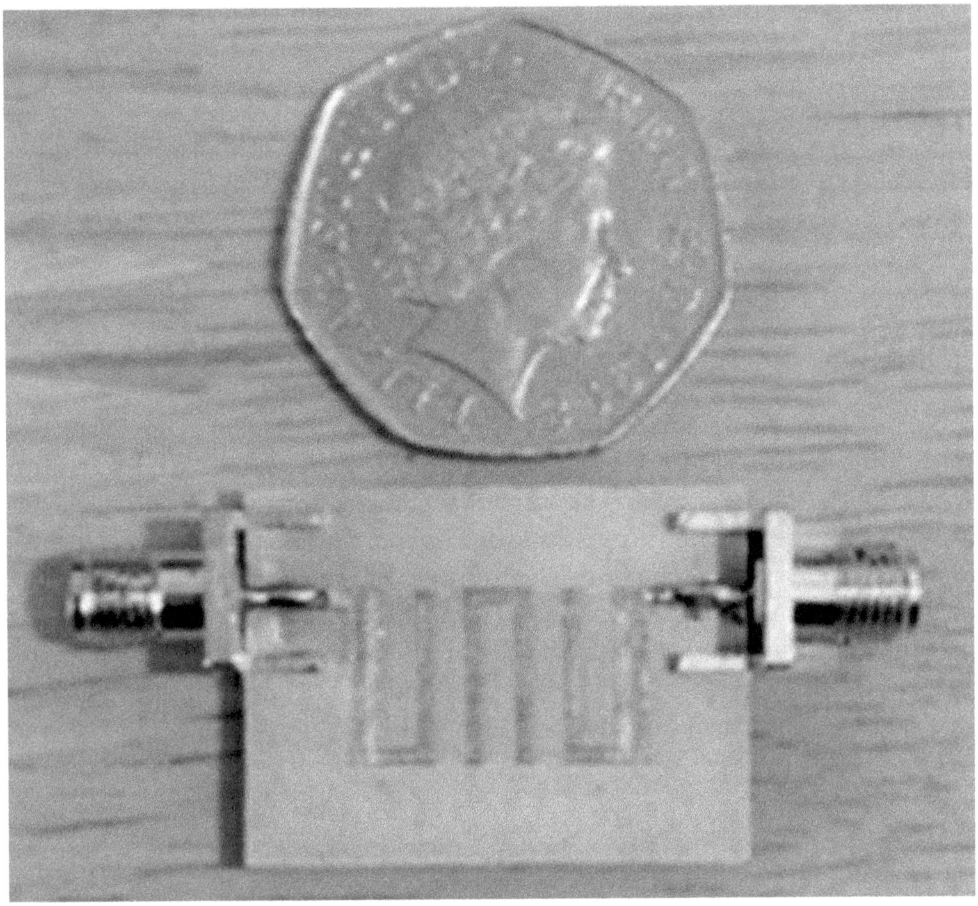

Figure 14. Comparison of the circuit model and the microstrip layout results of the proposed BPF.

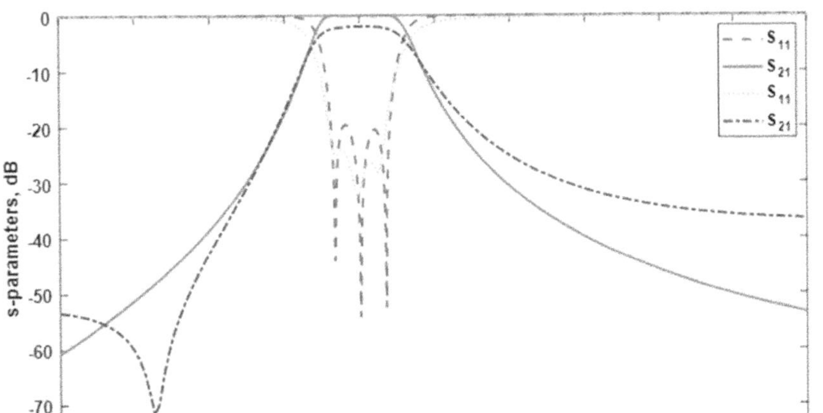

Table 3. Comparison of the proposed filter performances with related published works.

Ref.	f_0 (GHz)	Filter Order	Size (λ_g x λ_g)	IL[1] (dB)	RL[2] (dB)
(Zhang et al., 2020)	4.50	2	0.28 x 0.09	1.6	10.0
(Hu et al., 2020)	3.35	2	0.12 x 0.09	2.4	20.0
(Jung & Hwang, 2007)	2.45	2	0.15 x 0.13	2.4	20.0
This work	2.60	3	0.16 x 0.09	1.8	25.0

[1.] insertion loss,[2.] return loss.

6. CONCLUSION

A step-by-step approach to the design of bandpass filters has been presented in this chapter. A 3rd order (3-pole) bandpass filter has been proposed, designed, and implemented using the microstrip transmission line technology to illustrate the technique. The design was completely based on well-known formulations and relied on microstrip hairpin resonators for miniaturisation. The circuit model and microstrip layout results show good agreement; with centre frequency of 2.6 GHz, good selectivity and sharp roll-off. The simulation measured minimum insertion loss of less than 1.8 dB, and minimum in-band return loss of greater than 25 dB were achieved. The fabricated filter is of compact size with a small footprint of 18.0 mm by 10.7 mm (i.e. 0.16λg x 0.09λg), where λg is the guided wavelength of the 50 Ω microstrip line at the center frequency of the bandpass filter.

REFERENCES

AbuHussain, M., & Hasar, U. C. (2020). Design of X-bandpass waveguide Chebyshev filter based on CSRR metamaterial for telecommunication systems. *MDPI Electronics*, *9*(1), 101. doi:10.3390/electronics9010101

Azad, A. R., & Mohan, A. (2018). Substrate integrated waveguide dual-band and wide-stopband bandpass filters. *IEEE Microwave and Wireless Components Letters*, *28*(8), 660–662. doi:10.1109/LMWC.2018.2844103

Cai, C., Wang, J., Zhu, L., & Wu, W. (2016). A new approach to design a microstrip wideband balun bandpass filter. *IEEE Microwave and Wireless Components Letters*, *26*(2), 116–118. doi:10.1109/LMWC.2016.2516760

Dahle, R., Laforge, P., & Kuhling, J. (2017). 3-D printed customizable inserts for waveguide filter design at X-band. *IEEE Microwave and Wireless Components Letters*, *27*(12), 1080–1082. doi:10.1109/LMWC.2017.2754345

Dainkeh, A., Nwajana, A. O., & Yeo, K. S. K. (2016). Filtered power splitter using microstrip square open loop resonators. *Progress In Electromagnetic Research*, *C*, 64, 133–140. doi:10.2528/PIERC16042005

Dalmia, S., White, G., Sundaram, V., & Swaminathan, M. (2004). Design of quasi-lumped element diplexer for multiple wireless applications using liquid crystalline polymer (LCP) based substrates. *Proceedings of the 34th European Microwave Conference (EuMC)*, 737–740.

Hong, J.-S. (2011). *Microstrip filters for RF/microwave applications* (2nd ed.). John Wiley & Sons. doi:10.1002/9780470937297

Hou, Z., Liu, C., Zhang, B., Song, R., Wu, Z., Zhang, J., & He, D. (2020). Dual-/tri-wideband bandpass filter wit high selectivity and adjustable passband for 5G mid-band mobile communications. *MDPI Electronics*, *9*(2), 205. doi:10.3390/electronics9020205

Hu, S., Hu, Y., Zheng, H., Zhu, W., Gao, Y., & Zhang, X. (2020). A compact 3.3–3.5 GHz filter based on modified composite right–/left–handed resonator units. *MDPI Electronics*, *9*, 1.

Huang, F., Wang, J., & Zhu, L. (2016). A new approach to design a microstrip dual-mode balun bandpass filter. *IEEE Microwave and Wireless Components Letters*, *26*(4), 252–254. doi:10.1109/LMWC.2016.2538726

Jung, E.-Y., & Hwang, H.-Y. (2007). A balun-BPF using a dual-mode ring resonator. *IEEE Microwave and Wireless Components Letters*, *17*(9), 652–654. doi:10.1109/LMWC.2007.903442

Liu, Y.-T., Leung, K. W., & Yang, N. (2020). Compact absorptive patch antenna. *IEEE Transactions on Antennas and Propagation*, *68*(2), 633–642. doi:10.1109/TAP.2019.2938798

Nwajana, A. (2017). *Analysis and design of a substrate integrated waveguide multi-coupled resonator diplexer* (PhD Thesis). University of East London, London, UK.

Nwajana, A. O. (2020). Circuit modelling of bandpass/channel filter with microstrip implementation. *Indonesian Journal of Electrical Engineering, and Informatics*, *8*(4), 696–705.

Nwajana, A. O. (2020). *Dual-band microwave filter for WiMax application*. Lambert Academic Publishing.

Nwajana, A. O., Dainkeh, A., & Yeo, K. S. K. (2017). Substrate integrated waveguide (SIW) bandpass filter with novel microstrip-CPW-SIW input coupling. *Journal of Microwaves, Optoelectronics and Electromagnetic Applications*, *16*(2), 393–402. doi:10.1590/2179-10742017v16i2793

Nwajana, A. O., Ijemaru, G. K., Ang, K. L.-M., Seng, J. K. P., & Yeo, K. S. K. (2021). Unbalanced two-way filtering power splitter for wireless communication systems. *Electronics MDPI*, *10*(5), 617.

Nwajana, A. O., Okafor, P. U., & Arinze, S. N. (2018). T-junction diplexer for RF front end of a cellular base station. *International Journal of Microwave and Optical Technology*, *15*(6), 539–545.

Nwajana, A. O., & Yeo, K. S. K. (2016). Microwave diplexer purely based on direct synchronous and asynchronous coupling. *Wuxiandian Gongcheng*, *25*(2), 247–252. doi:10.13164/re.2016.0247

Nwajana, A. O., & Yeo, K. S. K. (2016). Multi-coupled resonator microwave diplexer with high isolation. *Proceedings of the 46th European Microwave Conference (EuMC)*, 1167–1170. 10.1109/EuMC.2016.7824556

Nwajana, A. O., Yeo, K. S. K., & Dainkeh, A. (2016). Low cost SIW Chebyshev bandpass filter with new input/output connection. *Proceedings of the 16th Mediterranean Microwave Symposium (MMS)*, 1–4.

Qui, L.-F., Wu, L.-S., Xie, B., Yin, W.-Y., & Mao, J.-F. (2018). Substrate integrated waveguide filter with flat passband based on complex couplings. *IEEE Microwave and Wireless Components Letters*, *28*(6), 494–496. doi:10.1109/LMWC.2018.2829349

Shi, Y., Zhang, J., Zhou, M., Feng, W., Cao, B., & Che, W. (2019). Miniaturized W-band gap waveguide bandpass filter using the MEMS technique for both waveguide and surface mounted packaging. *IEEE Transactions on Circuits and Wystems. II, Express Briefs*, *66*(6), 938–942. doi:10.1109/TCSII.2018.2873236

Wei, F., Zhao, X.-B., & Shi, X. W. (2019). A balanced filtering quasi-Yagi antenna with low cross-polarization levels and high common-mode suppression. *IEEE Access: Practical Innovations, Open Solutions*, *7*, 100113–100119. doi:10.1109/ACCESS.2019.2931141

Wong, S.-W., Wang, K., Chen, Z.-N., & Chu, X.-Q. (2014). Electric coupling structure of substrate integrated waveguide (SIW) for the application of 140-GHz bandpass filter on LTCC. *IEEE Transactions on Components, Packaging, and Manufacturing Technology*, *4*(2), 316–322. doi:10.1109/TCPMT.2013.2285388

Yeo, K. S. K., & Nwajana, A. O. (2013). A novel microstrip dual-band bandpass filter using dual-mode square patch resonators. *Progress In Electromagnetic Research, C*, 36, 233–247. doi:10.2528/PIERC12120312

Zhang, P., Liu, L., Chen, D., Weng, M.-H., & Yang, R.-Y. (2020). Application of a stub-loaded square ring resonator for wideband bandpass filter design. *MDPI Electronics*, *9*(1), 176. doi:10.3390/electronics9010176

Chapter 6
Non-Reciprocal Series-Fed Microstrip Patch Antenna Array Based on Graphene-Black Phosphorus for THz Applications Using the Iterative Method

Aymen Hlali
University of Carthage, Tunisia

Hassen Zairi
University of Carthage, Tunisia

ABSTRACT

Two-dimensional materials have already demonstrated their potential in electronic applications. In this chapter, the authors propose and investigate a non-reciprocal series-fed microstrip patch antenna array based on graphene-black phosphorus for THz applications. The modeling and simulation of this structure are made with the wave concept iterative process method, in which the 2D materials are incorporated as conductive surface boundaries. Combining the advantages of graphene and BP, the antenna exhibits both strong anisotropic and plasmon responses that are not available in either individual graphene or black phosphorus layer. The authors envision that this strategy of hybridization of graphene and BP may pave the way towards non-reciprocal plasmonic components with enhanced functionalities at THz band with important applications in imaging and communications.

DOI: 10.4018/978-1-7998-6992-4.ch006

INTRODUCTION

Two-dimensional materials have received the burgeoning amount of interest in recent years due to exciting electronic, optical, mechanical and thermal properties (F. Xia et al., 2014). Among these materials, graphene is the most popular one since its isolation by the Manchester group in 2004 (A. K et al., 2007). According to recent progress in graphene research, it has been shown that graphene exhibits anisotropic properties in the case of magnetic polarization (I. Crassee et al., 2011). Several non-reciprocal graphene-based devices to magnetic biasing were proposed in the past decades such as antennas, couplers, circulators, phase shifters and other nonreciprocal devices (D. C. Serrano et al., 2015). However, these devices with magnetic polarization are accompanied by several drawbacks, such as the requirement of bulky biasing devices and their incompatibility with semiconductor technology, which make their integration challenging (Nicholas Aaron Estep et al., 2016).

Black phosphorus, one of the newest members in the 2D material family, has attracted considerable attention due to its unique electrical properties (T. Low et al., 2014). Different from graphene, the atoms in BP are arranged to form a puckered honeycomb structure, which leads to strong in-plane anisotropic properties (H. Liu et al., 2014). Based on their excellent properties, many potential applications have been explored, including nano-antennas, SIW resonators, transistors, saturable absorbers, photodetectors and others. However, most of the Black phosphorus-based structures have the disadvantages of low doping concentration. Therefore, the resonance band cannot be tuned effectively. This is mainly because the plasmon resonances in individual doped and patterned BP are relatively weak.

Along with this research process, new methods are much in demand to model-based electronic devices based on the two-dimensional (2D) materials (M. Feizi et al., 2014). In most of the numerical methods, the 2D materials are usually incorporated in the algorithm by transforming the surface conductivity to a volumetric conductivity by dividing the thickness of the material, which might introduce a multiscale computational problem in simulation, require massive CPU time and memory, and make the optimization process tiresome. In order to overcome this problem, the wave concept iterative process method can be applied by treating the 2D materials as conductive surface boundaries, to yield not only precision in the result, but also reduces the time consumption and memory size (Aymen Hlali et al., 2020).

In summary, we investigate a plasmonic non-reciprocal series-fed microstrip patch antenna array based on hybrid graphene-black phosphorus over an ultrawide terahertz band. In particular, the response of the proposed array antenna can be controlled conveniently by adjusting the chemical potential of graphene and it is significantly decreased by adjusting the electron doping of black phosphorus. This type of hybrid architecture uses the graphene surface plasmons and the anisotropic plasmons in black phosphorus nanoribbons, which opens a new door for high performance 2D material non-reciprocal plasmonic devices.

The remaining part of this paper is organized as follows. In Section II, some theoretical aspects of the surface conductivity of graphene and black phosphorus, and a brief overview to the WCIP method are described. In Section III, details of the modeling and implementation of the boundary conditions of two-dimensional materials in the WCIP method are carefully demonstrated. Section IV investigates numerical results to demonstrate the capability of the proposed algorithm for the modeling and simulation of non-reciprocal plasmonic devices based on 2D materials, accuracy, and efficiency of the proposed algorithm is also examined. Some conclusions are presented in section V.

THEORY AND FORMULATION

1. Conductivity Model of the Monolayer Black Phosphorus

The anisotropy of the electric properties is the single characteristic distinguishing the BP from graphene and other 2D materials. The essential reason for this characteristic is that the atoms in 2D BP layer are organized to form a puckered hexagonal honeycomb structure, which leads to highly in-plane anisotropic properties (H. Liu et al., 2014).

Figure 1 shows the general schematic of the geometry under consideration, we consider a monolayer BP with conductivity tensor printed at the interface of isotropic material and air above.

Figure 1. Schematic representation of single-layer black phosphorus.

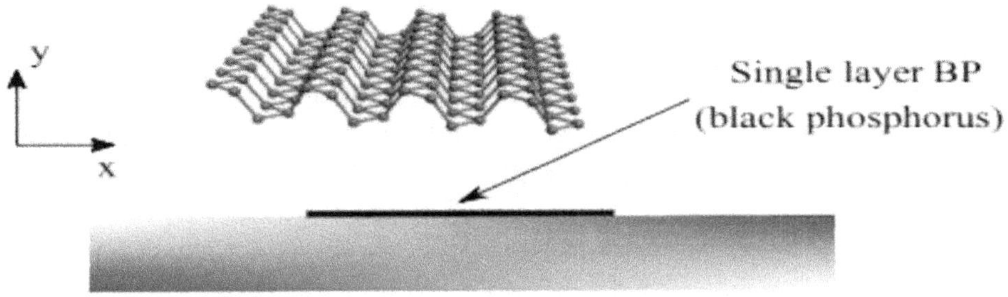

The monolayer BP was modeled as an ultra-thin metallic film with an anisotropic conductivity. The effective conductivities σ_{xx} and σ_{yy} are designated to be along the *x* and *y* directions, respectively (H. Liu et al., 2014)

$$\hat{\sigma} = \begin{pmatrix} \sigma_{xx} & 0 \\ 0 & \sigma_{yy} \end{pmatrix} \tag{1}$$

The surface conductivity σ_{jj} of a monolayer black phosphorus can be calculated via Drude model as (H. Liu et al., 2014)

$$\sigma_{jj} = \frac{iD_j}{\pi(\omega + \frac{i\eta}{\hbar})} \tag{2}$$

where ω is the frequency of incident light. η is the relaxation rate. D_j is the Drude weight and can be described by

$$D_{jj} = \frac{\pi e^2 n}{m_j} \tag{3}$$

where e and n are the electron charge and electron doping, m_j present the electron mass along the x and y direction, which can be given as

$$m_x = \frac{\hbar^2}{\frac{2\lambda^2}{\Delta} + \eta_c}, m_y = \frac{\hbar^2}{2V_c} \tag{4}$$

For a monolayer BP, we have $\Delta = 2$eV, $\eta_c = \hbar/0.4m_0$, $V_c = \hbar/1.4m_0$ and $\gamma = 4a/\pi$ eV m.

2. Conductivity Model of the Graphene Sheet

Graphene is modeled as a sheet conducting whose surface conductivity model includes of two parts, where the first part is corresponding to the intraband conductivity and the second part is due to the interband conductivity. The intraband term can be calculated via Kubo formula as (Aymen Hlali et al., 2020)

$$\sigma_{\text{intrabande}} = -j\frac{e^2 K_B T}{\pi\hbar^2(\omega - j2\Gamma)}\left[\frac{\mu_c}{K_B T} + 2\ln\left(e^{-\frac{\mu_c}{K_B T}} + 1\right)\right] \tag{5}$$

and the interband conductivity is

$$\sigma_{\text{interbande}} = -j\frac{e^2}{4\pi\hbar}\ln\left[\frac{2|\mu_c| - (\omega - j2\Gamma)\hbar}{2|\mu_c| + (\omega - j2\Gamma)\hbar}\right] \tag{6}$$

where $\hbar$ is the reduced Planck's constant, K_B is the Boltzmann constant, e is the electron charge, ω is the angular frequency, Γ is the scattering rate, μ_c is the chemical potential and T is the temperature.

The value of the interband conductivity is very small compared with that of the intraband conductivity and has no significant effect on the overall surface conductivity. However, as the frequency increases, the interband conductivity becomes comparable with the intraband conductivity, especially in the frequency range larger than 8 THz. In order to obtain an accurately modeling, both the interband and intraband conductivities are incorporated in the numerical simulation of graphene.

3. Formulation of WCIP Method

The formulation of the WCIP method has been described in different articles (Aymen Hlali et al., 2020). This method is based on the wave concept to solve the electromagnetic problem, which the wave concept is introduced by writing the incident $\mathbf{A}_i$ and reflected $\mathbf{B}_i$ waves as a function of the tangential electric field $\mathbf{E}_i$ and surface tangential current density $\mathbf{J}_i$. The relationship between these waves and the electromagnetic field is given by the following equation (Aymen Hlali et al., 2020)

$$\boldsymbol{A}_i = \frac{1}{2\sqrt{Z_{0i}}}\left(\boldsymbol{E}_i + Z_{0i}\boldsymbol{J}_i\right)$$
$$\boldsymbol{B}_i = \frac{1}{2\sqrt{Z_{0i}}}\left(\boldsymbol{E}_i - Z_{0i}\boldsymbol{J}_i\right) \tag{7, 8}$$

4. Modeling of two-dimensional Materials in the WCIP Method

a. Implementation of the Boundary Conditions of Graphene in WCIP Method

Graphene is implemented in the WCIP method as a boundary condition and is computed using the surface conductivity. After a mathematical resolution, we can derive the scattering matrix of graphene as (Aymen Hlali et al., 2018)

$$\begin{pmatrix} \boldsymbol{B}_1 \\ \boldsymbol{B}_2 \end{pmatrix} = \underbrace{\begin{pmatrix} S_{11}^G & S_{12}^G \\ S_{21}^G & S_{22}^G \end{pmatrix}}_{\Gamma_G} \begin{pmatrix} \boldsymbol{A}_1 \\ \boldsymbol{A}_2 \end{pmatrix} \tag{9}$$

where S^G_{ij} are defined by

$$S_{11}^G = \frac{Z_{02} - Z_{01} - \sigma Z_{02} Z_{01}}{Z_{02} + Z_{01} + \sigma Z_{02} Z_{01}}; S_{12}^G = \frac{2Z_{02}\sqrt{Z_{01}}}{\sqrt{Z_{02}}(Z_{02} + Z_{01} + \sigma Z_{02} Z_{01})}$$
$$S_{21}^G = \frac{2Z_{01}\sqrt{Z_{02}}}{\sqrt{Z_{01}}(Z_{02} + Z_{01} + \sigma Z_{02} Z_{01})}; S_{22}^G = -\frac{Z_{02} - Z_{01} + \sigma Z_{02} Z_{01}}{Z_{02} + Z_{01} + \sigma Z_{02} Z_{01}}$$

b. Implementation of the Boundary Conditions of BP in WCIP Method

As an analog to graphene, BP is introduced in the WCIP method as boundary condition via the surface current and the electric field using the surface conductivity. The electromagnetic boundary condition at monolayer BP can be written as (Aymen Hlali et al., 2020)

$$\boldsymbol{n} \times \left(\boldsymbol{H}_1 - \boldsymbol{H}_2\right) = \boldsymbol{J}_e^s = \hat{\sigma}\boldsymbol{E} \tag{10}$$

An equivalent model of single-layer black phosphorus in the WCIP method is presented in Figure 2.

Figure 2. Equivalent model of BP in the WCIP method.

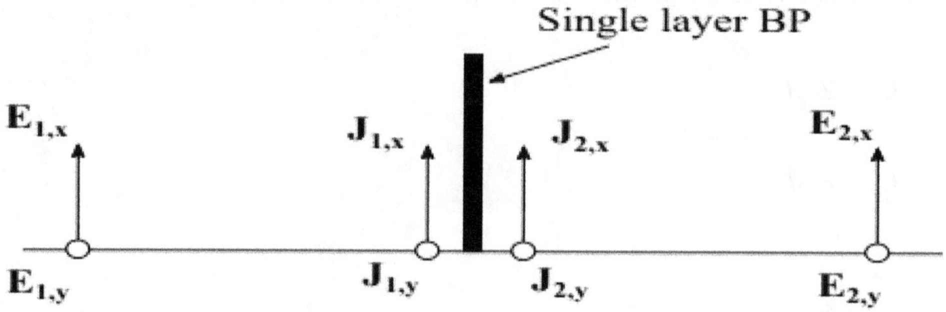

In general, the implementation of boundary conditions in the WCIP method consists of changing the magnetic field by the current density, the current matrix in the x and y-directions can be expressed as (Aymen Hlali et al., 2020)

$$\begin{aligned} \boldsymbol{J}_{1,x} + \boldsymbol{J}_{2,x} &= \hat{\sigma}_{xx}\boldsymbol{E}_{i,x} \\ \boldsymbol{J}_{1,y} + \boldsymbol{J}_{2,y} &= \hat{\sigma}_{yy}\boldsymbol{E}_{i,y} \end{aligned} \tag{11, 12}$$

The implementation of boundary conditions of BP in the WCIP method is applied by redefining the electric field and the current density in terms of waves. After some manipulation; we can determine the scattering operator of BP as

$$\begin{pmatrix} \boldsymbol{B}_{1,x} \\ \boldsymbol{B}_{1,y} \\ \boldsymbol{B}_{2,x} \\ \boldsymbol{B}_{2,y} \end{pmatrix} = \underbrace{\begin{pmatrix} S_{11}^{BP} & 0 & S_{13}^{BP} & 0 \\ 0 & S_{22}^{BP} & 0 & S_{24}^{BP} \\ S_{31}^{BP} & 0 & S_{33}^{BP} & 0 \\ 0 & S_{42}^{BP} & 0 & S_{44}^{BP} \end{pmatrix}}_{\Gamma_{BP}} \begin{pmatrix} \boldsymbol{A}_{1,x} \\ \boldsymbol{A}_{1,y} \\ \boldsymbol{A}_{2,x} \\ \boldsymbol{A}_{2,y} \end{pmatrix} \tag{13}$$

where S^{BP}_{ij} are defined by

$$S_{11}^{BP} = \frac{Z_{02} - Z_{01} - \sigma_{xx} Z_{02} Z_{01}}{Z_{02} + Z_{01} + \sigma_{xx} Z_{02} Z_{01}} ; S_{13}^{BP} = \frac{(1 - \sqrt{Z_{02}})(Z_{02}\sqrt{Z_{01}})}{\sqrt{Z_{02}}(Z_{02} + Z_{01} + \sigma_{xx} Z_{02} Z_{01})}$$
$$S_{22}^{BP} = \frac{Z_{02} - Z_{01} - \sigma_{yy} Z_{02} Z_{01}}{Z_{02} + Z_{01} - \sigma_{yy} Z_{02} Z_{01}} ; S_{24}^{BP} = \frac{(1 - \sqrt{Z_{02}})(Z_{02}\sqrt{Z_{01}})}{\sqrt{Z_{02}}(Z_{02} + Z_{01} - \sigma_{yy} Z_{02} Z_{01})}$$
$$S_{33}^{BP} = \frac{Z_{01} - Z_{02} - \sigma_{xx} Z_{02} Z_{01}}{Z_{02} + Z_{01} - \sigma_{xx} Z_{02} Z_{01}} ; S_{31}^{BP} = \frac{(1 - \sqrt{Z_{01}})(Z_{01}\sqrt{Z_{02}})}{\sqrt{Z_{01}}(Z_{02} + Z_{01} - \sigma_{xx} Z_{02} Z_{01})}$$
$$S_{44}^{BP} = \frac{Z_{01} - Z_{02} - \sigma_{yy} Z_{02} Z_{01}}{Z_{02} + Z_{01} + \sigma_{yy} Z_{02} Z_{01}} ; S_{42}^{BP} = \frac{(1 - \sqrt{Z_{01}})(Z_{01}\sqrt{Z_{02}})}{\sqrt{Z_{01}}(Z_{02} + Z_{01} + \sigma_{yy} Z_{02} Z_{01})}$$

c. Definition of the Diffraction Operator for an Anisotropic Medium

To create the overall diffraction operator, it is essential to establish the relationship between the waves with the anisotropic form while maintaining the isotropy of the graphene, metal, dielectric and source operators. In the isotropic case, the relations between the waves are written as follows (Aymen Hlali et al., 2020)

$$B_1 = S_{11}^{i} A_1 + S_{12}^{i} A_2, i = G, M, D, S$$
$$B_2 = S_{21}^{i} A_1 + S_{22}^{i} A_2, i = G, M, D, S \tag{14}$$

In the anisotropic case, these relationships are reformulated and take the following form

$$\begin{pmatrix} \boldsymbol{B}_{1,x} \\ \boldsymbol{B}_{1,y} \\ \boldsymbol{B}_{2,x} \\ \boldsymbol{B}_{2,y} \end{pmatrix} = \begin{pmatrix} S_{11,x}^{i} & 0 & S_{13,x}^{i} & 0 \\ 0 & S_{22,y}^{BP} & 0 & S_{24,y}^{i} \\ S_{31,x}^{i} & 0 & S_{33,x}^{i} & 0 \\ 0 & S_{42,y}^{i} & 0 & S_{44,y}^{i} \end{pmatrix} \begin{pmatrix} \boldsymbol{A}_{1,x} \\ \boldsymbol{A}_{1,y} \\ \boldsymbol{A}_{2,x} \\ \boldsymbol{A}_{2,y} \end{pmatrix} \tag{15}$$

Where

$$S_{11}^{i} = S_{11,x}^{i} = S_{11,y}^{i} ; S_{12}^{i} = S_{12,x}^{i} = S_{12,y}^{i}$$
$$S_{21}^{i} = S_{21,x}^{i} = S_{21,y}^{i} ; S_{22}^{i} = S_{22,x}^{i} = S_{22,y}^{i}$$

Lastly, it is possible to establish the total diffraction operator, consequently, the total diffraction operator in the spatial domain can be evaluated by the following matrix

$$\Gamma = \Gamma_G + \Gamma_{BP} + \Gamma_M + \Gamma_d + \Gamma_s \tag{16}$$

where Γ_G, Γ_{BP}, Γ_M, Γ_D and Γ_S are respectively the graphene, black phosphorus, metal, dielectric and source diffraction operators.

d. Determination of the Diffraction Operator for Hybrid Structure

In a hybrid structure as shown in Figure 3, the total diffraction operator assumes another form, where it takes the boundary conditions of graphene in the calculation of the relationship between the incident and reflected waves in the medium 1 and the boundary conditions of BP in medium 2.

Figure 3. Incident and reflected waves on a hybrid structure.

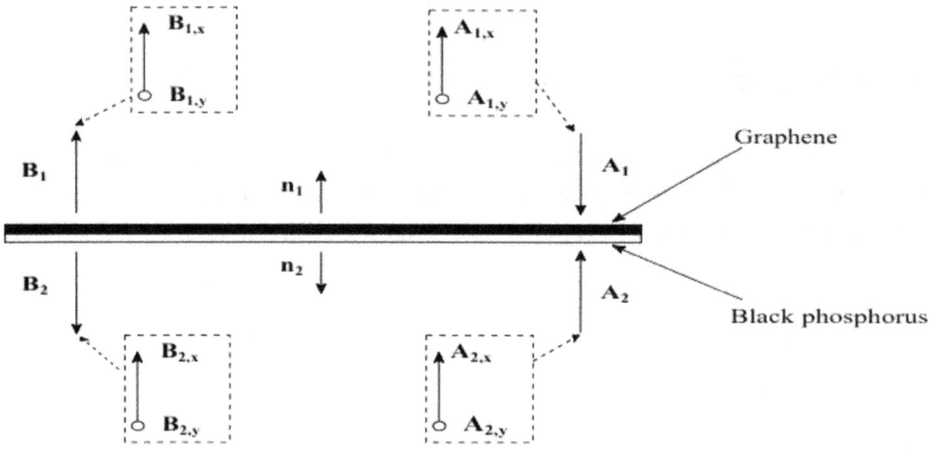

NUMERICAL RESULTS AND DISCUSSION

According to the conductivity, there are different scenarios for the properties of the material. The usual case is obtained when the properties are not dependent on the direction; this is the case of a graphene sheet, where it is a natural example of isotropic sheet. Contrary to the case of isotropic surfaces, the anisotropic case occurs when the parts of the conductivity components are different $\mathrm{Im}(\sigma_{xx}) \neq \mathrm{Im}(\sigma_{yy})$ and $\mathrm{Re}(\sigma_{xx}) \neq \mathrm{Re}(\sigma_{yy})$ (Xianglian Song et al., 2018). Figure 4 illustrates the real and imaginary parts of the conductivity of a monolayer BP versus frequency for two various values of electron doping.

Results confirm a high degree of anisotropy in the real and imaginary parts of the conductivity components of the monolayer BP with a tunability for the variations of the electron doping.

In this paper (Aymen Hlali et al., 2018), we demonstrated that the interband conductivity of graphene has a significant effect and cannot be neglected in the numerical simulation at the frequency band larger than 8 THz. For this, we consider this conductivity. Figure 5 illustrates the variations of the real and imaginary parts of the total conductivity versus frequency traced for different values of chemical potential for 0 to 0.4 eV, with $T = 300$ K and $\tau = 0.1$ ps.

Figure 4. Real (a) and imaginary (b) parts of BP conductivity versus frequency plotted for various values of electron doping.

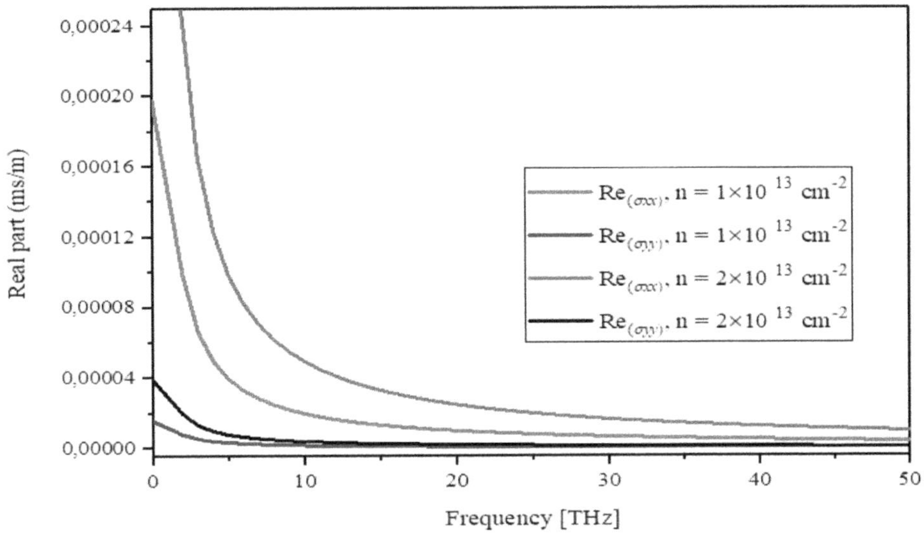

(a)

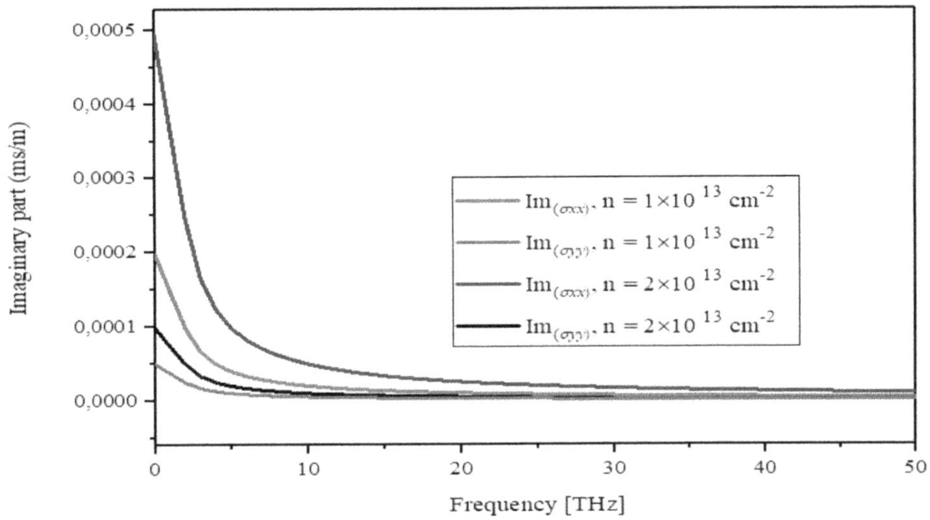

(b)

Results show that the graphene has a high degree of tunability for the variations of the chemical potential. Where, the chemical potential can be controlled in practice through an externally applied electrostatic bias.

Next, in order to demonstrate the accuracy of the proposed algorithm, we simulate a non-reciprocal series-fed microstrip patch antenna array for Terahertz applications. Figure 6 shows the schematic illustration of an eight-element series-fed microstrip patch antennas including the transmission lines to connect these elements, and the geometric parameters of our design.

Figure 5. Real (a) and Imaginary (b) parts of graphene conductivity versus frequency plotted for various values of chemical potential.

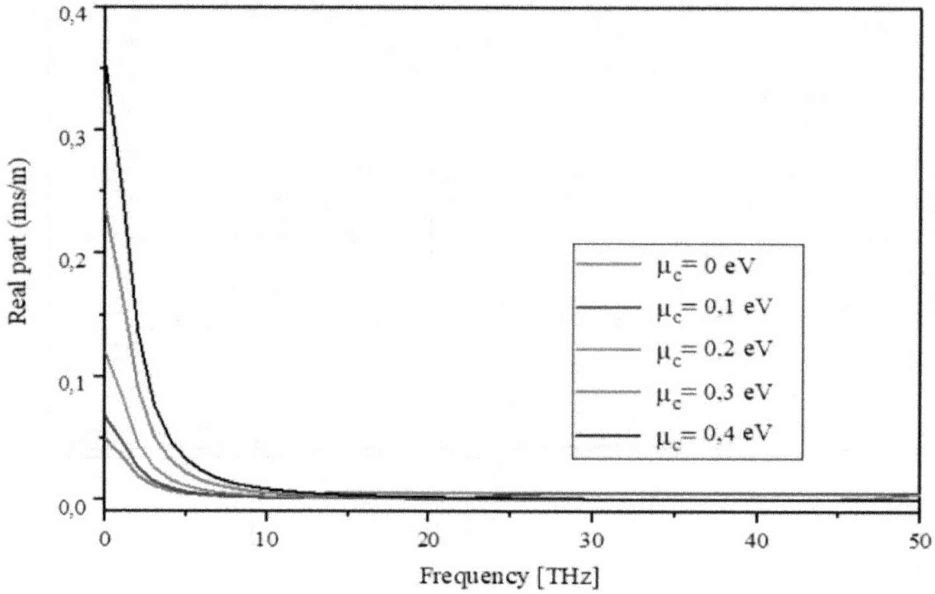

(a)

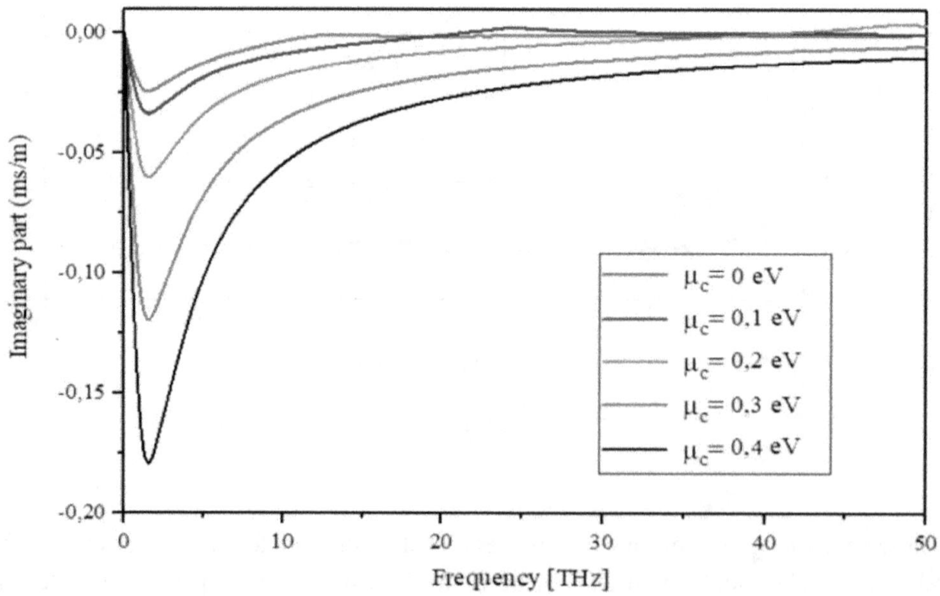

(b)

Figure 6. Schematic of the proposed array antenna. Parameters are $W_1 = 0.312\ \mu m$, $W_2 = 0.05\ \mu m$, $W_3 = 2\ \mu m$, $L_1 = 1.4\ \mu m$, $L_2 = 0.9\ \mu m$, $L_3 = 1.6\ \mu m$.

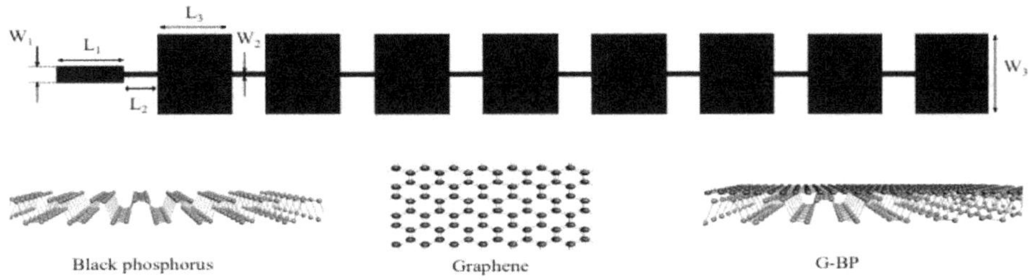

Figure 7. Convergence of the reflection coefficient.

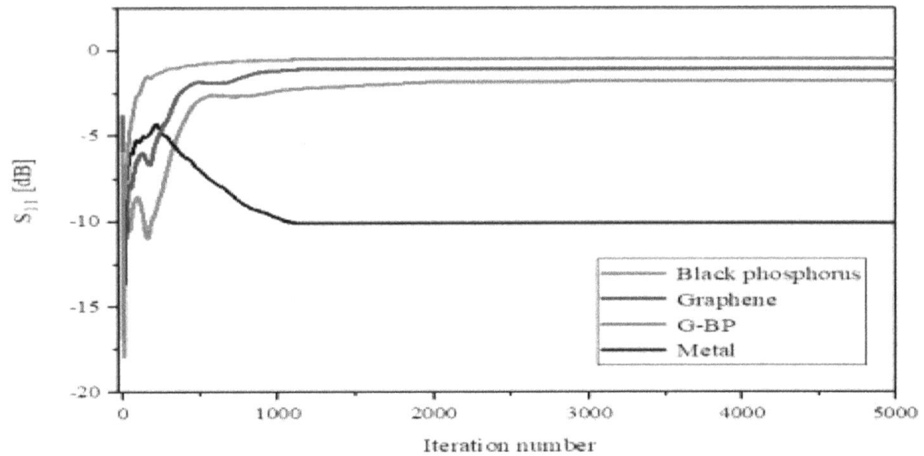

The antenna is designed on a Rogers substrate with dielectric constant of $\varepsilon_r = 2.6$. The width and length of the structure are assumed to be 5 µm and 25 µm, whereas the thickness is 0.2 µm. In order to examine the results obtained by the WCIP method, we begin by studying the convergence and the boundary conditions. Figure 7 illustrates the variation of the reflection coefficient versus the number of iterations.

We observe that the convergence of the metal-based structure is obtained from 1200 iterations, 1800 iterations for the graphene and black phosphorus structures and 1900 iterations for the G-BP. Figures 8 and 9 show the distribution of currents density and electric fields for each configuration.

It is found that the electrical field and current density verify the boundary conditions. The current density distribution is maximum on the graphene, BP and metal; and zero on the dielectric. The electric field is zero on graphene, BP and metal; and different from zero on the dielectric.

In order to validate the results obtained by the WCIP method, figure 10 illustrates the variation of the reflection coefficient for the four configurations, namely, black phosphorus, graphene, G-BP and metal obtained through the WCIP method and CST simulator.

It is found that the simulated result obtained by WCIP method is close to that simulated by the CST simulator. This validates the formulation developed by the WCIP method. Table 1 presents the efficiency of our proposed algorithm in comparison with CST simulator.

Figure 8. Distribution of the current density: (a) Black phosphorus, (b) graphene, (c) G-BP, and (d) metal.

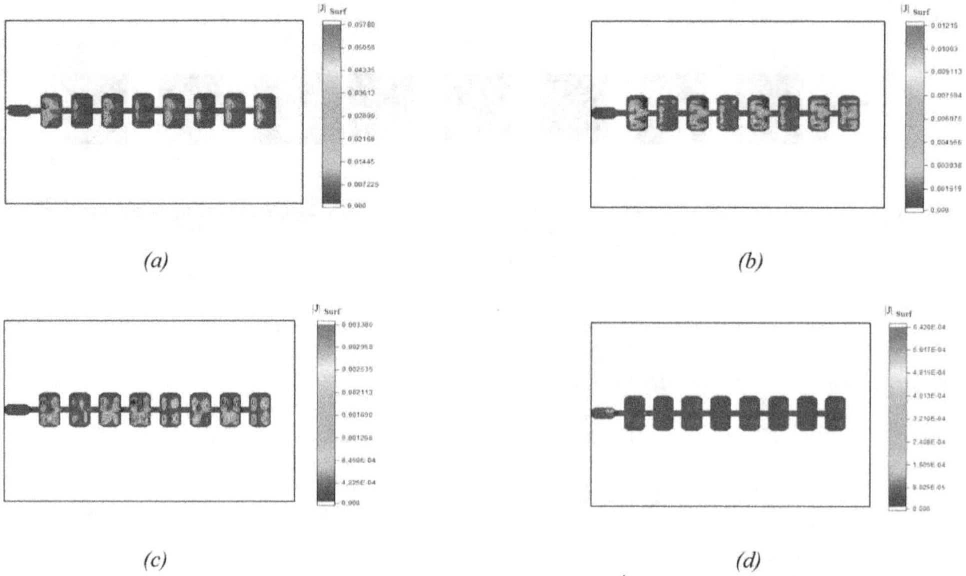

Figure 9. Distribution of the electric field: (a) Black phosphorus, (b) graphene, (c) G-BP, and (d) metal.

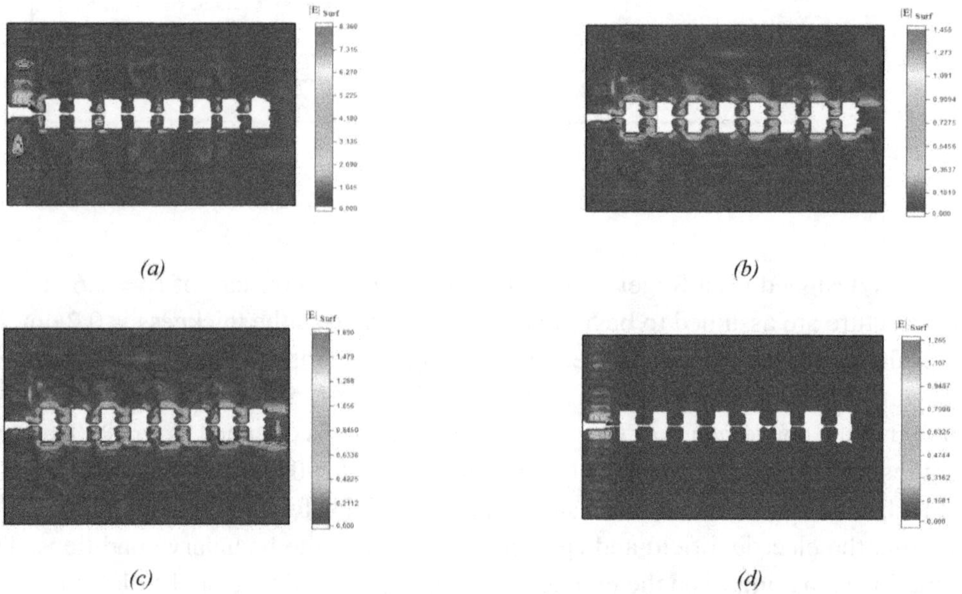

In this table, we present the time of simulation and memory size for the four configurations. For the CST simulator, the resolution of the problem requires a 3D mesh and a long computation time. As for the WCIP method, it transforms the 3D problem into a 2D problem.

So, the resolution becomes more straightforward with a reduced computation time and memory size. The computations of all structures were performed on Intel Core i7 CPU 2.5 GHz, 6 GB RAM.

Figure 10. Performance comparison of the reflection coefficient versus frequency: (a) Black phosphorus, (b) graphene, (c) G-BP, and (d) metal.

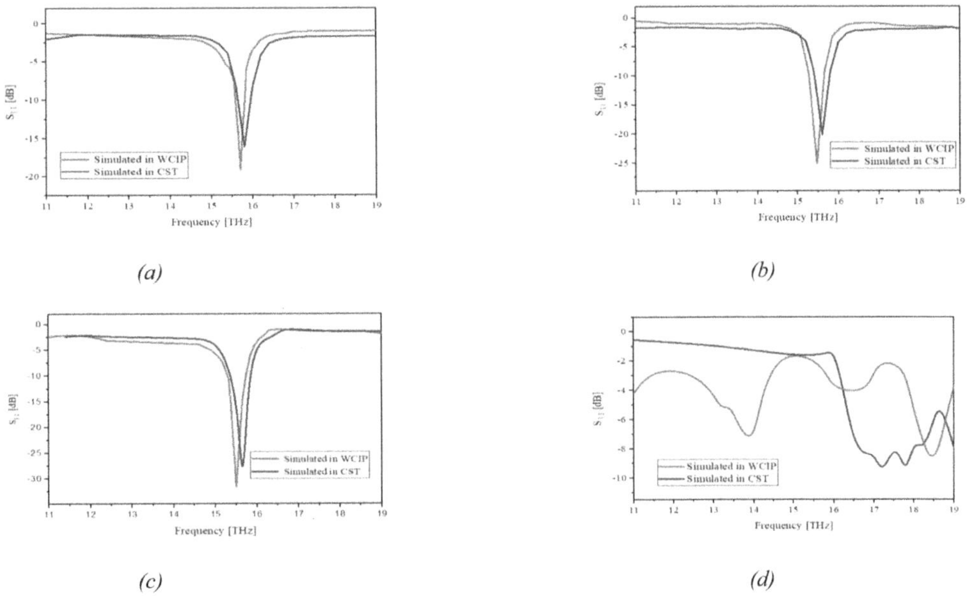

Table 1. Comparison between the WCIP results and the CST simulator regarding the computational time and memory size for the four configurations.

		Black phosphorus	Graphene	G-BP	Metal
Computation time	**In CST**	31 min 42 s	29 min 12 s	46 min 25 s	18 min 4 s
Computation time	**In WCIP**	24 min 3 s	20 min 3 s	21 min 34 s	10 min 45 s
Memory size	**In CST**	195.0 Mo	190.2 Mo	243.2 Mo	158.4 Mo
Memory size	**In WCIP**	13.900 Mo	12.864 Mo	13.150 Mo	8.390 Mo

According to figure 10 and table 2, the reflection coefficient of the metal-based antenna is wrong adapted. We also notice that, graphene and BP give more optimized results and the performance is enhanced compared to the metal-based configuration. But, the hybrid structure G-BP gives better adaptation and more efficient results than the other configurations.

So, we continue our research with this configuration. As we mentioned above, the chemical potential and the electron doping can affect the electromagnetic performances of the antenna because it directly determines the surface conductivity of the 2D materials.

Now, we investigate the influences of the electron doping of BP and the chemical potential of graphene on the antenna performances. The variation of the reflection coefficient of the antenna for different chemical potential of graphene is presented in figure 11.

Fig. 11(b) illustrates the tunability of the bandwidth Δf_r and the resonance frequency f_r with modification of the chemical potential. It is found that as the chemical potential increases from 0.1 to 0.4 eV, the value of f_r is increased from 15.486 to 16.95 THz and Δf_r is varied from 0.445 to 1.18 THz.

Table 2. Performance characteristics of the four configurations.

	Black phosphorus	Graphene	G-BP	Metal
Resonant frequency, THz	15.705	15.472	15.48	-
Return loss, dB	-19.21	-24.99	-31.53	-
Bandwith, THz	0.229	0.372	0.445	-
Gain, dBi	5.817	6.81	7.970	-

Figure 11. (a) Simulated results for different chemical potentials, and (b) variation of resonance frequency and bandwidth.

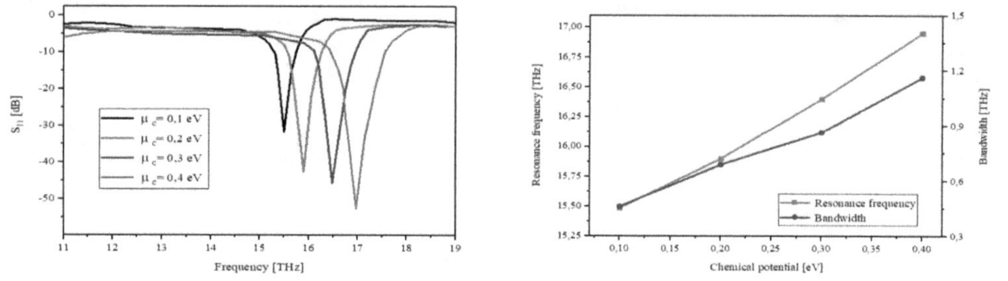

(a) *(b)*

Figure 12. (a) Simulated results for different values of electron doping, and (b) variation of resonance frequency and bandwidth.

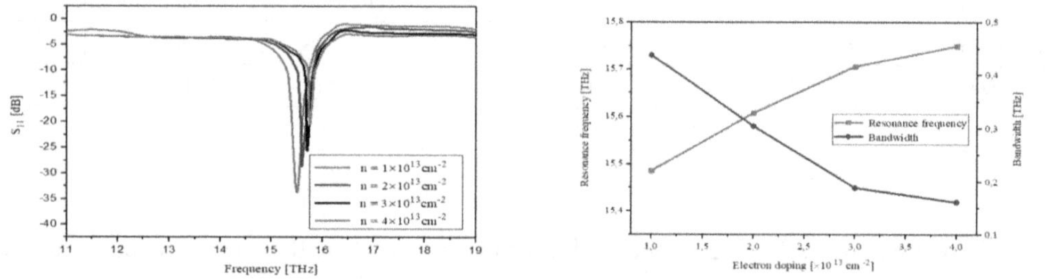

(a) *(b)*

Similar to graphene, the variation of the reflection coefficient of the antenna for different electron doping of BP is presented in figure 12.

Fig. 12(b) illustrates the tunability of the bandwidth Δf_r and the resonance frequency f_r with changing of the electron doping of BP. It is found that there is a small change in peak position, when n increases from 1 to 4 × 1013cm^{-2}, the resonance peak undergoes from 15.486 to 15.75 THz, and the intensity of the peak is gradually increased until 20.102 dB; and Δf_r is varied from 0.152 to 0.445 THz. It is exciting to note that the tunability characteristic is decreased in comparison with that of the change in chemical potential of graphene. It should be noted here that only the electron doping of BP and the chemical potential of graphene are changed, and the other parameters remain unchanged.

CONCLUSION

In conclusion, we have theoretically investigated anisotropic plasmonic responses of non-reciprocal series-fed microstrip patch antenna array based on graphene and black phosphorus for THz applications using the iterative method. The results suggest that by putting the graphene on the BP, the bandwidth and the return loss has been enhanced. Furthermore, by varying the electron doping of BP, the tunability characteristic is decreased in comparison with that of the change in chemical potential of graphene. The wave concept iterative process method has been numerically validated, and excellent agreement has been obtained between our numerical results and CST software. The proposed algorithm not only gives precision in the result, but also reduces the time consumption and memory size. The proposed strategy of hybridization of graphene and BP could be used to build 2D anisotropic plasmonic devices with enhanced functionalities at THz band, with important applications in imaging and communications.

REFERENCES

Crassee, I., Levallois, J., Walter, A. L., Ostler, M., Bostwick, A., Rotenberg, E., Seyller, T., Marel, D. d., & Kuzmenko, A. B. (2011). Two-dimensional material nano photonics. *Nature Physics*, *1*(7), 48–51.

Estep, N. A., Sounas, D. L., & Alu, A. (2016). Magnetless Microwave Circulators Based on Spatiotemporally Modulated Rings of Coupled Resonators. *IEEE Transactions on Microwave Theory and Techniques*, *1*(64), 502–518.

Feizi, M., Nayyeri, V., & Ramahi, O. M. (2014). Modeling Magnetized Graphene in the Finite-Difference Time-Domain Method using an Anisotropic Surface Boundary Condition. *IEEE Transactions on Antennas and Propagation*, *1*(66), 233–241.

Geim, A. K., & Ram Subramaniam, A. (2007). Two-dimensional material nano photonics. *Nature Materials*, *1*(3), 183–191.

Hlali, A., Houaneb, Z., & Zairi, H. (2018). Tunable filter based on hybrid metal graphene structures over an ultrawide terahertz band using an improved Wave Concept Iterative Process method. *International Journal for Light and Electron Optics*, *1*(181), 423–431.

Hlali, A., Houaneb, Z., & Zairi, H. (2020). Modeling of Magnetically Biased Graphene Coupler at Terahertz Frequency Through an Improved Anisotropic WCIP Method. *IEEE Transactions on Magnetics*, *1*(56), 1–9.

Hlali, A., Houaneb, Z., & Zairi, H. (2020). Terahertz microstrip circulator based on graphene. *Journal of Electromagnetic Waves and Applications*, *1*(34), 1–6.

Liu, H., Neal, A. T., Zhu, Z., Luo, Z., Xu, X., Tománek, D., & Ye, P. D. (2014). Phosphorene: An unexplored 2D semiconductor with a high hole mobility. *ACS Nano*, *1*(4), 4033–4041.

Low, T., Rodin, A. S., Carvalho, A., Jiang, Y., Wang, H., Xia, F., & Neto, A. H. C. (2014). Tunable optical properties of multilayers black phosphorus thin films. *Physical Review*, *1*(7), 075434.

Serrano, D. C., Diaz, J. S. G., Alù, A., & Melcón, A. Á. (2015). Electrically and magnetically biased graphene-based cylindrical waveguides: Analysis and applications as reconfigurable antennas. *IEEE Transactions on Terahertz Science and Technology*, *1*(5), 951–960.

Song, X., Liu, Z., Xiang, Y., & Aydin, K. (2018). Two-dimensional material nano photonics. *Optics Express*, *1*(26), 5469–5477.

Xia, F., Wang, H., Xiao, D., Dubey, M., & Ram Subramaniam, A. (2014). Two-dimensional material nano photonics. *Nature Photonics*, *1*(8), 89–907.

Chapter 7
Development of Solar-Powered Microcontroller-Relay-Based Control System Omnidirectional Wheelchair

Patrick Uche Okafor
Enugu State University of Science and Technology, Nigeria

Ndidi Stella Arinze
Enugu State University of Science and Technology, Nigeria

Osondu Ignatius Onah
Enugu State University of Science and Technology, Nigeria

Ebenezer Nnajiofo Ogbodo
Enugu State University of Science and Technology, Nigeria

ABSTRACT

A solar-powered omnidirectional wheelchair is implemented for physically challenged persons. The framework was mounted on the wheels that were connected with two direct current (DC) motors. The ratings of the battery and solar module were determined using system voltage (12V). A 7,805-voltage regulator was used to supply 5VDC to the AT89352 microcontroller. The microcontroller was programmed to provide a reference signal to the motor. The motor provides the needed torque to drive the wheels through interconnected relays. The relays are energized by the microcontroller and omnidirectional movement achieved through relays connected with microprocessor and micro switches, eliminating the need for joysticks and complex control mechanisms. System performance test result showed that the auxiliary solar power supply of the wheelchair increased the travel range by approximately 86% compared with that of a wheelchair powered by battery alone.

DOI: 10.4018/978-1-7998-6992-4.ch007

1. INTRODUCTION

The part of the society with disabilities (difficulties in walking) is excluded basically in all the day to day activities of the society. Apparently one can say that there is non-inclusive participation of the physically challenged in society mostly in developing countries like Nigeria. These categories of the population are more likely to experience adverse socioeconomic outcomes such as less education, poorer health outcomes, lower levels of employment, and higher poverty rates as stated by Worldbank.org 2020, not to mention disfranchising them from exercising their civic rights or being part of the congregations in various religion. This barrier must be bridged so that equality in all ramifications will exist among the people. One of the ways of bridging this gap is through the development of an affordable assistive technology that can help disabled persons to move around without external help and hindrances.

Wheelchairs are assistive technology for enhancing the mobility of disabled persons (Sharma et al, 2017). According to Worldbank.org 2020, one billion people representing 15% of the world's population, in one way or the other, experience some form of disability, and disability prevalence is higher for developing countries. Also, as the COVID-19 pandemic continues to have wide-reaching impacts across the globe, it is very vital to note how these people living with a disability are uniquely impacted by the pandemic, including health, education, and transport considerations. The current situation globally is the total and partial lockdowns imposed by different countries because the COVID-19 effectively reduced or totally stopped the public transport systems which the disabled persons rely on, for the accessibility of necessities or critical medical appointments.

In Nigeria and other developing countries, people with disabilities go through difficulties in integrating themselves into daily routines. Persons without access to a wheelchair are compelled to stay indoors, while crawling is their only means of moving around when they cannot get help. Another statistic reported by Physio-pedia.com (2020), showed that about 10% of the global population, representing about 650 million people, lives with disabilities and out of these, about 10% required a wheelchair. But Woods and Watson (2013), in a research report stated that only 2% of people with disabilities have access to a wheelchair and the reason could be as a result of economic situation of the people. This has become a global big challenge that needs urgent attention.

Reviews show that the primary challenge of the elderly, disabled, and those with accident-related defects is the dependence on external help towards movement. Their movements are often limited by those entrusted with taking care of them. Non-inclusive participation in society is another ugly menace posed to those living with disabilities especially in developing countries like Nigeria. On the other hand, existing wheelchairs have many limitations and could not serve to a greater extent as an assistive technology to this group of people. Some of the limitations include stress and pains in pushing the hand-rim, inability in moving to long distant locations, much needed extra human effort, high cost of foreign-made wheelchairs, etc. Though some of the existing wheelchairs are powered by solar, their maneuverability is achieved using complex control systems such as joysticks. Joysticks are known for their prone to calibration and freezing problems. Thus, there is need for a control system that can be developed using locally available materials for easy maintenance, usage and affordability.

2. THEORY OF WORK

The present work relates to electric-powered mobility devices in form of a wheelchair, providing assistive technology to enhance the mobility of physically challenged persons, providing omnidirectional movement particularly through relays connected with a microprocessor and micro switches, eliminating the need for joysticks and complex control mechanisms. Wheelchairs are assistive technology for enhancing the mobility of disabled persons (Sharma et al, 2017). There are conventional and smart wheelchairs in existence but, they have many limitations and are foreign-made that are quite expensive in developing countries. The conventional wheelchairs have bulkiness, extra-human support, enormous energy input, etc. as limitations. The smart ones are electrically powered but, one of the problems facing them is limitations in travel ranges because their source of energy cannot be replenished. On the other hand, some of the smart ones that can replenish their energy through solar and batteries can only navigate in two directions (forward or backward), which is a great limitation. Others that can navigate in all directions (omnidirectional) are designed to use joysticks as their navigation control.

Many researchers have proposed different ways of coming up with more appropriate wheelchairs to assist the needy. Wafa and Yusra (2020), in their work, used a decentralized algorithm to control the motion of omnidirectional wheelchairs. They incorporated three separated independent components of motion that can be controlled separately in the body frame of the omnidirectional wheeled wheelchair, which includes rotational motion, horizontal motion, and vertical motion. The authors also deployed two approaches of artificial intelligent-based controllers (PD-Fuzzy-P and GA-PID controllers) in their design to optimally enhance the manoeuvrability of wheelchairs. From the work, it is obvious that the authors considered a complex control system for having integrated the three separate controllers. Such a system cannot be affordable and easily maintained in a developing world.

Pacha, Srinivasan and Gnana, (2020) developed an android application for controlling the movement of wheelchairs using voice commands. Their design work showed that sensors are fixed in the wheelchair to detect and avoid obstacles when the chair is on the move but, the question is how far is the wheelchair able to travel without a solar system perhaps; travel distance is a limitation in their design.

Jagdish, Poonam, and Kedar (2019), designed and fabricated a lever-operated wheelchair for a disabled person with no legs. The wheelchair was designed in such a way that it requires less effort by an operator to operate, the researchers believed that the wheelchair is economically affordable as compared to other advanced wheelchair but the limitation is that it cannot achieve omnidirectionality in navigation. Budiharto and Gunawan (2018), in their work, used a voice recognition module to control the wheelchair to move forward, left, right, and stop, and additionally, used a joystick module to achieve the reverse mode. The researchers applied a speaker-dependent module to the wheelchair control systems by training the voice recognition module which is based on Arduino. The complexity of the system is not recommendable for users in developing countries.

Also, Ananda, Oishee, Prasanna, and Subhasis (2017), in their work developed 4 wheeled omnidirectional wheelchairs and their control using a myoelectric user intention interface. The developed system is driven by a holonomic drive system, it was obvious that their design achieved greater maneuverability compared to conventional powered wheelchairs but, it is expensive and not easy to maintain. Gurjot, Paramdeep, and Sandeep (2016) proposed a model that helps physically challenged people to make their journey through an electronic assisted module that works on signal processing over speech and image. The developed system works via speech or through recognition of hand gestures. Loss of signal due to real-time inputs is the main gap of the model but, was overcome using different signal enhancement techniques to achieve a high rate of accuracy and stability. Though the wheelchair has good stability, it is complex and expensive for people living in developing countries. Chi-Sheng, Tung-Yung, and Tze-Yuan (2014), in their study, utilized a quality function deployment method to develop a wheelchair with a user-selectable manual/electric propulsion mode and an auxiliary solar power supply system. The researchers stated that the auxiliary solar power supply of the wheelchair increased the travel range by approximately 26% compared with that of a wheelchair powered by battery alone but, the system did not achieve omnidirectional in its navigational control.

In essence, the existing solar-powered omnidirectional wheelchairs achieve their navigational control or maneuverability using joysticks. Joysticks are an expensive programmable device that is prone to calibration and freezing problems. The calibration problem arises when the system will continue in a particular direction and unable to change its course because the joystick could not find any of its coordinates. The freezing occurs in a situation where the joystick is cannot activate any of its inputs, in most cases, these problems lead to an accident. This calibration and freezing problems the joysticks has created a gap that needs to be bridged. That is the more reason this work is proposing a solar-powered omnidirectional wheelchair using micro-switches that are locally available in solving the calibration and freezing problems associated with joysticks. The micro-switches are cheap, not software-based, easy to maintain, and are readily and locally available. Summarily, some of the gaps left by conventional wheelchairs include bulkiness, low speed, limited travel ranges, single navigational direction, Calibration and freezing of navigational controllers, etc.

3. MATERIALS AND METHODS

The system units are related as shown in Figure1. The prototype of the invention was constructed using an existing wheelchair. From Fig 1, the 12V 50watts Solar Panel tracks the sun rays and then converts them to electrical energy for the recharging of the battery. The solar panel was sized using the system voltage rating of 12V DC, 50W. The current produced by the panel is 4.12A. The solar panel was connected to the 12V 20A Charge Controller that monitors the charging of the battery to make sure that the battery is not overcharged.

The charge controller is connected to the battery that supplies the needed system power. The battery specification is 12V, 26Ah. The energy produced by the battery is equal to 312Whr or 0.312Kwhr. Optimal charging of the system is considered using the depth of discharge (DOD) of 80%. This implies charging up to 312Whr ×80% is 249Whr. The energy required from the solar panel is 0.249KWhr. The time it takes the battery to be fully charged by the solar panel is given by:

$$\text{Time} = \frac{\text{Energy of the battery}}{\text{Power Rating of the solar panel}} \tag{1}$$

Figure 1. Brief Description of the System Components as They Relate to Each Order

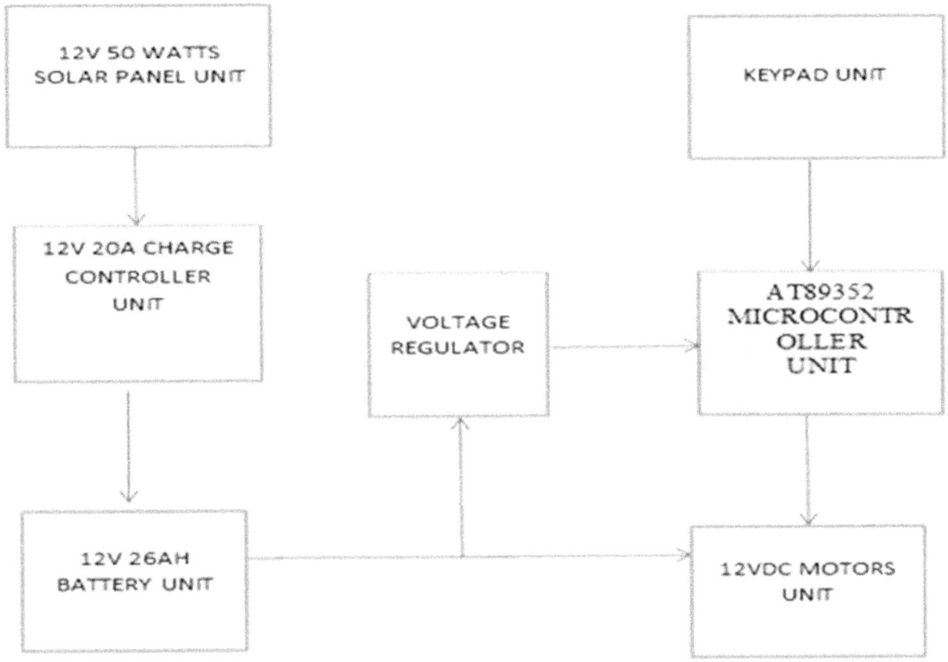

The time is computed to be 5hrs. The rated speed of the motor is 55rpm. The diameter of the back wheel is 0.38m and the perimeter of the wheel 1.194m thus, 1 rpm in km/hr = 0.072km/hr. Two DC motors were used. The motors are an electromechanical machine that provides the needed high torque that makes the wheelchair to move through the rotation of the wheels. The DC motor used was based on the system input voltage which is 12V. A 12V Operating voltage, 55 rpm speed, 2A free moving current, 10A stall torque, and 120W DC motor specification was used. The motor rotates at speed of 55rpm and the speed in km/hr. was calculated to be equal to 3.96km/hr. The speed was approximated to 4km/hr., which is 1.1m/s. It was assumed that the motor moves at 4km, with 120m steepness on the road then, the gradient was calculated to be equal to 3%. The system power (Watts) is equal to:

P = total weight * g * speed * gradient. (2)

The total weight of the wheelchair is a summation of the wheelchair frame (39kg), weight of solar panel (8kg), weight of the two DC motors (6kg), and the weight of the battery (34kg). The total weight of the wheelchair is (87kg). The motor power rating for the weight of 60kg (load), where the load is a human being, is the system's total weight of 147kg. The system power (Watts) is equal to the multiplication of total weight * g * speed * gradient. System Power (watt) equals 47.6 watts ≈55 watts (making room for tolerance). 110 watts is the approximated value for the two motors. Thus the wheelchair is rated 12volt; 110 watts with a Load Current of 9Amp. The two DC motors draw a current of 4.5A each. The battery lifespan is protected by applying a DOD of 80% to produce 20.8Ah (Amp-hr. charge). The brain of the wheelchair is the AT89352 microcontroller driven by instruction codes written in assembly language.

From Figure 2, the positive and negative terminals of the solar panel were connected to the input pins 1 and 2 of the Charge Controller. The positive and negative terminals of the battery were connected to the output pins 4 and 3 of the charge controller to deliver regulated 5v to the microcontroller through the 7805 voltage regulator. The voltage regulator receives 12V from the battery through its input (VI) and sends out a regulated output of 5V through its output pin (Vo) to the input pin 40 (Vcc) of the AT89352 microcontroller. The 5V is what powers the microcontroller.

Figure 2. Illustrates the omnidirectional wheelchair circuitry which controls the system operation.

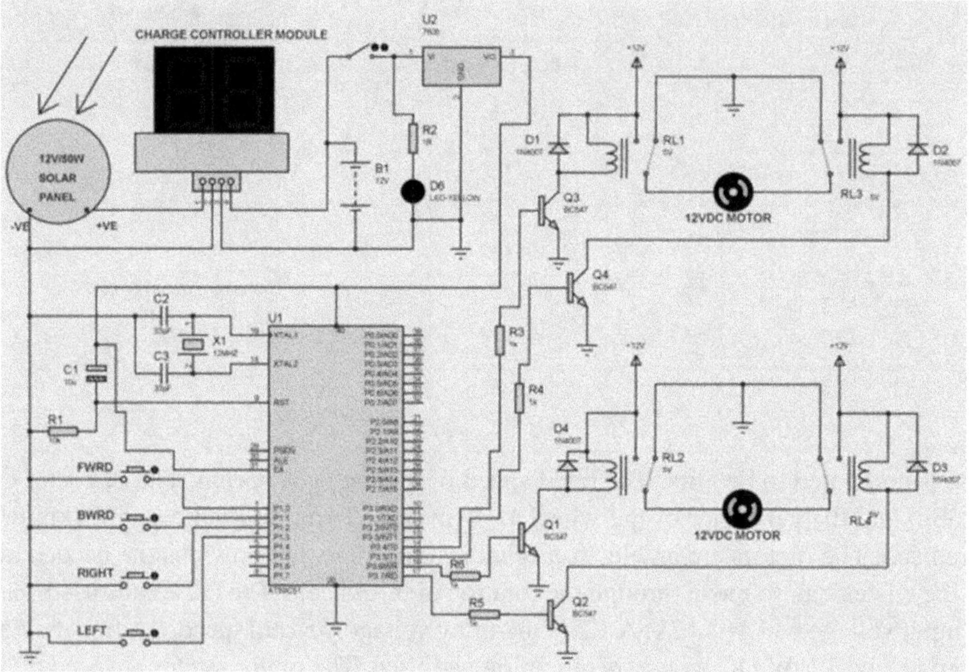

The AT89352 has four parts or thirty-two bits that can be assigned as inputs and outputs. It is the brain of the wheelchair, which receives the input voltage signal from the input (micro switches) and according to the reference input command, sends a corresponding output signal through the assigned port thereby initiating action in the actuator which is the DC motors (Okafor, Arinze, and Onah, 2020). The light-emitting diode positive terminal was connected to the positive terminals of R2 and battery, its negative terminal was connected to the negative terminals of R2 and the voltage regulator to indicate when the system is powered ON. Between the battery and the regulator exists an open circuit that can be controlled using a switch. From the datasheet, the LED 7 sinks 25.5mA, therefore, the value of the resistor can be calculated which is equal to 470 Ohms using Ohm's law. The four resistors: R3, R4, R5, and R6 of equal values were connected between the output pins 14, 15, 16 1and 17 of the microcontroller and the four BC547 transistors: Q1, Q2, Q3, and Q4.

The four resistors are the base resistors of the BC547 transistor that allows the needed current for the transistor biasing. From the datasheet, using the transistor as a switch means that it can be operated in the saturation and cut-off region with biasing current and voltage of IB = 5mA, and VBE=5V respectively.

The relationship between the base resistor *RB,* biasing current *IB* and base voltage *VBE* is yielded *RB* to be equal to 1kΩ. Transistors Q1 and Q2 were connected to the lower motor through relays RL2 and RL4. Both transistors were connected to the relays through 1N4007 PN junction diodes D3 and D4, for voltage rectification.

In likewise manner, transistors Q3 and Q4 were connected to the upper motor, through relays RL1 and RL3. Also, both transistors were connected to the relays through 1N4007 PN junction diodes D1and D2. The diodes are also for rectification. The relays are what switch the DC motor ON. The pushbuttons: FWRD, BWRD, RIGHT, and LEFT are the control input of the system. It is made up of four micro-switches which are used in achieving the navigation of the wheelchair in all directions with the switching of the relays. There is a half-duplex communication existing between buttons and the microcontroller unit. When the switch button is pressed, it sends a voltage signal to the microcontroller that instructs the microcontroller on which motor to energize through the relays.

Command inputs sent from the controller unit at the push of the buttons to drive the motor are received by the motor through the output pins 14, 15, 16 1and 17 of the microcontroller. The pushbuttons are connected to the input pins 1, 2, 3, and 4 of the microcontroller through R1. The resistor is a pull-down resistor which ensures that the input voltage to the switches is always in a known state. It holds the logic signal near zero volts when no other active device is connected. The value was taken to be 10k considering the voltage coming down from the power rail. Capacitors C2 and C3 are ceramic capacitors that fine-tune clock frequency and ground the electrical noise around the clock.

The parallel connection of the capacitors with the crystal oscillator X1 forms the crystal oscillator that provides a constant frequency output to the XTAL1 and XTAL2 input pins of the microcontroller under varying load conditions. The crystal stabilizes the operating frequency of the microcontroller and the value is taken as 12MHz according to the AT89S52 datasheet. Values of the capacitors were calculated to be 30 pF each using a supply voltage of 5V and a crystal value of 12MHz. It should be noted that during implementation, 33pf was used as a perfect substitute because of the unavailability of the 30pF. Capacitor C1 is the Polarized electrolytic capacitor that holds the reset power of the microcontroller; it also filters and smoothens the DC signal. The value was taken to be 10μF based on the datasheet and previous works recommendations. The capacitor was connected to the negative terminals of the pushbuttons and commonly grounded with the system.

The Omni-directionality of the wheelchair was simply achieved when the user through the micro switches sent a reference input to the microcontroller. The relays through the microcontroller receive the input signal and activate the DC motor following the directional specification made by the user. For example, to move the wheelchair in a forward or backward direction, the pushbuttons FWRD and BWRD are pressed and the corresponding voltage signal is received by the microcontroller. The microcontroller through its output pins energizes the four replays that are normally close to open. Torque is produced by the two motors in the forward or reverse direction to initiate movement. To move the wheelchair in the left or right direction, the RIGHT and LEFT buttons are activated as the case may be. The same process for forward and reverse directions **are** initiated but, in this case, it is either that the replays are opened and relays closed for the left direction or that relays are open while relays are closed for the right direction. With the opening and closing of the relays, movement in 360° direction can be achieved both in a clockwise and anticlockwise direction.

4. TEST AND RESULT

The Speed of the Wheelchair was determined using the rated DC motor speed and the diameter of the back wheel as follows:

Speed of the motor is 55rpm (rated), Diameter (D) of the back wheel = 0.38m and Perimeter of a circle $= \pi D$; 1.194m

$$\text{Then, } \frac{perimeter\ of\ the\ circle * 60}{1000} = \text{km/hr.} \quad (3)$$

$$\text{The speed in km/hr.} = 55rpm * \frac{1.194 * 60}{1000} = 1.1\text{m/s}$$

A photograph of the prototype is presented in Figure 3. The test was carried out in two phases using different loads (humans). In phase one, the wheelchair was test-run for hours using only the battery without charging it through the solar panel, and the maximum range it can travel before the battery dissipates to its 20% capacity was determined. In the second phase, the solar panel was used to recharge the battery and the maximum range it covered before the battery also dissipates to its 20% capacity was determined. The test results are tabulated in Tables 1 and 2. Table 1 shows the effect of recharging the wheelchair system using the solar panel.

From the table, it could be observed that the distance covered by the wheelchair when recharging through the solar panel is farther compared to when the wheelchair is running only on the battery. Results showed that the auxiliary solar power supply of the wheelchair increased the travel range by approximately 86% compared with that of a wheelchair powered by battery alone.

Table 2 shows the relationship between the different loads (humans), power rating, maximum current is drawn, travel range, and the total time of travel.

From the table, it can be deduced that at 50kg and a speed of 1.1m/s, a maximum current of 8.3A was drawn as the battery recharges through the solar panel at 4A/hr. Thus, as the load is increasing, the maximum current is drawn increases while the travel range covered decreases with a decrease in time.

Figure 3. Photograph of the Wheelchair Prototype

Table 1. Travel Ranges Covered

Parameter	Current drawn from the battery (Amp)	Time of test (Km)	Maximum travel range (Km)
Running with battery only	9	2.3	2.1
Running with both solar and battery	4.8	4.3	3.9

Table 2. Effect of different loads on current, travel range, and time of travel

Load (Kg)	Power Rating (W)	Max. Current Drawn (A)	Travel Range (Km)	Total Time of Travel (Hrs.)
50	100	8.3	4.8	5
60	110	9	3.9	4.3
70	116	9.7	3	3.7
80	122	10.2	2.7	3.4

5. CONCLUSION

The work utilized an omnidirectional control system equipped with micro-switches, microcontroller and relays to control the movement of the mobility device. The prototype system was achieved following the design specification. The dimensions of the framework were determined using an existing comfortable wheelchair and hence the skeletal system fabricated using locally available mild steel. The ratings of the battery, motor, and solar panel were determined using a system voltage of 12V. The device control system has micro switches as an input device for sending a signal to the microcontroller. The relays, through the microcontroller, receive the input signal and energize the relays. The relays activate the DC motor following the directional specification made by the user. System performance test carried out using different loads showed that the auxiliary solar power supply of the wheelchair increased the travel range by approximately 86% compared with that of a wheelchair powered by battery alone and as the load is increasing, the maximum current is drawn increases while the travel range covered decreases with decrease in time.

REFERENCES

Ananda, S. K., Oishee, M., Prasanna, K. L., & Subhasis, B. (2017). Omnidirectional Assistive Wheelchair: Design and Control with Isometric Myoelectric Based Intention Classification. *Procedia Computer Science*, *105*, 68–74. doi:10.1016/j.procs.2017.01.200

Budiharto, W., & Gunawan, A.A.S. (2018). *Design and algorithm for electric wheelchair using voice recognition and joystick*. . doi:10.24507/icicel.12.11.1171

Chi-Sheng, C., Tung-Yung, H., & Tze-Yuan, L. (2014). Design and development of solar power-assisted manual/electric wheelchair. JRRD, 51(9), 1411–1426.

Gurjot S. G., Paramdeep S. Sandeep K. A. (2016). *Electronic Wheelchair for Physically Disabled Persons*. . doi:10.17485/ijst/2016/v9i21/89753

Jagdish, P., C., Poonam, G. C., Kedar, M. K. (2019). *Design and Fabrication of Lever Operated Wheelchair for Disabled Person with No Legs*. doi:10.17148/IARJSET.2019.611

Okafor, P.U., Arinze, S.N., & Onah, O. I. (2020). Artificial Neural Network Inverse Model (AIM) Based Position Control Of DC Motor. *International Journal of Engineering, Science and Mathematics*, *9*(2).

Pacha, S., Srinivasan, S., & Gnana, S. (2020). Mobile controlled automated wheel chair for disabilities. *International Journal of Innovative Technology and Exploring Engineering, 9*(1S).

Physio-pedia.com. (2020). *Role of the Wheelchair*. Available Online: https://www.physio-pedia.com/Role_of_the_Wheelchair

Sharma, J., Anbarasu, M., Chandan, C., & Shanmugasundaram, M. (2017). *Iris movement based wheel chair control using raspberry pi.* . doi:10.1088/1757-899X/263/5/052049

Wafa, B., & Yusra, A. (2020). Decentralized Motion Control for Omnidirectional Wheelchair Tracking Error Elimination Using PD-Fuzzy-P and GA-PID Controllers. *Sensors (Basel), 3525*(12), 3525. Advance online publication. doi:10.339020123525

Woods, B., & Watson, N. (2013). The social and technological history of wheelchairs. *International Journal of Therapy and Rehabilitation, 11*(9), 407–410. doi:10.12968/ijtr.2004.11.9.19587

Worldbank.org. (2020). *Disability Inclusion*. Available online: https://www.worldbank.org/en/topic/disability

Chapter 8
Digital Terrestrial Television Broadcasting in 5G

Chandrakala K.
Noorul Islam Center For Higher Education, India

Jaya Kumari J.
Mar Baselios College of Engineering and Technology, India

Sreeja T. K.
Noorul Islam Center for Higher Education, India

ABSTRACT

Digital multimedia broadcasting is a technology for transmitting multimedia signals to devices like mobile phones, PC, laptops, etc. This allows everyone to watch TV on mobile devices. With the fifth-generation network, one can download a movie in a few seconds. Mobile TV offers round-the-clock participation in communication and broadcasting around the globe. This has become possible by watching a live TV program on handheld devices, as it facilitates real-time updates. Digital terrestrial television broadcasting aimed at transmitting two types of information: global and local content. The global content aims at broadcasting over the entire network, and the local content is location specific. Broadcasting the local content using multiple transmitters at different frequencies in multifrequency network requires additional spectrum. Global content can be transmitted through single frequency network. Therefore, if both global and local information are transmitted in a single frequency network (SFN), the system will become highly spectrally efficient and cost-effective.

1. INTRODUCTION

Broadcasting through a digital network can be done with a Multifrequency Network (MFN) or a Single Frequency Network (SFN). The analog broadcasting of the terrestrial television signal uses a different transmitter with different frequencies to avoid interference. At the receiver, the signal received from one transmitter interferes with the signal from another transmitter. This is caused due to reflection and

DOI: 10.4018/978-1-7998-6992-4.ch008

scattering of electromagnetic waves or due to various obstacles. If the two transmitters broadcast on the same channel, the receiving antenna receives the signal from both the transmitter simultaneously. Practically there is a time delay between the receptions of one signal to another signal. This phase shift may be constructive or destructive interference, which results in fading. Due to this, there is a ghost effect on the TV screen (Fisher, 2008). Since analog broadcasting uses amplitude modulation for video transmission, there is no way to avoid interference other than using a different frequency for different transmitters. The same frequency channel used in one area can be reused in another area separated by large distances to avoid co-channel interference. Such an arrangement is called MFN.

Figure 1. Multifrequency and Single frequency Network coverage

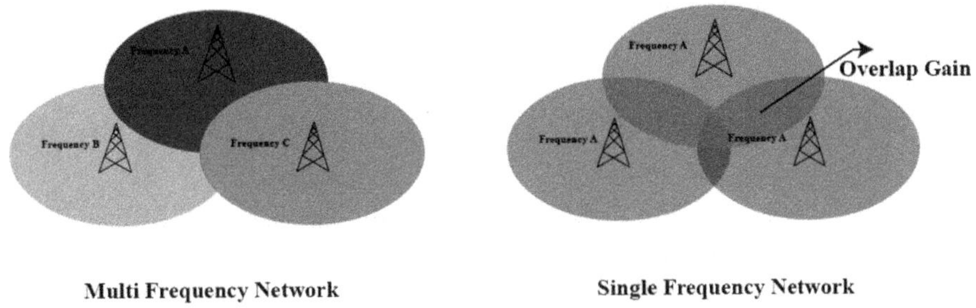

In the meantime, a Single-frequency network uses transmitters operating on the same frequency transmitting the same programs as illustrated in Figure 1. In other words, the broadcasting in the entire country is done with some frequency. A Single-frequency network provides wide coverage and network gain. Digital terrestrial television networks are designed to achieve better coverage in the confined topographical areas. The dimension of the coverage area can be as large as the entire country, state, or small region. This is the service area of the broadcast network. Digital standards like DVB-T, DVB-T2, Terrestrial Digital Audio Broadcasting (T-DAB), and DVB-NGH effectively utilize single frequency network benefits. The single frequency covers the entire service area, which improves the usage of the spectrum effectively which results in spectral efficiency (Dai *et al.,* 2012).

Within a given geographical area, it is desirable to integrate several smaller SFNs into a wide area network. i.e. Multiple SFN, to meet requirements of regional and local programs. Each SFN gets the benefits of overlap gain. It provides a homogeneous field strength distribution throughout its coverage area and the multi SFN gives the possibility for program diversity. In a multi SFN, the total coverage area is divided into several smaller areas, each served by a different frequency block. It increases the adaptability in the network design. Network gain or Overlap gain is the constructive interference of the signals from the transmitter, which improves the signal strength in a particular region. The network coverage can also be gracefully improved by adding more transmitters without the need for another planning of frequency. This is an added advantage. The constraints for SFN are to provide better time synchronization, coherency in frequency, and to behave identical and synchronized content over the entire transmitter. SFN offers many advantages over MFN in spectral efficiencies, quality coverage, and flexibility but still, there is a trade-off between its geographic size of SFN and its robustness, cost, and capacity. For the large size SFN, the cost of setup and the required capacity is high (Mattsson, 2005).

A Single-frequency network allows the Orthogonal Frequency Division Multiplexing (OFDM) technology due to its resistance to multipath fading. Most digital terrestrial television broadcasting uses multicarrier modulation like OFDM. The robustness is further improved by introducing the Forward Error Correction (FEC) to generate the Coded OFDM signal. OFDM takes a serial data stream of low data rate and modulate with a large number of carriers (ETSI, 2004). The number of carriers varies depends on the standard which ranges from 1K,2K,4K,8K, and so on. The carriers are spaced uniformly as the inverse of symbol duration. At the receiver, the carriers are orthogonal to each other, so that the influence of one carrier on the other carrier must be zero on demodulation. To avoid Inter Symbol Interference (ISI), a guard interval is introduced which copies a portion of the active symbol and appends at the front of the previous symbol.

2.LOCAL AND GLOBAL CONTENT IN DIGITAL TERRESTRIAL BROADCASTING

Digital terrestrial television broadcasting aimed at transmitting two types of information that is global and local content. The global content aims at broadcasting over the entire network and the local content is location-specific. Broadcasting of Global content is generally done using a single frequency Network. Broadcasting the local content using multiple transmitters i.e., using MFN with different frequencies requires additional spectrum. Therefore, if both the information is transmitted in a Single Frequency Network (SFN), the system will become highly spectrally efficient and cost-effective. To achieve this, Wavelet-based Hierarchical modulation Technique and selected subcarrier method is proposed in this chapter. These techniques are discussed in the following sections of this article. The section 3 explains about the Hierarchical modulation technique followed by its simulated results and discussion. Section 4 gives a detailed description on selected subcarrier method of broadcasting and its simulated results.

The conventional Fourier-based DVB system uses a cyclic prefix to overcome intersymbol interference, which consumes nearly 25% of the bandwidth. In this work, the wavelet OFDM based hierarchical modulation technique is proposed which replaces the conventional Fourier based OFDM with wavelet-based OFDM. The application of wavelet transform in communication systems is receiving attention in recent researches. The articles by Sharma and Singh (2014), Abdullah and Hussain (2007) analyze the performance of various wavelets like daubechies, symlet over the AWGN channel. This work proves that the wavelet OFDM is more bandwidth efficient than conventional OFDM and is less affected by the Doppler shift. The analysis has been done with diverse modulation schemes like BPSK, QPSK, 8-PSK, QAM, 8QAM, and 16QAM. The same analysis has been done on the multipath fading channel by Kumbasar and Kucur (2012), Lee and Ryu (2018). These articles also highlight the significance of wavelet OFDM over the conventional OFDM system. Recently, wavelet transform is applied in various research problems. Dibal *et al.,* 2018 have proposed the application of wavelets in cognitive radio for spectrum sensing. The main significance of the wavelet based OFDM system is that it does not use cyclic prefixes as in Fourier based system. Even without cyclic prefix wavelet, based OFDM provides good throughput in the UWB system as explained by Dilmaghani and Ghavami (2008). This work uses the Daubechies wavelet. Biag *et al.,* 2018 and cenikliglu *et al.,* 2018 discuss the role of a discrete wavelet transform in the 5G communication system. Wavelet OFDM also finds its significance in the MIMO system as discussed by Hasan M.M, (2012). This work proves that the wavelet OFDM based system overcomes the FFT based system in both AWGN and Fading channel environment.

3. HIERARCHICAL MODULATION

Hierarchical modulation finds its application in a digital video broadcasting system and wireless multimedia services. Jiang and Wilford (2005) propose that hierarchical modulation upgrades the existing digital broadcast system by adding the secondary constellation along with the basic information. The secondary constellation carries the additional information to be transmitted in the broadcast system. This system is backward compatible. The original system transmits only the basic information but the upgraded system carries both basic and secondary constellation. DVB standards like DVB S2 and DVB-SH also adopt Hierarchical modulation to improve spectral efficiency. Méric *et al.,* (2011) and Méric*et al.,* (2012) undergo research on satellite broadcast systems. O'Leary, (1997) states that time-sharing systems accompanied by hierarchical modulation improve spectral efficiency. Hierarchical modulation is a technique in which the two different signals are multiplexed together into a single complex signal. This multiplexing of the signals is done as a layered constellation. The base layer of the constellation consists of basic information that represents the quadrant of the constellation. The additional or secondary layer represents the enhancement layer that determines the location of the signal within the quadrant. The base layer stream always undergoes QPSK modulation and the modulation scheme of the secondary layer information is chosen based on the desired order of hierarchically modulated signal. Both the modulated signals are added to get the resultant Hierarchical modulated signal whose order of modulation depends on the modulation scheme preferred for the Secondary layer signal. At the receiver, the signal gives a different bit error rate performance for base and secondary layer information. The user with good signal strength will be able to receive both base layers and secondary layer signals perfectly. The transceiver model of the DVB system is shown in Figure (2). In the DVB system, the concatenated coding with BCH and LDPC coder is implemented. The forward error correction coder BCH forms the outer coder that is followed by the inner LDPC coder. An interleaver and modulator that construct the Bit Interleaved Coded Modulation (BICM) system follow the encoder. A mapper based on the order of the hierarchical modulated system modulates the channel-coded information.

In digital terrestrial television broadcasting, the base layer consists of information that is to be broadcasted to the entire network namely the Global content. The secondary layer comprises the information which is aimed at a specific region known as the local content. The mapping of binary information to the complex signal of a global stream is always done with a QPSK mapper since it represents the quadrant in which the resulting complex signal falls. This is done with two bits of four combinations with a phase shift of $(2i-1)\frac{\pi}{4}, i = 1,2,3,4$. The modulated complex global signal is given in equation (1)

$$G = \left|g\right| \angle e^{\varphi_g} \tag{1}$$

where, $\left|g\right|$ is the magnitude of the modulated complex global signal, φ_g represents the phase of the complex global signal.

Figure 2. System Model of Hierarchical Modulated DVB system

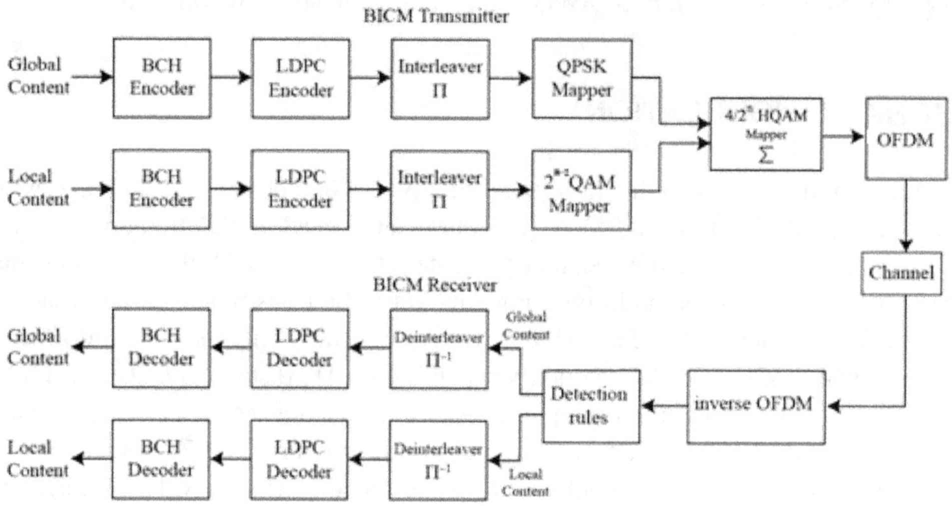

In accordance with the required order of resultant hierarchical modulated signal, the choice on the QAM mapper for local content is made. The complex QAM modulated local signal with 2^{n-2} message points frames a modulated signal that consists of 2^n message points by adding with QPSK modulated global signal. The modulated local signal is given in equation (2)

$$L = \left|l\right| \angle e^{\varphi_l} \tag{2}$$

where, $\left|l\right|$ is the magnitude of the modulated complex local signal, φ_l represents the phase of the complex QPSK signal. The resultant QPSK modulated complex of the global and local signal is summed up to obtain the hierarchically modulated signal as in equation (3)

$$X_{HQAM} = G + L \tag{3}$$

The resulting constellation of the hierarchically modulated signal for the 16HQAM and 64HQAM systems is given in Figures (3).

The first two bits represent the global bit, which is the same for all signal points in the quadrant. It varies with quadrant. The next two bits in 16HQAM and four bits in 64HQAM are the local bits, which differ within the quadrant. The Hierarchical modulated signal is applied to the OFDM block before transmitting into the channel. At the receiver, inverse OFDM has been applied.To detect the local and global contents different decision rules are used.

Figure 3. Constellation diagram of 64HQAM and 16HQAM signal respectively.

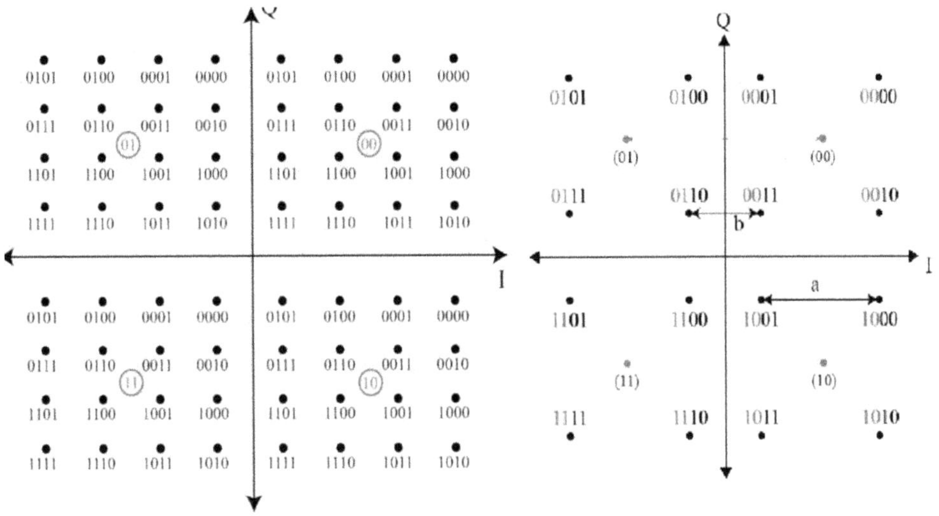

Figure 4. BER of global content in 16HQAM and 64HQAM system for 7/15 coder

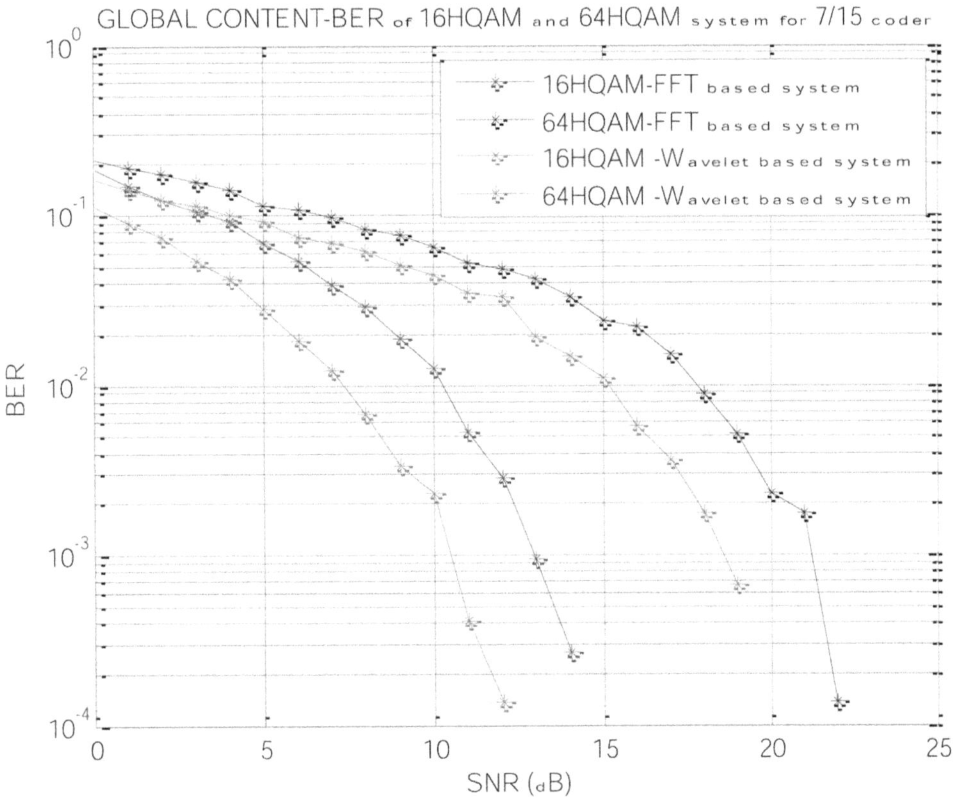

3.1 Simulated Results and Discussion:

Simulation of 16HQAM and 64HQAM systems with Fourier OFDM is done with BCH as the outer coder and LDPC as the inner coder. In the case of 16HQAM, the error rate reduces to 10^{-4} at 13dB for the local stream and 18dB for the global stream. With an increase in signal-to-noise ratio, the error rate reduces for both global and local streams. At 13dB, the error rate of the wavelet based 16HQAM system is 37% less than the Fourier based system for the global stream as shown in Figure 4 and 5. In the meantime, the error rate of 64HQAM is 27% greater than the 16HQAM system. It is clear from the figure that wavelet OFDM based DVB system shows an improved error performance for the global and local streams.

Figure 5. BER of Local content in 16HQAM and 64HQAM system for 7/15 coder

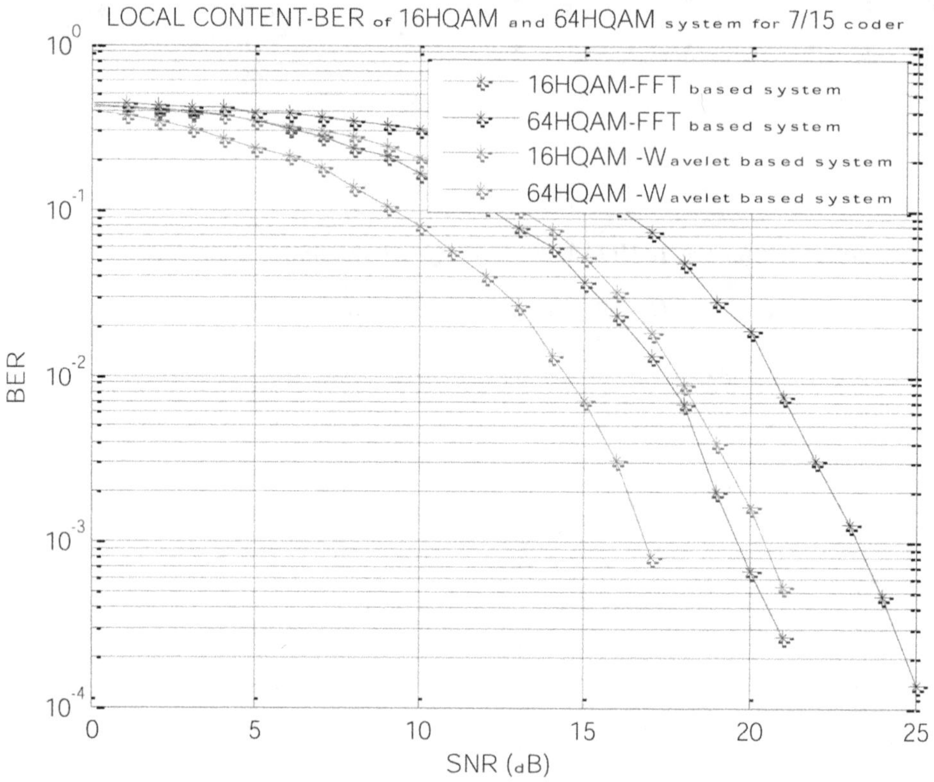

The wavelets used for multi-carrier modulation must be orthogonal like the carriers derived from FFT. Orthogonal wavelets like Daubechies, Symlet, and Coiflet are used in a wireless communication application. In this chapter, simulation using various wavelets has been done in Figure 6 and 7 for both global and local Information. All three wavelets Daubechies, Symlet, and Coiflet give almost the same performance for both information contents.

Figure 6. Performance Comparison of various wavelets for global content

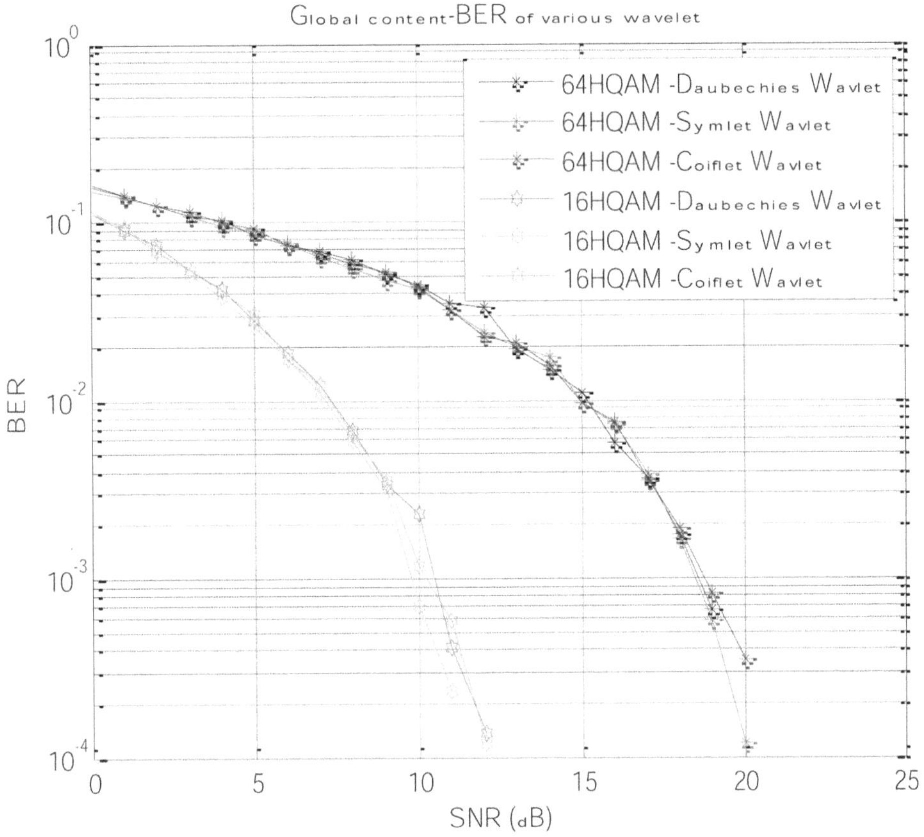

Simulation on Hierarchical Parameters (a and b **)**: Figure 8 shows the constellation of 16HQAM signal depicting the hierarchical parameters.

The performance of the system is highly affected by the selection of hierarchical parameters 'a', 'b' and alpha α. The separation between the message points in adjacent quadrant be 'b' and the distance between the message points within the quadrant be 'a'. The variation of the parameters 'a' and 'b' has a considerable impact on the bit error rate performance. The analysis is done based on the selection of these parameters with two cases on local and global content streams by varying 'a' and 'b' ranging from 1 through 4.The simulation is done for the 16HQAM signal in AWGN channel. Table 1 gives the simulation parameter for analysis.'

Case 1: Varying 'a' and Constant 'b': Figures 9 to 24 show the performance of the system for 'b' as constant and varying 'a'. The distance of the message points between the nearest quadrants is kept constant and the spacing between message points within the quadrant is varied. This condition is analyzed separately for both local and global contents.

Performance of Local Content: In case if b=1, to achieve an error rate of 10^{-4}, it requires SNR of minimum 20dB for local content in case of Fourier based system and 17dB in the case of wavelet based system which is much greater than the Fourier based system as given in Figure 9 and Figure 10. This happens because when 'b' reduces, the spacing between the message points between the quadrants reduces. This increases the interference. In case if 'a=b=1', the error rate remains constant after 15dB as

Figure 7. Performance Comparison of various wavelets for local content

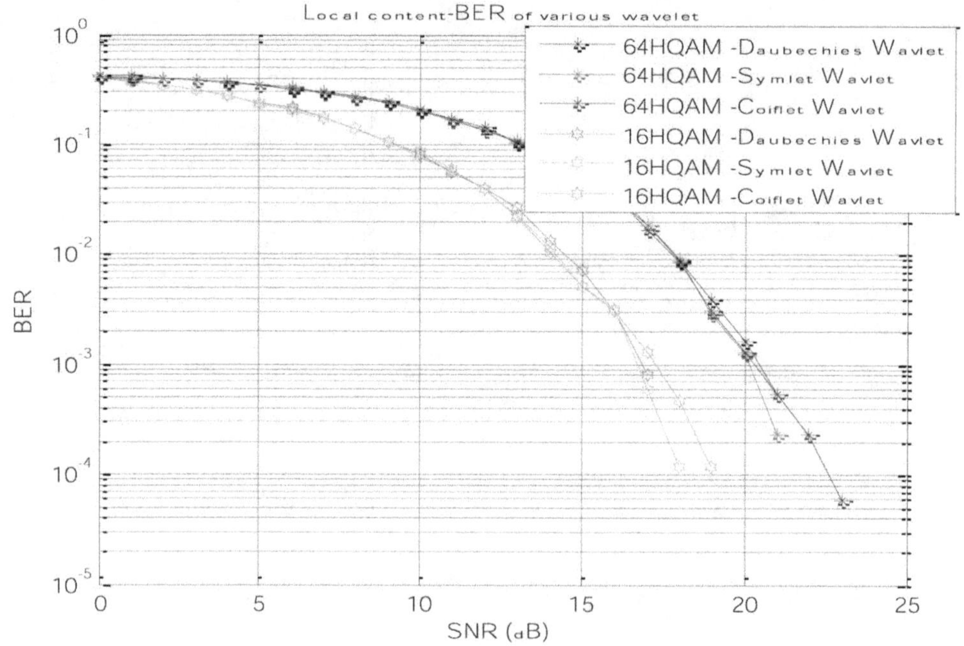

Figure 8. Constellation describing the Hierarchical parameters

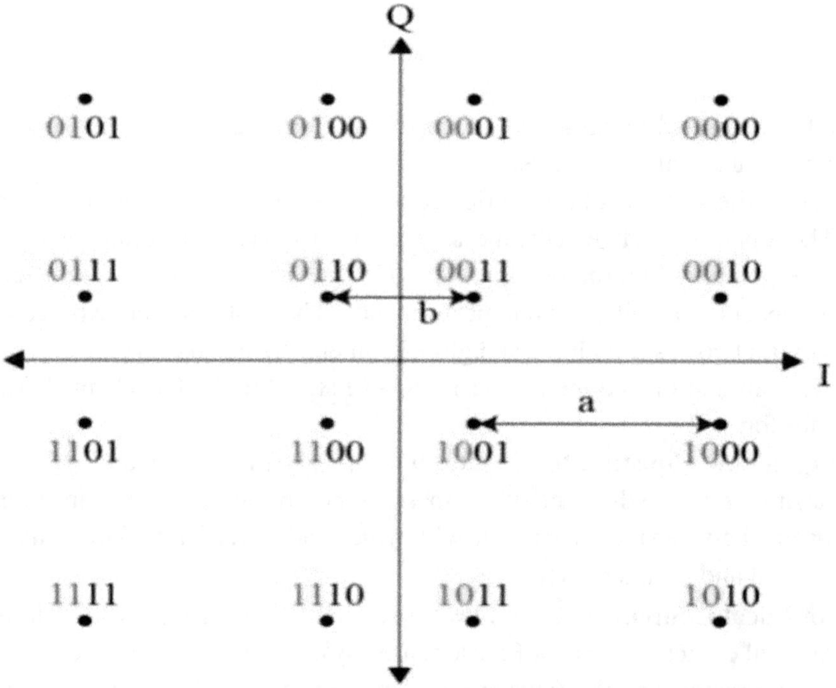

Table 1. Simulation Parameters for Analysis on Hierarchical parameters

Parameters	Specification
Code rate	7/15
Channel	AWGN
Outer/Inner coder	BCH/LDPC
Transform size	2K

10^{-1}. Thus the condition for different values of 'a' at 'b=1' comparatively suffers a high bit error rate and is not suggested.

When the value of b is 2 and 3 as in Figure 11, Figure 12, Figure 13 and Figure 14, the local content shows a non-sequential variation in performance for each case of 'b' with every increase in the value of 'a'. The minimum requirement of signal to noise ratio is nearly 17dB and it increases to the maximum of nearly 23dB in case of Fourier based system and it is 15dB and 20dB for wavelet based system respectively. The better results occur at a condition when 'a and 'b' have same value i.e., (a=3, b=3) and (a=2, b=2).

Figure 9. BER of local content for varying a and $b=1$ with wavelet based OFDM

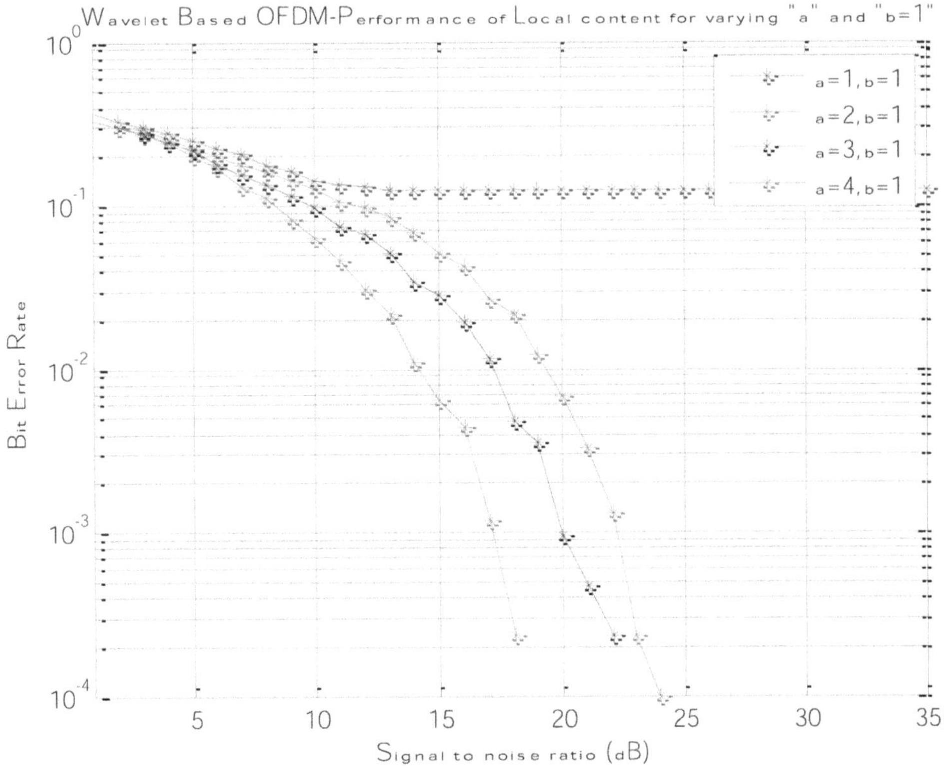

Figure 10. BER of local content for varying and with Fourier based OFDM

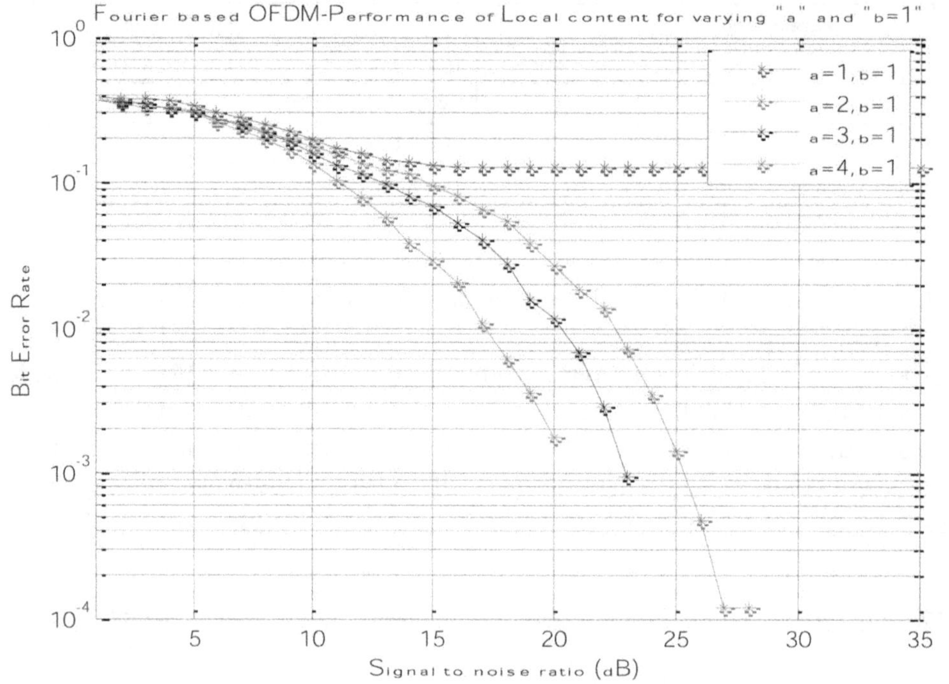

When b=4, Figure 15 and 16 show that as 'a' increases the performance of the system improves for both wavelet based and Fourier based system. This means if the spacing between messages points within the quadrant widens, the bit error rate reduces. This is because the discrimination between the message points increases due to an increase in the maximum likelihood distance. For every increase in the value of 'a ', there is nearly a drop-off in the requirement of signal to noise ratio by 2dB. To achieve 10^{-4} an error rate, the wavelet-based system requires 15dB SNR for $'a = 4'$ but in the case of Fourier based system; it requires 17dB SNR to reach an error rate of 10^{-3}.

Performance of Global Content: It is explicitly shown in the figures (17-24) that as 'a' increases the system performance for global content reduces. This means if the spacing between message points within the quadrant widens the SNR requirement increases. The value of bit error rate touches nearly to 10^{-4}. For 'a=1', to achieve an error rate of 10^{-4},requires 12dB SNR in case of Fourier based system and is 10dB for wavelet based system; and there is a requirement of 2dB SNR for every increase in the value of 'a' for any constant 'b'.

Case 2: Varying 'b' and Constant 'a'

Figures 25 to 40 show the performance of the system for 'a' as constant and varying 'b'. The distance of the message points between the nearest quadrants is varied and the spacing between message points within the quadrant is kept constant. This condition is analyzed separately for both local and global contents.

Performance of Local Content: In case if a=1 as shown in Figure 25 and Figure 26, to attain the error rate of 10^{-4}, it requires SNR of minimum 22dB for local content in case of Fourier based system

Figure 11. BER of local content for varying and with wavelet based OFDM

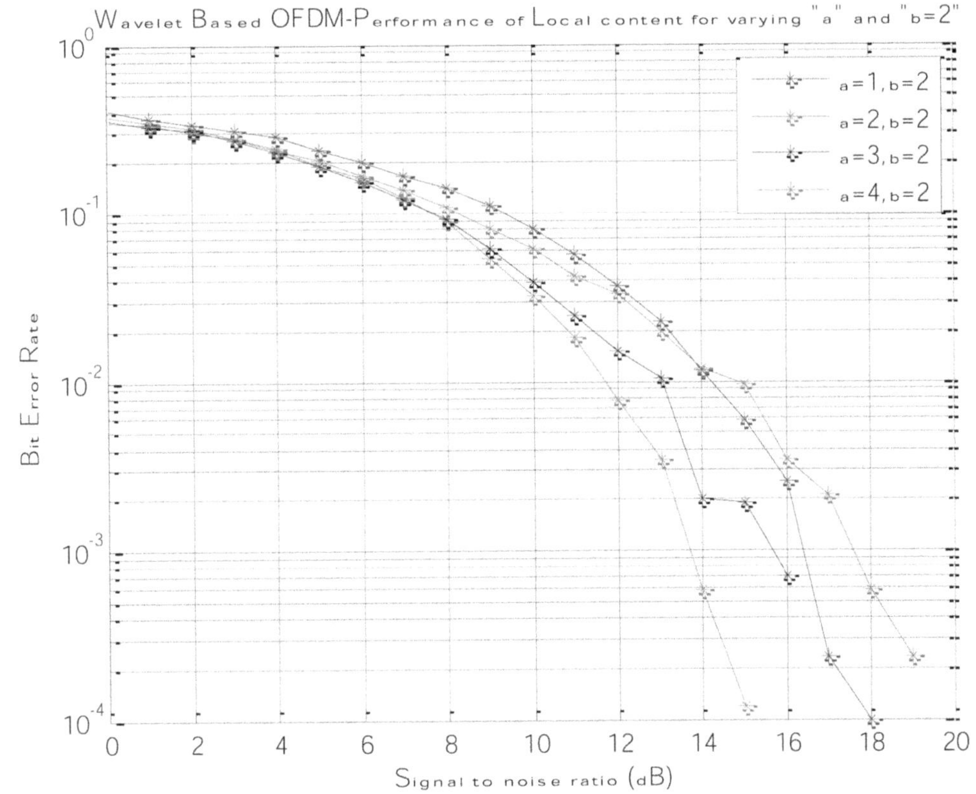

based and 17dB in case of wavelet based system which is very much greater compared to any other cases. In case if 'a=b=1', the error rate remains constant after 10dB for wavelet based system and 15dB for Fourier based system as 10^{-1}. Thus the condition for different values of 'b' at 'a=1' comparatively suffers high bit error rate and is not suggested.

From figure 27, Figure 28, Figure 29 and Figure 30, it is understood that when the value of 'a' is 2 and 3 the local content shows a non-sequential variation in performance for each case of 'a' with every increase in value of 'b'. The better results occurs at condition when 'a' and 'b' have same value i.e., (a=3, b=3) and (a=2, b=2).This means the constellation is uniform. Figure 27 and Figure 28 show that the SNR requirement to achieve the error rate of 10^{-4} is 15dB for wavelet based system and 17 dB for Fourier based system at a=b=2. Figure 29 and Figure 30 shows that that the SNR requirement to achieve the error rate of 10^{-4} is 15dB for wavelet based system and 17 dB for Fourier based system at a=b=3. The SNR requirement for the proposed system is 2db less than the existing Fourier based system.

When a=4, Figure 31 and Figure 32 show that as 'b' increases the performance of the system improves. This means if the spacing between messages points between the quadrants lengthens, the bit error rate reduces. This is because the discrimination between the message points increases due to increase in the maximum likelihood distance. For every increases in value of 'b 'there is nearly a drop off in the requirement of signal to noise ratio by 2dB.

Figure 12. BER of local content for varying and with Fourier based OFDM

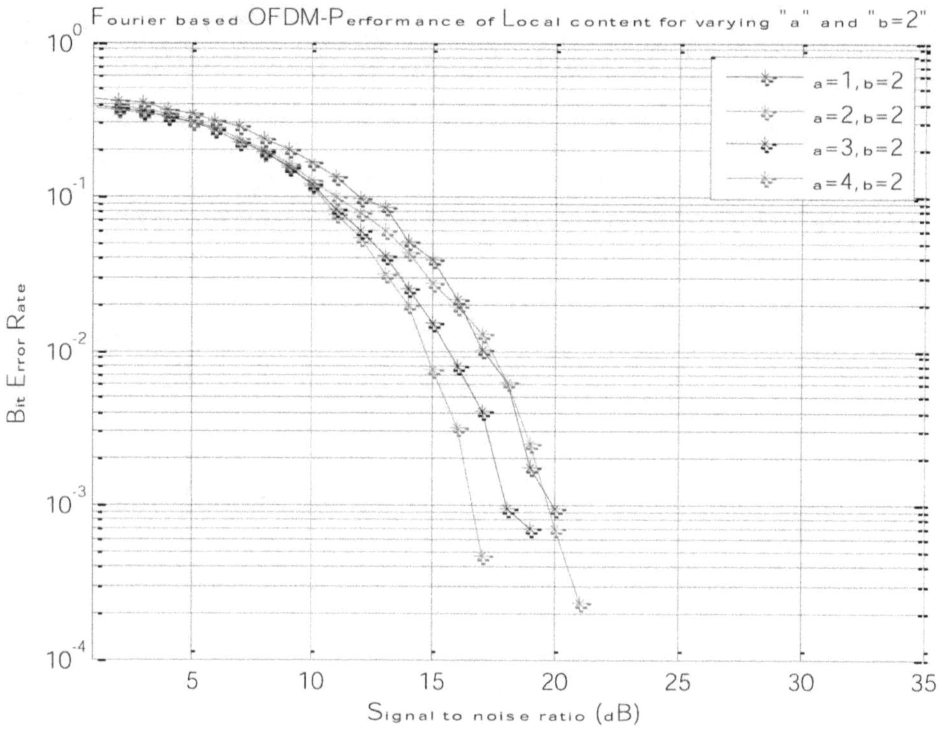

Performance of Global Content: It is presented in the figure (33-40) that as 'b' increases the performance of the system improves for all values of 'a'. This means, if the spacing between message points and the quadrants widen, the bit error rate reduces. The value of the bit error rate reaches nearly 10^{-4}. For 'a=1', the error rate is 10^{-4} at 13dB for Fourier based system and 9dB in case of wavelet based system, there is a requirement of 3dB SNR for every increase in the value of 'b' for any constant 'a'. For 'a=2, 3, 4', the minimum SNR requirement is greater than the previous case of 'a=1' to achieve the error rate of nearly 10^{-4}.

Analysis on Hierarchical Parameter Alpha: The parameter α denotes the ratio of the distance between the message point of adjacent quadrants and the spacing between the message point within quadrants .i.e., $\alpha = \frac{b}{a}$. If Hierarchical modulation is optioned in any DVB system to transmit two different signals by the single constellation in base and secondary layer, the α value has to be allocated. DVB allows α =1,2 and 4 in case of 16HQAM system and $\alpha = 1,3$ for 64HQAM system. If α =1, the constellation mapping of message point is uniform and it is nonuniform for α =2,4.The sensitivity to noise of both the global and local stream is affected by the hierarchical parameter α .

The bit error rate analysis for various values of alpha for local and global content is shown in figures 41,42 . In case if α=1, both the streams show equal susceptibility to noise as understood from the figures 41,42.Numerically the above statement can be vindicated by the observation which shows an SNR requirement of 17dB to achieve the error rate of 10^{-4} in case of fourier based system. This condition is analyzed in four cases with a and b as 1,2,3 and 4. For all cases except a=b=1,the error performance for both global and local stream is same. The exceptional condition shows a constant error rate of 10^{-2} for every increase in signal to noise ratio.

With the greater value of α, the global content is more robust in noise since the distance between the quadrant increases. This happens with the additional requirement of average transmission power. It is observed, from the figure 41and42 that there is not great variation in SNR for a global content stream but there is a notable increase in SNR requirement of the local content stream.

Figure 13. BER of local content for varying and with wavelet based OFDM

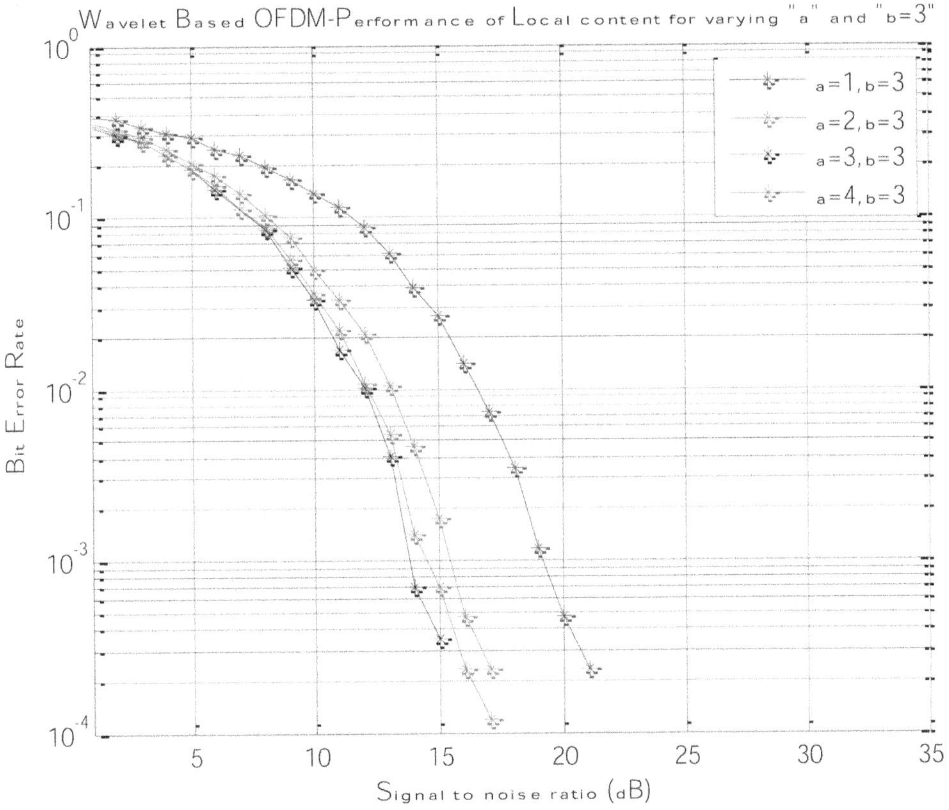

4. SELECTED SUBCARRIER METHOD FOR DIGITAL VIDEO BROADCASTING

Selected subcarrier method of Digital Video Broadcasting also focuses on transmitting both global and local content in a single frequency network. This technique is well suited in a scenario where the transmission of two different information is targeted over the entire globe at the same time. This methodology proficiently handles the concept of Orthogonality principles among the OFDM subcarrier. The advantage of selected subcarrier technique is that it provides local service across the entire network with same coverage as the global services. Therefore, this technique is suitable for transmission of local news or advertisement as the temporal window in the global services.

Figure 14. BER of local content for varying and with Fourier based OFDM

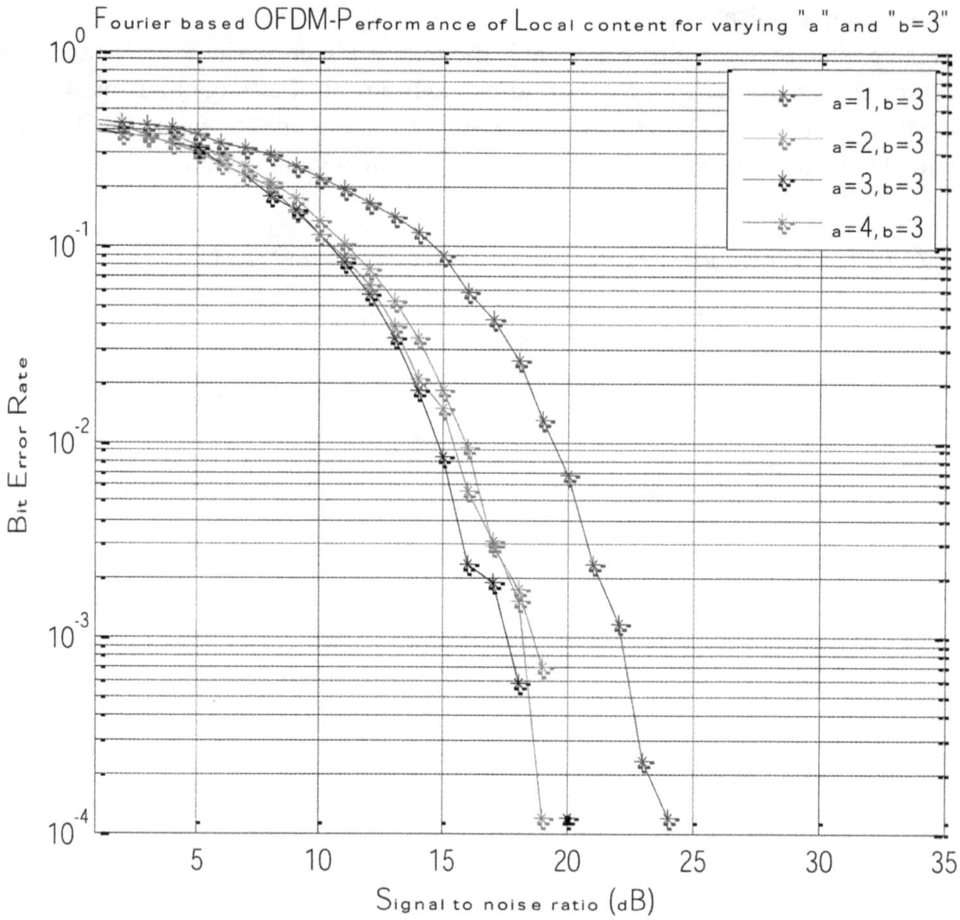

Figure 43 shows the coverage of transmitters for global and local service in selected subcarrier method of digital video broadcasting. The global information content is transmitted to the entire network. In the local service area, the information differs based on the requirement. This technique can be illustrated by an example. Say the global information is the one, which transmits to the entire state. The local information may be the advertisement, flash news or weather forecast to the individual districts. This information differs from one district to another district.

Sub Carrier Allocation Schemes: In this method, a set of dedicated OFDM subcarriers is reserved for local service transmission. Within the same OFDM symbol, different local service area transmitters use different sets of subcarriers. The same subcarrier used by other transmitter remains unused. The orthogonality of the subcarrier ensures no interference between the adjacent transmitters. With spatial reuse pattern, the Local service areas (LSA) separated by a long distance can reuse the same carrier. The local service has same coverage, as global service. The insertion of local service does not affect the global service coverage.

Figure 15. BER of local content for varying a and $b = 4$ with wavelet based OFDM

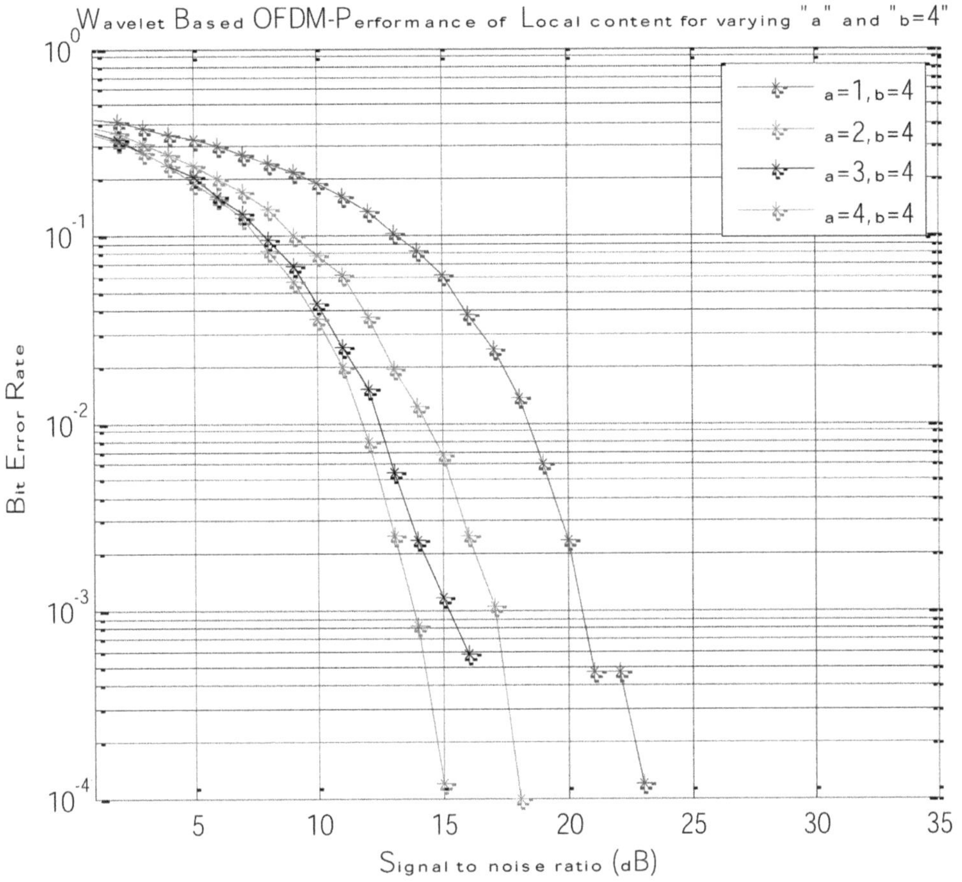

The above Figure 44 shows the illustration of assignment of subcarrier of local service. Each local transmitter allocates a specific set of OFDM sub-carriers to its local services. The subcarrier in one OFDM symbol used to transmit local service in one region may remain unused in other OFDM symbol of another region. This prevents the interference between local services

The block diagram representation of the selected subcarrier method of broadcasting is shown in figure 45 .The local content and global contents are separately encoded and modulated before it is subjected to multicarrier modulation .As the point of multicarrier modulation, where the carrier allocation for local and global content is done.

Simulation of Proposed System with FFT Based and Wavelet-Based OFDM: The simulation of the performance of the proposed system with FFT based and Wavelet-based OFDM is shown in figure (46) and (47). It is noted from the figure that the wavelet-based system provides good performance compared to FFT based system.

Case 1: The simulation result when the global and local contents are QPSK modulated is shown in Figure 46. To achieve an error rate of10^{-4}, the SNR requirement of FFT based system is 3dB greater than the wavelet-based system. The BER of both the streams shows the same variation, since the local content and global content are transmitted with same energy level.

Figure 16. BER of local content for varying and with Fourier based OFDM

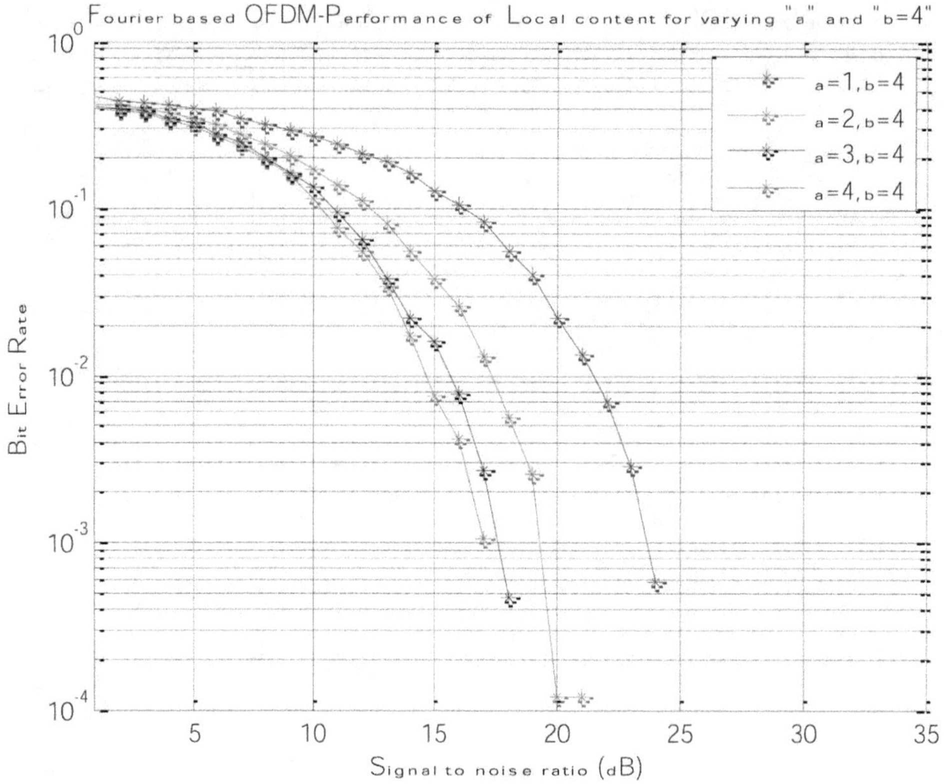

Figure 17. BER of Global content for varying a and b = 1 with wavelet based OFDM

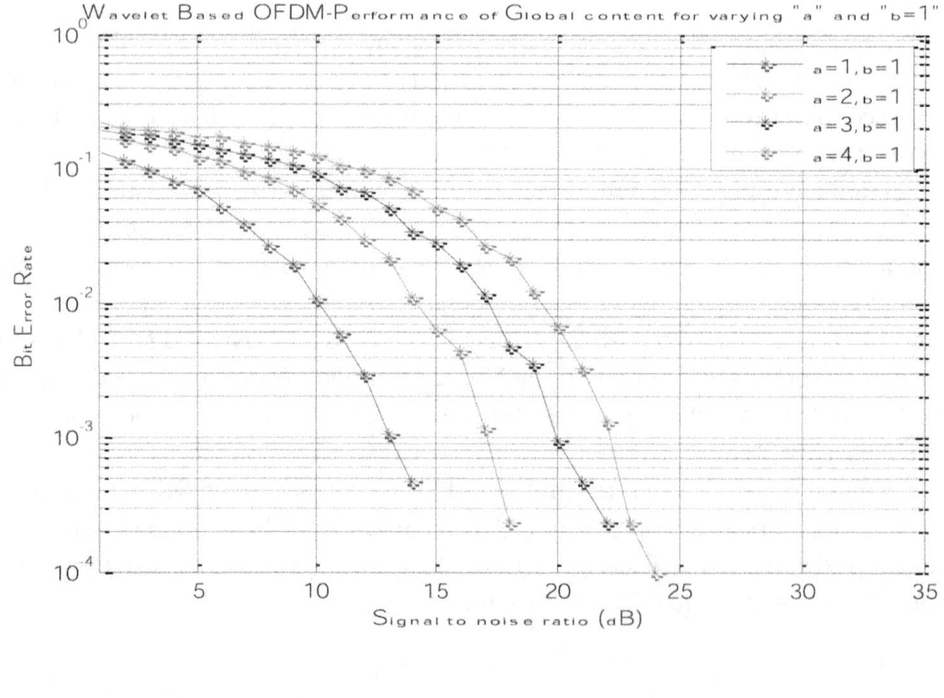

Figure 18. BER of Global content for varying and with Fourier based OFDM

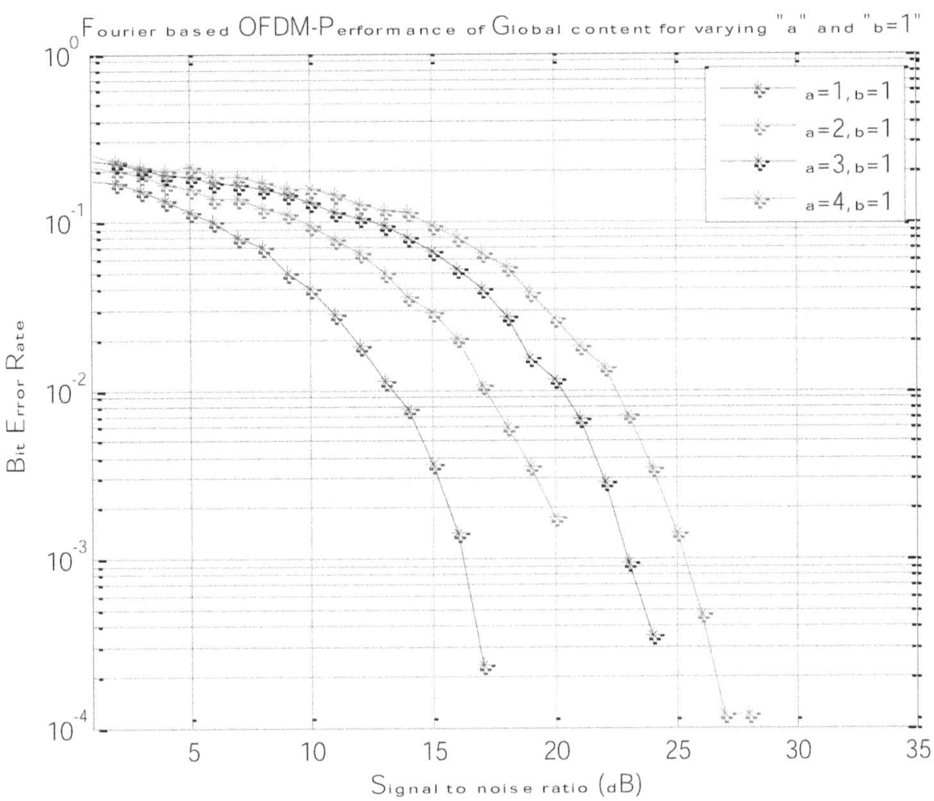

Case II: The simulation is extended to another case where the local content is QAM modulated and the global content is QPSK modulated. The Performance shown in figure (47) illustrates that the local content shows a higher error rate than the global content. This is because when modulation order increases, the data rate increases with sacrifice in error rate. The situation can be reversed by choosing the QAM for global content and QPSK for local content. The behavior of the system depends on the modulation scheme and power of the signal. Even with case II, the wavelet OFDM outperforms the FFT based system. To achieve 10^{-4} error rate, the local content of the FFT based system requires 13dB SNR and 11dB for wavelet-based system. There is a difference in SNR of 2dB. On the other hand, the global content of FFT based system requires 18dB SNR and for the wavelet-based system, it is 3dB less than that.

SUMMARY

This article aims to propose a technique to transmit both the local and global information through a Single Frequency Network by embedding both the information using a hierarchical modulation technique to contribute better spectrum efficiency. In this, the global information is always QPSK modulated and the local information is QAM modulated. Both modulated signals are added to get a new constellation where the global and local signals are transmitted. This chapter also proposes an algorithm namely the selected subcarrier method. Moreover, it also focuses on transmitting both global and local content in a Single Frequency Network. This technique is well suited in a scenario where the transmission of two different information is targeted over the entire network at the same time. In this, the orthogonality of the OFDM subcarrier is well utilized. Here the information is broken into small packets and modulated with orthogonal carriers. In this method, a set of dedicated OFDM subcarriers is reserved for local service transmission among the total subcarriers. Within the same OFDM symbol, different local service area transmitters use different sets of subcarriers. The same subcarriers used by other transmitters remain unused. The orthogonality of the subcarrier ensures no interference between the adjacent transmitters. With spatial reuse pattern, the Local service areas separated by a long distance can reuse the same carrier. The local service has the same coverage, as global service. Thus, both information does not interfere with each other. The insertion of local service does not affect global service coverage. Therefore, this technique is suitable for transmission of local news or advertisement as the temporal window in the global services. This research work includes the Digital Video Broadcasting Next-generation Handheld (DVBNGH) standard, which is the extension of DVB-T2. This standard upholds the globe towards 5G by dealing with requirements like flexibility, robustness, and spectral efficiency.

Figure 19. BER of global content for varying a and $b = 2$ with wavelet based OFDM

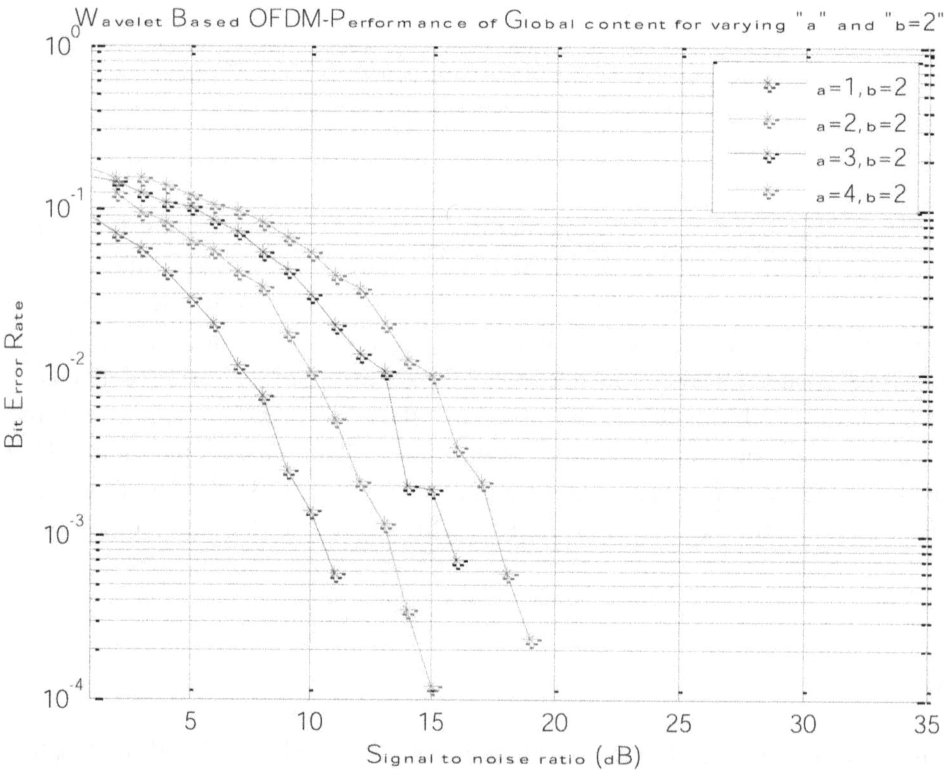

Figure 20. BER of global content for varying and with Fourier based OFDM

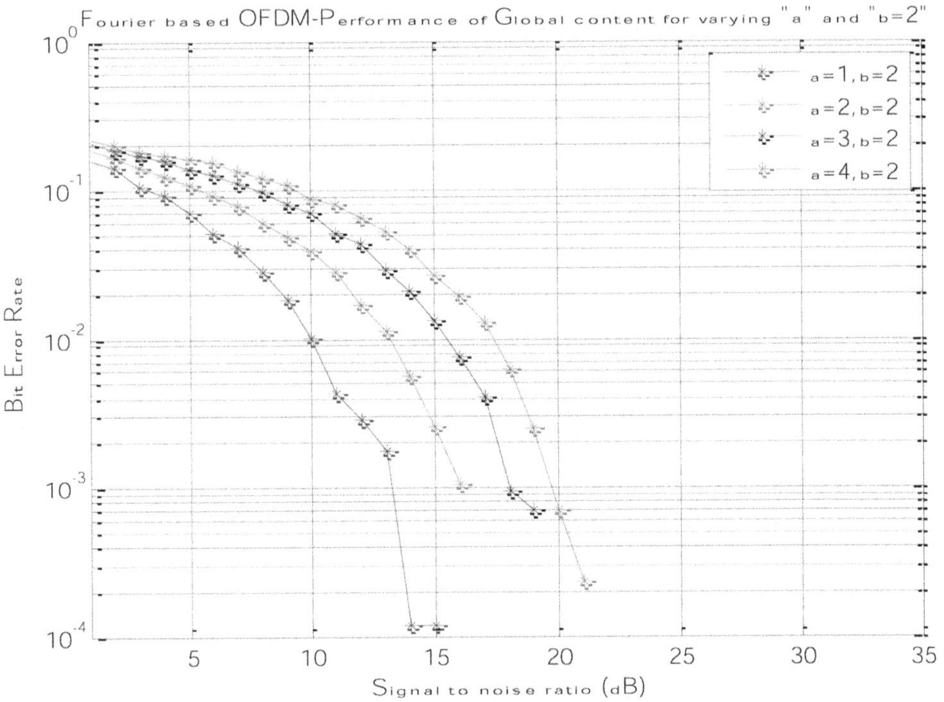

Figure 21. BER of global content for varying and with wavelet based OFDM

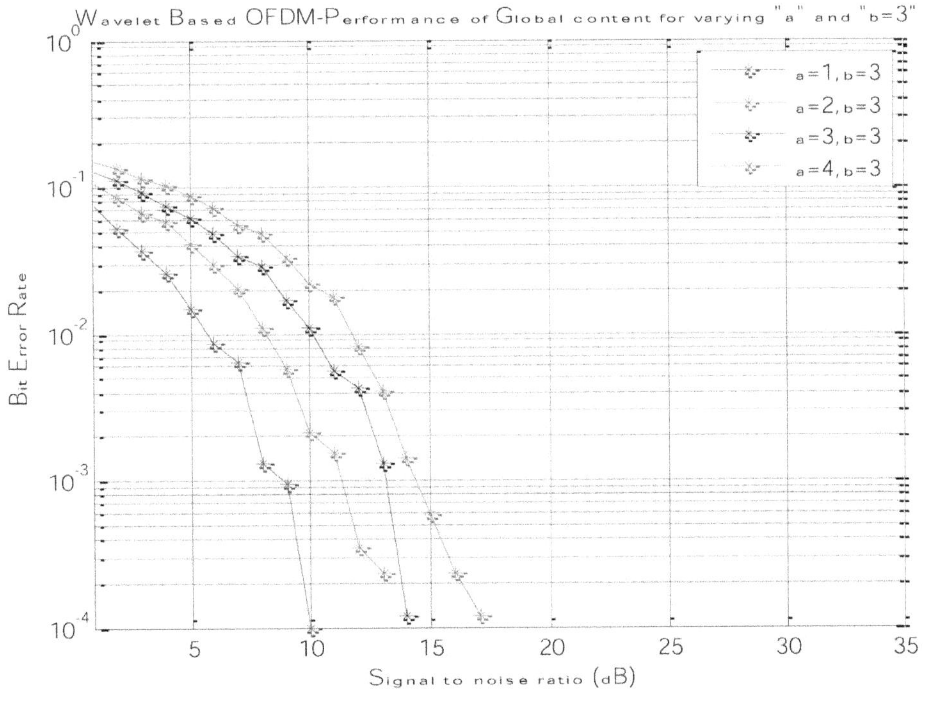

Figure 22 .BER of global content for varying and with Fourier based OFDM

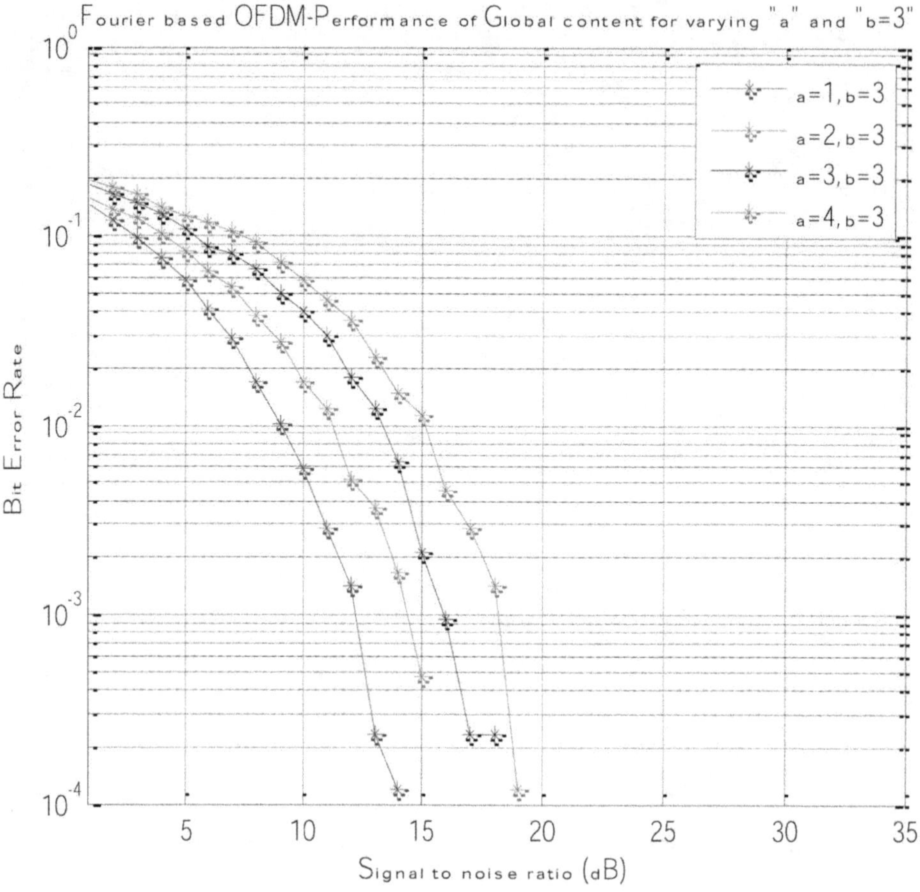

Figure 23. BER of global content for varying a and $b = 4$ with wavelet based OFDM

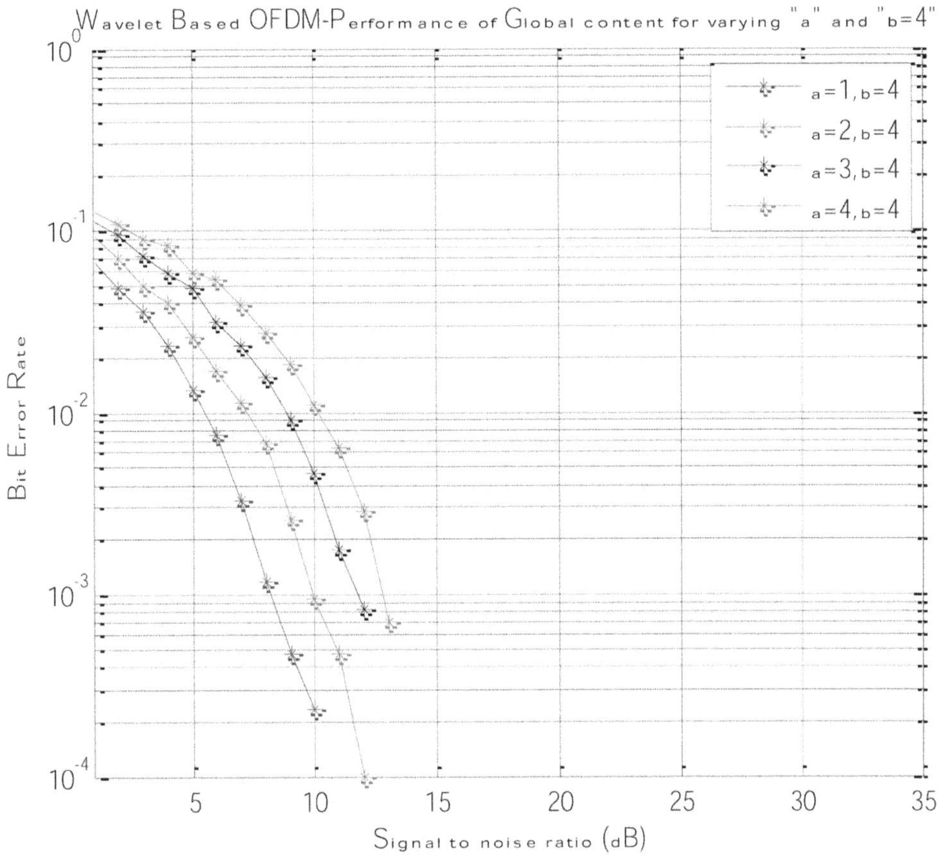

Figure 24. BER of global content for varying a and b = 4 with Fourier based OFDM

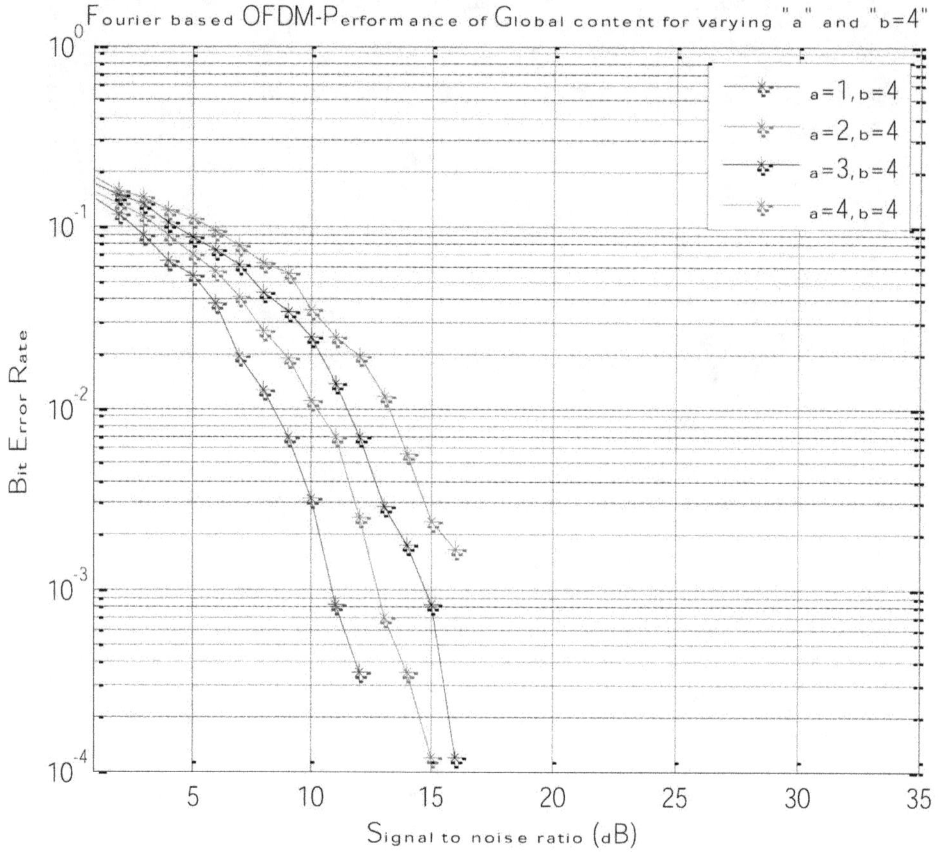

Figure 25. BER of local content for varying b and $a = 1$ with wavelet based OFDM

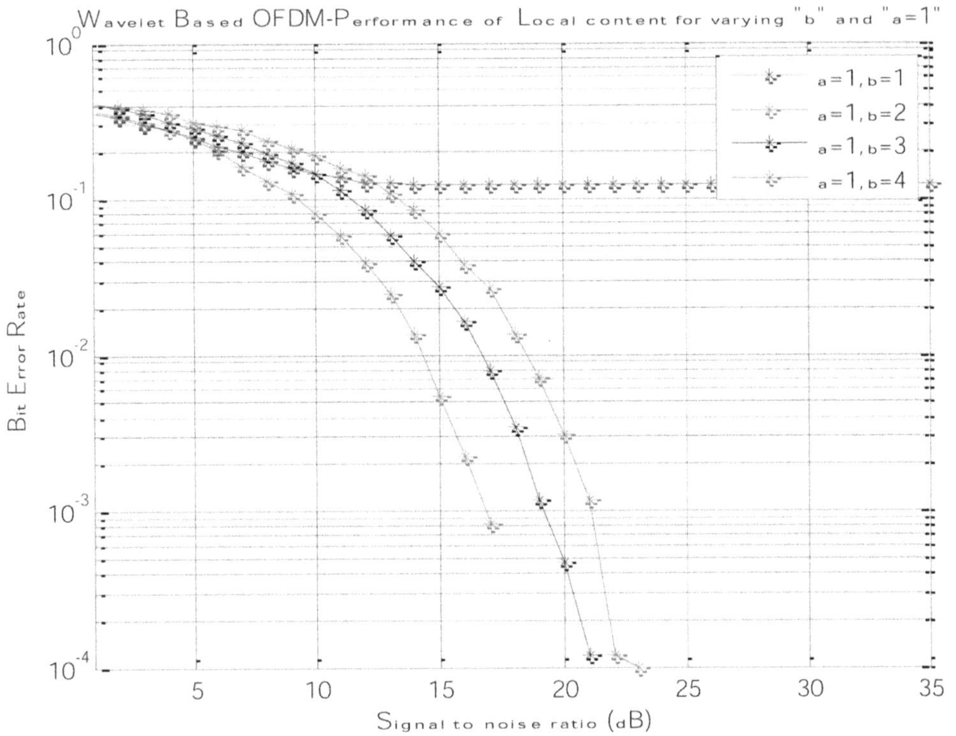

Figure 26. BER of local content for varying b and $a = 1$ with Fourier based OFDM

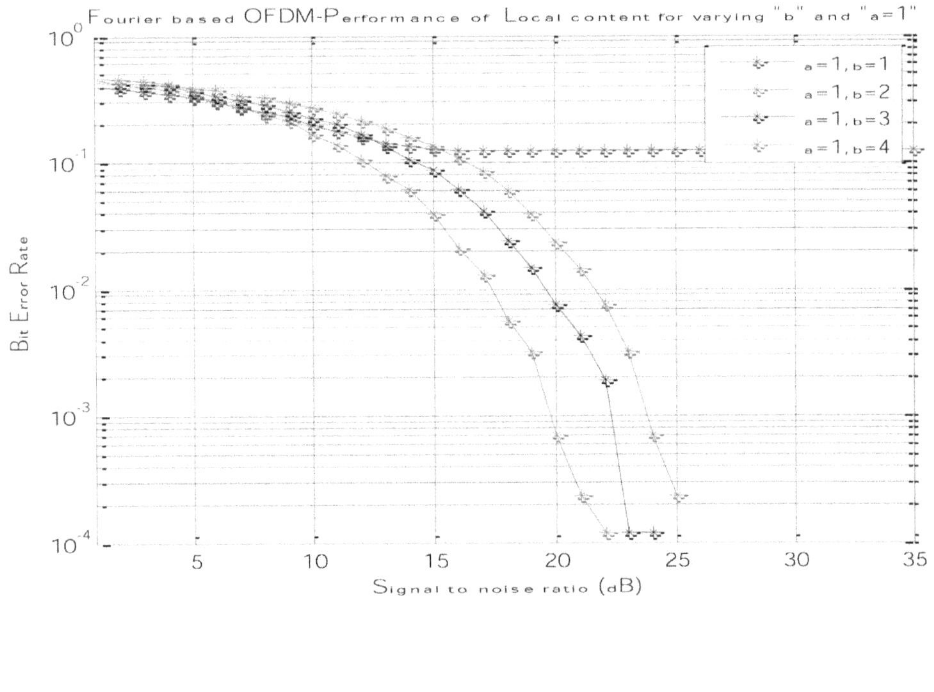

Figure 27. BER of local content for varying b and $a = 2$ with wavelet based OFDM

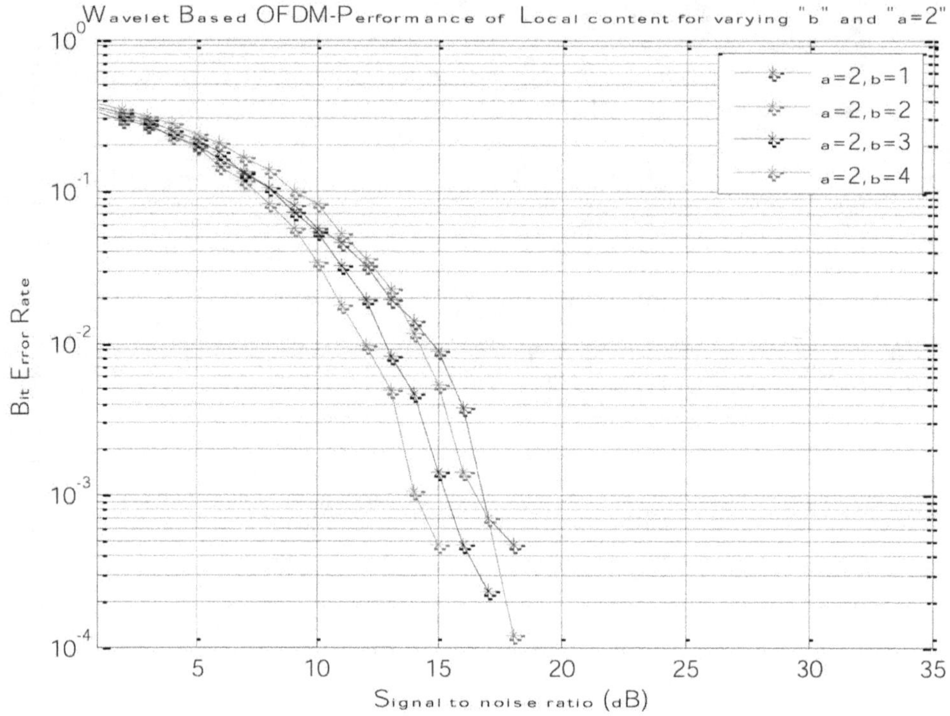

Figure 28. BER of local content for varying b and $a = 2$ with Fourier based OFDM

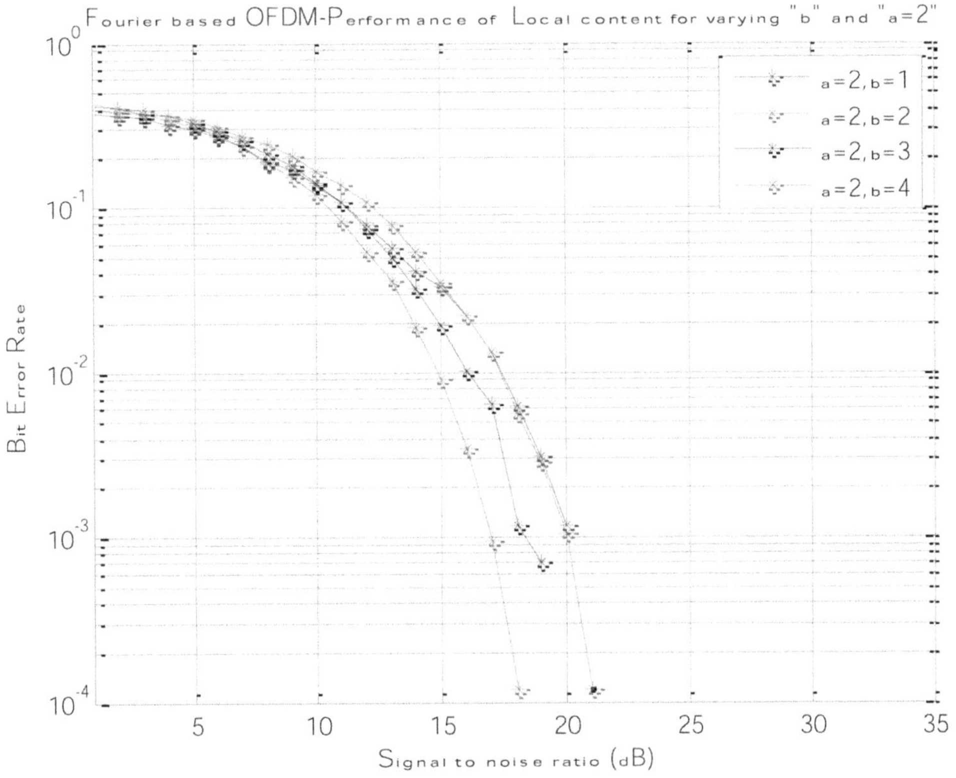

Figure 29. BER of local content for varying b and a = 3 with wavelet based OFDM

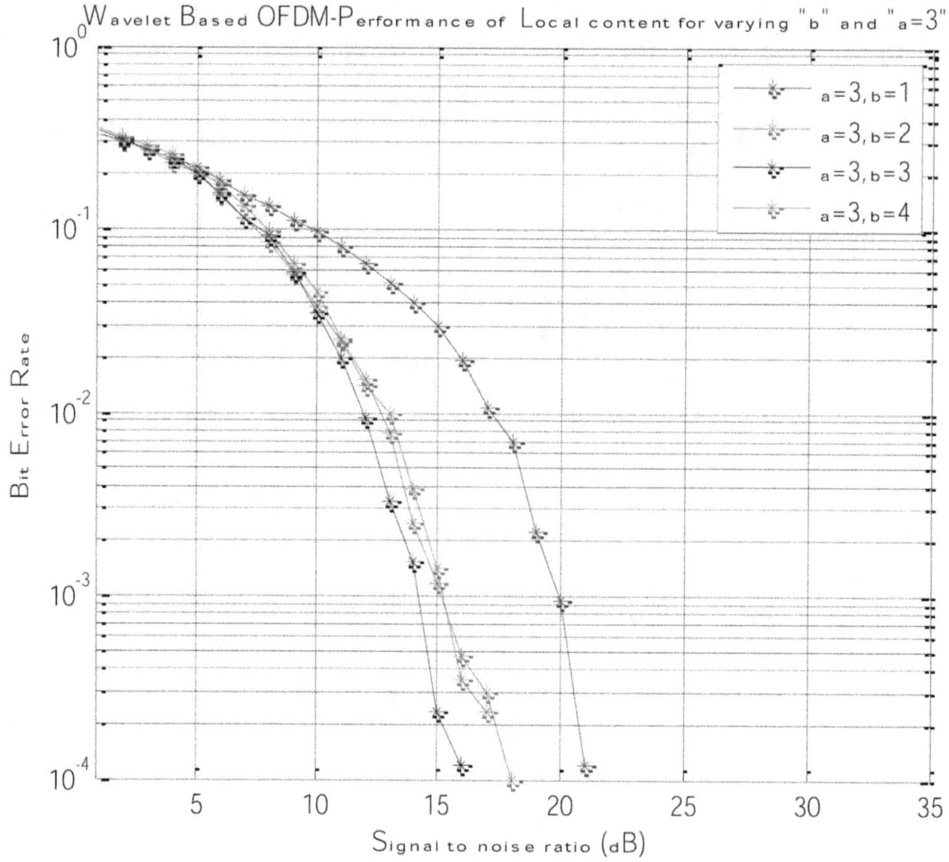

Figure 30. BER of local content for varying b *and* $a = 3$ *with Fourier based OFDM*

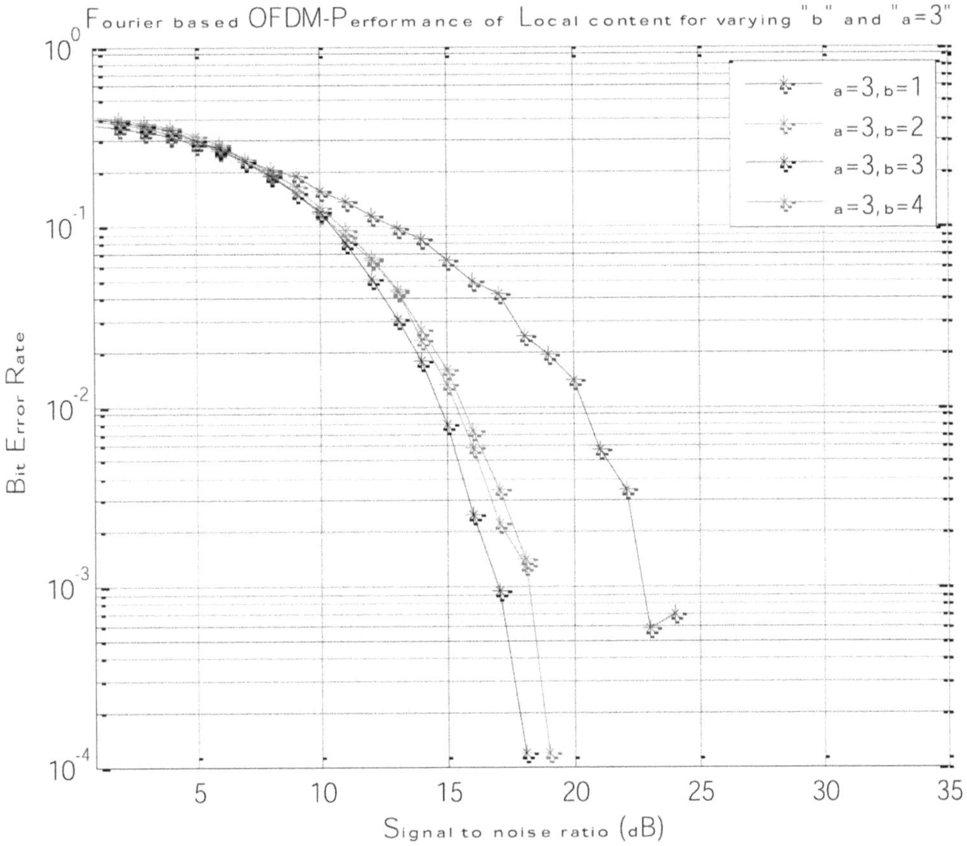

Figure 31. BER of local content for varying b and $a = 4$ with wavelet based OFDM

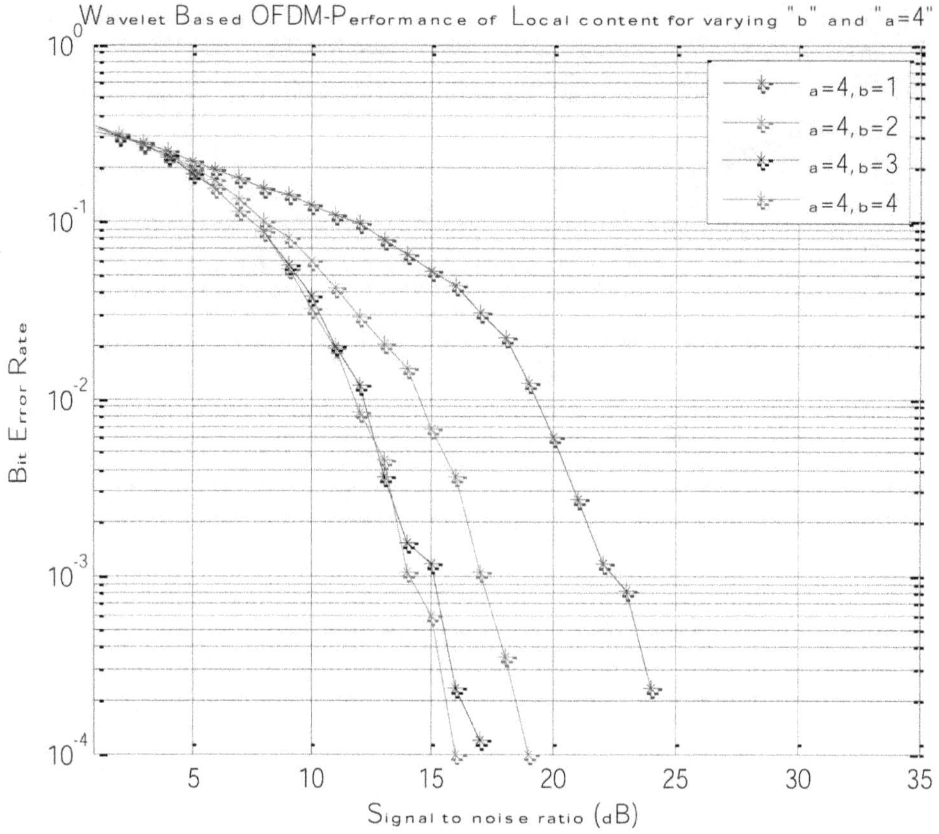

Figure 32. BER of local content for varying b and $a = 4$ with Fourier based OFDM

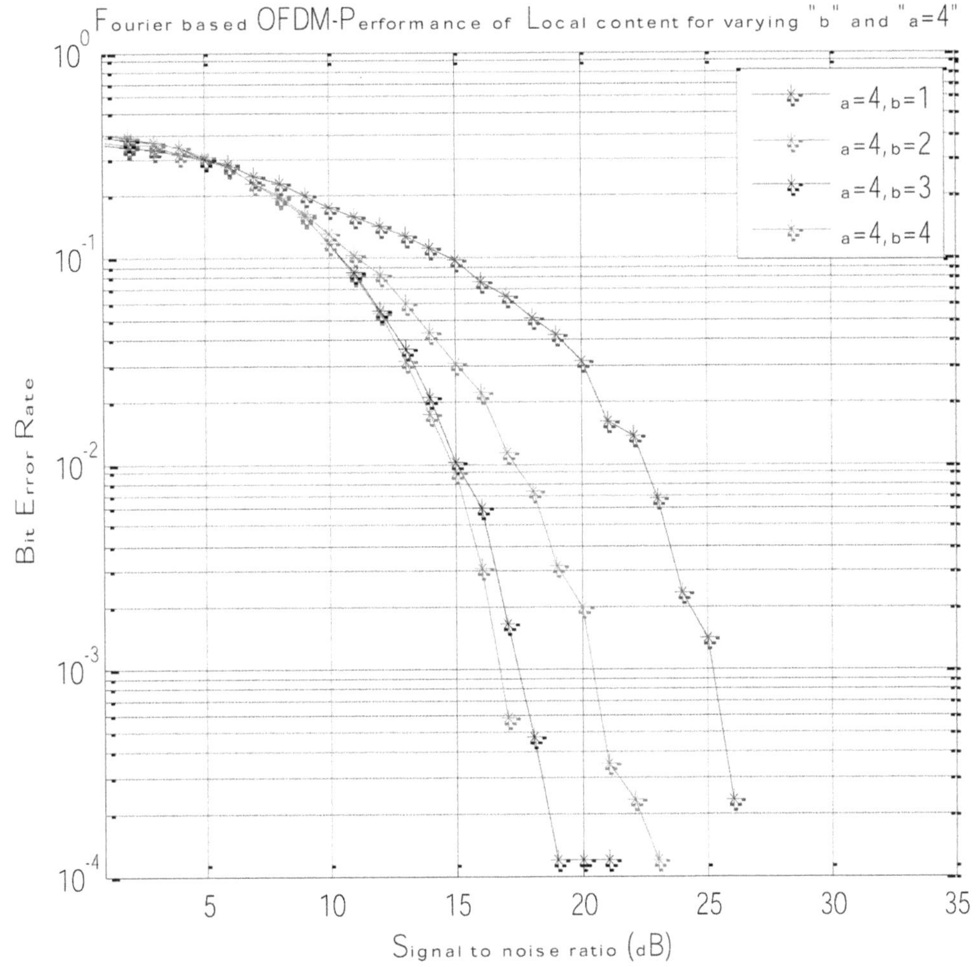

Figure 33. BER of global content for varying b and $a = 1$ with wavelet based OFDM

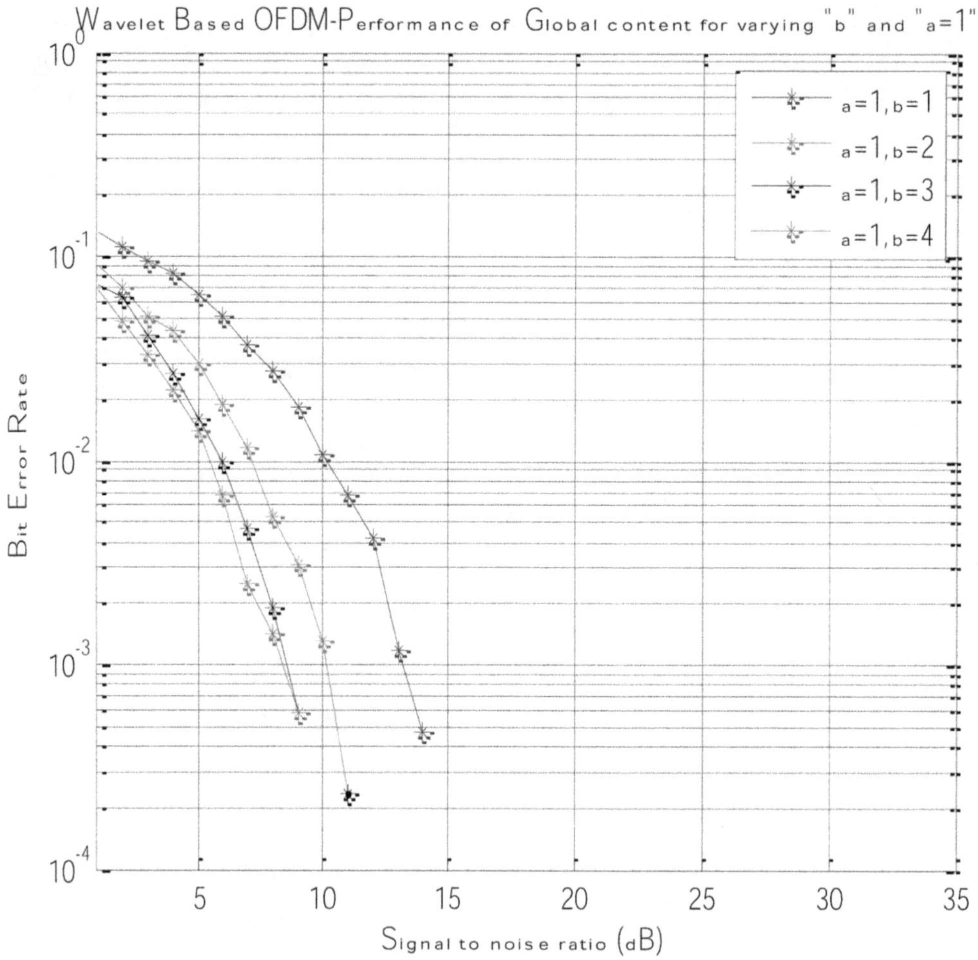

Figure 34. BER of global content for varying b and $a = 1$ with Fourier based OFDM

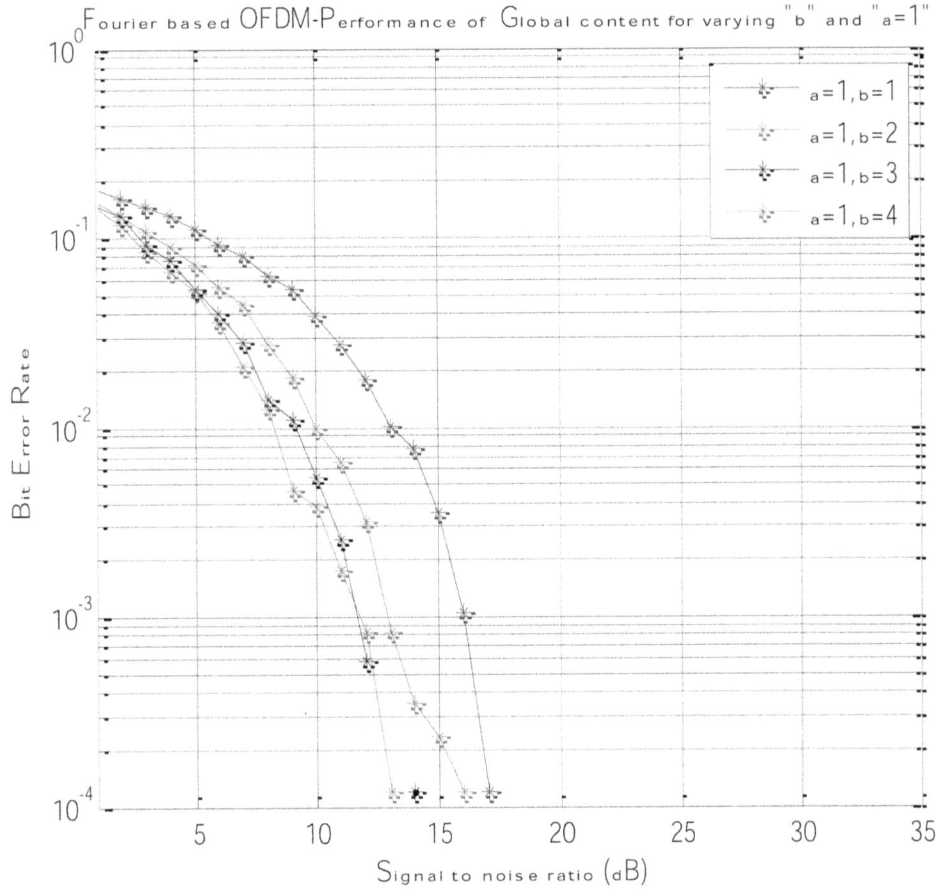

Figure 35. BER of global content for varying b and $a = 2$ with wavelet based OFDM

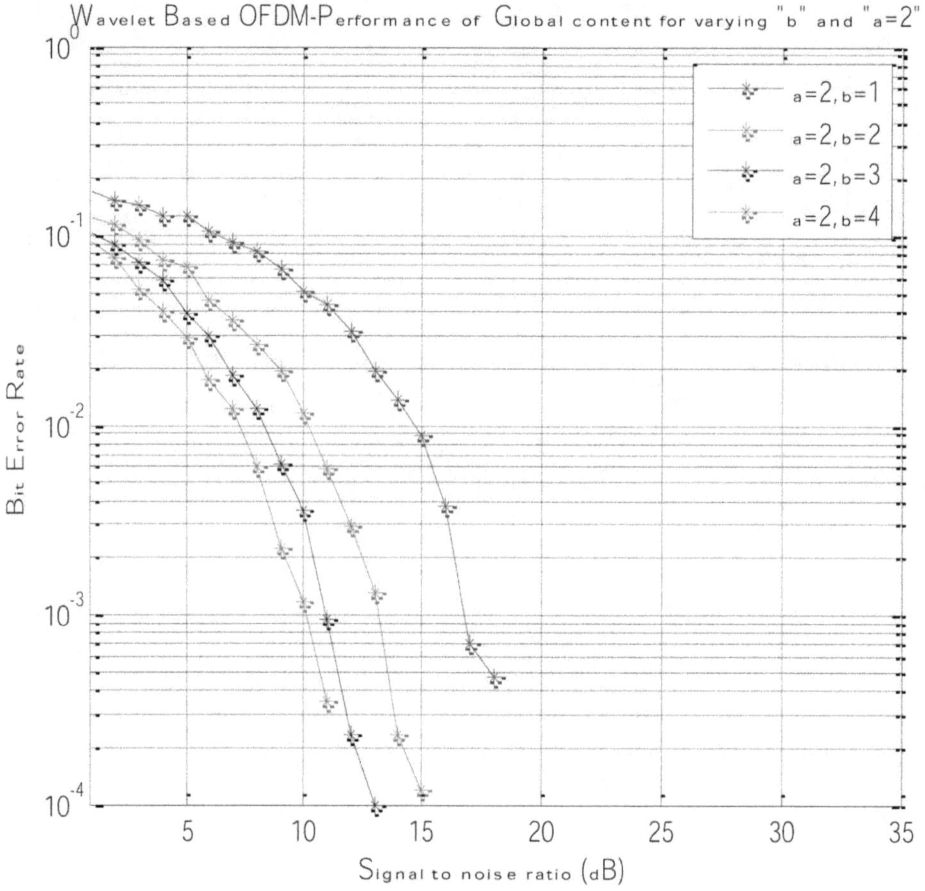

Figure 36. BER of global content for varying b *and* $a = 2$ *with Fourier based OFDM*

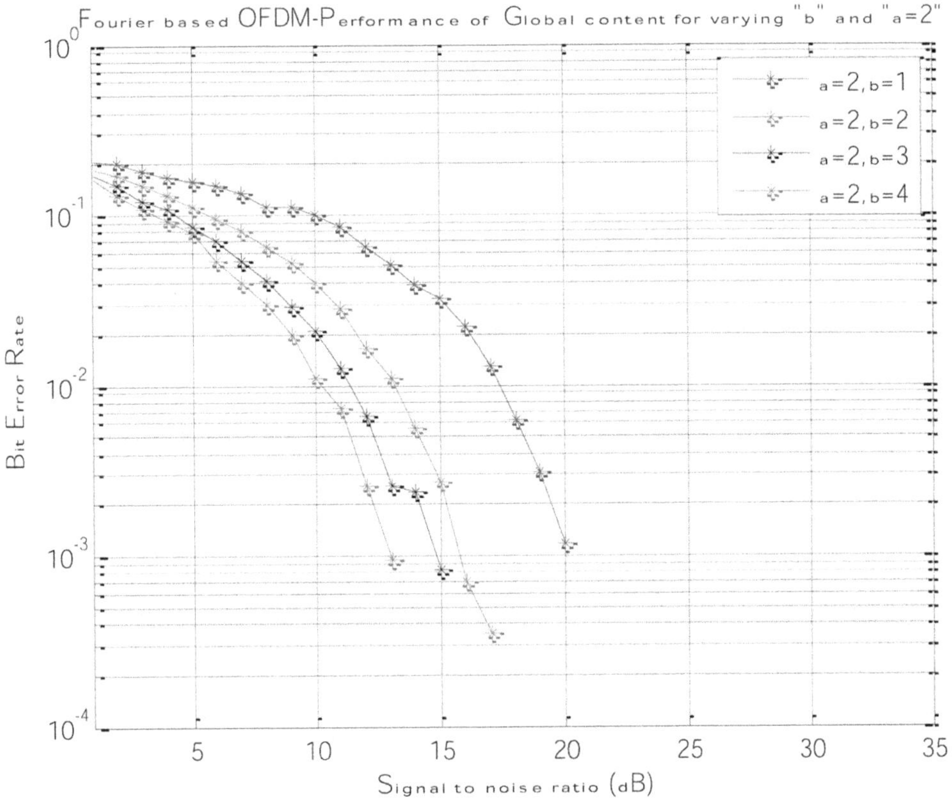

Figure 37. BER of Global content for varying b and $a = 3$ with wavelet based OFDM

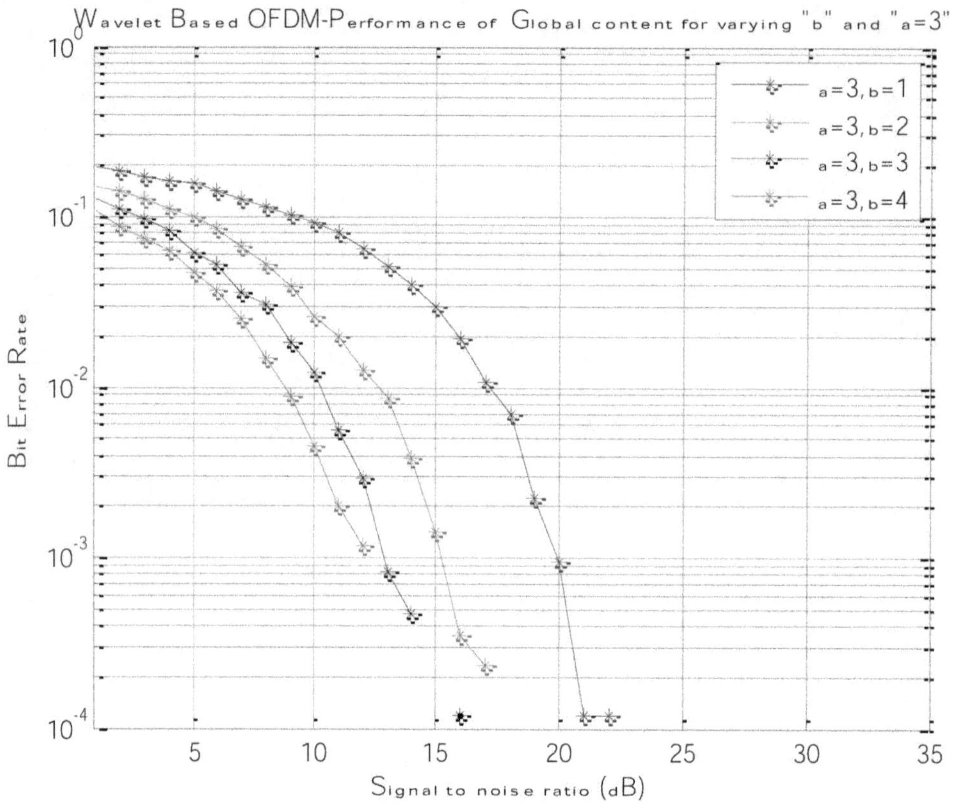

Figure 38. BER of global content for varying b and $a = 3$ with Fourier based OFDM

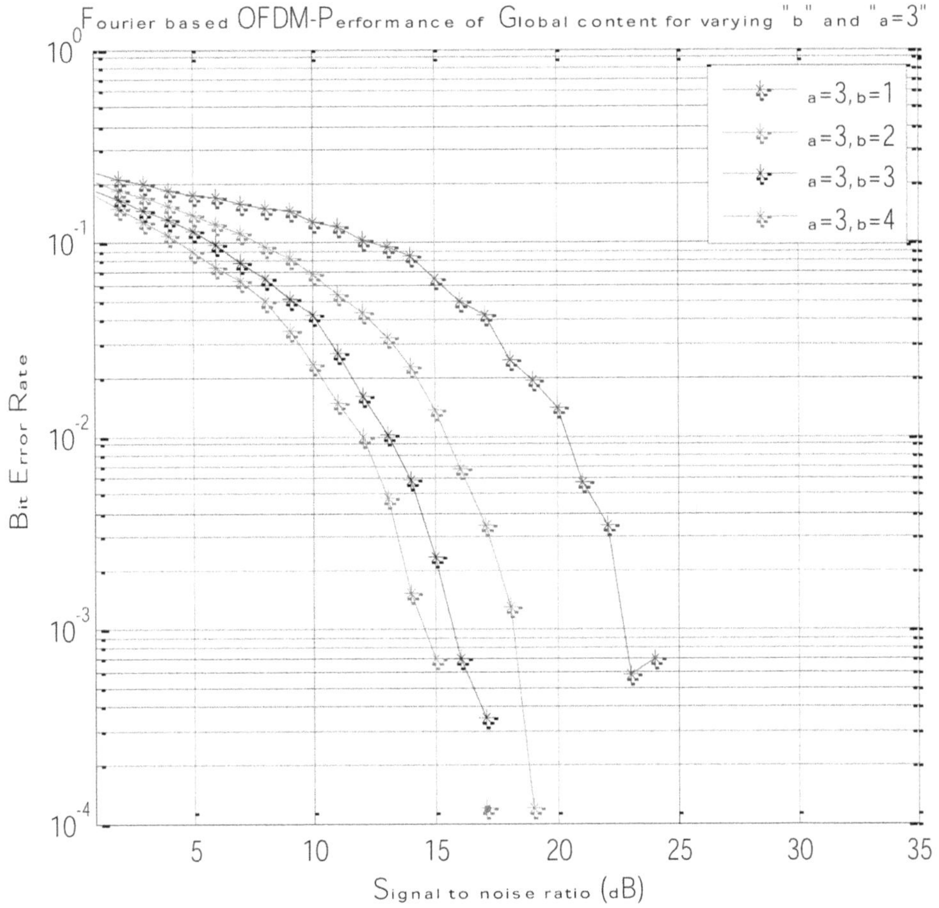

Figure 39. BER of Global content for varying b and a = 4 with wavelet based OFDM

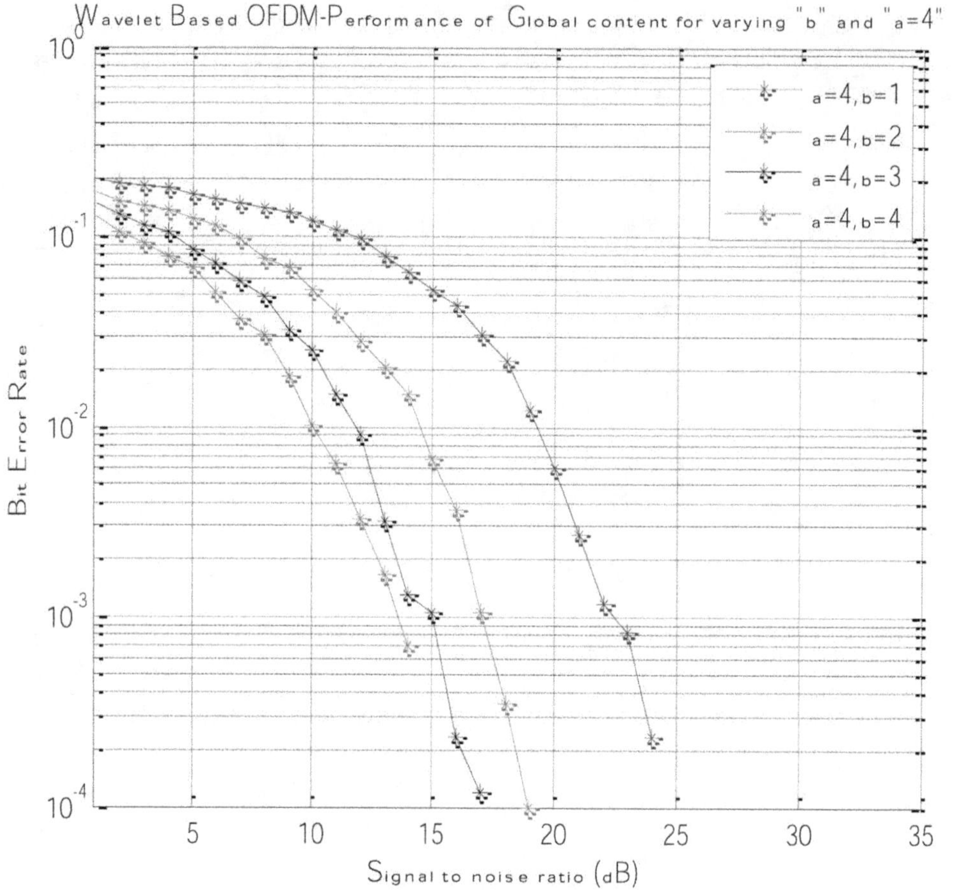

Figure 40. BER of global content for varying b and $a = 4$ with Fourier based OFDM

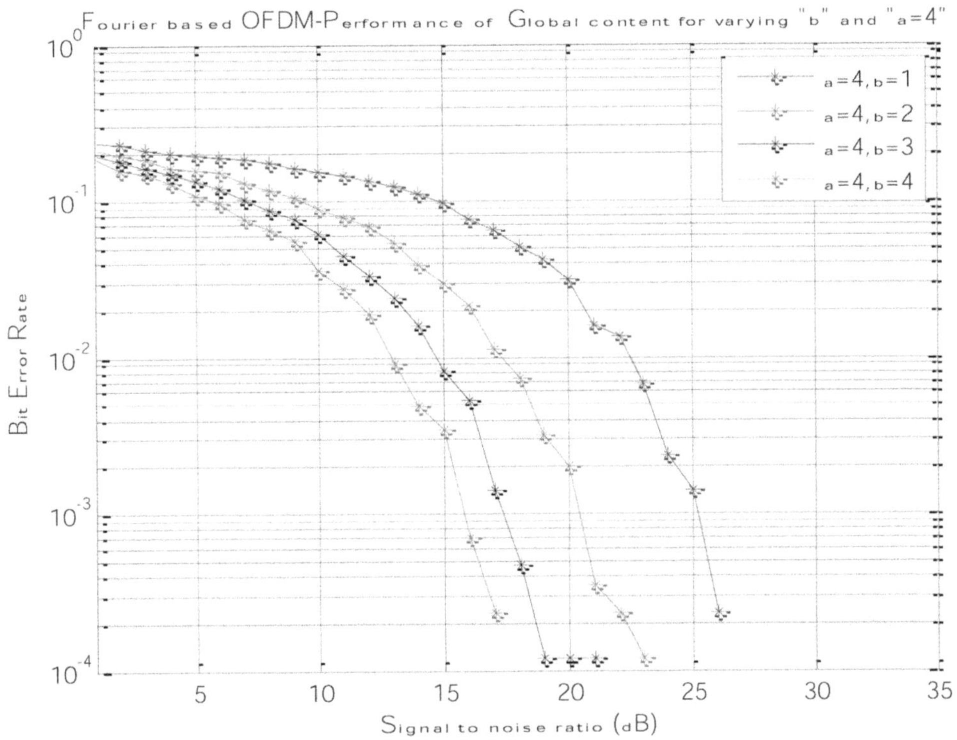

Figure 41. BER of local content for $\alpha = 1, 2$ and 4 *with Fourier based OFDM*

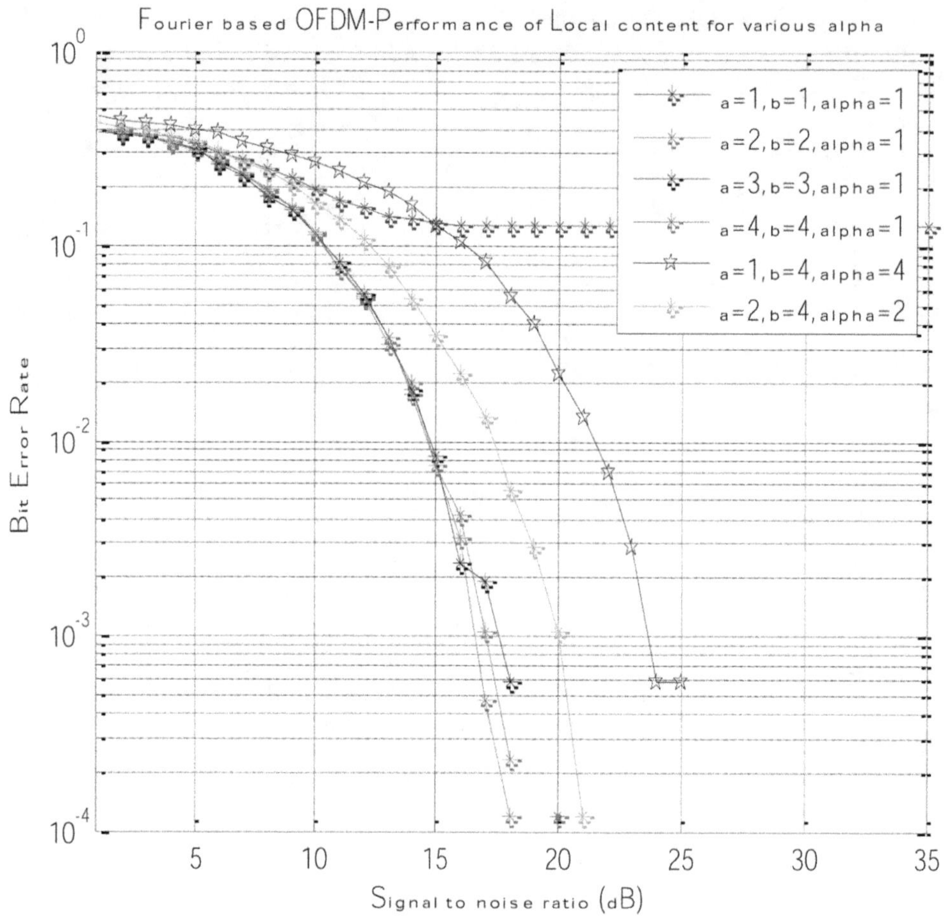

Figure 42. BER of global content for $\alpha = 1, 2$ and 4 *with Fourier based OFDM*

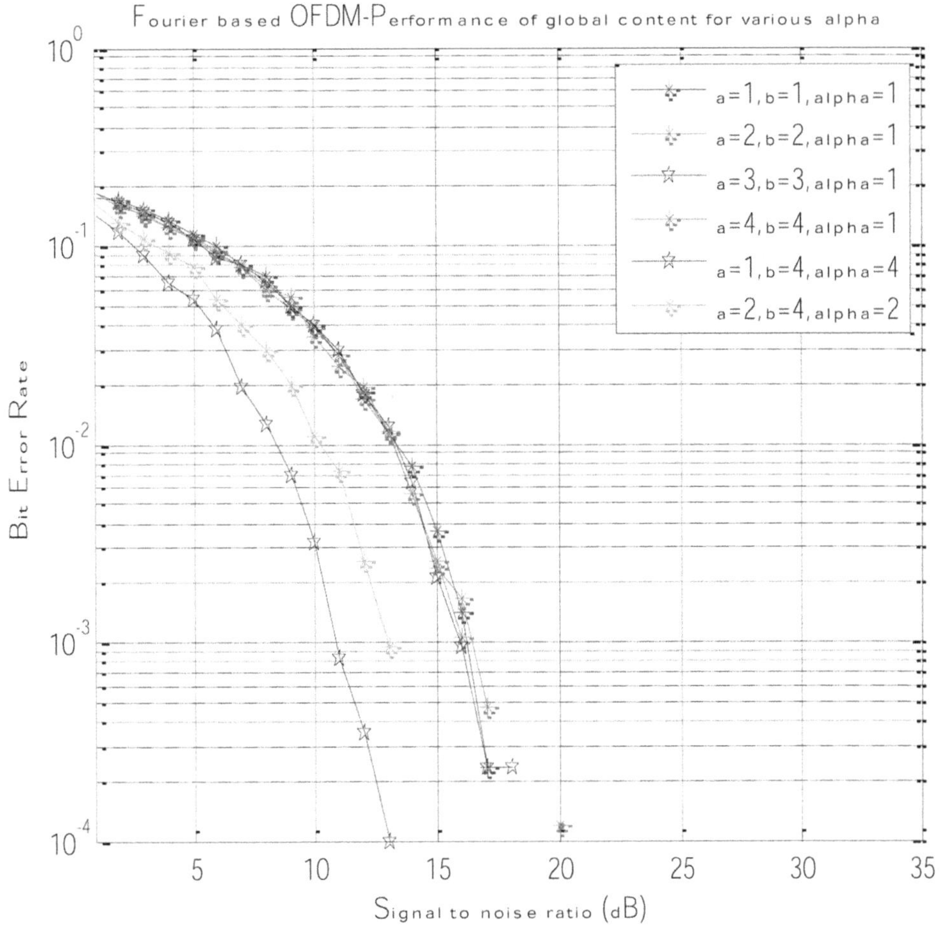

Figure 43. Coverage areas for global and local services in a SFN using selected subcarrier method.

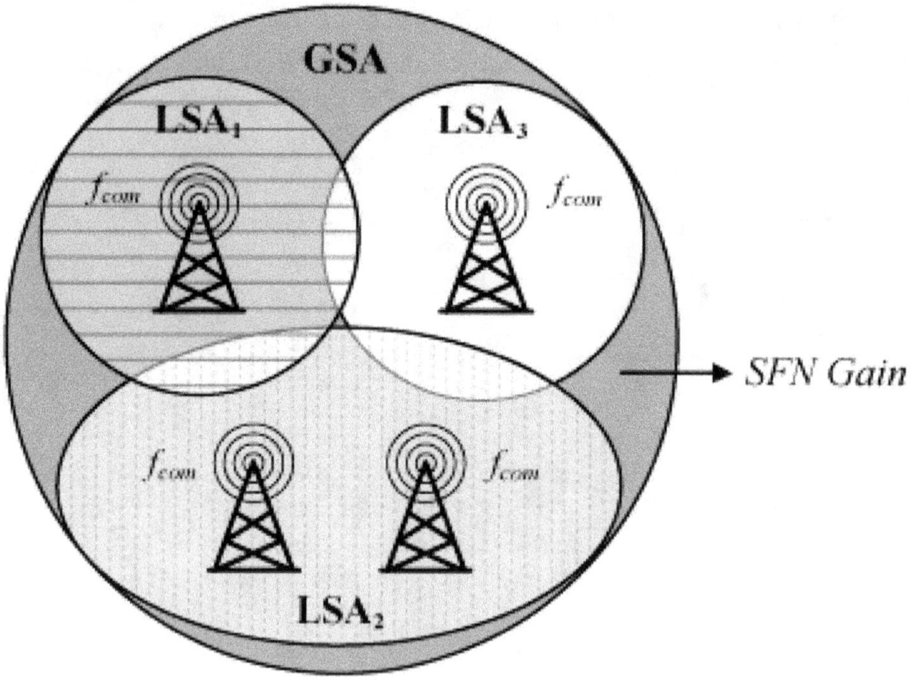

Figure 44 .Concept of the selected subcarrier technique.

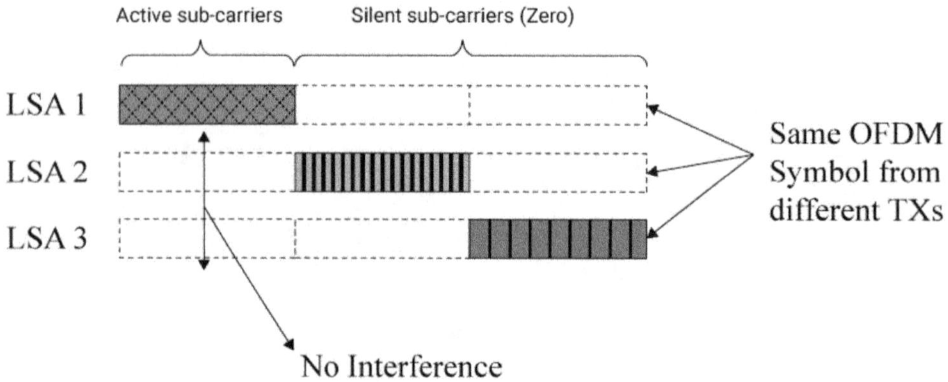

Figure 45. Selected subcarrier Method of Digital Video Broadcasting.

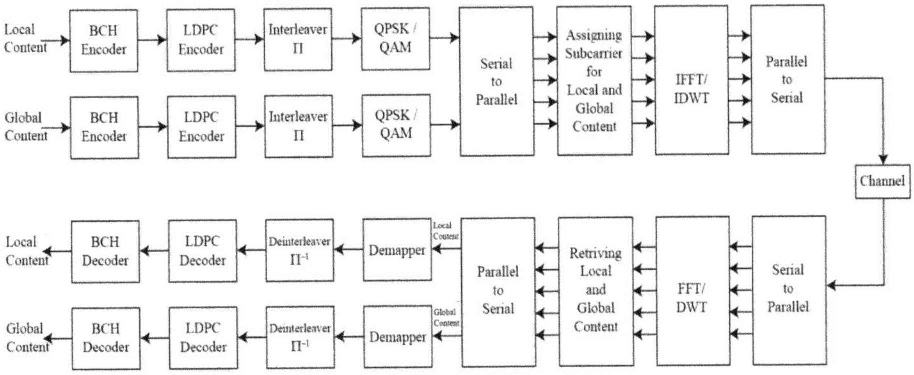

Figure 46. BER of QPSK Modulated global and local contents

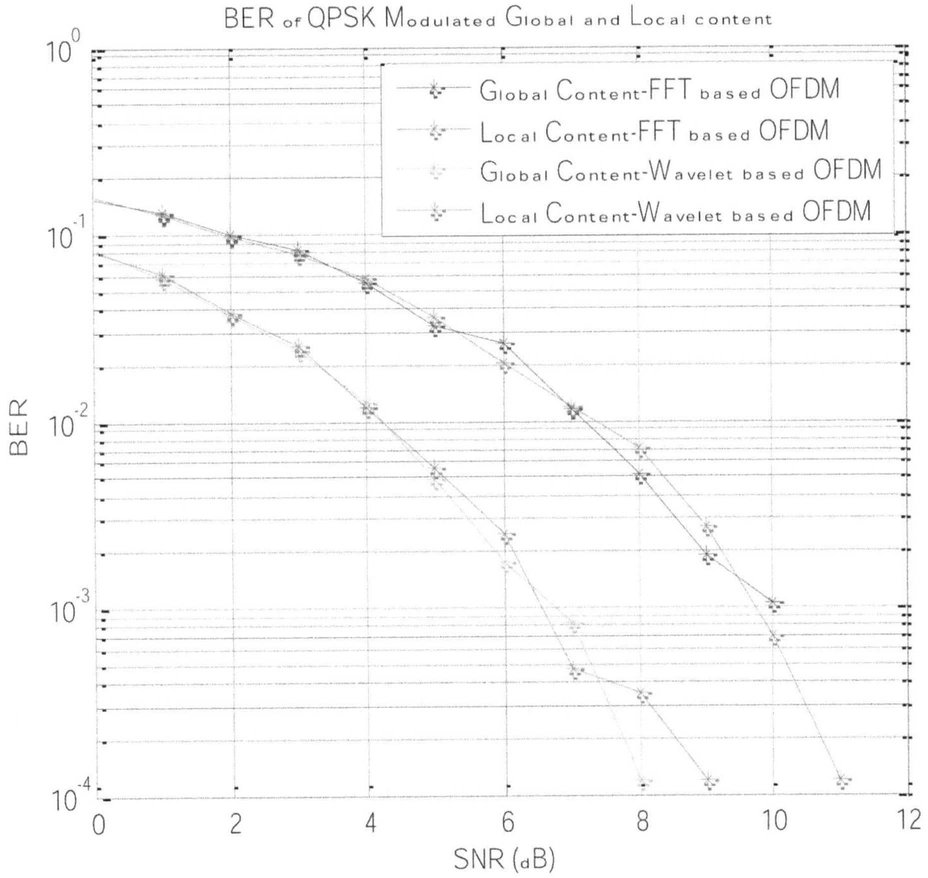

Figure 47. BER of QPSK modulated global content and QAM modulated local conten

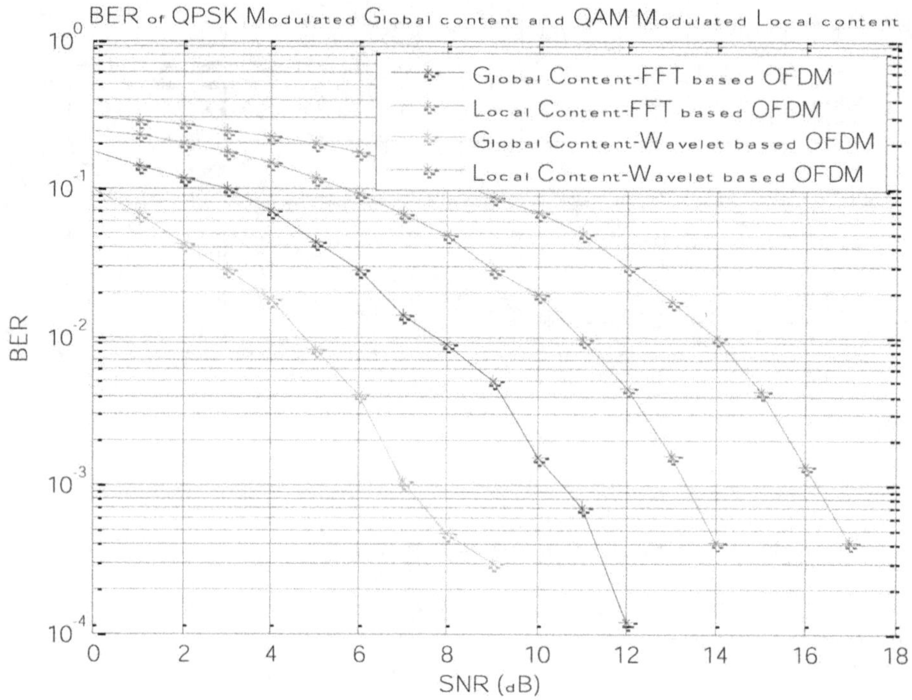

REFERENCES

Abdullah, K., & Hussain, Z. M. (2007, December). Performance of Fourier-based and wavelet-based OFDM for DVB-T systems. In *2007 Australasian Telecommunication Networks and Applications Conference* (pp. 475-479). IEEE. 10.1109/ATNAC.2007.4665248

Baig, S., Asif, H. M., Umer, T., Mumtaz, S., Shafiq, M., & Choi, J. G. (2018). High Data Rate Discrete Wavelet Transform-Based PLC-VLC Design for 5G Communication Systems. *IEEE Access: Practical Innovations, Open Solutions*, *6*, 52490–52499. doi:10.1109/ACCESS.2018.2870138

Ceniklioğlu, B., Özen, A., & Develi, I. (2018, May). Zero-tail discrete wavelet transform spread OFDM system for 5G and beyond. In *2018 26th Signal Processing and Communications Applications Conference (SIU)* (pp. 1-4). IEEE.

Dai, L., Wang, Z., & Yang, Z. (2012). Next-generation digital television terrestrial broadcasting systems: Key technologies and research trends. *IEEE Communications Magazine*, *50*(6), 150–158. doi:10.1109/MCOM.2012.6211500

Dibal, P. Y., Onwuka, E. N., Agajo, J., & Alenoghena, C. O. (2018). Application of wavelet transform in spectrum sensing for cognitive radio: A survey. *Physical Communication*, *28*, 45–57. doi:10.1016/j.phycom.2018.03.004

Dilmaghani, R., & Ghavami, M. (2007, September). Wavelet vs. Fourier based UWB systems. In *2007 IEEE 18th International Symposium on Personal, Indoor and Mobile Radio Communications* (pp. 1-5). IEEE. 10.1109/PIMRC.2007.4394653

ETSI. (2004). E. Digital Video Broadcasting (DVB); Framing structure, channel coding and modulation for digital terrestrial television. *ETSI EN*, *300*(744), V1.

Fisher, W. (2008). Digital Video and Audio Broadcasting. A Practical Engineering Guide. Springer.

Hasan, M. M. (2012). Performance comparison of wavelet and FFT based multiuser MIMO OFDM over wireless Rayleigh fading channel. *International Journal of Energy, Information and Communications*, *3*(4), 1–8.

Jiang, H., & Wilford, P. A. (2005). A hierarchical modulation for upgrading digital broadcast systems. *IEEE Transactions on Broadcasting*, *51*(2), 223–229. doi:10.1109/TBC.2005.847619

Kumbasar, V., & Kucur, O. (2012). Performance comparison of wavelet based and conventional OFDM systems in multipath Rayleigh fading channels. *Digital Signal Processing*, *22*(5), 841–846. doi:10.1016/j.dsp.2012.02.004

Lee, J. G., & Ryu, H. C. (2018, August). Wireless Communication System Based on OFDM System Using Wavelet Transform. In *2018 Progress in Electromagnetics Research Symposium (PIERS-Toyama)* (pp. 2041-2046). IEEE. 10.23919/PIERS.2018.8597769

Mattson, A. (2005). Single frequency networks in DTV. *IEEE Transactions on Broadcasting*, *51*(4), 413–422. doi:10.1109/TBC.2005.858419

Méric, H., Lacan, J., Amiot-Bazile, C., Arnal, F., & Boucheret, M. L. (2011, April). Generic approach for hierarchical modulation performance analysis: Application to DVB-SH. In *2011 Wireless Telecommunications Symposium (WTS)* (pp. 1-6). IEEE. 10.1109/WTS.2011.5960857

Méric, H., Lacan, J., Arnal, F., Lesthievent, G., & Boucheret, M. L. (2012, April). Improving broadcast channel rate using hierarchical modulation. In *European Wireless 2012; 18th European Wireless Conference 2012* (pp. 1-5). VDE.

O'Leary, S. (1997). Hierarchical transmission and COFDM systems. *IEEE Transactions on Broadcasting*, *43*(2), 166–174. doi:10.1109/11.598365

Sharma, V., & Singh, G. (2014). On BER assessment of conventional-and wavelet-OFDM over AWGN channel. *Optik (Stuttgart)*, *125*(20), 6071–6073. doi:10.1016/j.ijleo.2014.07.073

Chapter 9
Designing Filtering Antennas for 5G Applications

Adrian K. Stavrakis
https://orcid.org/0000-0002-9527-7103
Aston University, UK

Eugene Amobichukwu Ogbodo
Aston University, UK

ABSTRACT

This chapter aims to serve as an introduction to the integration process of filters and antennas into a single component, a filtering antenna, or filtenna, which is a topic of rising scientific interest and importance in recent years. As most receiving and transmitting devices nowadays call for a smaller footprint, better power management, and simplification of fabrication, this chapter showcases all the steps the reader must take to have a holistic approach to the integration of the aforementioned two components. Therefore, the fundamental principles of antenna, microstrip line, and filter theories are explored, and they are applied to a proof-of-concept patch filtering antenna operating in the S-Band, and at a frequency suitable for allocation to the 5G communications applications, while notable examples from other authors also demonstrate different ways of tackling this domain.

INTRODUCTION

The modern world relies greatly on systems of wireless communications for a variety of services, ranging from Wi-Fi to satellite communications to FM radio. All of these systems generally contain three discrete blocks, each with a specific functionality. A transmitter/receiver which is the antenna, a signal selection device which isolates the frequency of interest known as a filter, and a suitable amplification stage. Figure 1 presents a sample block diagram of the RF transceiver frontend.

Evidently, when the device is transmitting, the filter is responsible for rejecting out-of-band frequencies so that the signal will be amplified and reach the antenna as a last step. Reversely, while receiving the antenna only forwards the signal to the filter stage, which once again rejects all the unwanted frequencies

DOI: 10.4018/978-1-7998-6992-4.ch009

and forwards the signal to the amplification stage usually performed by an LNA, or low-noise amplifier. Aside from these fundamental blocks, the signal also undergoes frequency down- or up-conversion and digital processing at supporting circuit blocks, but the key outtake is that a proper impedance matching between the antenna and filter stages is of crucial importance, as any mismatches lead to severe performance degradation. In the conventional approach of circuit design, this comes with the unfortunate effect of increasing the overall circuit size.

Figure 1. Graphical representation of the RF transceiver block diagram and its fundamental blocks

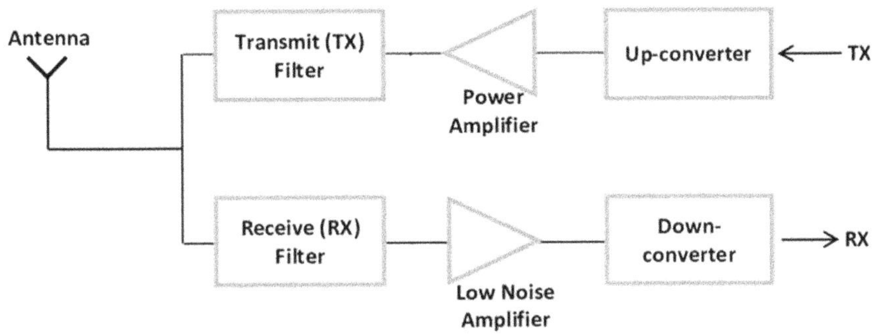

It is understandable, that in principle such systems are much more complex, but without these fundamental blocks wireless communication could not be established. However, the rapid rise of the Internet-of-Things (IoT) devices and in general the handheld consumer electronics, makes the need for miniaturization apparent. After years of scientific research and technological advances such as Micro-Electro-Mechanical Systems (MEMS) design which enable new fabrication techniques, the discrete blocks design tends to become obsolete, in favor of a seamless integration of the three blocks on a joint module.

The driving force behind all this research into the integration of the RF front end is simply owed to power constraints. The vast majority of the new devices have to rely exclusively on a battery for their functionality, sometimes non-replaceable and non-chargeable, in an effort to save space in the design. Therefore, its capacity is limited, and the entire system must work with the least possible power consumption to prolong its operational time. Even though there are multiple constraints to trying to minimize power consumption, one of the most straightforward ways is to reduce losses.

Attenuation -or loss- generally occurs due to four major causes. Thermal dissipation, path from transmitter to receiver, coupling imperfections and impedance mismatches. The possible elimination of the last two, is the reason why emphasis is being given to the codesign of multiple blocks using the same materials and techniques on the same substrate, and the general tendency is to integrate the filter and the antenna blocks, into a filtering antenna. By eliminating the path between these two stages, it is possible to remove, or significantly diminish, transition loss, insertion loss and external noise, which are primarily associated with the imperfections of physical coupling of the path to the components.

A filtering antenna is a structure that creates a single module, in contrast to the conventional fabrication of an antenna and a filter separately, matched to a pure resistive impedance, such as a 75Ω or 50Ω load, and a direct connection between the two as shown in Figure 2.

Figure 2. Design of antenna and filter (a) with the conventional approach and the use of a matching network and (b) with the new approach

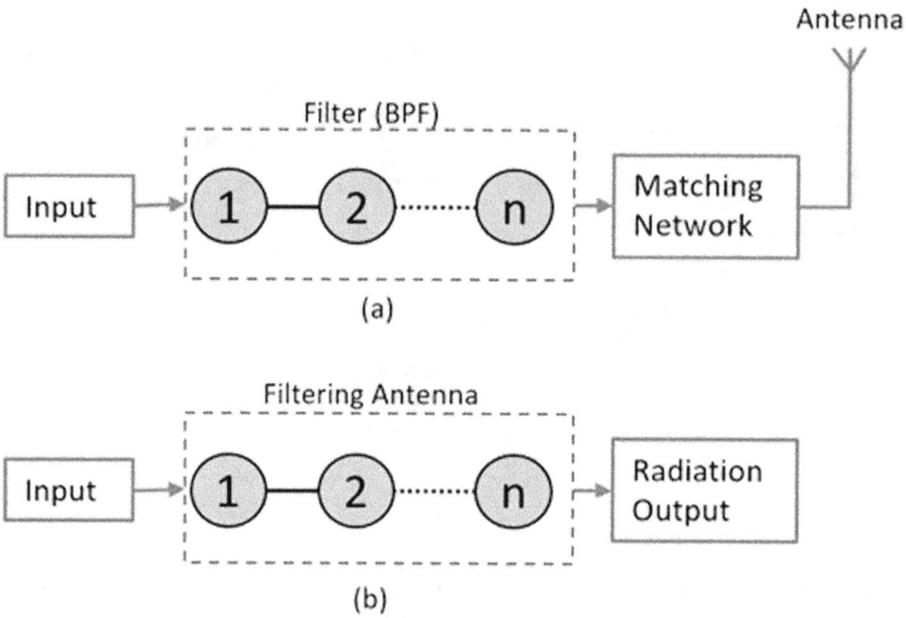

Theoretically, filtennas bring a number of benefits, such as reducing of the path of the signal transmission, overall system size, cost and complexity, along with minimizing the interference and losses associated with the antenna-filter transmission line. However, in reality, many difficulties and hurdles arise, on account of the mandatory trade-offs between a good performance of the filter (for instance maintenance of a good out of band rejection, a fast roll off and a high selectivity) and a good antenna structure (yielding a good directivity, polarization, and radiation pattern). Most importantly, the design needs to guarantee a reliable and sufficient isolation between the filtering parts and the radiating elements of the device. This can be achieved primarily in two ways; either by utilizing various filter synthesis techniques, or by mounting the filter directly on the feed line of the antenna.

The former way of approaching this task, allows oftentimes indirect coupling in order to achieve that good selectivity required, while the latter requires designing a filter which already possesses a good skirt selectivity, whilst ensuring an adequate match to the antenna at the integration step.

After having briefly touched down on the issues and requirements associated with the design and fabrication of a filtering antenna, next some notable design attempts will be presented. Following that, the chapter aims to introduce the mandatory theoretical background in both antenna theory and transmission line theory, and lastly touch on the necessary background on microwave filter design. Lastly, the sample filtering antenna design and simulation will be analyzed, so that the reader can gain knowledge on the field from start to finish, and opt for their own designs, simulations and fabrication.

BACKGROUND

It is of great value to understand the different approaches to designing a filtering antenna. As outlined in the previous section, usually the blocks for the antenna and the filter are designed by using a matching network cascaded between them. In recent years, the rise of many designs of filtering antennas have been observed, such as such as planar coaxial/linear antennas (Yu et al., 2012), resonator-based antennas (Chu et al., 2018; Liu et al., 2019; Pan et al., 2018), FSS (Frequency-Selective Surface) horn covered antennas (Luo et al., 2007), coupled resonator filter antennas with a connection to a monopole (Chuang & Chung, 2011b), patch (Zuo et al., 2009), or slot (Yusuf & Gong, 2011) configuration, as well as 3D printed filtering aperture antennas (He et al., 2018).

However, these designs are still victims of relatively large footprint, and during fabrication, complex techniques and steps are required (He et al., 2018), or utilize non-uniform synthetic materials (Chu et al., 2018; Liu et al., 2019; Pan et al., 2018), and are often deemed unsuitable for many applications such as flexible wearable devices. Therefore, the challenging task of creating an efficient implementation capable of satisfying the requirements of both the antenna and the filter parts using microstrips, has attracted a lot of scientific interest.

Many research groups opt to remove the 50Ω standardized impedance constraint, by codesigning the antenna and filter, (S. Chen et al., 2014; Chuang & Chung, 2009, 2011b; Zuo et al., 2009). These implementations are still challenging, because good filtering performance is generally inversely analogous to the level of compactness in these designs. In (Chuang & Chung, 2011b), a N number of parallel coupled microstrip line sections is utilized for the illustration of the synthetic principles, and the fabricated design operates around 2.45GHz with a footprint of 7cm by 3cm. A peak gain of 0.65dBi is observed as well, and it yields a better than 22dB suppression at this dimensionality, with a response similar to that of a 3rd order Chebychev filter.

A microstrip patch antenna is layered above a hairpin microstrip filter in (Zuo et al., 2009), but the impedance of the design is performance-based and not conformant to the standard 50Ω. The achieved overall footprint is 3cm by 2cm at a frequency of operation of 4.16±0.1GHz, but as a next step, additional elements that add complexity have been introduced to improve the bandwidth, and the measurements showed a 4.7% better bandwidth with a larger peak gain of 4.3dBi in comparison to the traditional one which has an impedance value of 50Ω.

A significantly smaller device measuring only 2.7cm by 2cm is presented in (He et al., 2018) but a PIN diode is utilised to achieve the combination of different working states and the enablement of a dual-band function at the 2.45/5.2GHz Wi-Fi frequency sets. The drawback of this attempt, however, is the extra requirement for a DC bias current in order to properly utilize the diode, and since it is a component not jointly fabricated on the same substrate, there is a whole new set of limiting factors jeopardizing the robustness of the design, such as spontaneous diode failure, and imperfections while soldering which can lead to fatigue and cracking.

Lastly, (Chuang & Chung, 2009) have created a system of a meander antenna and a shunt resonator, to act as the first and last resonators of a Chebychev bandpass filter. Even though the usage of a λ/4 resonator as shunt is a great step at minimizing the footprint of the design, it is no less than 6cm by 6cm. A positive trade-off is the nearly omnidirectional radiation pattern, with a peak gain of 2.78dBi, coupled with a sharp skirt selectivity. A drawback of this approach is that the antenna needs to exhibit a high inductance and a series RLC resonance, making the overall design incompatible with slight alterations for bandwidth tweaking.

A better performing approach, therefore, is to substitute the last resonator of the filter by an antenna, and various implementations exists for this. A circular patch antenna was created in (X. Chen et al., 2013) in principle by the combination of a Butterworth band pass filter with a traditional fan shaped antenna, by substituting the former's second port and resonator with the latter. Compared to the traditional approach, this design shows a wider bandwidth, good selectivity, and a flatter gain response while only measuring 3.5cm by 4.7cm. It yields a 2.3dBi peak gain and a 460MHz bandwidth at an operational frequency of 2.4GHz. Another work carried out (Ahsan, Islam, Habib Ullah, & Misran, 2014) which is designed to work at the same frequency shows a directivity that increases at higher frequencies. This is made possible by feeding a circular slotted patch by a meander microstrip and it demonstrates a secondary operating range between 600MHz and 1GHz.

The Γ -or inverted L- antenna, which is another design proposal by (Chuang & Chung, 2011a, 2011b), which relies on two perpendicular strips of microstrip line that overlap. In the first iteration, a parallel coupled microstrip line section is fed to the antenna, while in the second, a coupled line resonator is acting as one pole and the Γ antenna acting as the other pole of a second order band pass filter. The same group of authors comes up with a new idea in (Chuang & Chung, 2011a), and that is to include a miniaturized coupled line resonator, which acts as a provider of one pair of transmission zeros with tuneable frequencies, which enables in turn high selectivity at the band-edge. Among the advantages of the second proposal is an even smaller footprint.

A 5th order Chebychev filter with a T-shaped antenna as the last stage resonator is used in (Jadhav & Deore, 2017), with a central frequency at 2.45GHz. This approach achieves a great reduction to the overall size by utilizing a series of parallel coupled microstrip line segments, adding up to only 8cm by 4cm, while performing quite well, exhibiting a gain higher than 3.5dB, a fractional bandwidth of 12.24%, and high selectivity.

The approach in (Duan et al., 2016) is to have a feeding network that leads into a driven patch and a stacked patch on top. They are using two H-shaped microstrip feed lines orthogonal to each other, to achieve dual polarization, a sharp roll-off and a radiation zero created by the stacked patch. By varying the physical dimensions of the design, the frequency changes so this is a rather versatile filtering antenna proposal, while additionally, a greater than 40dB out of band suppression, a gain around 9dBi, a high isolation of 35dB between ports and lastly a low cross-polarization, around 29dB are among the overall very good characteristics of the design.

Slot antennas are also widespread, such as (Ahsan, Islam, Habib Ullah, Mahadi, et al., 2014; Deng & Feng, 2019; Jin et al., 2018; Li et al., 2018; Qian et al., 2018; Samsuzzaman et al., 2014; Zhang & Xue, 2018). (Qian et al., 2018) developed an asymmetrically loaded tri-band antenna which after experimentation produces three discrete operational bands at 5.21-5.45 GHz, 3.50-3.61GHz and 2.58-2.68GHz. The peak in-band gains achieved are 5.87dBi, 5.83dBi and 7.51dBi respectively. The out-of-band radiation suppression of the design is also notable.

(Zhang & Xue, 2018) manage to create a filtering antenna operating at a centre frequency of 2.3GHz, by creating extra cross-coupling paths between the source/load and any non-adjacent resonators. The first layer is that of the ground plane and two separate coupling slots, the second contains the microstrip feedline, the third the driven patch and the fourth the stacked patch. This antenna can create N radiation zeros with N resonators. The fractional bandwidth of the fabricated example is 15.2%, and the radiation zeros can be independently tuned.

A rather complex slot combination is discussed in (Samsuzzaman et al., 2014) where a patch antenna is slotted with a rotated square slot, excited by a microstrip line, which enhances the bandwidth. Additionally, four L-shaped slots are used for gain boost. The design achieves a bandwidth of 88%, and an average gain of 4.17dBi. At an overall footprint of 6.2cm by 6.2cm the antenna operates between 3.07GHz and 6.15GHz and exhibits an average radiation efficiency greater than 92.5%.

A different filtering patch antenna can be created by merging a slotted patch, and a ground plane only with metal pins between the two, as well as a metal strip under the patch, as described in (Li et al., 2018). The group achieved a consistent gain of 6.8dBi at the operating band of 3.0GHz to 3.8GHz, while the two radiation beams were directed at ±30°.

Embedded P shaped slots in a conventional patch are proposed in (Ahsan, Islam, Habib Ullah, Mahadi, et al., 2014) with the device resonating at 1.5GHz and 4GHz with a 0.2GHz and 0.3GHz bandwidth respectively. It would be a good candidate for applications such as GPS, since it also has an overall form factor of only 3cm by 3cm.

In (Ogbodo et al., 2018), the integration of a dual-mode patch antenna and a dual-mode band pass filter is demonstrated. To achieve this, a $\lambda/2$ square patch resonator is designed with a cut corner, so as to be perturbed into two new resonances, in this case 1.8GHz and 2.1GHz respectively. It is stacked back-to-back with a rectangular shaped patch antenna generating a vertical and a horizontal polarization. Overall, this proposal is characterized by high tunability, improved frequency selectivity, and a more compact form factor compared to the conventional design. By utilising a slot coupling technique, radiation interference is diminished and more freedom in controlling the channel bandwidths is achieved.

A U-shaped patch antenna is proposed in (Lin & Chung, 2011), once again substituting the last resonating element in a filter. Unconventionally, this approach creates a larger footprint than utilizing discrete stages for the filter and antenna, but it exhibits a twice as wide bandwidth with good frequency skirt selectivity, a flatter response in the passband, and an obtained -15dBi out-of-band rejection.

Perhaps the most interesting approach to filtering antennas design is the work of (Hassan et al., 2019). The group designed it to operate in the band between 6.0GHz and 8.5GHz, with maximum gain of 6dB in-band, a reflection coefficient below -15dB and a cross-polarisation level of -42dB, simply because its fabrication is treated as a topology optimisation problem, with pre-set requirements on directivity, filtering and impedance bandwidth. As this problem is tackled numerically, a powerful computer is required to constantly optimise the geometry after multiple iterations. This concept calls for a trade-off between antenna size and directivity/out-of-band suppression, and a very powerful piece of equipment to handle solving the complex topology problem.

To conclude, after having touched on all these different approaches to the design of a filtering antenna, the key outtake is that a successful integration demands trade-offs among bandwidth, gain, compactness, harmonic suppression, selectivity and polarization. The ideal filtennas would demonstrate a good harmonics rejection in the stop band, high frequency selectivity, while in the operating frequency range it would produce a relatively flat, high gain. Most importantly, the existence of appropriate isolation between the radiating and the feeding parts is required.

Four distinct methods of implementing a filtering antenna can be deduced from the existing literature. Firstly, direct integration between the antenna and the filter, which offers a wide bandwidth but introduces losses because of impedance mismatches and generally has a larger footprint. Secondly, circuit integration, which lacks in filtering performance and compactness but ensures the impedance matching. Thirdly, element co-design, which reduces the transition losses but suffers from a poor suppression level and selectivity. Lastly, the filter synthesis method which provides a better selectivity with reduced gain due to insertion losses and in general a larger footprint as a trade-off. One challenging remaining question is how to achieve a fast roll-off selectivity with a flat gain. In addition to that, multi band filtering antennas are generally less researched and there is still a lot of research potential in the field.

Therefore, for the scope of this chapter, a new antenna design is studied, capable of functioning in the frequencies required for in the S-Band and for potential future 5G transmissions, based upon global calls for spectrum liberation and allocation in underused frequencies such as in this band. This design is based on open loop resonators for the filtering stage and a patch for the radiating stage in order to create a functional antenna in the 3.2 GHz frequency, while trying to maintain the best possible characteristics. The fundamental description of the mechanism behind the selected resonator structures is found in the next subchapter.

THEORETICAL BACKGROUND FOR FILTERING ANTENNA DESIGN

Antenna Theory Fundamentals

An antenna is the passive device used in every wireless communication system to receive and/or transmit signals. In the receiving mode, it sends the EM wave from free space to the receiver and vice versa for the transmitting mode. The dipole antenna, essentially a piece of wire, is the most well-known antenna, while patch antennas, horn antennas and parabolic reflectors are other common types. The former is made of a flat conductor sheet on top of a dielectric substrate on top of a ground plane, connected at the feed line. It can be easily fabricated using conventional PCB technology. It is relatively low cost, and simple to design.

The main characterization parameters for antennas and their performance are their directivity, efficiency, gain, radiation pattern and bandwidth. Radiation patterns are generally showing lobes, which in turn can be grouped into main, minor, back and side as shown in Figure 3. The maximum radiation intensity direction of the antenna is very easily understood there. In addition to that, if one marks the points of the half power in the main lobe another parameter can be extracted, the half-power beam width (HPBW). The first null (zero) in the radiation pattern is introduced at the zero point of the main lobe. Side lobes are generally the ones close to the main lobe while the back lobe is the one directly opposite. Any remaining ones are minor lobes, and generally are considered a representation of unwanted radiation directions (Balanis, 2005).

The antenna directivity, denoting its radiation performance, can be found as (Balanis, 2005):

$$D = \frac{U}{U_0} = \frac{4\pi U}{P_{rad}}. \tag{1}$$

where U is the intensity of the radiation (W/steradian), U_0 the intensity of the radiation of an isotropic source (W/steradian) and P_{rad} the total radiated power (W). The antenna gain, which describes the power transmitted in the direction of maximum radiation, can be expressed as (Balanis, 2005):

Figure 3. Generic representation of a directional antenna radiation pattern

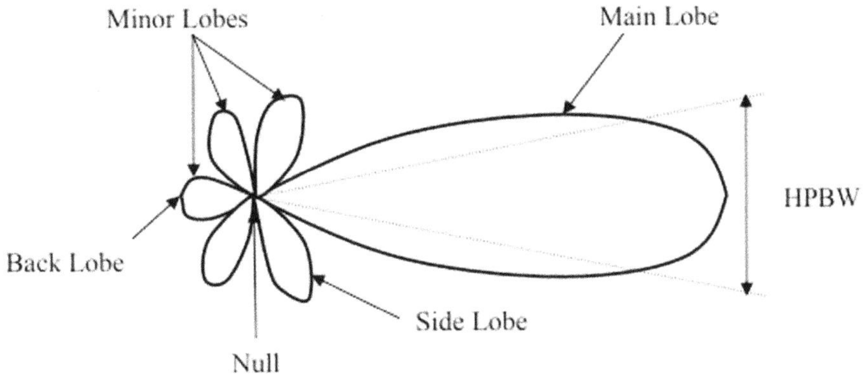

$$G = e_0 D \,. \tag{2}$$

with e_0 being the total antenna efficiency expressed as:

$$e_0 = e_r e_c e_d \,. \tag{3}$$

Where $e_r = 1 - |\Gamma|^2$ the reflection efficiency and e_c, e_d being the conductor and dielectric efficiency, respectively.

In terms of bandwidth, for antennas this can be derived as the specific frequency range at each side of the central frequency, for which a given performance characteristic, such as the return loss is within acceptable values. That value tends to be at least 10 dB (Balanis, 2005). This can be either expressed as the absolute (4) or fractional (5) bandwidth:

$$ABW = f_H - f_L \tag{4}$$

$$.\,FBW = \frac{f_H - f_L}{\sqrt{f_H f_L}} \times 100 \,. \tag{5}$$

f_H and f_L represent the upper and lower frequency range limit at a 10 dB return loss.

It is important to remember that radiation patterns are used to describe the properties of radiation of the antenna as a function of coordinates in space. Therefore, they are usually normalized power values with respect to the maximum ones and are almost always checked in the far-field where the distance does not intervene with the angular distribution of radiated power.

As the replacement of the third filter resonator in this design is a patch antenna, it is good to delve a bit further into the specifics of its design. In its simplest form, a patch antenna is a sandwich of two conductive layers and a dielectric. The second conductive layer acts as a ground plane, while any possible shape can be etched on the first conductive layer to form the patch. However, the tendency is to use some common shape, which simplifies both the fabrication and the analysis of the antenna, as shown in Figure 4. Rectangular patches are normally fabricated with a length L being between a third and a half of the free space wavelength. There are numerous substrates available according to the design needs, but normally their height is $0.003\lambda_0$£h£$0.003\lambda_0$ (Balanis, 2005). A typical range of the dielectric constant (ε_r) is between 2 and 12.

Figure 4. A typical patch antenna

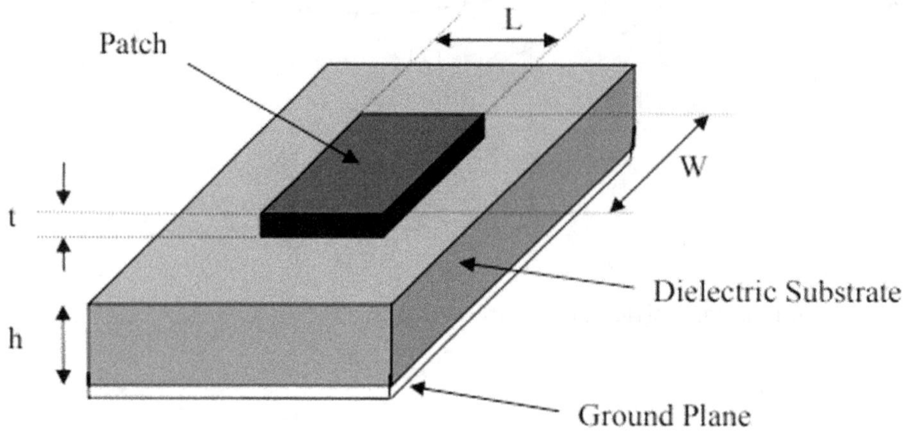

The mechanism of the patch antenna radiation is owing to the existence of fringing fields between the patch edge and the ground plane. The thicker the substrate material, the lower its ε_r values, which in turn improves the antenna efficiency and bandwidth, but at an increased size. Thus, a trade-off is almost always required, achievable by using substrates with higher values of dielectric constant. Patches have several advantages as low fabrication cost, volume and weight, easy integration and capability to handle multiple polarisations and multi-mode frequency operations which make them very popular. However, they are generally suffering from low capacity to handle power, narrow bandwidths, low overall efficiency and gain. They also generally exhibit a high antenna quality factor (Q), representative of the losses related to the antenna, which can prove problematic. Even though Q is inversely proportional to the thickness of the substrate, with thicker substrates a growing proportion of the power delivered to the antenna forms a surface wave. This is unwanted and can be also considered as a power loss.

Figure 5 showcases the five ways to feed a patch antenna, the most common of which is a microstrip line directly attached to the patch. Sometimes, inset gaps are formed on the patch, with the purpose of impedance matching. When using the probe fed approach, the position of attachment will guarantee impedance matching. The drawback is that since a hole needs to be drilled and there is a protrusion for the connection on the ground side, it is not entirely planar. In addition to that, applications in thicker substrates lead to mismatches due to the high inductive impedance component (Kumar & Ray, 2013). Another approach is by using a centred coupling aperture below the antenna, which determines how strong the coupling from the feed line is. The disadvantage of this method is the multi-layered setup which increases the overall thickness and the fabrication complexity. Lastly, through the proximity coupled feed, or alternatively electromagnetic coupling, a feed line is stacked between two dielectric layers and the patch is on top. This approach diminishes spurious radiation and provides a much higher performance controllability as the two substrates may differ, giving much better control over the two individual components, and is many times preferred.

Figure 5. Patch antennas fed through (a) a microstrip line, (b) a probe, (c) proximity coupling, (d) aperture coupling

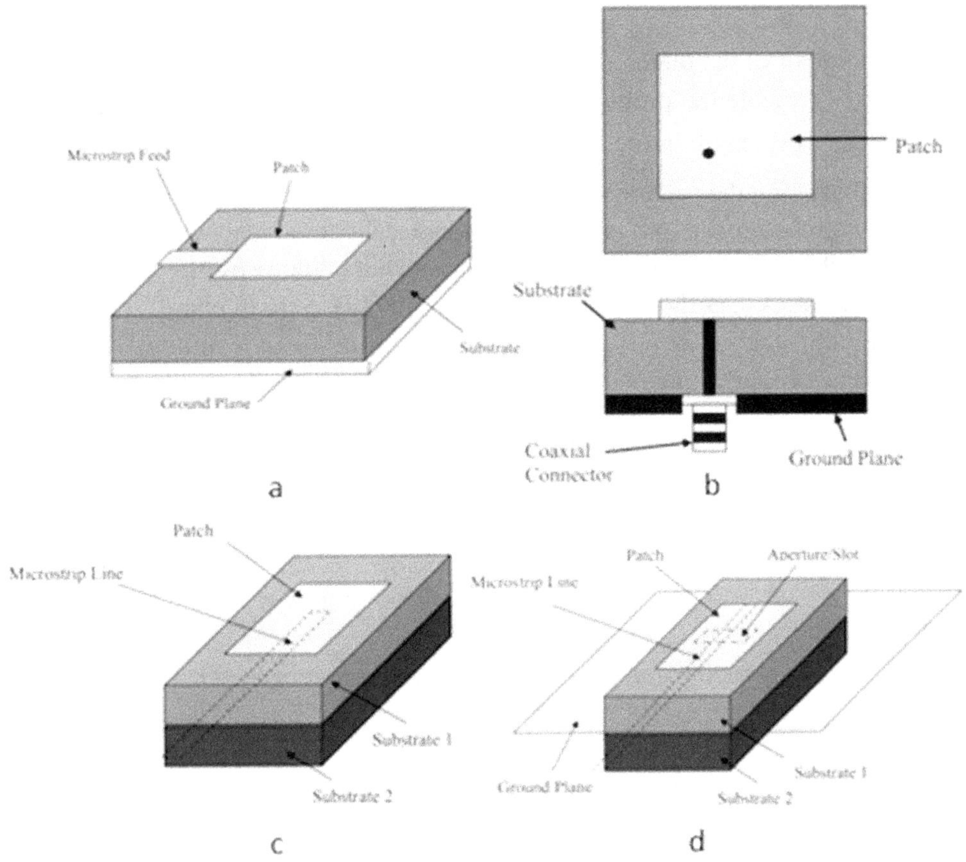

There are three models to analyze the antenna performance that are standard and widespread. In the first one, the microstrip antenna can be considered as a transmission line, and it is the simplest, at an expense of accuracy. The second one is the cavity model and it is quite complex approach but more accurate and gives good estimations, while the full wave model is extremely precise, and can take into account a number of arrays, elements, couplings and arbitrary shapes (Balanis, 2005).

In the case of the first model, it is common knowledge that most of the electric field lines are confined in the substrate, with the remaining few parts in the air around the patch. This results in the line not being capable of supporting a pure transverse electric magnetic (TEM) transmission mode, due to different phase velocities in the dielectric material in the substrate and the air. Therefore, the calculation of an effective dielectric constant is necessary to account for the inline propagation and fringing effects, which will be a little less in value than the nominal ε_r. (Balanis, 2005) presents the formula for the calculation of ε_{reff} as:

$$\varepsilon_{r_{eff}} = \frac{\varepsilon_r + 1}{2} + \frac{\varepsilon_r - 1}{2}\left[1 + 12\frac{h}{W}\right]^{1/2}. \tag{6}$$

with h being the height of the dielectric material and W being the width of the patch. It is also described how in order to enable the propagation of the fundamental mode, the patch length must be a little less than half the wavelength in the dielectric medium, which in turn equals to $\lambda_0 / \sqrt{\varepsilon_{r_{eff}}}$. It can simply be deduced that along the patch width there is maximum voltage and minimum current, while at the edges, the fields can be broken into tangential and normal components with respect to the plane of ground. This analysis shows that the normal components are out of phase by effect of the λ/2 patch length and cancel each other out. As per the tangential components, these are in phase, so they give the maximum radiation normal to the patch surface since their resulting fields combine. Due to fringing, the antenna is perceived to be longer than what it is in reality, and this can be expressed as:

$$\Delta L = 0.412h \frac{\left(\varepsilon_{reff} + 0.3\right)\left(\frac{W}{h} + 0.264\right)}{\left(\varepsilon_{reff} - 0.258\right)\left(\frac{W}{h} + 0.8\right)}. \tag{7}$$

With respect to the cavity model, the general difference is that it considers the slight field variations occurring along the radiating edges. It owes its name to the modelling of the interior of the dielectric substrate as a region of a cavity bound by conducting walls at the top and bottom edges. This is valid for thin substrates (h<<λ). In principle, the cavity walls and material are lossless, leading to a non-radiating cavity, but if one includes a radiation and a loss resistance, R_r and R_L respectively, the radiation and loss mechanism considered, and the new lossy cavity can now represent an antenna, by the means of the effective loss tangent $\delta_{eff} = 1/Q_T$. with Q_T being the total antenna quality factor, expressed as (Pozar, 2005):

$$\frac{1}{Q_T} = \frac{1}{Q_d} + \frac{1}{Q_c} + \frac{1}{Q_r}. \tag{8}$$

with Q_d, Q_c, Q_r, being the quality factors of the dielectric, conductor and radiation, respectively.

The last performance analysis tool is the Method of Moments, where the microstrip patch is modelled by using surface currents, and volume polarisation currents are utilised in modelling the fields in the slab of the dielectric. This method requires software capable of handling full wave EM solving, such as COMSOL Multiphysics, Keysight ADS, Matlab, CST etc. The working mechanism is that an integral equation is produced from the unknown currents and then the whole set of these equations is converted into matrix equations (Balanis, 2005; Harrington, 1968).

A practical approach to calculating patch antenna dimensions can be widely found in bibliographical references and online, and makes use of the following two formulas to calculate the patch antenna length L and width W for a known resonating frequency and material:

$$W = \frac{c}{2f_0\sqrt{\frac{\varepsilon_r + 1}{2}}} \tag{9}$$

$$L = \frac{c}{2f_0\sqrt{\varepsilon_{r_{eff}}}} - 0.824h\left(\frac{\left(\varepsilon_{reff} + 0.3\right)\left(\frac{W}{h} + 0.264\right)}{\left(\varepsilon_{reff} - 0.258\right)\left(\frac{W}{h} + 0.8\right)}\right). \tag{10}$$

Microstrip Theory Fundamentals

After already having addressed the fundamentals of antenna theory, it is worth exploring the principles behind microstrip transmission lines, as given the scope of the chapter, not only the patch antenna can be modelled as one, but also the resonators of the filter will be essentially pieces of microstrip. As already mentioned, microstrip lines are made of two conductive sheets of metal separated by a layer of dielectric material. The conductor properties, along with the substrate specifics define the signal transmission through the line. Figure 6 below makes it clear to understand that the EM fields not only are not entirely confined within the substrate, but also propagate slower than the air above given the greater than 1 value of dielectric permittivity.

Similarly to (6), the effective dielectric constant for narrow ($\frac{W}{h} < 1$. microstrip lines a can be defined as:

$$\mu_{r_{eff}} = \frac{\mu_r + 1}{2} + \left[\frac{\mu_r - 1}{2}\left(1 + 12\frac{h}{W}\right)^{-\frac{1}{2}} + 0.04\left(1 - \frac{W}{h}\right)^2\right]. \tag{7}$$

Another useful expression that can be defined for both wide (8) and narrow (9) lines is the characteristic impedance Z_0. The accuracy of the formulas is about 1% while W, h are the width of the line and height of dielectric respectively, and η is the free-space wave impedance, equal to 120πΩ.

Figure 6. A cross-section of a microstrip transmission line demonstrating the electrical and magnetic field distribution

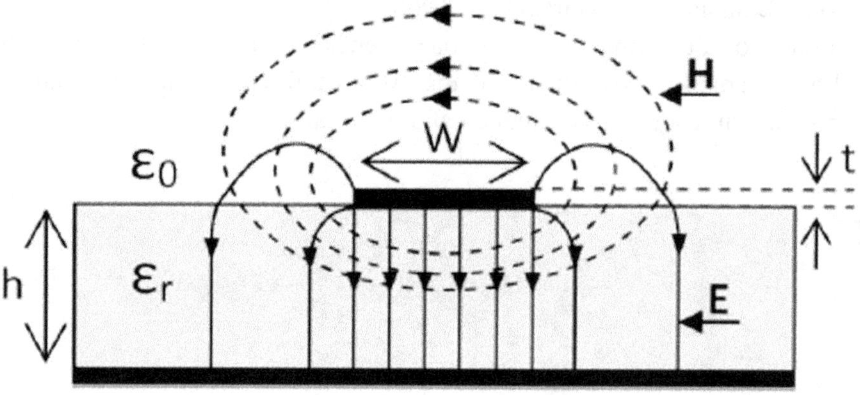

$$-_0 = \frac{\cdot}{\sqrt{\mu_{reff}}}\left[\frac{W}{h}+1.393+0.677\,ln\left(\frac{W}{h}+1.444\right)\right]^{-1}. \tag{8}$$

$$-_0 = \frac{\cdot}{2\grave{A}\sqrt{\mu_{reff}}}\left[ln\left(\frac{8h}{W}+\frac{W}{4h}\right)\right]. \tag{9}$$

Overall, microstrip transmission lines, can be considered highly lossy. These losses are due to the conductor due to power dissipation as heat owing to the finite metallisation, radiation because the semi-open structure nature of the line itself fails to restrict it from propagating away and dielectric because the existence of applied fields causes atoms vibration within the material and lead to rise in temperature. The overall performance is the least impacted by radiation losses, however the conductor loss is greatly varying with frequency due to skin effects. In principle, this means that in the RF domain any current is not any more evenly distributed along a given cross-section of the trace but instead it tends to concentrate near the edges, practically vanishing from the conductor centre. Therefore, at higher frequencies, the cross-sectional area is reduced, and the attenuation constant of the transmission line increases, due to the rise of the resistance of the conductor trace.

Although calculating the attenuation constant are lengthy and complex, thanks to the use of full-wave EM software, one can now fairly easily after simulations generate these constants from a matched microstrip line. The sum of the attenuation constants is useful for determining the resonator unloaded quality factors and is inversely proportional to their power dissipation.

Since the proposed filter here is fabricated using microstrip lines, consideration should be given next to the effect of discontinuities in these lines. Generally, they introduce parasitic reactance, which in turn perturbs the field distributions along the lines. Among the best-known discontinuities are width changes, gaps, bends, junctions and open ends. In Figure 7 these can be seen graphically along with their circuit equivalent. Overall, an equivalent capacitance represents a transformed electric field distribution while an equivalent inductance characterises a transformed magnetic field distribution.

Figure 7. Common types of microstrip line discontinuities and their circuit equivalents: (a) open end, (b) gap, (c) width change, (d) junction and (e) bend (Pozar, 2005)

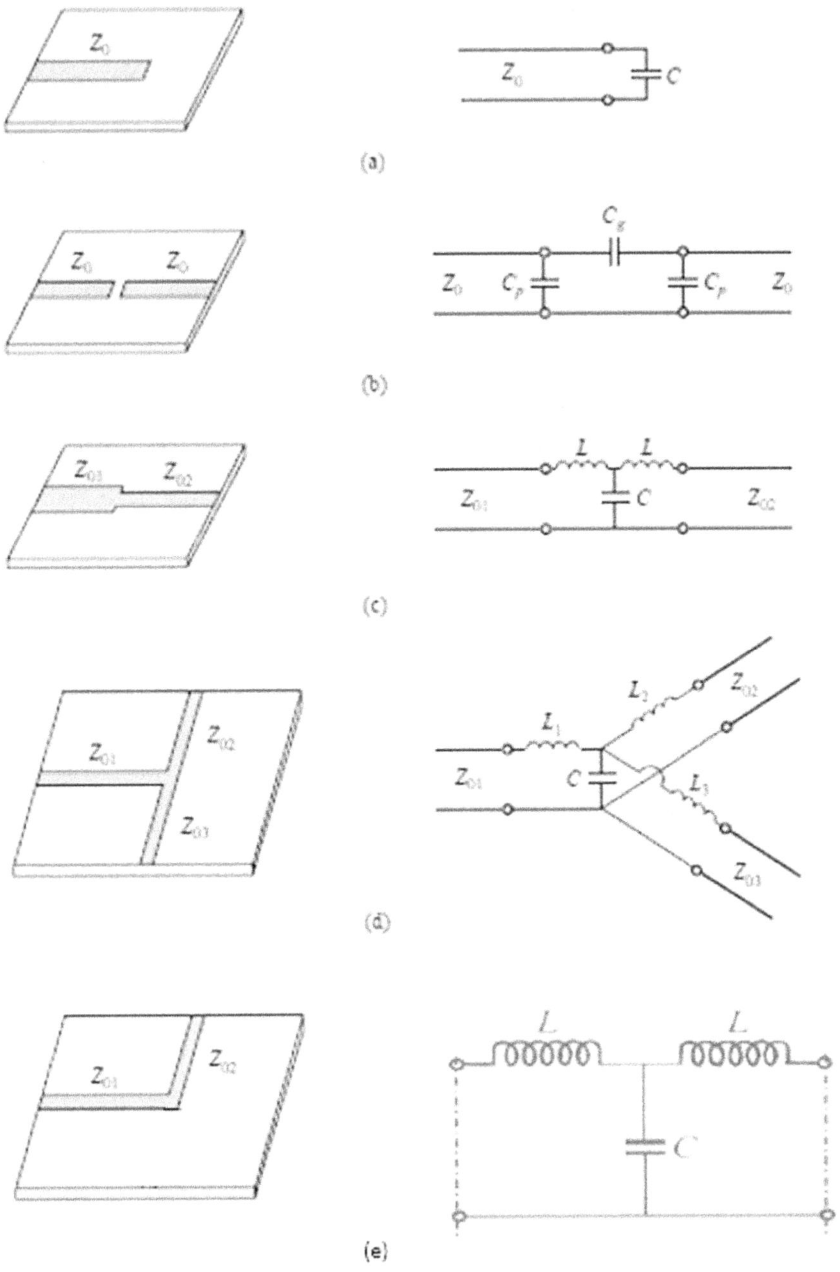

In an open end, a transmission line is open-circuited, which gives rise to fringing, and leads to an increase in the effective length of the transmission line. While $0.2 \le \frac{W}{h}$ .nd $50 \ge \varepsilon_r \ge 2$ .the length change and capacitance can be calculated as (Edward, 2000):

$$\Delta L = 0.412h \times \frac{\varepsilon_{reff} + 0.3}{\varepsilon_{reff} - 0.258} \times \frac{\frac{W}{h} + 0.264}{\frac{W}{h} + 0.8} \tag{10}$$

$$C_p = \Delta L \times \frac{\sqrt{\varepsilon_{reff}}}{cZ_0} \tag{11}$$

A gap is a self-explanatory discontinuity between two microstrip lines. A Π-network consisting of three capacitances C_p, C_p, C_g leads to the deduction of the corresponding suspectances. In the case a plane of symmetry exists, i.e. two lines are identical, an odd and even mode analysis method can be utilised. In the former, the incident waves are identical but, in the latter, they are 180^0 out of phase. When the values for the odd and incident mode capacitances C_o and C_e are known, the formulas below can lead to the calculation of the required values (Edward, 2000):

$$C_p = 0.5C_e \tag{12}$$

$$C_g = 0.5C_o - 0.25C_e \tag{13}$$

$$C_e = 12W \times \left(\frac{\varepsilon_r}{9.6}\right)^{0.9} \times \left(\frac{s}{W}\right)^{m_e} \times e^{k_e} \tag{14}$$

$$C_o = W \times \left(\frac{\varepsilon_r}{9.6}\right)^{0.8} \times \left(\frac{s}{W}\right)^{m_o} \times e^{k_o} \tag{15}$$

$$m_o = \frac{W}{h} \times 0.619 log\left(\frac{W}{h}\right) - 0.3853 \Big|_{0.1 \le s/W \le 1} \tag{16}$$

$$k_o = 4.26 - 1.453log\left(\frac{W}{h}\right)\Bigg|_{0.1\leq s/W\leq 1} \tag{17}$$

$$m_e = 0.8675\Big|_{0.1\leq s/W\leq 0.3} \tag{18}$$

$$m_e = \frac{1.565}{\left(\frac{W}{h}\right)^{0.16}} - 1\Bigg|_{0.3\leq s/W\leq 1} \tag{19}$$

$$k_e = 2.043\times\left(\frac{W}{h}\right)^{0.12}\Bigg|_{0.1\leq s/W\leq 0.3} \tag{20}$$

$$k_e = 1.97 - \frac{0.03h}{W}\Bigg|_{0.3\leq s/W\leq 1} \tag{21}$$

In the case of a width change the equivalent circuit is comprised of a shunt capacitance C at the junction, and a series inductance L_1,L_2 at each side of the capacitance representing the two microstrip lines, and in case of symmetry can be expressed as (Edward, 2000):

$$C = 0.00137h\times\frac{\sqrt{\varepsilon_{reff1}}}{Z_{01}}\times\left(1-\frac{W_2}{W_1}\right)\times\frac{\varepsilon_{reff1}+0.3}{\varepsilon_{reff1}-0.258}\times\frac{\frac{W_1}{h}+0.264}{\frac{W_1}{h}+0.8} \tag{22}$$

$$L_1 = L\times\frac{L_{w1}}{L_{w1}+L_{w2}} \tag{23}$$

$$L_2 = L\times\frac{L_{w2}}{L_{w1}+L_{w2}} \tag{24}$$

$$L = 0.000987h \times \left[1 - \frac{Z_{01}}{Z_{02}}\sqrt{\frac{\varepsilon_{reff1}}{\varepsilon_{reff2}}}\right]^2 \tag{25}$$

$$L_{wi} = \frac{Z_{0i}\sqrt{\varepsilon_{reffi}}}{c}. \tag{26}$$

with h being the height of the dielectric, W_i the transmission line width, ε_{reffi} the effective dielectric constant, Z_{0i} the characteristic impedance and L_{wi} the inductance per unit length of the corresponding transmission line denoted by the i value, while c being the propagation speed of light in free space.

The last discontinuity which is relative to this design is the bend, where a transmission line direction is changed, regularly by 90^0. The equivalent circuit of a bend is a T-network of a shunt capacitance C, and two inductances L in series on both sides. An approximation of these values can be found below (Hong & Lancaster, 2001):

$$L = 100h \times \left(4\sqrt{\frac{W}{h}} - 4.21\right) \tag{27}$$

$$C = \left.\frac{\frac{W}{h}(14\varepsilon_r + 12.5) - 18\varepsilon_r + 2.25}{\sqrt{\frac{W}{h}}} + \frac{0.02\varepsilon_r}{\frac{W}{h}}\right|_{\frac{W}{h}<1} \tag{28}$$

$$C = W\left[\frac{W}{h}(9.5\varepsilon_r + 1.25) + 5.2\varepsilon_r + 7\right]\Bigg|_{\frac{W}{h}\geq 1} \tag{29}$$

Before proceeding to the filter design fundamentals, it is worth exploring the skin effect. This describes a behaviour when alternating current mostly flows towards the outer boundaries of an electrical conductor, increasing the current density near the surface. This is occurring at a given depth below the surface, termed skin depth, and the majority of the current flow, will be confined within a layer about 3 times this skin depth deeper into the material. A direct current on the other hand will remain evenly distributed across a given conductor cross-section. With the increase of frequency, this phenomenon increases the effective resistance of the conductor, and reduces its effective cross-section. Therefore, one should check the metallisation thickness of the selected board before proceeding to the fabrication. The skin depth δ can be expressed as:

$$\delta = \sqrt{\frac{2\rho}{\omega\mu_r\mu_0}} \tag{30}$$

with ρ being the metallisation material resistivity, μ_r its relative magnetic permeability, ω the angular frequency and μ_0 equal to $4\pi*10^{-7}$ H/m. As the aim of this chapter is to design a filtering antenna at 3.2GHz, and the substrate board selected uses copper metallisation at a 17 μm thickness, this effect will not be further discussed as the skin depth for the working range of frequencies is 0.96μm to 1.33μm, well below the metallisation thickness.

Filter Synthesis Fundamentals

As the project aims to integrate an antenna with a filter, it is time to tackle some basic principles of filter theory. The scattering, or S-Parameters are a common descriptor of a filter performance. For any two-port network they are related to the reflected and incident waves in terms of the following matrix (Balanis, 2005):

$$\begin{bmatrix} b_1 \\ b_2 \end{bmatrix} = \begin{bmatrix} S_{11} & S_{12} \\ S_{21} & S_{22} \end{bmatrix} \begin{bmatrix} \alpha_1 \\ \alpha_2 \end{bmatrix} \tag{31}$$

as commonly known, S_{11} is the input reflection coefficient, S_{22} is the output reflection coefficient with the respective ends properly terminated, S_{21} and S_{12} are the forward and reverse transmission coefficients. Filters, however, are also characterised by the transmission loss (L_A) and return loss (L_B), which are assumed positive and obtained by the S-parameters as:

$$L_A = -20\log_{10} S_{21}\, dB \tag{32}$$

$$L_B = -20\log_{10} S_{11} dB \tag{33}$$

When designing a filter, an initial assumption is made that the inside of the network is a lumped element circuit, also known as a prototype filter circuit and demonstrated in Figure 8 below. The g values of this filter can be extracted from its response and are presented in (34)-(39). Commonly, a Chebychev filter is used, as it has a maximally flat stop band and an equiripple passband response.

$$g_0 = 1.0 \tag{34}$$

$$g_1 = \frac{2}{\gamma}\sin\frac{\pi}{2n} \tag{35}$$

Figure 8. A lowpass prototype filter in a pi (a) and T configuration (b)

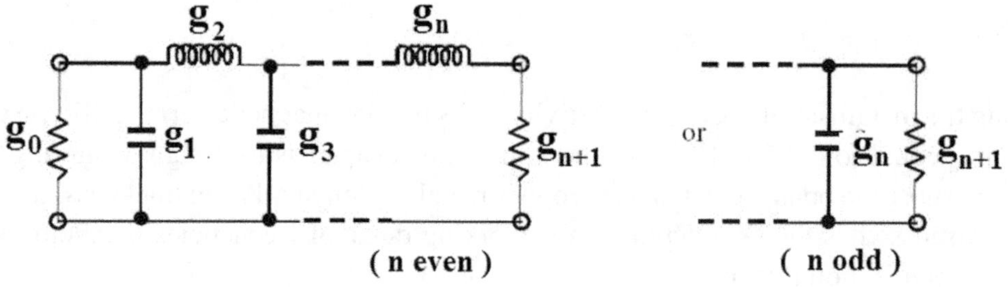

a

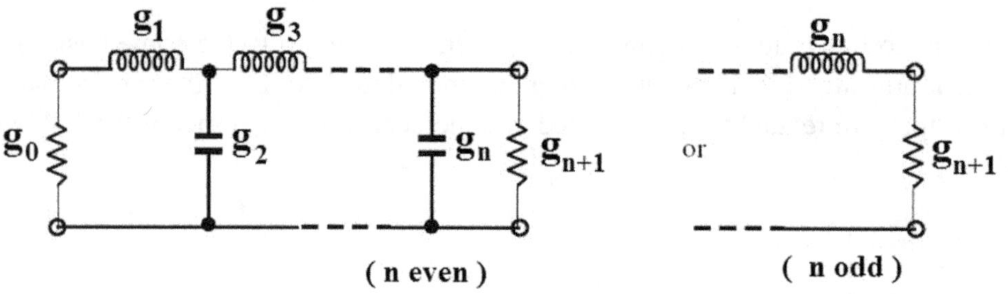

b

$$g_i = \frac{1}{g_{i-1}} \frac{4sin\left(\frac{(2i-1)\pi}{2n}\right)sin\frac{(2i-3)\pi}{2n}}{\gamma^2 + sin^2\left(\frac{(i-1)\pi}{n}\right)} \; for\, i = 2,3\ldots n \tag{36}$$

$$g_{n+1} = \begin{cases} 1.0 \; for\, n\, odd \\ \coth^2 \frac{\beta}{4} \; for\, n\, even \end{cases} \tag{37}$$

$$\beta = \ln(\coth \frac{L_{Ar}}{17.37}) \tag{38}$$

$$\gamma = \sinh \frac{\beta}{2n} \tag{39}$$

with L_{Ar} the specified passband ripple level.

The above values can be utilized in a real filter to find the required inductances and capacitances for the realisation of different cut-off frequency and source impedance as shown in (Zhou, 2010):

$$C_k = \frac{1}{Z_s}\frac{g_k}{\omega_c^{'}} \tag{40}$$

$$L_k = Z_s \frac{g_k}{\omega_c^{'}} \tag{41}$$

$$Z_L = \begin{cases} \dfrac{g_{n+1}}{Z_s}(n\,even) \\ g_{n+1}Z_s(n\,odd) \end{cases} \tag{42}$$

with ω_c¢ being the new cut-off frequency, Z_S, Z_L the new source and load impedance respectively. It should be noticed that, in the prototype filter, the impedances of the source and load could be different for a Chebyshev filter, while they are the same for the Butterworth case.

In a similar way, we can study the transformation from lowpass to bandpass. By using lowpass to bandpass mapping (Figure 9) we can relate the BPF response to the one of the lowpass as (Hong & Lancaster, 2001; Matthaei et al., 1980):

$$\omega = \frac{\omega'}{\omega'_c}\frac{\omega_2 - \omega_1}{2} \pm \frac{1}{2}\sqrt{\left(\frac{\omega'}{\omega'_c}\right)^2 (\omega_2 - \omega_1)^2 + 4\omega_1\omega_2} \tag{43}$$

In order to obtain the transformed BPF response, the shunt capacitance in the prototype lowpass filter will have to be changed to a new susceptance:

$$jB_k = j\frac{\omega'_c\,\omega_0}{\omega_2 - \omega_1}\left(\frac{\omega}{\omega_0} - \frac{\omega_0}{\omega}\right)g_k = j\omega\frac{\omega'_c\,g_k}{\omega_2 - \omega_1} + \frac{1}{j\omega\dfrac{\omega_2 - \omega_1}{\omega'_c\,\omega_1\omega_2 g_k}}(k\,odd) \tag{44}$$

which is equivalent to a shunt tuned resonator with a shunt inductance and capacitance given as:

$$C_k = \frac{\omega'_c\,g_k}{\omega_2 - \omega_1} \tag{45}$$

Figure 9. The circuit equivalent of a Chebychev type BPF after transformation from a prototype lowpass filter (a) and its typical response (b))

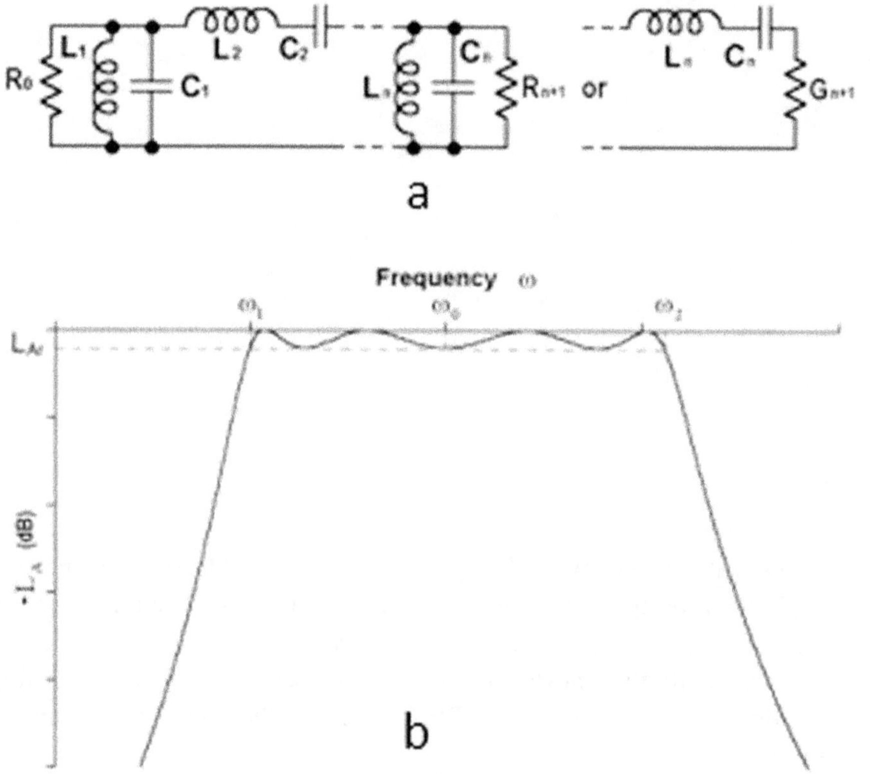

$$L_k = \frac{\omega_2 - \omega_1}{\omega'_c \omega_1 \omega_2 g_k} (k\, odd) \tag{46}$$

In a similar way, the series inductance can be changed to a new reactance, equivalent in turn to a series tuned resonator with series inductance and capacitance given as (Zhou, 2010):

$$L_k = \frac{\omega'_c g_k}{\omega_2 - \omega_1} \tag{47}$$

$$C_k = \frac{\omega_2 - \omega_1}{\omega'_c \omega_1 \omega_2 g_k} (k\, odd) \tag{48}$$

Microwave filters are commonly created using microwave resonators. These devices can be cavity resonators, microstrip resonators, dielectric etc., and can be modelled similarly to a lumped element one. In this case a resonant circuit is a typical RLC configuration either in series or in parallel. The losses in a resonant circuit can either be described by R in the lumped element representation, or by the unloaded quality factor (Q_u), defined as $Q_u = \frac{\omega L}{R}$ .n series and $Q_u = \frac{\omega C}{G}$ .in parallel circuits. Similar to the antenna analysis, the total unloaded quality factor is once again dependent on the conductor, dielectric and radiation losses and is expressed as (Pozar, 2005):

$$\frac{1}{Q_u} = \frac{1}{Q_d} + \frac{1}{Q_c} + \frac{1}{Q_r} \tag{49}$$

Many a time, these losses in resonators, can degrade the passband response of the filter, and the Q_u can be used for the insertion loss calculation, of a BPF design using resonators with finite value of $Q_{u.}$ The increase in this passband insertion loss in dB is (Hong & Lancaster, 2001):

$$\Delta L_{A0} = 4.343 \sum_{i=1}^{n} \frac{\Omega_c}{FBW \times Q_{ui}} g_i \tag{50}$$

A coupled resonator circuit is generally used to design a filter, by combining accordingly all the resonators. For a coupled resonator BPF of nth order, the external quality factor Q_{e1} and Q_{en} and the coupling coefficient ($M_{i,\,i+1}$) are the necessary parameters (Hong & Lancaster, 2001).

$$Q_{en} = \frac{g_n g_{n+1}}{FBW} \tag{51}$$

$$M_{i.i+1} = \frac{FBW}{\sqrt{g_i g_{i+1}}} \ for\ i = 1, 2, \ldots, N-1 \tag{52}$$

For the scope of this project, the resonators used are microstrip square open loop resonators which can be obtained by folding a straight open resonator. Due to the corners and the fringing capacitance between the open ends, it is impractical to calculate the EM fields in a square resonator. However, a qualitative analysis can be performed, as an analogy to the behaviour of the straight one. As a next step, by utilising full wave EM software a comparison can be made against the actual EM field distribution.

It is simple to see that the resonant frequency for the above resonator can be found by its input admittance at any given point in its length. If its total length is $\alpha_t = \alpha_1 + \alpha_2$ the resonator can maintain a standing wave whenever the input admittance Y_{in} is zero, which produces infinite resonating points at $\alpha_t = n\pi$ or $\alpha = n\lambda/2$.

When the input admittance is infinite, or in other words the input impedance is zero, a voltage null can be expressed. At the first resonant frequency of the resonator such a null exists only at the position π/2, at the second resonant frequency we can find two at π/2 and 3π/2 and so on. This means in principle, that once we know the position of these voltage nulls, we know where we cannot excite the resonator. Therefore, we can also select this way to excite only the odd or even modes etc. as seen in Figure 10.

Figure 10. Resonator excitation (a) in a null of the fundamental mode and (b) symmetrically with respect to both open ends. Both ways only target the even modes.

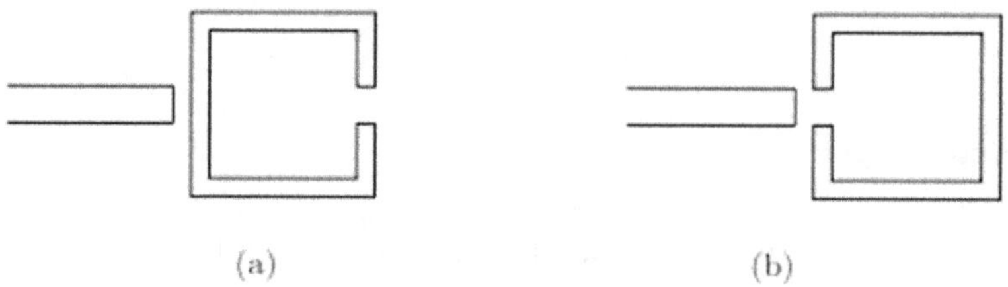

After having briefly seen the way a resonator can be excited from a feed, it is time to delve deeper into the main interaction between resonators, proximity coupling. The ratio of coupled energy to stored energy, is coupling coefficient k, and can be expressed as (Hong & Lancaster, 2001):

$$k=\frac{\int \varepsilon \mathrm{E}_1\cdot \mathrm{E}_2 dv}{\sqrt{\int \varepsilon \mathrm{E}_1{}^2 dv \int \varepsilon \mathrm{E}_2{}^2 dv}}+\frac{\int \mu H_1\cdot H_2 dv}{\sqrt{\int \mu H_1{}^2 dv \int \mu H_2{}^2 dv}} \tag{53}$$

The E, H represent the electric and magnetic fields of the resonator corresponding to their subscript, while the first term describes the coupling due to the electric fields interaction and the second the coupling due to the magnetic fields interaction. Hence, the coupling can be electric, magnetic, or mixed, depending on the position of the resonators, as demonstrated in Figure 11.

An alternative expression to (53) can be obtained based on the fact that two coupled resonators resonate together at f_1 and f_2 generally different from their fundamental resonant frequency f_0 with this difference increasing following an increase in their coupling.

$$k=\frac{f_2^2-f_1^2}{f_1^2+f_2^2} \tag{54}$$

This is a very useful formula as by using of full wave EM simulation software and loosely coupling a resonator to two identical feed lines, one can plot for different separation S the different coupling coefficient.

Figure 11. A representation of different coupling modes due to different resonator orientations. Electric (a) magnetic (b) and mixed (c), (d).

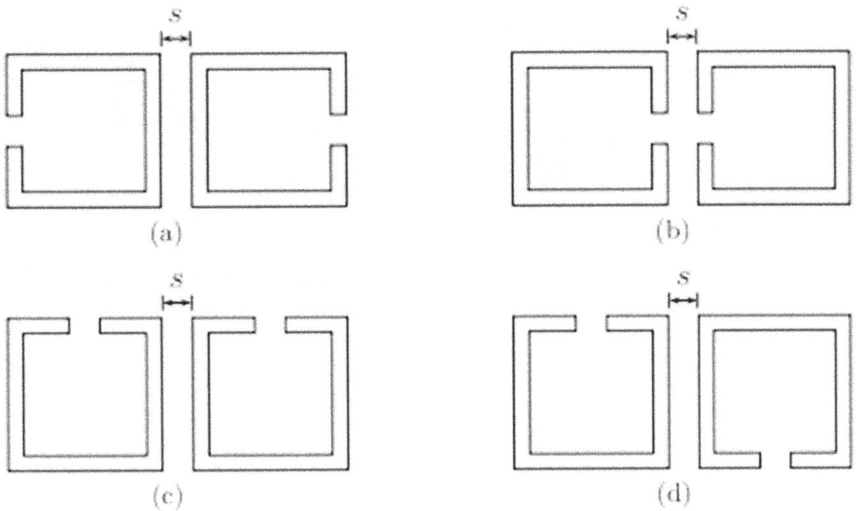

Filtering Antenna Simulation Setup

An overview of the basic operational principles of an EM full wave simulation software like Keysight ADS could be presented here, but instructions are readily available online, and setting up the program parameters is trivial. However, it is worth mentioning the steps of the simulation to be carried out to successfully model and characterize a filtering antenna, based on the fundamental principles presented earlier. Simply, the designer should first model the filter and antenna separately, to extract and optimize the necessary parameters for their integration, and then move on to merging the two stages.

Initially, one should model a single -in this case open loop- resonator, loosely coupled to two identical 50Ω microstrip lines. Software like ADS have parameter tuning capabilities, so when for example the operating frequency is 3.2GHz, but the simulated response has a peak at 3.197GHz, the software can be programmed to tweak the parameters automatically to fix such issues. When the final parameters are satisfactory, two resonators should be put together, loosely coupled to two identical 50Ω microstrip lines. This step will allow the estimation of the coupling coefficient k between them, and it is good practice to get this value with respect to the inter-resonator spacing to observe any anomalies. That process is described in Figure 12, where the layout of the two resonators is shown, and the coupling coefficient is plotted as a function of resonator spacing.

Figure 12. Layout view of the two resonators layout for the estimation of the coupling coefficient k at the top, and the coupling coefficient versus the resonator spacing at bottom.

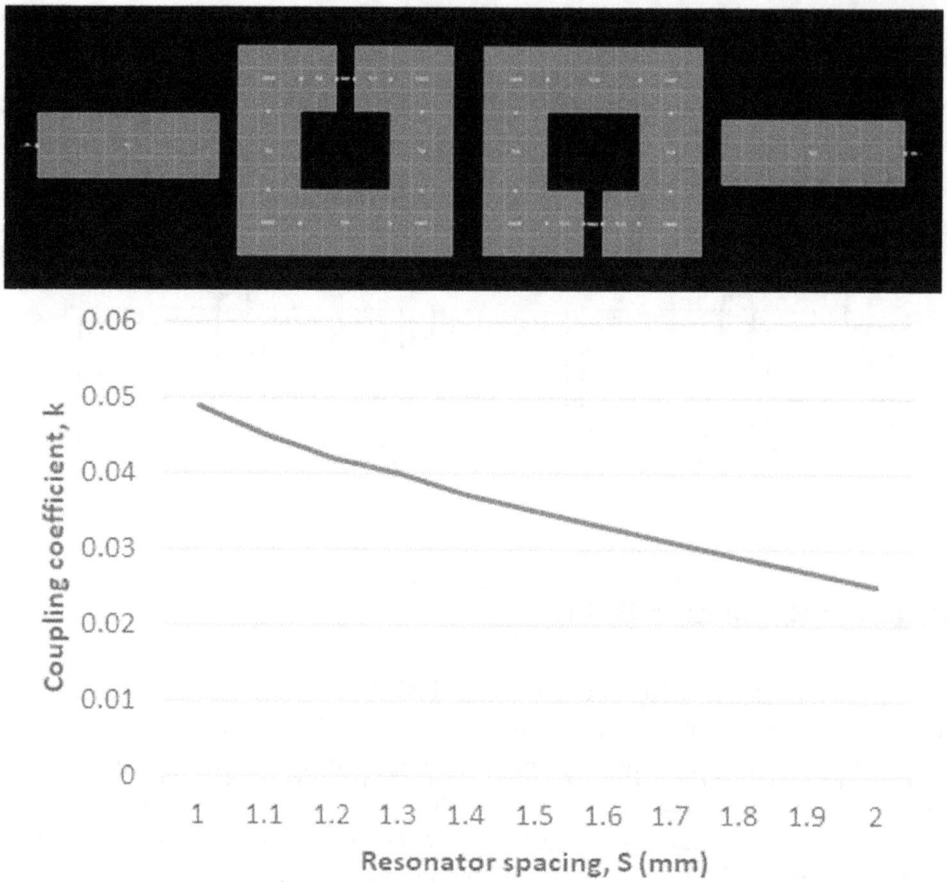

Next, a single resonator once again, should be coupled to the actual feeding design of the filtering antenna and at the same 50Ω transmission line on the other side, to estimate the external quality factor of the input resonator (Figure 13). This value, which models the losses related to the resonator excitation system can be found by plotting the S(2,1) magnitude response. It should generally be plotted as a function of the spacing between the input resonator and the input feed line. Following that, the full filter stages should be assembled, and some parameter optimization should be performed to tweak the final filter response.

Figure 13. Matching circuit for the estimation of Q_{ext} and its variation versus the change of the feed line-resonator spacing

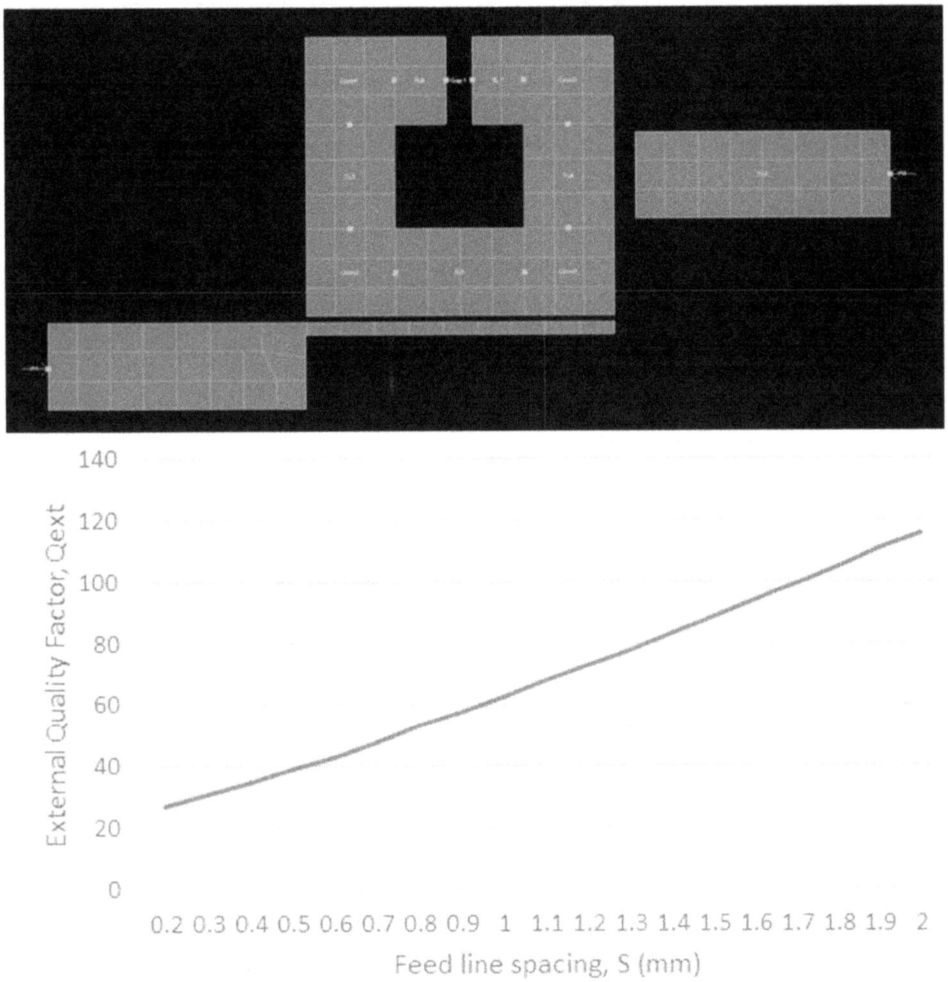

Once the steps for characterizing the filter stage are complete, the patch antenna should be designed, to a geometry deduced by the formulas discussed earlier. It should be noted that extra care should be taken in maintaining a similar Q_{ext} value, to ensure proper matching with the rest of the filter, since there is no direct feed line towards it from the filter, and all signal propagation will reach it solely by means of electromagnetic coupling.

When the components are integrated, the designer should expect a very large amount of fine-tuning of the individual parameters, such as geometries, distances and positioning, to achieve the desired response, but once again, albeit time-consuming, this task can be automated by the software. Table 1 below summarizes these key parameters, both in the theoretical calculation stage, and after optimization by the software.

Table 1. Key design parameters of the proposed filtering antenna

Parameter	Original Value	Final value
Feed Line Width, Outer	3.437mm	3.437mm
Feed Line Width, Inner	0.5mm	0.62mm
Feed Line Length, Outer	10mm	10mm
Feed Line Length, Inner	11.718mm	12.214mm
Resonator Width	3.437mm	3.437mm
Resonator Gap	1mm	1mm
Resonator Height	11.718mm	10.694mm
Resonator 1 Length	11.718mm	12.214mm
Resonator 2 Length	11.718mm	12.275mm
Coupling Gap, Resonators	1.7mm	1.32mm
Coupling Gap, Feed	0.2mm	0.082mm
Coupling Gap, Patch	0.75mm	0.81mm
Patch Length	24.39mm	23.307mm
Patch Width	31.06mm	37mm

The end product in this example is shown in Figure 14 below where from left to right are displayed the overall design and footprint, the return loss of the filtering antenna, and its gain and directivity. The design demonstrates a fractional bandwidth of 3.62% (or 116MHz), a high radiation efficiency, a return loss better than -15dB in the operating band, as well as the fact that there exist three reflection zeros, indicative that the design indeed functions as a third order bandpass filter, cementing the assumption that the patch can substitute the third filter resonator. It can also be noticed that the response has no insertion loss, due to the radiating antenna co-designed with the filter.

Figure 14. Layout view of the designed filtering antenna, its return loss response, and its gain and directivity.

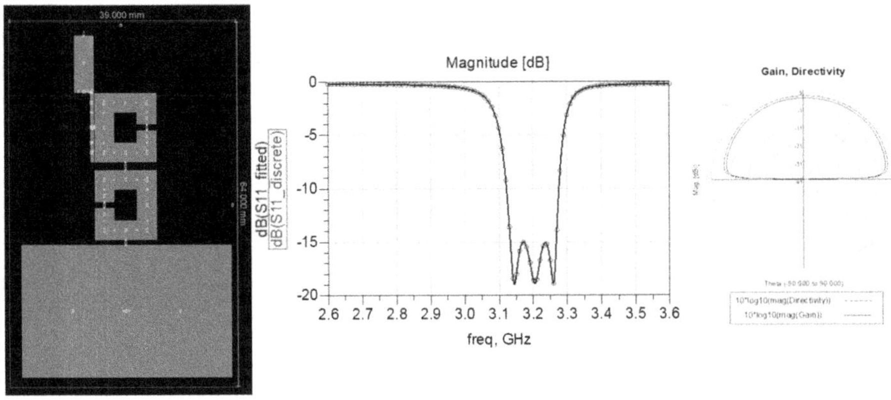

SOLUTIONS AND RECOMMENDATIONS

An important question to ask when designing a filtering antenna is what the maximum possible size is the design could take up, and based on that select a suitable substrate, as based on its properties, the dimensions will change. Sometimes, a substrate should be thin, flexible, or have some other requirement, that will restrict the selection process.

Careful consideration should also be given to the operating frequency of the filtering antenna, as very high frequencies start introducing problems of their own, and degrade the performance of the devices, to the point that microstrip line technology is no longer suitable for these applications.

Generally, one could verify the filtering antenna by comparing its filtering characteristics to a similar order filter, and its radiating characteristics to an equivalent antenna. If there is no severe performance difference such as in this case, the device is an adequate substitution of the two discrete ones.

Lastly, another method of verifying the device performance, is to compare it with that of the lumped element model equivalent circuit, but that requires another set of analysis to create.

FUTURE RESEARCH DIRECTIONS

Even though the focus of this chapter has been to provide the reader with understanding of the basic process of a successful filtering antenna design through introducing the basic theory required, as well as the various parameters and issues to be taken into consideration, this research area is very accommodating to future work, particularly since the practical applications of filtering antennas are rising and will continue to rise for the foreseeable future.

A great direction to be explored, would be the addition of another set of resonators, oriented on the other axis of the antenna, which will also have it as the last element, effectively creating a dual-band filtering antenna, which can of course be tuned according to the elemental geometry to two distinct frequencies. A great advantage is that by doing so, a single antenna can service devices needing connectivity to different technologies, not necessarily operating in the same bands, such as 4G and Bluetooth, or ZigBee and 5G.

Another potential research interest would be the miniaturization of the individual components, which would lead to a further reduction in the filtering antenna footprint, without significantly compromising performance. This could be achieved through various ways, such as incorporating the resonators within the patch, using complementary resonators, or even 3D stacking.

Lastly, one of the trending research directions is the integration of wireless devices, and by extension filtering antennas, to unconventional hosts, for example in textiles. The fabrication is substantially different, as in this case the substrates, layers and geometries are all based on conductive or insulating threads, but the general principles remain the same.

CONCLUSION

The effort of the miniaturization of the RF frontend by means of integrating the antenna and filter blocks is still an active research area. This is owing to the rising demand of utilization in the sub-6GHz bands by a number of wireless communication devices, and the associated large physical dimensions of the traditional resonators. Many techniques have already been briefly discussed in this chapter, but the most prominent design incorporates open loop resonators, due to their ease of fabrication, generally easy geometrical calculation, versatility, and compactness.

This chapter analyzed an integration technique for the filter and antenna, through which the final stage resonator of a band pass filter is replaced by a patch antenna. As each analysis step demonstrated, this patch antenna can prove an adequate replacement for the resonator without a significant compromise in the individual component performance in either filtering or radiating. The overall dimensionality of the proposed design is 3.8cm by 6.4cm and its resonant frequency is at 3.2GHz, if fabricated on a substrate of similar dielectric and thickness as to the one of the calculations, such as Rogers RO4003c, with a thickness of 1.524mm.

EM simulation solver software is a great aid in predicting the performance of the filtering antenna, and to avoid going through the expensive fabrication process before having an understanding of the expected behaviour, while the capabilities and applications for filtering antennas are practically limitless.

ACKNOWLEDGMENT

This research was supported by Erasmus+ Programme Key Action 1: Erasmus Mundus Joint Master Degrees [Ref. Number 2017 – 2734/001 – 001, Project number - 586686-EPP-1-2017-1-UK-EPPKA1-JMD-MOB].

REFERENCES

Ahsan, M. R., Islam, M. T., Habib Ullah, M., Mahadi, W. N. L., & Latef, T. A. (2014). Compact Double-P Slotted Inset-Fed Microstrip Patch Antenna on High Dielectric Substrate. *The Scientific World Journal, 2014*, 1–6. Advance online publication. doi:10.1155/2014/909854 PMID:25165750

Ahsan, M. R., Islam, M. T., Habib Ullah, M., & Misran, N. (2014). Bandwidth enhancement of a dual band planar monopole antenna using meandered microstrip feeding. *TheScientificWorldJournal, 2014*, 1–8. Advance online publication. doi:10.1155/2014/856504 PMID:24723832

Balanis, C. (2005). *Antenna theory: Analysis and design.* Wiley-Interscience.

Chen, S., Zhao, Y., Peng, M., & Wang, Y. (2014). A codesigned compact dual-band filtering antenna with PIN loaded for WLAN applications. *International Journal of Antennas and Propagation, 2014*, 1–6. Advance online publication. doi:10.1155/2014/826171

Chen, X., Zhao, F., Yan, L., & Zhang, W. (2013). A compact filtering antenna with flat gain response within the passband. *IEEE Antennas and Wireless Propagation Letters, 12*, 857–860. doi:10.1109/LAWP.2013.2271972

Chu, H., Hong, H., Zhu, X., Li, P., & Guo, Y. X. (2018). Implementation of Synthetic Material in Dielectric Resonator-Based Filtering Antennas. *IEEE Transactions on Antennas and Propagation, 66*(7), 3690–3695. doi:10.1109/TAP.2018.2819891

Chuang, C. T., & Chung, S. J. (2009). New printed filtering antenna with selectivity enhancement. *European Microwave Week 2009, EuMW 2009: Science, Progress and Quality at Radiofrequencies, Conference Proceedings - 39th European Microwave Conference, EuMC 2009,* 747–750. 10.1109/EUMC.2009.5296120

Chuang, C. T., & Chung, S. J. (2011a). A compact printed filtering antenna using a ground-intruded coupled line resonator. *IEEE Transactions on Antennas and Propagation, 59*(10), 3630–3637. doi:10.1109/TAP.2011.2163777

Chuang, C. T., & Chung, S. J. (2011b). Synthesis and design of a new printed filtering antenna. *IEEE Transactions on Antennas and Propagation, 59*(3), 1036–1042. doi:10.1109/TAP.2010.2103001

Deng, J., & Feng, L. (2019). Dual-band microstrip filtering antennas with symmetrical slots. *Progress in Electromagnetics Research Letters, 86*(May), 13–19. doi:10.2528/PIERL19050404

Duan, W., Zhang, X. Y., Pan, Y. M., Xu, J. X., & Xue, Q. (2016). Dual-Polarized Filtering Antenna with High Selectivity and Low Cross Polarization. *IEEE Transactions on Antennas and Propagation, 64*(10), 4188–4196. doi:10.1109/TAP.2016.2594818

Edward, T. C. (2000). *Foundation of Interconnect and Microstrip Design* (3rd ed.). Wiley. doi:10.1002/9781118894514

Harrington, R. F. (1968). *Field Computation by Moment Methods.* Macmillan.

Hassan, E., Martynenko, D., Wadbro, E., Fischer, G., & Berggren, M. (2019). Compact differential-fed planar filtering antennas. *Electronics (Switzerland), 8*(11), 1–15. doi:10.3390/electronics8111241

He, X., Zhang, Y., Du, M., & Xu, J. (2018). Lightweight and compact high-gain filtering aperture antenna fabricated by three-dimensional printing technology. *IEEE Antennas and Wireless Propagation Letters*, *17*(7), 1141–1144. doi:10.1109/LAWP.2018.2830815

Hong, J. S., & Lancaster, M. J. (2001). *Microstrip Filters for RF/microwave applications*. Wiley & Sons. doi:10.1002/0471221619

Jadhav, J. B., & Deore, P. J. (2017). Filtering antenna with radiation and filtering functions for wireless applications. *Journal of Electrical Systems and Information Technology*, *4*(1), 125–134. doi:10.1016/j.jesit.2016.10.007

Jin, J. Y., Liao, S., & Xue, Q. (2018). Design of Filtering-Radiating Patch Antennas with Tunable Radiation Nulls for High Selectivity. *IEEE Transactions on Antennas and Propagation*, *66*(4), 2125–2130. doi:10.1109/TAP.2018.2804661

Kumar, G., & Ray, K. P. (2013). Broadband Microstrip Antennas. In *Microstrip and Printed Antenna Design*. Artech House Inc. doi:10.1049bew048e_ch4

Li, J. F., Chen, Z. N., Wu, D. L., Zhang, G., & Wu, Y. J. (2018). Dual-Beam Filtering Patch Antennas for Wireless Communication Application. *IEEE Transactions on Antennas and Propagation*, *66*(7), 3730–3734. doi:10.1109/TAP.2018.2835519

Lin, C. K., & Chung, S. J. (2011). A compact filtering microstrip antenna with quasi-elliptic broadside antenna gain response. *IEEE Antennas and Wireless Propagation Letters*, *10*, 381–384. doi:10.1109/LAWP.2011.2147750

Liu, Y. T., Leung, K. W., Ren, J., & Sun, Y. X. (2019). Linearly and Circularly Polarized Filtering Dielectric Resonator Antennas. *IEEE Transactions on Antennas and Propagation*, *67*(6), 3629–3640. doi:10.1109/TAP.2019.2902670

Luo, G. Q., Hong, W., Tang, H. J., Chen, J. X., Yin, X. X., Kuai, Z. Q., & Wu, K. (2007). Filtenna consisting of horn antenna and substrate integrated waveguide cavity FSS. *IEEE Transactions on Antennas and Propagation*, *55*(1), 92–98. doi:10.1109/TAP.2006.888459

Matthaei, G., Young, L., & Jones, E. M. T. (1980). Microwave Filters, Impedance-matching Networks and Coupling Structure (A. H. Inc, Ed.). Academic Press.

Ogbodo, E. A., Aly, M. G., Ali, A. E. T., Wu, Y., & Wang, Y. (2018). Dual-band filtering antenna using dual-mode patch resonators. *Microwave and Optical Technology Letters*, *60*(10), 2564–2569. doi:10.1002/mop.31396

Pan, Y. M., Hu, P. F., Leung, K. W., & Zhang, X. Y. (2018). Compact Single-/Dual-Polarized Filtering Dielectric Resonator Antennas. *IEEE Transactions on Antennas and Propagation*, *66*(9), 4474–4484. doi:10.1109/TAP.2018.2845457

Pozar, D. M. (2005). *Microwave Engineering* (3rd ed.). Wiley & Sons.

Qian, J. F., Chen, F. C., & Chu, Q. X. (2018). A Novel Tri-Band Patch Antenna With Broadside Radiation and Its Application to Filtering Antenna. *IEEE Transactions on Antennas and Propagation*, *66*(10), 5080–5585. doi:10.1109/TAP.2018.2853160

Samsuzzaman, M., Islam, M. T., Mandeep, J. S., & Misran, N. (2014). Printed wide-slot antenna design with bandwidth and gain enhancement on low-cost substrate. *TheScientificWorldJournal*, *2014*, 1–10. Advance online publication. doi:10.1155/2014/804068 PMID:24696661

Yu, C., Hong, W., Kuai, Z., & Wang, H. (2012). Ku-band linearly polarized omnidirectional planar filtenna. *IEEE Antennas and Wireless Propagation Letters*, *11*, 310–313. doi:10.1109/LAWP.2012.2191259

Yusuf, Y., & Gong, X. (2011). Compact low-loss integration of high-Q 3-D filters with highly efficient antennas. *IEEE Transactions on Microwave Theory and Techniques*, *59*(4 Part 1), 857–865. doi:10.1109/TMTT.2010.2100407

Zhang, B., & Xue, Q. (2018). Filtering antenna with high selectivity using multiple coupling paths from source/load to resonators. *IEEE Transactions on Antennas and Propagation*, *66*(8), 4320–4325. doi:10.1109/TAP.2018.2839968

Zhou, J. (2010). Microwave Filters. In *Microwave and Millimeter Wave Technologies from Photonic Bandgap Devices to Antenna and Applications*. IntechOpen. doi:10.5772/9052

Zuo, J., Chen, X., Han, G., Li, L., & Zhang, W. (2009). An integrated approach to RF antenna-filter co-design. *IEEE Antennas and Wireless Propagation Letters*, *8*, 141–144. doi:10.1109/LAWP.2009.2012732

ADDITIONAL READING

Balanis, C. (2005). *Antenna theory: Analysis and design*. Wiley-Interscience.

Edward, T. C. (2000). *Foundation of Interconnect and Microstrip Design* (3rd ed.). Wiley. doi:10.1002/9781118894514

Harrington, R. F. (1968). *Field Computation by Moment Methods*. Macmillan.

Hassan, E., Martynenko, D., Wadbro, E., Fischer, G., & Berggren, M. (2019). Compact differential-fed planar filtering antennas. *Electronics (Switzerland)*, *8*(11), 1–15.

Hong, J. S., & Lancaster, M. J. (2001). *Microstrip Filters for RF/microwave applications*. Wiley & Sons. doi:10.1002/0471221619

Jin, J. Y., Liao, S., & Xue, Q. (2018). Design of Filtering-Radiating Patch Antennas with Tunable Radiation Nulls for High Selectivity. *IEEE Transactions on Antennas and Propagation*, *66*(4), 2125–2130. doi:10.1109/TAP.2018.2804661

Pozar, D. M. (2005). *Microwave Engineering* (3rd ed.). Wiley & Sons.

KEY TERMS AND DEFINITIONS

Far-Field Radiation: A region away from a source of EM radiation, where the EM fields are dominated by radiating fields, the E and H-fields are orthogonal to each other and to the direction of propagation as with plane waves.

Filtering Antenna: A composite RF device, integrating the filter and antenna stages found at the front-end of wireless communication systems.

Full Wave EM Simulator: A commercially available piece of software, capable of running the complex descriptive math and produce solutions in terms of electromagnetic parameters such as S-Parameters and amplitude response.

Open Loop Resonator: A piece of microstrip line bent in a certain way that can be used in the filter design to create the desired filter response.

Patch Antenna: A piece of microstrip line of a calculated geometry which can emit or receive electromagnetic radiation to and from free space.

Skin Effect: A phenomenon occurring in solid conductors, by which alternating currents tend to avoid travelling through it but limiting themselves to conduction near the surface.

Chapter 10
The Adverse Effect of Electromagnetic Radiation From Cellular Base Stations in Nigeria

Ndidi Stella Arinze
Enugu State University of Science and Technology, Nigeria

Patrick Uche Okafor
Enugu State University of Science and Technology, Nigeria

Osondu Ignatius Onah
Enugu State University of Science and Technology, Nigeria

ABSTRACT

On a global scale, the telecommunication industry is experiencing tremendous growth in mobile phones. Mobile phones communicate with base stations that are erected by the telecommunication industry. The base station produces radio frequency and exposes the people near the base stations to radiation. The effect of electromagnetic radiation from four base stations located at the residential area was studied by measuring their electric field strength and calculating their magnetic field strength and power density at different distances covering a frequency range of 900MHz to 2100MHz. The obtained values showed that the four cellular base stations are operating above the standard values of the International Commission on Non-Ionizing Radiation Protection Electromagnetic Field Radiation. The specific absorption rate was measured to determine the amount of radio frequency electromagnetic radiation absorbed by the human body. The result which is in the range of 3.22-3.70 W/kg is higher than the acceptable 2 W/kg for localized specific absorption rate.

INTRODUCTION

The global communication industry in recent years has seen unprecedented growth that increased the number of wireless devices. In 2011, according to the Nigerian Communications Commission (NCC),

DOI: 10.4018/978-1-7998-6992-4.ch010

the National Bureau of Statistics report showed that 84% of the urban population and 58.5% of the rural population have access to mobile phones (NCC, 2012). In 2014, cell phone subscribers in Nigeria numbered 125 million with four or more wireless service providers. The number of cellular base station towers estimated as of 2014 was a figure of 28,340 (Microlab, 2014). Presently, 3G (third-generation), 4G (fourth generation), and broadband services are available in Nigeria, and all this capacity requires an extensive telecommunication infrastructure by the service providers. The Association of Licensed Telecommunications Operators of Nigeria (ALTON) estimated that the market needs another 50,000 to 60,000 towers for optimum mobile network coverage across Nigeria. Next-generation services are continuing to drive the build-out of both new infrastructure as well as the adaptation of preexisting sites. Unfortunately, limited space availability and the lack of strict invigilation by regulatory authorities have led to the mounting of cellular base station towers on residential buildings. People are exposed to electromagnetic radiation emitted from these towers. Electromagnetic radiation, also known as electrosmog, is in the form of waves of electric and magnetic energy. The radiations are invisible and its potential harm over long periods of exposure is not realized until it manifests in the form of biological disorders. The electromagnetic radiations are of two types: Ionizing radiations (such as gamma rays and X-rays) and Non-ionizing radiations (such as magnetic and electric fields, radio waves, microwaves, infrared, visible and ultraviolet radiation). The International Agency for Research on Cancer (IARC) of the World Health Organization in 2011 classified RF radiation from base station towers as a "possible human carcinogen" Class 2B (WHO, 2011). Studies from several researchers have reported the ill-effects of radio-frequency electromagnetic field (RF-EMF) radiation. Meo et al. (2019), investigated the exposure impact of radio frequency electromagnetic field generated from base station towers. They used a total of 217 students from two different schools (school 1 was 124 students and school 2 was 91 students). Both schools were located within 200m from the cell tower. They discovered that the RF-EMF in School 1 was 2.010 $\mu W/cm^2$ (20.1 mW/m^2) and in School 2 was 10.021 $\mu W/cm^2$ (100.2 mW/m^2) with a frequency of 925 MHz. The students were exposed to electromagnetic frequency radiation for six hours a day, five days a week for a total period of two years. Chiaramello et al. (2019), presented research efforts on radio frequency radiation exposure in indoor environments over the last ten years. The authors considered different RF-EMF sources that can cause exposure in indoor environments and reported that the maximum mean levels of the exposure were found in offices (1.14 V/m) and public transports (0.97 V/m), while the homes and apartment had the minimum levels of exposure with mean values that in the range 0.13–0.43 V/m. Hardell et al. (2018), measured the RF radiation level of an apartment located at 6m away from the base station tower. The total mean RF radiation level was found to be 3,811 $\mu W/m^2$. The authors concluded that due to the level of radiation, the apartment is not suitable for long-term living. Also, the children are at high risk of biological disorder since they are more sensitive than adults. Gallastegi et al. (2018), studied the effect of RF-EMF exposure in children. They used 104 children of 8 years old and identified the locations (homes, school playground) and sources that contributed to a high level of vulnerability. Results showed that median exposure levels ranged from 29.73 $\mu W/m^2$ (0.11 V/m, 0.38% of the ICNIRP reference level) in children's bedrooms to 200.10 μW/m2 (0.27 V/m, 0.88% of the ICNIRP reference level) in school playgrounds were higher at outdoors than indoors. The authors discovered that mobile phone downlink and FM radio are sources that contributed most to the exposure. Kushpal et al. (2016) presented a study on the effect of electromagnetic radiations (EMRs) from mobile phone base station on general health and salivary function. The authors used twenty individuals residing very close to selected four mobile base stations in India. Questions related to sleep disturbances and other health problems were measured using the Pittsburgh Sleep Quality Index (PSQI). A Chi-square test was used for the statistical analysis. The result showed that the majority of the people complained of sleep disturbances, dizziness, headache, hypertension and concentration difficulties. This paper studied the effect of electromagnetic radiation from base stations located at the residential areas in Nigeria as well as their compliance with all aspects of the radiation guidelines of International Commission on Non-ionizing Radiation Protection.

THEORY OF WORK

Cellular Base Station

It is a cellular-enabled device site where antennae and communications equipment are placed on a tower to create adjacent cells in a cellular network as shown in Figure 1.

In a cellular network, cellular devices communicate with telephone network systems deployed throughout the world by transmitting and receiving the electromagnetic waves that are in the radiofrequency (RF) range (Arinze et al, 2020). Cellular base station comprises of Antennas for transmission and reception of signals, communication room for storage of electronic equipment, a computer system for monitoring the transmitters and receivers, a power system for power backup in case of power failures. The antennas of a cellular base station which may be directional or adaptive are placed on top of the tower. The antennas are the source of radiofrequency radiation that is transmitted through the volume of the cell that it serves.

Radiation

Radiation is the emission or release of energy in the form of waves (such as radio waves, microwaves, ultraviolet, visible light, infrared, x-rays and gamma rays) or particles (such as alpha radiation (α), beta radiation (β), and neutron radiation) through space or medium. Radiation is made up of electric and magnetic energy that are moving together through space at the speed of light. Human beings are exposed to natural radiation (ultraviolet light from the sun) and human-made radiation (X-Rays from x-ray machines, radio waves from television broadcasts and radio). Two categories of radiation are ionizing and non-ionizing radiation.

1. Ionizing Radiation: It is radiation with sufficiently high energy that removes electrons from the atoms. Its process strips electrons from atoms and molecules. X-rays, Gamma rays and the higher energy range of ultraviolet light forms the ionizing part of the electromagnetic waves while the neutrons, beta and alpha constitute the ionizing part of particles. Ionizing radiation's interaction with matter can change chemical reactions in the body that leads to damage in biological tissues. The practical use of ionizing radiation is in medicine. Still, it presents a health hazard if misused because its interaction with matter can change chemical reactions in the body and causes damage in biological tissues, genetic damage, cancer, organ failure and death.

Figure 1. Cellular Base Station

2. Non-Ionizing Radiation: It has insufficient energy that breaks molecules instead of atoms. It consists of radio waves, microwaves, infrared, visible light and lower frequencies of ultraviolet light. They

are not energetic enough to detach electrons from atoms but cause chemical changes and molecular damages that are similar to ionization; for example sunburn caused by solar ultraviolet causes vibrations in the bonds which are sensed as heat (National safety council, 2019). Although they can technically ionize by raising the temperature of a body through the process of thermal-ionization. Non-ionizing radiation sources are telecommunications, broadcasting antennas, microwave ovens, lasers, light bulbs, ultraviolent lamps, infrasound and ultrasound.

Radiofrequency (RF) radiation is electromagnetic waves emitted from wireless communication devices. Low levels of RF radiation are not considered hazardous, according to the Center for Construction Research and Training. Still, exposure to RF radiation levels from telecommunications equipment poses a considerable health risk (National safety council, 2019). High exposure risk is relative to the proximity and amount of time spent near the telecommunications equipment. Sometimes, symptoms such as overheating, reddening of skin and burns from an overexposed person can be delayed. One has to be careful since the International Agency for Research on Cancer (IARC, 2013) has classified exposure to RF waves as carcinogenic to humans.

Specific Absorption Rate

Specific Absorption Rate is the rate at which energy is absorbed by the human body when exposed to radio frequency (RF). The RF exposure limits used are expressed in terms of Specific Absorption Rate (SAR) which is a measure of the electric or magnetic field strength and power density for transmitters operating at frequencies from 300 kHz to 300 GHz.

SAR is described as the power absorbed per mass of tissue or power level per body mass. SAR unit is of watts per kilogram (W/kg). It can also be defined as a part of the distribution of the mass density and conductivity throughout the exposed tissue and can be calculated from the electric field within the tissue as

$$SAR = \frac{\sigma |E|^2}{\rho}. \tag{1}$$

{\displaystyle {\text{SAR}}={\frac {1}{V}}\int _{\textrm {sample}}{\frac {\sigma (\mathbf {r})|\ mathbf {E} (\mathbf {r})|^{2}}{\rho (\mathbf {r})}}d\mathbf {r} }

ere

. {\displaystyle \sigma } σ s the conductivity of the tissue in Siemens/meter (S/m)
. {\displaystyle E} E s the RMS electric field strength (V/m)
. is the tissue mass density (Kg/m^3)

The power density (S . can be written as

$$S = \frac{|E|^2}{Z_o} = Z_o |H|^2 \left(\frac{w}{m^2}\right) \tag{2}$$

if the electric-field vector, the magnetic field vector and the direction of propagation are in plane-wave condition. E .(V/m) is the electric field strength, H .(A/m) is the magnetic field strength and Z_o .s the impedance (377 Ω) in free space.

The Federal Communications Commission (FCC) and the International Commission on Non-Ionizing Radiation Protection (ICNIRP) are the two most popular standards that are widely used for RF radiation exposure limits. These standards are designed to protect humans against established adverse health effects of RF radiation. For the FCC guidelines, SAR levels of 1.6 W/kg in 1 g of tissue in the head and trunk (1-g SAR) and 4 W/kg in 10 g of tissue in limbs (10-g SAR) were adopted as the localized SAR limitations for the general public. The ICNIRP guidelines are very much alike to the FCC guidelines with few exceptions. The maximum localized SAR limitation of the ICNIRP guidelines is 2 W/kg in any 10 g of tissue in the head and trunk (10-g SAR) and 4 W/kg in any 10 g of tissue in the limbs (10-g SAR) for the general public (NCC, 2018). The summary of ICNIRP general public exposure limitation is shown in Table 1. NCC in Nigeria adopts the ICNIRP standard.

Table 1. ICNIRP General Public Safety Exposure

Frequency(Hz)	E-field(V/m)	H-field(A/m)	Power density(W/m^2)	Whole body average SAR (W/kg)	Localized SAR (head and trunk) (W/kg)	Localized SAR (limbs) (W/kg)
400-2000MHz	1.375f$^{1/2}$	0.0037f$^{1/2}$	f/200	0.08	2	4
2-300GHz	61	0.16	10	0.08	2	4

MATERIALS AND METHOD

Four base stations within Enugu residential metropolis shown in Figure 2 were used for the RF radiation measurement.

To obtain low uncertainty and good reproducibility, the material used for the measurement of the electric field strength of the cellular base station is the TS-EMF system by Rohde & Schwarz. The system comprised of FSH 3 spectrum analyzer with isotropic sensor and S-RFEX software for data acquisition and settings management on the laptop. With the isotropic sensor, the measurement was done independently from the direction or polarization of the emitter. The cabling of the TS-EMF system was done according to the diagram shown in Figure 3.

The tri-axis isotropic sensor has two connection cables which are fixed mounted to the probe. The RF coaxial cable with an N-type connector is plugged into the RF input of the spectrum analyzer. The control cable with 9-pin Sub-D plug is connected to the USB converter box that enables the switching of the three-axis X, Y and Z of the probe. The converter is also connected to the USB interface of the computer

The interface cable that comes with the FSH3 was used to connect the spectrum analyzer to the COM port of the computer. The spectrum analyzer was controlled via the GPIB Bus. The S-RFEX software was installed on the computer. The selected measurement settings on the S-RFEX software portal was average & peak measurements, pre-defined packet, Max Hold for trace mode, dwell time DT of 10000

Figure 2. Four Cellular Base Station used for measurement

ms, measurement period of 10 minutes. The distance was varied during the measurement on the four base stations. Magnetic Field measurement was obtained by calculation using equation 2 where;

$$E = H * Z_0, S = H^2 * Z_0 .$$

SAR was also calculated using equation 1

Figure 3. Rhode & Schwartz TS-EMF Measurement System (a) Connection Diagram of TS-EMF System (b) TS-EMF System

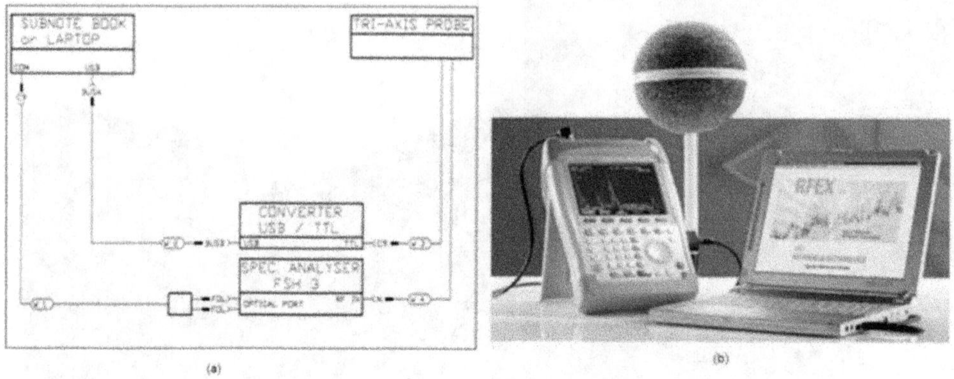

RESULTS

The result of electric field strength obtained from the TS-EMF System and the calculated result of the magnetic field strength and power density for four cellular bases stations are shown in Tables 2, 3, 4 and 5.

Table 2. EMF of Cellular Base Station 1

Distance(m)	Frequency range	E-field(V/m)	H-field (A/m)	Power density (W/m²)
20	900MHz	43.52	0.115	4.98
50	1800MHz	60.23	0.160	9.65
100	2100MHz	65.12	0.173	11.28

Table 3. EMF of Cellular Base Station 2

Distance(m)	Frequency range	E-field(V/m)	H-field (A/m)	Power density (W/m²)
20	900MHz	43.63	0.116	5.07
50	1800MHz	60.12	0.159	9.53
100	2100MHz	65.22	0.173	11.28

Table 4. EMF of Cellular Base Station 3

Distance(m)	Frequency range	E-field(V/m)	H-field (A/m)	Power density (W/m²)
20	900MHz	43.52	0.115	4.99
50	1800MHz	60.23	0.160	9.65
100	2100MHz	65.41	0.174	11.41

Table 5. EMF of Cellular Base Station 4

Distance(m)	Frequency range	E-field(V/m)	H-field (A/m)	Power density (W/m²)
20	900MHz	43.67	0.116	5.07
50	1800MHz	60.25	0.160	9.65
100	2100MHz	65.48	0.174	11.41

The EMF values of four cellular base station obtained in Tables 2-5 were compared with the reference values of ICNIRP General Public Safety Exposure and are shown in Figure 4

Figure 4. Comparison of EMF values of the four cellular base stations with ICNIRP reference values

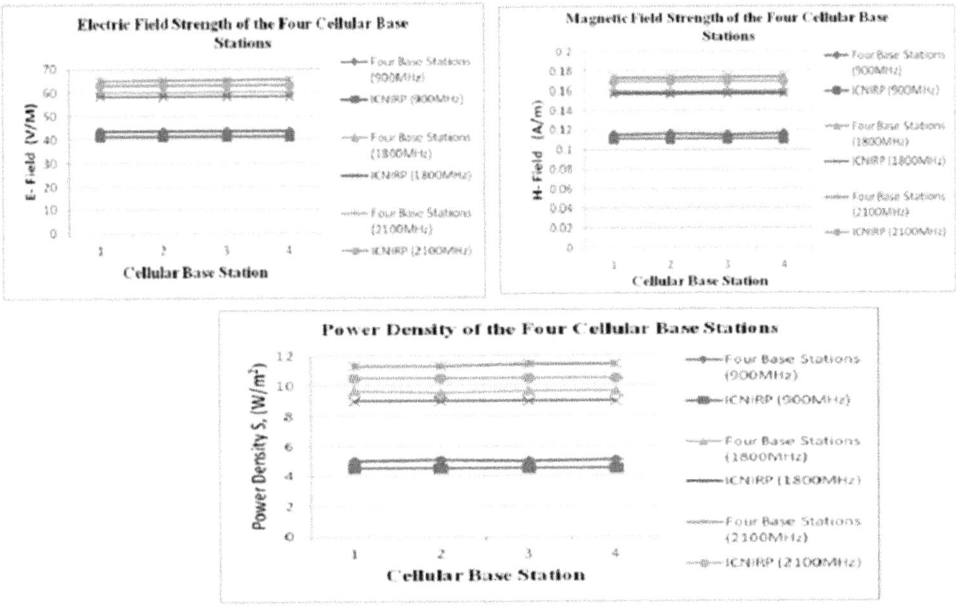

The result showed that all four central base stations are operating above the reference level of the ICNIRP general public safety guidelines for limiting radiation exposure. Therefore the operators of the cellular base stations are not in compliance with all aspects of the guidelines.

The specific absorption rate gives an RF radiation level and the result is shown in Table 6

SAR values of the four cellular base stations under measurement was compared with the ICNIRP SAR recommended value shown in Figure 5

The comparison of SAR values of four cellular base stations with ICNIRP recommended value showed that the energy absorbed by the human body is high. As the distances increase, the SAR level decreases, showing that the farther away one is to the cellular base station, the less the energy is absorbed by the human body. Furthermore, the closest distance to the cellular base station was investigated to

Table 6. SAR Values of the Four Cellular Base Stations

Cellular Base Station 1					
Distance (m)	**Frequency range**	**E-field(V/m)**	**Mass density (Kg/ m³)**	**Conductivity (S/m)**	**SAR (W/kg)**
20	900MHz	43.52	500	0.97	3.67
50	1800MHz	60.23	1550	1.40	3.28
100	2100MHz	65.41	2380	1.80	3.24
Cellular Base Station 2					
Distance (m)	Frequency range	E-field(V/m)	Mass density (Kg/ m³)	Conductivity (S/m)	SAR (W/kg)
20	900MHz	43.63	500	0.97	3.69
50	1800MHz	60.12	1550	1.40	3.26
100	2100MHz	65.22	2380	1.80	3.22
Cellular Base Station 3					
Distance (m)	Frequency range	E-field(V/m)	Mass density (Kg/ m³)	Conductivity (S/m)	SAR (W/kg)
20	900MHz	43.52	500	0.97	3.67
50	1800MHz	60.23	1550	1.40	3.27
100	2100MHz	65.41	2380	1.80	3.24
Cellular Base Station 4					
Distance (m)	Frequency range	E-field(V/m)	Mass density (Kg/ m³)	Conductivity (S/m)	SAR (W/kg)
20	900MHz	43.67	500	0.97	3.70
50	1800MHz	60.25	1550	1.40	3.28
100	2100MHz	65.48	2380	1.80	3.24

Figure 5. Comparison of SAR values of the four cellular base stations with ICNIRP reference values

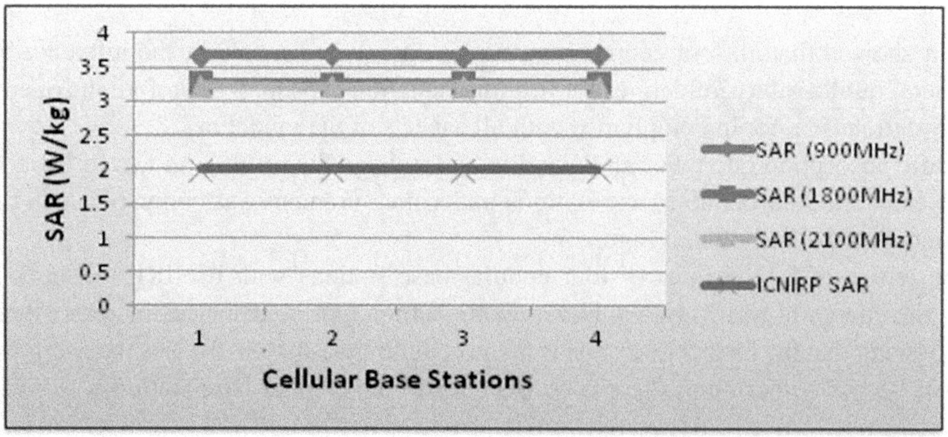

find the level of exposure to the children and adults. The calculation of the body density was carried out by incorporating measurements of limb circumferences, height and weight. The result obtained for the SAR in children and adults exposed to the radiation from the cellular base station at a distance of 20 meters is shown in table 7

Table 7. SAR Values for the Children and Adult exposed to Radio Frequency

Age(years)	Body Density (Kg/m³)	E-field(V/m)	Conductivity (S/m)	Whole Body Average SAR (W/kg)
Under 10	900	43.52	0.97	2.04
10-15	1060	43.52	0.99	1.76
16-25	1030	43.52	1.02	1.88
26-40	1010	43.52	1.04	1.95
41-60	1004	43.52	1.06	1.99
Above 60	1001	43.52	1.03	1.94

The result showed that the whole-body averaged SAR is higher than the ICNIRP general public exposure limitation. Also, the risk of radiation exposure for children and adolescent are greater than the adults.

CONCLUSION

Human beings are exposed to radio frequency radiation every day due to the presence of radio frequency signals around them. This study evaluated the electromagnetic radiation and specific absorption rate of human body exposure to four cellular base stations located in a residential area. It was discovered that the safe exposure limit is not complied by the operators of the cellular base station. The significant health risk is high. Nigerian Communications Commission should ensure that cellular base stations should not be erected very close to residential houses.

REFERENCES

Alnajjar, H., Alnaser, N. W., & Alnaser, W. E. (2019). Level of Electromagnetic Radiation Emitted from Mobile Phone Base Station during F1 Race in the Kingdom of Bahrain. *International Journal of Physics: Study and Research*, *2*(1), 65–70. doi:10.18689/ijpsr-1000109

Arinze, S. N., Onoh, G. N., & Abonyi, D. O. (2020). Performance of Light Fidelity and Wireless Fidelity Networks in a WLAN. *International Journal of Research in Engineering and Science*, *4*(1), 10-20.

Arinze, S. N., Onoh, G. N., & Abonyi, D. O. (2020). Development of A Light Fidelity System for Improving Network Performance in a Wireless Local Area Network. *International Journal of Emerging Trends in Engineering and Development*, *2*(10), 31-41.

Buckus, R., Strukčinskienė, B., Raistenskis, J., Stukas, R., Šidlauskienė, A., Čerkauskienė, R., Isopescu, D., Stabryla, J., & Cretescu, I. (2017). A Technical Approach to the Evaluation of Radiofrequency Radiation Emissions from Mobile Telephony Base Stations. *International Journal of Environmental Research and Public Health, 14*(3), 244–254. doi:10.3390/ijerph14030244 PMID:28257069

Chiaramello, E., Bonato, M., Fiocchi, S., Tognola, G., Parazzini, M., Ravazzani, P., & Wiart, J. (2019). Radio Frequency Electromagnetic Fields Exposure Assessment in Indoor Environments: A Review. *International Journal of Environmental Research and Public Health, 16*(6), 955–984. doi:10.3390/ijerph16060955 PMID:30884917

Gallastegi, M., Huss, A., Santa-Marina, L., Aurrekoetxea, J. J., Guxens, M., Birks, L. E., Ibarluzea, J., Guerra, D., Röösli, M., & Jiménez-Zabala, A. (2018). Children's exposure assessment of radiofrequency fields: Comparison between spot and personal measurements. *Environment International, 118*, 60–69. doi:10.1016/j.envint.2018.05.028 PMID:29803802

Hardell, L., Carlberg, M., & Hedendahl, L. K. (2018). Radiofrequency radiation from nearby base stations gives high levels in an apartment in Stockholm, Sweden: A case report. *Oncology Letters, 15*(5), 7871–7883. doi:10.3892/ol.2018.8285 PMID:29725476

Kushpal, S., Anup, N., Asif, Y., Shravani, G., Sonia, P., & Preeti, V. (2016). Effect of electromagnetic radiations from mobile phone base stations on general health and salivary function. *Journal of International Society of Preventive & Community Dentistry, 6*(1), 54–59. doi:10.4103/2231-0762.175413 PMID:27011934

Meo, S. A., Almahmoud, M., Alsultan, Q., Alotaibi, N., Alnajashi, I., & Hajjar, W. H. (2019). Mobile Phone Base Station Tower Settings Adjacent to School Buildings: Impact on Students' Cognitive Health. *American Journal of Men's Health, 13*(1), 1–6. doi:10.1177/1557988318816914 PMID:30526242

Microlab. (2014). Nigeria Migrates to the Independent Tower Company Business Model. *Inside Towers.*

National Safety Council. (2019). *Understanding RF radiation.* https://www.safetyandhealthmagazine.com/articles/18168-understanding-rf-radiation

Sivani, S., & Sudarsanam, D. (2013). Impacts of radio-frequency electromagnetic field (RF-EMF) from cell phone towers and wireless devices on biosystem and ecosystem. *Biology and Medicine (Aligarh), 4*(4), 202–216.

Venkatesulu, S., Vijai Bhaskar, P., & Vijayasree, R. (2018). Mathematical Analysis Of Electromagnetic Radiation Emit From A Cellular Base Station. *International Journal of Advance Engineering and Research Development,* 5(2).

Venkatesulu, V. S., Giri, P. M. N., & Venkata, R. P. (2014). Monitoring Of Electromagnetic Radiation for Cellular Base Stations Using Arm Processor. *International Journal of Innovative Research in Computer and Communication Engineering, 2*(6), 4603–4609.

Chapter 11
Implementations of ACJ Technique in Different Four-Port Filter Networks

Eugene A. Ogbodo
https://orcid.org/0000-0002-9798-723X
Aston University, UK

ABSTRACT

This chapter proposes the use of asynchronously coupled-resonator junctions (ACJ) in the design of a multi-input multi-output (MIMO) filtering network and a masthead combiner (MHC). By employing the resonator junctions, miniaturised circuits are achieved without using any transmission-line-based impedance matching circuits. The main challenge in the designs is the control and implementation of the external couplings at the common ports of this all-resonator-based MIMO filtering network and MHC. Both devices are four ports-based with the MIMO filtering network operating at 1.8, 2.1, and 2.6 GHz, while the MHC operates at the two channels of 1.8 and 2.1 GHz. The demonstrated designs achieved fractional bandwidths of 1.764 GHz to 1.836 GHz, 2.058 GHz to 2.142 GHz, and 2.548 GHz to 2.652 GHz, respectively. Good agreements have been achieved between the measurements of the prototype devices and the simulations.

INTRODUCTION

In this work, the design concept of a multi-input multi-output (MIMO) filtering network is demonstrated using two different types of four-port networks. A multi-input multi-output filter and diplexer, and a mast head combiner (MHC) is designed and presented. The designs are concentrated on a miniaturised circuit size by applying an all-resonant-structure design technique in the implementations. The asynchronously coupled junction (ACJ) technique is utilised in all the designs. The design of the MIMO filter and diplexer was based on the idea of having a filtering network with the capability to function as a multi-purpose filter with different channel response. In this case, it functions as a diplexer with different channel responses, depending on the port(s) of excitation. The technic employed in the design made it a

DOI: 10.4018/978-1-7998-6992-4.ch011

suitable device that can save cost during the purchases of different filtering devices. The size achieved for the design contributes to the reduction of carbon footprint. The MHC device also utilised the ACJ technique in its implementation, the result of the utilisation of the ACJ allowed the embedding together of three different filters of which two have the same channel response while the other have a different channel response. From the two designs presented in this chapter, it is obvious that the ACJ can play a vital role in the design of different types of components for 5G applications. It is worth iterating the importance and capability of the ACJ for next generation of designers, besides the presentation of these two designs. Here, the step-by-step procedure of how to achieve a suitable ACJ that operates at 1.8 GHz and 2.1 GHz as well as 1.8GHz and 2.6 GHz is properly demonstrated. This ACJ operating frequencies are the capable operating frequencies of the designs presented.

IMPLEMENTATION OF ASYNCHRONOUS COUPLED JUNCTION (ACJ)

To achieve the ACJ operating at 1.8 GHz and 2.1 GHz, a resonator of 1.8 GHz is coupled to 2.1 GHz resonator. This technique is referred to as asynchronous coupling; this is because it comprises two resonators of different inherent resonant frequencies. Fig. 1 presents the frequency responses of the two coupled resonators against their separation, $d_{1,4}$. It is worth mentioning that the type of coupling response a designer might want, be it magnetic coupling or electric coupling, it can be implemented at this stage. An example is that if an ACJ is implemented to act as a signal distributor, a designer might choose to improve selectivity with the adjacent channel. After achieving the required coupling, the next stage is to achieve the Q-factor. The properties of the ACJ were determined using the general filter synthesis.

To achieve the Q_{ext} factor for the ACJ, a coupled feed line technique (Ogbodo, Wu, & Wang, 2017; Son and Pyo, 2005; Mansour, Lancaster, Hall, Gardner, & Nugoolcharoenlap, 2014) was used to feed the resonators at port-1 as shown in Fig. 2. Port-2 and port-3 were weakly coupled to the resonators while the coupling gaps, width and length at port-1 were adjusted. Fig. 2(a) presents the typical frequency response. Fig. 2(b) presents the simulated current distribution of the ACJ at 1.8 GHz and 2.1 GHz. The current distribution presented shows the level/concentration of currents at various sections of the ACJ during operations. It can be used to trace the directions of the current shown here. It is evident that at the operating frequencies of 1.8 GHz and 2.1 GHz, the currents are concentrated on the corresponding resonators, this shows that the ACJ directed the signal to different channels. The ACJ does the signal splitting and combining in the diplexer circuitry.

The same procedure was followed in creating the second ACJ which is done by coupling together of resonators 1.8 GHz and 2.6 GHz. After realising the ACJ, the next step in the designing of the proposed MIMO filter and diplexer or MHC can then be done by coupling the corresponding filter-elements to the ACJ. A detailed step-by-step procedure on how to realise the proposed designs are presented below.

Figure 1. Resonant frequencies of the asynchronous coupled resonators with an inset of topology used

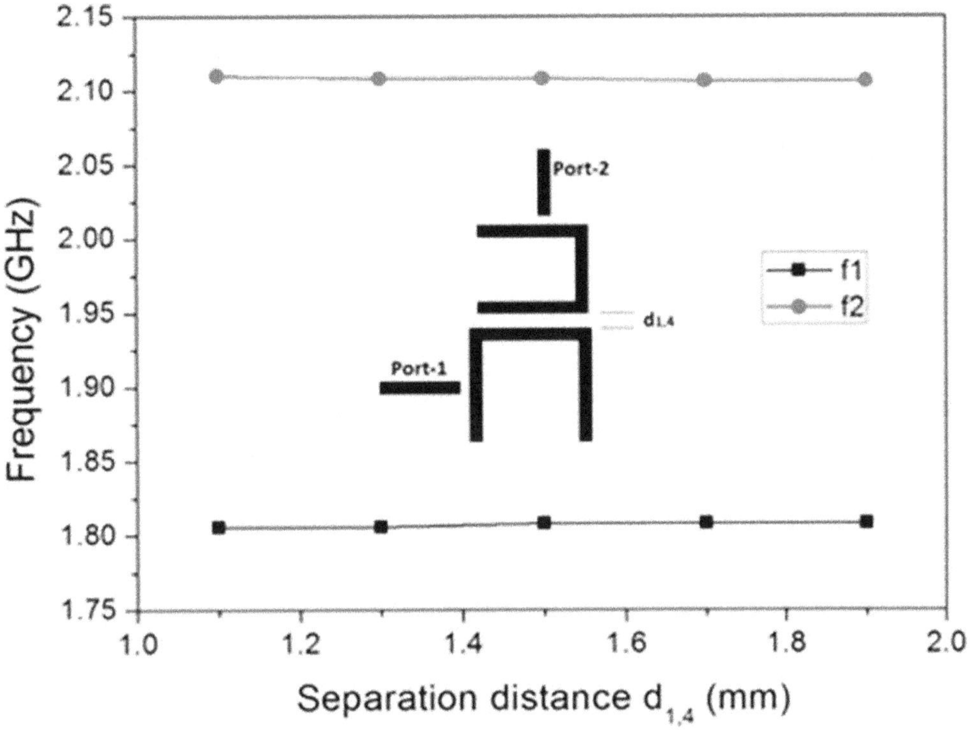

Figure 2a. Simulated typical response of the configuration used to extract the Q_{ext};

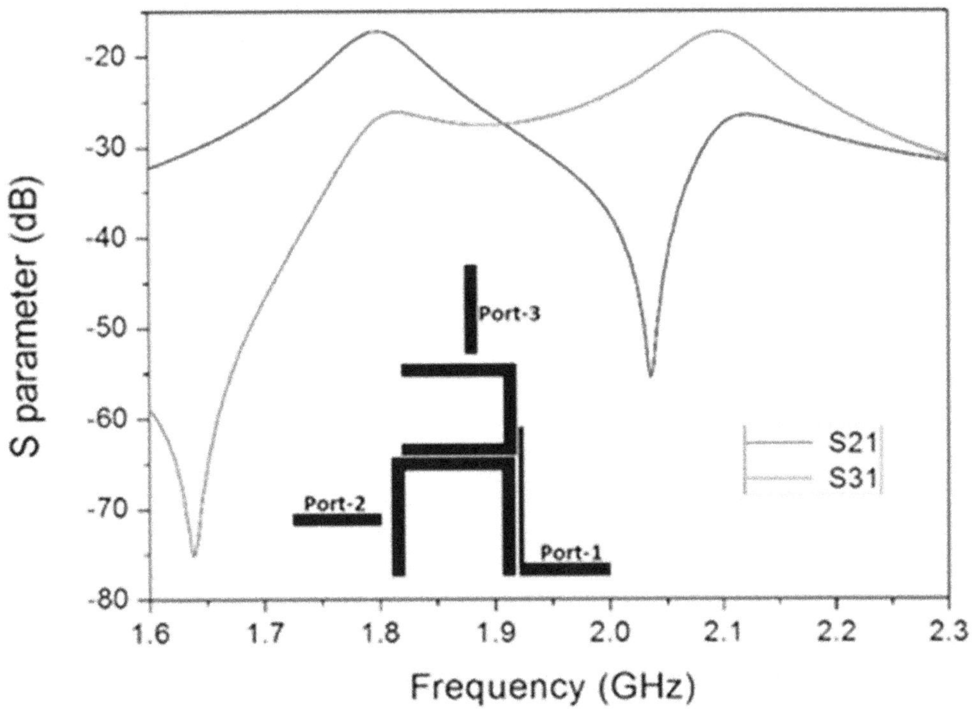

Figure 2b. Current distribution of the ACJ

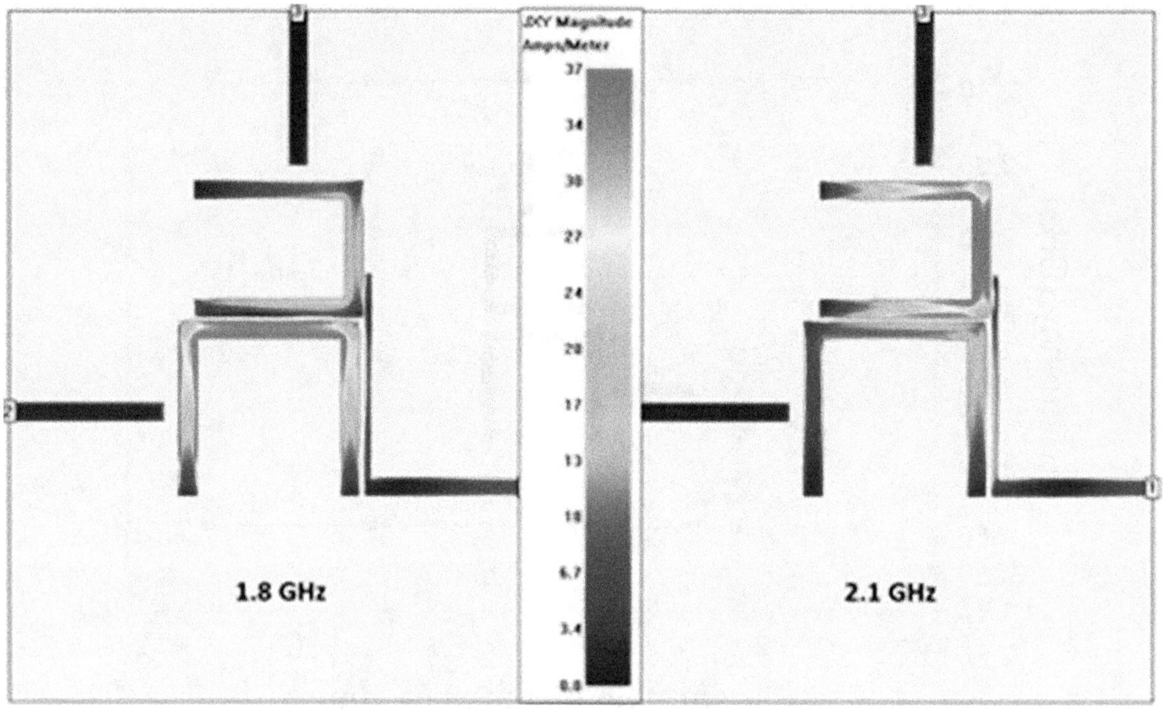

MULTI-INPUT MULTI-OUTPUT (MIMO) FILTER AND DIPLEXER

As explained in the introduction, this design is proposed to have a multi-purpose function. It can be used as a filter and/or diplexer. Its operating frequencies are 1.8 GHz, 2.1 GHz and 2.6 GHz, all having a 20 dB return loss and 4% FBW. By using all resonant structures in the design, the general filter synthesis technique is applied using (1), (2), and (3) where (1) is used in determining the coupling coefficients between the resonators, (2) is used for determining the quality external factors of the designs, and (3) for determining the geometrical positioning of the input feedline which corresponds to the impedance matching of the designs with external devices. The coupling topology used for the design is presented in Fig. 3. The resonators 1, 2 and 3 represents the 1.8 GHz filter channel. Resonators 4, 5 and 6 represents the 2.1 GHz filter channel while resonators 7, 8 and 9 represents the 2.6 GHz filter channel.

$$m_{i,i+1} = \frac{FBW}{\sqrt{g_i g_{i+1}}}, for\; i = 1,2....N-1. \tag{1}$$

$$Q_{ex1(2)} = \frac{g_0 g_1}{FBW}. \tag{2}$$

Where FBW is the fractional bandwidth and can be expressed as,

$$FBW = \frac{\omega_2 - \omega_1}{\omega_0} \cdot Q_{ex1(2)} = \frac{f}{\Delta f} \cdot \quad (3)$$

Where f .is the centre frequency, Δf .is the 3dB bandwidth of the resonance curves.

Figure 3. Multi-input multi-output filter and diplexer coupling topology

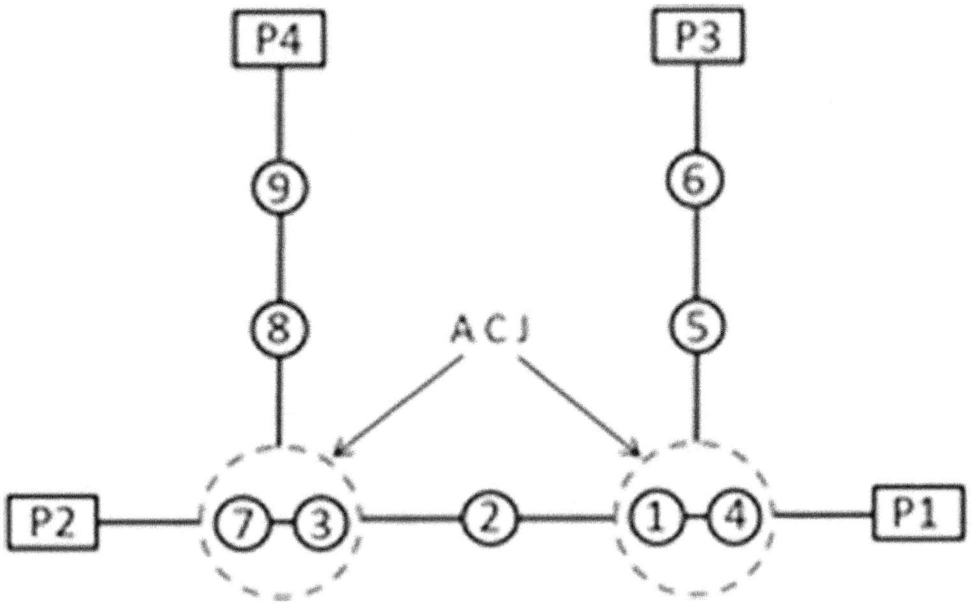

At this stage, the two required ACJs designed in the previous section is then used to synchronously couple the remaining resonators of the filter channels to their respective path. Here, the design has the first resonator of 1.8 GHz filter asynchronously coupled to the first resonator of 2.1 GHz filter. That is resonator 1 coupled to resonator 4 as shown in Fig. 3 where it represents the ACJ. As the ACJ comprises the first resonators of both channel filters which were of different inherent resonant frequencies, the corresponding filters are now joined to the ACJ by replacing its first resonators with the ACJ which does the signal splitting and combining in the diplexer circuitry. The next step in the design process is optimisation to achieve the required response properties. The design layout is presented in Fig. 4(a) with optimised parameter values achieved while Fig. 4(b) shows the block diagram of the circuit. It can be seen that a coupled feedline is utilised in the design.

Figure 4a. Proposed design layout;

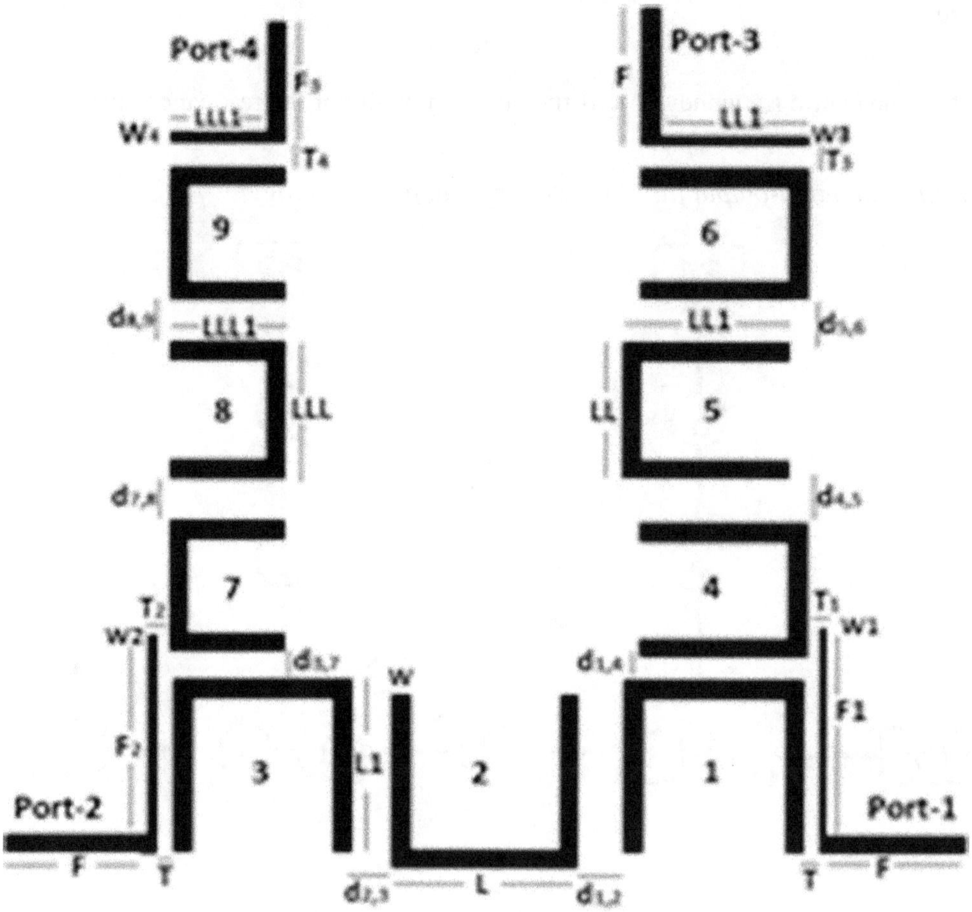

The simulated responses of the Multi-Input Multi-Output filter and diplexer are presented in Fig. 5 and Fig. 6, where Fig. 5 illustrates the response of a diplexer operating at 1.8 GHz and 2.1 GHz at the chosen port-1 as the common port. Fig. 6 illustrates that of 1.8 GHz and 2.6 GHz diplexer at the chosen port-2 as the common port. Fig. 5 and Fig. 6 are the two diplexers that can be selected from this circuit. It can be seen that these selections are through the choice of excitation port.

Fig. 7 presented a situation where the MIMO network is selected to function as two filters. This situation allowed the two possible channel filters to operate at the same time, that is 2.1 GHz filter and 2.6 GHz filter. In this case, port-1 and port-3, as well as the port-2 and port-4, are utilised for this purpose. In Fig. 5, 6 and 7, the measured responses are also presented for comparison with the simulations.

From the responses presented, all the return losses are over 20 dB. Measurements and simulations show reasonably good agreement except for the frequency shift of 10 MHz. This is believed to be due to the thinning of the substrate from milling as well as the tolerance of the dielectric constant of high permittivity anisotropic substrate material used for prototyping. From Fig. 5, it is notice that the return losses of the simulated response are at about 2 dB for both frequencies while it is about 4 dB and 5 dB for the 1.8 GHz and 2.1 GHz respectively. In the Fig. 6, the return losses of the simulated response also remained at about 2 dB for both frequencies while it is about 4 dB and 6 dB for the 1.8 GHz and 2.6 GHz respectively. This shows that as the frequency increases, the return loss also increases, and it is expected due to the same resonator width been used throughout the filtering channels. The similarity of the differences in the return loss is also witnessed in Fig. 7 where the device is operating as dual filter. These responses presented in the Fig. 5, Fig 6, and Fig. 7 has shown good agreement between the simulated and measured responses. Fig. 8 presents the isolation response between port-1 and port-4, port-2 and port-3 as well as port-3 and port-4. The solid lines represent the measured responses while the dashed lines represent the simulated responses.

Figure 4b. Circuit block diagram. F = 10mm, F1 = 15mm, F2 = 15mm, T = 0.4mm, T1 = 0.2mm, T2 = 0.38mm, T3 = 1 mm, T4 = 0.4 mm, W = 1.2mm, W1 = 0.4mm, W2 = 0.6mm, W3 = 0.6mm, L = 9.6mm, L1 = 11.5mm, LL = 6.6mm, LL1 = 11mm, d1,2 = d2,3 = 1.46mm, d1,4 = d3,7 = 0.22 mm, d4,5 = 1.65 mm, d5,6 = 1.7 mm, d7,8 = 1.84 mm, d8,9 = 1.82 mm

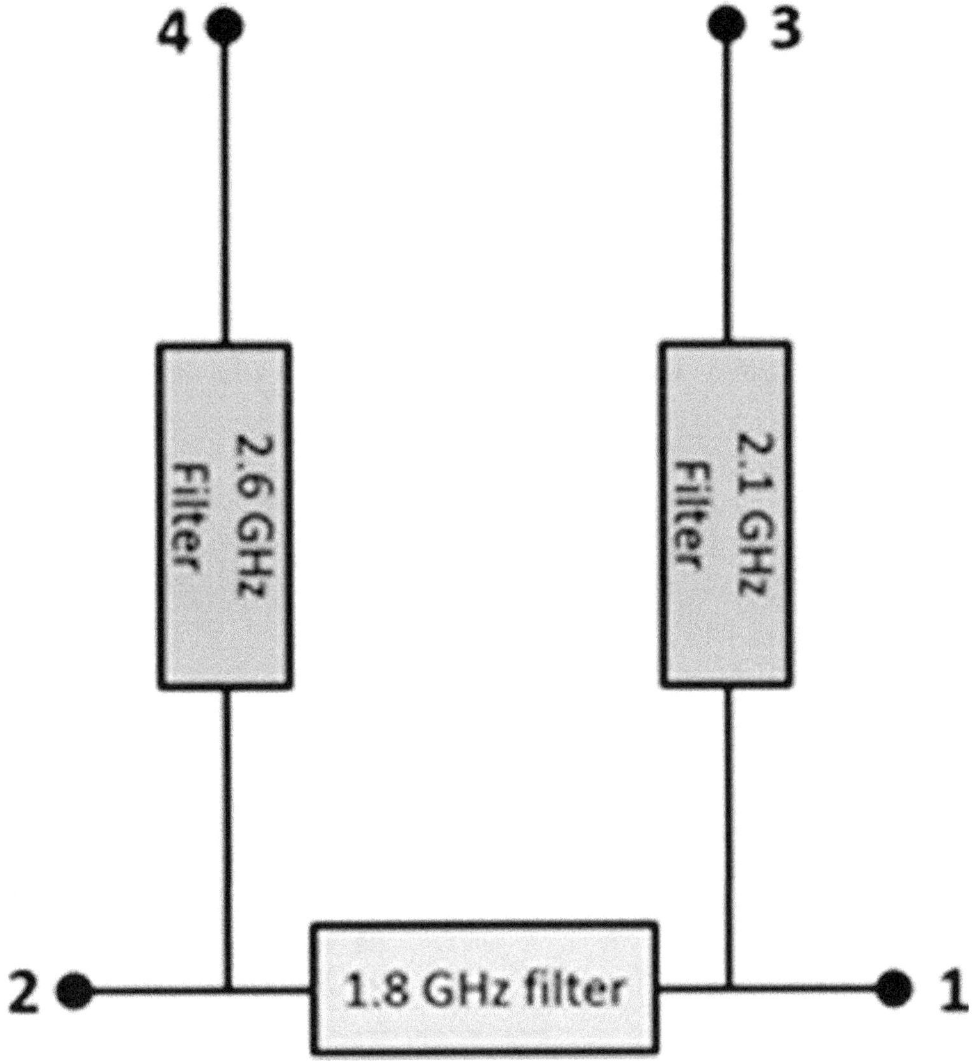

Figure 5. The duplexing responses when Port-1 is excited as the common port.

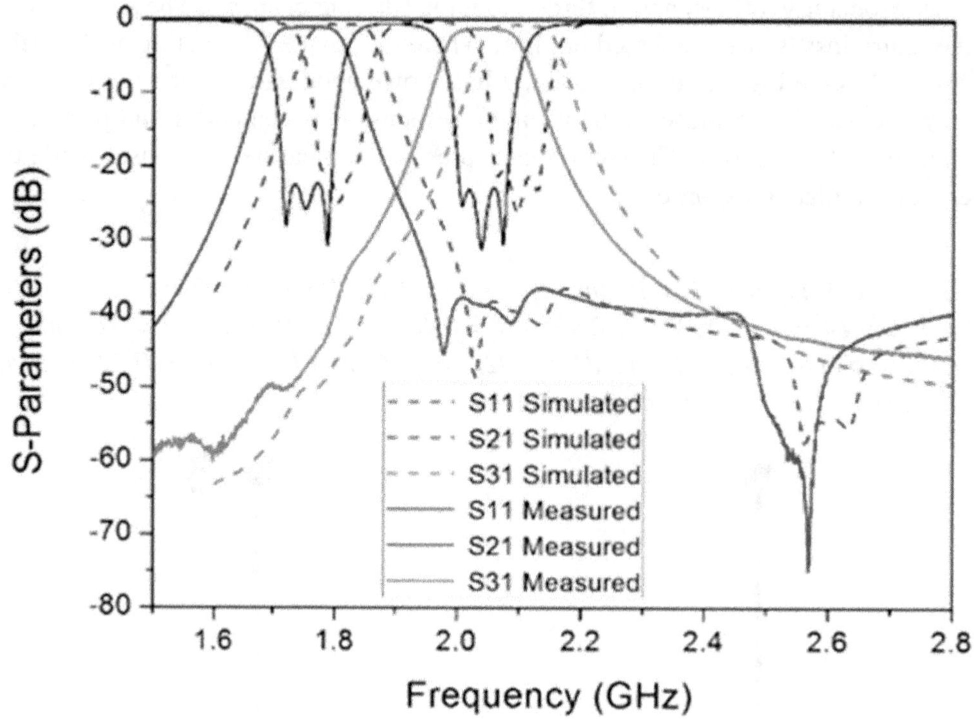

Figure 6. The duplexing responses when Port-2 is excited as the common port.

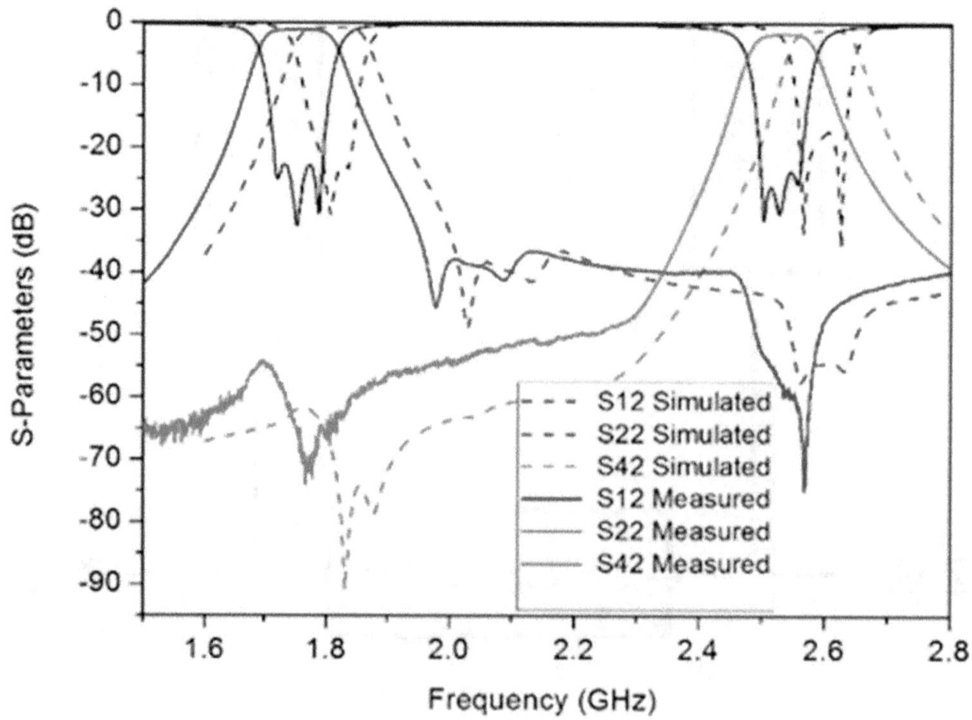

Figure 7. Dual filter operation mode

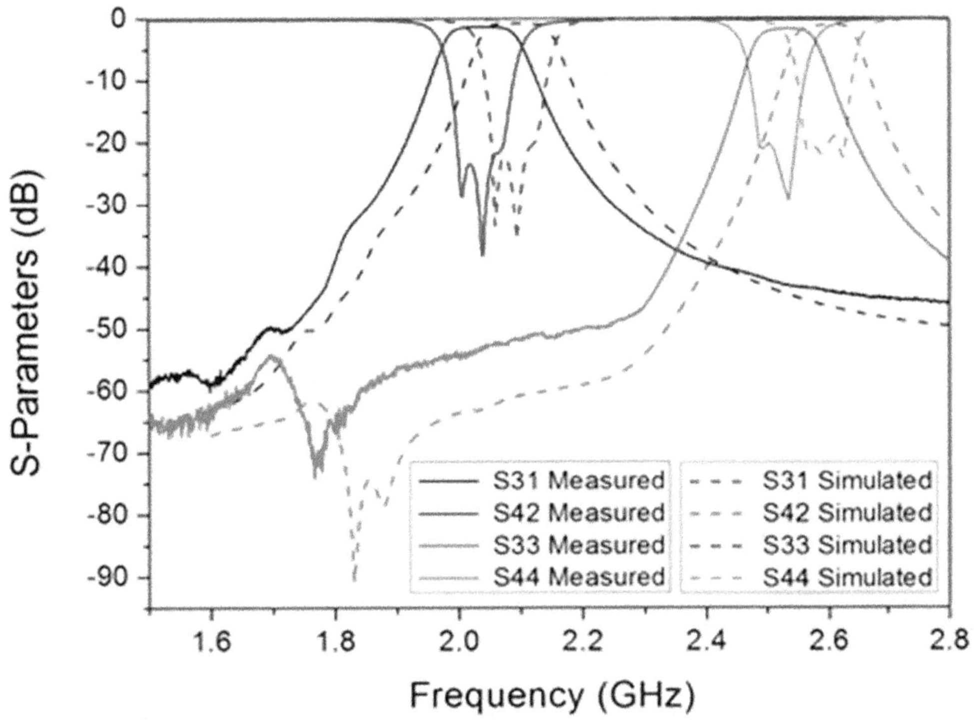

Figure 8. Measured and simulated isolation responses

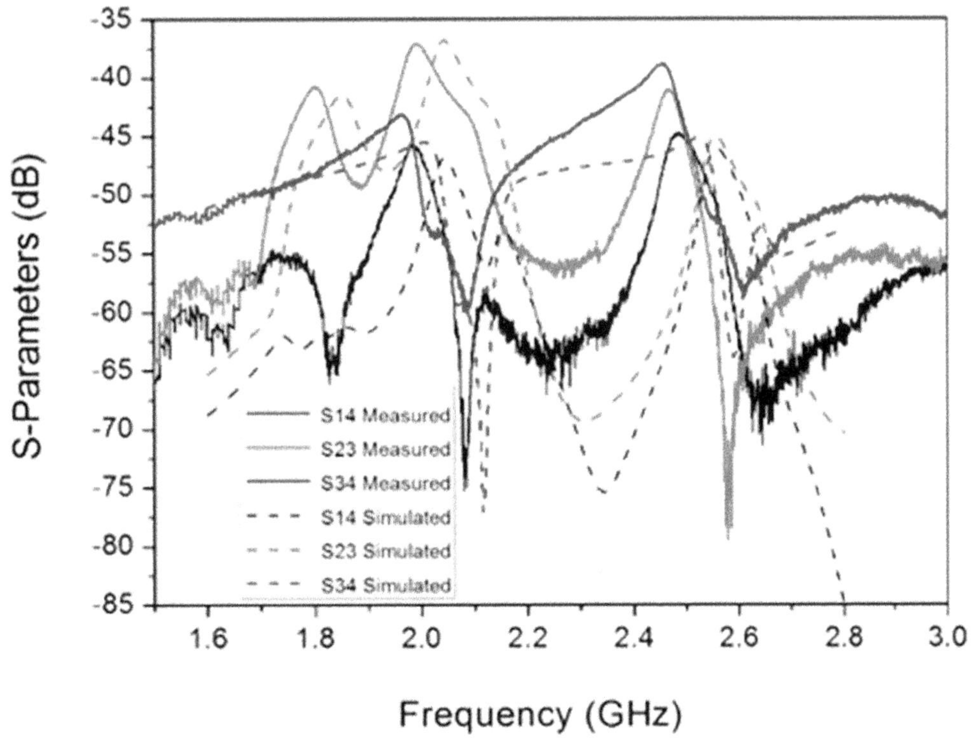

Figure 9. Prototyped multi-input multi-output filter and diplexer

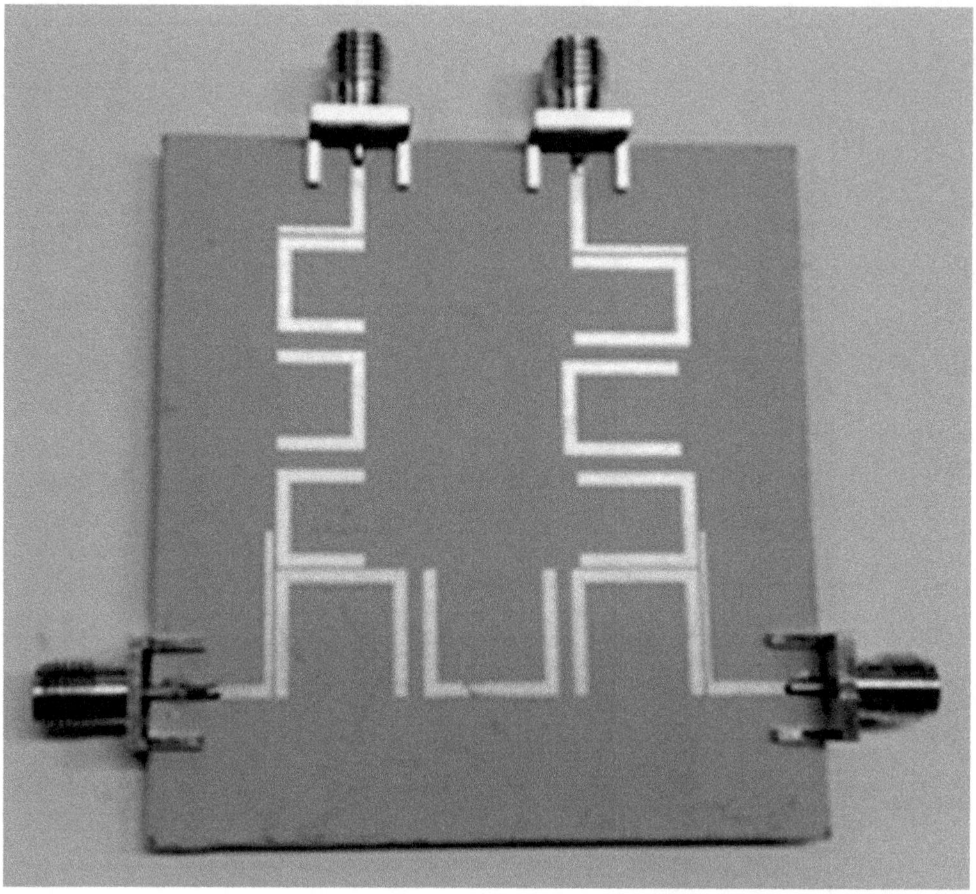

The designed MIMO filter and diplexer were prototyped using the Rogers 3010 substrate with a thickness of 1.27 mm, a relative permittivity of 10.8 and a loss tangent of 0.0022 and is shown in Fig. 9. Measurements were taken using the Agilent Network Analyser N5230A and compared with the simulated responses.

The current distribution of the channels during operation is presented in Fig. 10 using the same current density scale. A good illustration can be seen when the 1.8 GHz frequency channel is in operation, the current density distribution at 1.8 GHz channel is at its peak. At 2.1 GHz and 2.6 GHz channels, the current density distribution is at its lowest level. If the current is then considered at 2.1 GHz and 2.6 GHz frequencies, it will be noticed that at 1.8 GHz frequency, the current density distribution will be at its lowest level.

Figure 10. Current density distribution of the MIMO filtering network

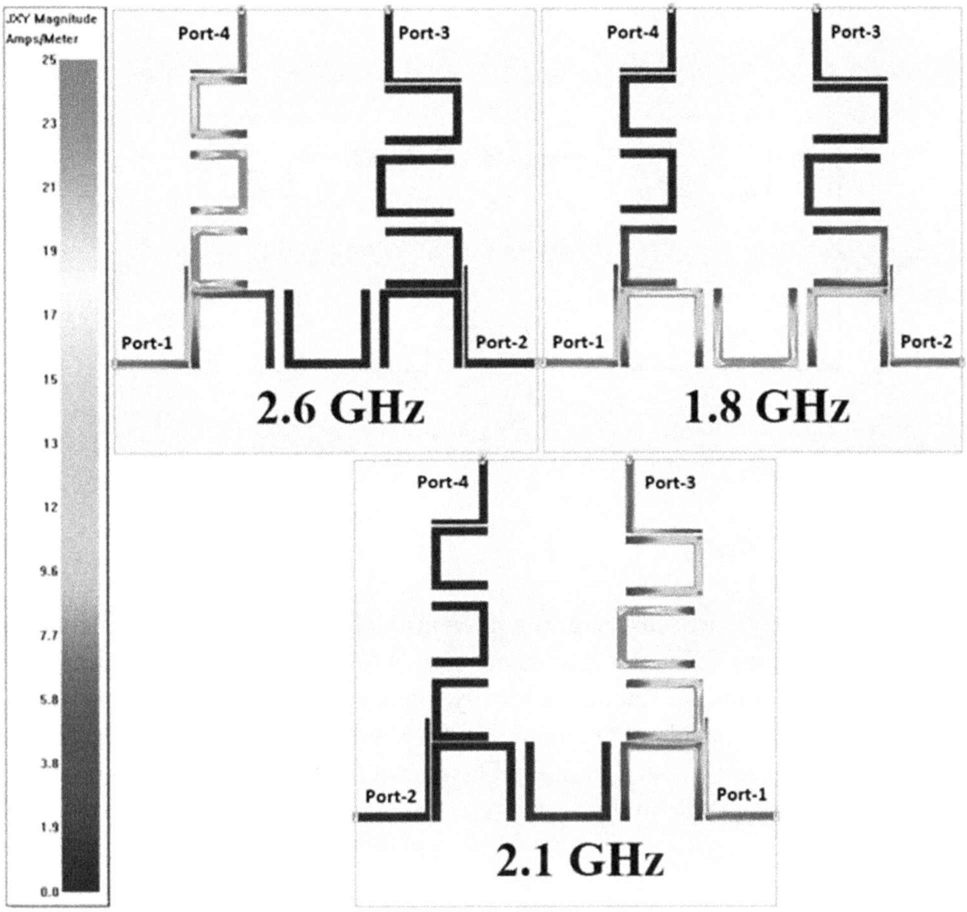

Figure 11. Classical connection in an antenna and transceiver module

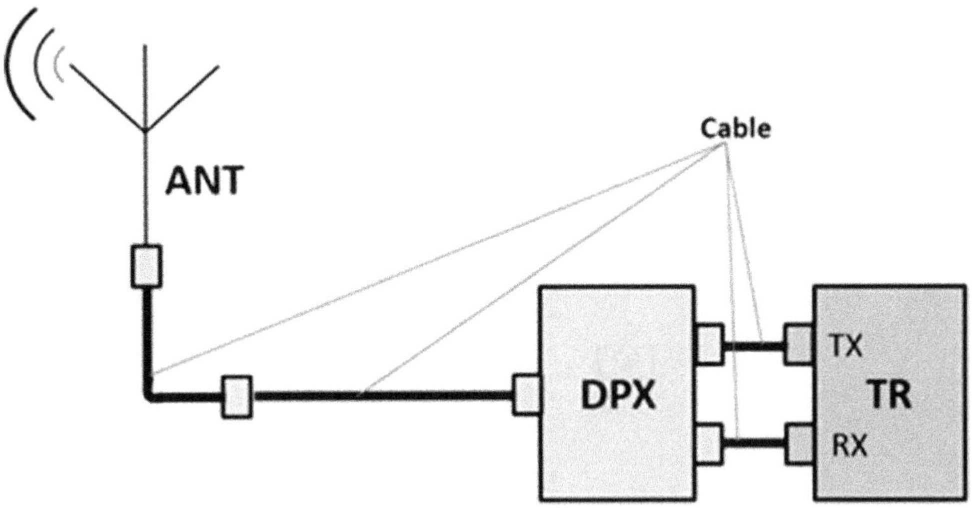

Figure 12. Masthead combiner allowing the introduction of an LNA at the antenna output

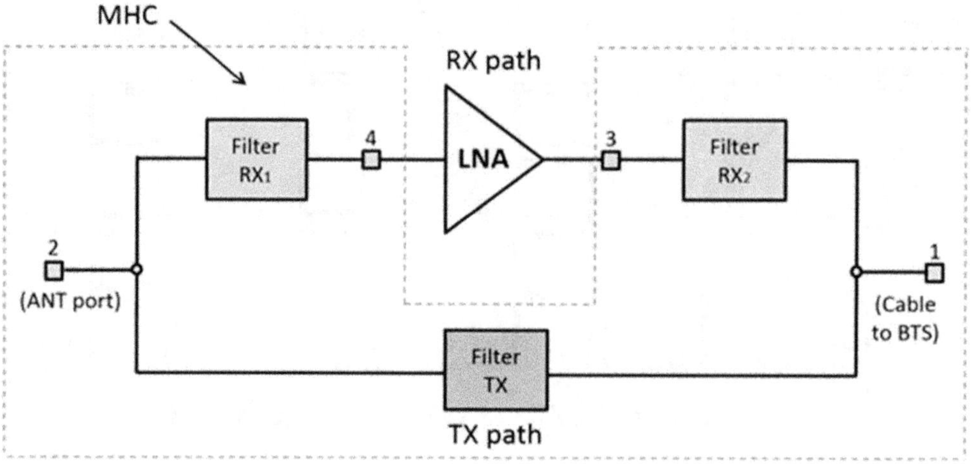

MastHead Combiner (MHC)

In the communication system, the transceiver output is normally connected to an antenna using selective diplexer which permits the TX and RX to share a common antenna; this is common in most communications systems. This common antenna is mostly mounted on the top of a communication mast, so a transmission line with a considerable length will be required for the transceiver connection. The transmission line will in effect introduce additional losses which worsen the overall noise figure of the receiving system. This scenario is illustrated in Fig. 11 where the TX and RX (transceiver) are connected to a diplexer close by.

Figure 13. Masthead combiner coupling topology with an LNA connected

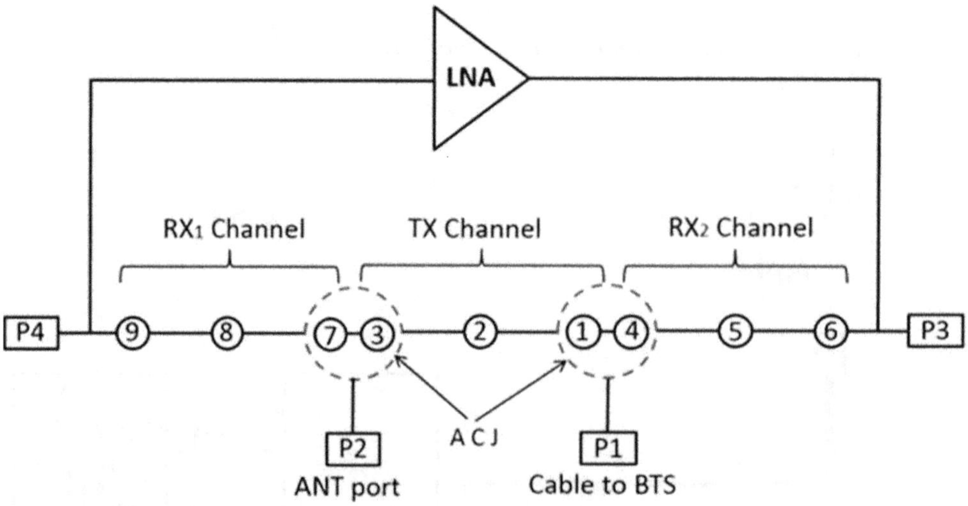

Figure 14a. Proposed microstrip MHC: (a) design layout

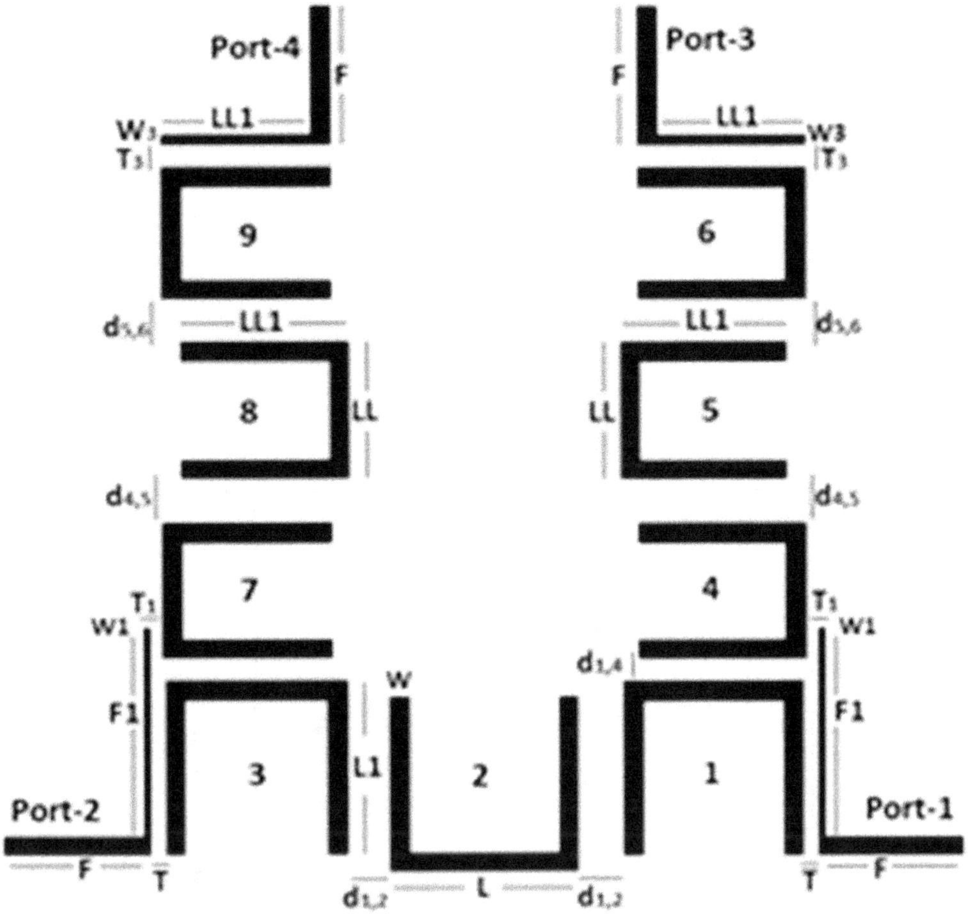

In recent time, it was suggested that the noise figure degradation caused by the cable losses could be compensated by introducing a Low Noise Amplifier (LNA) at the receiver chain while preserving the connection of TX to the antenna (Ogbodo, Wu, Callaghan, & Wang, 2018; Tamiazzo and Macchiarella, 2013; Macchiarella and Tamiazzo, 2015). The MastHead Combiner (MHC) is the device which allows the addition of the LNA as shown in Fig. 12.

As can be seen from Fig. 12, the MHC is the passive device with four ports. It is made up of three filters. The RX_1 and RX_2 filters of the MHC provides the protection of LNA input and output against the TX signal. The losses of the RX_1 filter should be at the lowest possible while the TX filter provides the TX signal path. It is worth mentioning that these filters (RX_1, RX_2 and TX) may possess multi-band response depending on the application requirement (Ogbodo EA, 2017; Ogbodo, Wang, & Yeo, 2016; Ogbodo, Wu & Wang, 2017; Ogbodo, Ngobigha, Obadiah, & Atuba, 2021).

The approach to designing the proposed MHC presented here uses the Asynchronously Coupled Junction (ACJ) as described in Fig. 1 and 2 to represent the two junctions of the MHC. The three channel filters required for this design was separately designed to have a three-pole Chebyshev ripple factor of 0.043 dB and to meet a 4% FBW each with a 20 dB return loss. The conventional hairpin resonators were used for the design of the filters for the MHC. They operate at 1.8 GHz for the TX channel, 2.1

GHz for the RX_1 and RX_2 channels. Fig. 13 presents the coupling topology of the design with a block diagram of an LNA (not included in the design). The circuit layout is presented in Fig. 14(a) with the optimised dimension parameters achieved and listed in the caption while Fig. 14(b) shows the block diagram of coupling topology used.

Figure 14b. coupling topology. F = 10mm, T = 0.4mm, T1 = 0.2mm, T3 = 0.4mm, W = 1.2mm, W1 = 0.4mm, W3 = 0.6mm, L = 9.85mm, L1 = 11.5mm, LL = 6.6mm, LL1 = 11mm, d1,2 = 1.7mm, d1,4 = 0.2mm, d4,5 = 1.6mm, d5,6 = 1.6mm

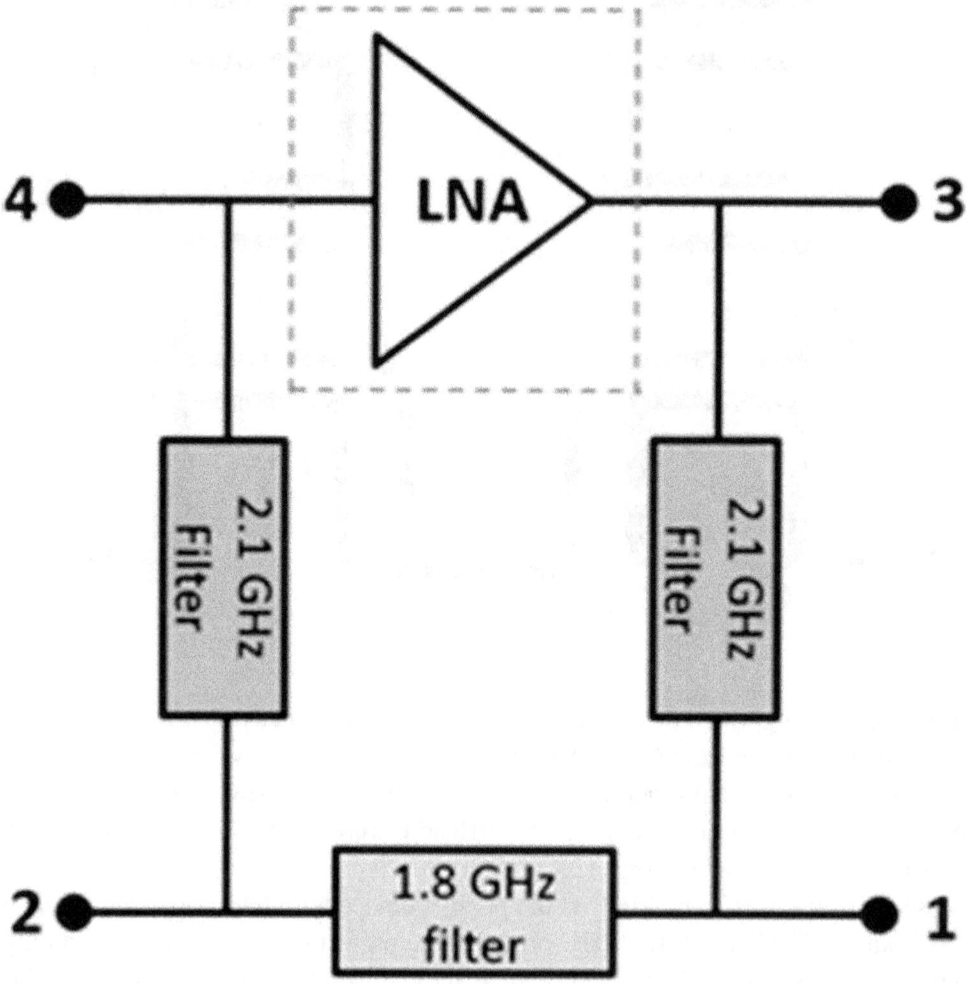

From Fig. 13, the resonators 1, 2 and 3 represents the TX filter while the resonators 7, 8 and 9 represents the RX_1 filter and resonators 4, 5 and 6 represents the RX_2 filter. The two required ACJs are identified using dashed ring circles where resonators 3 and 7 forms the ACJ or the node linking the TX filter and RX_1 filter. The resonators 1 and 4 forms the second ACJ or the node linking the TX filter and the RX_2 filter. This makes the MHC look more like a double diplexer sharing one TX filter in a branch, just as that of Fig. 4 for the MIMO filter/diplexer. In fact, the MHC represents a special case where two branches (7-8-9 and 4-5-6) are of the same frequency as against Fig. 4 where all the branches are of different frequencies.

Figure 15. Simulated response of MHC showing RX channel, TX channel and isolation response

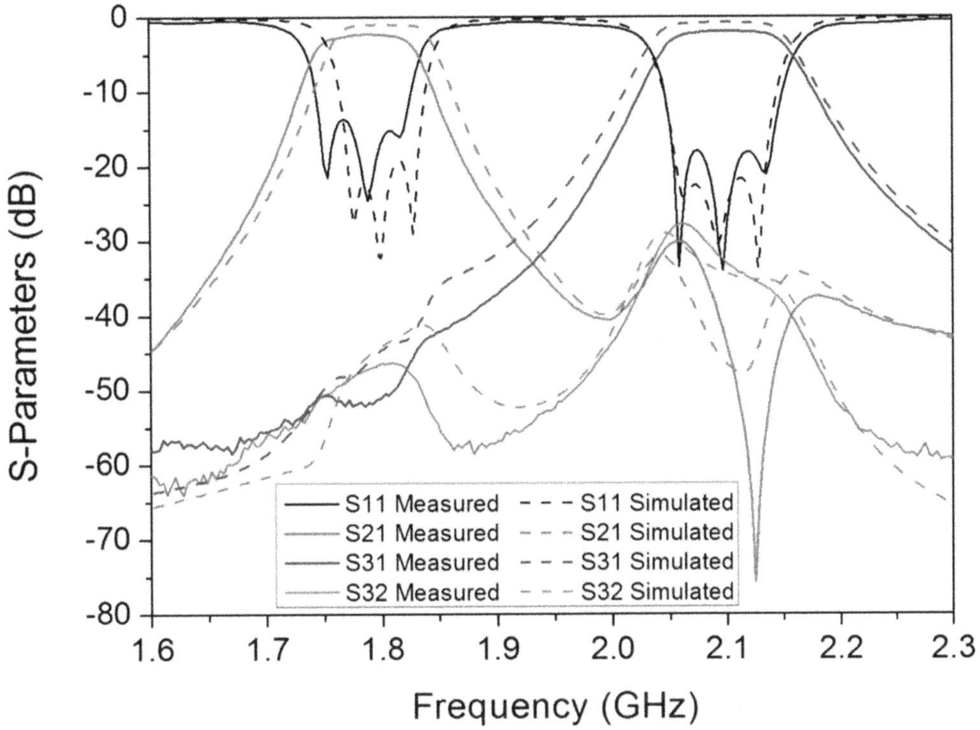

The external Q-factors of the three filters and the two nodes used for the MHC are calculated using (2). (1) is used for obtaining the coupling coefficient between the coupled resonators of the three filters as well as the two coupled resonators that make up the ACJ. After obtaining the physical dimensions of the ACJ, the ACJ are then used in joining the three designed filters of 1.8 GHz and two 2.1 GHz. The individually designed two 2.1 GHz filters had their first resonator removed and replaced with the ACJ while the 1.8 GHz filter, had its first and third resonator removed and replaced with the ACJ. The removing of the resonators and replacing them with the ACJ makes the circuit more compact.

Figure 16. Current density distribution at TX, and RX1 channels

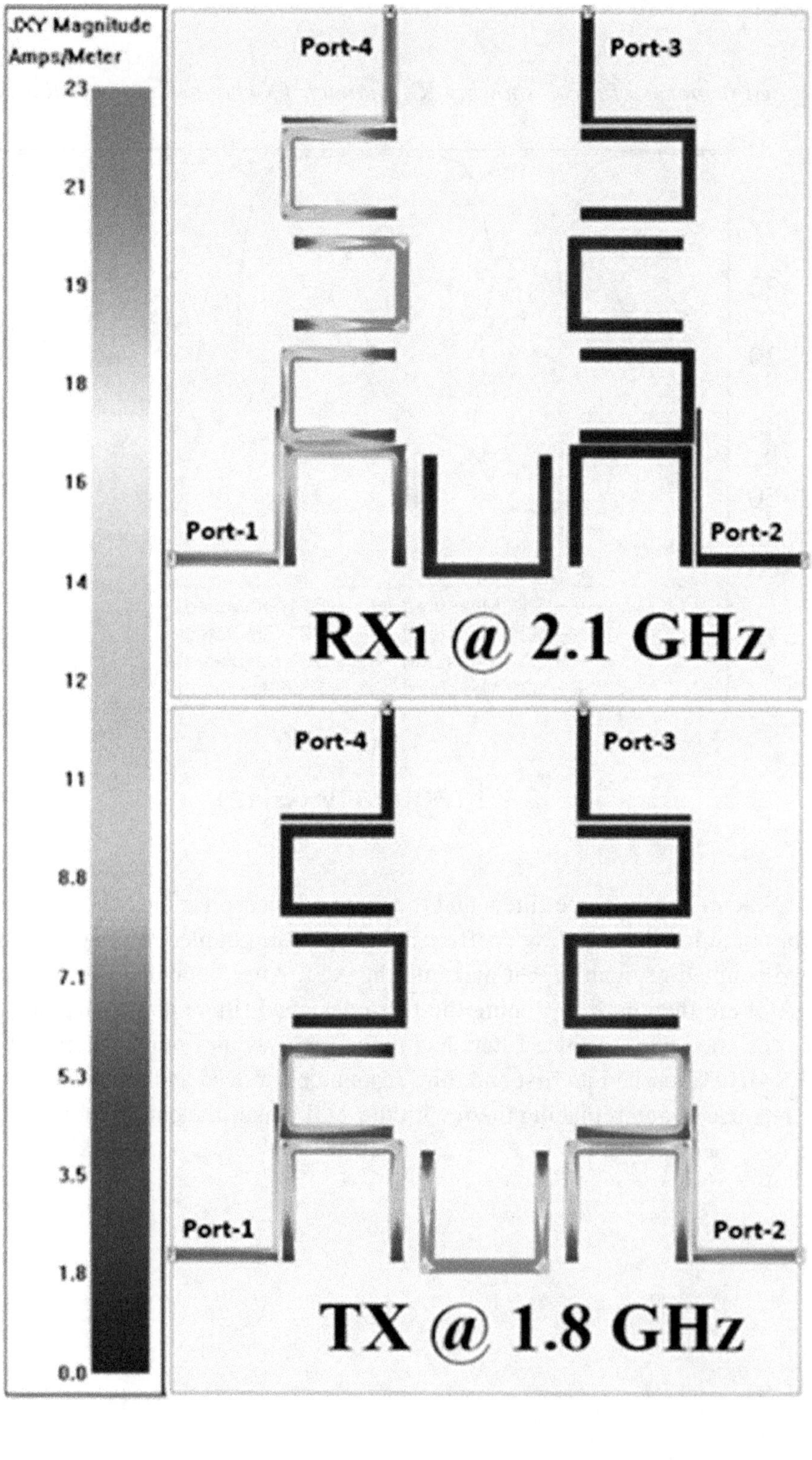

Figure 17. Prototyped MHC circuit

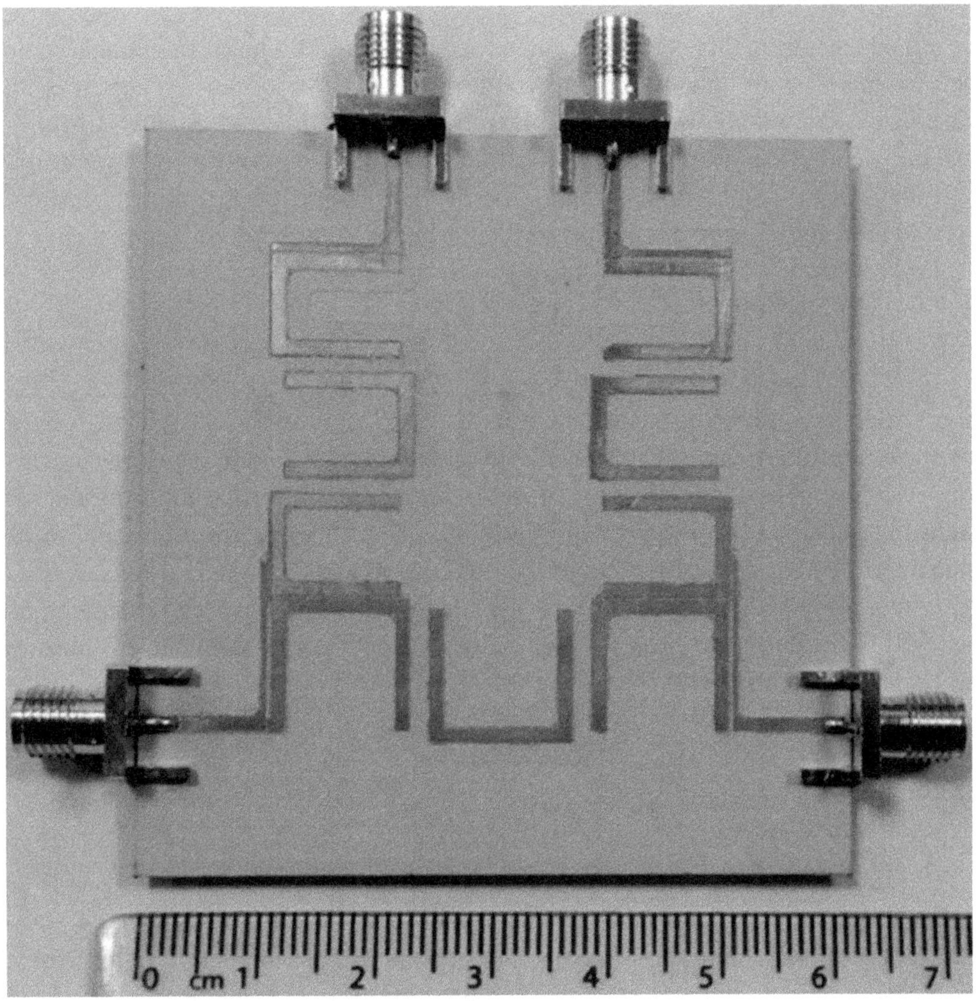

Fig. 15 presents the simulated responses of the MHC results using dashed lines for the RX_1 filter and the TX filter. Due to symmetry, the frequency response of the RX_2 filter and the TX filter will be identical to that of the RX_1 and TX filter. It can be seen that the responses achieved in the simulation met the specification of 4% FBW and 20 dB return loss at both passbands with an isolation around 30 dB on the RX_1 channel, while the TX channel has about 43 dB. The Fig. 15 also presents the measured results using thick lines in comparison with the simulated response. The comparison indicates that the measured responses of the TX channel have a frequency shift of about 10 MHz to the left with a bandwidth of 5 MHz more than the simulated response. The return loss is about 15 dB with an insertion loss of about 2.5 dB. The measured RX_1 channel have approximately the same bandwidth with the simulated response. A return loss of 18 dB and insertion loss of 2 dB is also recorded from the measured result. The measured isolation response is at about 28 dB for the RX_1 channel and about 48 dB on the TX channel. The current distribution of the MHC during operation at the TX and RX_1 channels are presented in Fig. 16. Due to symmetry, the pattern of current distribution on the RX_1 channel will be the same on the RX_2 channel. The prototyped design is presented in Fig. 17 measuring around 62 mm in length.

FUTURE WORK

From the presented designs in this chapter, it can be seen that the ACJ helps in miniaturisation of these devices when compared with their traditional counterparts. This is because it removes the need for traditional T-junction commonly used in multiplexing and power dividing circuits. Its miniaturisation capability can be extended to other devices in 5G application for satellite and base station communication systems where applicable. Suitable applications are in the designing of power splitters, power combiners, different types of multiplexers and demultiplexers, filtering antenna arrays, switchplexer, etc.

SUMMARY

In this chapter, two four-port devices based on the MPFN concept is presented. All the designs are based on extending the capability of the Asynchronous Coupled Resonant Junction (ACJ). The designs demonstrated the feasibility of realising multi-input multi-output (MIMO) filter/diplexer whose frequency selectivity relies on the choice of the excitation ports. At this stage, no specific application about the proposed multi-port devices is identified apart from the Masthead combiner. However, this represents a new design opportunity offered by the coupled-resonator filtering networks for future development. The designed MIMO filter/diplexer has the capability to function as a switchable diplexer and filters based on the choice of chosen ports.

REFERENCES

Macchiarella, G., & Tamiazzo, S. (2015). Design of "masthead" combiners. *2015 European Microwave Conference (EuMC)*, 686-689. 10.1109/EuMC.2015.7345856

Mansour, G., Lancaster, M. J., Hall, P. S., Gardner, P., & Nugoolcharoenlap, E. (2014). Design of Filtering Microstrip Antenna Using Filter Synthesis Approach. *Progress in Electromagnetics Research*, *145*, 59–67. doi:10.2528/PIER14011405

Ogbodo, E. A. (2017). *Novel microwave devices based on multi-port filtering networks*. University of Greenwich.

Ogbodo, E. A., Ngobigha, F., Obadiah Ali, N., & Atuba, S. A. (2021). diplexer with direct coupled resonant junction. *Microwave and Optical Technology Letters*, 1–5.

Ogbodo, E. A., Wang, Y., & Yeo, K. S. K. (2016). Microstrip dual-band bandpass filter using u-shaped resonators. *Progress in Electromagnetics Research Letters*, *59*, 1–6. doi:10.2528/PIERL15072303

Ogbodo, E. A., Wu, Y., Callaghan, P., & Wang, Y. (2017). Asynchronous Coupling Resonant Junctions for Compact Diplexers and Multi-Port Filtering Networks. *Microwave and Optical Technology Letters*, *59*(12), 3046–3051. doi:10.1002/mop.30877

Ogbodo, E. A., Wu, Y., Callaghan, P., & Wang, Y. (2018). Masthead Combiner Employing Asynchronously Coupled Resonant Junctions. *48th European Microwave Conference (EuMC)*. 10.23919/EuMC.2018.8541754

Ogbodo, E. A., Wu, Y., & Wang, Y. (2017). *Dual-Path Dual-Band Filters Based On Patch Resonators.* International Workshop on Electromagnetism (iWEM), London, UK. 10.1109/iWEM.2017.7968779

Son, H., & Pyo, C. (2005). Design of RFID tag antennas using an inductively coupled feed. *Electronics Letters*, *41*(18), 994. doi:10.1049/el:20051536

Tamiazzo, S., & Macchiarella, G. (2013). Design and realization of filtering units in a triple-band mast-head amplifier system for BTS of mobile communications. *IEEE MTT-S Int. Microwave Symp. Dig*.

Chapter 12
Investigating the Effects of Temperature, Paste Height, and Time on Solder Paste to Its Printing Performance

Patrick Uche Okafor
Enugu State University of Science and Technology, Nigeria

James Eke
Enugu State University of Science and Technology, Nigeria

Ndidi Stella Arinze
Enugu State University of Science and Technology, Nigeria

Osondu Ignatius Onah
Enugu State University of Science and Technology, Nigeria

ABSTRACT

Due to pressure to meet consumer requirements, miniaturization, reliability, and increased functionality have become more important. To achieve good assembly of electronic devices, there must be smooth deposition of tiny and consistent paste deposits on the board which must be up to an acceptable height and volume based on desirable process parameters in the output. The printing performance of the solder paste is dependent on conditions such as paste deformation behavior, good paste roll, complete aperture filling, and paste release onto the substrate pads. Eighteen tests were performed using two types of lead-free solder paste. Three temperatures were investigated under three different time intervals. Results analysis showed that variability of the paste distribution was lowest and more uniform for paste stored at 25°C for 48 hours. From the analysis, it is obvious that temperature has tremendous effects on the printing performance of solder paste. This study recommends 25°C as a good temperature for solder paste printing.

DOI: 10.4018/978-1-7998-6992-4.ch012

1. INTRODUCTION

The electronics manufacturing industry has turned out to be one of the largest industries in the world with more than $1trillion investment (Lau et al, 2002). It is a vibrant, attractive, and important area of manufacturing. The industry has had a steady growth rate since the early 1990s with personal computers being one of the most lucrative products in the market as shown in Figure 1.

Figure 1. Sales of PCs from the US according to the National Safety Council (Lau et al, 2002)

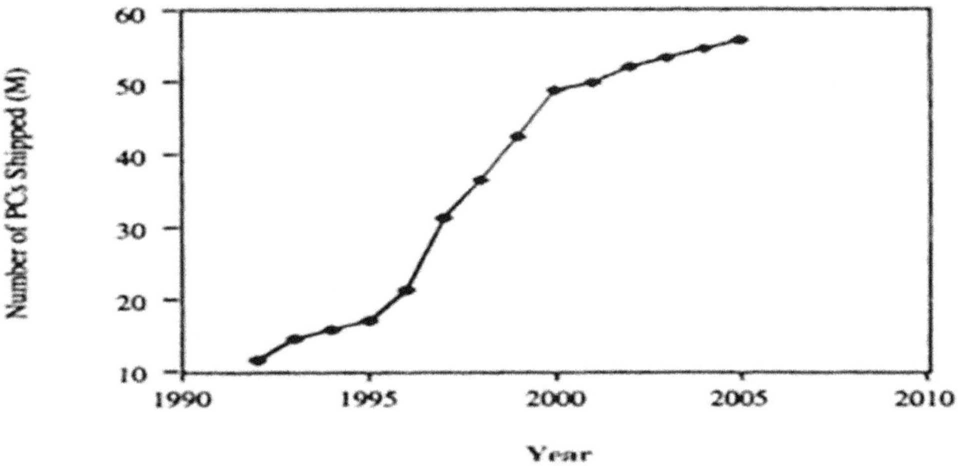

In the past few years, the electronic industry has been facing an impending change in light of upcoming lead-free technology legislation. This is because the electronic industry has relied on tin-lead (SnPb) solders for its products (Lau et al, 2002). In 1998, the European Commission introduced two draft proposals called the Waste Electrical and Electronic Equipment (WEEE) and Reduction of Hazardous Substance (ROHS) directives. The main objective of these complementary proposals is to minimize the risks and impacts that the production, use, treatment, and disposal of waste electrical and electronic equipment have on human health and the environment.

There has been a continuous increase in worldwide interest in lead-free solder paste, not only for environmental or regulatory reasons but also because of market differentiation. The demand for better, cheaper, more reliable, and more portable products have forced international competition in all manufacturing sectors. This has compelled the Electronic industry to look for innovative methods to decrease cost, weight, and volume of products while making resolute efforts towards achieving higher levels of productivity and reliability (Eke, Okafor, and Arinze, 2020). There has been an increase in complexity and reduction in the size and weight of the electronic component. In the electronic industry today, the newest approach to compete is to make products more personal by making them smarter, lighter, smaller, thinner, shorter, and faster, while at the same time making them more user friendly, functional, powerful, reliable, robust, innovative, creative, and less expensive (Durairaj *et al.* 2001). Surface mount technology (SMT) makes it possible to produce more reliable assemblies at a reduced weight, volume, and cost.

As the trend toward miniaturization and the compact product continues (Okafor and Eneh, 2013), the assembling process of the electronic component or surface mount devices (SMDs) becomes more complex and there is a need for some form of automation. The automated process of assembling the electronic component or surface mount devices is known as *Surface Mount Technology (SMT)* by using solder paste as interconnecting material to provide an electrical, thermal and mechanical function (Lau *et al.* 1996). Figure 2 shows a typical example of surface mount technology using Solder paste to attach components unto PCB.

Several interconnection technologies have been developed within the electronics manufacturing industry. The effort to develop reliable interconnection solutions is to meet up with the rising consumer demand for better, faster, and cheaper products. Broadening traditional surface mount technology (SMT) infrastructure (process capabilities, equipment, and soldering materials), stencil printing has been proven to be the most economical solution for flip-chip interconnection (Huang et al, 2002, and Nguty et al, 2001). Although this technique is vital to producing high volume, low-cost joints, it also accounts for about 60% of assembly defects within the SMT process (Durairaj et al, 2002).

Figure 2. Surface Mount Technology (google.com)

This work focuses on the use of stencil printing in surface mount technology. According to Eke, Okafor, and Arinze (2020), the majority of the components used on a printed circuit board assembly (PCBA) are based on surface mount technology (SMT) being assembled using solder paste printing (SPP) and then fixed by the reflow soldering (RS) process (Lau and Yeung, 1996). There are three major challenges in the fine pitch stencil printing process. These challenges are the solder paste formulation, the stencil manufacturing process, and the optimization of the process parameters in stencil printing. The solder paste which is primarily used as a bonding medium is typically deposited using the stencil printing process. In the SMT industry, the solder paste deposition operation is already recognized as a major contributor to end-of-line defects and there is wide agreement in the industry that the paste printing process accounts for the majority of assembly defects. Most defects originate from a poor understanding of the effect of printing process parameters on printing performance and the nature of their interactions.

Good paste printing performance depends on conditions such as good paste roll, complete aperture filling, and paste release from aperture onto the substrate pads. The implication of this is that the paste flow and its rheology are paramount in defining the printing performance of any solder paste. The paste flow and rheology are functions of viscosity. Since viscosity is affected by temperature, it is, therefore, important to study the effects of temperature on the solder paste. This will help to understand the solder paste behavior before, during, and after the stencil printing process and therefore develop better interconnection solutions.

2. THEORY OF WORK

Solder paste is a homogeneous, stable suspension of solder powder particles suspended in a flux binder. The metal content of a typical solder paste is between 88% - 91% by weight, and about 30% - 70% by volume. The main constituent of the flux medium is a naturally occurring rosin or resin. Solder paste is one of the most important process materials today in surface mount technology (Nguty and Ekere 2000). The characteristics and flow properties of solder pastes depend primarily on their chemical and physical composition. The solder paste printing process is responsible for common causes of assembly defects in the electronic manufacturing industry.

Stencil printing of solder paste onto PCBs constitutes an important stage in the reflow soldering of surface mount devices (Nguty and Ekere, 2000). The quality of solder paste printing is very sensitive to environmental conditions such as temperature and humidity. Different investigation has shown that printed circuit board assembly (PCBA) defective derived out from reflow soldering inspection, in-circuit testing, and function testing up to final quality inspection are heavily induced by poor solder paste printing (Lau and Yeung 1997). This is likely to continue with the trend toward miniaturization and the implementation of die-size packages.

To achieve repeatable solder deposits from board-to-board and pad-to-pad requires an understanding of the paste rheology. One of the key factors that influences solder paste rheology is temperature. A temperature change will cause the viscosity of the solder paste to change. This change could be ambient or from the stencil printing process itself. This is likely to impact the performance of the solder paste. Current models show a single variable dependence of viscosity with temperature (Lau and Yeung 1997). The shape and size of the metal particles in combination with the rheological properties of the binder are tailored to match the application method of the paste onto the PCB (Nguty and Ekere, 2000).

The Solder Paste Stencil Printing Process: Lead-free solders have higher surface tension and so do not spread as well as SnPb solders. Therefore, they tend to leave the edge of the PCB pads exposed after reflow. This may be more prominent on PBC with an OSP surface finish. To minimize this occurrence, the stencil apertures could be widened to allow more solder paste coverage on the pad maximizing solder spread during reflow (Bath, 2007).

Stencil Printing: Stencil printing is one of the techniques used in surface mount technology. The solder paste is concurrently deposited across the entire wafer making the process very fast. According to Clements et al (2007) with the recent advances in stencil manufacturing, especially in the reproducibility of micron-size aperture shapes, stencil printing is rapidly gaining importance in direct-chip attach technology due to its inherent low cost. Stencil printing continues to be a leading process used for the deposition of solder paste onto printed circuit boards (PCBs) in the volume production of electronic assemblies, despite problems in achieving a consistent print quality at an ultra-fine pitch. (Durairaj et al, 2002).

Solder Paste Printing Performance: The printing process for lead-free assembly is similar to that of the SnPb assembly process. Nevertheless, the use of lead-free solder paste may entail some minor adjustments to the process such as modifying printing pressure and print speed (Bath, 2007). The examination of rheological properties is critical when investigating the suitability of a solder paste for the printing process, especially in those instances where a developmental paste is to be considered; hence the need to understand the properties of new formulations of paste is suitably justified. An ideal solder paste should be viscous enough to flow through the stencil openings at a high shear rate, while also be elastic enough to recover itself when the shear is stopped (Durairaj et al, 2006).

Thermal Effect on Solder Paste and its Print Performance: Stencil printing of solder paste is a critical stage in the surface mount assembly line where the solder paste is deposited onto the printed circuit board (PCB). Components are placed on the deposits and the entire assembly is heated to complete the formation of the joints. In the case of solder paste, a high temperature (higher than 25°C) will cause the solder paste to evaporate faster (Ekere and Mallik, 2008; Mallik, 2009). Though current research has provided new temperature stable solder paste, LOCTITE GC 1, the author of the work stated that the lead-free, zero halogen solder paste is stable at 26.5°C for one year and at 40°C for one month (Ian, 2018) but, there has not been reviews as regards to the effect of the LOCTITE GC 1 on the joints after some years has elapsed. The formations of thermally, mechanically, and electrically reliable solder joints depend mostly on the quality of solder paste deposition in the stencil printing process (Ekere and Mallik, 2008). High processing temperatures increase the dissolution rates of metal finishes and/or conductor metals into the solder and can increase the rate the intermetallic compound formation in bulk solder. This reduces compliance. High lead-free processing temperature can affect the assembly yields of particular components due to thermal mass and the size or pitch of a component.

3. MATERIALS AND METHODS

The solder paste can be identified by its properties such as viscosity, normal force, density, and surface tension when considered as a homogeneous and kinetically stable mixture (Hwang, 1989). According to Hwang (1989) solder paste are made up of three components;

1. Solder alloy powder melts to form the bond between the components and the substrate.
2. The flux system helps the alloy powder and the surface to be joined to maintain a clean and metallic state, so that good wetting and metallic continuity between solder alloy and the surface joined can be formed.
3. The vehicle carrier system is primarily a carrier for solder powder and provides desirable rheology.

Lead-Free Solder Paste used for the Experiment

Two commercially available lead-free solder pastes P1 and P2 prepared from fluxes F1 and F2 as shown in Figure 3 were used in the experimental studies reported in this study (Eke, Okafor and Arinze, 2020). The solder particles for all the paste samples are made of the same tin-silver-copper alloy with a melting point of 217°C. Both P1 and P2 have a percentage metal loading of 88.5%. P1 is a Type 3 solder paste and P2 is Type 4. For this experiment, the flux medium makes up for 11.5% of the solder paste weight. The flux medium contains a stable resin system and slow evaporating solvents with minimal odour. The formulation meets the requirements of the Telcordia (formerly known as Bellcore) GR-78-CORE and ANSI/J-STD-004 for a type ROL0 classification.

Figure 3. Lead-free solder paste LF318 and LF328

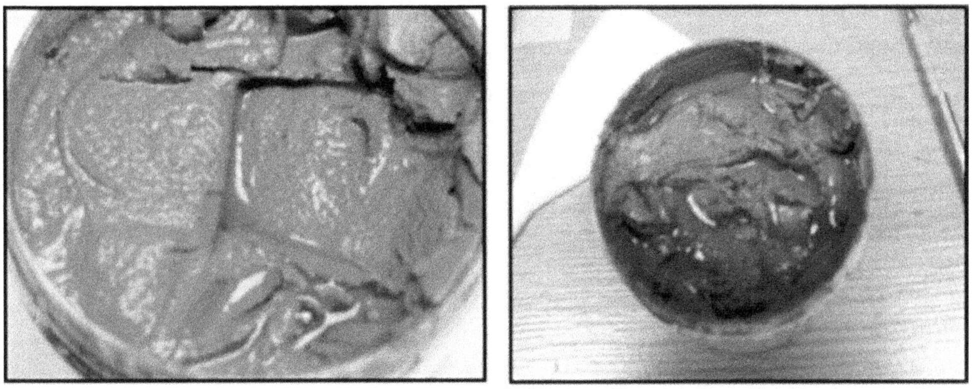

DOE4 DOE8

First Part: Solder Paste Separation and Thermal Conditioning

The first part of the experimental procedure is explained here to provide an insight into how the experiment was designed. The two lead-free solder pastes LF318 and LF328 used in this experiment were separated into nine different jars respectively, making it eighteen jars as a whole. Each jar contained one hundred grams of lead-free solder paste. This separation was to allow for the number of experiments that were carried out in the study. For quick and easy identification, LF318 was named P1, and LF328 was named P2. Since oxidation is a major concern of solder pastes, the eighteen jars were stored in a refrigerator where they were taken out in batches for the experimental process. The storage in refrigerator was done to keep the solder paste airtight and maintain low temperature to reduce the rate of oxidation and flux degradation to the barest minimum. This is important because,

When each of the solder paste is taken out from the refrigerator, it is allowed to return to room temperature before putting it into the thermal chamber for storage under constant temperature. Both P1 and P2 went through nine different processes of varying time and temperature, one jar for one process. The temperatures under investigation were 15°C, 25°C, and 35°C. As shown in Table 1, two solder pastes P1 and P2 were stored at each of these temperatures for 24hours, 48hours, and 72hours respectively, making it nine experiments for each solder paste.

Table 1. Storage parameters for P1 and P2 at the thermal chamber

Solder paste	Temperature (°C)	Storage time (minutes)	Storage time (minutes)	Storage time (minutes)
P1	15(°C)	1440	2880	4320
P2	15(°C)	1440	2880	4320
P1	25(°C)	1440	2880	4320
P2	25(°C)	1440	2880	4320
P1	35(°C)	1440	2880	4320
P2	35(°C)	1440	2880	4320

After thermal conditioning, the solder paste was printed using the stencil printer. For each printing process, room temperature and humidity are recorded since they could not be controlled. Each solder paste was printed about six times before taking the seventh print like the one to be investigated. This procedure was adopted to maintain standard printing practice.

Second Part: Process Parameters

The printing parameters used for the experiment are outlined in table 2. Previous work reported by Marks et al (2007) was used as a benchmark.

Table 2. Printing parameters

Printing parameters	Values used
Printing/Squeegee speed	20mm/sec
Squeegee loading (or pressure)	8kg
Separation speed	100% (3mm/sec)
Snap-off/Print gap	0.0mm (On-contact printing)

After the printing parameters have been set, the procedures used in this part of the experiment are outlined below (Okafor and Eneh, 2013):

- The stencil was placed in the stencil holder compartment of the printer. The stencil is firmly held in the compartment by touching the 'table lock' and 'stencil lock' button in the computer interface.
- The stencil was then cleaned using Iso-Propanol (IP) solvent which readily evaporates.
- The squeegees were cleaned using IP before being placed on the squeegee holder. The screw on the squeegee easily secures the squeegee. Double squeegee was used in this experiment for forward and reverse strokes respectively.
- The solder paste was then applied to the centre of the stencil. This is to allow paste elongation during the printing process.
- A copper board was then placed onto the tooling pins. The copper board represents the Printed Circuit Board (PCB).

- The machine was set to 'step-mode'. This mode allows the user to input required data step by step through a series of questions. In this mode, the input data can be changed/altered before the printing commences.
- The 'print stroke' (distance traveled by the squeegee during the printing) was set to the dimension of the wafer pattern.
- Once the paste was supplied on the stencil, the experiment begins.
- The room temperature and humidity were recorded.
- In each experiment, there were 7 runs. In each run, the settings in the printer machine remained constant throughout the experiment. The seventh run of each experiment was used for inspection.
- After every print, the upper and bottom sides of the stencil is cleaned with iso-propanol. Since the apertures are small, clogging is a major problem. Therefore this step is very important to ensure a good paste deposit.
- Eighteen boards were printed as a whole; the solder paste deposit on each board was measured and inspected after printing.
- Sixteen different paste heights were measured each time from two different apertures.

A total of eighteen experiments (seven runs for each experiment) were conducted during this part of the experiment with two different solder pastes.

Third Part: Print Performance Assessment

In assessing the print performance of the solder paste, the following procedure was followed (Okafor and Eneh, 2013):

- The solder paste was methodically stirred (for about 2 minutes) before placing it on the stencil.
- Print parameters were set to standards: Pressure- 8 Kg, Speed-20 mm/sec, and separation speed- high-100%.
- A sufficient amount of solder paste was placed on the stencil for each of the experimental runs.
- At the end of the seventh run of each experiment, the average heights of the selected apertures were measured.
- After measuring for average heights, bridging was also checked and noted at a selected location on the board.

This procedure was followed for both Type 3 and Type 4 solder paste. A total of eighteen experimental assessments were conducted during this part of the experiment with two different solder pastes.

Equipment Used

Thermal Chamber: The Weiss Technik test systems with modular programmable controller MOPS/CTC-C was used for the conditioning of the solder paste. The storage of the solder paste was an isothermal process whereby the temperature of the system was kept constant: $\Delta T = 0$. The temperatures that were studied are; 15°C, 25°C, and 35°C. **Stencil Printer:** The DEK 260 semi-automatic printer was used. The printer principally consists of the PC interface, stencil frame holder, the squeegee system, the motor, the stencil/substrate alignment system, and the substrate holder. **Stencil:** The printing was done using a laser-cut Nickel-plated stencil with a design wafer. The stencil has a thickness of 125 microns (0.125mm). The stencil is designed to match the Benchmarker II test board design. **Squeegee system:** A squeegee is a tool fixed to the printer. Its main function is to mechanically transfer the solder paste from the stencil into the aperture. **Stereomicroscope:** One of the inspection devices used in the examination of the print performance in this project was the Leica S6D zoom stereomicroscope. This stereomicroscope was used to examine the solder paste deposits for bridging/slumping and spread behaviors. **Vision master measurement device:** The vision master is a three dimensional (3-D) measurement device. It was used to measure the height of the solder paste deposits.

4. IMPLEMENTATION

Section 1: Gauge R&R: Gauge R&R stands for Gauge Repeatability and Reproducibility. It is a quantitative assessment of gauge effectiveness. Before experiments were carried out in this study, a Gauge R&R was conducted to determine if the measuring system is capable of measuring the solder paste height. A repeatability test was carried out by operator A to estimate variability using the same operator and samples. Ten samples were measured twice, making a total of twenty measurements. Variability was measured by operator B using the same gauge and samples. This is called the reproducibility test. All the results recorded were put into the Gauge calculator. The R&R% got was 19.43%. This is well below 30% which is the acceptable standard. This means that the measurement device is capable of the task.

Section 2: Plotting the Average: In the formation of solder joints, print thickness determines the volume of solder in the joint. The thickness of the paste print is determined by the thickness of the metal mask of the stencil. Though stencil thickness controls the paste thickness, other variables such as snap-off height and the condition of the printing equipment determine the print height. The reflow solder height is not just a factor of solder paste height only, but also the metal content of the paste. Since the print gap for the experiment is 0.0mm, the snap-off height becomes the thickness of the stencil which is 125μm. Therefore, at this stage, we compute the average of the average heights got from the regions in the chosen locations on the stencil. This will make it easier to get a better picture of the behavior of the solder paste in terms of paste height, hence, help to determine the printing performance.

In each of the two locations under investigation, four regions were measured. In each region, two deposits were measured. The averages of the two deposits were taken. Subsequently, the averages of the locations were then computed based on the four regions. A sample of the 3-D image of the measured heights is shown in Figures 4 and 5. Figure 4 shows the height of P1 while Figure 5 shows that of P2. The blue base indicates the reference point, which is the surface of the substrate. The red colour indicates the maximum height. The results from the computed averages are shown in Tables 3 and 4, while Figures 6 and 7 show the graphs of the computed average height against the solder paste storage/aging time to the storage temperatures.

Figure 4. The 3-D image of the solder paste deposit for location 1

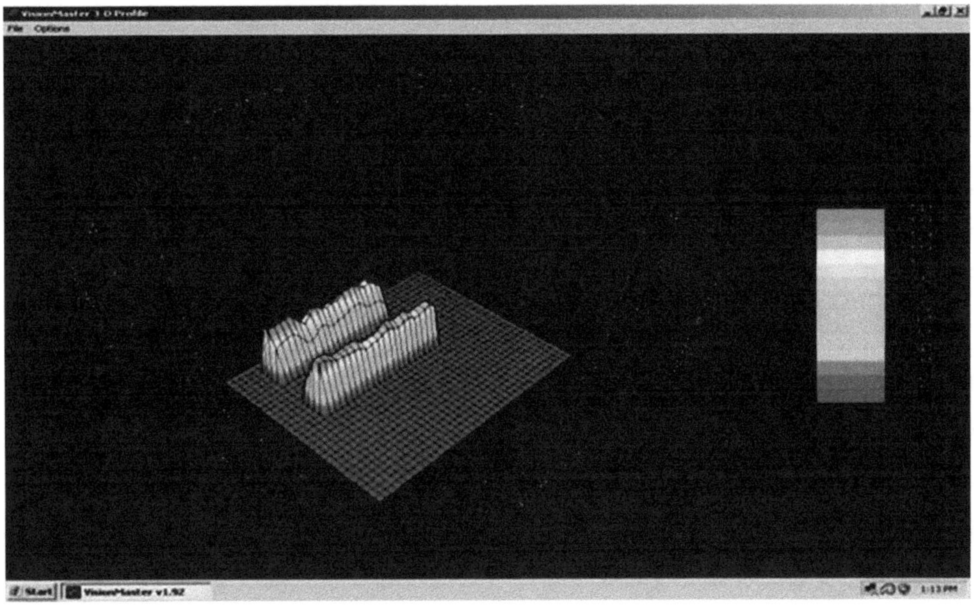

Figure 5. The 3-D image of the solder paste deposit for location 2

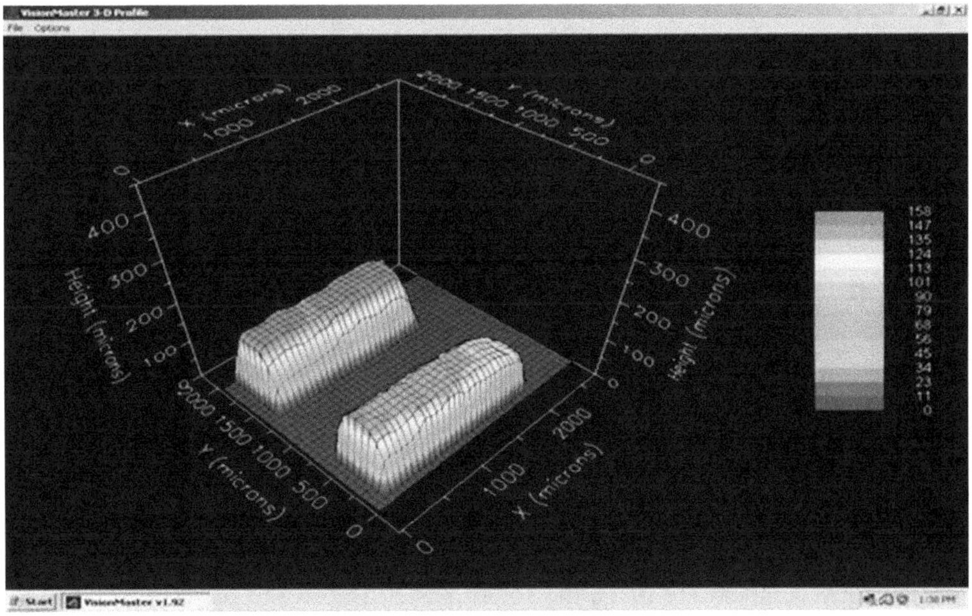

Table 3. Results for solder paste deposit heights for LF318

Storage temperature (°C)	Storage time (Hours)	Target height (µm)	Height for Location 1 (µm)	Height for Location 2 (µm)	Average height (µm)
15	24	125	103.36	89.5	96.43
15	48	125	91.63	109.66	100.65
15	72	125	92.24	115.9	104.07
25	24	125	89.15	115.58	102.37
25	48	125	83.48	122.74	103.11
25	72	125	102.08	116.79	109.44
35	24	125	100.25	120.15	110.2
35	48	125	95.26	118.69	106.98
35	72	125	112.02	119.43	115.74

Figure 6. Temperature effect results for solder paste deposit height for LF318

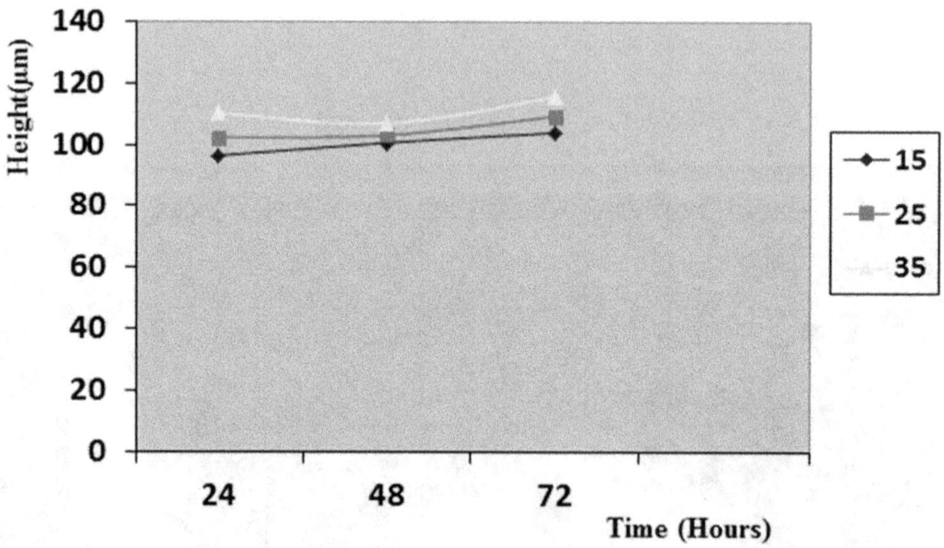

During the printing process for LF318 aged at 15°C for 24 hours, the solder paste just rolled minimally. The temperature of the solder paste during the printing was low, and that had a great effect on the solder paste viscosity. The low temperature made the solder paste viscosity to be too high. The temperature was not conducive for the rheological properties of the solder paste to be at their best. Therefore, the paste could not role, giving rise to skipping and low aperture filling which in effect caused ragged edges. This was one of the contributing factors to the low height seen after the printing. This is evident from the points of the height for 15°C in the graph (Figure 4).

Table 4. Results for solder paste deposit heights for LF328

Storage temperature (°C)	Storage time (Hours)	Target height (μm)	Height for Location 1 (μm)	Height for Location 2 (μm)	Average height (μm)
15	24	125	120.83	103.06	111.95
15	48	125	89.63	109.66	99.93
15	72	125	92.24	108.31	96.76
25	24	125	108.5	126.26	117.38
25	48	125	98.49	102.08	100.29
25	72	125	100.21	11538	107.80
35	24	125	99.63	114.48	107.06
35	48	125	110.15	133.9	122.03
35	72	125	117.14	129.48	123.31

It can be observed that as the temperature increased, the height against time also increased. This trend was seen for both P 1 and P 2, except for P 2 stored at 35°C for 24 hours. Normally, it would be expected that an increase in temperature would give rise to a decrease in height, given that higher temperature would cause the solder paste to slump. Rather than decrease the height as the temperature increased, the height was increasing. This is because, at low temperature, the paste's viscosity increases causing the solder paste flow to slow down or not to flow at all. But as the temperature increases, the material flow rate increases making it possible for the squeegee to fill in the aperture effectively. On the other hand, when the temperature becomes too high, the effect will be slumping of the solder paste deposits which will lead to bridging.

Figure 7. Temperature effect results for solder paste deposit heights for LF328

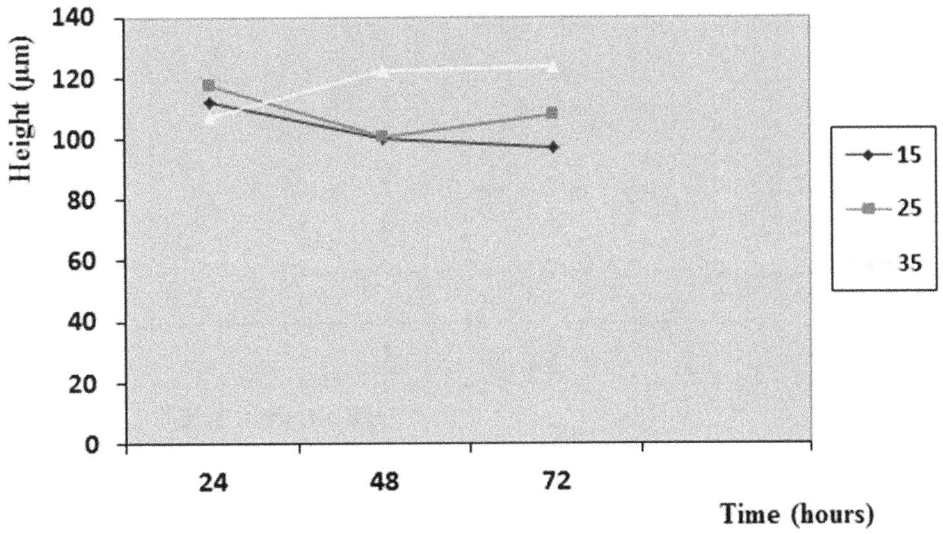

Figure 8. Comparing solder paste deposit heights of P 1 and P 2 for 24 hours aging

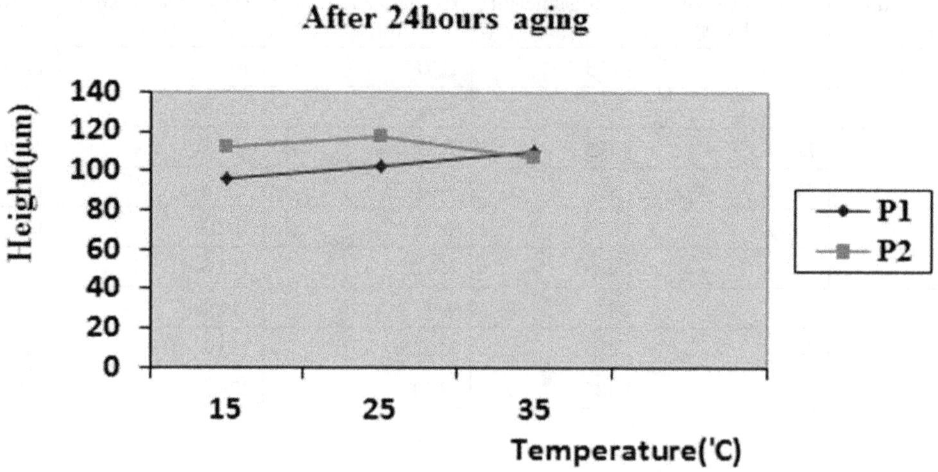

The height increased as the time of storage and/or conditioning increased. This could be as a result of paste particles and the flux system having more time to align to a certain structure. For the fact that temperature affects the structural alignment/formation of the solder paste, time plays a major role because the alignment and re-alignment of pastes happen at a slow rate. Therefore, the more time that is given for this to happen, the more the structures become stable at that temperature.

Figure 9. Comparing solder paste deposit heights of P 1 and P 2 for 48 hours aging

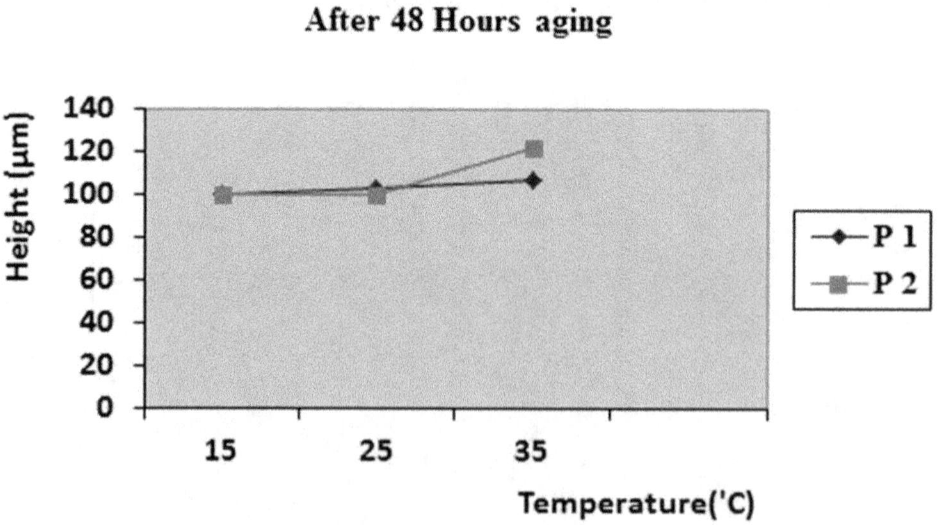

Figure 10. Comparing solder paste deposit heights of P 1 and P 2 for 72 hours aging

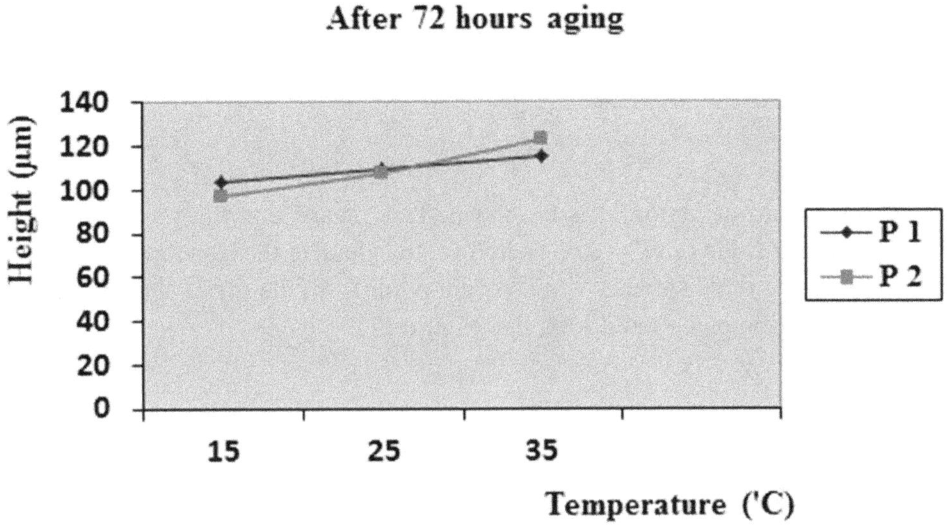

When the solder pastes were compared, it was found as shown in Figures 8-10, that the height increases as the temperature increases. This trend was seen as earlier discussed, for all the experiments except for P2 aged/conditioned at 35°C for 24 hours. This shows that an increase in height with an increase in temperature and or storage time is common for both P1 and P2. Judging from the common trend of increased height with increased temperature, the behavior of P2 conditioned at 35°C for 24 hours may be a special cause of variation. Considering the relationship between paste heights, the ability of the solder paste to roll during printing, viscosity, temperature, and time, the results got in this experiment were compared with those got by Nguty and Ekere (2000). They investigated the effects of storage conditions and duration on the rheological properties of the solder pastes. They conducted experiments on solder paste samples kept at both room temperature and refrigerated temperature over a period of five months. From the results that were got, it was shown that there was an increase of 60% in paste viscosity for samples kept at room temperature. This shows that like the results that were got in this study, longer time and higher temperature increases viscosity thereby increasing the tendency of the solder paste to roll. However, the temperature must not be excessive.

5. VARIABILITY ANALYSIS

Standard Deviation: In probability theory and statistics, the standard deviation is a measure of the variability or dispersion of a statistical community, a data set, or a probability distribution. A low standard deviation indicates that the data points tend to be very close to the mean, whereas a high standard deviation indicates that the data are spread out over a large range of values. When variability is reduced, the cost of production goes down, errors and mistakes become fewer, and hence, quality increases. Calculating the standard deviation will help to find out if the data is a normal distribution which means that most of the points in the data are close to the average. It is desirable to have a normal distribution. This entails consistency of performance which Taguchi has taken a broad view of it as the definition of quality. The following mathematical equation is used to calculate the standard deviation.

$$\sigma = \sqrt{\frac{\left(x_1 - \mu\right)^2 + \left(x_2 - \mu\right)^2 + \cdots + \left(x_N - \mu\right)^2}{N}} \tag{1}$$

Where σ is the standard deviation, μ is the mean. This means that the standard deviation σ is the square root of the average value of $(X - \mu)^2$. Therefore, to calculate the standard deviation of the solder paste heights, we compute the difference of each data point from the mean and then square the result. A normal standard deviation plot should look like Figure 11.

Figure 11. Normal standard deviation diagram

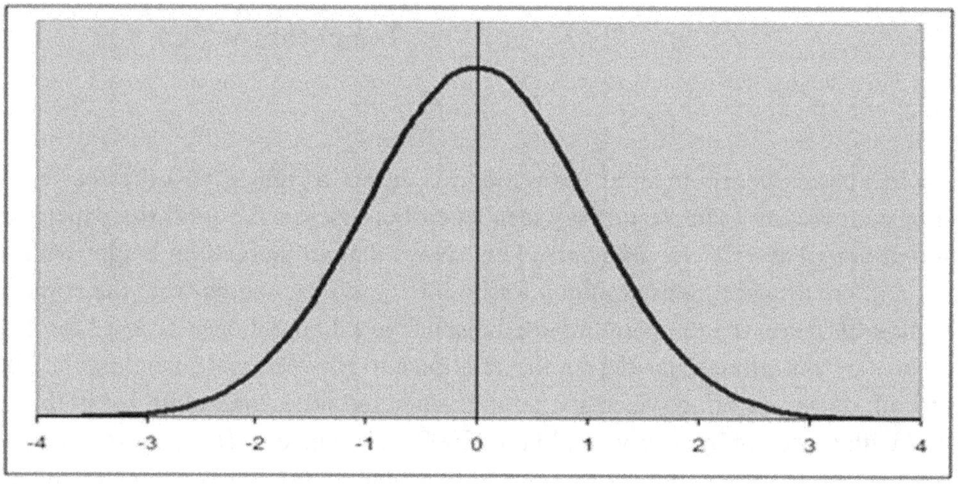

Mean-Square Deviation

Mean-Square deviation (MSD) is a quantity that is used to measure the performance of a population's test samples in terms of their consistency. Since consistent performance is obtained when performance is closer to the target more often, the MSD is computed by adding squares (to avoid the effect of the sign) of the deviations of all the sample results from the target.

Assuming $y_1, y_2, y_3, \ldots\ldots\ldots y_n$ are n data points, the MSD for it can be computed as

$$MSD = \sum(Y_i - Y_0)^2 = (Y_1 - Y_0)^2 + (Y_2 - Y_0)^2 + (Y_3 - Y_0)^2 + K \tag{2}$$

n n

It can also be shown that:

$$MSD = \sigma^2 + (Y_{avg} - Y_0)^2 \tag{3}$$

Where, Y_0 is the target value and σ is the standard deviation (population, using n as the divisor). MSD can be seen as a deviation from the origin. For the fact that like other forms of deviations, MSD is always preferred to be a smaller quantity in all three quality characteristics, in the case of bigger quality characteristics, the inverse of the deviation squares are used as shown below. In this way, despite the quality characteristics of the original results, the desirability of MSD is always retained as smaller is better.

$$\text{Nominal: } MSD = (Y_1 - Y_0)^2 + (Y_2 - Y_0)^2 + (Y_3 - Y_0)^2 + (Y_i - Y_0)^2 \tag{4}$$

n

$$\text{Smaller: } MSD = Y_1^2 + Y_2^2 + Y_3^2 + K \tag{5}$$

n

$$\text{Bigger: } MSD = 1/Y_1^2 + 1/Y_2^2 + 1/Y_3^2 + K \tag{6}$$

n

Percentage Variation

Percentage variation is a measure of the variability or dispersion of a statistical area or a pair of data by expressing the result in terms of a number as a fraction of 100. In this case, the variability of the two deposits measured in each region was being calculated in terms of percentage. It is desirable to obtain a very low percentage variation. This is because getting a low percentage variation means that by expressing the variance between a pair of data or statistical areas in terms of percentage, the value is close to each other. This, in other words, shows consistency of performance, not with regards to a set target, but with regards to each other.

Assuming that y_1 and y_2 is a pair of data, the percentage variation between them will be expressed by Equation 8;

$$\text{Percentage variation} = y_1 - y_2 \times 100\% \tag{7}$$

y_1

For this study analysis, standard deviation and percentage variation were used to ascertain the level of variability at different levels of performance. While standard deviation was used to study the variation at the location level, a more in-depth study of variation was conducted at the region level using percentage variation.

Applying Standard Deviation

By applying equation 2 to the results, the results for 15°C at 24 hours, we have that;

$$\mu = \frac{89.85 + 106 + 113.3 + 104.3}{4}. = 103.36$$

$(89.85 - 103.36)^2 = (-13.51)^2 = 182.52$

$(106 - 103.36)^2 = (2.64)^2 = 6.97$

$(113.3 - 103.36)^2 = (9.94)^2 = 98.80$

$(104.3 - 103.36)^2 = (0.94)^2 = 0.88$

$$\sigma^2 = \frac{\sqrt{(-13.5)2 + (2.64)2 + (9.94)2 + (0.94)2}}{4} \text{ Therefore, } \sigma = \sqrt{72.29} = 8.5$$

This means that for the lead-free solder paste aged before printing at 15°C for 24 hours, the standard deviation (σ), of the printed deposits for location 1 is equal to 8.5. The same process was used to compute the standard deviation of all the locations measured for both P 1 and P 2. The computed results are shown in Table 5.

Table 5. Result of standard deviation for LF318 (P 1) and LF328 (P 2)

Temperature (°C)	Time (Hours)	Standard deviation (P 1)		Standard deviation (P 2)	
		Location 1	Location 2	Location 1	Location 2
15	24	8.5	9.9	8.9	7.6
15	48	4.0	5.2	9.2	6.1
15	72	12.2	3.7	9	3.8
25	24	10.4	8.1	5.3	5
25	48	7.6	6.1	6.6	8.6
25	72	12.3	3.7	7.4	6.9
35	24	8.7	8.6	13.1	2.9
35	48	8.6	1.9	8.8	6
35	72	13.4	1.2	12.6	1.8

Table 6. Percentage variation result of deposited solder paste height for LF318

Location	Region	Time (Hours)	Percentage variation (%) for P1		
			15°C	25°C	35°C
Loc 1	1	24	7.2	- 0.8	5.2
	2	24	- 10.7	1.2	10.4
	3	24	8.6	18.0	1.6
	4	24	16.2	- 11	4.1
Loc2	5	24	3.4	4.0	17.0
	6	24	9.5	7.9	5.9
	7	24	2.6	- 5.2	11.0
	8	24	4.3	5.7	19.3
Loc 1	1	48	8.9	5.2	- 0.6
	2	48	3.3	8.8	4.6
	3	48	7.5	1.1	- 2.2
	4	48	1.2	3.7	4.5
Loc 2	5	48	1.2	- 2.2	20.0
	6	48	13.8	8.9	- 4.2
	7	48	16.7	- 2.5	14.1
	8	48	8.3	13.6	11.1
Loc 1	1	72	5.3	- 0.1	0.1
	2	72	24.2	6.6	4.9
	3	72	14.8	1.7	2.0
	4	72	- 3.3	7.1	8.9
Loc 2	5	72	3.6	17.3	6.9
	6	72	15.0	18.0	3.1
	7	72	4.5	7.4	11.3
	8	72	20.5	7.8	13.6

From Table 5, it can be seen that the lowest standard deviation is at (P 1), location 2 of the paste aged at 35°C for 72 hours. This means that the solder paste deposits tend to be very close to the mean. In other words, the printing was able to produce a normal distribution at this location, which means that most of the apertures in that particular location had deposits that are close to their average. When this result (standard deviation) is compared to the average height of the solder paste deposit for the same paste at the same location which is 119.43µm, it will be seen that the value is relatively close to the target value 125µm.

Generally, location 2 seems to have a normal distribution than location 1. This is most likely because the aperture sizes at location 2 are larger than that of location 1, therefore enabling the squeegee to deposit more solder paste at a balanced level. As the paste rolls in front of the squeegee, filling the apertures in the stencil some distance ahead of the squeegee, the apertures with larger surface area absorbed more solder paste as the squeegee sheared off the paste in the apertures.

Applying Percentage Variation

For P1 location 1, region 1, the percentage variation was calculated as

$$\% \text{ variation} = \frac{93.2 - 86.5 * 100}{93.2} . 7.2\%$$

The same procedure was used to calculate for all the regions measured.

Table 7. Percentage variation result of deposited solder paste height for LF328

Location	Region	Time (Hours)	Percentage variation (%) for P2		
			15°C	25°C	35°C
Loc 1	1	24	18	10.5	0.2
	2	24	5.1	3.9	- 5.0
	3	24	13.3	- 4.9	8.8
	4	24	4.8	- 3.3	27.8
Loc 2	5	24	4.6	6.7	8.2
	6	24	0.4	19	10.3
	7	24	- 6.9	5.1	19.7
	8	24	3.3	9.1	14.4
Loc 1	1	48	5.1	0	7.6
	2	48	4.4	- 4.8	10.6
	3	48	9.5	5.3	4.9
	4	48	29.4	1.9	4.3
Loc 2	5	48	10.6	8.1	18.9
	6	48	8.7	2.0	9.5
	7	48	- 5.8	2.2	3.6
	8	48	0.7	14.2	9.1
Loc 1	1	72	3.6	0.5	12.3
	2	72	7.6	11	6.1
	3	72	3.4	1.3	- 6.8
	4	72	3.3	3.9	13.3
Loc 2	5	72	5	10.6	- 14.4
	6	72	13.4	16	- 6.7
	7	72	4.7	11.5	- 4.0
	8	72	16.5	3.9	3.7

According to Prasad (1997), the maximum allowable variation in paste height should be about 20%. Figures 12 to 17 show the graph of the percentage variation.

Figure 12. Percentage variation plot for P1 at 24 hours

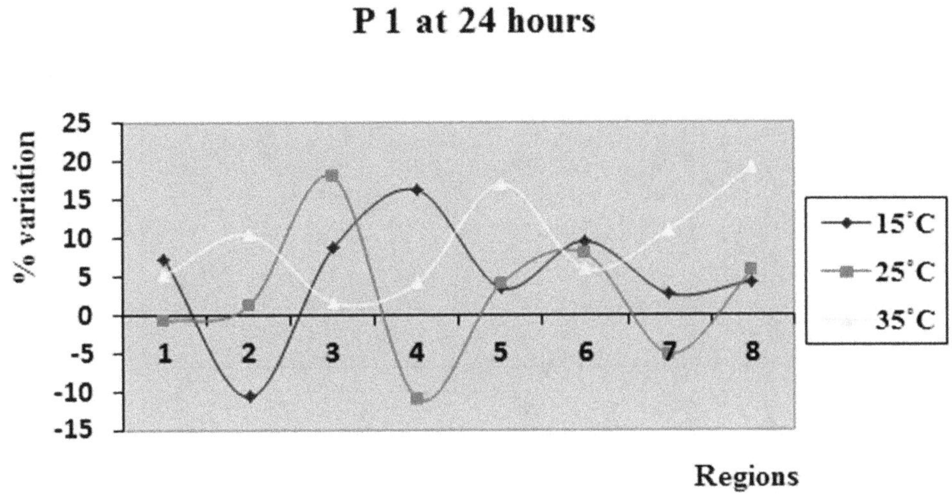

Figure 13. Percentage variation plot for P1 at 48 hours

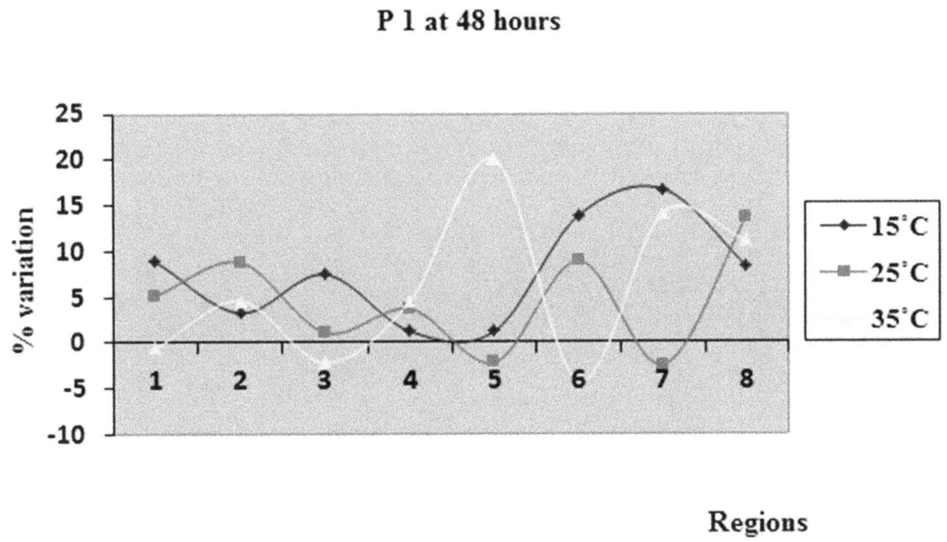

Figure 14. Percentage variation plot for P1 at 72 hours

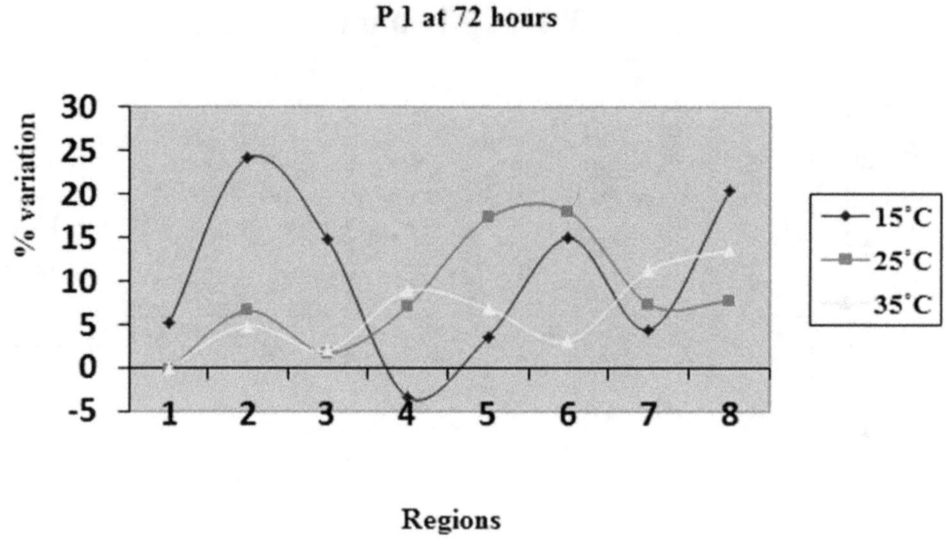

Figure 15. Percentage variation plot for P2 at 24 hours

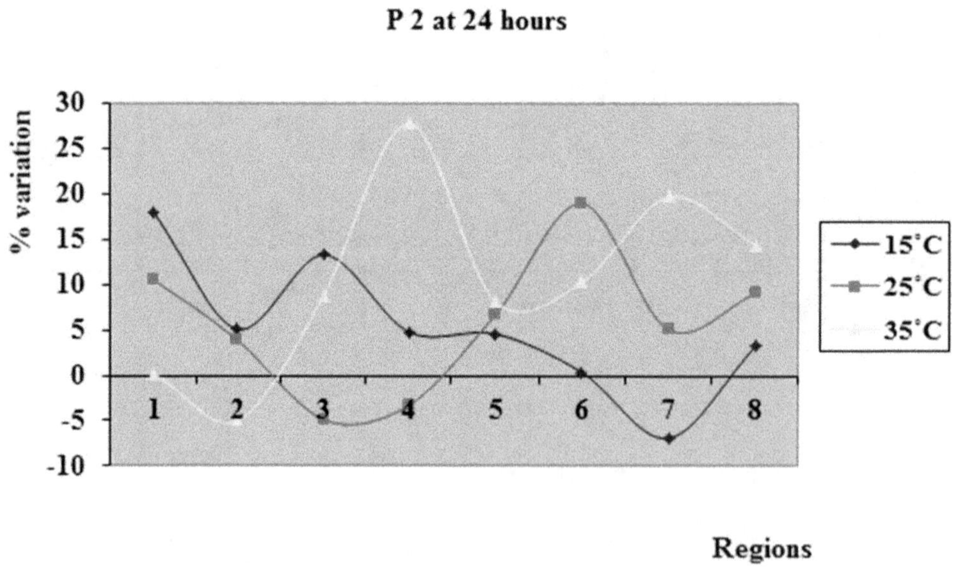

Figure 16. Percentage variation plot for P2 at 48 hours

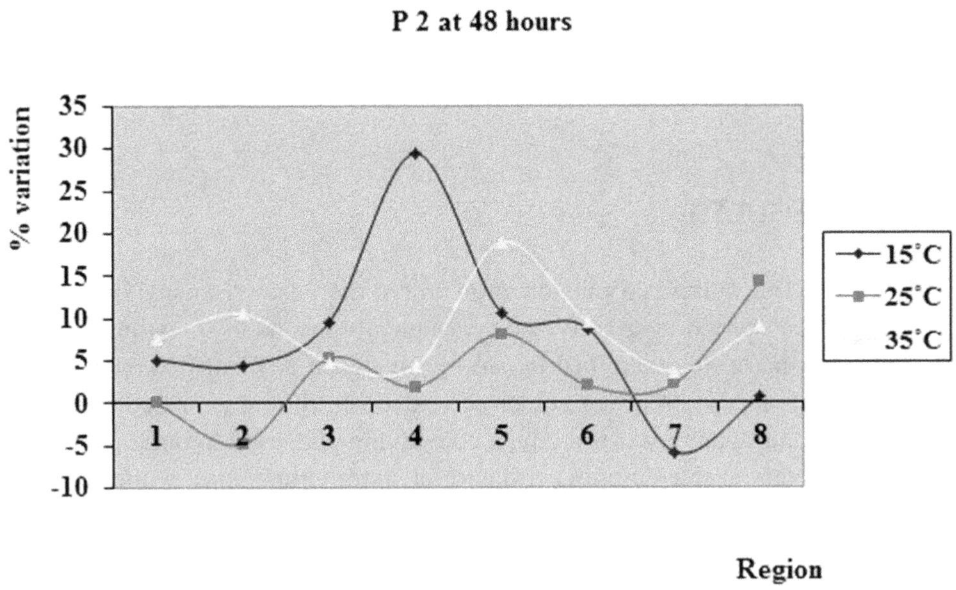

Figure 17. Percentage variation plot for P2 at 72 hours

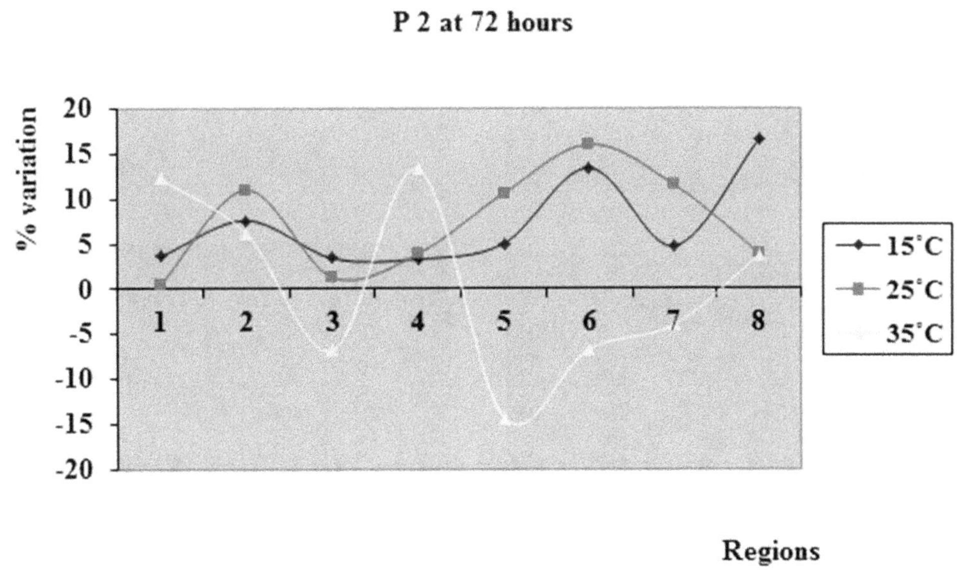

Since 20% is the maximum allowable percentage variation, the upper control limit (UCL) and lower control limit (LCL) were set at 20% and -20% respectively for this experiment. Therefore, these control limits were used to check for out of control conditions in the printing process. For the whole process/ experiment, only four out of control conditions (freak) were found at the region which is an infinitesimal part of the deposits.

6. SLUMP TEST RESULTS

According to Lee (2002), a slump is a phenomenon where the paste viscosity is not high enough to resist the collapsing force exerted by gravity and consequently results in spreading beyond the area to be deposited. The solder paste was examined for slump which can be identified by a decrease in height and an increase in length and width of the solder paste deposit. Slump is categorized as cold and hot slump to temperature conditions. For this study, a cold slump was used to determine the solder paste behavior. Cold slump refers to the slumping behavior at room temperature while hot slump refers to slumping during reflow. Figure 18 shows the slump location that was visually observed using a Leica S6D zoom stereomicroscope.

Figure 18. Four edges of the slump location for P1

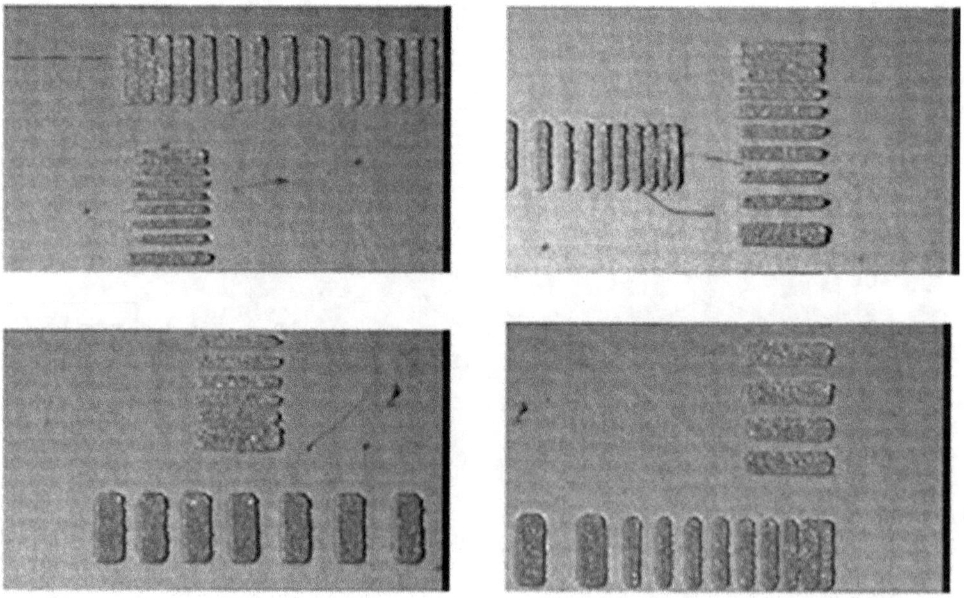

From the observations, paste stored at 35°C had the highest level of slumping. For all the experiments, most of the slumping occurred within the areas with the least pitch sizes.

CONCLUSION

From the experiments, it was established that stencil opening does not obstruct solder paste flow. Unlike other printing methods, the stencil provides a 100% open area for the paste to be printed through, compared to the screen which provides only about 50%. The printing parameters also play a major role in determining the performance of the print. The stencil print process is a critical process step for controlling the quality of finished solder joints. From the measured heights of the solder paste deposit, the paste stored at 35˚C for 72 hours, produced the height closest to the target. However, it was found that a lot of slumping occurred, making it unfavorable.

Though more slumping occurred for the paste stored at 35˚C, it was not enough to get the height lower than that stored at 25˚C or 15˚C. Because of the slumping behavior of solder paste, it was believed that the height of the solder paste would decrease as the temperature increased. This was not exactly the case, the height rather increased slightly as the temperature increased. This behavior was a result of several factors. First, the paste was not able to role at low temperatures such 15˚C. Secondly, as the temperature increased, the paste rolled and was able to fill the apertures effectively. Therefore, considering this slumping phenomenon, the paste stored at 25˚C for 48 hours produced the best results of good height with minimal slumping. The percentage variation showed zero variation in one of the regions for P2. With the results of the variability analysis, 25˚C showed less variability. From the analysis, it is obvious that temperature has tremendous effects on the printing performance of solder paste. This study recommends 25˚C as a good temperature for solder paste printing.

REFERENCES

Bao, X., Lee, N., Raj, R. B., Rangan, K. P., & Maria, A. (1998). Engineering solder paste performance through controlled stress rheology analysis. *Soldering & Surface Mount Technology, 10*(2), 26–35. doi:10.1108/09540919810219949

Bath, J. (2007). *Lead-free soldering. XIV*. Hardcover. doi:10.1007/978-0-387-68422-2

Clements, D., Desmulliez, M. P. Y., & Abraham, E. (2007). The evolution of paste pressure during stencil printing. *Soldering & Surface Mount Technology, 19*(3), 9–14. doi:10.1108/09540910710843720

Colin, C. J., & Joseph, K. (1989). *Solder paste technology: principles and applications. Edition: Illustrated*. TAB Professional and Reference Books.

Durairaj, R., Ekere, N. N., & Salam, B. (2004). Thixotropy flow behaviour of solder and conductive adhesive pastes. *Journal of Materials Science Materials in Electronics, 15*(10), 677–683. doi:10.1023/B:JMSE.0000038923.62923.1e

Durairaj, R., Mallik, S., & Ekere, N. N. (2008). Solder paste characterisation: towards. *Soldering & Surface Mount Technology, 20*(3), 34-40.

Durairaj, R., Mallik, S., Marks, A., Winter, M., Bauer, R., & Ekere, N.N. (2006). *Rheological Characterisation of New Lead-Free Solder Paste Formulations for Flip-Chip Assembly*. Electronics Manufacturing Engineering Research Group, Medway School of Engineering, University of Greenwich Chatham Maritime.

Durairaj, R., Nguty, T. A., & Ekere, N. N. (2001). Critical Factors Affecting Paste Flow During Stencil Printing of Solder Paste. *Soldering & Surface Mount Technology*, *13*(2), 30–34. doi:10.1108/09540910110385239

Durairaj, R., Ramesh, S., Mallik, S., Seman, A., & Ekere, N. N. (2009). Rheological characterisation and printing performance of Sn/Ag/Cu solder pastes. *Materials & Design*, *30*(3), 3812–3818. doi:10.1016/j.matdes.2009.01.028

Eke, J., Okafor, P. U., & Arinze, S. N. (2020). Investigating the Effects of Temperature, Paste Height and Time on the Rolling of Solder Paste. *Innovative Systems Design and Engineering*. www.iiste.org

Ekere, N. N., & Mallik, S. (2008). A methodology for Characterising New Lead-Free Solder Paste Formulations used for Flip-chip Assembly Applications. *Smart Processing Technology*, *2*(1), 59–64.

Greene & Srihari. (2008). *A procedure for determining the high-speed stencil printing performance of solder pastes in an electronic service provider's environment.* Emerald Group Publishing Limited.

Harper, C. A. (2000). *High performance printed circuit boards. Edition: illustrated.* Published by McGraw-Hill Professional.

Howard, H. M. (1995). *Soldering handbook for printed circuits and surface mounting: Design, materials, processes, equipment, trouble-shooting, quality, economy, and line management* (2nd ed.). Published by Springer.

Hwang, J. S. (1989). *Solder Paste in Electronics Packaging*. Van Nostrand Reinhold. doi:10.1007/978-94-011-6050-6

Hwang, J. S. (2004). *Lead-free implementation and production: A manufacturing guide*. McGraw-Hill Professional.

Ian, W. (2018). *First-Ever Temperature Stable Solder Paste Unveiled.* Available online: https://www.henkel-adhesives.com/com/table_images/Technical_Article_--_Temperature_Stable_Solder_Paste_Unveiled_final.pdf

John, H. L., Wong, C. P., Ning-Cheng, L. S., & Ricky, L. V. (2002). *Electronics manufacturing: with lead-free, halogen-free, and conductive-adhesive materials. Edition: illustrated.* McGraw-Hill Professional.

Lau, F. K. H., & Yeung, V. W. S. (1997). A hierarchical evaluation of the solder paste printing process. *Journal of Materials Processing Technology*, *69*(1-3), 79–89. doi:10.1016/S0924-0136(96)00045-3

Lee, N. C. (2002). *Reflow Soldering Processes: SMT, BGA, CSP and Flip-Chip Technologies*. Newnes Publication. Available Online http://books.google.co.uk/books?id=Tv_AuIPrZmEC

Mallik S., (2009). *Study of the Time-dependent Rheological Beehaviour of Lead-free Solder Pastes and Flux Mediums used for Flip-Chip Assembly Applications*. Academic Press.

Mallik, S., Ekere, N. N. R., Durairaj, R., & Marks, A. E. (2008). An investigation into the rheological properties of different lead-free solder paste for surface mount application. *Soldering & Surface Mount Technology,* 3-10.

Marks, A., Mallik, S., Durairaj, R., & Ekere, N. N. (2007). Effect of Abandon Time on Print Quality and Rheological Characteristics for Lead-Free Solder Pastes used for Flip-Chip Assembly. *32nd International Electronics Manufacturing Technology Symposium (IEMT)*. 10.1109/IEMT.2007.4417048

Nguty, T. A., & Ekere, N. N. (2000). Modelling the effects of temperature on the rheology of solder paste and flux system. *Journal of Materials Science Materials in Electronics*, *11*(1), 39–43. doi:10.1023/A:1008904220118

Nguty, T. A., & Ekere, N. N. (2000). The rheological properties of solder and solar pastes and the effect on stencil printing. *Rheologica Acta*, *39*(6), 607–612. doi:10.1007003970000117

Ogata, S. (1991, August). Rheology and Printability of Solder Paste. *IEICE Transactions*, *E74*(8), 2378–2383.

Okafor, P. U., & Eneh, I. I. (2013, September-October). Investigation of Printing Performance of Solder Paste at Different Temperatures. *International Journal of Engineering Research and Applications*, *3*(5), 818–829. www.ijera.com

Shea, C., Arora, S., & Brown, S. (2007). *Wave Soldering Flux Selection for Pb-Free Assembly*. Circuit Assembly. Available Online: http://circuitsassembly.com/cms/cms/content/view/4511/95/

Chapter 13
Comparative Study of Transmission Line Junction vs. Asynchronously Coupled Junction Diplexers

Eugene A. Ogbodo
https://orcid.org/0000-0002-9798-723X
Aston University, UK

ABSTRACT

A conventional diplexer with a T-junction transmission line is designed in this work and compared with a new approach of designing a diplexer using the asynchronously coupled resonant junction (ACJ). The use of ACJ is to explore its features associated with enhancing the output response of the diplexer as well as its ease in control and implementation of the external couplings at the common port of an all-resonator-base diplexer. The measured and simulated responses of the diplexers were compared, and they agreed with each other.

INTRODUCTION

This work presents a comparison of a diplexer with a conventional transmission line junction and a diplexer with a resonant junction based on the multi-port filtering network (MPFN) concept. The comparison between the two diplexers is drawn between their topology, sizes, rejection, isolation, current density distribution, and power handling capability.

DESIGN OF THE CONVENTIONAL DIPLEXER

A conventional non-resonant T-junction diplexer (Kordiboroujeni and Boenemann, 2014; Zhang and James, 1998) and a diplexer with a resonant junction (Macchiarella, 2011; Zhao and Wu, 2014; Wu,

DOI: 10.4018/978-1-7998-6992-4.ch013

Wang, Ogbodo, 2016; Ogbodo, Wu and Wang, 2016; Ogbodo, Wu, Callaghan, and Wang, 2017; Chuang and Wu, 2011; Ogbodo, Ngobigha, Ali and Atuba, 2021) is designed and compared. Both diplexers are designed to operate at 1.8 GHz and 2.1 GHz with 4% fractional bandwidth and 20 dB return loss. To achieve the designs, two channel filters operating at 1.8 GHz and 2.1 GHz were first separately designed to specifications using hairpin resonators. For the conventional diplexer, the two channel filters were joined using a T-junction transmission line followed by optimisation to meet the required specifications. Fig. 1 presents the coupling topology used for the design.

Figure 1. Coupling topology of a diplexer with a transmission-line T-junction

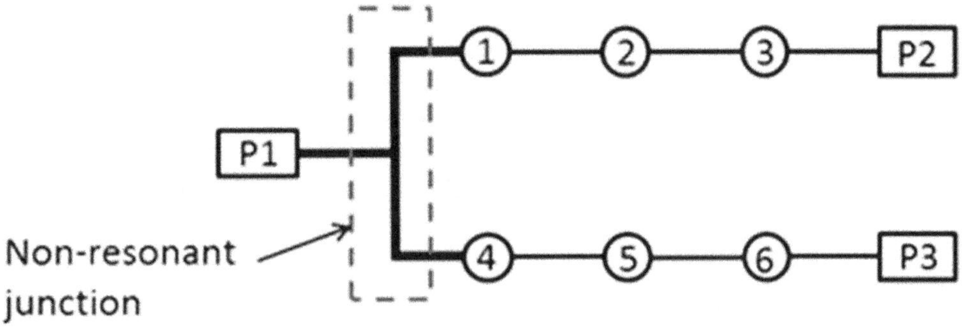

To design the non-resonant T-junction transmission line, two asymmetric line sections are used. The line section connected to one channel filter is around quarter-wavelength long at the centre frequency of the opposite channel and vice versa. The T-junction is essentially an impedance matching network. Fig. 2 shows the layout and dimensions of the T-junction and the diplexer after optimisation.

Figure 2. Conventional T-junction diplexer. F = 10mm, F1 = 13.7mm, X1 = 13.8mm, X2 = 17.6mm, T = 0.3mm, T1 = 0.4mm, W = 1.2mm, W1 = 0.6mm, W2 = 0.6mm, L = 11.5mm, L1 = 9.95mm, LL = 11mm, LL1 = 6.59mm, d1,2 = 1.77mm, d2,3 = 1.77mm, d4,5 = 1.75mm, d5,6 = 1.75mm, TH = 25.2mm, TL = 124.28mm.

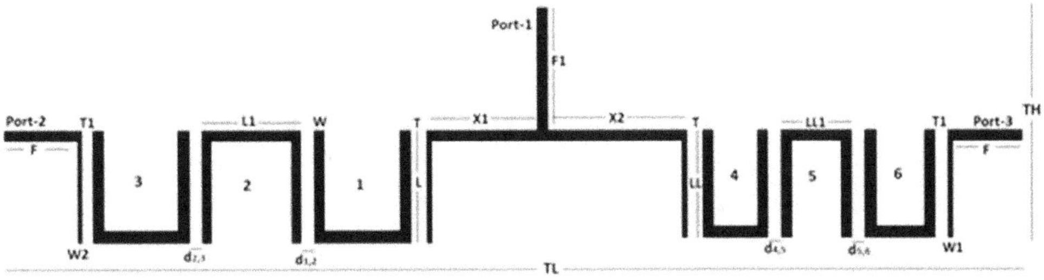

DESIGN OF THE RESONANT JUNCTION DIPLEXER

To achieve the diplexer with a resonant junction, the two individually designed channel filters were asynchronously coupled together using the first resonators of each channel filter, and their external Q factor was determined using the general filter synthesis presented in (1) while (2) was used for the physical positioning of the layout. In the process, one feed line was removed to allow both channel filters share one input feed line for receiving and transmitting functions. The technique used resulted in a compact circuit topology and the diplexer having an all-resonant structure as presented in Fig. 3.

Figure 3. Coupling topology of a diplexer with an asynchronously coupled resonant junction (ACJ)

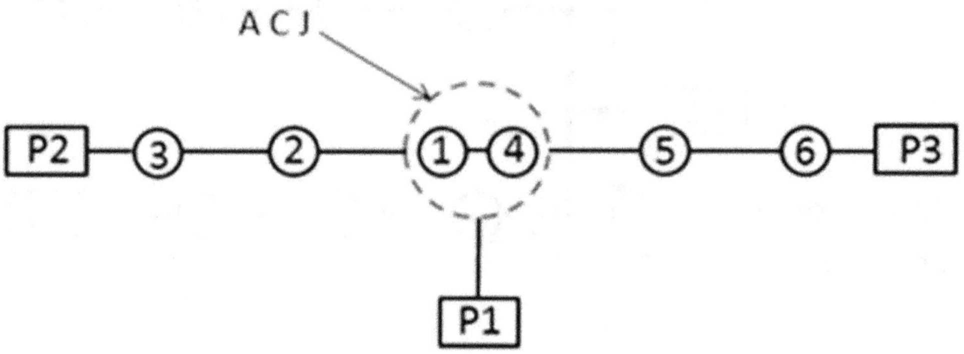

$$Q_{ex1(L)} = Q_{ex1(H)} = Q_{ex2} = Q_{ex3} = \frac{g_1 g_0}{FBW} = 21.29. \tag{1}$$

Where $Q_{ex1(L)}$ .and $Q_{ex1(H)}$ .are the external Q-factors from the common port to the low and high channel respectively.

$$Q_{ex1(L)/(H)} = \frac{f_{1(L)/(H)}}{\Delta f_{1(L)/(H)}}. \tag{2}$$

Where $f_{1(L)/(H)}$ .and $\Delta f_{1(L)/(H)}$.

are the centre frequencies and the 3dB bandwidths of the resonance curves corresponding to the low and high channels.

The dashed circle marks out the asynchronously coupled junction (ACJ) which comprises the first resonators of both channel filters. The 'asynchronous' in the name refers to the two coupled resonators with different inherent resonant frequencies, corresponding to the two channels respectively. The ACJ does the signal splitting and combining in the diplexer circuitry. Fig. 4 presents the frequency responses of the two coupled resonators against their separation, $d_{1,4}$.

Figure 4. Resonant frequencies of the ACJ with an inset of topology used

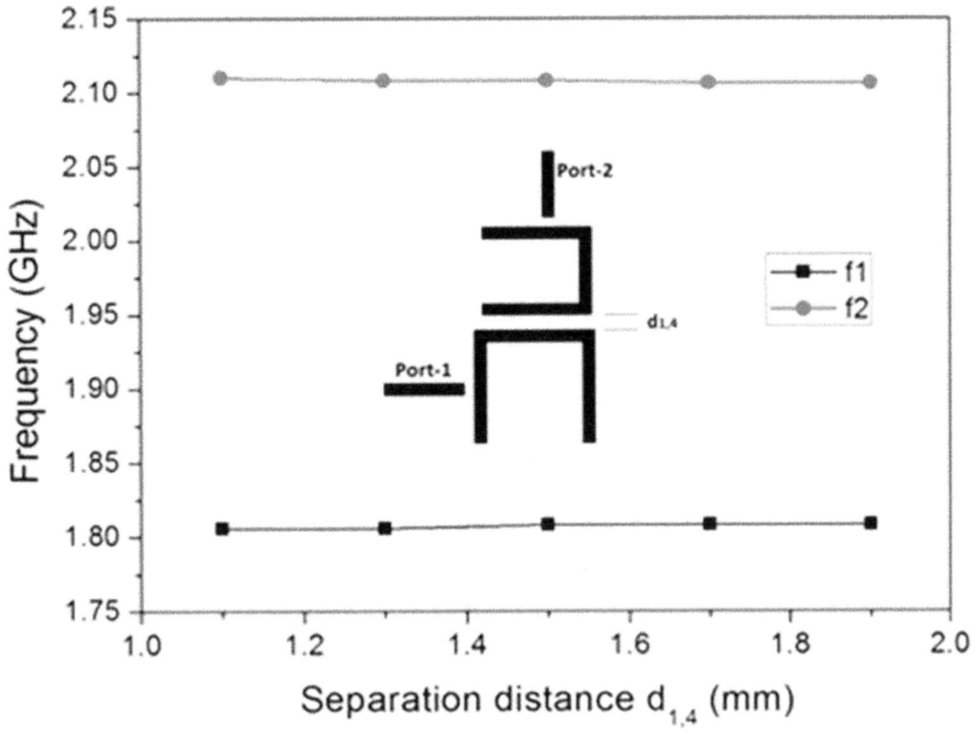

To achieve the Q_{ext} factor for the ACJ, a coupled feed line technique (Ogbodo, Wu, & Wang, 2016; Son and Pyo, 2005; Mansour, Lancaster, Hall, Gardner, & Nugoolcharoenlap, 2014) was used to feed the resonators at port-1 as shown in Fig. 5. Port-2 and port-3 were weakly coupled to the resonators while the coupling gaps, width and length at port-1 were adjusted. Fig. 5(a) presents the typical frequency response. Fig. 5(b) presents the simulated current distribution of the ACJ at 1.8 GHz and 2.1 GHz. It is evident that at the operating frequencies, the ACJ directed the signal to different channels.

At this point, the corresponding channel filters are coupled to their associated resonant pole of the ACJ. Layout optimisation is performed to meet the design specifications.

It is worth mentioning that the coupling coefficients was calculated using (3) while (4) was used in determining the positions of the resonators. This is also applicable to the conventional non-resonant T-junction diplexer design.

$$M_{1,2} = M_{2,3} = M_{4,5} = M_{5,6} = \frac{FBW}{\sqrt{g_1 g_2}} = 0.041 . \quad (3)$$

$$M_{1,2} = M_{2,3} = M_{4,5} = M_{5,6} = \frac{f_2^2 - f_1^2}{f_2^2 + f_1^2} . \quad (4)$$

Figure 5. (a) Simulated typical response of the configuration used to extract the Q_{ext}; (b) Current distribution

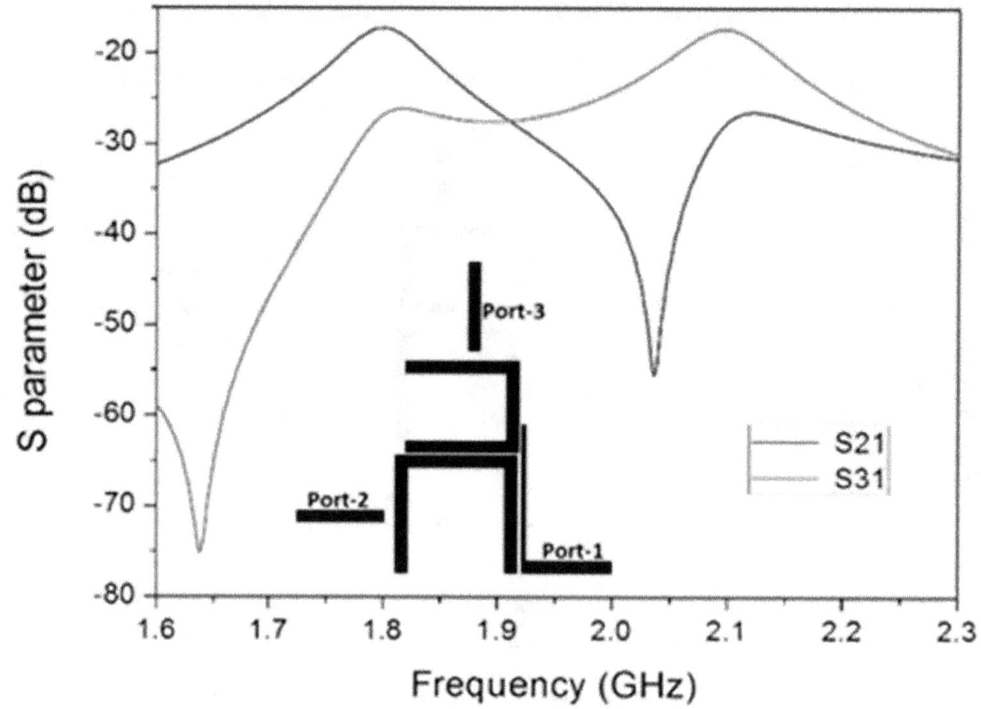

COMPARISON

Size

After matching the simulated responses of the non-resonant T-junction diplexer and the resonant junction diplexer, a comparison between them can be drawn. Fig. 2 and Fig. 6 present the two diplexer layouts respectively. The physical dimensions achieved after optimisation are given in the caption. It can be seen that the diplexer with the resonant junction is more compact than the diplexer with a non-resonant T-junction when their dimensions are compared. This is due to the elimination of the non-resonant T-junction.

Rejection and Isolation

The simulated frequency response of the conventional diplexer is shown in Fig. 7 in comparison with the simulated frequency response of the diplexer using the resonant ACJ. The bandwidths are seen to match each other. The insertion losses are the same while the return losses meet the 20 dB mark.

The rejections at the high passband S_{31} frequency in both diplexers are very similar. The ACJ offers slightly higher rejection due to a transmission zero at 2.02 GHz. At the low passband frequency S_{21}, the T-junction offers better rejection mainly due to a transmission zero generated at 1.85 GHz.

Figure 6. Resonant junction diplexer. F = 10mm, T = 0.4mm, T1 = 0.2mm, W = 1.2mm, W1 = 0.4mm, W2 = 0.6mm, W3 = 0.6mm, L = 9.6mm, L1 = 11.5mm, LL = 6.6mm, LL1 = 11mm, d1,2 = 1.7mm, d1,4 = 0.2mm, d2,3 = 1.7mm, d4,5 = 1.65mm, d5,6 = 1.7mm, TH = 65mm, TL = 61.5mm.

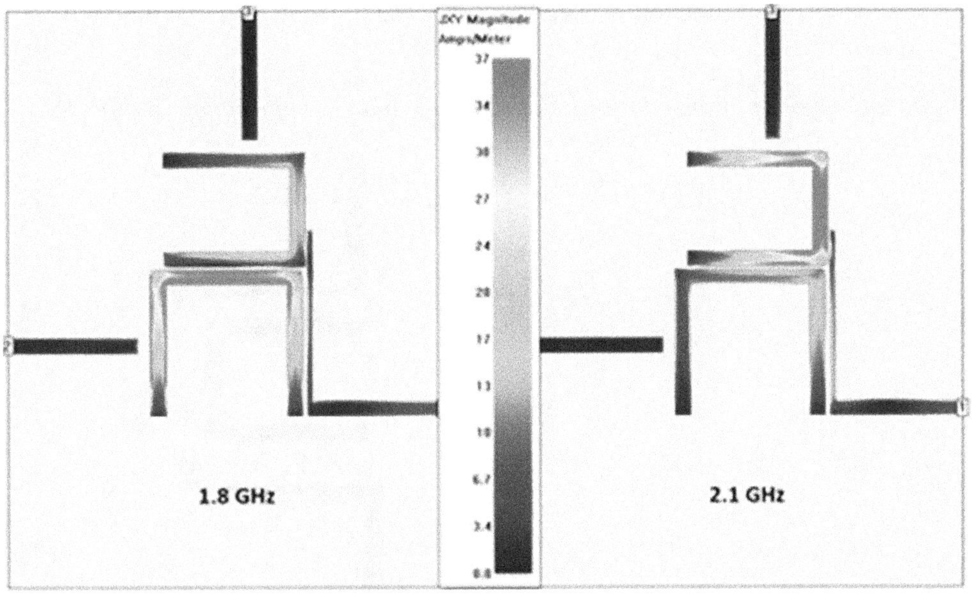

Fig. 8 presents the isolation S_{32} for both diplexers. It indicates that both isolations are better than 43 dB. At the low pass band, the conventional diplexer is at about 55 dB while the resonant ACJ has a lower isolation of 43 dB.

To realise the design, both diplexers were fabricated using the Rogers 3010 substrate with a thickness of 1.27mm, a relative permittivity of 10.8 and a loss tangent of 0.0022 and are presented in Fig. 9. Measurements were taken using the Agilent Network Analyser N5230A and compared with each other. Fig. 10 presents the measured responses. It can be noticed that a good agreement is achieved between the two diplexers. Measurements also agree well with the simulations. The frequency shifts between the measured and simulated responses are due to fabrication errors and the variation of the dielectric constant.

Current Density Distribution and Implication on Power Handling

The current distributions of both diplexers are simulated and presented in Fig. 11. By comparing the current density, the power handling capability of the diplexers was investigated. To carry out this investigation, the 1.8 GHz channels of both diplexers were excited and the current density distribution across the microstrip line collected from the middle of the channels at the circled area in Fig. 11. The current density distribution at 2.1 GHz was also collected from the circled areas in the high channels. A plot of the distribution at 1.8 GHz of the diplexer with ACJ is presented in Fig. 12(a) in comparison with that of the conventional T-junction diplexer. It shows that the T-junction diplexer has slightly higher current density. The same observation can be made at 2.1 GHz as shown in Fig. 12(b). This seems to indicate the ACJ diplexer may have slightly higher power handling capability. It is worth mentioning that the limitation imposed by the simulation software used for this investigation could not allow for accurate values of the current density as it is highly dependent on the size of the subsections. In this investigation, the smallest subsection used is 0.01 mm and the comparison were made when the same subsection is used for both diplexers. Although it is difficult to draw a definitive conclusion about the current density and the implied power handling, the comparison does show, at least, there is a slight difference in the current distribution when the resonant junction is used to replace the transmission line junction.

Figure 7. Simulated response of the conventional diplexer and resonant junction diplexer

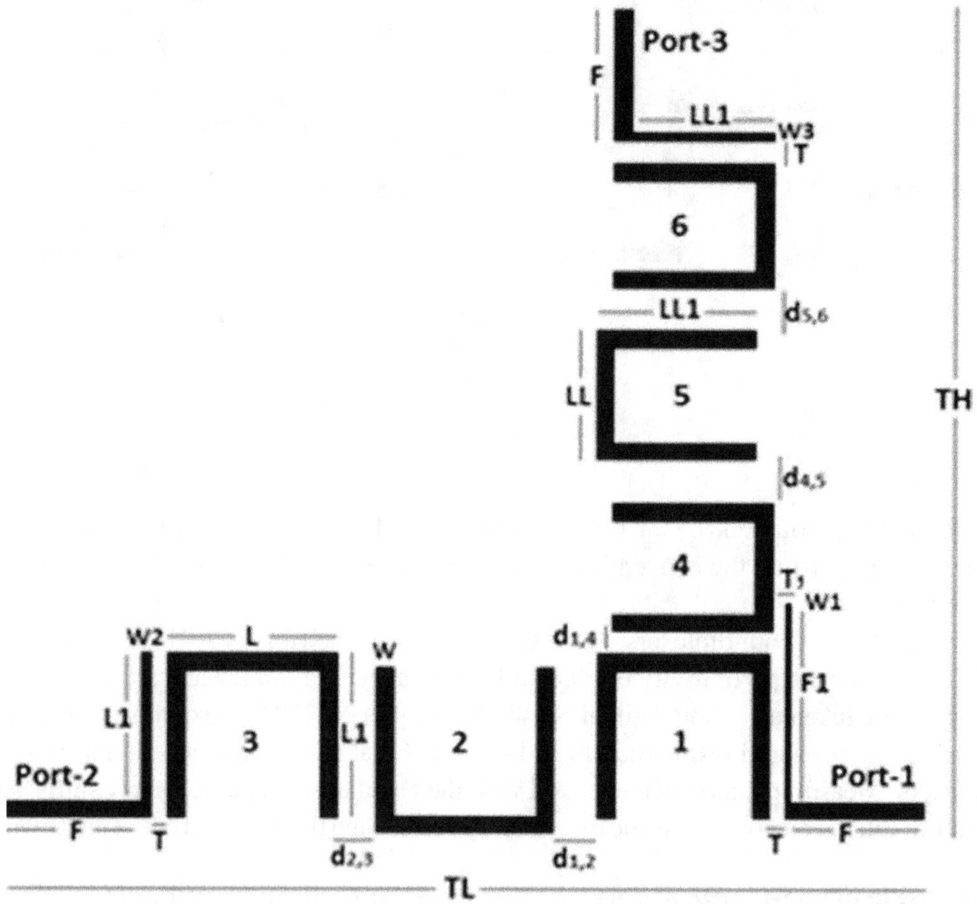

SUMMARY

In this chapter, the conventional diplexer based on a transmission-line T junction and the resonant junction diplexer based on the MPFN concept has been compared in terms of their physical dimensions, current distribution and the frequency responses. It is evident the ACJ based diplexer has a more compact structure by the elimination of the transmission line junction. However, in this case study, this is at a cost of rejection and isolation at the low band as shown in Fig. 7 and Fig. 8. An improved rejection and isolation may be achievable by further optimisation of the coupling and the structure around the junction. Although the finding of this comparison does not dictate a limitation of the junction resonator in the diplexer, it is plausible to highlight the potential challenge about the rejection and isolation in the design of diplexers and multiplexers based on the MPFN concept with junction resonators.

Figure 8. Simulated isolation response of the conventional diplexer and resonant junction diplexer

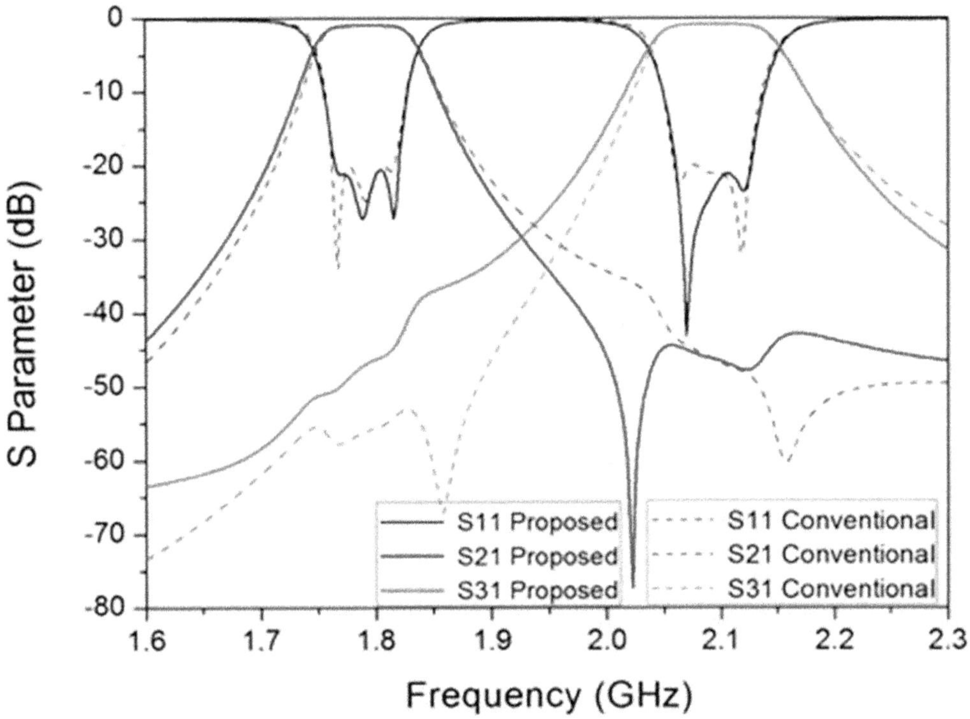

Figure 9. Fabricated designs: (a) Conventional T-junction diplexer; (b) Coupled resonant Junction diplexer

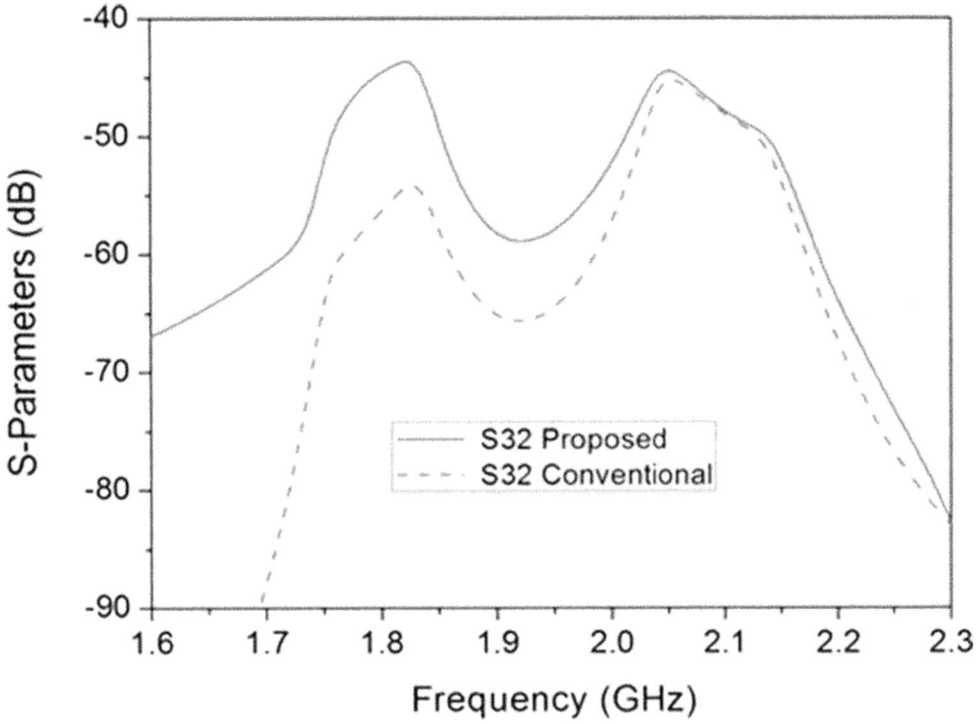

Figure 10. Measured frequency responses of the two diplexers

Figure 11. Simulated current distribution of the two diplexers, plotted using the same current density scale.

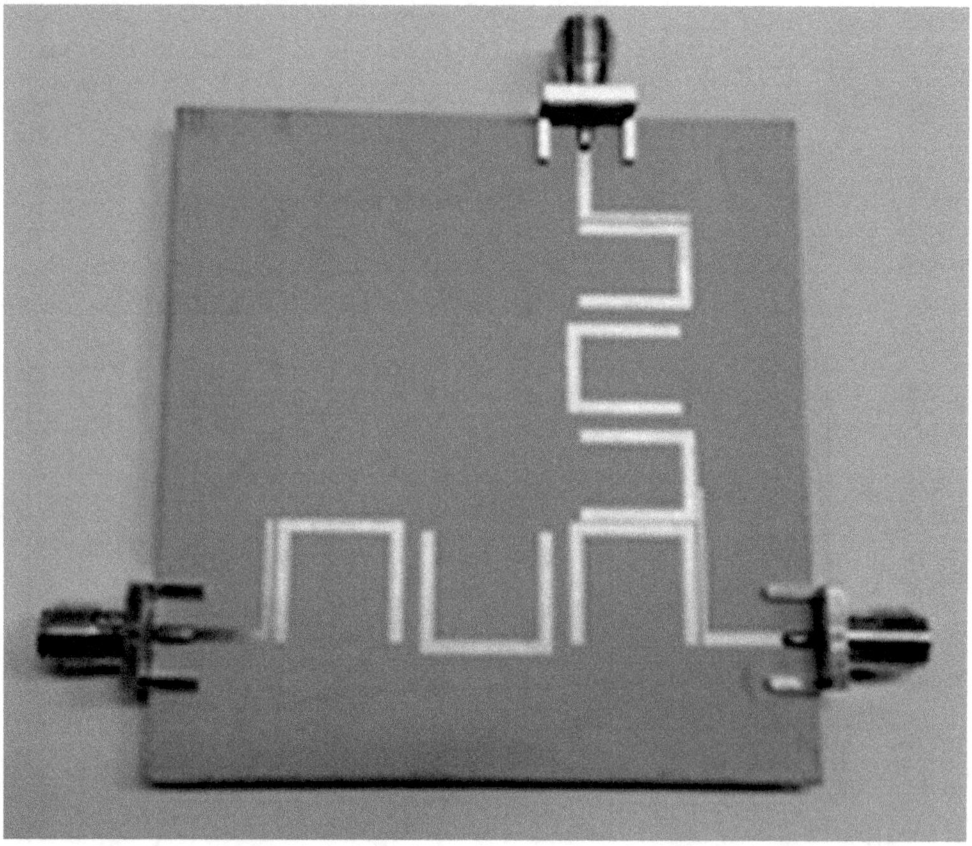

Figure 12. Simulated current density distribution across the microstrip lines at the circled areas of Fig. 11: (a) 1.8 GHz; (b) 2.1 GHz. The range of the X-axis is proportional to the width of the microstrip line.

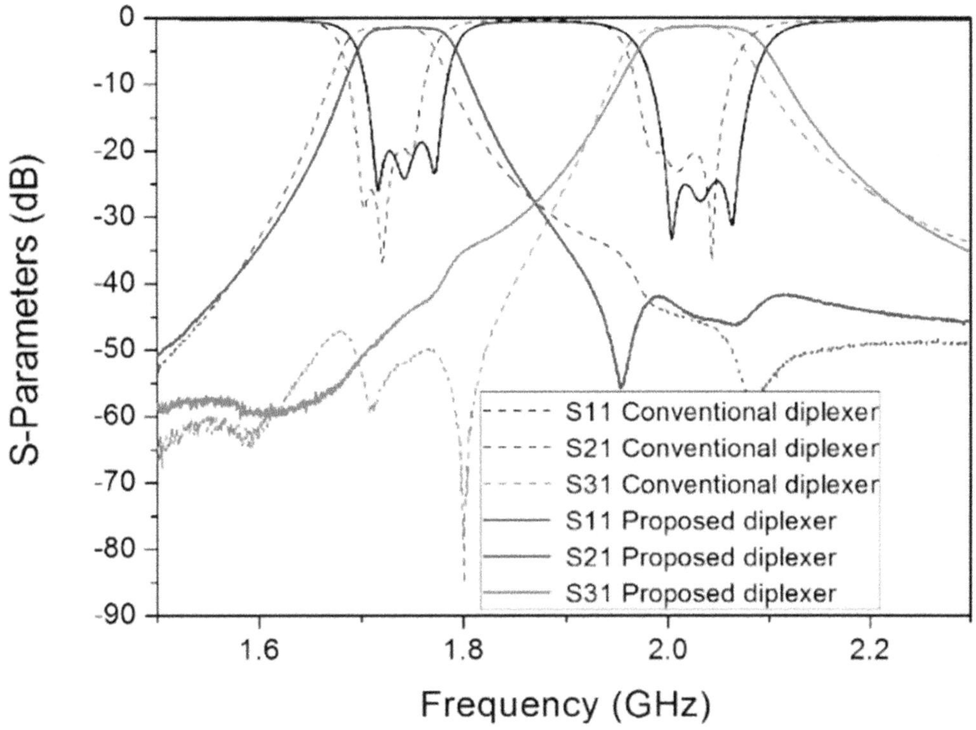

REFERENCES

Chuang, M., & Wu, M. (2011). Microstrip Diplexer Design Using Common T-Shaped Resonator. *IEEE Microwave and Wireless Components Letters*, *21*(11), 583–585. doi:10.1109/LMWC.2011.2168949

Kordiboroujeni, Z., & Bornemann, J. (2014). Mode-matching analysis and design of substrate integrated integrated waveguide T-junction diplexer and corner filter. International Journal of Numerical Modelling: Electronic Networks. *Devices and Fields*, *28*(5), 497–507.

Macchiarella, G. (2011, October). Synthesis of Star-Junction Multiplexers. *IEEE Microwave Magazine*, *12*(6), 101–109. doi:10.1109/MMM.2011.942011

Mansour, G., Lancaster, M. J., Hall, P. S., Gardner, P., & Nugoolcharoenlap, E. (2014). Design of Filtering Microstrip Antenna Using Filter Synthesis Approach. *Progress in Electromagnetics Research*, *145*, 59–67. doi:10.2528/PIER14011405

Ogbodo, E. A., Ngobigha, F., Ali, N. O., & Atuba, S. (2021). A Diplexer with Direct Coupled Resonant Junction. Microwave and Optical Technology Letters. *Microstrip diplexers with dual-mode patch resonant junctions. 46th European Microwave Conference (EuMC)*, 1155-1158.

Ogbodo, E. A., Wu, Y., Callaghan, P., & Wang, Y. (2017). Asynchronous Coupling Resonant Junctions for Compact Diplexers and Multi-Port Filtering Networks. *Microwave and Optical Technology Letters*, *59*(12), 3046–3051. doi:10.1002/mop.30877

Son, H., & Pyo, C. (•••). Design of RFID tag antennas using an inductively coupled feed. *Electronics Letters*, *41*(18), 994.

Wu, Y., Wang, Y., & Ogbodo, E. A. (n.d.). Microstrip wideband diplexer with narrow guard band based on all-resonator structures. *46th European Microwave Conference (EuMC),* 1163-1166. doi:10.1049/el:20051536

Zhang, H., & James, G. (1998, January). A broadband T-junction diplexer with integrated iris filters. *Microwave and Optical Technology Letters*, *17*(1), 69–72. doi:10.1002/(SICI)1098-2760(199801)17:1<69::AID-MOP19>3.0.CO;2-6

Zhao, P., & Wu, K. (2014, December). An Iterative and Analytical Approach to Optical Synthesis of a Multiplexer with aStar-Junction. *IEEE Transactions on Microwave Theory and Techniques*, *62*(12), 3362–3369. doi:10.1109/TMTT.2014.2364222

Chapter 14
Enhanced Simulation Tools for Analyzing Signal Processing Techniques in the Optical Transmission Path

Rastislav Róka
https://orcid.org/0000-0001-9767-9547
Slovak University of Technology, Slovakia

ABSTRACT

With emerging applications and needs of ever-increasing bandwidth, it is anticipated that 5G networks with much higher bandwidth present a natural path forward to satisfy these demands. For 5G network operators, it is interesting and profitable to develop valuable broadband optical networks at the same time. Broadband optical networks naturally utilize the optical transmission medium, and for increasing transmission rates of broadband applications and services in 5G networks, it is necessary to characterize and investigate some specific features of advanced signal processing techniques utilized in the optical transmission path. For this case, advanced techniques must also be analyzed and evaluated for real long-distance, metropolitan, and access optical networks as an integral part of 5G networks. In this chapter, enhanced simulation tools for analyzing advanced signal processing techniques in the optical transmission path are presented together with possibilities for their practical implementations in real single-channel and multi-channel optical transmission systems.

INTRODUCTION

For new multimedia broadband services and advanced modern applications provisioned in 5G networks to both residential and business customers, a reliable and safety transmission medium - the optical fiber - must be necessarily used in a certain form. One of the prominent technologies is the broadband optical network. In advanced forms of broadband optical networks, active elements should be considered for including in optical network terminals. Architectures of broadband optical networks must be reliable

DOI: 10.4018/978-1-7998-6992-4.ch014

and cost effective; therefore, no switching and control elements in the optical distribution network are preferable against active ones. Moreover, optical network terminals on the subscriber side must be cheap and simple. The optical line terminal on the central office side can be more sophisticated because it represents the main managing equipment of broadband optical networks.

The broadband optical network is a bi-directional multipoint-to-multipoint system that contains passive (optical fibers, passive optical splitters, couplers, connectors) optical elements in a distribution part and active (optical line and network terminals) optical elements placed in end terminating points of the access, metropolitan and long-distance network. The transmission path between optical transmitters and receivers is presented by single-mode optical fibers. It can be divided into the three main parts:

1. A transmitting part is responsible for generating optical signals using the laser and for modulating generated signals according to relevant input data into a form suitable for transmitting in the environment of the optical fiber.
2. A transmission channel represents the optical transmission medium with linear and nonlinear effects with negative influences on transmitted signals. A main attention is focused on the attenuation, various types of the dispersion and different nonlinearities. Because these negative influences expressively interfere into the communication and represent its main limiting factors, they present a critical part of the simulation model and, therefore, it is necessary exactly to recognize and express their characteristics by correct transmission parameters.
3. A receiving part is responsible for demodulating received optical signals and for calculating of the bit error rate. Moreover, graphical presentations (eye diagrams, signal constellation diagrams) of optical signals in any part of the optical transmission path can be utilized.

Nowadays, an interest in the signal transmission through optical fibers rapidly increases due to demand of 5G networks. Constructing new optical transmission paths can be time consuming, expensive and sometimes not available solution. In the electric domain, utilizing new advanced signal processing can lead to increasing of the transmission capacity. Such solutions can be easily integrated. With increasing of modulation rates, linear and nonlinear influences in the optical transmission medium on transmitted optical signals are growing and by this way additional bit errors in information signals are generating. Therefore, it is important to design and analyze advancements in the optical domain with respect to the optical transmission medium. The simulation gives transmission boundaries of each designed advanced signal processing techniques for the optical transmission path and allows comparing all the solutions before their deployment and implementation into real optical transmission systems. The simulation allows increasing the data rate and the transmission range of deployed optical transmission systems using advanced signal processing techniques and allows designing a new optical transmission system with different optical fibers. Due to increased demands for transmission rates in the optical transmission medium, it is very important to avoid expensive practical demonstration and testing. Therefore, simulation platforms are increasing their role. However, it is important to search for a simulation platform that will be able to accurately describe behavior and limitations of the optical transmission system under various working conditions. After becoming acquainted with several various commercial simulation software products, we decided for designing and creating our own simulation platform based on the Matlab Simulink programming environment.

The beginning phase of our work was focusing on an optical transmission medium, primarily on linear and non-linear effects present in the single-mode optical fiber and on analyzing environmental negative influences on transmitted optical signals. Its summary presents the base of simulation tools for broadband optical networks prepared in (Róka & Čertík, 2015a). The developing phase of our work is focusing on another significant components of the optical transmission path with an emphasis placed on their specific features influencing transmitted optical signals. This chapter presents a complex of enhanced simulation tools for analyzing advanced signal processing techniques in the optical transmission path. These are representing an upgrade and enhancement the base of simulation tools including specific optical components focused on advanced optical signal processing techniques suitable for transmitted optical signals associated with new multimedia broadband services and advanced modern applications provisioned in 5G networks.

This chapter is consisting of three main contributions. In the first, features of basic functional blocks in the optical simulation program are presented with more details. This contribution presents specific objectives focusing on detailed block schemes created in the base of simulation tools. Also, examples of the optical signal presentation are graphically displayed at various points of the optical transmission path. In the second contribution, specific objectives focusing on possibilities for specific optical components - laser, interferometer - are discussed and their inclusion in the enhanced simulation tool is introduced. Simultaneously, time variations of laser output characteristics acquired by numerical simulations are displayed. In the third contribution, specific objectives focusing on principles of systems utilizing coarse wavelength division multiplexing (CWDM) and dense wavelength division multiplexing (DWDM) techniques - are discussed in detail. Subsequently, the simulation model for multiple channel transmission systems is presented and environmental influences of the optical transmission path on optical signals are analyzed using the frequency spectrum.

DESIGN AND SIMULATION OF THE OPTICAL TRANSMISSION PATH

Ultra-capacity and high-speed optical communications have become essential technologies for backbone, long-distance, metropolitan and access transmission networks. Due to increased demands for the data transmission speed (100 Gbit/s per one transmission channel), it is very important to avoid expensive practical demonstration and testing. Therefore, simulation platforms are increasing their role. There exist various approaches in this situation. In (Ahmed & Ali, 2007), an optical link simulator was designed like a platform on which the transmission and optimization of communication systems could be carried out. The simulator has been developed under the Simulink environment in the form of modules. In (Del Río Bellisco, Hermosilla, Monteroy & Espí, 2004), a software tool developed for simulating of optical communication systems was presented. The simulation allows determination of the spectral and frequency responses of blocks and the whole system. The tool has been developed using the Matlab programming environment. The accurate determination of fiber nonlinearities is an important issue in the design of optical communication systems. In (Batagejl, 2002), the need of knowing fiber non-linear coefficients in global optical networks was presented.

However, it is important to search for a simulation platform that will be able to accurately describe behavior and limitations of different components in the optical transmission system under various working conditions. Such a simulation platform must be based on verified measurement parameters to predict a performance of the optical transmission system. Commercial simulation software products are often expensive and not easily adaptive. Many of them are not based on measurements parameters and a data extraction for further processing is difficult if even possible. Therefore, we decided for creating our own simulation platform based on the Matlab Simulink programming environment that has become a universal modelling tool in most universities and research laboratories around the world. As a comparison, simulation models created in Matlab Simulink and VPIphotonics for a specific optical component are presented in (Čučka, Šalík, Róka, Műnster & Filka, 2018).

A design and simulation of the optical transmission medium is performed in Matlab Simulink and Matlab GUI. The simulation program starts with a main screen that control whole simulations. The main screen allows performing adequate operations to users with required only basic knowledge about characteristics of optical fibers. The Matlab GUI is used to create a main screen with the drag and drop system. The main screen can run and control the simulation model after the calculation performed for the simulation model. As we can see in Figure 1, the main screen contains basic input optical fiber parameters that characterize its features and can be provided by optical fiber manufactures. A transmission rate of optical signals, different modulation formats and a shape of optical signals are included in this main screen. Our simulation program includes 1 Gbit/s, 10 Gbit/s and 100 Gbit/s transmission rates. The design of optical transmission systems allows selecting from different modulation formats and various signal types.

Figure 1. The main screen of the optical simulation program

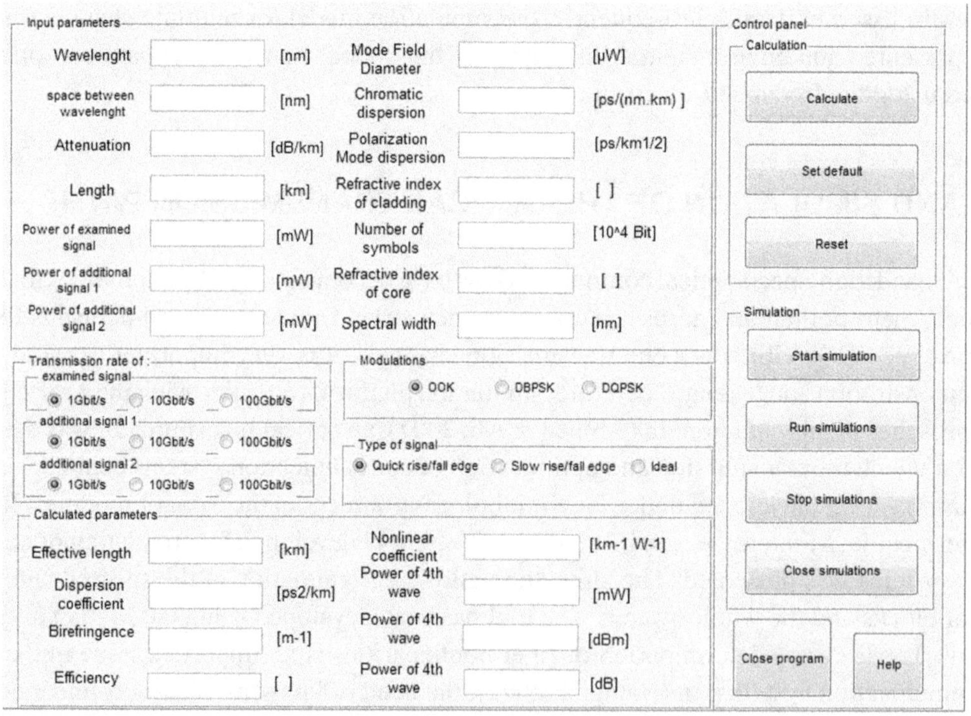

A simulation program has two main functions (calculation and simulation) representing by two independent systems. The simulation part simulates an environment of the optical transmission medium with linear and nonlinear effects influenced on optical signals transmitted with different modulation and encoding techniques. The simulation part uses the Communication Blockset and Communication

Toolbox tools to simulate transmission of optical signals through the optical transmission path. In this simulation program, except already created blocks - modulators, generators, operation functions and scope blocks, new designed and prepared blocks for simulating some of linear or nonlinear effects in the optical fiber are also used.

For modeling of the optical transmission path, we used the software program Matlab Simulink together with additional libraries like Communication Blockset and Communication Toolbox. The realized model (Figure 2) represents the signal transmission in the environment utilizing optical fibers for very high-speed data signals in both directions. Optical communication technologies will always be facing the limits of high-speed signal processing and modulation, which is an important factor to consider when discussing advanced optical modulation formats. The main task of the simulation model is the analysis of various modulation and encoding techniques.

Basic functional blocks realized in the optocommunication simulation model can be divided into the three main parts:

1. A transmitting part - it is responsible for the generating and for modulating of generated signals according to required information inputs into a form suitable for the transmission channel. The modulation block contains a basic set with the on-off keying (OOK) modulation and other variations of modulation techniques.
2. A transmission channel (the optical fiber) - this part of the model realizes negative influences on the transmitted signal. Because these negative influences expressively interfere into the communication and represent its main limiting factors, they present a critical part of the model and, therefore, it is necessary exactly to recognize and express their characteristics by correct parameters. At the present time, the chromatic dispersion (CD) and the polarization mode dispersion (PMD) blocks are realized, where the group velocity dispersion (GVD) and differential group delay (DGD) parameters are used to characterize and simulate the linear CD and PMD effects. From Kerr nonlinearities, the four wave mixing (FWM) block realizes this effect that mainly depends on the power of adjacent-channel neighbors, the wavelength channel spacing and the dispersion only in WDM systems. The self-phase modulation (SPM) block changes a signal phase and frequency depending on signal intensity changes. The cross-phase modulation (XPM) block is like the SPM and depends on signal intensity changes of adjacent-channel neighbors. For scattering nonlinearities, an influence of the signal intensity caused by these effects realized in the stimulated Brillouin scattering (SBS) and stimulated Raman scattering (SRS) blocks change the signal envelope and impact on the optical signal-to-noise ratio OSNR. The environment of fixed transmission media and negative influences in the simulation are described in detail in (Róka, 2015a), (Róka, 2015b).
3. A receiving part - it is conceptually inverted in a comparison with the transmitter. At the receiver side, a signal is demodulated by appropriate demodulator and the BER ratio is calculated. Also, blocks for graphical presenting of transmitted optical signals can be utilized.

Figure 2. The Simulink model for analyzing the optical transmission path

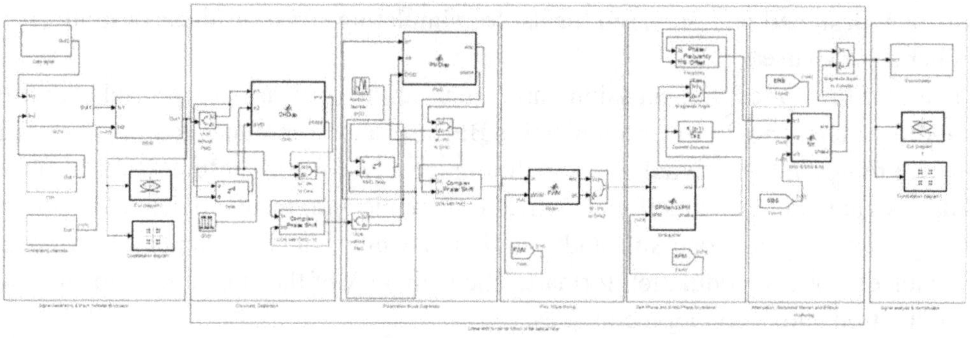

Features of Basic Functional Blocks in the Optical Simulation Program

A design of the optical transmission system for considered digital modulation techniques is shown in Figure 3. Within this subchapter, each block in the optical simulation program will be discussed. For demo purposes, we can consider a standard single-mode fiber (SMF) with these system parameters: three optical source generators with the power 1 mW at wavelengths of 1550,5 nm, 1551 nm and 1551,5 nm, the optical fiber length 10 km, the *CD* coefficient 18 ps/(nm.km), the *PMD* coefficient 0,4 ps/$\sqrt{\text{km}}$ and the $\alpha_{specific}$ attenuation coefficient 0,21 dB/km.

Figure 3. The block scheme of the OOK modulation technique

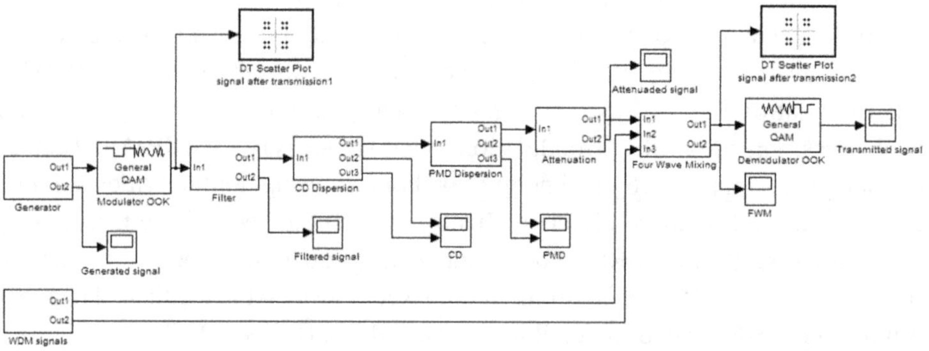

As a bit source, we used the Bernoulli generator that generates two pulses - the logic “1” and the logic “0”. As an alternative generator, a random integer generator can be used. Additional parameters for selected generator must be inserted: a probability of zero and an initial speed that represents the system transmission rate. If the random integer generator is preferred, a number of states must be inserted. We use the interconnection between Matlab GUI and Matlab Simulink to insert these input parameters via the main screen. The Bernoulli generated signal diagram with the appropriate block scheme is shown in Figure 4. The complete generator scheme involves the Bernoulli generator that generates signals and the discrete generator that changes samples per 1 bit.

Figure 4. The Generator block scheme and the example of the ideal electrical signal

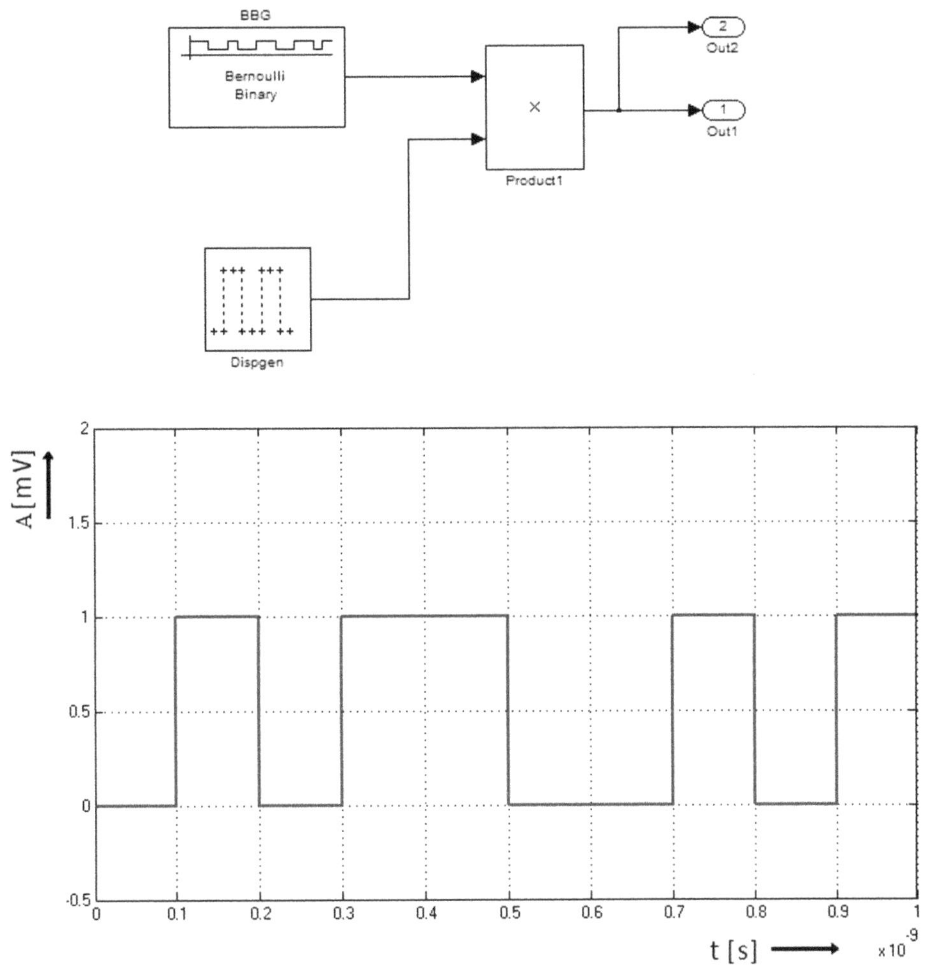

After being generated, the ideal signal is modulated by appropriate modulations that are shown in Figures 5 and 6. Modulation blocks are included in the Matlab Simulink Communication Blockset. As we can see from figures, the OOK modulation changes the signal magnitude to 1 and 0. Differential binary phase-shift keying (DBPSK) and differential quadrature phase-shift keying (DQPSK) modulations have constant signal envelope and change only the signal phase, therefore constellation diagrams are presented. The aim of this optical simulation program is to simulate a behavior of signal transmission in the environment of real optical fibers utilized in applied optical transmission systems without infinity bandwidth. Assuming a finite bandwidth, we restrict modulated signals with filtering and the appropriate filter block scheme is shown in Figure 7 and the OOK filtered signal is shown in Figure 8. The filter block uses a discrete filter for reshaping an ideal pulse into the super-Gaussian pulse and few switch blocks for keeping optical signals in boundaries (between the logic "0" and "1"). We use super-Gaussian pulses to simulate optical pulses from directly modulated laser diodes. The magnitude of DQPSK and DBPSK modulated signals is the same and equals to one, therefore there is no special reason for showing it after filtering.

Figure 5. The example of the ideal optical signal modulated with the OOK modulation

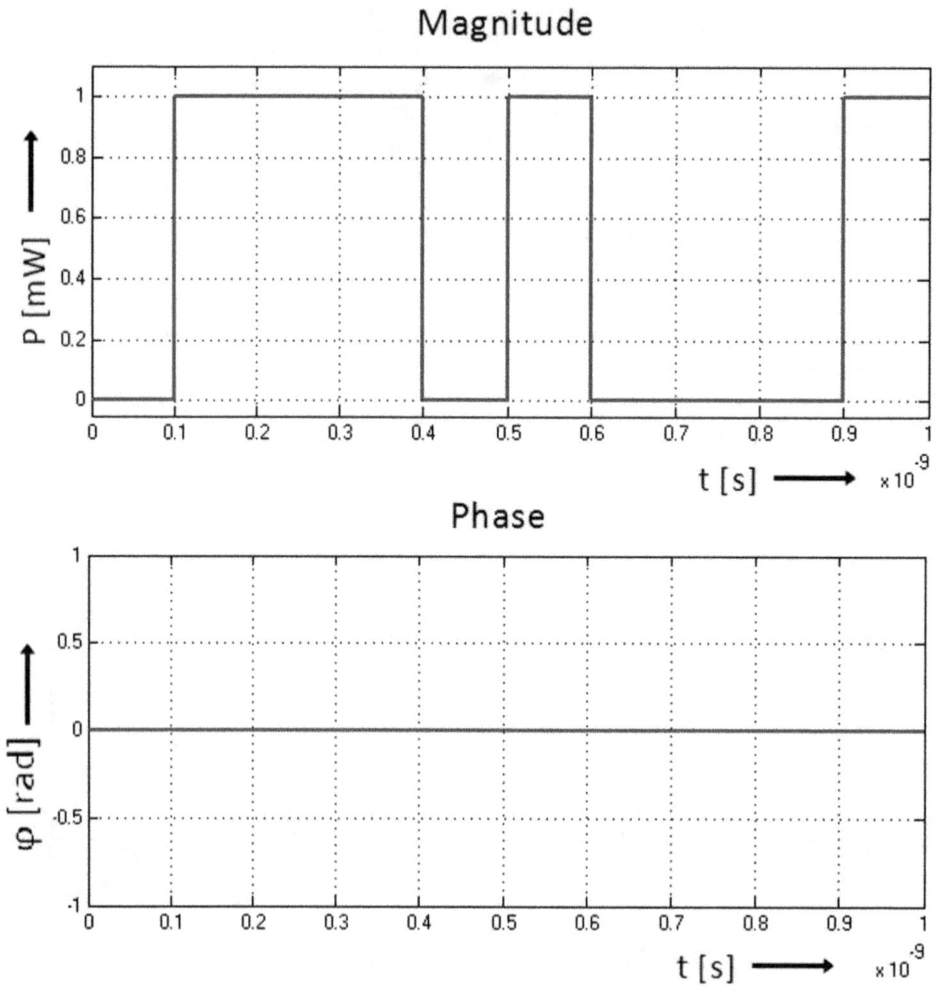

Figure 6. Constellation diagrams for the ideal a) DBPSK b) DQPSK optical signal

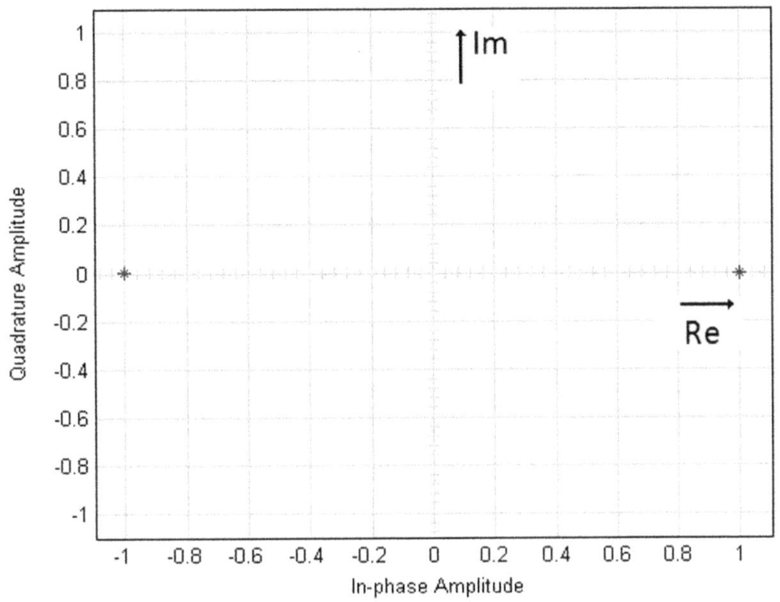

a)

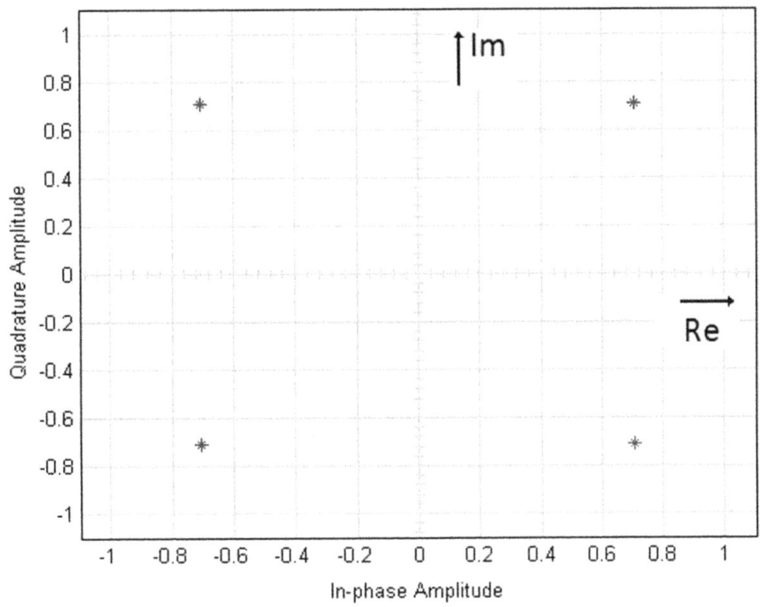

b)

Figure 7 The Filter block scheme

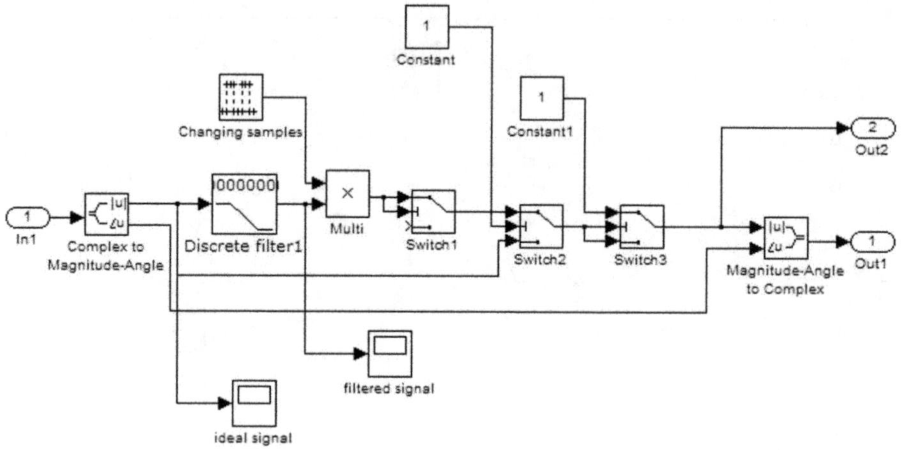

Figure 8 The example of the OOK modulated signal after filtering

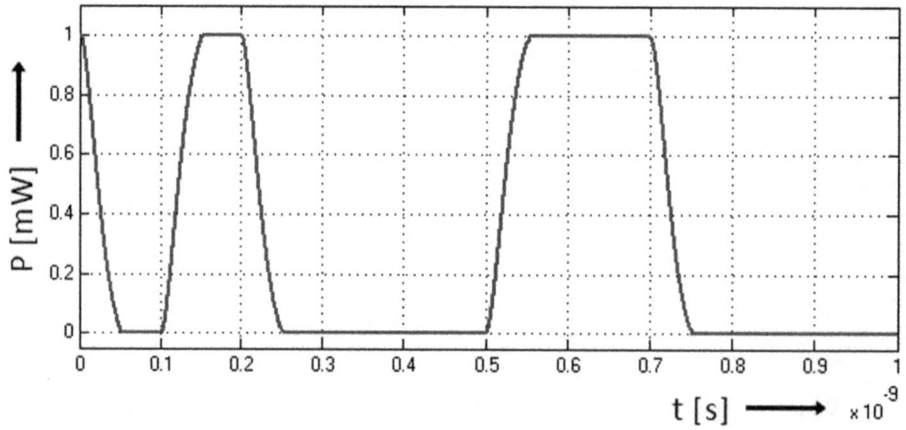

Figure 9. The CD Dispersion block scheme

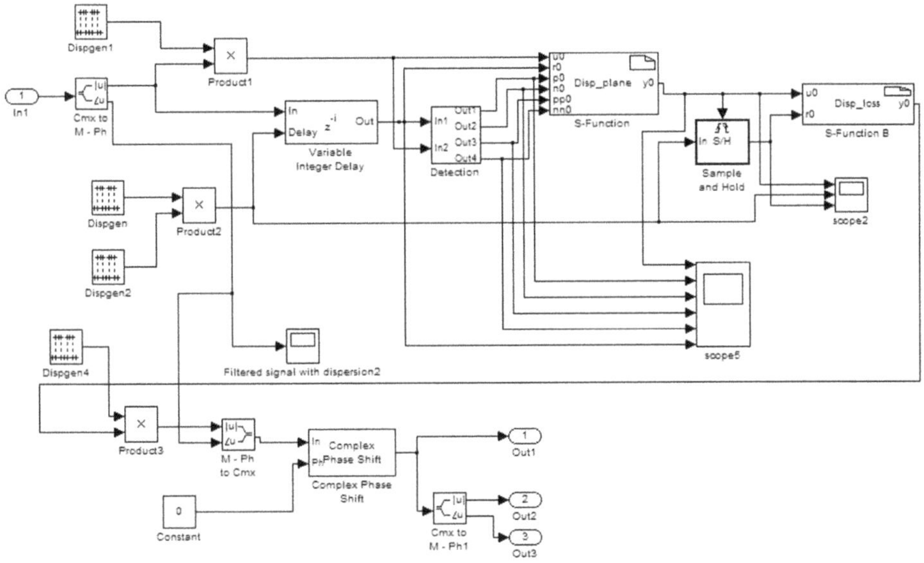

After signal filtering, the CD and PMD blocks are integrated. We assume that standard single-mode fibers are utilized and therefore we are not considering simulating a multimodal dispersion. CD and PMD dispersion blocks are expanding an original signal in the time domain and the phase shift occurs due to this signal broadening. The block scheme for simulating of the chromatic dispersion effect is shown in Figure 9. In both CD and PMD schemes, we use the discrete generator, where a signal magnitude is set to the GVD value (constant for the CD effect), respectively the DGD value (for the PMD effect). This original signal inputs to the variable delay block that delays the modulated signal by the signal magnitude values. Modulated and delayed modulated signals are brought to the disp_plane block where a new signal is created from the interference of input signals. This signal is input to the disp_loss block, where is attenuated with appropriate values (a law of the energy conservation). After this block, the complex phase shift block is used, where the modulated signal is phase shifted with a signal driven from the discrete generator. The appropriate value is calculated at start of the simulation. The PMD block is very similar to the CD block, includes a dynamic generator with the normal distribution instead of a discrete generator to generate the dynamic DGD parameter. Disp_plane and disp_loss blocks are programmed in s-function that uses C++ coding. First, we can demonstrate the influence of presented blocks on optical signals modulated with OOK, DBPSK and DQPSK modulations with the 10 Gbit/s transmission system rate for the 10 km fiber length. For better signal representation of the OOK modulation, it is showed its magnitude in time. For DBPSK and DQPSK modulations, constellation diagrams are presented. As we can see in Figure 10, the chromatic dispersion broadens the optical signal with the value of 180 ps and therefore causes the ISI generation. As we mentioned before, the dispersion broadening limits must be 1/10 of a bit time, otherwise signal boundaries are broken, and a receiver is not able to recover an original form of the received signal. This is the main reason for using of dispersion shifted fibers or dispersion compensators. DBPSK and the DQPSK modulations have a constant magnitude during whole transmission, therefore the signal broadening does not arise and only phase shift occurs as we can see in Figure 11.

Figure 10. Examples of the OOK filtered signal a) without b) with CD and PMD influences

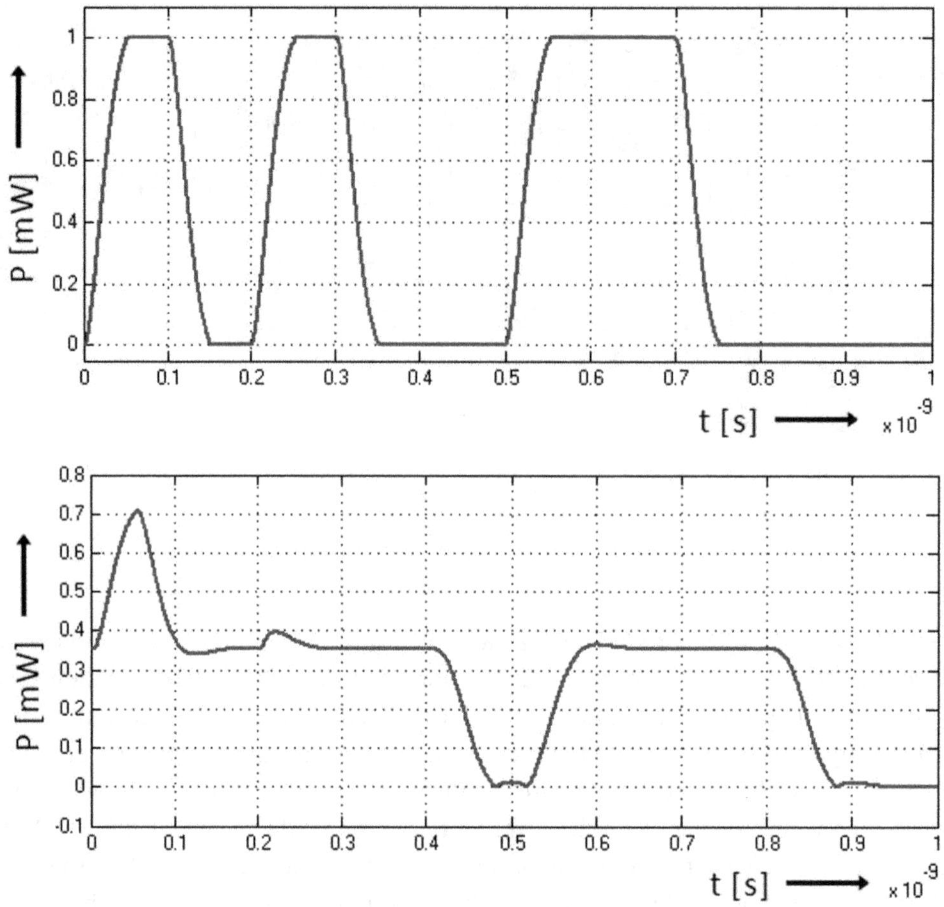

For demonstration of the CD compensation, we can consider a SSMF with the CD coefficient reduced to 0,1 ps/(nm.km) value and with the extended 100 km optical fiber length. If we change the fiber length from 10 km up to 100 km, the CD can broaden the optical signal by 10 ps. After CD compensating, the transmission system errors are mainly resulting from the PMD effect, where the signal broadening exceeds the limit as shown in Figure 12. However, DBPSK and DQPSK modulated signals are more influenced by the dynamic phase shifting as shown in Figure 13.

The 10 Gbit/s transmission system with the 100 km fiber length and with the attenuation coefficient 0,21 dB/km causes the total signal attenuation on the transmission path equal to 21 dB. The attenuation block scheme and the OOK modulated signal after attenuation is shown in Figure 14. We are using the dB gain block to simulate signal attenuation. First, signal attenuation is calculated in the calculation part, then an acquired value is converted into decibel units and consequently a result is inserted into the dB gain block. A negative gain value causes loss of the signal power.

Figure 11. Constellation diagrams for the a) DBPSK b) DQPSK modulated signal with CD and PMD influences

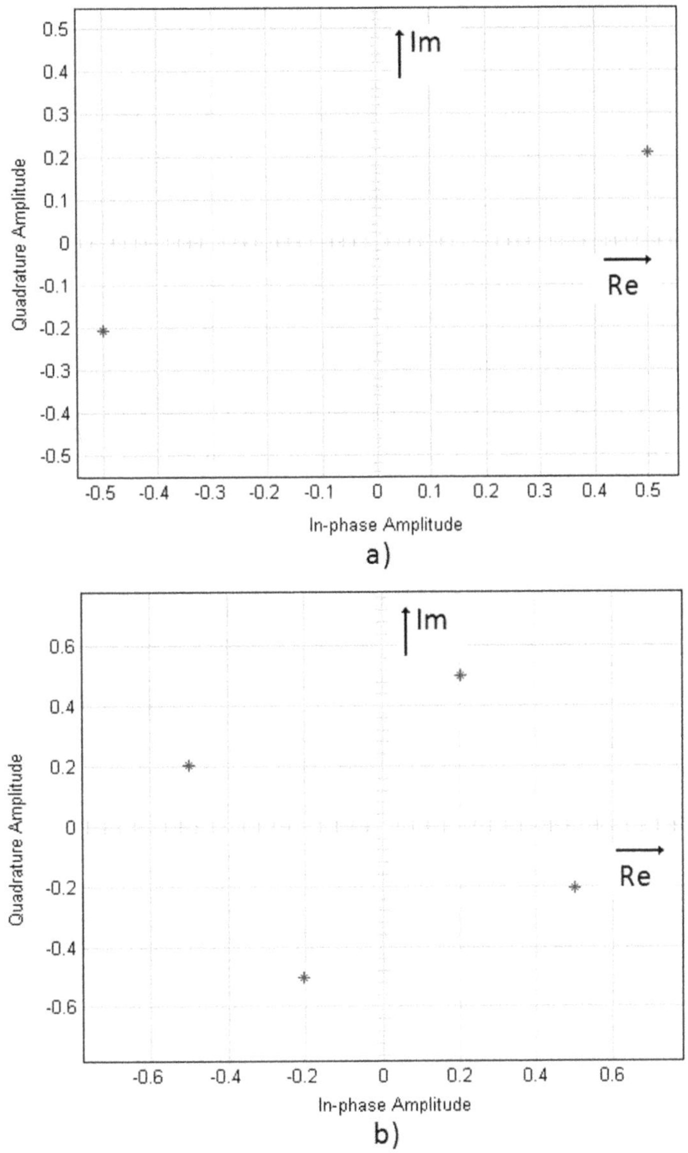

Another source of bit errors can be caused by the FWM effect. This FWM effect occurs only in WDM transmission systems and therefore we must generate additional optical signals with the same modulation technique on different wavelengths. Three originated signals are mixed (in the Four Wave Mixing block) with the Sum block and a new generated optical signal is created with the adequate amplitude that is ensured by the dB gain block and calculated in the calculation part. This 4th wavelength signal interferes with transmitted signals. The FWM block scheme is shown in Figure 15. The FWM effect between various wavelengths in WDM transmission system mainly depends on optical signal powers,

Figure 12. Examples of the OOK filtered signal a) without b) with CD and PMD influences on the 100 km optical fiber length after CD compensating

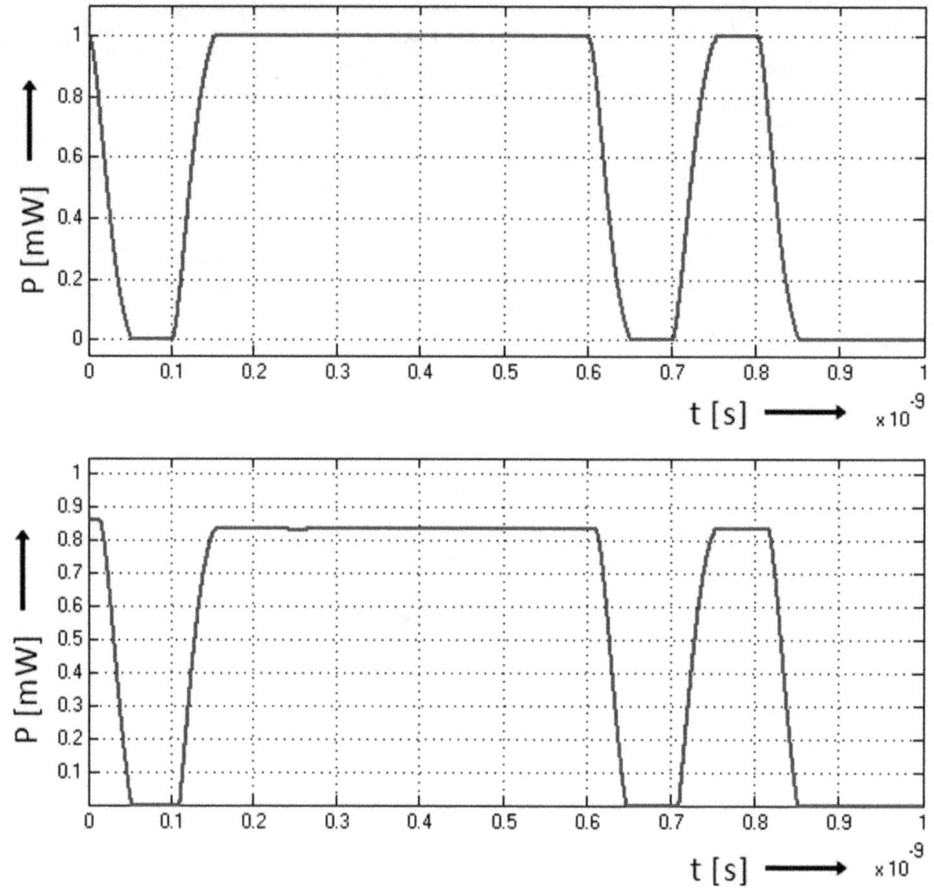

dispersion and attenuation parameters, so we can present the FWM influence on the OOK modulated signal with the 10 Gbit/s transmission rate with the CD compensation in Figure 16.

For analyzing and comparing different modulation techniques in the optical transmission system and for evaluating various environmental influences of the optical transmission medium, the bit error rate (BER) parameter can be calculated. The BER measurement can be accomplished by bit-by-bit comparing original input data sequences with received output ones. This option can be set in the main screen of the simulation program. As we can see in Figure 17, the BER calculation block contains displays showing the error rate calculation, a number of error bits and a number of transmitted bits.

Figure 13. Constellation diagrams for the a) DBPSK b) DQPSK modulated signal with CD and PMD influences on the 100 km optical fiber length after CD compensating

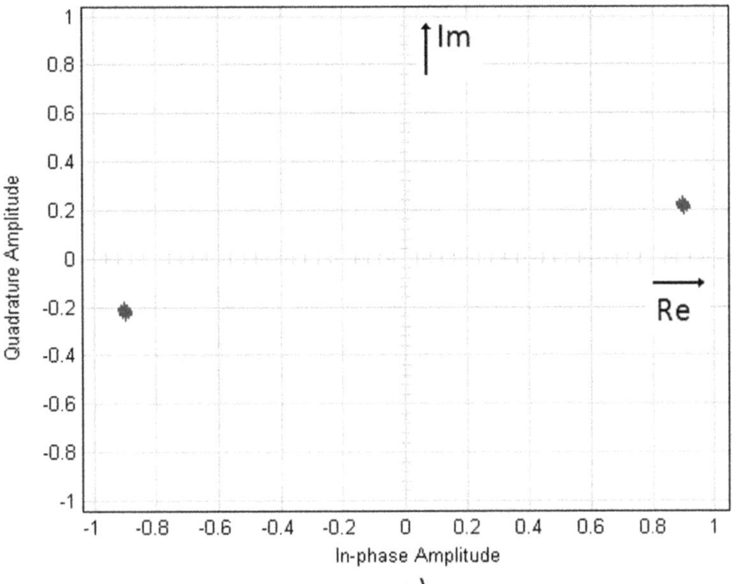

a)

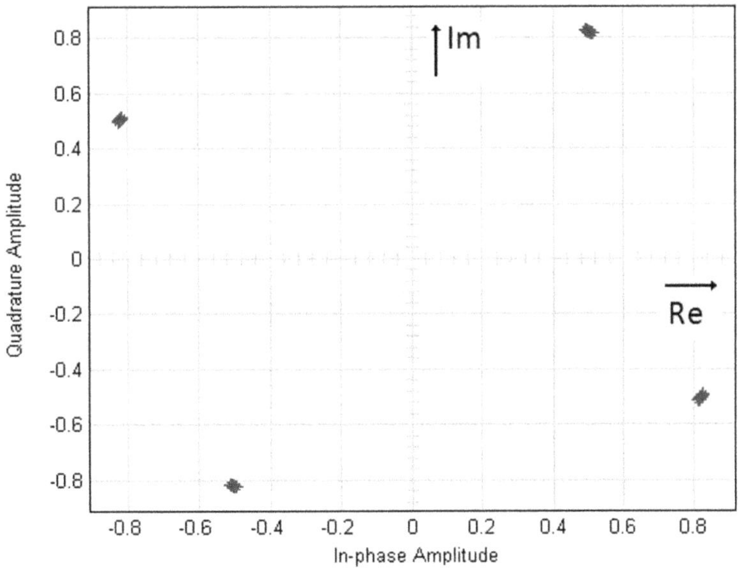

b)

Figure 14. The Attenuation block scheme and the example of the OOK modulated signal after attenuation

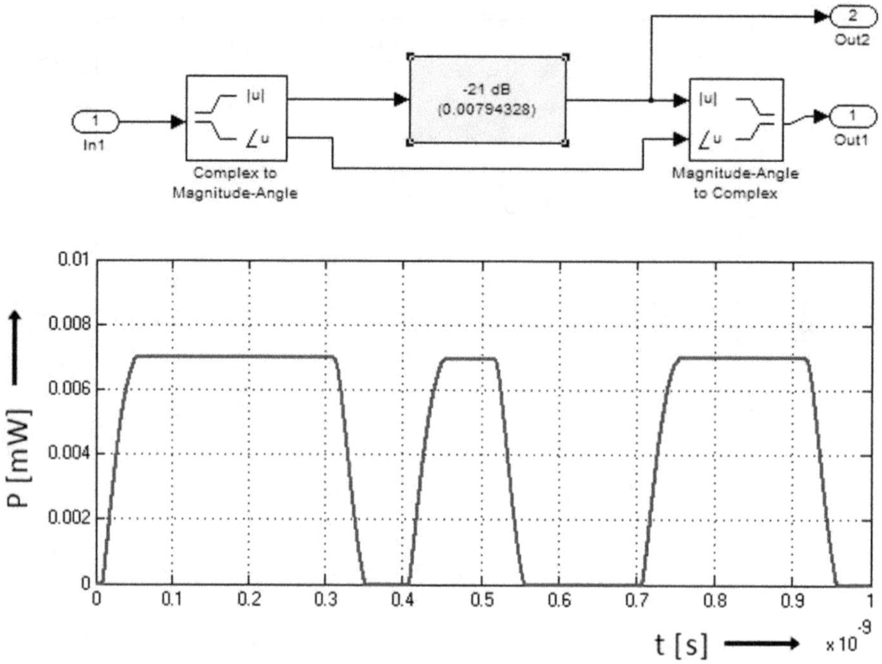

Figure 15. The Four Wave Mixing block scheme

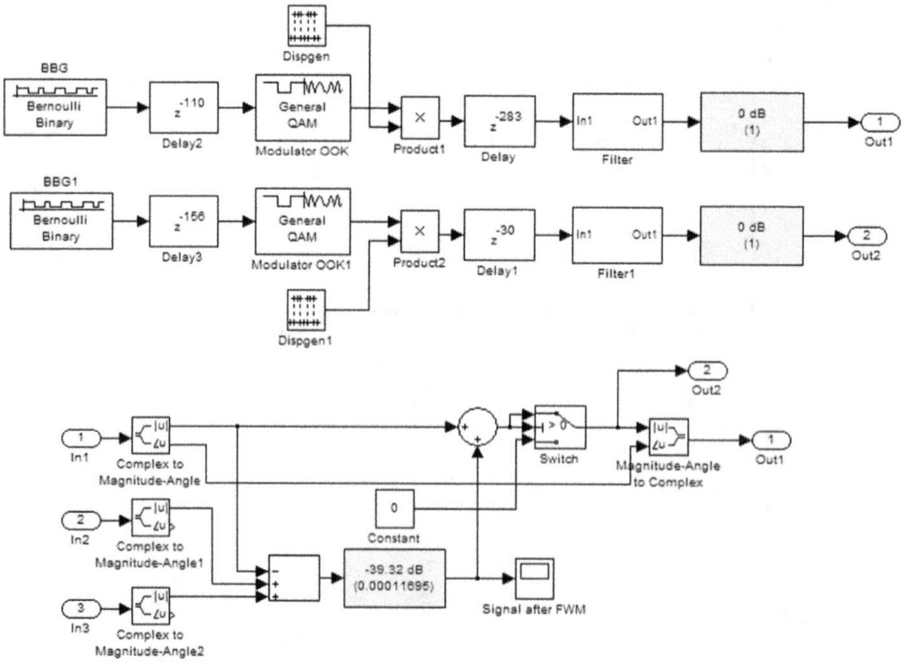

Figure 16. The example of the 10 Gbit/s OOK modulated signal after the FWM influencing

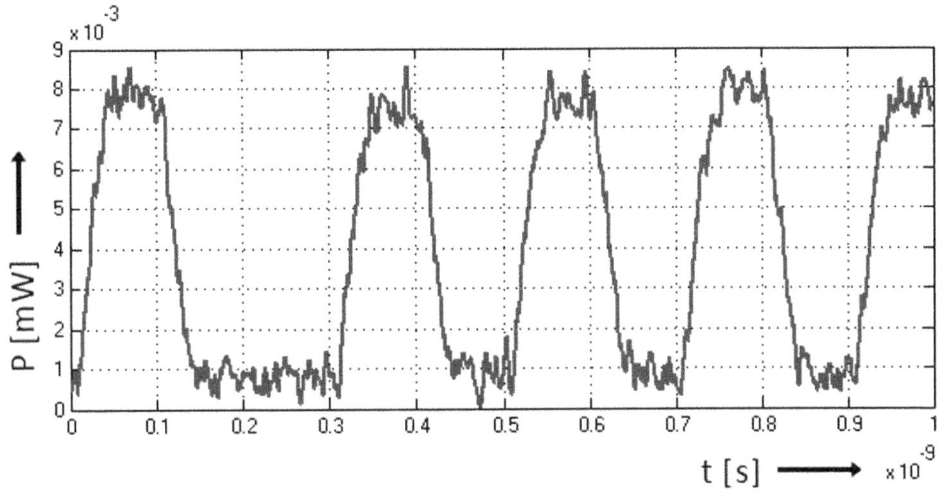

Figure 17. The Bit Error Rate block scheme

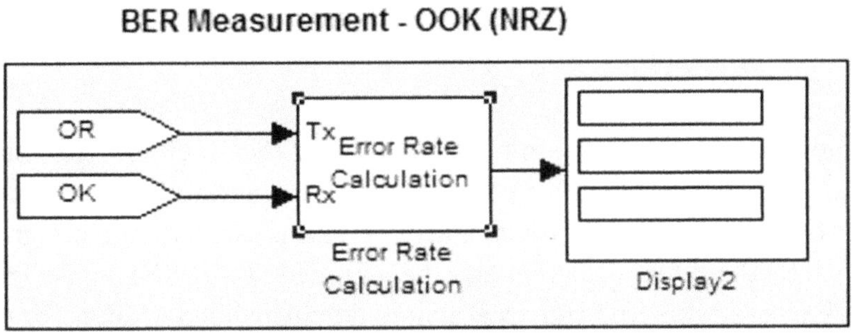

For this simulation model, specific parameters and power levels of concrete linear and nonlinear effects are setting suitable for analyzing their influences on optical signals in WDM transmission systems (Čertík & Róka, 2012), (Čertík & Róka, 2014), (Róka & Šalík, 2017), (Róka & Šalík, 2020). The simulation model is realized for the single-mode optical fiber with parameters satisfying the specification (ITU-T, 2009a), (ITU-T, 2009b), (ITU-T, 2010). Overall, partial contributions of environmental influences are relatively very small comparing to the total signal degradation. For presenting functionalities of the optical transmission path, a graphical representation called the eye diagram can be successfully utilized. The eye diagram is an oscilloscope displaying a received signal repetitively sampled with the bit interval. The simulation of the signal transmission in the optical transmission medium will be demonstrated using the initial on/off keying OOK. Assuming the OOK modulation is noncoherent, the demodulation process changes the intensity of optical signals into the electric domain directly.

In Figure 18, eye diagrams corresponding to OOK optical signals at the 10 Gbit/s transmission rate before transmitting into the optical path (left) and after transmitting in the 60 km optical fiber length influenced by all mentioned environmental effects (right) are presented. Signals are received in the Demodulator block by the direct detection. For modeling a noise in this block, the additive white Gaussian noise (AWGN) channel with the 20 dB SNR level is assumed. For this case, the output BER value is more than 10^{-12}.

Figure 18. Eye diagrams for OOK optical signals before (left) and after (right) transmitting in the 60 km the optical transmission path

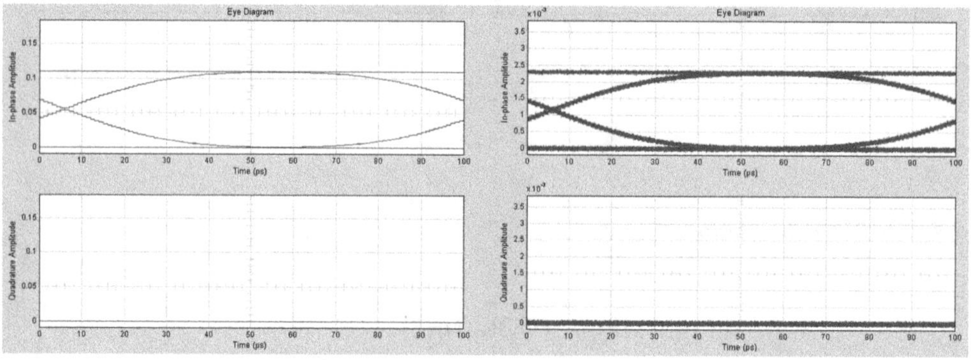

In Figure 19, eye diagrams corresponding to optical signals at the 10 Gbit/s transmission rate after transmitting in the 90 km optical fiber length (left) and after transmitting in the 160 km optical fiber length (right) influenced by all mentioned environmental effects are presented. Again, signals are received in the Demodulator block by the direct detection and the assumed 20 dB SNR noise level is the same as in the previous case. However, received signals are more attenuated and noisier. For these cases, the output BER values with more than 10^{-9} are insufficient.

Figure 19. Eye diagrams for OOK optical signals after transmitting in the 160 km (left) and 90 km (right) the optical transmission path

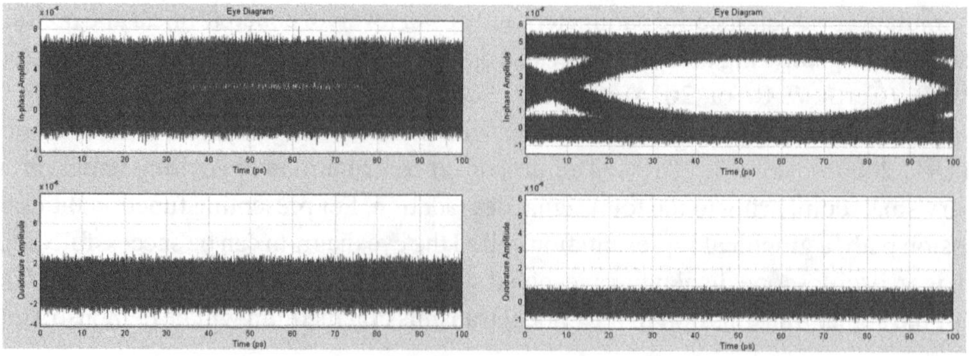

Figure 20. Constellation diagrams for OOK optical signals before (left up) and after transmitting in the 60 km (right down), 90 km (right up) and 160 km (left down) optical transmission path

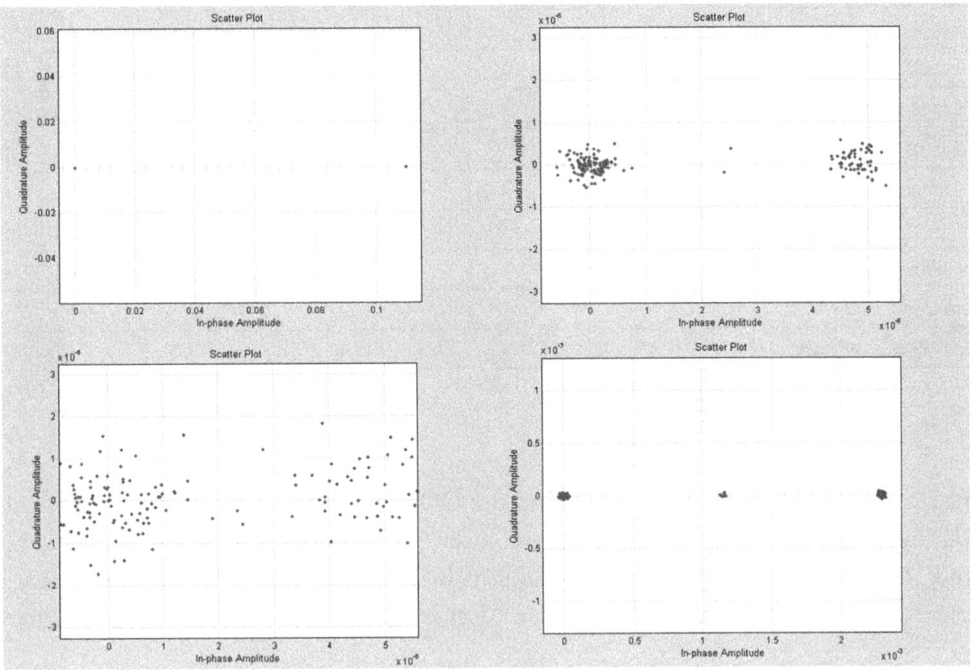

Figure 21. The frequency spectrum of 80 optical transmission channels (red) and the analyzed 193,4 THz frequency channel (blue)

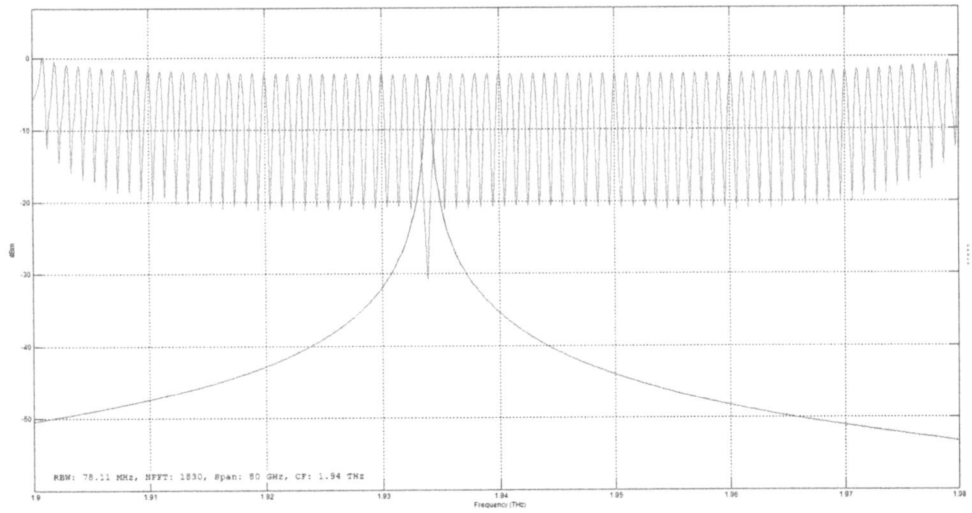

Table 1. Extracted parameters for laser rate equations

Parameter name	*Parameter*	*Value*
electron lifetime	τ_n	0.33×10^{-9} s
photon lifetime	τ_p	3.0×10^{-12} s
optical confinement factor	Γ	0.8
gain compression ratio	ε	2.5×10^{-24} m^3
active region volume	V_a	1.5×10^{-16} m^3
group velocity	v_g	7.5×10^{7} ms^{-1}
carrier density at transparency	N_0	1.0×10^{-12} m^{-3}
spontaneous emmision factor	β	3.0×10^{-5}
total differential quantum efficiency	η	0.042
central wavelength	λ	1550×10^{-9} m

In Figure 20, constellation diagrams related to the OOK modulation technique for optical signals at the 10 Gbit/s transmission rate before (left up) and after transmitting in the 60 km (right down), 90 km (right up) and 160 km (left down) optical fiber lengths influenced by all linear and nonlinear effects are presented. As can be seen, a spreading of constellation points is more expressive with increasing fiber lengths.

In Figure 21, a complete frequency spectrum of 80 optical transmission channels (red) in the simulation model is presented together with the analyzed channel at the 193,4 THz frequency (blue).

Using this simulation model, a possible utilization of advanced signal processing techniques in optical transmission systems can be simulated. By this way, different modulation and/or encoding techniques can be independently or in combinations analyzed.

Possibilities for Specific Optical Components in the Optical Simulation Program

Possibilities for Laser Numerical Simulations

Dynamic properties of the distributed feedback laser (DFB) can be modelled via coupled rate equations in (1)-(3). These equations describe inner interactions between the electron number *N(t)*, the photon number *S(t)* and the optical phase of photons *Φ(t)*. This approach considers fluctuations of parameters described below and, by using the coupled rate equations, it is possible to create the efficient numerical model for the DFB laser source (Šalík & Róka, 2017), (Šalík & Róka, 2018):

$$\frac{dN(t)}{dt} = \frac{I(t)}{q} - \frac{N(t)}{\tau_n} - g\frac{N(t) - N_0}{1 + \varepsilon S(t)} S(t) + F_N(t) \quad (1)$$

Figure 22. Light current curves for different central wavelengths

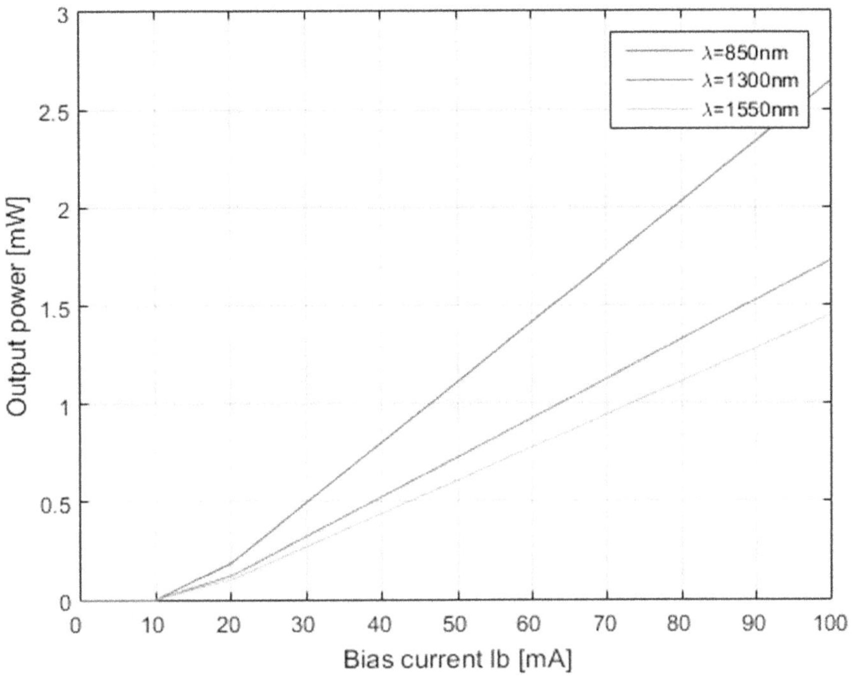

Figure 23. Output power curves for different bias currents

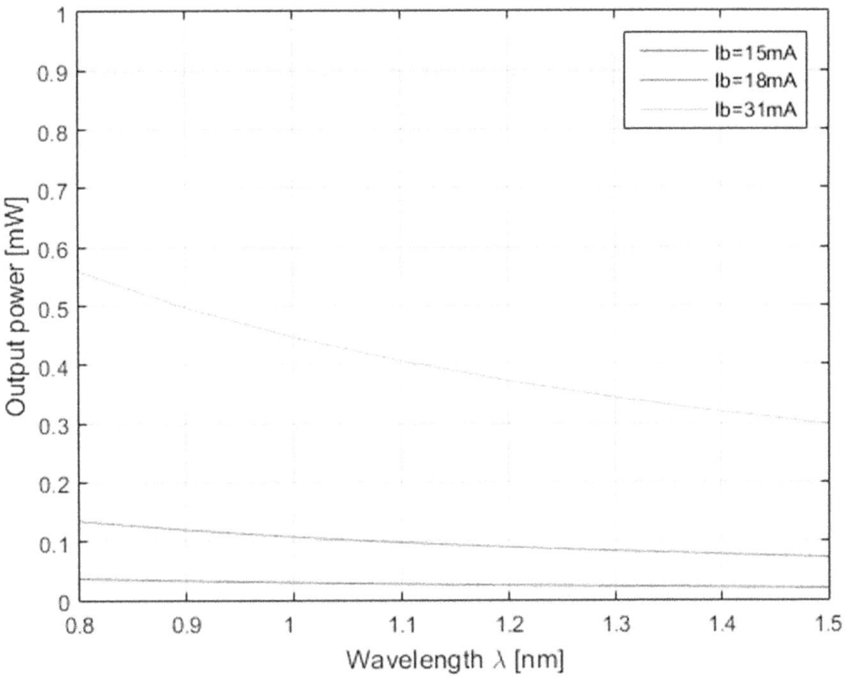

Figure 24. Time variations of DFB laser output characteristics for the steady-state mode a) the electron number b) the photon number c) the lasing frequency fluctuation

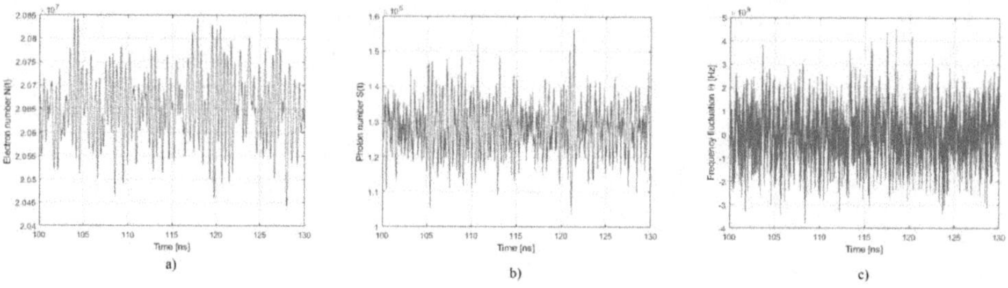

$$\frac{dS(t)}{dt} = g\frac{N(\mathrm{t}) - N_0}{1 + \varepsilon S(\mathrm{t})} S(t) - \frac{S(\mathrm{t})}{\tau_p} + \frac{\beta N(\mathrm{t})}{\tau_n} + F_S(t) \tag{2}$$

$$\frac{d\theta(t)}{dt} = \frac{\alpha}{2} g\left(N(\mathrm{t}) - N_{avg}\right) + F_\theta(t) \tag{3}$$

where g is the gain slope constant coefficient, N_0 is the carrier density at transparency, ε is the gain compression factor, τ_p is the photon lifetime, β is the spontaneous emission coupled into a lasing mode, $I(t)$ is the injection current, q is the electron charge, τ_n is the electron lifetime, α is the linewidth enchancement factor, N_{avg} is the time averaged carrier number. For simulation purposes, it is necessary to consider noise parameters that are representing intensity phase and frequency fluctuations. These fluctuations are caused by the spontaneous emission and the electron hole recombination and can be assumed to be Langevin noise sources $F_i(t)$, where i can be substituted with s, n or θ depending on the equation in that it is used. More about a noise source modelling for the DFB laser can be found in (Ahmed, Yamada & Saito, 2001), (Gao, Li, Flucke & Boeck, 2004), (Fatadin, Ives & Wicks, 2006).

Figure 25. Time variations of DFB laser output characteristics for the steady-state mode a) the output power fluctuation b) the phase fluctuation c) the power spectrum

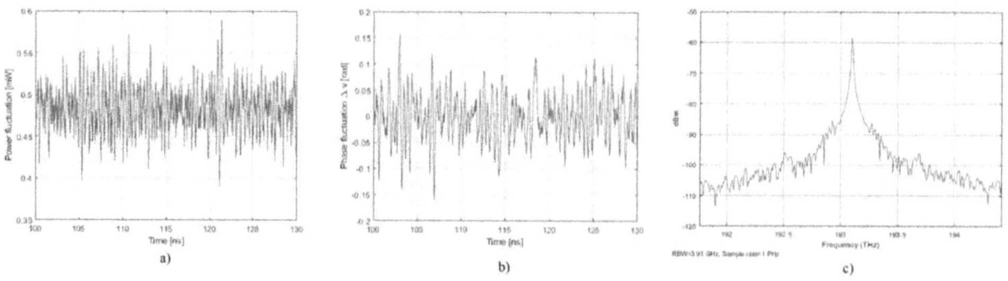

Figure 26. The principal structure of the optical intensity modulator based on the MZI structure

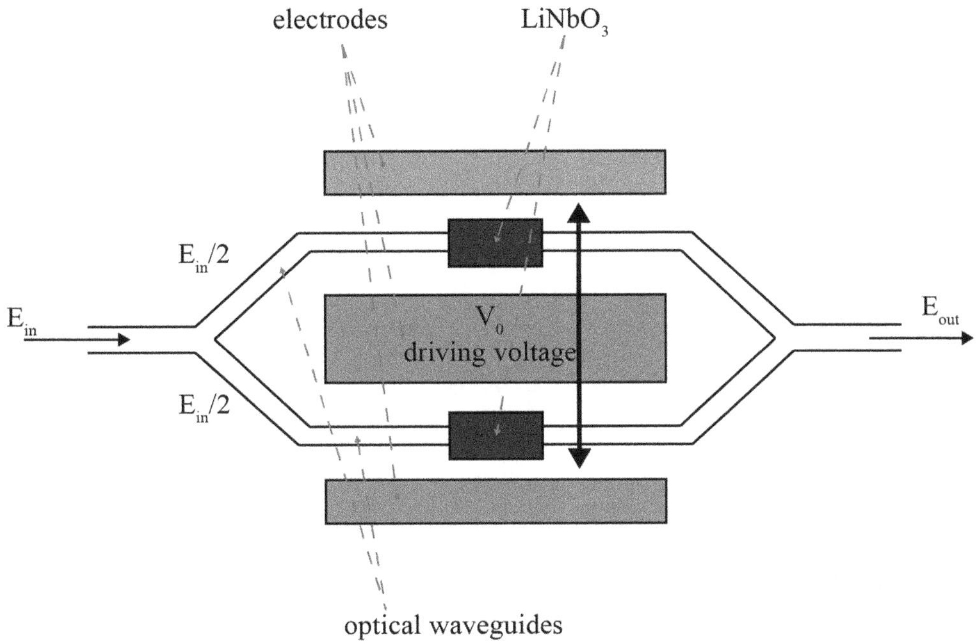

Numerical results of DFB laser output parameters are obtained by means of the fourth-order Runge-Kutta algorithm using the short interval $\Delta t = 10$ ps. Rate equations in the laser model are presented in a form containing only parameters that were extracted from measurements. More information an extraction process and parameters used in this simulation can be found in (Šalík & Róka, 2017). More detailed description of the steady-state equations can be found in (Agrawal & Dutta, 1993).

The Simulation Model for Analyzing a Laser Behavior

The simulation is performed in the Matlab environment. Two simulation models are designed for analyzing a behavior of the DFB laser. Simulations are based on rate equation models mentioned above. These models only simulate the first two operation regions of the laser, which are carrier and photon density equations. For a purpose of the simulation, extracted parameters (can be seen in Table 1) are used (Šalík & Róka, 2018).

Numerical results for the carrier number N(t), photon number S(t) and output power P(t) are obtained using the fourth-order Runge-Kutta method using a short interval of $\Delta t = 10$ ps. In Figure 22, light current (L-I) curves are shown for different central wavelengths. We can see that if the bias current level is below the threshold current level, the output power is zero. By increasing the bias current, the maximum output power level can be reached.

In Figure 23, dependencies of the output optical power on the central wavelength are shown for different bias currents. The time evolution of laser output characteristics for the carrier number *N(t)*, photon number *S(t)* and output power *P(t)* without and with noise function are presented in (Šalík, Róka & Tomáš, 2018). It is readily seen that adding noise operators causes a fluctuation of these physical quantities around their steady-state values.

Figure 27. Examples of transmitted lightwaves in the MZI structure

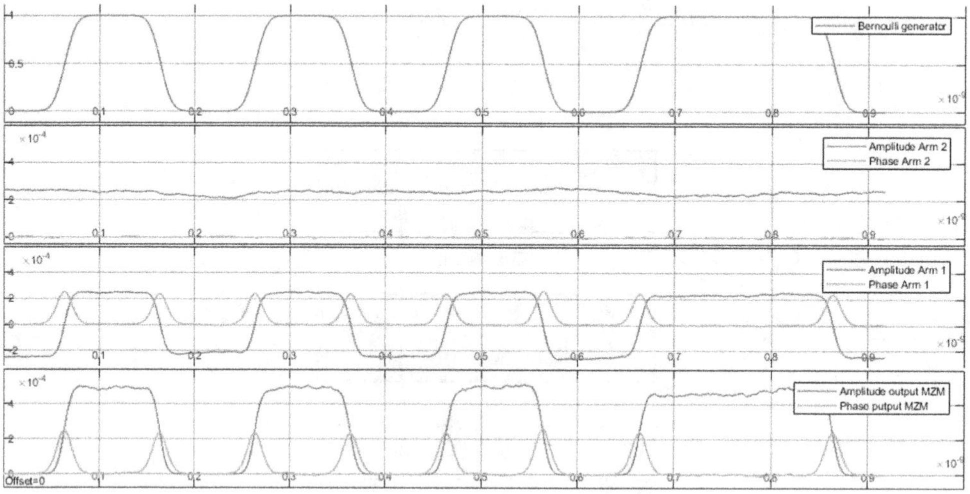

Time variations of DFB laser output characteristics for the steady-state mode (continues wave lasing mode) (*N(t), S(t), θ(t)*) can be seen in Figure 24.

Nonlinear and diffusion effects are lumped together as the field-dependent optical gain compression. The term *1/(1+ε.S(t))* used in equations (1)-(2) is the best option (according to (Tomita & Suzuki, 1991)) for a numerical solution for the DFB laser with the high photon density. The simulation platform is focused on the external modulation, therefore the optical feedback representing the light reflection on a connector connected to the optical fiber can be neglected. If anyone wants to consider light reflections, the equations (1)-(3) will be no longer suitable.

For accurate results, it is important to keep the condition (according to (Šalík & Róka, 2017), (Šalík & Róka, 2018)) that the bias current *(Ib = 12,9 mA)* must be greater the threshold current (I_{th} = *9,93 mA)*.

The output power fluctuation, the phase fluctuation and the output power spectrum of the steady state of the DFB laser is displayed in Figure 25. The central lasing frequency equals to *193,1 THz* corresponding to the *1550 nm* wavelength. It is better to use a transmitted lightwave in a form of the complex envelope in baseband than the high frequency signal, if it meets the condition that the optical signal's frequency is much larger that the optical signal's spectral bandwidth. The frequency used for a purpose of the optical transmission is too high to fulfil sampling theorem, therefore is necessary to adjust the optical carrier interpretation.

Figure 28. a) The detailed block scheme of the Mach-Zehnder Interferometer block with internal connections b) The detailed block scheme of the Phase Shifter block with internal connections

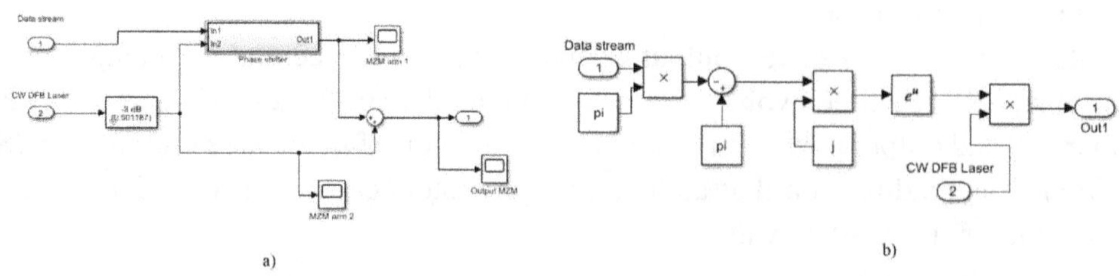

Figure 29. The WDM system scheme

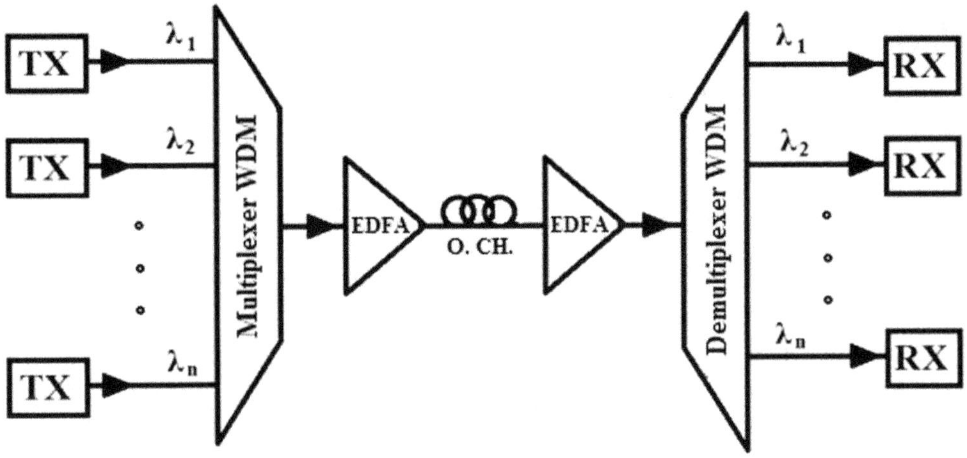

Results of the Analysis

A detailed mathematical methodology, based on laser rate equations of the DFB laser to describe its behavior under different conditions, has been presented. The first simulation is based on the laser rate equation model that can describe electron and photon population fluctuations. The rate equations in the laser model are presented in a form containing only parameters that were extracted from measurements. This approach can be also used to simulate a direct modulation of the DFB laser. The second simulation is based on steady-state conditions to obtain light current curves for different central wavelengths. The shape of output power curves has proven a functionality of selected numerical approach in theory. In the future, this approach will be tested via comparison with measured data from real DFB laser diodes. The comparison is focused on the relative intensity noise and phase noise comparisons.

Possibilities for the Interferometer Numerical Simulations

The Mach-Zehnder interferometer (MZI) is the optical intensity modulator that operates on a principle of the optical field interference between two lightwave components. The waveguide structure splits the input lightwave in two arms. The attenuation in each waveguide is normally 3 dB. Each modulator arm contains $LiNbO_3$ crystal that employs a phase modulation due to a driving current. Lightwaves from two arms are coupled at the MZI output. For a simulation platform, the single-drive MZI option is used, i.e. it means there is only single RF voltage driving on modulator arms. The principal structure of the single-drive optical intensity modulator is shown in Figure 26.

The transmitted optical field *E(t)* can be written as:

$$E(\mathrm{t}) = \frac{E_{iRMS}}{2} e^{j\omega_c t}\left(e^{j\phi_1(t)} + e^{j\phi_2(t)}\right) \tag{4}$$

where ω_c is the carrier frequency, *E(t)* is the transmitted optical field, E_{iRMS} is the root mean square value of the carrier and $\theta_1(t)$, $\theta_2(t)$ are temporal phase changes in $LiNbO_3$ crystal.

Figure 30. The example of the four wave mixing spectrum

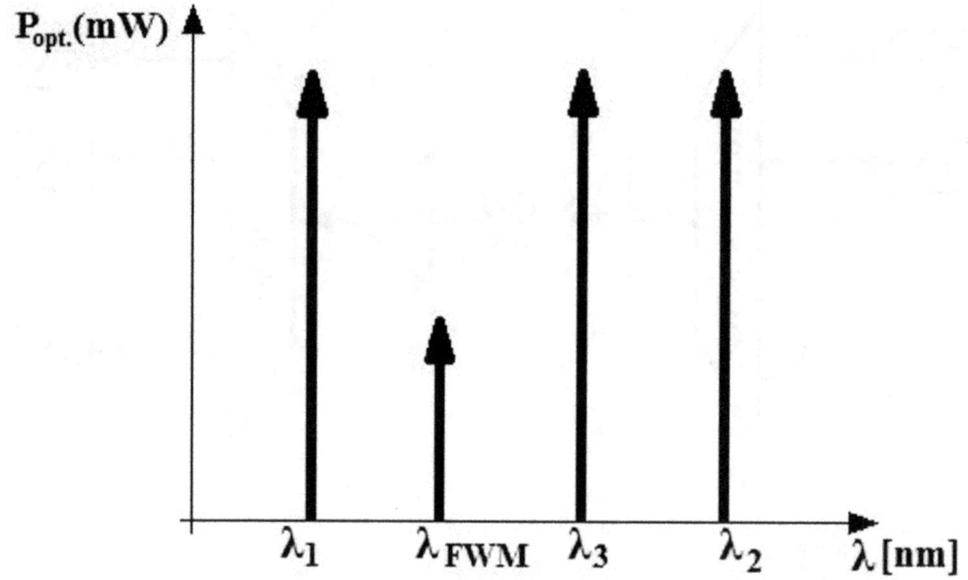

As can be seen in Figure 27, a data stream is shaped as the Gaussian pulse. In the second arm of the MZI, there is no phase change in a transmitted lightwave, on the other hand, a driving voltage in the first MZI arm causes the phase shift π when the logic “0” occurs. Finally, two lightwaves from arms are coupled together and produce the output modulated signal. A signal from the DFB laser is interpreted as the complex envelope. Phase chirps are caused by a single-drive constellation and can be removed by the MZI push-pull constellation.

In Figure 28, detailed schemes related to the MZI block and its phase shifter with internal connections can be found. First, the lightwave must be attenuated by 3 dB to simulate a splitter at the MZI input. Then, lightwaves are processed in two separate MZI arms. A lightwave in the lower arm remains unchanged, while a lightwave in the upper arm has a phase change according to data stream. The Phase Shifter block as a component of the MZI block works as follows. Randomly generated Gaussian pulses from data stream are multiplied by π. This operation converts a data stream to a range of radian values (π for the logic “1” and zero for the logic ”0”). The MZI block generates the 180-degree phase shift only if a zero is required at the modulator output. This causes that the zero-degree phase shift is generated ($-\pi + \pi = 0$) for the logic “1”and the 180-degree phase shift is generated ($0 + \pi = \pi$) for the logic “0”.

More detailed description of numerical simulations for specific components is in (Šalík, Róka & Tomáš, 2018).

Figure 31. The Simulink model of the 4-channel CWDM optical transmission path

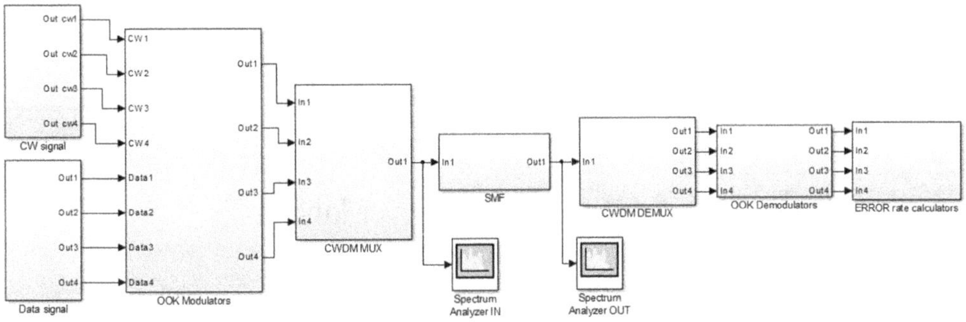

Principles of Wavelength Division Multiplexing Systems

Negative environmental effects play an important role in a transmission of optical pulses through the optical fiber. Knowing which fundamental linear and nonlinear interactions dominate is helpful to conceive techniques that improve a transmission of optical signals, including multiplexing, advanced modulation formats, optical signal processing and a distributed optical nonlinearity management. Basic transmission parameters of the standard optical single-mode fiber (SMF) utilized in telecommunications are determined by linear and nonlinear effects (Mukherjee, 2006), (Binh, 2010), (Róka, 2012), (Róka & Čertík, 2012).

Linear effects represent most losses at the optical signal transmission signal through the optical fiber. These linear effects are mainly caused by the attenuation and the dispersion. The attenuation limits the optical signal transmission and the dispersion influences transmission rates of optical signals. Nonlinear effects in the optical fiber may potentially have a significant impact on the performance of optical communication systems utilizing the wavelength division multiplexing (WDM) technique. In the WDM system, these effects place constraints on the spacing between adjacent wavelength channels and they limit the maximum power per channel, the maximum bit rate and the system reach.

Each channel in the WDM system, which is transmitted through a shared optical fiber, is represented by a different wavelength of the light radiation (Bao & Shieh, 2007), (Sharma & Kaur, 2015). To the WDM multiplexer enters several channels, which are coupled into one resulting optical signal at the fiber input using the optical coupler. On the other fiber end, the WDM demultiplexer splits one coupled signal to corresponding individual optical channels using the optical splitter and a set of optical filters. Wavelengths, which pass through optical filters, should correspond to wavelengths, which are used for individual channel transmissions. A scheme with basic system principles of the WDM transmission is shown in Figure 29.

The transmission capacity of WDM systems is dependent on the number of transmitted channels and on the channel spacing. If the wavelength distance between optical channels is smaller, then more channels can be transmitted in optical transmission windows. The wavelength band is strongly dependent on the operation band of utilized optical amplifiers, concretely the EDFA.

Figure 32. The block scheme of the Data signal source

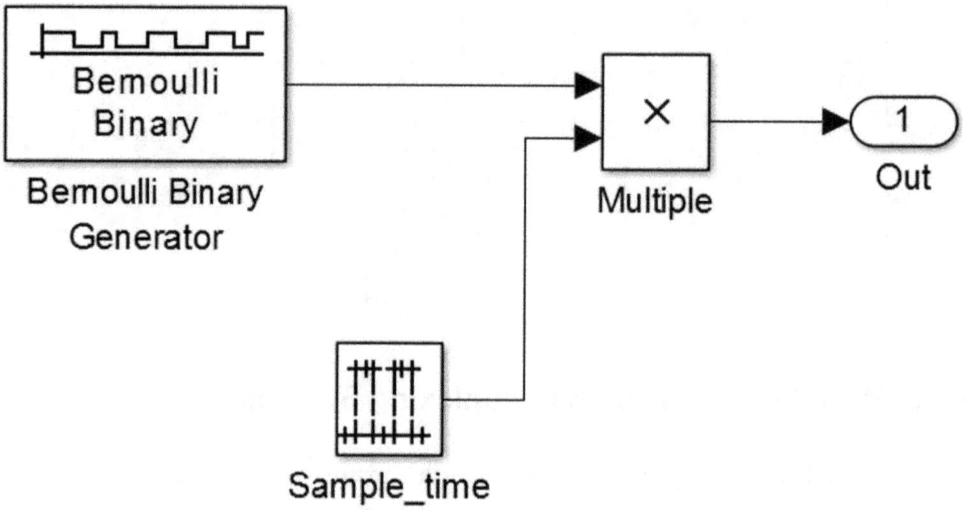

The Coarse Wavelength Division Multiplexing

The coarse wavelength division multiplexing (CWDM) is a simpler version of the WDM multiplexing technique, where no optical amplifiers are considered (ITU-T, 2003a, 2003b). Therefore, its simulation and analyzing is much easier than the dense wavelength division multiplexing (DWDM).

For the CWDM system, there is used a larger channel spacing that allows utilizing simpler and therefore cheaper system components, e.g. non-cooled lasers with a larger wavelength tolerance and optical filters with a wider passband. The CWDM technology uses II. and III. optical transmission windows (concretely, wavelengths from 1270 nm to 1610 nm) with the 20 nm channel spacing and 18 transmitted channels. The CWDM technology is used for short and medium distance network applications (Basak, Talukder & Islam, 2013).

Figure 33. The block scheme of the CW signal source

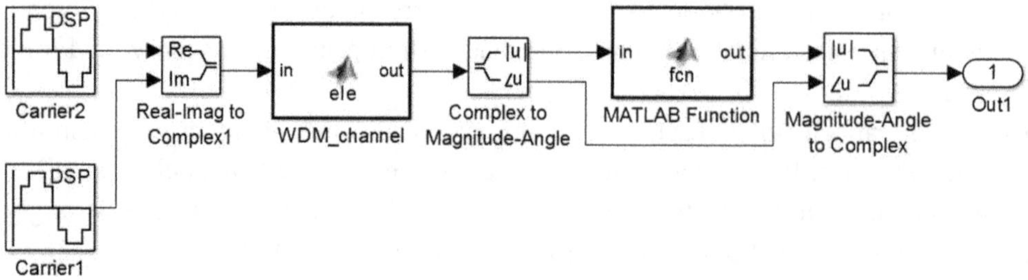

Figure 34. The detailed block scheme with internal connections of the OOK Modulators block

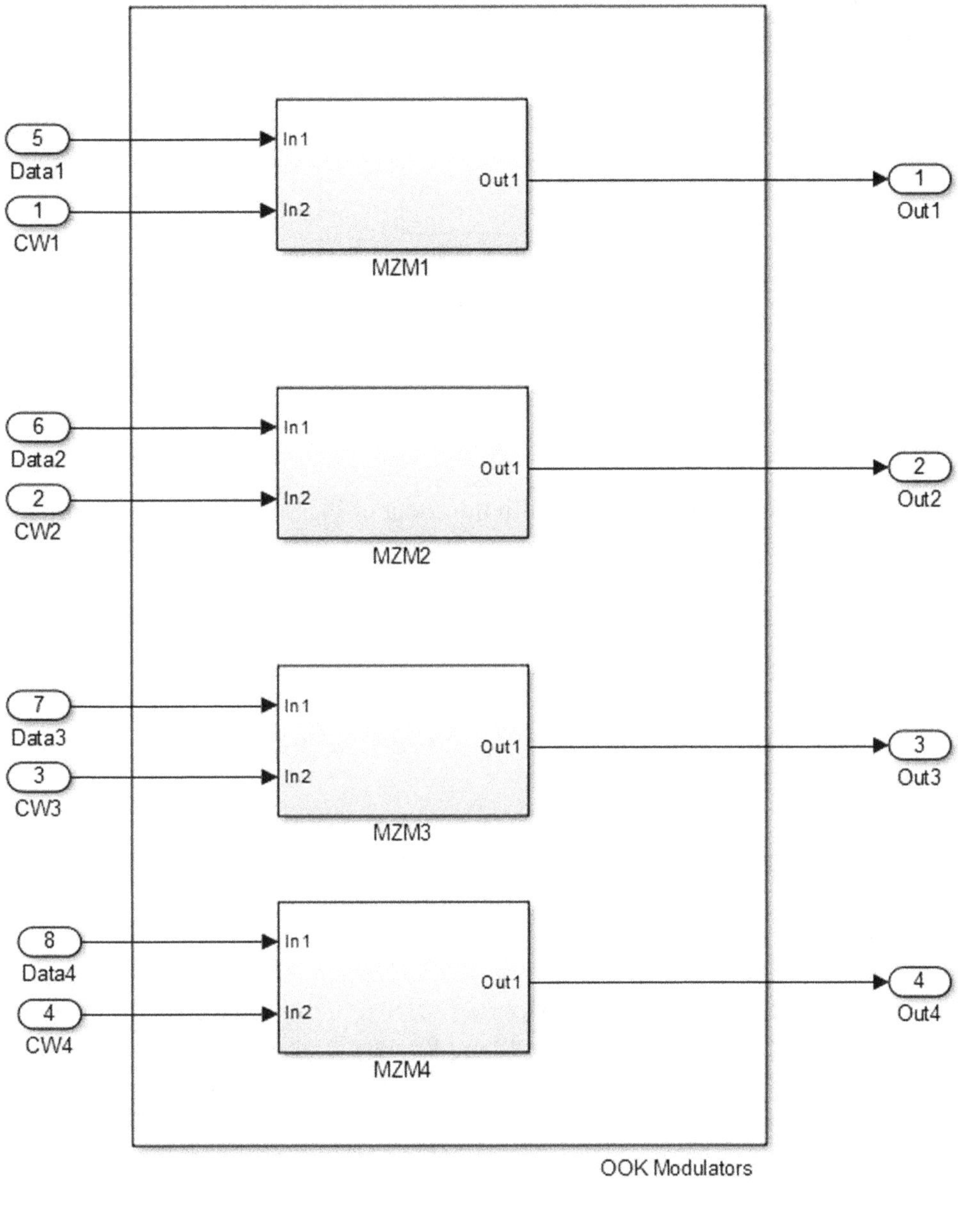

As was mentioned before, the four wave mixing (FWM) is the nonlinear effect with one of the strongest impacts on WDM signals. Basically, FWM arises when optical signal of three different wavelengths enters the fiber and form a new false wave on another wavelength. A schematic spectrum of the FWM effect in WDM system is shown in Figure 30, where λ_1, λ_2 and λ_3 are the transmitted waves and λ_{FWM} is the new generated false wave. The presence of this false wave (λ_{FWM}) can lead to interferences between channels, generating of additional noise that decreases a system performance and a power of required signals lost to false signals.

Figure 35. The block scheme for generating multiplexing signals in the VWDM MUX block

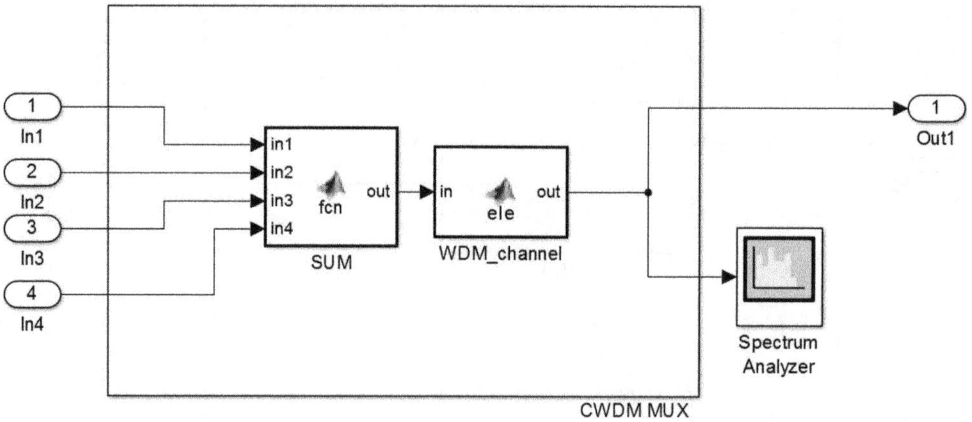

The count of mixing components increases with the count of WDM channels. The number of mixing components M is defined as

$$M = \frac{1}{2}.N^2.\left(N-1\right) \tag{5}$$

where N is the number of transmitted WDM channels (Cisco, 2006), (Deskmukh & Jagtap, 2013), (Kibria & Austin, 2015).

The Simulation Model for Analyzing the CWDM Transmission

For modeling of the optical transmission path, we used the software program *Matlab Simulink* together with additional libraries like *Communications System Toolbox* and *DSP System Toolbox*. The realized model (Róka, Mokráň & Šalík, 2017), (Róka & Mokráň, 2018) represents the CWDM signal transmission in the environment utilizing optical single-mode fibers for very high-speed data signals transmitted by four different wavelengths (Figure 31). For more accurate simulation of negative effects on each channel, this model of optical transmission path is divided into four paths (each channel has its own path) that are interconnected with each other. This interconnection interacts the transmitted channel using adjacent channels (i.e. the 1. channel is affected by 2., 3. and 4. channels, the 2. channel is affected by 1., 3. and 4. channels, etc.). Since effects on the channels are simulated individually, the CWDM Multiplexer block is connected after the SMF block of the optical transmission path.

Optical communication technologies will always be facing the limits of high-speed optical signal processing and advanced optical modulation formats, which are important factors to consider when discussing more effective utilization of possible transmission capacities in the optical fiber. Therefore, a main task of the CWDM simulation model is an analysis of various multiplexing, modulation and encoding techniques.

For the CWDM optical transmission path, the channel wavelengths are selected as follows corresponding to the 20 nm CWDM channel spacing:

Figure 36. The detailed block scheme with internal connections of the CWDM DEMUX block

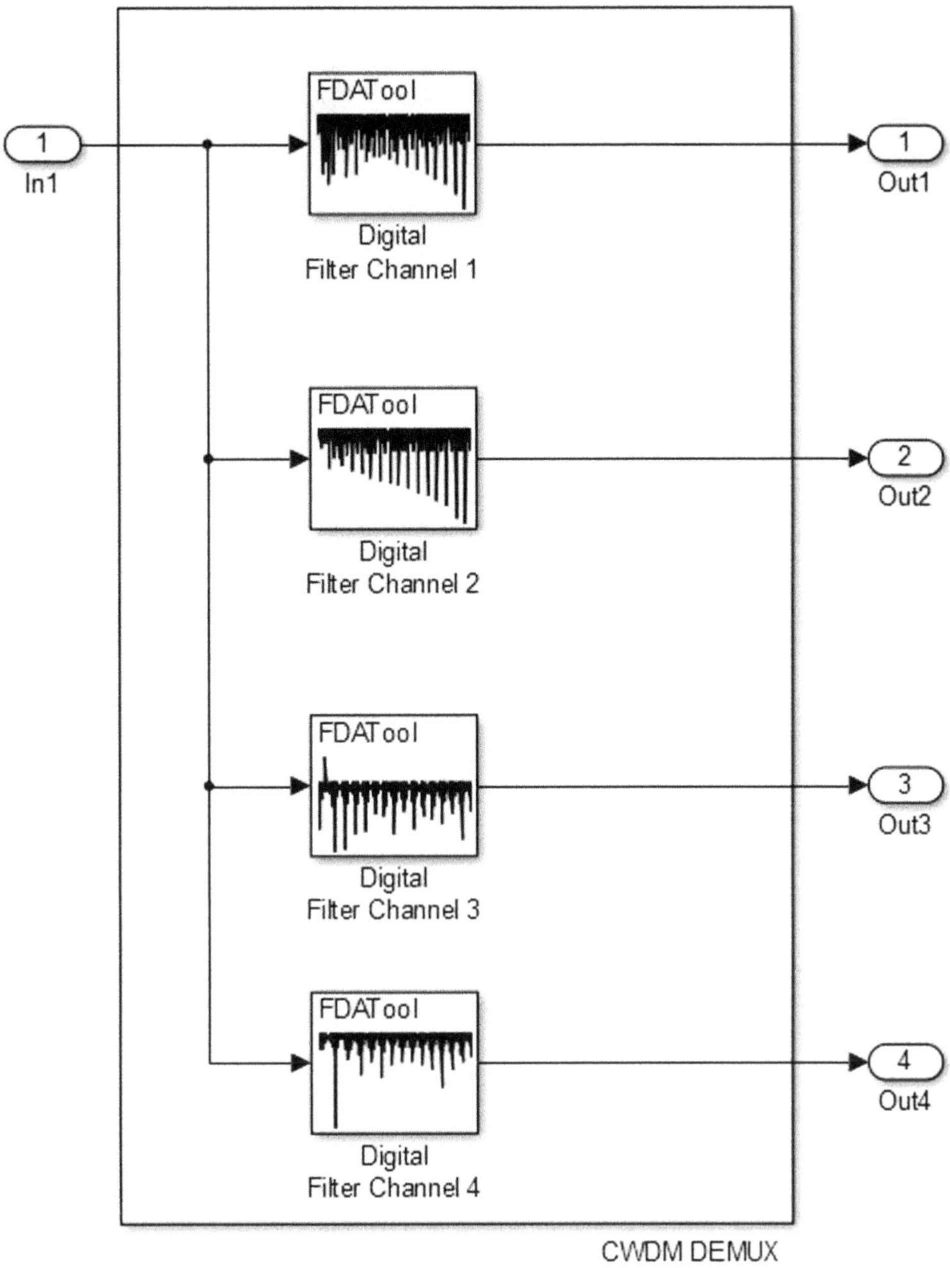

1. channel: λ_1=1571 nm => f_1= 190,83 THz
2. channel: λ_2=1551 nm => f_2= 193,29 THz
3. channel: λ_3=1531 nm => f_3= 195,81 THz
4. channel: λ_4=1511 nm => f_4= 198,41 THz

This Simulink model of the CWDM optical transmission path consists of next fundamental block:

Figure 37. The eye diagram for OOK optical signals after transmitting in the 80 km optical transmission path using the 4-channel CWDM system

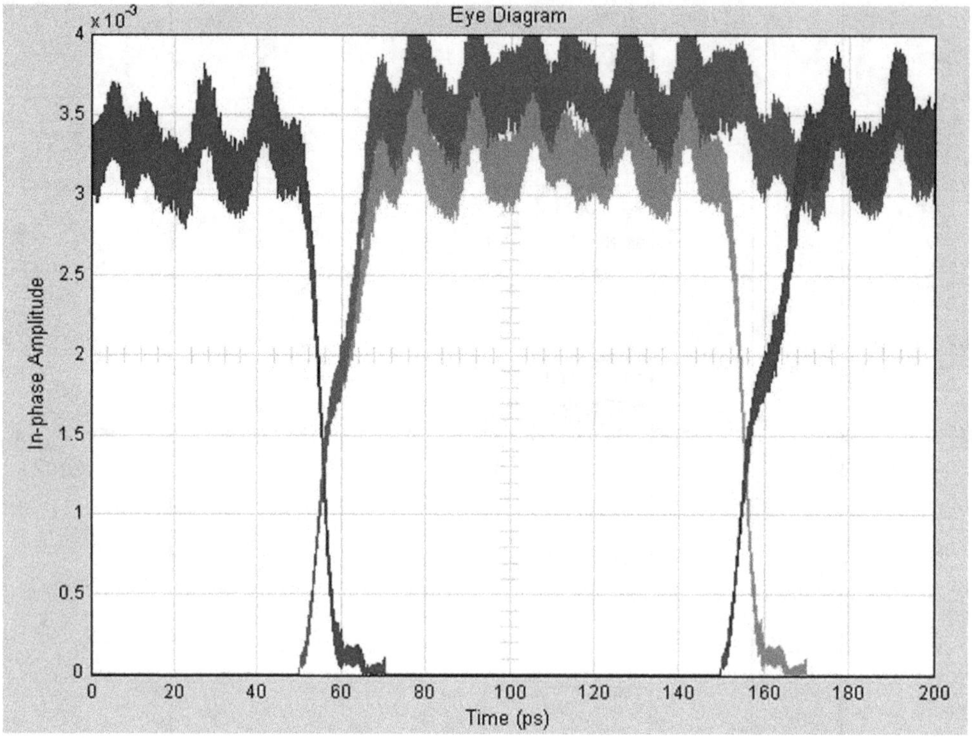

- Data signal
- CW signal
- OOK Modulators
- CWDM Multiplexer
- SMF - model of the optical transmission path
- CWDM Demultiplexer
- OOK Demodulators
- ERROR rate calculators

The Data signal block consists of four (Bernoulli) binary generators, which represent informative flows modulated and multiplexed into four outgoing signals. One of the data signal sources is shown in Figure 32.

The continuous wave (CW) signal block represents a set of carrier signal sources, which enters the OOK Modulators block. It is the basic block for advanced optical modulation techniques and its output simulates optical signals needed to information transmission. One carrier signal source consists of several sine generators, which are set to generate many continuous wave signals at the same time, since a real source of the optical radiation is not monochromatic (i.e. it has not only one carrier wavelength (λ), but there are more wavelengths ($\Delta\lambda$)). The internal connection of the CW signal source is shown in Figure 33.

Figure 38. The detailed block scheme with internal connection of the Nonlinear effects block for the 4-channel CWDM system

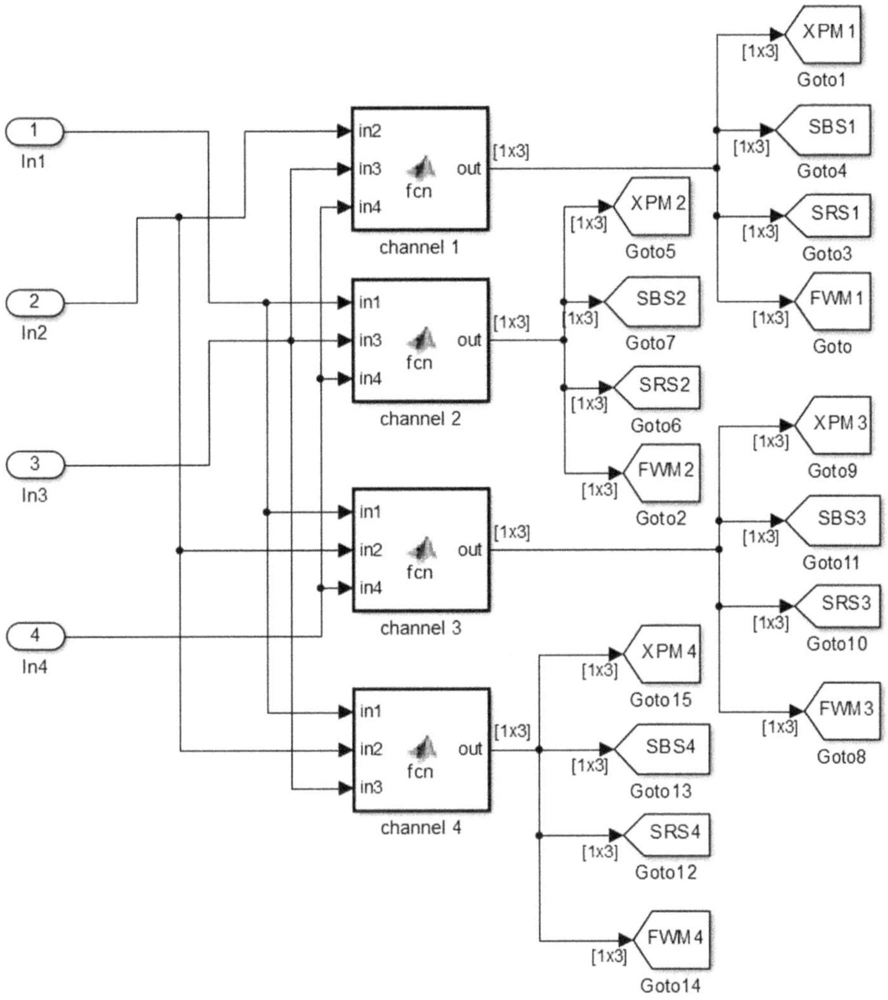

Outputs of the Data signal and CW signal blocks are connected to the OOK Modulators block. This block consists of four MZM OOK modulators whereby internal connections of the OOK modulator are described in (Mokráň, Róka & Čertík, 2015), (Róka & Mokráň, 2015). The detailed scheme of the OOK Modulators block is shown in Figure 34.

Optical modulated signals are going into the CWDM MUX block where are coupled into one originating signal before transmission in the optical SMF medium. Internal connections of the CWDM multiplexer are shown in Figure 35.

The outgoing signal from the CWDM MUX block goes into the SMF block that simulates the optical transmission path with negative environmental effects on transmitted optical signals like dispersions, an attenuation and non-linear negative effects. Details of the SMF block are available in (Róka & Čertík, 2012), (Róka & Čertík, 2015b), (Róka & Čertík, 2015c).

Figure 39. The frequency spectrum of four CWDM channels a) entering IN and b) outgoing OUT the SMF block

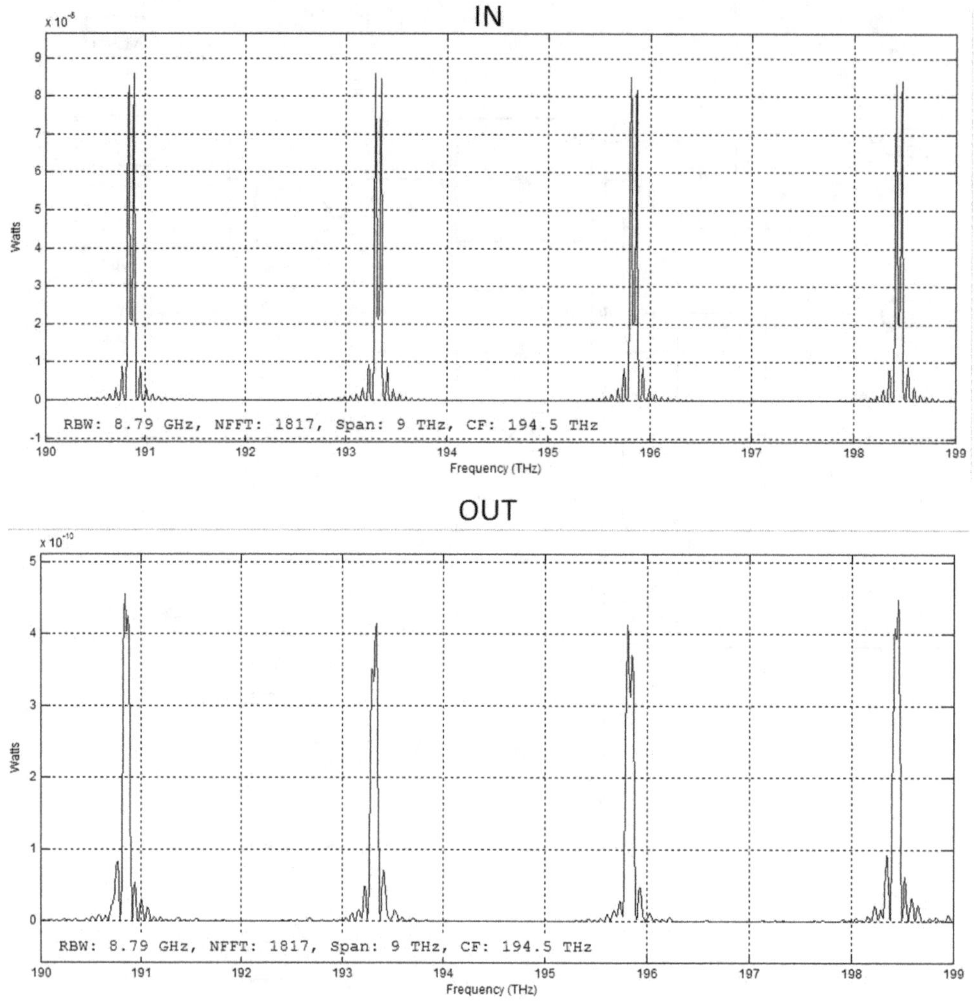

After passing through the SMF block, the one transmitted signal enters the CWDM DEMUX block. The task of this demultiplexer is splitting the incoming signal into individual outgoing signals, whereby each of them contains only one data channel. The CWDM DEMUX block consists of a set of filters (for each outgoing channel one), which are represented with band-pass filters. Filters are designed using *the Digital Filter Design* block that is contained in the *Matlab Simulink DSP System Toolbox*. Internal connections of the CWDM demultiplexer is shown in Figure 36.

Using this simulation model the 4-channel CWDM optical transmission path, it is possible to analyze a distance between logical levels with an eye diagram. The eye diagram for second simulated channel (λ_2 = 1551 nm) transmitted through the SMF block is shown in Figure 37.

Figure 40. The detailed frequency spectrum of the 2nd CWDM channel a) entering IN and b) outgoing OUT the SMF block

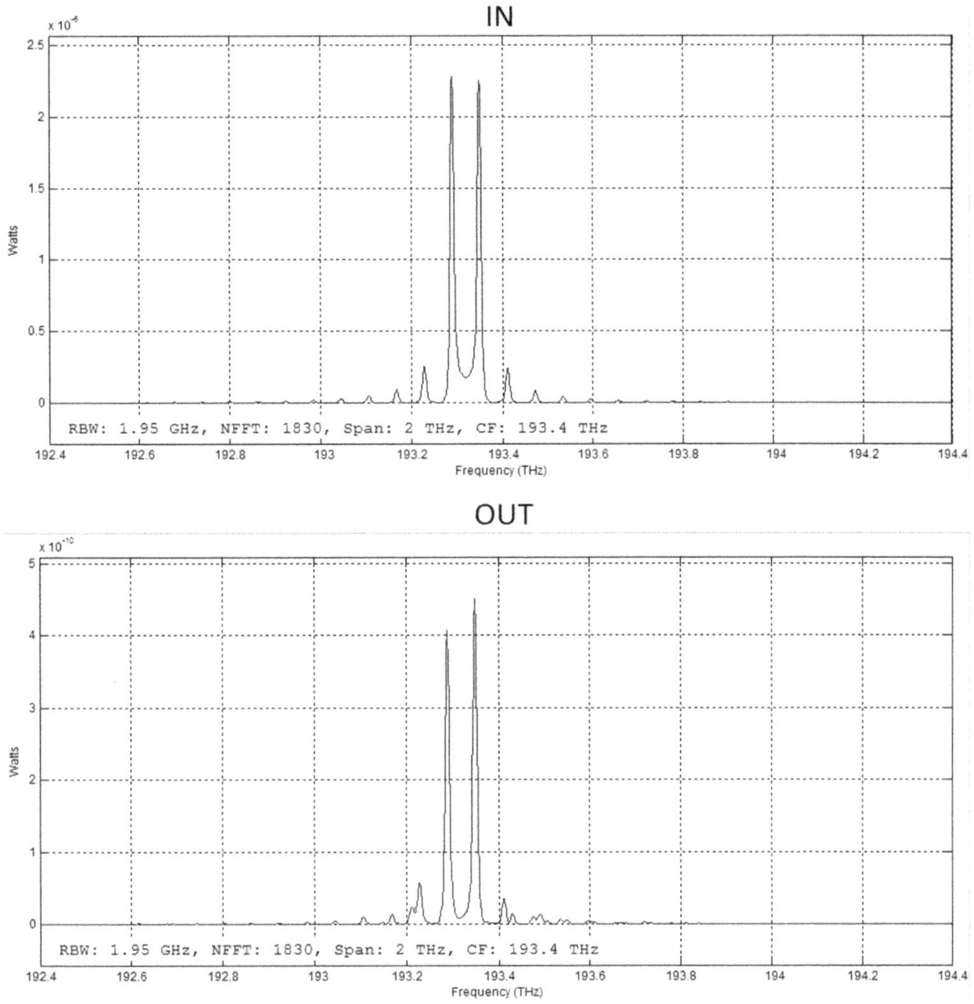

In Figure 38, the internal connection of nonlinear effect channels is shown. The task of this auxiliary block is preparing WDM channels outgoing from OOK modulators for using in the Nonlinear effects block of SMF optical transmission paths for the 4-channel CWDM system. For each transmitted channel, three other channels are multiplexed and used as input for the Nonlinear effects block. E.g., 2., 3. and 4. channels are sent to the Nonlinear effects block for specifying their influence on the 1. channel.

Results of the Analysis the CWDM Transmission

The CWDM signal transmission's simulation is performed in Matlab Simulink and consists of implementation of four wavelength channels using the CWDM technique to the simulation model of negative environmental influences at the signal transmission in the optical transmission path (Róka, 2012), (Róka & Čertík, 2012), (Róka & Čertík, 2015a), (Róka & Čertík, 2015c). In our simulation, it is assumed the fiber length of L = 80 km and 4 different wavelengths λ_1 - λ_4. Parameters of the optical single-mode fiber are set up particularly for each wavelength. For example, the total attenuation a_{total} = 16,8 dB (i.e. $\alpha_{specific}$ = 0,21 dB/km) and other specific values PMD = 10 ps/$\sqrt{}$km, CD = 10 ps/(nm.km) are valid for the wavelength λ_2 = 1551 nm. For a simulation, the OOK technique is used with the same 10 Gbit/s transmission speed for each channel.

Figure 41. The detailed frequency spectrum of the 2nd CWDM channel for individual negative environmental influences

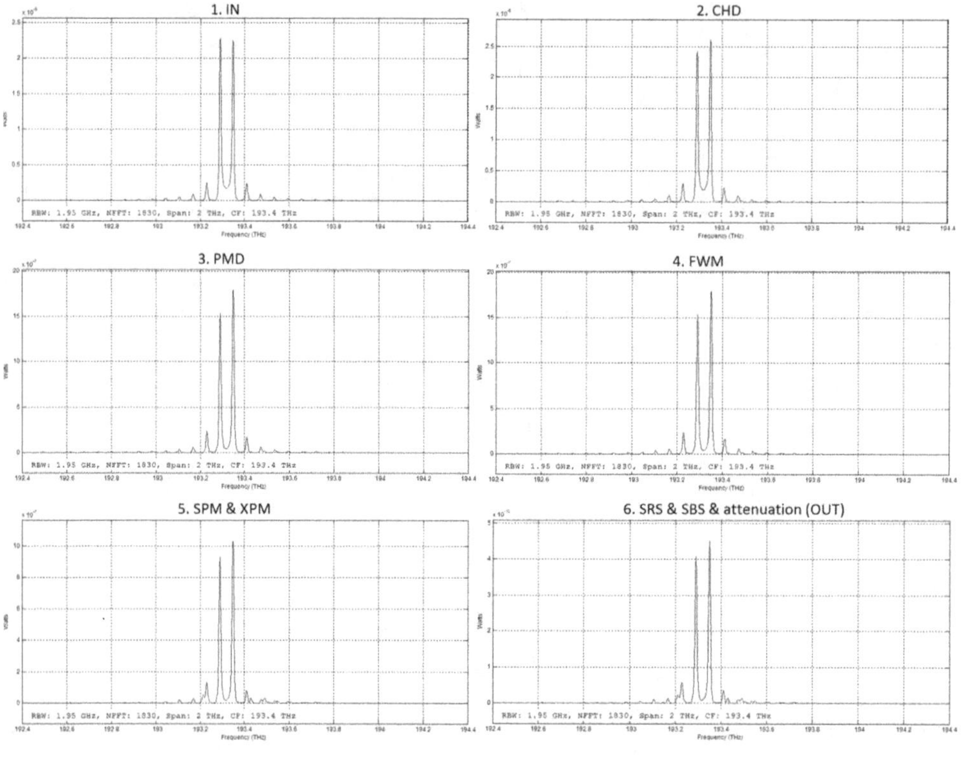

The simulation model of the 4-channel CWDM optical transmission path also allows analyzing an impact of the optical transmission path on utilized signal spectrum (Róka & Mokráň, 2018), (Róka & Mokráň, 2019). The spectrum for all four CWDM simulated channels is shown in Figure 39, where the spectrum of signals entering the SMF block is marked as "IN" and the spectrum of outgoing signals is marked as "OUT".

In Figure 40, a particular spectrum of the second simulated channel (λ_2 = 1551 nm) entering and outgoing the SMF block is presented.

In Figure 41, concrete spectra of the second simulated channel (λ_2 = 1551 nm) for individual negative environmental influences are presented. In the simulation model of the CWDM optical transmission path, following negative influences considered for the optical transmission medium included (Róka, 2012), (Róka & Čertík, 2012), (Róka & Čertík, 2015a), (Róka & Čertík, 2015c) are included:

Figure 42. The DWDM frequency grid with fixed 50GHz channel spacing

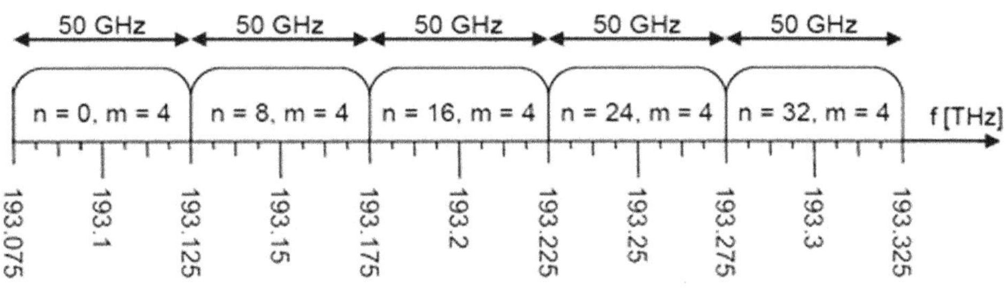

- chromatic dispersion (CD)
- polarization mode dispersion (PMD)
- four-way mixing (FWM)
- self-phase modulation (SPM) & cross-phase modulation (XPM)
- stimulated Raman scattering (SRS) & stimulated Brillouin scattering (SBS) & attenuation

As can be seen in Figure 41, the signal spectrum after transmission is changed due to negative influences of the optical medium. Shapes of signal sidebands are deformed and unsymmetrical. Environmental effects with the strongest impact are dispersions (CD and PMD) and FWM.

This subchapter deals with main features, characteristics and the simulation of negative environmental influences in the optical transmission medium at the CWDM signal transmission and is focused especially on the FWM effect. Also, basic principles of optical communication systems utilizing the coarse WDM technique are introduced.

Figure 43. The DWDM frequency grid with flexible channel spacing

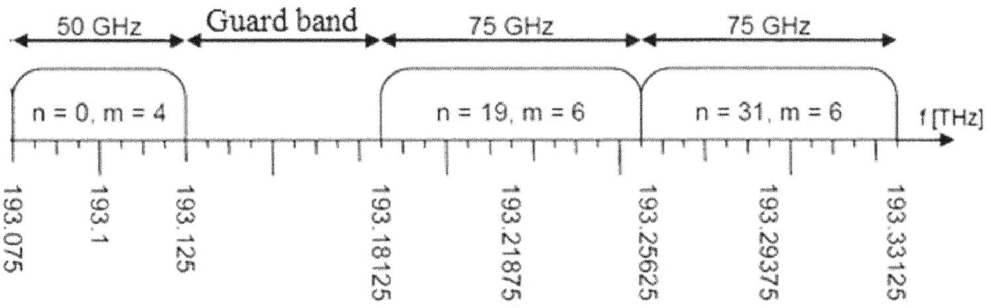

Figure 44. The frequency spectrum of four DWDM channels a) entering IN and b) outgoing OUT the SMF block

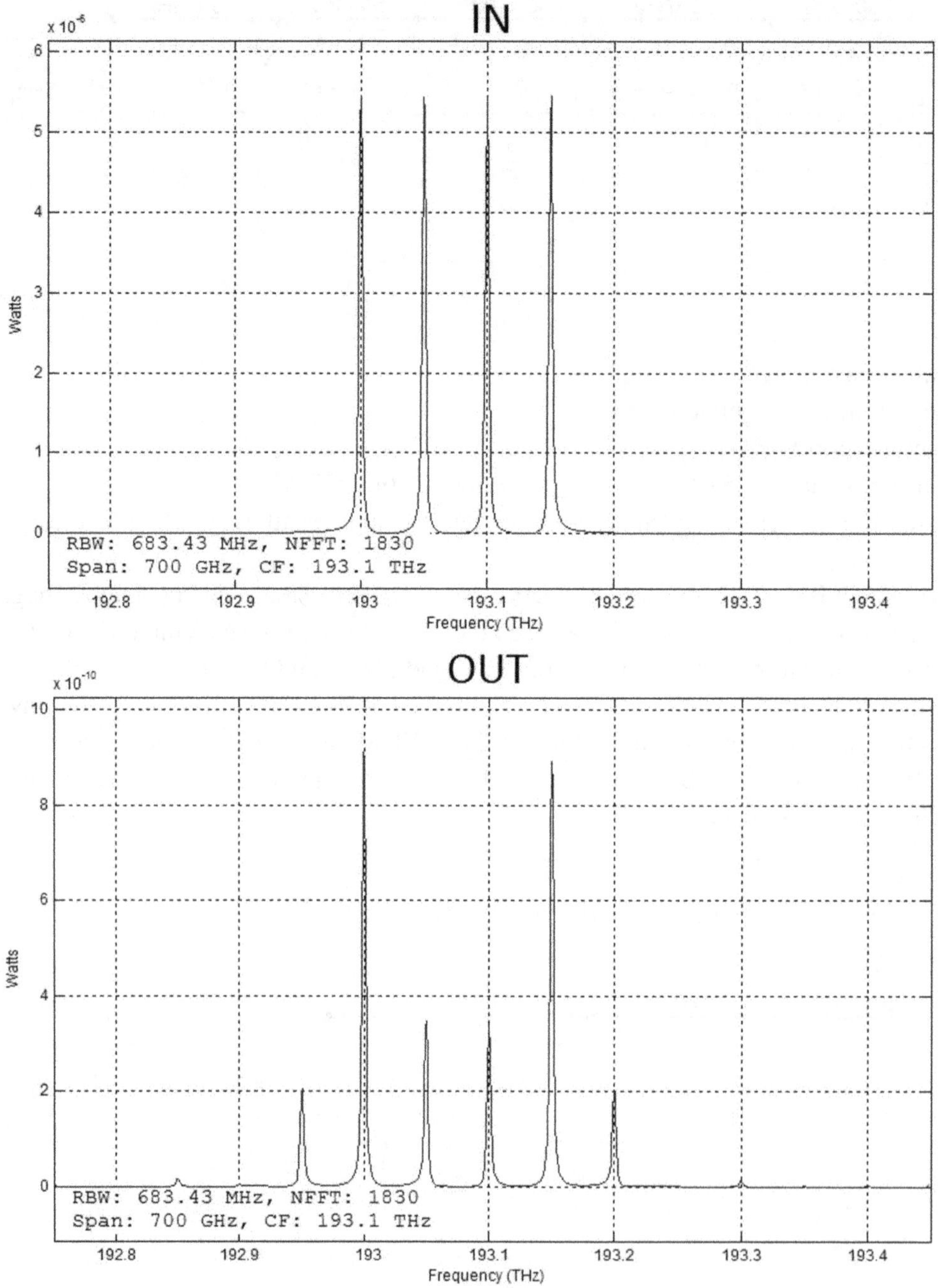

Modeling of the CWDM multiplexing technique with four wavelength channels in the optical transmission medium using Matlab Simulink environment is presented in (Róka & Mokráň, 2018). The Simulink model of the 4-channel CWDM optical transmission path allows executing different requested analyses for advanced optical signal processing techniques. The original model used in (Róka, Mokráň & Šalík,

2017) is modified to achieve more accurate simulation of negative influences (especially the FWM) in the optical transmission environment. In this case, negative effects are simulated individually for each channel with interconnections between each other. These interconnections simulate interferences between wavelength channels. The impact of this optical transmission path on transmitted optical signals is presented using eye diagram and frequency spectrum.

In the eye diagram, the larger distance between logical levels can be caused by a simpler implementation of the CW source (with low noise level) for the logic "0". In the frequency spectrum, it can be seen the strongest impact of negative environmental influences, which affects the symmetry and shapes of optical signal sidebands. The strongest impact on spectrum is observable for the first channel (Figure 39) where a new false part of sideband is generated.

In a future, the model of the 4-channel CWDM optical transmission path will be enhanced by increasing the number of CWDM channels to maximum (18 channels) and by improving the CW signal source. Moreover, the extension of this model for the DWDM signal transmission will be prepared in the next step (Róka & Mokráň, 2019).

The Dense Wavelength Division Multiplexing

The dense wavelength division multiplexing (DWDM) uses for transmission wavelengths from 1530 nm to 1565 nm for C optical transmission band resp. from 1565 nm to 1625 nm for L optical transmission band with channel spacing from 0,1 nm to 0,8 nm. The maximum number of transmitted channels can be varied in dependence of channel spacing from 45 to 360 channels. The DWDM can be further divided according to channel spacing into:

- DWDM with fixed channel spacing,
- DWDM with flexible channel spacing.

The DWDM with fixed channel spacing has the equal channel spacing for each transmitted channel. Channel frequencies are defined by (ITU-T, 2003a) by equations:

$$193{,}1 + n_{p12,5} \cdot 0{,}0125 \; [GHz] \text{ for the 12,5 GHz channel spacing} \tag{6a}$$

$$193{,}1 + n_{p25} \cdot 0{,}025 \; [GHz] \text{ for the 25 GHz channel spacing} \tag{6b}$$

$$193{,}1 + n_{p50} \cdot 0{,}05 \; [GHz] \text{ for the 50 GHz channel spacing} \tag{6c}$$

$$193{,}1 + n_{p100} \cdot 0{,}1 \; [GHz] \text{ for the 100 GHz channel spacing} \tag{6d}$$

where $n_{p12,5}$, n_{p25}, n_{p50} and n_{p100} are integers (positive, negative or zero). An example of the DWDM frequency grid with fixed 50 GHz channel spacing is shown in Figure 42.

The DWDM with flexible channel spacing allows to multiplexing channels with different transmission speeds. For DWDM with flexible channel spacing, nominal central frequencies are defined by the equation:

$193,1+n_f.0,00625\ [GHz]$ (7a)

where n_f is an integer (positive, negative or zero) and a slot width is defined by:

$$\Delta f=12,5.m\ [GHz] \tag{7b}$$

where m is a positive integer. The recommendation (ITU-T, 2003a) allows any combinations of frequency slots if no two frequency slots overlap. In Figure 43, an example of frequency grid with one 50 GHz channel and two 75 GHz channels is shown. The frequency slot between 193,125 THz and 193,18125 THz is unallocated and left as guard band between two sets of channels. Instead of this guard band can be this slot used for another 50 GHz channel (Cisco, 2006).

Because the 0,4 nm channel spacing is pre-determined for the DWDM system, the selected wavelengths are:

1. channel: λ_1 = 1553,3288 nm => f_1 = 193,00 THz
2. channel: λ_2 = 1552,9265 nm => f_5 = 193,05 THz
3. channel: λ_3 = 1552,5244 nm => f_3 = 193,10 THz
4. channel: λ_4 = 1552,1225 nm => f_4 = 193,15 THz

The simulation model of the 4-channel DWDM optical transmission path also allows analyzing an impact of the optical transmission path on utilized signal spectrum (Róka & Mokráň, 2019). In the Figure 44, spectral characteristics of environmental influences on transmitted DWDM signals present in the optical transmission medium are shown. "IN" is the spectrum of input signals entering in the SMF model and "OUT" is the spectrum of outgoing signals. As can be seen, there are new optical signals on 192,95 THz and 193,2 THz wavelengths. These additional signal frequency components are generated because of the FWM nonlinear influence on the transmitted signals utilizing DWDM multiplexing.

FUTURE RESEARCH DIRECTIONS

Broadband optical systems present optical infrastructures in long-distance, metropolitan and access networks to support various applications of many service providers. Technological improvements of wireless broadband and broadcast technologies will require a reliable support in optical networks. Therefore, appropriate advanced converged technologies utilizing the optical transmission medium and wireless transmission can be expected in a near future and more reasons begin effect on utilization and implementation of the WDM multiplexing technique in converged broadband optical fiber-wireless networks.

On the other hand, the broadband optical technology must be able to protect the investment of legacy passive and active optical networks. Therefore, a strong interest for utilization of advanced signal processing techniques in the optical transmission medium in converged broadband optical fiber-wireless networks is present. Naturally, a combination of advanced modulation and encoding techniques seems to be a practical solution. However, an area of research directions in converged broadband optical fiber-wireless networks is very wide and can include e.g. a mutual convergence of technologies between optical metropolitan and optical/wireless access networks, a utilization of power saving methods for subscriber terminals, an expansion of dynamic bandwidth and wavelength allocation mechanism, etc.

CHALLENGES AND OPEN ISSUES

In the near future, a convergence of optical and wireless technologies will be continued. Mobile base stations can be backhauled through the hybrid passive optical network and they can be thought of as common optical network terminal elements. Between the central office and the remote base station will be introduced another intermediate node responsible for generation and transmission of control data streams over longer fiber distances. Broadband optical networks can be deployed by various scenarios to address different cases with diverse degrees of consolidation - the urban, the passive extended reach and the active extended reach. With these, also various converged broadband optical fiber-wireless deployments could be included into possible future extensions. Moreover, a variety of users - residential, corporate/business and telco - must be considered. And, therefore, more flexible broadband optical architectures including very promising and different wavelength division multiplexing techniques should be considered.

For practical implementation, there are some open issues. A high importance is dedicated to the power saving in broadband optical long-distance, metropolitan and access networks and energy efficient efforts on both terminal sides in converged broadband optical fiber-wireless networks. From this viewpoint, possibilities of broadband optical systems with reliable traffic protection provisioning can also play an important role. For enhancement of broadband optical networks, approaches for extending reach of optical metropolitan and access systems and for flexible per-wavelength rate upgrades based on dynamic bandwidth and wavelength allocation mechanisms can be utilized. Within this context, other specific components in future broadband optical systems must be improved and optimized.

CONCLUSION

This chapter introduces enhanced simulation tools for analyzing advanced optical signal processing techniques in the optical transmission medium as an enhancement of the base of simulation tools for broadband passive optical networks prepared in (Róka & Čertík, 2015a).

The simulation program analyzes a pulse propagation inside optical fibers and influences of linear and nonlinear effects on information signals transmitted in the optical transmission path. This part presents features of basic functional blocks in the optical simulation program with more details together with specific block schemes created in the base of simulation tools. Also, examples of the optical signal presentation are graphically displayed at various points of the optical transmission path. Consequently, possibilities for some specific optical components are discussed and their inclusion in the enhanced simulation tool is introduced. Simultaneously, time variations of laser output characteristics acquired by numerical simulations are displayed for visualization. Finally, an attention is focused on principles of optical systems utilizing wavelength division multiplexing techniques. Belonging to this domain, the simulation model for multiple channel transmission systems is presented in details and environmental influences of the optical transmission path on optical signals are analyzed using the frequency spectrum.

A design of the optical transmission path offers a possibility for visual displaying of optical signals transmitted through the optical transmission medium and affected by its negative environmental influences by means of constellation diagrams, waterfall curves for bit error rate comparison and eye diagrams. The simulation program of the optical transmission path can be used for analyzing of optical modulation and encoding techniques at the signal transmission in the single-channel and multi-channel optical transmission paths.

ACKNOWLEDGMENT

This work is a part of research activities conducted at Slovak University of Technology Bratislava, Institute of MICT, within the KEGA agency project - 034STU-4/2021 "Utilization of Web-based Training and Learning Systems at the Development of New Educational Programs in the Area of Optical Wireless Technologies".

REFERENCES

Agrawal, G. P., & Dutta, N. K. (1993). *Semiconductor Lasers*. Springer.

Ahmed, G. H., & Ali, O. (2007). Design and conception of optical links simulator for telecommunication applications under Simulink environment. *WSEAS International Conference on Applied Electromagnetics, Wireless and Optical Communications*, 52-57.

Ahmed, M., Yamada, M., & Saito, M. (2001). Numerical modeling of intensity and phase noise in semiconductor lasers. *IEEE Journal of Quantum Electronics*, *37*(12), 1600–1610. doi:10.1109/3.970907

Bao, H., & Shieh, W. (2007). Transmission simulation of coherent optical OFDM signals in WDM systems. *Optics Express*, *15*(8), 4410. doi:10.1364/OE.15.004410 PMID:19532688

Basak, A., Talukder, Z., & Islam, R. (2013). Performance analysis and comparison between Coarse WDM and Dense WDM. *Global Journal of Research in Engineering - Electrical and Electronics Engineering*, *13*(6).

Batagejl, B. (2002). Need of knowing non-linear coefficient in optical networks. *WSEAS International Conferences*.

Binh, L.-N. (2010). *Optical fiber communications systems: Theory, practice and Matlab Simulink models*. CRC Press.

Čertík, F., & Róka, R. (2012). The nonlinear FWM effect and its influence on optical signals utilized different modulation techniques in the WDM transmission systems. *Optical Communications Conference*.

Čertík, F., & Róka, R. (2014). Nonlinear SPM and XPM effects and their influence on optical signals utilized different modulation techniques in WDM transmission systems. *Optical Communications Conference*.

Cisco. (2006). *ONS 15454 - DWDM Engineering and Planning Guide*. Author.

Čučka, M., Šalík, P., Róka, R., Műnster, P., & Filka, M. (2018). Simulation models of pulse generator for OTDR in Matlab and VPIphotonics. *International Conference on Telecommunications and Signal Processing*. 10.1109/TSP.2018.8441274

Del Río Bellisco, J., Hermosilla, J., Monteroy, R., & Espí, P. (2004). Education tool for optical communication engineering teaching. *WSEAS International Conferences*.

Deshmukh, G., & Jagtap, S. (2013). Four wave mixing in DWDM optical system. *International Journal of Computational Engineering Research*, *3*(6).

Fatadin, I., Ives, D., & Wicks, M. (2006). Numerical simulation of intensity and phase noise from extracted parameters for CW DFB lasers. *IEEE Journal of Quantum Electronics*, *42*(9), 934–941. doi:10.1109/JQE.2006.880117

Gao, J., Li, X., Flucke, J., & Boeck, G. (2004). Direct parameter-extraction method for laser diode rate-equation model. *Journal of Lightwave Technology*, *22*(6), 1604–1609. doi:10.1109/JLT.2004.829211

Kibria, R., & Austin, M. W. (2015). All optical signal-processing techniques utilizing four wave mixing. *Photonics*, *2*(1), 200–213. doi:10.3390/photonics2010200

Mokráň, M., Róka, R., & Čertík, F. (2015). Analysis of possible utilization of phase shift keying modulations for long-haul optical transmission systems. Applied Physics of Condensed Matter.

Mukherjee, B. (2006). *Optical WDM networks*. Springer Science Business Media Inc.

Róka, R. (2012). Fixed transmission media. In *Technology and engineering applications of Simulink*. InTech. doi:10.5772/37442

Róka, R. (2015a). The simulation of negative influences in the environment of fixed transmission media. *International Conference on Applied Physics, Simulation and Computers*, 58-68.

Róka, R. (2015b). The environment of fixed transmission media and their negative influences in the simulation. *International Journal of Mathematics and Computers in Simulation*, *9*, 190–205.

Róka, R., & Čertík, F. (2012). Modeling of environmental influences at the signal transmission in the optical transmission medium. *International Journal of Communication Networks and Information Security*, *4*(3), 146–162.

Róka, R., & Čertík, F. (2015a). Simulation tools for broadband passive optical networks. In Simulation technologies in networking and communications: Selecting the best tool for the test. CRC Press, Taylor and Francis Group.

Róka, R., & Čertík, F. (2015b). Modeling of negative influences at the signal transmission in the optical medium. Applied Physics of Condensed Matter.

Róka, R., & Čertík, F. (2015c). Simulation and analysis of the signal transmission in the optical transmission medium. *International Conference on Simulation and Modeling Methodologies, Technologies and Application*, 219-226. 10.5220/0005569602190226

Róka, R., & Mokráň, M. (2015). Modeling of the PSK utilization at the signal transmission in the optical transmission medium. *International Journal of Communication Networks and Information Security, 7*(3), 187–196.

Róka, R., & Mokráň, M. (2018). Effect of the FWM influence on the CWDM signal transmission in the optical transmission media. *International Journal of Circuits. Systems and Signal Processing, 12*, 42–47.

Róka, R., & Mokráň, M. (2019). Simulation platform for analyzing environmental effects in multi-wavelength transmission systems in Matlab Simulink. *Przeglad elektrotechniczny, 95*(6), 9-14.

Róka, R., Mokráň, M., & Šalík, P. (2017). Simulation of negative influences on the CWDM signal transmission in the optical transmission media. *International Journal of Circuits. Systems and Signal Processing, 11*, 75–80.

Róka, R., & Šalík, P. (2017). Impact of environmental influences on multilevel modulation formats at the signal transmission in the optical transmission medium. *International Journal of Communication Networks and Information Security, 9*(1), 76–87.

Róka, R., & Šalík, P. (2020). An effective integration of the PM 16-QAM modulation in enhanced metropolitan networks with the EDFA amplification. *Journal of Electrical Engineering - EČ, 71*(5), 317-325.

Šalík, P., & Róka, R. (2017). Analysis of possibilities for numerical simulations of continues wave DFB laser. *International Congress on Ultra-Modern Telecommunications and Control Systems.*

Šalík, P., & Róka, R. (2018). Analysis and simulation of dynamic properties for the DFB laser. *Przeglad elektrotechniczny, 94*(7), 17-20.

Šalík, P., Róka, R., & Tomáš, G. (2018). Simulation platform of optical transmission system in Matlab Simulink. *Procedia Computer Science, 134*, 196–203. doi:10.1016/j.procs.2018.07.162

Sharma, V., & Kaur, D. (2015). Review on multiplexing techniques in optical communication systems. *European Scientific Journal*, 2.

ITU-T Telecommunication Standardization Sector. (2003a). *Recommendation G.694.1 - Spectral grids for WDM applications: DWDM wavelength grid.*

ITU-T Telecommunication Standardization Sector. (2003b). *Recommendation G.694.2 - Spectral grids for WDM applications: CWDM wavelength grid.*

ITU-T Telecommunication Standardization Sector. (2009a).: *Recommendation G.652- Characteristics of a single-mode optical fiber and cable.*

ITU-T Telecommunication Standardization Sector. (2009b). *Recommendation G.657 - Characteristics of a bending-loss insensitive single-mode optical fiber and cable for the access network.*

ITU-T Telecommunication Standardization Sector. (2010). *Recommendation G.656 - Characteristics of a fibre and cable with non-zero dispersion for wideband optical transport.*

Tomita, A., & Suzuki, A. (1991). A new density matrix theory for semiconductor lasers, including non- Markovian intraband relaxation and its application to nonlinear gain. *IEEE Journal of Quantum Electronics, 27*(6), 1630–1641. doi:10.1109/3.89987

Section 2

Computing and Computer Science

Chapter 15
Exploiting Patterns of Object Use for Human Activity Recognition

Isibor Ihianle
https://orcid.org/0000-0001-7445-8573
Nottingham Trent University, UK

Syed Islam
University of East London, UK

Usman Naeem
Queen Mary University, UK

Solomon Henry Ebenuwa
University of East London, UK

ABSTRACT

The accurate recognition of activities of daily living (ADL) is fundamental in the support and provision of assistance for the elderly and cognitively impaired. Current ontology-based techniques and knowledge-driven model object concepts form assumptions and everyday knowledge of objects used for activities. Activities modelled from assumptions and everyday knowledge can lead to incorrect recognition results of routine activities and possible failure to detect abnormal activity trends. A significant step to the accurate recognition of activities of daily living is the discovery of the object usage for specific routine activities. This chapter presents an approach that discovers object usage for routine activities using latent Dirichlet allocation (LDA) topic modelling. The object usage discovery augments an activity ontology that enables recognition of simple activities of daily living in the home environment. The proposed approach is evaluated and validated using the Kasteren and Ordonez datasets.

DOI: 10.4018/978-1-7998-6992-4.ch015

1 INTRODUCTION

Activity recognition is an emerging area of research in ubiquitous and pervasive computing with significant importance to the provision of support and assistance to the elderly, disabled and cognitively impaired. It is involved in the detection, identifying and recognition of what an individual is doing, e.g. sleeping, showering, making food. Recent efforts so have focused on the use of video (Akdemir, Turaga, & Chellappa, 2008; Leo, Medioni, Trivedi, Kanade, & Farinella, 2017), wearable sensors (D. Huynh, 2008; Labrador & Yejas, 2013) and wireless sensor net works (Ordonez, Toledo, & Sanchis, 2013; Kasteren, Englebienne, & Krose, 2010; I. K. Ihianle, Naeem, & S. Islam, 2018; I. Ihianle et al., 2020) to monitor simple human activities. Video based activity recognition captures and monitors images which are segmented and then classified using context-based analysis. Sensor based activity recognition use sensors to monitor object usage from the interactions of users. Unlike sensor-based activity recognition, video-based approaches suffer from accurate segmentation of captured images which affects classification process and

also raises privacy concerns. The process of recognising activities via sensor-based activity recognition can be data driven, knowledge driven or a combination of both techniques. Data-driven techniques employ machine learning and statistical methods to discover the data patterns for activities to be inferred. Although data driven techniques have been used by the authors in (Ordonez et al., 2013; Kasteren et al., 2010; Chen, Nugent, & Wang, 2011) demonstrating its strengths in the recognition of activities, inferences in most cases for data driven approaches are hidden and or latent. Recognised activities are however needed to be expressed in an understandable format for the end user. Data driven techniques, in addition, are unable to integrate context aware features to enhance activity recognition.

Conversely, ontology based and knowledge driven techniques model activities as concepts. These concepts are associated to the everyday knowledge of object usage in the home environment through a knowledge engineering process. The modelling process involves associating low level sensor data to the relevant activity to build a knowledge base of activities in relation to sensors and object. The recognition of activities is achieved by logical inference and or inclusion of subsumption reasoning. Unlike data driven techniques, knowledge driven techniques are expressively clear and activity recognition results are in the format easily understood by the end user (Chen, Nugent, Hoey, Cook, & Yu, 2012). Modelling activity ontology concepts for Knowledge driven techniques largely depend on everyday knowledge of activities and object use to build and construct activity ontologies. Knowledge of object use are generic and mostly by assumptions regular everyday knowledge of the objects are used for routine activities or even wikiknowhow[1] (Okeyo, Chen, Wang, & Sterritt, 2012a).

This paper follows a sensor-based approach to capture object use as the result of the interactions of objects in the home environment. These sensor-based object use and interactions are regarded as atomic events leading to activities. Activity situations should be a result of a set of observed objects. There may be cases of same object sets resulting in different activities, making the process of identifying activity from object use complex. This paper proposes an approach which follows object use as events and entries to recognise activities. This is in line with what is obtainable in real world situations and most home environments. Interactions of objects as object use in the home environment result in activities. These objects use and interactions as atomic events result in particular routine activities. This paper's motivation stems from the premise that knowledge driven activity recognition constructed from the everyday knowledge of object use may not fit into certain activity situations or capture routine activities in specific home environments. If an activity model has been developed based on the generic and or assumed knowledge of object use, the recognition model may fail due to activities and objects fittings which differ with individuals and home environments. Generic ontology models have been designed and developed as in Chen et al. (Chen et al., 2011) to emphasise reusability and shareability, however, they share accuracy.

As a way forward, it is essential to accommodate object concepts that are specific to the routine activities with regards to the individual and the contextual environment. To provide assistance and support to the elderly and cognitively impaired, the recognition of their ADLs must be accurate and precise with regards to the object use events. In this context, the major aim of this paper is to present an approach for activity recognition by extending knowledge driven activity recognition techniques to include a process of acquiring knowledge of object use to describe contexts of activity situations. This approach also addresses the problem of object use specification for routine activities, especially where the object use has not been predefined. So, the challenge becomes discovering specific object use for particular routine activities as a way to emphasise the object set for the routine activities. This paper focuses on the process of the recognition of ADL as a step towards the provision support and assistance to the elderly disabled and cognitively impaired. It also emphasises on knowledge driven techniques and shows that the activity discovery process augments the activity ontology, thus extending it beyond the use of generic and common knowledge of object use to build activity ontology. Given these, the work in this paper harnesses the complementary strengths of data and the knowledge driven techniques to provide solutions to the limitations and challenges highlighted above.

In particular, this paper makes the following contributions:

- Ability to deal with streaming data and segmentation using a novel time-based windowing technique to augment and include temporal properties to sensor data.
- Automated activity object use discovery in context for the likely objects use for specific routine activities. We use Latent Dirichlet Allocation (LDA) topic modelling through activity object use discovery to acquire knowledge for concepts formation as part of an ontology knowledge acquisition and learning system.
- Extend the traditional activity ontology to include the knowledge concepts acquired from the activity object use discovery and context description which is vital for recognition especially where object use for routine activities have not been predefined.
- The evaluation and validation of the proposed approach to show that it outperforms the current stateoftheart knowledge-based recognition techniques.

The remainder of the paper is organized as follows. Section 2 provides an overview of the related works, while Section 3 describes the proposed activity recognition approach. Section 4 provides experimental results for Kasteren and Ordonez datasets to validate the proposed approach and finally, Section 5 concludes this paper.

2 RELATED WORK

Activity recognition approaches can be classified in two broad categories data and knowledge driven approaches. This classification is based on the methodologies adopted, how activities are modelled and represented in the recognition process. Data driven approaches can be generative or discriminative. According to (Chen et al., 2012), the generative approach builds a complete description of the input (data) space. The resulting model induces a classification boundary which can be applied to classify observations during inference. The classification boundary is implicit, and a lot of activity data is required to induce it. Generative classification models include Dynamic Bayes Networks (DBN) (Modayil, Bai, & Kautz, 2008; Tapia et al., 2007) Hidden Markov Model (HMM) (Patterson, Liao, Fox, & Kautz, 2003; Patterson, Kautz, & Philipose, 2005), Naive Bayes (NB) (Langley, Iba, & Thompson, 1992; Tapia, Intille, & Larson, 2004), Topic model Latent Dirichlet Allocation (LDA) (Katayoun & GaticaPerez, 2011; T. Huynh, Fritz, & Schiele, 2008). Discriminative models, as opposed to generative models, do not allow generating samples from the

joint distribution of the models (Chen et al., 2012). Discriminative classification models include nearest neighbour (T. Huynh, Blanke, & Schiele, 2007; Lee & Mase, 2002), decision trees (Bao & Intille, 2004), support vector machines (SVMs) (Cao, Masoud, Boley, & Papanikolopoulos, 2009; Qian, Mao, Xiang, & Wang, 2010), conditional random fields (CRF) (Liao, Fox, & Kautz, 2007), multiple eigenspaces (T. Huynh & Schiele, 2006), and kmeans (T. Huynh et al., 2007). Data driven approaches have the advantage of handling incomplete data and managing noisy data. A major drawback associated with data driven approaches is that they lack the expressiveness to represent activities.

Topic modelling inspired by the text and natural language processing community have been applied to discover and recognise human activity routines by (Katayoun and GaticaPerez, 2011) and Huynh et al (D. Huynh, 2008). Huynh et al (D. Huynh, 2008) applied the *bag of words* model of the Latent Dirichlet Allocation (LDA) to discover activities like dinner, commuting, office work etc. Whilst (Katayoun and GaticaPerez, 2011) discovered activity routines from mobile phone data, Huynh et al (D. Huynh, 2008) used wearable sensors attached to the body parts of the user. Activities discovered in both works were latent, lacked expressiveness and minimal opportunities to integrate context rich features. The approach presented in the paper also significantly differs from (Katayoun and GaticaPerez, 2011) and Huynh et al (D. Huynh, 2008) with the modelling using an ontology activity model.

Knowledge driven ontology models follow web ontology language (OWL) theories for the specification of conceptual structures and their relationships. Ontology based activity recognition is a key area within knowledge driven approaches. It involves the use of ontological activity modelling and representation to support activity recognition and assistance. Ontology uses the formal and explicit specification of a shared conceptualization of a problem domain (Gruber, 1993). Vocabulary for modelling a domain is provided by specifying the objects and concepts, properties, and relationships. This then uses domain and prior knowledge to predefine activity models to define activity ontologies (Chen et al., 2012). Chen et al. (Chen et al., 2012; Chen & Nugent, 2009) proposed an ontology-based approach to activity recognition in which they constructed context and activity ontologies for explicit domain modelling. Sensor activations over a period of time are mapped to individual contextual information and then fused to build a context at any specific time point. Subsumption reasoning were used to classify the activity ontologies, thus inferring the ongoing activity. Knowledge driven ontology models also follow web ontology language (OWL) theories for the specification of conceptual structures and their relationships (Yamada et al., 2007). OWL has been widely used for modelling human activities for recognition, which most of the time involves the description of activities by their specifications using their object and data properties (Chen & Nugent, 2009). In ontology modelling, domain knowledge is required to encode activity scenarios, but it also allows the use of assumptions and common sense domain knowledge to build the activity scenarios that describe the conditions that drive the derivation of the activities (Okeyo, Chen, Wang, & Sterritt, 2012b). Recognising the activity then requires the modelled data to be fed to the ontology reasoner for classification. The authors of (Chen et al., 2012) and (Okeyo et al., 2012b) followed generic activity knowledge to develop an ontology model for the smart home users. Whilst these approaches to model activities depending on common sense domain knowledge and its associated heuristics, which is commendable however they often lead to incorrect activity recognition due to lack of specificity of the object use and contexts describing the activity situations. The problem arises to the inability of the generic models to capture specific considerations to object use for routine activities, home settings and

Figure 1. An Overview of the Activity Recognition Approach.

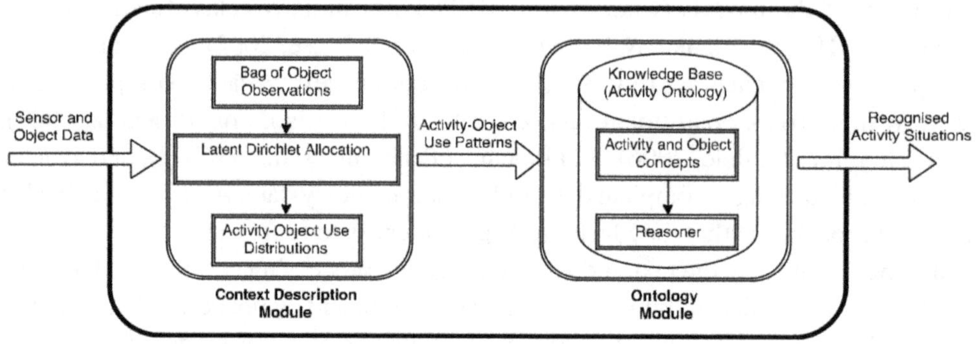

individual object usage between differing users. They also do not follow evidenced patterns of object usage and activity evolution as they rely on generic know-hows and hows-to to build ontology models. In view of these limitations, an LDA enabled activity discovery technique as applied to discover likely object use for specific routine to augment the generic ontology modelling pro cess. The approach presented in this paper, also extends the works in (I. Ihianle, Naeem, & Tawil, 2015, 2016) by the inclusion of the ontology activity model.

3 OVERVIEW OF OUR ACTIVITY RECOGNITION APPROACH

The proposed approach (Figure 1) supports objects as contexts descriptors of activity situations by activity object use discovery and information fusion of activity and object concepts through activity ontology design, development and modelling to achieve activity recognition.

This proposed approach is composed of two component modules the context description module and the ontology module. The object use for the specific routine activities are discovered using the LDA topic model as activity object use distributions in the context description module. In this paper, we refer the object use for the specific routine activities as the contexts describing the activity situation. These context descriptors discovered for the routine activities then form concepts to be modelled and encoded in an activity ontology which the ontology module is com prised of. Activities are recognised by mapping observed objects on the activity ontology using an object query to retrieve the relevant activity situation. The two component modules context description and ontology function uniformly to provide a seamless activity recognition progressively with observed sensors/objects representing atomic events of activities as inputs. These component modules are further described in the subsections below.

3.1 Context Description Module

The context description module augments the activity ontology for the specification of the object set used for the routine activity very much unlike the traditional knowledge driven activity recognition approaches (Chen et al., 2011; Chen & Nugent, 2009; Riboni, Pareschi, Radaelli, & Bettin 2011). Its function is the discovery of the activity object use distributions which forms the knowledge of object use for specific routine activities and context descriptors. To provide the basis for an activity recognition, the knowledge of object use for respective activity concepts are required. The necessity of these object use knowledge is such that activities as high-level events are a result of low-level tasks or atomic events of object interactions. Traditional ontology driven and knowledge driven activity recognition approaches (Chen et al., 2011; Okeyo et al., 2012a) model activity ontologies from assumptions of object use and the knowledge of the everyday use of objects in the home environment. However, activity ontologies modelled traditionally from generic knowledge and assumptions may not fit into every home setting or environment and this may lead to erroneous object descriptions of low-level tasks and eventually incorrect activity recognition. An accurate process of determining the object set or context descriptors which describe activities specifically as a higher level of events must be employed to design and model activity situations ontologically. The context description module functions by discovering the activity object use discovery and activity context description based on the activity object use discovery enabled a topic model process. The modular process is described below:

3.1.1 Activity Object use Discovery by Latent Dirichlet Allocation

The activity object use discovery uses the LDA approach introduced by Blei et al (Blei, Ng, & Jordan, 2003). The LDA generatively classifies a corpus of documents as a multinomial distribution of latent topics. It takes advantage of the assumption that there are hidden themes or latent topics which have associations with the words contained in a corpus of documents. It then requires the *bag of words* (documents) from a corpus of documents as input and number of topics as a key parameter. In the context of the activity object use discovery process, the activity topic number and *bag of sensors* observations corresponds to topic number and *bag of words* respectively of the LDA. The activity topic number and *bag of sensors* observations are explained below.

- **Activity Topic Number:** A key parameter needed by the LDA process is the topic number. In the context of the activity recognition proposed in this paper, the number of activities corresponds to the topic number. The activities to be recognised adds up to the activity topic number. For experimental purposes, this paper uses the number of activities specified in the dataset.
- **Bag of Sensors Observations:** The *bag of objects* observation is analogous to the *bag of words* used in the LDA text and document analysis. In text and document analysis, a document (bag) in a corpus of texts can be represented as a set of words with their associated frequencies independent

of their order of occurrence (Griffths & Steyvers, 2004). This paper follows the LDA *bag of words* approach to representing discrete observations of objects or sensors of specific time windows generated as events in the use or interaction of home objects. Observed objects in the time windows forms object segments which correspond to the LDA documents (bags). In this regard, it is referred to as the *bag of objects* observations. To achieve bagging of the objects accordingly, the stream of observed sensors or objects data are partitioned into segments of suitable time intervals. The objects and the partitioned segments then respectively correspond to the words and documents of the *bag of words*. If a dataset is given by D made of $x_1...x_n$ objects, D can be partitioned

using suitable sliding time window intervals into $d_1...d_D$ segments similar to the schema 1.

$$D = \begin{bmatrix} \{x_1 \ldots, x_{n1}\} objects\ in\ segment\ d_1 \\ \{x_1 \ldots, x_{n2}\} objects\ in\ segment\ d_2 \\ \ldots.. \\ \{x_{N1} \ldots, x_N\} objects\ in\ segment\ d_N \end{bmatrix} \tag{1}$$

The observed objects $x_1...x_n$ in each of the segments $d_1...d_D$ are then represented with their associated frequencies f to form a segmentobjectfrequency matrix similar to the schema given in 2.

$$Bag\ of\ Sensor\ Observation = \begin{bmatrix} d_1\ x_1\ f_1 \\ d_1\ x_1\ f_1 \\ \ldots.. \end{bmatrix} \tag{2}$$

The LDA takes advantage of the assumption that there are hidden themes or latent topics which have associations with the words contained in a corpus of documents. It also involves the use of *bag of words* in the corpus of documents which are generatively classified to latent themes or topics and word distributions. We conversely apply this assumption to the activity object pattern discovery context that latent activity topics would have associations with the features of object data in the partitioned segments of the *bag of sensors* observations discussed above. The documents are presented in the form of objects segments $d_1...d_D$ composed of cooccurring object data observations. With D composed of object segments $d_1...d_D$ similar to schema 1 above, d_i would be made of objects represented as $x_{i1} ...x_{in}$ from X objects of $x_1...x_n$. The LDA places a dirichlet prior $P(\theta_d|\alpha)$ with parameter α on the document topic distributions $P(z|\theta_d)$. It assumes a Dirichlet prior distribution on the topic mixture parameters θ and ϕ, to provide a complete generative model for documents D. θ describes D x Z matrix of document specific mixture weights for the Z topics, each drawn from a Dirichlet(α) prior, with hyperparameter α. ϕ is an X x Z matrix of word specific mixture weights over X objects for the Z activity topics, drawn from β which is a Dirichlet prior. The probability of a corpus or observed object dataset, is equivalent to finding parameter α for the Dirichlet distribution and parameter β for the topicword distributions $P(x|z, \beta)$ that maximize the likelihood L of the data for documents $d = d_1...d_D$ as given in the equation 3 following Gibbs sampling for LDA parameter estimation (Blei et al., 2003).

$$L(\alpha,\beta) = \prod_{d=1}^{D} \int P(\theta_d|\alpha) \left(\prod_{x=1}^{X_d} \sum_{z=1}^{Z} P(X_{xi}^d|z,\beta) P(z|\theta_d) \right) d\theta_d \tag{3}$$

Given the assumptions above, the activity object patterns can be calculated from the equation 4 below:

$$P(x|z,\phi) = \prod_{z=1}^{Z} \prod_{x_i=1}^{X} \left(\phi_z^{x_i}\right)^{n_z^{x_i}} \tag{4}$$

where $n^x_z i$ is the number of times an object x_i is assigned to a topic z.

In the context of the activity recognition, the context descriptors to be modelled as ontology concepts would rely on the activity object use distributions. Eq 4 above computes the activity object use distributions that are linked to specific activity topics. The activity object use distributions given the specific activity topics form the object use and the respective activity context descriptions. These context descriptors are then modelled and encoded as object concepts and the respective activity concepts in the activity ontology model. Given the number of activities in a dataset D as activity topic number and the constructed *bag of objects* observations as inputs, the activity object use distributions determined could be expressed similarly to the schema 5. The process continues by annotating and linking of the objects to the specific activities using the activity object use distribution as the contexts describing the activities.

$$P(x|z) = \begin{bmatrix} \{x_{k1a}, x_{k1b}, \ldots, x_{k1n}\}\ Assigned\ to\ k_1 \\ \{x_{k2a}, x_{k2b}, \ldots, x_{k2n}\}\ Assigned\ to\ k_2 \\ \{x_{Ka}, x_{Kb}, \ldots, x_{KN}\}\ Assigned\ to\ K \end{bmatrix} \tag{5}$$

3.2 Ontology Module

The ontology module is comprised of the knowledge base which is made of activity ontology concepts, data and encoded rules to support activity recognition. The knowledge base also functions as an ontology repository of modelled concepts organized, shared and can be searched. The activities and the objects concepts from the context description module are modelled and encoded following Description Logic (DL), knowledge representation and formalism to populate the knowledge base. This process is such that the encoded concepts in the knowledge forms the TBox (Terminological Box), ABox (Assertional Box) and linked by properties. The terminological box is a collection of ontology concepts modelled as class concepts, and in this case, the different activities and objects. The class concepts could also be modelled hierarchically to reflect the relation between them given their properties and attributes. The class concepts and instances carry with evidence of semantics which does not only adds sense to the concepts and how they are related but also enhances their expressiveness and clarity. The activity ontology development process gradually populates the TBox by encoding the activities and context descriptors from the context description module as ontology concepts. The assertional box is the collection of the ontology concepts as instances and individuals. They are asserted through properties that may be object or data properties. For the class concepts in the TBox, instances and individuals of these are asserted through different properties to populate our ABox. In addition to the activities and context descriptors (objects) class concepts and instances, this paper introduces the use of temporal class concepts and instantiated them following the 4D fluent approach for a realistic reflection activity evolution and transition. The 4D fluent approach allows ontology concepts to be associated with time as an attribute. In addition, the activity ontology is further extended by modelling activities as either static or dynamic activities using their temporal properties as the basis. The resultant activity ontology is created with the fusion of likely object use and behavioural information from the activity context descriptions making it possible for activity inferencing. Also, in the ontology module is the reasoner which checks the relationships between the concepts in the TBox and the consistencies in the ABox for the individuals and instances to perform activity recognition by information retrieval. The eventual results of the information retrieval are the activities or activities situations.

Table 1. Ontology concepts used

Ontology Concepts	Descriptions
ADL	Activity of Daily Living and super activity of all activity concepts
Activity or Activity situation	Concepts representing type of ADL examples include *Breakfast, Dinner, Drink, Toileting* etc.
Objects or Resources	Concepts representing concepts of object used
	in the home environment examples include *Microwave, Plates Cupboard, Fridge, Cup* etc.
TimeSlice	Time slice of an object
TimeInterval	Time interval of an activity
Ontology Properties	
hasUse	An object property used to associate usage of an
	object for an activity
hasStartTime	A data property used to associate time reference when an object was observed
tsTimeSliceOf	Fluent property for time slice of an object.
tsTimeIntervalOf	Fluent property for the interval describing range
	of time an activity.

Table 2. Ontology notations used

Notations	Descriptions
$\sqsubseteq$	subclass of
$\exists$	Property of
$\sqcap$	Intersection
$\sqcup$	Union
$\equiv$	is an equivalent class of
$\rightarrow$	Implication

This paper models activity concepts from a set of objects and context descriptors by asserting all the objects concepts and with their times to be encoded to represent the activity situation (See Tables 1 and 2 for a list of concepts and notations used). If an activity situation *Breakfast* is the result of *Microwave_On* and *Fridge_On* at times *t1* and *t2* respectively, then *Breakfast* can be asserted with the properties hasUse and hasStartTime as hasUse(*Microwave_On*, hasStartTime *t1*), hasUse(*Fridge_On*, hasStartTime *t2*). The activity is therefore modelled as a list of the objects and with their times ordered temporally. The example of *Breakfast* from *Microwave_On* and *Fridge_On* at *t1* and *t2* can then be encoded by the expression 6 below.

$$Breakfast \subseteq ADL \sqcap \exists hasUse\left(\left(Microwave\_On \sqcap \exists hasStartTime\left(=, t1\right)\right) \sqcap \exists\left(Fridge\_On \sqcap \exists hasStartTime\left(=, t1\right)\right)\right) \quad (6)$$

Typically, the activities performed in the home environment are a result of specific objects use. To model activities accurately, it is important to extend the traditional activity ontology modelling to include specific resources and the contexts describing the routine activity. The activity context descriptors resulting from the activity object use discovery process forms the resources and the context concepts to be modelled in the activity ontology for the specific routine activities such that:

Activities: The activity topics are annotated as activity concepts analogous to the activity situations in the home environment. This represents a class collection of all types of activities set as $z_1 \ldots z_k$.

Objects: These represents class collection of all objects as activity context descriptors in the home environment set as $x_1 \ldots x_n$.

Recall P(x|z) computes the likely objects use for each of the activity topics, then z_i from $z_1 \ldots z_k$ would have a subset of likely objects defined as $x_{1z} \ldots x_{iz}$. The function f maps the activity context descriptors $x_{1z} \ldots x_{iz}$ objects for activity z_i as the likely objects use for this activity such that:

$$f : z_i \rightarrow x_{1z} \ldots x_{iz} \quad (7)$$

3.2.1 Static and Dynamic Activities

Activities are performed differently, in different ways and times within the 24-hour day path in the home environment. Some of these activities can be performed at specific times or time intervals of the day making them have static times (Static activities). On the other hand, dynamic activities do not have specific times or time interval they are performed. Also, activities in some cases may have same or share similar object usage. This paper makes analogy using *Breakfast*, *Lunch*,

Dinner, *Toileting* and *Showering* as examples of activities in the home environment. *Breakfast*, *Lunch* and *Dinner* are examples of different activity concepts which can be performed with same or share similar object usage given that they are food related activities. In contrast, they are performed within a specific time interval of the day making them static activities.

Considering they share the same or have similar object usage, they can be modelled as sub classes of the activity *Make Food*, however, they differ with regards to their respective temporal properties. Whilst they inherit all the properties of *Make Food* by subsumption, they can be easily confused in the recognition process if modelled without the specification temporal properties. Dynamic activities like *Toileting* and *Showering* do not have specific times and intervals they are performed making them less dependent on their temporal properties. So, dynamic activities are not constrained within any time interval. Applying the 4Dfluent approach (Baratis, Petrakis, Batsakis, Maris, & Papadakis, 2009), this paper extends the traditional activity ontology to include static and dynamic activities as class concepts requiring the temporal class concepts *Timeslice* and *TimeInterval* to be specified using the relational properties tsTimesliceOf and tsTimeIntervalOf respectively. The time intervals *Interval1* and *Interval2* holds the temporal information of the time slices for the static and dynamic activities respectively. An instance of a *TimeSlice* of an activity whether static or dynamic is linked by the property tsTimeSliceOf and property tsTimeInterval which then links this instance of *TimeSlice* with an instance of the class *TimeInterval*.

Modelling Static Activities: Ontology modelling of static activities require that the activity concepts are encoded with the specification of their temporal attribute. In this case, the temporal attribute specification of the times and the time interval they are performed can be encoded by using the class concepts *TimeSlice*, *TimeInterval* and with the context descriptors for that activ ity. The hasUse object property is also used to encode the object usage for an activity and then specifying static activity as the domain class concept as well as ranging to all the object classes which describe the context descriptors. The tsTimeInterval property of a static activity is also encoded so that its domain are *TimeSlice* and *Resources* and it ranges to the class *TimeInterval* to capture specific time interval of the day through *Interval* (a sub class of *TimeInterval*). Further, the time instants of the static activity are captured through the tsTimeSliceOf with domain *TimeSlice* and *Resources* and it ranges to *TimeSlice*. The expression 11 encodes a static activity so that *Interval1* asserts the time interval of the day the static activity is performed using the object x_{1z}.

$$z_i hasUse\, x_{1z} \ldots x_{iz} \tag{8}$$

Modelling a Dynamic Activity: Dynamic activities are also modelled in the similar manner as the static activities. An instance of a *TimeSlice* of a dynamic activity is linked by the property tsTimeSliceOf and property tsTimeInterval and then links this instance of the class *TimeSlice* with an instance of class *TimeInterval* which may be *Interval2*. Unlike the *Interval1* for the static

activities, *Interval2* of the dynamic activities ranges to cover the full 24-hour cycle of the day as asserted by expression 13.

$$z_i hasUse\left\{\left(x_{1z}\_On, hasStartTime \ldots t_{1z}\right)\ldots\left(x_{iz}\_On, hasStartTime\, t_{i-}\right)\right\} \quad (9)$$

3.3 Activity Recognition by Object Use Query

Activity recognition is facilitated by the Algorithm 1 and achieved by an observed object use query by mapping this query on to the activity ontology to retrieve to closest activity situation from the activity ontology. A typical object use query uses constructs adapted from the Temporal Ontology Querying Language (TOQL) (Baratis et al., 2009) on the knowledge base to retrieve activity situations fitting the requirements of the query. As an advantage, sensor states and status of object use as implemented in the activity ontology can be used in queries to reflect real situations of object usage in the home environment. A typical query is comprised of SQL like construct (SELECT FROM WHERE) for OWL which treats the ontology classes and properties like database tables and columns. An additional AT construct in the query compares the time interval for which a property is true with a time interval or instant. The inputs are observed sensors or objects along their timelines as $x_1,\ldots x_n$ from segments $d_1\ldots d_D$. The process maps the sensors or objects x_i on to the ontology and by reasoning and inference rules to determine that x_i is a context descriptor for a static or dynamic activity. If an activity is returned, then it is reported as the activity recognised for the observed object x_i.

4 EXPERIMENTS AND RESULTS

The approach presented in this paper is validated using the Kasteren A (Kasteren et al., 2010) and Ordonez A (Ordonez et al., 2013) datasets captured in two different home settings with similar events and activities (see tables 3 and 3 for an overview of the home setting descriptions and activity instances). The choice of these datasets was due to the fact that the Kasteren and Ordonez dataset contains a lot of sensor activations as object use with dense sensing applied. Multiple sensor types (e.g. pressure sensors, magnetic sensors tagged to home objects like microwave, dishes, cups) were used to capture object interactions for different activities. Also, the activities performed in varied ways and accurately annotated with the ground truth. Experiments were performed on the datasets following steps mentioned in the sections above and then validated using a 4-fold cross validation on the datasets. The criterion for evaluation is to compare recognised activities with the ground truth provided with the dataset based on the average true positives TP, false positives FP and false negatives FN per activity. The results are further evaluated based on the metrics precision, recall and FScore.

Table 3. Home Setting and Description

	Kasteren A	Ordonez A
Setting	Apartment	Apartment
Rooms	3	4
Duration	22 Days	14 Days
Sensors	14	12
Activities	Kasteren A	Ordonez A
Sleeping	25	14
Toileting	114	44
Leaving	36	14
Showering	24	14
Grooming	Na	51
Breakfast	20	14
Lunch	Na	9
Dinner	10	Na
Drink	20	Na
Snack	Na	11
Spare Time	Na	11

Algorithm 1: Activity recognition algorithm.

```
Input: Observed Sensors X = {x1,...xn} in Partition of Sensor Segment D =
{d1...dD}, ADL ontology (ADL)
Result: Static Activity (SA) or Dynamic Activity (DA).
Begin:;
while data stream is active do
Extract observed object xi ∈ X = {x1... xn}, from di ∈ d1...dD for each di ∈ D.
Create Activities Z ≡ Z(x1) ⊔...Z(xn)
for each x ∈ A do;
Map xi to a Static Activity (SA) or Dynamic Activity (DA)
So that;
(        )
Z ≡ ADL ⊓ ∃ hasUse.  xi_On ⊓ ∃ hasStartTime.(=, t)
Activity Inference
if an activity Zi is returned then
Report (SA) or (DA) as recognised.;
for all Repeat process for next x.
```

```
end
end
```

Table 4. Activity Concepts and the discovered context descriptors for Kasterens House A

Activities	Context Descriptors
Leaving	Front Door.
Toileting	Hall Toilet Door, Toilet Flush.
Showering	Hall Bathroom Door.
Sleeping	Hall Bedroom Door.
Make Food	Fridge, Plates Cupboard, Cups Cupboard,
	Groceries Cupboard, Microwave, Freezer.
Make Drink	Fridge.

4.1 Activity Object Use Discovery and Context Description Evaluation

The context descriptors for the routine activities were discovered following the process described in subsection IIIA above. The number of activities as given in the dataset was used as the topic number which is an input requirement for the LDA activity discovery process. The datasets were partitioned using 60 seconds sliding windows to construct the *bag of objects* observations also to be used as input requirement for the LDA activity discovery process. The activity object use distributions are discovered using the input requirements (activity topic number and *bag of objects* observations) and also by setting the Dirichlet hyperparameters α to $50/K$ and β to 0.01 as recommended by Steyvers and Griffiths (Griffths & Steyvers, 2004). The activity context descriptors for the specific routine activities were then annotated and linked from the distributions discovered as given in tables 5 and 6. The idea is that for an object to be a context describing an activity topic, it must have been assigned to an activity topic as in the activity object use distribution discovered.

Table 5. Activity Concepts and the discovered context descriptors for Ordonez House A

Activities	Context Descriptors
Leaving	Main Door.
Toileting	Toilet, Basin.
Showering	Shower.
Sleeping	Bed.
Make Food	Cupboard, Fridge, Microwave, Toaster.
Spare Time	Seat.
Grooming	Basin, Cabinet.

Set 6 activity topics were discovered (from the 7-activity set in the Kasteren A ground truth) in this process and annotated as *Leaving*, *Toileting*, *Showering*, *Sleeping*, *Make Food* and *Drink*. *Make Food* activity in this case represents *Breakfast* and *Dinner* due to same and similar context descriptors or object usage. However, these activities were distinguished modelled as static activities and encoding their temporal attributes. This is similar for Ordonez A with 7 activities annotated as *Leaving*, *Toileting*, *Showering*, *Sleeping*, *Make Food*, *Spare Time* and *Grooming*. Our *Make Food* topic maps to Ordonez House A's activity sets *Breakfast*, *Lunch* and *Snack*.

4.2 Activity Ontology and Recognition Performance

To this point, context descriptors for the activities for Kasteren A and Ordonez A houses have been discovered. The activities and context descriptors are modelled in an activity ontology to

Figure 2. Common object concepts for Kasteren and Ordonez Houses

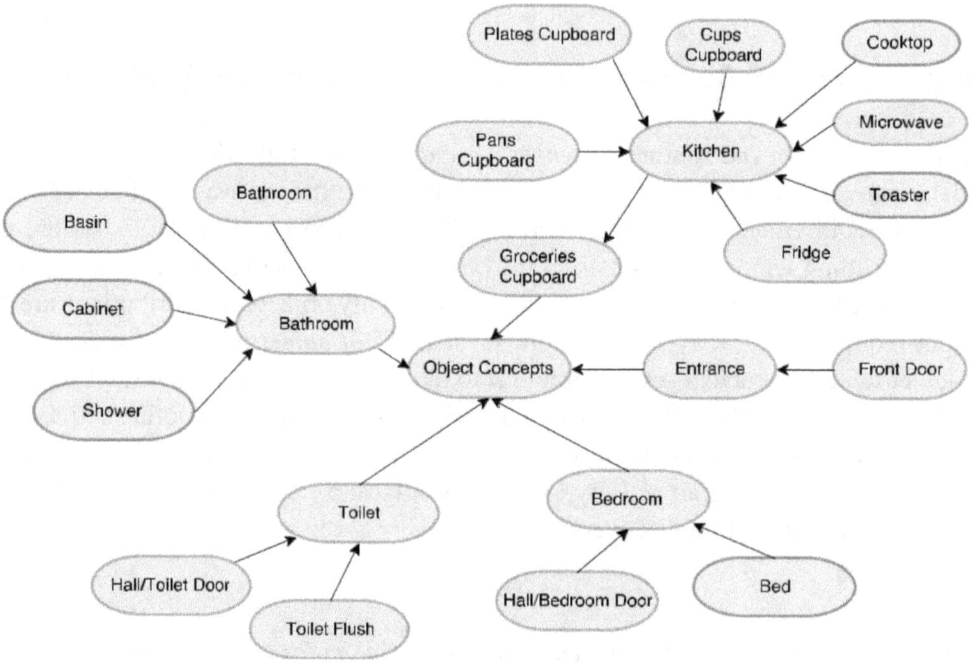

facilitate activity recognition. The activity and object concepts are further modelled to support a unified and reusable activity ontology model of commonly shared concepts for the Kasteren and Ordonez datasets as illustrated in Figure 2. The green coloured rounded rectangles represent common object concepts in both homes. The blue rounded rectangle has been used specifically Ordonez concepts which are not shared in Kasteren concepts. This unified ontology model can also be extended and adapted further for similar homes thus reducing the amount of time taken to construct and develop activity ontologies.

The activity concepts were also modelled accordingly, but due consideration was given to *Make Food* which represented a group of activities. *Make Food* correspond to a group of activities involving making of food and ranges from *Breakfast* and *Dinner*. For Kasteren dataset, *Breakfast* and *Dinner* are modelled as static activities with *Make Food* as superclass. The share same or similar context descriptors and are performed at specific times of the day. To enhance shared ontology concepts and reuse, the Kasteren and Ordonez activity concepts are harmonised to form a set of unified activity concepts as illustrated in Figure 2. The activity concepts have been colour coded so that the green rounded rectangle represents common activity concepts and blue rounded rectangle as activity concepts in the Ordonez house and not in the Kasteren House. In addition to this, instances and individuals have been added to the object concepts making the model more expressive (assertions used in populating the ABox) for example instantiating *Microwave* with *Microwave_On* to suggest the state of the object or sensor when in use. The ABox was further populated with assertions using object and data properties as explained in subsection 3.2 incorporating the context descriptors for the activity situations preparatory for activity recognition. The modelled activity situations or concepts are then linked to their respective context descriptors as object states through the properties as assertions added to the ABox. Activity recognition is achieved using observed sensors or objects use queries to retrieve activity situations as events representing the observed sensors or objects.

Table 6. Activity recognition performance for Kasteren A and Ordonez A

Activities	True Positives (%)		False Positives (%)		False Negatives (%)	
	Kasteren A	Ordonez A	Kasteren A	Ordonez A	Kasteren A	Ordonez A
Leaving	100	100	0.00	0.00	0.00	0.00
Toileting	100	100	0.00	0.00	0.00	0.00
Showering	94.7	63.2	3.6	0.00	1.7	0.00
Grooming	Na	68.7	Na	26.6	Na	4.8
Sleeping	100	100	0.00	0.00	0.00	0.00
Spare Time	Na	100	Na	0.00	Na	0.00
Breakfast	72.5	81.4	17.3	17.5	10.2	1.1
Lunch	Na	61.3	Na	29.5	Na	9.2
Dinner	69.2	Na	23.5	Na	7.3	Na
Drink	64.5	Na	29.8	Na	5.6	Na
Snack	Na	58.7	Na	31.6	Na	9.7

Data Set	Precision	Recall		FScore
Kasteren A	88.60	95.48		91.91
Ordonez A	84.15	94.54		88.73

The experimental results were compared with the ground truth and evaluated for performance. *Leaving Toileting*, *Sleeping* and *Showering* for both datasets were recognised with significantly high results as in Table 7. This performance can be attributed to the discovered object use and context descriptors for these activities. For these activities, the context descriptors were accurately specific with minimal false positives. The process of discovering likely object use for routine activities significantly ensured that these activities were associated with the objects used to perform them. *Breakfast* and *Dinner* for Kasteren House and *Breakfast* and *Lunch* for Ordonez House showed lower performance due to confusions from same and similar object use with *Drink* and *Snack* respectively. These activities have shared same and similar object interactions as observed with context description process hence been classed under the super activity *Make Food*. Recall that to further distinguish *Breakfast* and *Dinner*, they were modelled as subclasses of *Make Food* and as static activities given the specific time of the day they are performed. To enhance their recognition, time interval properties and concepts enabled by 4D fluent approach were included. They were often recognised concurrently and led to high false positives in the process. Overall, the average precision, recall and FScore with the datasets as illustrated in Table 8 show impressive performance.

4.3 Learning Performance of the Approach

Activities can be performed in different ways or object use for activities may differ. An activity recognition approach should be able to recognise activities irrespective of the object use and interactions. This paper evaluates the learning capability at the activity level further using the ground truth as the basis for evaluation. It should be able to return almost the same number

Table 7. Summary of correctly recognised activity instances for Kasteren and Ordonez datasets

Activities	Ground Truth		Instances Recognised		Differences	
	Kasteren A	Ordonez A	Kasteren A	Ordonez A	Kasteren A	Ordonez A
Leaving	36	14	36	14	0	0
Toileting	114	44	114	44	0	0
Showering	24	14	22	9	2	5
Grooming	Na	51	Na	35	Na	16
Sleeping	25	14	25	14	0	0
Spare Time	Na	11	Na	11	Na	0
Breakfast	20	14	14	11	6	3
Lunch	Na	9	Na	6	Na	3
Dinner	10	Na	7	Na	3	Na
Drink	20	Na	13	Na	7	Na
Snack	Na	11	Na	6	Na	5

of activity traces as in the ground truth. The results as presented in Table 9 shows good performance of the learning process. *Leaving*, *Toileting*, *Sleeping* and *Showering* were recognised with significantly high instances and very little difference with the ground truth for both datasets. *Breakfast*, *Lunch* and *Dinner* showed lower performance due to high false positives and confusions from same and similar object use with *Drink*. The number of correctly recognised instances in comparison to the ground truth is also high.

5 SUMMARY AND CONCLUSION

The proposed activity recognition approach provides the basis to learn and recognise activity situations. This paper evaluated the performance of the proposed approach using the results from experiments on two publicly available datasets. Given the performance and evaluations, it can be said that *i)* The discovered activity object use distributions provide accurate context descriptors for routine activities which are then modelled as object concepts and the latter as activity concepts in an activity ontology. *ii)* Extending the activity ontology to include their temporal attributes of the objects and activity concepts as a reflection of activities evolving with time enhanced activity recognition. *iii)* Comparing the performance of the proposed approach with results published using the same datasets show it is significantly good and encouraging. The experimental and evaluation process using these datasets suggests that the features, components and the entire activity recognition process have been fully verified. The future work would involve extending the activity ontology to include more contextual features from multiple sensing devices. It should also consider extending the temporal entities and features for progressive activities as they evolve. In the future, we also intend to conduct a largescale empirical study with other data sets to establish the generality of the results reported in this paper.

REFERENCES

Akdemir, U., Turaga, P., & Chellappa, R. (2008). An ontology-based approach for activity recog nition from video. *Proceedings of 16th ACM international conference on Multimedia (MM, 08).*

Bao, L., & Intille, S. (2004). Activity recognition from user annotated acceleration data. *2nd International Conference Pervasive Computing (PERVASIVE, 2004).*

Baratis, E., Petrakis, E., Batsakis, S., Maris, N., & Papadakis, N. (2009). *Toql: Temporal ontology querying language*. Advances in Spatial and Temporal Databases.

Blei, D., Ng, A., & Jordan, M. (2003). Latent dirichlet allocation. *Machine Learning.*

Cao, D., Masoud, O., Boley, D., & Papanikolopoulos, N. (2009). Human motion recognition using support vector machines. *Computer Vision and Image Understanding*, *113*(10), 113. doi:10.1016/j.cviu.2009.06.002

Chen, L., & Nugent, C. (2009). Ontology based activity recognition in intelligent pervasive environments. *International Journal of Web Information Systems*, *5*(4), 410–430. doi:10.1108/17440080911006199

Chen, L., Nugent, C., Hoey, J., Cook, D., & Yu, Z. (2012). Sensor based activity recognition. *IEEE Transactions on Systems, Man and Cybernetics. Part C, Applications and Reviews*, *42*(6), 790–808. doi:10.1109/TSMCC.2012.2198883

Chen, L., Nugent, C., & Wang, H. (2011). Knowledge driven approach to activity recognition in smart homes. *IEEE Transactions on Knowledge and Data Engineering.*

Griffths, T., & Steyvers, M. (2004). Finding scientific topics. *Proceedings of the National Academy of Sciences.*

Gruber, T. (1993). A translation approach to portable ontology specifications. *Knowledge Acquisition, 5*(2), 5. doi:10.1006/knac.1993.1008

Huynh, D. (2008). *Human activity recognition with wearable sensors* (PhD thesis). Technische Universit Darmstadt.

Huynh, T., Blanke, U., & Schiele, B. (2007). Scalable recognition of daily activities with wearable sensors. *3rd International Conference on Location and Context awareness (LOCA 07).*

Huynh, T., Fritz, M., & Schiele, B. (2008). Discovery of activity patterns using topic models. *Ubiquitous computing (UbiComp 08).*

Huynh, T., & Schiele, B. (2006). Unsupervised discovery of structure in activity data using multiple eigenspaces. *2nd International Workshop on Location and Context Awareness (LoCA, 2006).*

Ihianle, I., Naeem, U., & Tawil, A. (2015). A dynamic segmentation based activity discovery through topic modelling. *IET International Conference on Technologies for Active and Assisted Living (TechAAL 2015).*

Ihianle, I., Naeem, U., & Tawil, A. (2016). Recognition of activities of daily living from topic model. *Procedia Computer Science.*

Ihianle, I., Nwajana, A., Ebenuwa, S. H., Otuka, R. I., Owa, K., & Orisatoki, M. O. (2020). A deep learning approach for human activities recognition from multimodal sensing devices. *IEEE Open Access, 8.*

Ihianle, I. K., Naeem, U., Islam, S., & Tawil, A.-R. (2018). A hybrid approach to recognising activities of daily living from object use in the home environment. *Informatics (MDPI), 5*(1), 5. doi:10.3390/informatics5010006

Kasteren, T., Englebienne, G., & Krose, B. (2010). *Human activity recognition from wireless sensor network data: Benchmark and software.* Atlantis Press Book.

Katayoun, F., & GaticaPerez, D. (2011). Discovering routines from largescale human locations using probabilistic topic models. *ACM Transactions on Intelligent Systems and Technology, 2*(1).

Labrador, M., & Yejas, O. (2013). *An ontology based approach for activity recognition from video. In Human Activity Recognition: Using Wearable Sensors and Smartphones.* CRC Press.

Langley, P., Iba, W., & Thompson, K. (1992). An analysis of bayesian classifiers. *10th National Conference on Artificial Intelligence.*

Lee, S., & Mase, K. (2002). Activity and location recognition using wearable sensor. *IEEE Pervasive Computing.*

Leo, M., Medioni, G., Trivedi, M., Kanade, T., & Farinella, G. (2017). Computer vision for assistive technologies. *Computer Vision and Image Understanding, 154*, 115. doi:10.1016/j.cviu.2016.09.001

Liao, L., Fox, D., & Kautz, H. (2007). Extracting places and activities from gps traces using hierarchical conditional random fields. *The International Journal of Robotics Research, 26*(1), 119–134. doi:10.1177/0278364907073775

Modayil, J., Bai, T., & Kautz, H. (2008). Improving the recognition of interleaved activities. *10th International Conference on Ubiquitous Computing (UbiComp 2008).*

Okeyo, G., Chen, L., Wang, H., & Sterritt, R. (2012a). Dynamic sensor data segmentation for realtime activity recognition. *Pervasive and Mobile Computing.*

Okeyo, G., Chen, L., Wang, H., & Sterritt, R. (2012b). A hybrid ontological and temporal approach for composite activity modelling. *IEEE 11th International Conference on Trust, Security and Privacy in Computing and Communications.*

Ordonez, J., Toledo, P., & Sanchis, A. (2013). *Activity recognition using hybrid genera tive/discriminative models on home environments using binary sensors.* MDPI Sensor Open Access.

Patterson, D., Kautz, D., & Philipose, M. (2005). Fine-grained activity recognition by aggregating abstract object usage. *9th IEEE International Symposium on Wearable Computers.* 10.1109/ISWC.2005.22

Patterson, D., Liao, L., Fox, D., & Kautz, H. (2003). Inferring high level behavior from low level sensors. *5th Conference on Ubiquitous Computing (UbiComp, 03).*

Qian, H., Mao, Y., Xiang, W., & Wang, Z. (2010). Recognition of human activities using svm multiclass classification. *Pattern Recognition Letters, 31*(2), 100–111. doi:10.1016/j.patrec.2009.09.019

Riboni, D., Pareschi, L., Radaelli, L., & Bettini, C. (2011). Is ontology based activity recognition really effective? *IEEE International Conference on Pervasive Computing and Communications Workshops (PERCOM Workshops).* 10.1109/PERCOMW.2011.5766927

Tapia, E., Intille, S., & Larson, K. (2004). Activity recognition in the home using simple and ubiquitous sensor. *2nd International Conference in Pervasive Computing (PERVASIVE 2004).* 10.1007/978-3-540-24646-6_10

Tapia, E., Intille, S., Larson, K., Haskell, W., Wright, J. A. K., & Friedman, R. (2007). Realtime recognition of physical activities and their intensities using wireless accelerometers and a heart rate monitor. *11th IEEE International Symposium on Wearable Computers.*

Yamada, N., Sakamoto, K., Kunito, G., Yamazaki, K., Isoda, Y., & Tanaka, S. (2007). Applying ontology and probabilistic model to human activity recognition from surrounding things. *IPSJ Digital Courier, 3.*

Chapter 16
The Changing Global Landscape With Emerging Technologies and Their Implications for Smart Societies

Patrice Seuwou
Northampton University, UK

Vincent F. Adegoke
London South Bank University, UK

ABSTRACT

The opportunities offered by digital technology are enormous. The global social and economic system is being reconfigured at an incredible rate. Connectivity is increasingly reshaping our world and redefining the way we interact with our environment. The rise of digital technologies is transforming almost every aspect of modern life. More and more of our interactions are mediated by machines. Along with the rapid evolution comes the risks, threats, and vulnerabilities in the system for those who plan to exploit it. In this chapter, firstly, the authors explore the role of 5G, big data, the internet of things (IoT), artificial intelligence (AI), autonomous vehicles (AV), and cloud computing play in the context of smart societies; secondly, they analyse how the synergy between these technologies will be used by governments and other stakeholders around the world to improve the safety of citizens albeit increasingly relinquishing privacy rights and encouraging mass surveillance at the expense of liberty.

1. INTRODUCTION

Digital technologies have advanced more rapidly than any other innovative technology in history reaching around 50% of the world's population in the last twenty years, transforming societies, accessibility to public services and businesses (UN, 2020a, 2020b). This has led to continual move of people to the big cities globally. The mandatory legislation to solve related migration problem smartly without deple-

DOI: 10.4018/978-1-7998-6992-4.ch016

tion in ecosystem is a major issue that many urban policy makers are currently facing. Several concepts such as smart city, smart society, smart region, smart home, smart car, etc. are therefore emerging in an attempt to provide intelligent and environmentally friendly solutions. The components of smart city among other are smart economy, smart environment, smart governance, smart citizen, smart transportation, and smart wellbeing of the citizen.

Adequate implementation of these conceptual components will enhance suitable provision of healthcare, waste management, cyber control and essential social services such as policing, housing and education to the citizens. Likewise, the advent of digital technologies and connection of technological devices are suddenly changing our method of communication with our environment and other citizens (Arts, Wal, & Adams, 2015). As of 2019, almost half of the world population are digitally connected by social media and internet users are growing by an average of more than one million new users every day (Kemp, 2019). Review shows that as of October 2020 almost 4.66 billion people were active internet users encompassing 59 percent of the global population (Clement, 2020). Despite these researches, review further shows that robots and artificial workers could make 800 million real-workers (BBC, 2017; Manyika, et al., 2017) redundant by 2030.

The internet technology that was originally meant for military has suddenly accelerated the pace of industrial revolution and a major key player in global unification and citizenship. Advanced development of AI technologies, chains of several mathematical programmes, simulations and integration of big data has made it possible for machines to impersonate human intelligence in several areas: autonomous vehicles, industrial automated systems and Industrial Internet of Things.

The society is moving toward big brother state; UK has more CCTV cameras per person than any other country in the western world, the citizen faces surveillance by their own government. USA, China, EU and other countries are following similar trend. It is becoming more intrusive than the domestic spying carried out by Big Brother in George Orwell's '1984' (Adams, 2019). Big brother ideology is no longer a fiction but few steps from being a reality due to availability of big data and digitization technology.

The emerging technology affects the way people interact with each other on daily basis, it enhance efficiencies of processes, products, and services. It therefore plays important role in today's society. However, it also has negative effects on critical infrastructure and come with threats, risks and other susceptibilities that are associated with their deployment. Therefore, there is need for robust policies and regulations to optimize the positive benefits and control the negative impacts on the society.

The purpose of this paper is to explore the emergence of smart-contemporary technologies, their benefits and consequences to the societies. This will provide an important opportunity to advance our knowledge and understanding of current development in Industrial Revolution 4.0 (Renjen, 2019) and how science and technology is digitally realizing one-society-one-world, at the same time progressively eroding our liberty and changing ways of life.

This chapter is organized as follows. In Section 2, we briefly discuss state of the art: smart cities and concepts. Their potentials, economic benefits, challenges and global citizenship. In Section 3, we present summaries of emerging technologies: Internet of Things (IoT), Smart homes, big data, 5G-6G wireless technologies and cryptocurrency/blockchain technology. Section 4 discuss journey towards a big brother state, a framework that uses voluntary surveillance technology that is pervasive and invasive more than ever. Finally, Section 5 provides possible open research issues and concludes this chapter.

2. STATE OF THE ART REVIEW

Smart Cities

For much of the 20th century, the idea that a city could be smart was a science fiction that was pictured in the popular media but quite suddenly with the massive proliferation of computable devices across many scales and with a modicum of intelligence being embedded into such devices, the prospect that a city might become smart, is now a reality. Some of the top smart cities' governments are located in London, Singapore, Seoul, New York & Helsinki. Making a city "smart" is emerging as a strategy to mitigate the problems generated by the urban population growth and rapid urbanization.

The convergence of information and communication technologies is producing urban environments that are quite different from anything that we have experienced previously. More than half of the planet's population now lives in urban areas (Dirks S. G., 2010) (Dirks S. &., 2009) (Dirks S. K., 2009). This shift from a primarily rural to a primarily urban population is projected to continue for the next couple of decades (see http://www.unfpa.org). It is predicted that more than two-thirds of the world's population will be urbanized by 2050 (Braun, 2018). Such enormous and complex congregations of people inevitably tend to become/create messy and disordered places.

Cities, megacities, generate new kinds of problems. Difficulty in waste management, scarcity of resources, air pollution, human health concerns, traffic congestions, and inadequate, deteriorating and aging infrastructures are among the more basic technical, physical, and material problems (Borja, 2007) (Marceau, 2008) (Toppeta, 2010) (Washburn, 2010) (Allam & Dhunny, 2019). These cities will feature countless smart devices, each of which will be communicating with the smart network by sending, copying, and processing data. This process will generate an immense amount of data, some of which will be confidential and should be secured. By nature of smart city interconnectivity, data will be transferred and utilized throughout the smart city processes, with multiple parties communicating and gaining access to information. From the makers of the smart sensors, to the city's transportation authority, to individuals accessing the smart city through their smartphones, each organization contributing to the smart city will uniquely use and handle data in ways that may endanger personal privacy (Allam & Dhunny, 2019). Furthermore, since each smart city stakeholder will have different priorities, there will exist gaps between the different stakeholders' privacy standards.

THE CONCEPT OF SMART SOCIETIES

There are different possibilities of smart society's frameworks depending on the conceptual approach and subjectivity. Therefore, several authors have tried to formulate the definition of smart society with digital society, wired society, information society and sensing society (Winkowka, Szpilko, & Pejic, 2019). However, smart societies could be viewed as ecosystems which are communities of interacting organisms and their environment that are typically described as complex networks formed as a result (Gretzel, Werthner, Koo, & Lamsfus, 2015) of resource interdependencies. British Standards Institute defines smart city as: the effective integration of physical, digital and human systems in the built environment to deliver sustainable, prosperous and inclusive future for its citizens (BIS, 2014). In terms of data-driven, it is a scalable solution that take advantage of information and communications technology (ICT) to increase efficiencies, reduce costs, and enhance quality of life (Falconer & Mitchell, 2012).

The concept comprises of numerous elements, interconnections, collaborations and functions that are intelligent, autonomous and adaptive agents that form communities that can adapt to addition or removal of elements. It is "an interdependent social system of actors, organizations, material infrastructures, and symbolic resources" (Maheshwari & Janssen, 2014).

Dimensions of Smart Society

Communication, innovation, government, infrastructure, transport, social economy, natural resources, societal-well-being, emerging technologies, environmental sustainability, etc. are some of the landscapes that could be used to describe the concept of smart societies. Technically, six dimensions of smartness have been identified. These are smart economy, smart mobility, smart environment, smart people, smart living and smart governance (Zanella & et.al., 2014; Caragliu, Del-Bo, & Nijkamp, 2011) and several sub-layers as illustrated in Figure 1.

Critical Factors and Goals of Smart Society

The need to be smart ready in contemporary society cannot be sidestepped. It is a roadmap that requires endurance and challenge. Some of the critical factors for the creation of smart societies are smart marketing, digital democracy, innovation capacity, knowledge-based workforce and broadband communication infrastructure (ICF, 2015). It is essential to note that most smart initiatives are rooted in contemporary emerging technologies. Though, in some cases it might result in disruption or extinction of existing systems, products, structures or technologies. However, the ultimate goal is the improvement and better use of resources to improve the quality of services offered to citizens (Zanella & et.al., 2014) at a competitive cost. At the same time enhancing societal value creation, innovation and sustainability of healthier society.

POTENTIALS AND ECONOMIC BENEFITS OF SMART SOCIETIES

The economic potentials that the concepts of smart society can offer through technological innovations are enormous. However, it requires constant economic investments to its implementation and constant advancement in technologies. Investments are thriving, it was $342.4 billion in 2016 and projected to reach $774.8 billion in 2021. North America alone invested $188.5 billion in 2016 and is expected to invest $244.5 billion in 2021 (SNG, 2017). This is because smart technology can leverage connected technologies and help cities to operate more proficiently and improve service provision to citizens and businesses.

Review shows that different advantages and disadvantages have been reported to support the application of smart concept to city, region or country (Angelidou, 2014). In this report we took a holistic view of some its potentials and economic benefits.

Figure 1. Dimensions and sub-layers of Smart City

1. **Reliable data driven decision**-The availability of big data, well-designed analysis strategy and connection to other devices enable policy makers to analysis data, make real-time decisions.
2. **Enhanced e-government collaboration**–Access to real time government information through digital services, collaboration tools, KPI boards in communities enhance closer government engagement.
3. **Smart and safer policing**–Smart community is a safe community, application of emerging technologies such as: plate number recognition, movement detection, gunshot detectors, security cameras, etc. can enhance law enforcing agents to be proactive in handling crimes.
4. **Smart environmental ecosystems**–Air quality control, reduced carbon emission footprints, and cleaner cities. This increases standard of living and wellbeing of citizens.
5. **Smart infrastructure/transportation**–Smart buildings, renewable energy sources, modern transportation systems, etc. that are digitally connected with each other enhance sustainability and efficient provision of utilities and public services to citizens.
6. **Smart digital equity**-Accessibility to high speed internets and placement of public Wi-Fi at hotspots offer reliable IoT services and affordable communication cost to the citizens and visitors.
7. **Economic opportunities**–Smart cities attracts global investments, businesses, reduces unemployment and generates revenues that can be invested in community welfare and social service projects to control crime.
8. **Smart connectivity**-It enables connection of people and devices through cloud to each other. It ensures that citizens stay connected, productive, and informed. It enables city operators and government agents to monitor assets, online-self-service, and traffic- control, remote-working and respond to real-time emergencies.
9. **Smart economic prosperity** – Emerging smart city technologies could save enterprises, governments and citizens worldwide over US$5 trillion annually by 2022 (Ismail, 2017). In 2014 the city

Table 1. Challenges and implication of emerging technologies in smart societies

Potential applications	Challenges	Implications
Governance/ Politics	• Political instability • Lack of trust between government and contractors that are implementing the concept.	• Collision with foreign government agents. • Exposure and stolen of sensitive national information and secrets.
Economic	• Uncertainty and volatile global economy. • Cost of implementation and maintenance.	• Delay in addressing government targets and citizen demands. Rise cost of tools. • Increased in health issues because of pollutions.
Social	• The absence of participation and involvement among citizens. • Social and geographical exclusion due to political and unethical reasons.	• Flexibility and scalability of demand problem. • Unbalanced geographical development. • Inequality in social citizenship and inclusion
Technology	• The absence of technological knowledge and constraints. • Coverage, security, and demand capacity issues	• Citizens privacy are no longer private. • Hackers and issues of stolen identity as a result of security and privacy problems.
Environmental	• The absence of environmental ecological policy • Inadequate carbon emission policy and guidelines	• Rise in activists & civil unrests Delay in making progress decisions. • Absence of public safety that required increased policing.
Ethical & Legal	• Absence of regulatory laws • Lack of policies, standardization and directions • Personal data protection issues	• Unequal citizen inclusion • Unbalanced predictive policing and unfairness. • Unending legal challenges Lack of regulatory constraints

of Barcelona saved over €75 million by adopting IoT-driven smart initiatives and creating 47,000 new jobs related to smart city developments (Bolos, Tsoros, & Holopainen, 2015).

10. **Smart quality life**– Smart city initiatives: smart parking, digitized government services, car charging points, etc. make citizens happier and attracts global inflow of talents, visitors and increase GDP.

CHALLENGES OF SMART SOCIETIES IN A DIGITAL WORLD

Emerging smart technologies such as artificial intelligence, big data, analytics, and robotics can offer potential advantages of creating sustainable smart societies (Kitchin R., 2015b) and improve the well-being of the citizens. Despite the potential benefits (Rana & et.al., 2019; Ghosal & Halder, 2018) of several smart society projects, there are barriers and challenges (Kitchin R., 2013) that come with their implementation, deployment and maintenance. This include issues with privacy and impact of legal and ethical problems of applying the concept. Some of the potentials and challenges of the concept among others are summarised in Table 1.

GLOBAL CITIZENSHIP

The contemporary interconnectedness among people, countries and unified worldwide economies through big data, AI, other evolving and advanced smart technologies have led to increased homogeneity and unification among diverse societies. That were once unconnected and divided culturally and ethnically in the past. Global-societal connectivity and recent political developments is making citizens' identity to surpass geography or political borders. Present-day digitisations and development in the global arena which were not achievable few decades ago have led to the emergence of universal oneness and global citizenship concepts. Review shows that education and tourism play a huge impact in establishing the concept of global citizenship (UNICEF, 2013; Oxfam, 2020). The theory of globalization has been demarcated in various ways, however the concept is unquestionably contested among researchers and government agents. Oxfam define it as 'a global citizen is someone who is aware of and understands the wider world - and their place in it. They take an active role in their community, and work with others to make our planet more equal, fair and sustainable' (Oxfam, 2020). On the other hand, 'Global citizenship can be defined as a moral and ethical disposition that can guide the understanding of individuals or groups of local and global contexts and remind them of their relative responsibilities within the various communities' (Haydey, Levy, & Thompson, 2015). In order to address these three theoretical frameworks have emerged to explain and argue the positions of global citizenship. These theories are: the political theory of global citizenship, educational theory of global citizenship, and the social theory of global citizenship (Pittman, 2017). The impact of emerging technologies on global citizenship are enormously positive however it comes with its undesirable detriments. Though it has changed the institution of citizenship sense of belonging and state membership.

Advantages and Disadvantages

Review shows that in some cases the law enforcement agencies are employing AI, big data and machine learning technologies to create algorithmic identities that uses online activity to predict "digital citizenship", to understand ancestry, migration history and form genetic citizenship (EUI, 2020). Such could have monitoring consequences on citizens that challenges or try to evade the perception of the state. A classic example is the use of R-app to monitor and control the spread of COVID-19 and movement of UK citizens, people in the England could face fines up to £10,000 for refusing to self-isolate when asked to do so (Page & et.al., 2020), however some saw it as an invasion of privacy. On the other hand, it encourages people to develop the knowledge, skills and values they need to engage with the world. Therefore, global citizenship helps people to: build their own understanding of world events, think about their values and what's important to them. It takes learning into the real world, challenge ignorance and intolerance, get involved in their local, national and global communities, develop an argument and voice their opinions, see that they have power to act and influence the world around them (Oxfam, 2020; Stoner & et.al., 2013).

THE SOCIAL CREDIT SYSTEM (CHINA)

The Social Credit System (SCS) is a national reputation system being developed by the Chinese Communist Party (CCP), under CCP General Secretary Xi Jinping's administration (Shahin, 2018)The program initiated regional trials in 2009, before launching a national pilot with eight credit scoring firms in 2014. Under the SCS, the Chinese government proposes to assign a social credit rating to every citizen representing their "trustworthiness" based on their everyday social and economic activities. The idea started being discussed in official and non-official circles in the early 2000s and draft planning outlines were proposed. The State Council of China's planning outline for the construction of an SCS (2014–2020), issued in 2014, states that the SCS will focus on four major areas: honesty in government affairs, commercial integrity, societal integrity, and judicial credibility. The goal is to "raise the awareness for integrity and the level of credibility within society" (China Copyright and Media, 2014). To implement it, the government will track and evaluate what you buy at the shops and online; where you are at any given time; who your friends are and how you interact with them; how many hours you spend watching content or playing video games; and what bills and taxes you pay (or not). (Wired, 2017). In essence, SCS will allow the Chinese government, and the technology companies that work with it, to monitor every social, economic, and political action taken by every citizen as well as most of their private activities. The proposal stems from a lack of social and economic trust that has engulfed Chinese society since its turn to market economics in the 1980s. Among other things, it prevents individuals and small businesses from getting loans from banks. But "rather than promoting the organic return of traditional morality to reduce the gulf of distrust, the Chinese government has preferred to invest its energy in technological fixes" (Hawkins, 2017). The idea has also raised the hackles of data protection and privacy advocates, who have called it "an excuse to implement surveillance and control" (para. 17).

AADHAAR IN INDIA

Aadhaar (foundation or base) is a 12-digit unique identity number that can be obtained voluntarily by residents or passport holders of India, generated after recording their demographic as well as biometric data such as fingerprints and iris scans. The data is collected by the Unique Identification Authority of India (UIDAI), a statutory authority established in January 2009 by the government of India, under the jurisdiction of the Ministry of Electronics and Information Technology. Aadhaar is the world's largest biometric ID system. World Bank Chief Economist Paul Romer described Aadhaar as "the most sophisticated ID programme in the world". The Indian authorities started registering residents for Aadhaar in 2010 and it is reported that 99% of the residents were registered by the end of 2017. This Big Data project was referred to by the government as a strategic policy tool for social and financial inclusion, public sector delivery reforms with the ability to manage fiscal budget, increase convenience and promote hassle-free people-centric governance (Unique Identification Authority of India, n.d.)(UIDAI.gov, n.d., para. 4). Officially, the main aims of the project were to limit corruption in the transfer of benefits to the public by removing the middlemen and providing the opportunity to the population, particularly the poorer people to have direct access to a variety of welfare schemes.

Although it was claimed that participation was entirely voluntary, it was mandatory for Indians if they wished to access a range of social and financial services. Limited legislative systems were implemented for the protection of citizen's data being collected by technology companies working on the government behalf or the authorities (Dixon, 2017). This ambitious Biometric system promise to consolidate 12 billion of its citizens' fingerprints, 2.4 billion iris scans, and 1.2 billion photographs (Sarkar, 2014). While the West already has existing biometric identity systems, it is nowhere near India's scale and scope. This project is the brainchild of the technology entrepreneur Nandan Nilekani, who cofounded and built the multibillion-dollar outsourcing company Infosys prior his appointment by the government to head this project. It has received much media attention, with discourse leaning toward the empowerment of the marginalized. R. S. Sharma, secretary of the Department of Electronics & Information Technology in India, declared that "digital India is not for rich people . . . it is for poor people" (Toness, 2014). The project has raised fears of privacy violations and mass surveillance. Soon after it was launched, Indian intelligence officials said its purpose was to flush out illegal immigrants in the country—the reason why residents and not just citizens were asked to register for Aadhaar—and the claim that Aadhaar would help deliver welfare schemes was just a ruse.

3. EXPLORING THE IMPACT OF EMERGING TECHNOLOGIES IN THE CONTEXT OF SMART SOCIETIES

Artificial Intelligence and Evolving Smart Solutions

Currently, about 55% of the world's population lives in urban areas this proportion will increase to 68% by 2050 and add another 2.5 billion people to the already densely populated area by the same period (United, 2018). This development requires solution to several related problems and service provisions such as: population, infrastructures, housing, pollution, transportation/congestion, crime, security, social poverty, etc. Artificial intelligence (AI), machine learning, Internet of Things, and other emerging technologies concept offers holistic and smart interconnected solutions to address some of these challenges. It also boosts economic performance, environmental sustainability and improve the wellbeing of the people. Even though there is no generally accepted definition for AI. However, AI is a chain of mathematical programmes and simulations that try to impersonate human intelligence, behaviour and action. In many areas of science and technology and AI has been applied such that it smartly and efficiently predicts events (Adegoke, Chen, Banissi, & Barsikzai, 2020) that outperform man in learning and handling difficult tasks.

Emerging AI Solutions

The emerging of AI technologies could offer endless solutions to smart society challenges in several ways: improve performance efficiency of systems, processes, management, etc. However, AI must be adequately integrated to enhance economic and environmental sustainability that benefits societal transformation. Some of the smart city areas and related contemporary AI-technologies solutions are illustrated in Table 2.

Table 2. AI-technology and possible solutions for smart societies

Smart City Area	AI-technology	Possible solution
Smart Economy	E-business, e-commerce, e-tourists.	Urban development, sustainability and investment attractiveness (Hattingh & et.al., 2020). Attract investors and tourists.
Smart Agriculture	Agricultural robots	Innovative solution - Crop, soil and weather monitoring: predictive analytics (Malapela, 2017)
Smart Governance	AI for planning and administration	AI technologies work patterns, policy designs, tax collections, social credits, etc.
Smart ecosystem	AI for smart home and buildings	Smart cyber ecosystem, smart home/office ecosystem and other smart devices.
Smart Healthcare	AI in healthcare and pharmaceutical services	Delivery of predictive services, remote consolations/ robotic assisted surgery, automated disease/image analysis and smart IoT devices for monitoring patient health (OECD, 2015).
Smart Mobility / Transportation	AI-autonomous vehicles	Self-driving, autonomous vehicles and delivery services (Ohnsman, 2018).
Smart Energy	Intelligent energy management software, IoT devices	Smart metering, automated energy storage and diversion, smart grid (Rathor & Saxena, 2020)
Smart Security/Justice	e-security and CCTV; Intelligence justice	Fraud detection and prevention, biometric/image surveillance and detection (Giles, 2018; Kleinberg & al., 2017). Predictive algorithms in the legal system (Wyllie, 2013)
Smart Education	e-education, virtual meeting	Virtual advanced training, universities, virtual meetings and facilitators, remote working/office (Williamson, 2015)
Smart Citizen	Smart state	Engagement of citizens: communities, developers, researchers, etc. to collaborate and address environmental issues in cities (Noveck, 2015; Hemment & Townsend, 2013).

INTERNET OF THINGS AND SMART HOMES

Sustainability and economic survival of smart society requires the integration of essential, contemporary ICT tools and connections of billions of sensors and emerging smart devices. The concept of Internet of Things (IoT) enables the connection of several physical and remote devises using integrated sensors over the internet network. This enable seamless communication, exchange of data and the ability to carry out automated tasks. Report shows that by the end of 2020, over 50 billion devices that worth $19 trillion opportunity will be connected (Forbes, 2020) by IoT and related technology. IoT is tremendously changing the way we live, move and work. It has got to a point that many people cannot do without essential IoT devices like smart-phones, speakers, light-bulbs, thermostats, wearables, payment/cashless devices, virtual assistants and essential smart devices. In major cities and modern homes, IoT is making the use of technology smarter to control other devices and proactively react to events. City leaders are taking advantage of available real-time data to understand traffic and weather patterns, heating/cooling systems, landscaping and move of citizens from one part of the city to other part of city (Boulos & Al-Shorbaji, 2014; Fork & Hines, 2015) therefore able to control various aspects citizen's life. Development in the Industrial sector reveals that Industrial Internet of Things (IIoT) is the next generation of IoT. The technology will enable companies to use IIoT to capture growth and performance. Therefore, boost revenues by increasing production/creating new hybrid business models, exploit intelligent technologies to fuel innovation, and transform their workforce. Thus, enabling different industries leveraging a network of embedded sensors and devices connected to many appliances, vehicles, equipment, warehouses and logistics, operations. It will enable industries to benefit from real-time process monitors and control that are connected via IoT (Daugherty & et.al., 2015). Some of the potential benefits and barriers of implementing and deploying IoT technology are summarised in Table 3

Table 3. Benefits and Barriers of implementing and deploying IoT

Applications/Benefits	Barriers/Threats
• Remote working options • Business opportunities/expansion • Efficient services and products development • Predictive analytics and Market understanding • Customer self-service and personalisation • Environmental sustainability • Smart management /Ease parking & transportation • Healthier citizens • Data Sharing and Decision Making • Inventory Tracking and Management • Reduce operating costs • Ability to track and monitor things **• Industrial IoT: Smart manufacturing, maintenance, power grids, logistics/supply chains.**	• Lack privacy and protection of sensitive data • Low acceptance, trust and security • Initial cost and maintenance • Standardisation issues • Unexpected failure of IoT • Upgrade and compatibly issues • Yielding unemployment • Technology dependability/overreliance • Fear of major failure • Complexity and specialist requirement

BIG DATA

In the past few decades, businesses, governments, and individuals have continued to generate vast amounts of data than they have been able to cope with and utilize. According to IBM, we create around 2.5 quintillion bytes of data on a daily basis, and it is estimated that the size of data will double in every 2 years (IBM, 2016). In today digital society, data is now the world's most valuable commodity, the citizen as user are now digital human interacting massively with the technology and generating tremendous amount of data. Indeed, our data is the new oil, the new gold, the new currency. Big data is the term commonly applied to massive data sets that are generated through a variety of sources, including environmental and body sensors, mobile devices, electronic medical records, imaging and laboratory studies, and administrative claims data. Although big data is considered a new form of capital along with the traditional financial and intellectual capitals for businesses, many firms fail to fully utilize this kind of capital due to their limited capabilities of dealing with the technologies (e.g., advanced big data analytics tools) and their limited understanding of the fit and the governance of these new technologies with respect to their specific business goals. For the retail business, big data is utilized for getting a prominent understanding of customer shopping propensities and how to draw in new customers. Organizations are being enabled and empowered by big data analytics to create customer proposals dependent on their buying history, resulting in customized shopping experiences. These big data analytics solutions additionally help in anticipating patterns and taking strategic decisions that are based on an analysis of the market. Big data offers the potential for vastly enhanced data analytics. Used properly, organizations can employ big data to spot entirely new trends, to segment customers to an astonishing degree of accuracy, and to allow unprecedented levels of innovation in technology and product design. While it's easy to get caught up in the opportunities big data offers, it's not necessarily a cornucopia of progress. If gathered, stored, or used wrongly, big data poses some serious dangers. Big data comes with security issues; Bad players can abuse big data—if data falls into the wrong hands, big data can be used for phishing, scams, and to spread disinformation. Insights are only as good as the quality of the data they come from—bad, noisy, or 'dirty' data (or applying poor best practice) can lead to poor insights, which can be risky in the wrong situations. There are ethical issues—as a new field, the ethics of big data is still evolving. The battle between big data's potential and its dangers remains ongoing. However, identifying and acknowledging its potential risks goes a long way to resolving them (Hillier, 2020).

Table 4. Advantages and controversial threats of 5G/6G technologies

Applications/Enablers	Threats and disadvantages
• Device cost mobility & lower latency • Massive density /Increased bandwidth • Reduced signalling overhead & energy consumption • Integrate public infrastructure network within network slices • Support for pub/sub message-oriented communication • Adaptation of big-data and input interfaces • Enhanced capacity/game technology • Greater transmission speed and coverage • Autonomous car-technology • Robotic medical advancements • IoT and IIoT evolution	• Security and privacy issue • Development, deployment and maintenance cost • Standardization and legislation of cyber law • Lack of trust in 5G • Issues surrounding conspiracy theories about Radiofrequency (RF) radiation and health concerns. • Threats to citizens job security • Overdependence on machines and technology

5G AND 6G WIRELESS GENERATIONS TECHNOLOGY

The 5G and 6G cellular technology refer to the fifth and sixth generations of mobile wireless networks respectively. The 5G technology is known to have data transmission frequency at least 10 times greater than 4G. On the other hand, 6G is the successor to the 5G technology that is set to have data transmission speed as 100 times faster than 5G. According to Ericsson, an expert in the field of cellular technology "5G evolution will move us into an era of ubiquitous, high-capacity radio. Taking us to ever lower levels of latency and extensive Gbps capacity. The evolution of 5G technology will pave the way for highly intelligent haptic IoT technologies at hyper scale, leveraging new forms of vehicle-to-infrastructure, vehicle-to-vehicle, vehicle-to-pedestrian, and person-to-person connectivity (Ericsson, 2020). Among the notable features of the 5G technology are the IIoT, ultra-reliable low latency communication (URLLC), intelligent transportation systems (ITS) and vehicle-to-anything (V2X) communications and positioning (Peisa & et.al., 2020). Firstly, the IIoT bring together network of a multitude of industrial devices connected by communications technologies that can monitor, exchange, collect, analyse and deliver valuable insights. Such insights can enable industrial productivity (GE, 2020) and profitability through enhancements of smarter and faster business decision making. Secondly, ITS technology will transform application of transportation systems and infrastructures concept. Therefore, ITS technology will revolutionize traffic management and make road safer, enabling buses, autonomous vehicles, trains and transportation to be more reliable, run efficiently and economically than ever before. However, despite the tremendous benefits the technology offers it also come with a number, disadvantages and controversial issues some are illustrated in Table 4.

AUTONOMOUS VEHICLE (AV)

As transportation technology evolves, more and more car manufacturers are announcing their will to introduce new kinds of vehicles into the automobile market. Such vehicles include Autonomous (AV) and Connected Vehicles (CV) as well as Electric Vehicles. According to the Centre for Advanced Automotive Technology (Seuwou, Banissi, & Ubakanma, 2019)(Centre for Advanced Automotive Technology, 2018) Connected Vehicles are considered ''vehicles that use any of a number of different communication technologies to communicate with the driver, other cars on the road (vehicle-to-vehicle [V2V]), roadside infrastructure (vehicle-to-infrastructure [V2I]), and the ''Cloud'' [V2C]'' (Seuwou, Patel, & Ubakanma, 2014). The implementation of Connected and Autonomous Vehicles (CAVs) as well as the use of alternative fuel as gas or electric power to empower vehicles, are supposed to have an impact on the amount of the GHG emissions originating from traffic. Autonomous vehicle technology may be able to provide certain advantages compared to human-driven vehicles (CAAT, 2018) . One such potential advantage is that they could provide increased safety on the road – vehicle crashes cause many deaths every year, and automated vehicles could potentially decrease the number of casualties as the software used in them is likely to make fewer errors in comparison to humans. A decrease in the number of accidents could also reduce traffic congestion, which is a further potential advantage posed by autonomous vehicles. Autonomous driving can also achieve this by the removal of human behaviours that cause blockages on the road, specifically stop-and-go traffic (Seuwou, Chrysoulas, Banissi, & Ubakanma, 2020). Another possible advantage of automated driving is that people who are not able to drive – due to factors like age and disabilities – could be able to use automated cars as more convenient transport systems. Additional advantages that come with an autonomous car are elimination of driving fatigue and being able to sleep during overnight journeys. Like other producers, developers of autonomous vehicles are legally liable for damages that stem from the defective design, manufacture and marketing of their products. The potential liability risk is great for driverless cars because complex systems interact in ways that are unexpected. Manufacturers want to minimize the number of liability claims made against them. One way is to reduce the chance of their product being misused by educating consumers about how it works and alerting them to safety concerns. The safety and efficiency benefits of autonomous cars rely on computers making better, quicker decisions than people. Users input their desired destination and thereafter cede control to the computer. Full autonomy has — deliberately — not yet been adopted in transportation. People are still perceived as being more flexible, adaptable and creative than machines, and better able to respond to changing or unforeseen conditions. Pilots are able, therefore, to wrest control from flyby-wire technology when key computers fail. The public is right to remain cautious about full automation. Manufacturers need to explain how a car would protect passengers should crucial systems fail. A driverless car must be able to stop safely if its hazard-avoidance algorithms malfunction, its cameras break, or its internal maps die. But this is hard to engineer: for example, without cameras, such a car cannot see where it is going.

CRYPTOCURRENCY AND BLOCKCHAIN TECHNOLOGY: A NEW GLOBAL DIGITAL CURRENCY

A cryptocurrency is a digital currency form of virtual payment that can be used as a medium of exchange in business transactions like the US dollar or the European euro. The currency is securely created and stored electronically by cryptography concept using blockchain technology to verify transaction and to avoid counterfeit or double-spend (WallSteet, 2018). Consequently, cryptocurrency is a new financial paradigm of the global society with the purpose of providing peer-to-peer network and universal transaction that ensures financial security and privacy. Bitcoin (BTC) on the other hand is the best known category that was invented cryptocurrency using blockchain technology, other crypto currencies among other are Ethereum (ETH), Litecoin (LTC) and Ripple (XRP). Cryptocurrency is another contemporary technology that is shaping the financial markets (Giudici, Milne, & Vinogradov, 2020) and therefore affecting the future of global currency and financial transactions.

Currently, Bitcoin is the most widely used cryptocurrency with a total value at around US$63 billion and has topped US$300 billion. Additionally, over 300,000 transactions in bitcoin take place and around 2000 new bitcoin are created every day (Tredinnick, 2019). Review shows that whether it is independent of government control or underwritten by central banks or major financial organizations cryptocurrency has the potential of becoming the way of transaction in the future (Tredinnick, 2019; Houben, 2018). The currency is gaining global attention and the EU policy makers has proposed a regulatory framework. They are concerned about its phenomenon from a legal perspective, therefore focusing on the use of cryptocurrencies to avoid financial crime, money laundering and tax evasion (Houben, 2018) and misuse of virtual currencies that is currently estimated to exceed EUR 7 billion worldwide.

Table 5. Advantages and Limitation/Disadvantages of Cryptocurrency/Blockchain technology

Advantages/benefits	Disadvantages/Limitations
• Lower transactions cost	• Rising costs of mining
• Faster processing of transaction	• Power consumption/environmental consequences
• Worldwide acceptance/Cost reduction	• Market manipulation: pump/dump trading
• Mitigation of fraudulent transactions	• Regulatory concerns/implications
• Secure and private	• Scalability and cybersecurity issues
• Decentralization/Increased transparency	• No protection in case of loss
• Unparalleled transparency	• Challenges of market fluctuations
• Instant accessibility	• Difficulty to understand
• Absolute anonymity	• Complex technology
• Permanent ledger	• Competing platforms
• Accurate tracking	• Implementation challenges
• Supply is central bank independent	• Has no physical form or intrinsic value

Blockchain Technology

A blockchain technology is a decentralized ledger technology (DLT) that is based on a peer-to-peer (P2P) topology, it securely records information that is impossible to alter or hack. The technology enables users to confirm transactions without a need for a central clearing authority as required in the traditional non-virtual banking system. It also has a wide spectrum of potential applications such as financial services, settling trades, automotive, voting and healthcare (PWC, 2020), asset registries, supply chain management, DRM and intellectual property, document and records management, and smart contracts (Tredinnick, 2019). Blockchain possesses a great potential in empowering the citizens of the developing countries if widely adopted by e-governance applications for identity management, asset ownership transfer of precious commodities such as gold, silver and diamond, healthcare and other commercial uses as well as in financial inclusion. However, like any other emerging technologies there are numerous limitations and disadvantages some of these are summarized in Table 3.

4. TOWARDS A BIG BROTHER STATE

Before September 11, 2001, the United States wasn't officially engaged in any wars. Few of us had ever heard of al-Qaeda or Osama bin Laden, and ISIS didn't even exist. In US like in the UK, our surveillance state was a fraction of its current size. And perhaps hardest to believe, we didn't have to take off our shoes to go through airport security. America's involvement in the War on Terror — prompted by the 9/11 terrorist attacks. Two of the planes were flown into the twin towers of the World Trade Centre in New York City, a third plane hit the Pentagon just outside Washington, D.C., and the fourth plane crashed in a field in Shanksville, Pennsylvania. Almost 3,000 people were killed during the 9/11 terrorist attacks. In the UK, the 7 July 2005 London bombings, often referred to as 7/7, were a series of coordinated Islamist suicide attacks in London, England, that targeted commuters travelling on the city's public transport system during the morning rush hour — resulting in a dramatic change in our nation's attitudes and concerns about safety, vigilance and privacy. It ushered in a new generation of policies like the counter terrorism laws in the UK, USA Patriot Act, prioritising national security and defence, often at the expense of civil liberties (Richard C. Leone and Jr. Greg, 2003). These changes continue to have ripple effects across the globe, particularly in the Middle East, where American-led military operations helped foment rebellions and ongoing warfare throughout the region. In the wake of revelations that the United States and the United Kingdom had conducted mass surveillance programs of their own and others' citizens, and shared much of the data with select allies and cooperating intelligence agencies ('the Snowden revelations'), (1) it was reported that sales of George Orwell's classic examination of the surveillance state, *1984*, had surged. This novel was published in 1949, but after more than 70 years its accuracy with today digital society has been recognised as astonishing. (2) The Big Brother of science fiction had taken contemporary form and significance.

In the future ubiquitous-computing environment, RFID tags will be attached to all kinds of products and other physical objects, even to people, and could become a fundamental technology for ubiquitous services where the tags are used to identify things and people automatically. However, despite this promise, the possible abuse (or just excessive use) by retailers and government agencies of RFID's tracking capability raises questions about potential violations of personal privacy. Companies and governments have plans to launch approximately 50,000 satellites to provide 5G and Wi-Fi services everywhere on Earth. Elon Musk's SpaceX is leading these efforts with its planned deployment of 42,000 Starlink satellites and its goal of establishing a global network (Tachover, 2020). The most powerful forces in our rapidly digitizing world are technologies that most of us don't know much about yet. Alone and in combination, new innovations in artificial intelligence, blockchain, natural language processing and 5G telecommunications will usher a decade of change that will make the last 10 years look pretty laid back. Many believe that the government is at least 40 years ahead of the public in technology (West, 2014). 5G brings a range of benefits to the IoT, which are not available with 4G or other technologies. These include 5G's ability to support a massive number of static and mobile IoT devices, which have a diverse range of speed, bandwidth and quality of service requirements (Opiah, 2020). Along with this literary revival, the response to the National Security Agency ('NSA') surveillance scandal has been accompanied by a renewal of interest in privacy as a human right. This prompts the question whether a right to privacy can help us analyse and regulate against the incursions on liberty which have been increasingly routinized in our daily ritual usage of information and communications technologies and in the harvesting, accumulation and analysis of big data by our governments, communities, corporate actors and employers. We used to think that Big Brother would always take the form of the state in Orwellian terms. Then slowly commentators pointed to the dangers posed by transnational actors and companies: why were we worried about our governments when our local supermarket or our internet service or phone provider held incredibly sensitive personal information about us? The Snowden revelations point to yet another variation on this theme: that public and private actors now both act in ways which are potentially invasive and detrimental to our liberty, and often do so together.

5. CONCLUSION

The world is changing massively due to new communications facilities and technological development. With the rise of terrorist attacks in major cities around the globe, most governments in the developed world are setting measures to control their citizen following an exponential population growth and the need to expand their influences. It appears from this paper that privacy is a myth, it doesn't exist anymore, and this concept is reinforced through literature exploring how the Chinese government has developed its Social Credit System (SCS) and the Indian government has also introduced their Aadhaar system to have a better control of their respective populations. It is just a question of time before major countries in the west develop similar systems. According to Edward Snowden revelations in 2013, it appears that secret services both in the UK and in the US have given themselves powers to make the global surveillance a reality. The social and policy implications of our study are wide-ranging. Both bottom-up and top-down efforts are needed to alter the nature of structural relations that normalize surveillance and violations of privacy although such efforts will have to be aligned with the political, economic, and cultural particularities of each nation. First, data protection and digital rights activists need to recognize the importance of educating citizens about why privacy matters and how Big Data projects undermine it. This article frames the privacy concerns of the SCS and the Aadhaar users through a contextual and networked privacy lens that considers power differentials and user agency, the rise of emerging technologies such as AI, RFID, 5G, 6G network, the smart grid, big data, blockchain, AV and others changing the nature of our cities and normal day to day activities. It also provides a wide range of perspectives drawing on data and opinions from citizens as well as insights that will inform future debates about the collection, use, interpretation, and linking of 5G/6G networks, big data by businesses, governments, and other institutions. It is also evident that privacy can only be understood in the context of the relations within which it operates; in the case described in this article, the most important component of this context, perhaps, is the population's relationship with the state.

REFERENCES

Adams, J. (2019). *Mail Online - Britain faces Big Brother surveillance society WORSE than George Orwell's 1984: Lip-reading cameras that will force people to talk 'with hands over mouths' are latest attack on privacy, government watchdog warns.* Retrieved 09 December, 2020, from https://www.dailymail.co.uk/news/article-7399467/Britain-faces-Big-Brother-surveillance-society-WORSE-George-Orwells-1984-watchdog-says.html

Adegoke, V., Chen, D., Banissi, E., & Barsikzai, S. (2020). Enhancing Ensemble Prediction Accuracy of Breast Cancer Survivability and Diabetes Diagnostic Using Optimized EKF-RBFN Trained Prototypes. In A. Madureira, A. Abraham, N. Gandhi, C. Silva, & M. Antunes (Eds.), *Proceedings of the Tenth International Conference on Soft Computing and Pattern Recognition(SoCPaR 2018) Advances in Intelligent Systems and Computing*. Cham: Springer. doi:10.1007/978-3-030-17065-3_6

Allam, Z., & Dhunny, Z. (2019). On big data, artificial intelligence and smart cities. *Cities (London, England)*, *89*, 80–91. doi:10.1016/j.cities.2019.01.032

Allam, Z., & Dhunny, Z. (2019). *On big data, artificial intelligence and smart cities*. Academic Press.

Angelidou, M. (2014). Smart city policies: A spatial approach. *City*, S3–S11. doi:10.1016/j.cities.2014.06.007

Arts, K., Wal, R., & Adams, W. (2015). Digital technology and the conservation of nature. *Ambio*, *44*(S4), 661–673. Advance online publication. doi:10.100713280-015-0705-1 PMID:26508352

BBC. (2017). *BBC -Robot automation will 'take 800 million jobs by 2030' - report.* BBC. Retrieved November 26, 2020, from https://www.bbc.co.uk/news/world-us-canada-42170100

BIS. (2014). Smart cities framework – Guide to establishing strategies for smart cities and communities. *PAS*, *181*, 2014.

Bolos, N., Tsoros, A., & Holopainen, A. (2015). Social, innovative and smart cities are happy and resilient': Insights from the WHO EURO 2014 International Healthy Cities Conference. *International Journal of Health Geographics*, *14*(3). PMID:25588543

Borja, J. (2007). Counterpoint: Intelligent cities and innovative cities. *E-Journal on the Knowledge Society*. Retrieved from http://www.uoc.edu/uocpapers/5/dt/eng/mitchell.pdf

Boulos, M., & Al-Shorbaji, N. (2014). On the Internet of Things, smart cities and the WHO Healthy Cities. *International Journal of Health Geographics*, *13*(10). PMID:24669838

Braun, T. F., Fung, B. C. M., Iqbal, F., & Shah, B. (2018). Security and privacy challenges in smart cities. *Sustainable Cities and Society*, *39*, 499–507. doi:10.1016/j.scs.2018.02.039

CAAT. (2018, October 21). *Center for Advanced Automotive Technology*. Retrieved from http://autocaat.org/Technologies/Automated_and_Connected_Vehicles/

Caragliu, A., Del-Bo, C., & Nijkamp, P. (2011). Smart Cities in Europe. *Journal of Urban Technology*, *18*(2), 65–82. doi:10.1080/10630732.2011.601117

China Copyright and Media. (2014). *Planning outline for the construction of a social credit system (2014-2020)*. Retrieved from https://chinacopyrightandmedia.wordpress.com/2014/06/14/planning-outline-forthe-construction-of-a-social-credit-system-2014-2020/

Clement, J. (2020). *Worldwide digital population as of October 2020*. Statista. Retrieved November 26, 2020, from https://www.statista.com/statistics/617136/digital-population-worldwide/

Daugherty, D. (2015). *Driving Unconventional Growth through the Industrial Internet of Things*. Accenture.

Dirks, S. (2009). *A Vision of Smarter Cities: How Cities Can Lead the Way into a.* IBM Global Business Services.

Dirks, S. G. (2010). *Smarter Cities for Smarter Growth: How Cities Can Optimize*. IBM Global Business Services. Retrieved from ftp://public.dhe.ibm.com/common/ssi/ecm/en/gbe0334 8usen/GBE03348USEN.PDF

Dirks, S. K. (2009). *How Smart is Your City? Helping Cities Measure Progress*. IBM Global Business Services. Retrieved from ftp://public.dhe.ibm.com/common/ssi/ecm/en/gbe0324

Dixon, P. (2017). A failure to do no harm—India's Aadhaar biometric ID program and its inability to protect privacy in relation to measures in Europe and the U.S. *Health and Technology*, *7*(4), 539–567. doi:10.100712553-017-0202-6 PMID:29308348

Ericsson. (2020). *5G evolution – the latest from 3GPP*. Retrieved October 21, 2020, from Ericsson.com: https://www.ericsson.com/en/future-technologies/5g-evolution

EUI. (2020). *CitTech: Global Citizenship and Technology*. Retrieved from European University Institute: http://global-citizenship.eui.eu/research/global-citizenship-technology/

Falconer, G., & Mitchell, S. (2012). *Smart City Framework A Systematic Process for Enabling Smart+Connected Communities.* Cisco. Retrieved October 15, 2020, from https://www.cisco.com/c/dam/en_us/about/ac79/docs/ps/motm/Smart-City-Framework.pdf

Forbes. (2020). *Intenet of Things*. Retrieved October 17, 2020, from Forbes: https://www.forbes.com/internet-of-things/#d37e04476e53

Fork, W., & Hines, G. (2015, December 8). *The Internet Of Things: Making Cities -- And The Way They Use Technology -- Smarter*. Retrieved from Forbes: https://www.forbes.com/sites/uhenergy/2015/12/08/the-internet-of-things-making-cities-and-the-way-they-use-technology-smarter/#5a507e8f6f94

GE. (2020). Retrieved October 2020, 2020, from Everything you need to know about the Industrial Internet of Things: https://www.ge.com/digital/blog/everything-you-need-know-about-industrial-internet-things

Ghosal, A., & Halder, S. (2018). In Z. Mahmood (Ed.), *Building Intelligent Systems for Smart Cities: Issues, Challenges and* (pp. 107–125). Smart Cities. Computer Communications and Networks. doi:10.1007/978-3-319-76669-0_5

Giles, M. (2018). Cybersecurity's insidious new threat: workforce stress. *MIT Technology Review.* Retrieved October 18, 2020, from https://www.technologyreview.com/2018/08/07/141139/cybersecuritys-insidious-new-threat-workforce-stress/

Giudici, G., Milne, A., & Vinogradov, D. (2020). Cryptocurrencies: Market analysis and perspectives. *Economia e Politica Industriale*, *47*(1), 1–18. doi:10.100740812-019-00138-6

Gretzel, U., Werthner, H., Koo, C., & Lamsfus, C. (2015). Conceptual foundations for understanding smart tourism. *Computers in Human Behavior*, *50*, 558–563. doi:10.1016/j.chb.2015.03.043

Hattingh, M. (2020). SMART City and Economy: Bibliographic Coupling and Co-occurrence occurrence. In E-business, E-services, and E-society (pp. 102-113). Skukuza.

Hawkins, A. (2017). Chinese citizens want the government to rank them. *Foreign Policy*. Retrieved from https://foreignpolicy.com/2017/05/24/chinese-citizens-want-the-government-to-rank-them/

Haydey, M., Levy, J., & Thompson, J. (2015). *The SAGE Handbook of Research in International Education* (2nd ed.). SAGE. doi:10.4135/9781473943506

Hemment, D., & Townsend, A. (2013). Future Everything. Academic Press.

Hillier, W. (2020). *Is Big Data Dangerous?* Retrieved October 25, 2020, from https://careerfoundry.com/en/blog/data-analytics/is-big-data-dangerous/

Houben, R. (2018). *Cryptocurrencies and blockchain*. European Parliament - Policy Department for Economic, Scientific and Quality of Life Policies.

IBM. (2016, October 16). *What Is Big Data?* Retrieved from https://www-01.ibm.com/software/data/bigdata/what-is-big-data.html

ICF. (2015). *What is an Intelligent Community?* Retrieved October 12, 2020, from ICF: https://www.intelligentcommunity.org/what_is_an_intelligent_community

Ismail, N. (2017, December 5). *Smart cities could lead to cost savings of $5 trillion - report suggests*. Retrieved from Information Age: https://www.information-age.com/smart-cities-lead-cost-savings-5-trillion-123469863/

Kemp, K. (2019). *Digital 2019: Global Intenet use accelerates*. We are social. Retrieved November 26, 2020, from https://wearesocial.com/blog/2019/01/digital-2019-global-internet-use-accelerates

Kitchin, R. (2013). Big data and human geography: Opportunities, challenges and risks. *Dialogues in Human Geography*, *3*(3), 262–267. doi:10.1177/2043820613513388

Kitchin, R. (2015b). The Promise and Perils of Smart Cities. *Society for Computers & Law*, *26*(2), 1–5.

Kleinberg, J. (2017). *Human decisions and machine predictions*. NBER Working Paper.

Maheshwari, D., & Janssen, M. (2014). Reconceptualizing measuring, benchmarking for improving interoperability in smart. *Government Information Quarterly*, *31*, S84–S92. doi:10.1016/j.giq.2014.01.009

Malapela, T. (2017, December 19). *Can artificial intelligence help improve agricultural productivity?* Retrieved October 18, 2020, from e-agriculture: http://www.fao.org/e-agriculture/news/can-artificial-intelligence-help-improve-agricultural-productivity

Manyika, J., Lund, S., Chuin, M., Bughin, J., Woetzel, J., Batra, P., . . . Sanhvi, S. (2017). *McKinsy & Company - Jobs lost, jobs gained: What the future of work will mean for jobs, skills, and wages*. Retrieved November 25, 2020, from https://www.mckinsey.com/featured-insights/future-of-work/jobs-lost-jobs-gained-what-the-future-of-work-will-mean-for-jobs-skills-and-wages

Marceau, J. (2008). Introduction: Innovation in the city and innovative cities. *Innovation: Management. Policy*, (10), 136–145.

Noveck, B. (2015). *Smart citizens, smarter state the technologies of expertise and the future of governing*. Harvard Univ. Press. doi:10.4159/9780674915435

OECD. (2015). *Data-Driven Innovation: Big Data for Growth and Well-Being*. Paris: OECD. Retrieved from https://www.oecd-ilibrary.org/science-and-technology/data-driven-innovation_9789264229358-en

Ohnsman, F. (2018, May 31). *Forbes*. Retrieved October 18, 2020, from Waymo dramatically expanding autonomous taxi fleet, eyes sales to individuals: https://www.forbes.com/sites/alanohnsman/2018/05/31/waymo-adding-up-to-62000-minivans-to-robot-fleet-may-supply-tech-for-fca-models/

Opiah, A. (2020). *The 5G and IoT Revolution*. Retrieved October 28, 2020, from https://www.capacity-media.com/articles/3825285/the-5g-and-iot-revolution

Oxfam. (2020). *What is global citizenship?* Retrieved October 17, 2020, from Oxfam: https://www.oxfam.org.uk/education/who-we-are/what-is-global-citizenship/

Page, M. (2020, October 16). *Covid-19 news: Remdesivir has little effect on survival, finds WHO*. Retrieved October 17, 2020, from NewScientist: https://www.newscientist.com/article/2237475-covid-19-news-remdesivir-has-little-effect-on-survival-finds-who/

Peisa, J. (2020). *5G evolution - 3GPP reases a6 & 17 overview*. Stockholm: Ericsson. Retrieved October 18, 2020, from https://www.ericsson.com/49bdd9/assets/local/reports-papers/ericsson-technology-review/docs/2020/5g-nr-evolution.pdf

Pittman, J. (2017). Exploring Global Citizenship Theories to Advance Educational, Social,Economic and Environmental Justice. *Journal of Tourism & Hospitality (Los Angeles, Calif.)*, *6*(6). Advance online publication. doi:10.4172/2167-0269.1000326

PWC. (2020, October 25). *Making sense of bitcoin, cryptocurrency and blockchain*. Retrieved from PWC - United States: https://www.pwc.com/us/en/industries/financial-services/fintech/bitcoin-blockchain-cryptocurrency.html

Rana, N., Luthra, S., Mangla, S. K., Islam, R., Roderick, S., & Dwivedi, Y. K. (2019). Barriers to the Development of Smart Cities in Indian Context. *Information Systems Frontiers*, *21*(3), 503–525. doi:10.100710796-018-9873-4

Rathor, S., & Saxena, D. (2020). Energy management system for smart grid: An overview and key issues. *International Journal of Energy Research*, *44*(6), 4067–4109. doi:10.1002/er.4883

Renjen, P. (2019). *Harvard Business Review - How Leaders Are Navigating the Fourth Industrial Revolution*. Retrieved March 26, 2020, from https://hbr.org/sponsored/2019/03/how-leaders-are-navigating-the-fourth-industrial-revolution

Leone. (2003). *The War on Our Freedoms: Civil Liberties in an Age of Terrorism*. The Century Foundation.

Sarkar, S. (2014). The unique identity (UID) project, biometrics and re-imagining governance in India. *Oxford Development Studies*, *42*(4), 516–533. doi:10.1080/13600818.2014.924493

Seuwou, P., Banissi, E., & Ubakanma, G. (2019). The Future of Mobility with Connected and Autonomous Vehicles in Smart Cities. *Digital Twin Technologies and Smart Cities, Internet of Things (IoT)*.

Seuwou, P., Banissi, E., & Ubakanma, G. (2020). The Future of Mobility with Connected and Autonomous Vehicles in Smart Cities. In Digital Twin Technologies and Smart Cities. Internet of Things (Technology, Communications and Computing (pp. 37-52). Springer. doi:10.1007/978-3-030-18732-3_3

Seuwou, P., Chrysoulas, C., Banissi, E., & Ubakanma, G. (2020). Measuring Consumer Behavioural Intention to Accept Technology: Towards Autonomous Vehicles Technology Acceptance Model (AV-TAM). *Advances in Intelligent Systems and Computing*, 507-516.

Seuwou, P., Patel, D., & Ubakanma, G. (2014). Vehicular ad hoc network applications and security: A study into the economic and the legal implications. *International Journal of Electronic Security and Digital Forensics*, *6*(2), 115–129. doi:10.1504/IJESDF.2014.063107

Seuwou, P., Patel, D., & Ubakanma, G. (2014). Vehicular ad hoc network applications and security: A study into the economic and the legal implications. *International Journal of Electronic Security and Digital Forensics*, *6*(2), 115–129. doi:10.1504/IJESDF.2014.063107

Seuwou, P., Patel, D., & Ubakanma, G. (2014). Vehicular Ad hoc Network Applications and Security: A Study into the Economic and the Legal implications. *International Journal of Electronic Security and Digital Forensics*.

Shahin, S. (2018). Big data and the illusion of choice: Comparing the evolution of India's Aadhaar and China's Social Credit. *Social Science Computer Review*. Advance online publication. Retrieved October 10, 2020, from. doi:10.1177/0894439318789343

SNG. (2017, July 13). *Report: Smart city investment will double by 2021*. Retrieved October 13, 2020, from StateScoop: https://statescoop.com/report-smart-city-investment-will-double-by-2021/

Stoner, L. (2013). The Importance of Global Citizenship to Higher Education: The Role of Short-Term Study Abroad. *British Journal of Education Society & Behavioural Science*, *3*(2), 184–194. doi:10.9734/BJESBS/2013/2910

TachoverD. (2020). Retrieved from https://childrenshealthdefense.org/news/what-you-should-know-about-5g-satellites-how-musks-sci-fi-dreams-are-becoming-our-living-nightmare

Toness, R. V. (2014). *India building database to unite records for 1.2 billion*. Retrieved from https://www.bloomberg.com/news/articles/2014-12-10/india-building-database-to-unite-records-for-12-billion

Toppeta, D. (2010). The Smart City Vision: How Innovation and ICT Can Build Smart, "Livable". *The Innovation Knowledge Foundation*. Retrieved from http://www.thinkinnovation.org/file/research/23/en/Top peta_Report_005_2010.pdf

Tredinnick, L. (2019). Cryptocurrencies and the blockchain. *Journal Indexing and Metrics, 36*(1).

U.N. (2020b). *The Age of Digital Interdependence*. United Nations.

U.N. (2020a). *The Impact of Digital Technologies*. Retrieved November 26, 2020, from United Nations: https://www.un.org/en/un75/impact-digital-technologies

UN. (2020a). *United Nations - The Impact of Digital Technologies*. Retrieved November 26, 2020, from https://www.un.org/en/un75/impact-digital-technologies

UN. (2020b). *The Age of Digital Interdependence*. United Nations.

UNICEF. (2013). *Global citizenship: a high school educator's guide (grades 9-12)*. UNICEF. Retrieved October 13, 2020, from https://www.gcedclearinghouse.org/resources/global-citizenship-high-school-educators-guide-grades-9-12

Unique Identification Authority of India. (n.d.). *About Aadhaar*. Retrieved from https://uidai.gov.in/youraadhaar/about-aadhaar.html

United, N. (2018). *Revision of World Urbanization Prospects*. United Nations. Retrieved October 17, 2020, from https://www.un.org/development/desa/publications/2018-revision-of-world-urbanization-prospects.html

WallSteet. (2018, October 26). *Bit coin the next world currency*. Retrieved from WallStreet: https://wall-street.com/bit-coin-the-next-world-currency/

Washburn, D. S. (2010). *Helping CIOs Understand "Smart City" Initiatives: Defining the Smart City, Its Drivers, and the Role of the CIO*. Forrester Research, Inc. Retrieved from http://public.dhe.ibm.com/partnerworld/pub/smb/smart

West, A. (2014). *In US, government hacking is '40 years ahead' of the language in privacy laws*. The World. Available at https://www.pri.org/stories/2014-03-28/us-government-hacking-40-years-ahead-language-privacy-laws

Williamson, B. (2015). Educating the smart city: Schooling smart citizens through computational urbanism. *Big Data & Society*, *2*(2), 1–13. doi:10.1177/2053951715617783

Winkowka, J., Szpilko, D., & Pejic, S. (2019). Smart city concept in the light of the literature review. *Internationa Society for Manufaturin, service and Management Engineering*, 70-86.

Wired. (2017). *Big data meets big brother as China moves to rate its citizens*. Retrieved from https://www.wired.co.uk/article/chinese-government-social-credit-score-privacy-invasion

Wyllie, D. (2013, August 20). *How 'Big Data' is helping law enforcement*. Retrieved October 19, 2020, from Police1.com: https://www.police1.com/police-products/software/data-information-sharing-software/articles/how-big-data-is-helping-law-enforcement-LLLcFsngeq28DKVz/

Zanella, A., Bui, N., Castellani, A., Vangelista, L., & Zorzi, M. (2014). Internet of things for smart cities. *IEEE Internet of Things Journal*, *1*(1), 22–32. doi:10.1109/JIOT.2014.2306328

Chapter 17
The Role of Artificial Intelligence and Fifth-Generation (5G) Technology in Mitigating Cyber Physical Attacks on Power Systems

Kehinde Oluwafemi Olusuyi
Federal University Oye Ekiti, Nigeria

Paul Kehinde Olulope
Ekiti State University, Ado Ekiti, Nigeria

Abiodun Ernest Amoran
Federal University Oye Ekiti, Nigeria

Eno Edet Peter
Federal University Oye Ekiti, Nigeria

ABSTRACT

The present-day electric power system is an evolving cyber-physical system. Researchers and industry players in the energy world continue to deploy new technologies towards making the electric power system a smarter grid. This involves the integration of information, communication, and control technologies into the existing power grid in order to improve its stability, security, and operational efficiency. Reliance of the modern power system's applications such as state estimation, sequential control and data acquisition (SCADA) systems, phasor measurement units (PMUs), etc. on open communication technologies including the internet has exposed the smart grid to various vulnerabilities, threats, and cyber-physical attacks. This chapter seeks to exploit the robust synergy which exists between artificial intelligence (AI) and fifth-generation (5G) technology to mitigate these challenges. A comprehensive review of techniques which have hitherto proven efficient and/or effective in mitigating identified challenges was carried out with a view to availing researchers of future directions.

DOI: 10.4018/978-1-7998-6992-4.ch017

1. INTRODUCTION

The electric power system at generation, transmission and distribution levels comprises interconnection of various smart, intelligent electronic devices together with other components which help in the efficient utilization of electrical energy. The integration of Renewable Energy (RE) and other non-conventional sources to the power grid has made it a complex network. The resulting smart grid has both cyber and physical components of various monitoring, measuring and communication infrastructure. It is thus a good example of a cyber-physical system (CPS). Reliance of the modern power system's applications such as state estimation, Supervisory Control and Data Acquisition (SCADA) systems, Phasor Measurement Units (PMUs) on open communication technologies including the internet has exposed the smart grid to various vulnerabilities and threats (Alimi et al, 2020).

Since the early 20th century, utilities have increasingly relied on automation to keep up with exponential increases in electricity demand and consumption, as well as reducing need for manpower. Utilities have steadily adopted increasing levels of system protection, automation and control capability to ensure the highest levels of reliability. As reliability levels and energy efficiencies have increased so has the demand for real time information and the expectancy of reliability. These advancing data needs and system requirements have continued to push for even broader automation and control capabilities (Mission Support Center Analysis Report, 2016).

The electric system of today relies on an advanced transmission system, market operations, independent power producers, system operators, as well as the traditional vertically integrated utilities to ensure reliability. The engineering of the power system and how the system is operated has been a very dynamic environment and in many of the new automation and control elements the potential of cyber threats was unrecognized at the time the systems were adopted. As utilities have upgraded Industrial Control Systems (ICS) and electric grid technology to meet present-day needs, the practicality of digital automation and data transfer has become necessary. A variety of vulnerabilities have emerged, particularly related to greater accessibility as a result of advanced communication means and Internet connectivity (Mission Support Center Analysis Report, 2016). These vulnerabilities include:

1. Various point-of-entry and paths that can be exploited by threat actors, sometimes remotely.
2. An ever-increasing number of new vulnerabilities introduced to ICS via system additions or integration.
3. Easier/broader access to systems and networks due to greater connectivity, including devices directly connected to the Internet.
4. Greater requirements for sensitive data collection and exchange (equipment operating status, power production and consumption, electricity pricing, etc.) among utilities, market coordinators, and customers create additional network paths and connections.

A primary purpose of the smart grid is to deploy digital communication networks (e.g. Ethernet, cellular service and satellite signal) to enable data acquisition and remote control between control centres and the large number of power grid facilities (e.g., substations and power plants). Due to the installation of Intelligent Electronic Devices (IEDs) on power grids, power system operators are able to monitor and control a power system from a remote control centre. These remote control and monitoring technologies are based on Information and Communications Technology (ICT). As a result, vulnerabilities with respect to cyber intrusions also become a serious concern (Sun, et al, 2016).

The rest of the paper is divided into various sections and sub-sections. Section 2 focuses on the electric power system as a cyber-physical system, section 3 discusses artificial intelligence applications for power system security and stability, section 4 is on mobile wireless technology, fifth-generation (5G) technology, the technical attributes of 5G, role of artificial intelligence (AI) in 5G deployment, and 5G networks security, while section 5 discusses mitigation techniques of cyber-physical power systems using AI and 5G technology, and then, the last section is the conclusion in that order.

2. THE ELECTRIC POWER SYSTEM AS A CYBER-PHYSICAL SYSTEM

The grids today have evolved from a network of transmission lines into a cyber-physical system where bi-directional information flow plays an important part in the decision making process. Integration of high speed communication links to the existing power network has made it more dynamic in terms of demand response and other energy management capabilities. However, increased reliance on information networks has rendered the system vulnerable to threats associated with security breaches in communication networks. In the paradigm of smart grids, security threats can come from malicious users violating communication protocols attempting to delay, block or deny the availability of information. These are classified under Distributed Denial of Services Attacks (DDoS) (Chatterjee et al, 2017).

The recent development of advanced information and communication technology such as fifth-generation (5G) technology has made modern power systems evolve into more complicated cyber-physical systems. 5G promises a dramatic increase in capabilities over earlier generations, particularly in its ability to support Internet of Things (IOT) applications requiring almost instantaneous data flows at very high data bandwidths (INSA Cyber Council, 2019).

Sridhar et al (2012) highlights the significance of cyber infrastructure security in conjunction with power application security to prevent, mitigate, and tolerate cyber-attacks on power systems. These attacks may be physical or cyber. According to Watts (2016), physical attacks on the power system can be in three dimensions:

1. Attacks upon the Power System – Here the target is the electric infrastructure. For example, terrorists could attack simultaneously two substations or key transmission towers in order to cause a blackout in a big area of grid. Another example could be an attack to the electric market.
2. Attacks by the Power System – Terrorists could use some installation, for example, using power plant cooling towers to dispense chemical or biological agents.
3. Attacks through the Power System – Terrorists could use some installations of the power system to attack civil infrastructure, for example, terrorists could couple an electromagnetic pulse through the grid to damage computer or telecommunication infrastructure (Watts, 2016).

2.1 Common Control System Vulnerabilities/ Weaknesses

2.1.1 Software/product Security Weaknesses

1. Improper input validation
2. Poor code quality
3. Permissions, privileges, and access controls

4. Improper authentication
5. Insufficient verification of data authenticity
6. Cryptographic issues
7. Credentials management
8. Configuration and maintenance

2.1.2 Configuration Weaknesses

1. Permissions, privileges, and access controls
2. Improper authentication
3. Credentials management
4. Security configuration and maintenance
5. Planning/ policy/procedures
6. Audit and accountability configuration

2.1.3 Network Security Weaknesses

1. Common network design weaknesses
2. Weak firewall rules
3. Network component configuration (implementation) vulnerabilities
4. Audit and accountability

Khan et al (2019) categorized threats and attacks on the 5G network into six categories. These are as stated in figure 1 below:

Figure 1. Types of Evolved Security threats for 5G Networks
Source: Khan, et al (2019)

A massive cyber-attack occurred on Ukraine's power system in December 2015. More than ten thousand homes and facilities experienced a power outage for hours, even days. This attack was enabled by a malware called "BlackEnergy" installed on the control centre's computers. This cyber intrusion event shows that attackers can damage a large-scale ICT network in a short time (Sun et al, 2016). Similarly, in Ukraine, researchers have found the first real-world malware that attacks physical infrastructure since Stuxnet. The researchers described that malware, which they alternately named "Industroyer" or "Crash Override," as only the second-ever known case of malicious code purpose-built to disrupt physical systems. The first, Stuxnet, was used by the US and Israel to destroy centrifuges in an Iranian nuclear enrichment facility in 2009 (Greenberg, 2017). An attack campaign known as Dragonfly 2.0 targeted Western energy companies with a variety of infection vectors. The series of attacks constitutes the latest push from Dragonfly, a threat actor which has been around since at least 2011 but then reemerged in 2014. Available evidence suggests the Dragonfly 2.0 attack campaign has been underway since December 2015 (Bisson, 2017). The foregoing are examples of cyber-physical attacks on power systems as listed in Figure 2 below.

Figure 2. Examples of Cyber-attacks on the Electric Grid Source: Resilient Energy Platform (2020)

Cyberattacks on the Electric Grid: Recent Examples

Year	Malware used	Target	Goal	Reference
2015	Laziok	Energy companies worldwide	Information-gathering	(Paganini 2015)
2015	BlackEnergy 3	Ukrainian electric distribution company	Power outage to 225,000 customers	(Lee, Assante, and Conway 2016)
2015-2017	Dragonfly 2.0	Energy companies in the Western United States	Information-gathering, potential access to operational systems	(Bisson 2017)
2016	Crash Override	Ukrainian electric transmission substation	Power outage to one-fifth of Kiev	(Greenberg 2017)
2019	Basic hacking toolkits	An electric utility in the Western United States	Disruption of internet-based communication	(Marks 2019)

3. ARTIFICIAL INTELLIGENCE APPLICATIONS FOR POWER SYSTEM SECURITY AND STABILITY

Owing to its flexibility and strong computational power, Artificial Intelligence (AI) has been extensively applied in the area of power system security and stability. Machine Learning (ML) and Computational Intelligence (CI), subsets of AI, have been deployed by researchers to mitigate challenges of the modern power system. Different computational strategies, agglutinated under the perspective of Artificial Intelligence (AI), have played significant roles in the development of smart grid systems (Santofimia-Romero et al, 2011). The smart grid is enhanced by AI techniques such as Complex event processing, qualitative reasoning, common-sense reasoning, multi-agent systems, and planning techniques in the areas of data filtering and processing diagnosis context-awareness as well as energy management and control. Optimization, modelling and monitoring of the smart grid have been realized with computational intelligence techniques such as fuzzy logic, evolutionary computing and artificial neural networks.

Wehenkel (1997) developed a framework that uses machine learning and other automatic-learning methods to assess power-system security. The framework exploits simulation models in parallel to screen diverse simulation scenarios of a system, yielding a large database. Using data-mining techniques, the framework extracts synthetic information about the simulated system's main features from this database.

Artificial Intelligence (AI) has played a significant role in the realization of the modern power grid. AI has helped in the deployment of renewable energy sources, real-time and dynamic stability assessment, wide-area monitoring and control, and dynamic energy management, to mention but a few. However, due to limitations in the state-of-the-art methods of AI employed, the smart grid is further open to vulnerabilities, threats and other cyber-physical attacks. These "AI attacks" are fundamentally different from traditional cyber-attacks (Comiter, 2019). One major challenge that gives researchers concern on the smart grid is how to intelligently manage the vast amount of information generated on it so that conclusions and inferences could be drawn to support the decision making process. This challenge is being mainly addressed from the perspective of Complex Event Processing (CEP) techniques. CEP techniques address event filtering to seek for relevant patterns (Santofimia-Romero et al, 2011).

Harley and Liang (2011) opined that Computational Intelligence (CI) could smarten a grid in many ways. Offline CI-based methods could assist the operator with anticipatory information. Furthermore, when the grid is stressed, CI-based multi-variable nonlinear optimal controllers can control the grid more reliably and more rapidly than humans, without requiring a traditional model of the grid.

Gao and Shi (2020) proposed an attack-mitigation dynamic game-theoretic scheme for security vulnerability analysis on transmission lines in which recovery time delay and Denial-of-Service (DoS) attacks were investigated. An improved Particle Swarm Optimization (PSO) approach combined with sequential quadratic programme was leveraged in order to solve the proposed optimization model. Similarly, Jiang et al (2019) developed a method employing particle swarm optimization to search for the critical attack sequence that can trigger significant line-switching attacks thereby resulting in cascading failure on the power systems. This method proved to be more efficient than its reinforcement learning counterpart when tested on the IEEE 24-bus reliability test system as well as the Washington 30-bus dynamic test system. In recent times, machine learning techniques (MLTs) have proven to be effective in numerous applications including power system studies. In the literature, various MLTs such as artificial neural networks (ANN), Decision Tree (DT), support vector machines (SVM) etc, have been proposed, resulting in effective decision making and control actions in the security and stable operations of the power system (Alimi et al, 2020).

Historically, when SCADA systems were first deployed, the major threat was sabotage through the physical destruction of the utility's hardware as the old SCADA systems had private and dedicated networks that are secured by traditional air-gapped separations. However, over the past two decades, SCADA networks have been equipped with IoT devices that sometimes communicate over open channels which expose the networks to numerous vulnerabilities and network-based cyber-attacks. The Industrial Control Systems Cyber Emergency Response Team (ICS-CERT) announced that, out of the 245 recorded cyber incidents on Critical Infrastructure (CI) in 2014, 79 were targeted at the energy sector (Alimi et al, 2020). Ten et al (2008) proposed a vulnerability assessment framework for cybersecurity in SCADA systems. According to the authors, vulnerability assessment of the SCADA system is performed in three levels: system, scenarios, and access points. The proposed framework is composed of two aspects: 1) cyber-net model, and 2) power flow simulation. A cyber-net defines the intrusion scenarios and its events and status. Power flow is the most basic model of the steady state behavior of a power system. The integration of these two models makes it possible to quantify the impact caused by a potential cyber-attack.

Monitoring of power systems plays an important role in security analysis and control of power systems. Traditionally, a security assessment is performed using the measurements obtained from the SCADA systems. However, these data were not sufficient to calculate the measurements across the whole power system network as it was not time-sychronized (Thirugnanasambandam and Jain, 2008). In recent years, various methods of security assessment using phasor measurements have been reported. In order to increase the prediction accuracy and reduce the false misclassification rate attributable to support vector machines (SVMs) used for security assessment and classification, Thirugnanasambandam and Jain (2008) proposed an Adaptive Boost (AdaBoost) SVM classifier that combines the AdaBoost algorithm and the SVM. The AdaBoost algorithm combines the learning ability of many weak classifiers to design a strong classifier.

In a research carried out by Diamantoulakis et al (2015), big data issues and challenges faced by dynamic energy management (DEM) employed in smart grid (SG) networks were highlighted. Energy management systems (EMSs) in SGs include 1) real-time wide-area situational awareness (WASA) of grid status through advanced metering and monitoring systems, 2) consumers' participation through home EMSs (HEMS), demand response (DR) algorithms, and vehicle-to-grid (V2G) technology, and 3) supervisory control through computer-based systems (Diamantoulakis et al, 2015).

Secure and reliable electric power grid operation is important to social wellbeing. In recent years, many blackouts due to natural disasters have been recorded such as the 2005 hurricane Katrina blackouts, 2011 Japan Earthquake blackouts, and 2012 Hurricane Sandy blackouts (Wang et al, 2015). Researchers have employed artificial intelligence techniques to harden the grid and increase the resilience of the power grid under natural disasters. At the same time, new technologies such as smart grid, micro grid, and wide area monitoring applications that could increase situational awareness as well as enable faster restoration of the system have been developed (Wang et al, 2015).

4. MOBILE WIRELESS TECHNOLOGY

Mobile technology has developed over the years from First-generation (1G) systems introduced in the 1980s (Gawas, 2015) all through to the present day fifth-generation (5G) mobile wireless technology. 5G is the next-generation of wireless technology that represents a complete transformation of telecommunication networks. Combining new and legacy technology and infrastructure, 5G will build upon previous generations in an evolution that will occur over many years, utilizing existing infrastructure and technology (CISA, 2019), (Fonyi, 2020). Table 1 below gives a basic comparison among 3G, 4G, and 5G mobile technologies.

Table 1. Basic comparison among 3G, 4G, and 5G technology

Technology/ features	3G	4G	5G
Data Bandwidth	2Mbps	2Mbps to 1Gbps	1Gbps & Higher (as demand)
Frequency Band	1.8 - 2.5 GHz [16]	2 - 8 GHz [16]	3-300GHz [16],[18]
Standards	WCDMA CDMA-200 TD-SCDMA [19]	All access convergence including:OFMDA,MC-CDMA Network-LMPS [19]	CDMA & BDMA
Technology	Broad bandwidth CDMA,IP technology [19]	Unified IP And seamless combination of broadband LAN/WAN/ PAN and WLAN [19];	Unified IP and seamless combination of broadband, LAN/WAN/PAN/WLAN [19] and technologies for 5G new deployment (could be OFDM etc.);
Service	Integrated high quality audio, video and data	Dynamic information access, wear-able devices, HD streaming; global roaming;	Dynamic information access, wear-able devices, HD streaming; any demand of users; upcoming all technologies; global roaming smoothly;
Multiple Access	CDMA	CDMA	CDMA & BDMA
Core Network	Packet Network	All IP Network	Flatter IP Network & 5G Network Interfacing(5G-NI)
Definition	Digital Broadband, packet data	Digital Broad band, Packet data, All IP	Digital Broadband, Packet data All IP, Very high throughput
Hand off	Horizontal	Horizontal & Vertical	Horizontal & Vertical
Start from	2001 [12]	2010 [12]	2015 [12]

Source: Hossain, (2013)

4.1. Fifth-Generation (5G) Technology

Defined by the International Telecommunication Union (ITU), 5G systems will have three major scenarios. The first is enhanced mobile broadband (eMBB), which targets extremely high data rate to fulfil the high-speed data access requirement of emerging services such as 3D and ultra-high-definition (UHD) video transmission, and virtual reality (VR) applications. The second is massive machine-type communications (mMTC), which aims to provide high connection density (up to 200, 000 devices/km^2) and low data rate (1 to 100kb/s per device) at low power consumption (up to 15 years battery life) to fulfil requirements of sensor networks used for smart city, the Internet of Things, and wearable device networks. Third, the ultra-reliable low-latency communication (URLLC) scenario aims to provide extremely high reliability (999.999 percent) and low latency ($<$ 1 ms) wireless services used for control networks such as smart meters, high-speed train control, transport safety control, remote medical surgery services, and industrial robotic control (Fu, et al, 2018), (Huawei Technologies, 2019). Figure 3 below illustrates the three major use-case scenarios of the 5G technology.

5G is supported by LAS-CDMA, OFDM, MC-CDMA, UWB, Network-LMDS, IPv6. It can be called the perfect real wireless world or World Wide Wireless Web (WWWW) as it has no limitations. The basic protocol for running on both 4G and 5G is IPv6. The 5G mobile is all-IP based for mobile and wireless network interoperability. Figure 4 shows the comparison of the frequency spectrum for 4G and 5G.

Figure 3. 5G usage Scenarios
Source: Joint radio Company Ltd (2019)

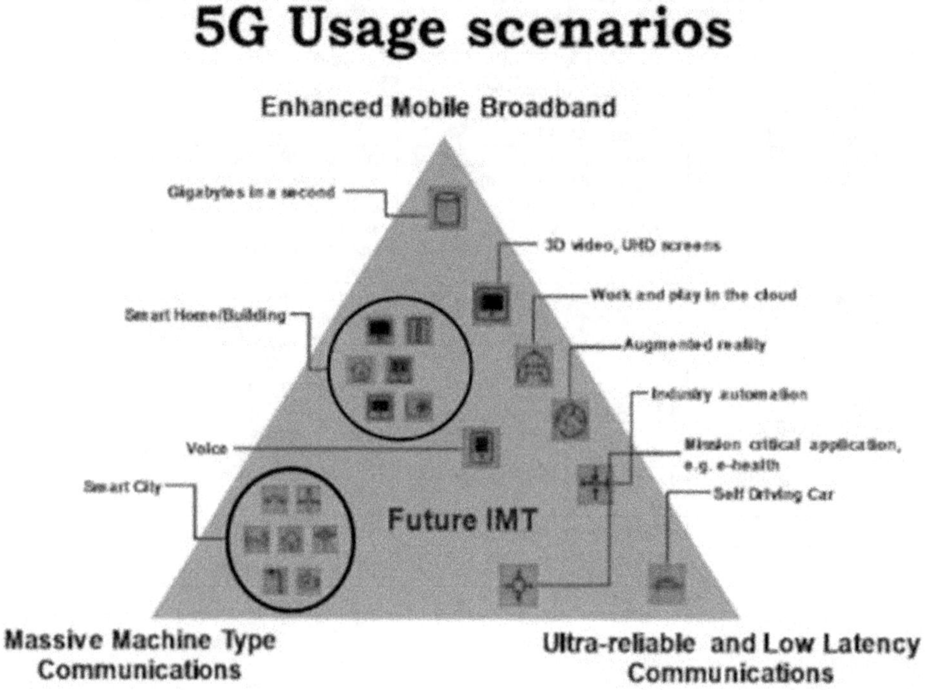

Figure 4. 4G and 5G Frequency Spectrum
Source: (CISA, 2019)

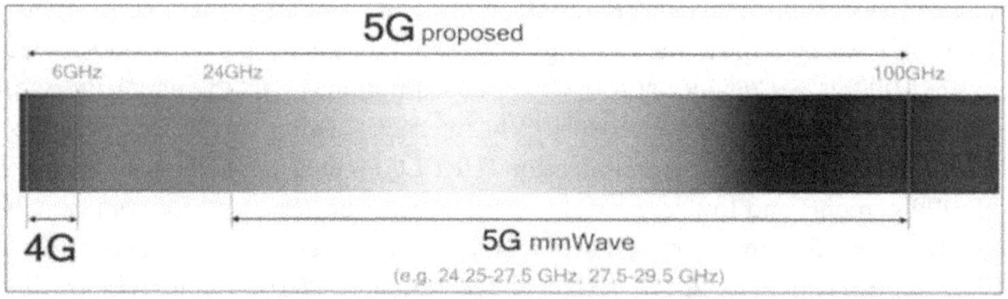

According to GSMA, “5G is more than just an evolutionary step to a new generation of technology; it represents a fundamental transformation of the role that mobile technology plays in society. As demand for continuous connectivity grows, 5G is an opportunity to create an agile, purpose-built network tailored to the different needs of citizens and the economy” (GSMA, 2019). It is a change in direction of the future of communication as incoming generations technologies will evolve from 5G especially with its instantaneous, high-powered connectivity to billions of devices. Among other benefits of 5G enables machines communication without human input in an Internet of Things (IoT) capable of driving a near-endless array of services. It facilitate safer, more efficient and cost-effective transport networks. It will offer improved access to medical treatment, reliably connecting patients and doctors all over the globe.

4.2. The Technical Attributes of 5G

Some of the technical attributes of 5G include (Pwc, 2020):

1. Network slicing, which provides a way for service providers to enable network-as-a-service (NaaS) to specific subscriber groups, giving them the flexibility to manage their own devices and services according to specific needs.

Figure 5. 5G Network Slicing Graphic
Source: Joint radio Company Ltd (2019)

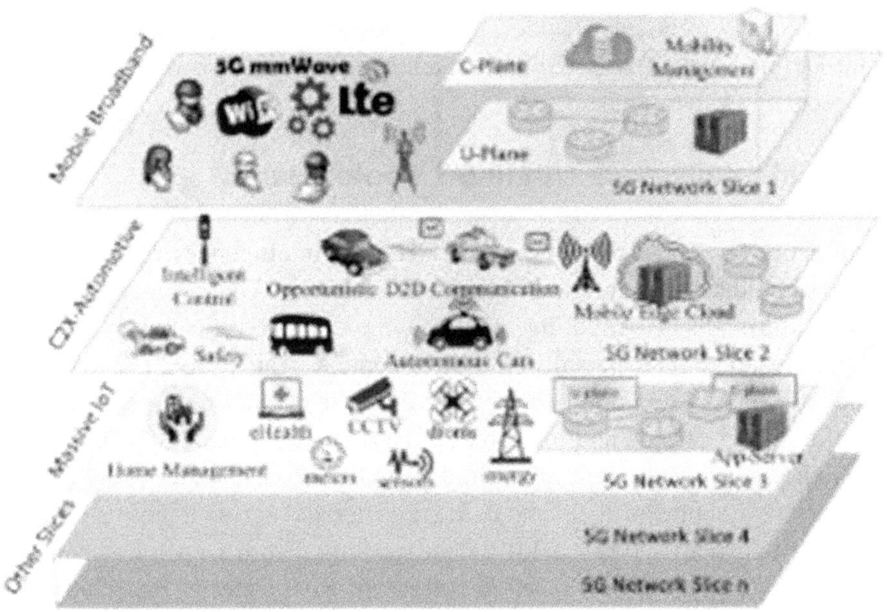

2. Enhanced mobile broadband (1-20 Gps), which supports applications such as 3D video transmission with 4k or 8k resolution screens, online gaming and so on.
3. Ultra-low latency (<1 ms), which is important for mission-critical services such as augmented reality (AR) and virtual reality (VR), telemedicine and healthcare, intelligent transportation, and industry automation.
4. Massive device connectivity for vehicles, mobile subscribers' enterprises, IoT and the like.
5. High availability and dense coverage, which will make it capable of providing unlimited connectivity for billions of different subscribers.

6. Low energy consumption, with up to ten-year battery life for M2M (machine-to-machine) communications.

The world is gradually moving from Internet of Things (IoT) to Industrial Internet of things (IIoT) which is practically based on the application of communication technologies to more complex industrial processes and the advantages of 5G is more suitable for this. According to Mockel & Makala, (2019), "economies that lacked 5G will eventually find themselves at a clear disadvantage."

4.3 5G Networks Security

The security architecture, procedures and requirements for 5G systems will be focusing on security standards for massive Machine Type Communication and Ultra-Reliable Low Latency Communications. The new security features aim to provide E2E security along with flexibility of incorporating multiple authentication frameworks, and higher-layer security protocols to support security for Service Based Architecture (SBA) in 5G. The SBA and network slicing in 5G networks allow higher modularity in the design measure of security protocols (Haider, 2020). 5G networks support virtualized architecture along with Software-Defined Networking (SDN), which can greatly improve security by introducing additional layers of security (Mavropoulos, 2020).

4.4. Role of Artificial Intelligence (AI) in 5G Deployment

Many researchers have conducted studies on the role of Artificial Intelligence in 5G technology development and deployment. Candidate AI techniques include deep learning (DL), deep reinforcement learning (DRL), etc. Recent breakthroughs in computing and AI have encouraged researchers in the field of wireless communications to leverage AI especially in the 5G context (Yao et al, 2019). An AI-defined 5G network provides the base stations (BSs)/Cloud with capability to create a cognitive and comprehensive data repository by splitting, processing and interpreting the operational data. (The BS is defined in this article as the remote radio head plus the centralized and virtualized baseband unit pool of a single mobile network operator.) Massive amounts of real-time data are generated across a large number of users and ranges from channel state information (CSI) to IoT device readings. From a machine-centric communication perspective, big data analytics are leveraged to extract massive patterns, especially at the physical (PHY) and medium access control (MAC) layers and enable self-organizing operations (Yao et al, 2019)

Current research on 5G network traffic management has already driven conventional approaches purely based on communication theory to the limit. It will be extremely challenging to solve the traffic management problem for networks and to achieve global optimal performance for the whole network. A primary direction to tackle the challenges is to adapt AI technologies to analyze and manage the traffic of 5G networks from network data. Machine learning (ML) and deep learning (DL) are two advanced AI methodologies that have attracted lots of interest to overcome the challenges of managing 5G network traffic (Fu et al, 2018).

5. MITIGATION TECHNIQUES OF CYBER-PHYSICAL ATTACKS ON POWER SYSTEMS USING AI AND 5G TECHNOLOGY

The primary task of mitigation actions is to ensure the reliability and stability of a power system. Mitigation actions are activated as soon as IDSs report an attack event. In general, mitigation methods are categorized into two parts: cyber and physical systems. For the cyber system, the mitigation techniques are aimed at verifying the legitimacy of on-line users and network packets. For the physical system, the mitigation control strategies are applied to maintain the stability of power systems (Sun et al, 2016).

AI and 5G technologies have shown good performance in the mitigation of cyber-physical threats and attacks on the power system. Many companies offering services in mobile communication technology especially the newly deployed 5G have embedded security features that can effectively tackle most of the cyber threats and vulnerabilities. Companies such as Verizon and Huawei Technologies have developed the 5G tech using security-by-design approach (Verizon, 2019). Other companies offer security as a service on their 5G infrastructure to utilities.

Opportunities for electric utilities to use 5G technology include but not limited to the following:

1. Unmanned aerial vehicle (VAVs) or drones for routine inspections and to discover and assess damage after storms. Such developments may radically enhance inspection capability and reduce restoration times following faults, generating a real business case for 5G deployment.
2. Enhanced and virtual reality. When understanding repairs and maintenance, demonstrations of virtual reality safety helmets are impressive. The helmets have head-up displays and voice activation to enable operatives to call up detailed work constructions, manuals, videos, network status information and infrared images whilst leaving their hands free for work. Such information enhances productivity and safety but requires high data bandwidth. 5G would be an essential facilitator for this enhanced method of working.

Utilities will also wish to participate in the 5G world by acquiring spectrum in order to have the option to construct their own private 5G networks and interface them into a 5G world. These private 5G network will take a variety of forms but will need to be able to integrate and interwork with commercial 5G infrastructure operated by telecommunications providers (JRC, Ltd, 2019). Electric utilities might want to operate private 5G networks for the following reasons:

1. Networks will be able to operate for extended periods in the absence of primary power.
2. Utilities will have access to greater security than offered by commercial networks.
3. Deterministic low latency services.
4. Coverage into areas not served by commercial operators being either remote rural areas, industrial sites with poor coverage, underground locations, tunnels, etc.
5. Redundant telecommunications provision.

The "Internet of things" (IoT) is an extension of the Internet and other network connections to different sensors and devices- or "things". (Attaran, 2021). The smart grid is one example of the application of IoT where components of the electric grid from transformers to power lines to home electric meters have sensors and are capable of two-way communication. The electric company can use the smart grid to manage distribution more efficiently, be proactive about maintenance, and respond to outages faster.

Smart grids integrate traditional power systems with information, communication, and control technology to improve the power system's stability, security, and operating efficiency. The new digitized power generation facility attempts to improve the efficiency of power systems by building a high capacity high-speed, real-time, secure, and stable communications networks. 5G greatly enhances the amount of spectrum used to send and receive data. It can act as an integrator and support the diverse requirements of smart grids. 5G is more efficient and faster than fibre optic and short-range wireless communications technology, supports over-the-air wireless connectivity, and has excellent disaster recovery capabilities. Leveraging cyber-physical systems and striving towards ever more automation and autonomous decisions requires substantial resources to deal with the resulting amount of data and needs to be gathered, analyzed, and transferred. 5G has the potential to bring the reliability, latency, scalability, mobility, and security that is required for mission-critical services in the IoT ecosystem (Attaran, 2021). The introduction of 5G technology will unleash a new wave of smart grid features and improve efficiency tenfold by allowing many devices that are currently unconnected, to be monitored for their energy usage. An important point in term of reliability is the investigation of cross dependencies between Energy Grid and 5G Networks (Energy suppliers need 5G network to monitor their infrastructure and 5G Networks need energy to run their infrastructure) (Mavropoulos, 2020).

There are a variety of specific technical and administrative cyber security measures that electric utilities are currently implementing. These include network firewalls, antivirus software, application control software, encryption of communication data, securing smart grid technology upgrades, intrusion detection systems, etc (Mission Support Center, 2016). AI and 5G technologies are vigorously deployed in all of these in order to improve power system security and stability. The advancements in the ICT has led to the possibility of new technologies such as 5G, edge and cloud computing to be integrated into the SCADA systems. The decisions from the SCADA control room are implemented in the power systems with the help of field controllers such as Intelligent Electronic Devices (IED) and Remote Terminal Units (RTU). Software architecture and protocols form the foundation for SCADA architecture as well as for different devices to communicate with each other (Narayan, 2019). Some utilities currently use encryption software to encrypt SCADA communications to prevent hackers from collecting information in control processes for the electric grid (Mission Support Center, 2016).

Another cyber security mitigation technique is IP fast hopping mechanism. Two categories are defined for these approaches: a) Attack prevention methods, which include refining protocol and overall system security level, resource allotment & accounting, firewalls, etc, and b): Attack detection methods, which include attack source recognition and appropriate reaction. IP fast hopping mechanism is a prevention mechanism that can be proposed to refine even huge malicious streams. It is a new method that can be applied to hide details and endpoints of user's communication period to counteract exhausting of servers resources initiated by Distributed Denial of Service (DDoS) attacks (Pour et al, 2017).

6. CONCLUSION

The 5G technology is still at its early stage of development and the current versions are the non-standalones which are built on top of the 4G LTE infrastructure. Their offerings include URLLC and mMMC (massive Machine-to-Machine Communications). It is not clear that there is a business case for utilities to become engaged in processing the volume of data generated by mMMC to better inform electricity network operations, and it is difficult to understand how this new URLLC technology will lead to higher levels of reliability since that is a function of network design and operation, not technology.

In the area of application of artificial intelligence to manage network traffic of 5G wireless networks, existing reported research has only focused on the core network, and applied ML to solve the routing problem in the core network; there is little research on traffic control with respect to the 5G network. For traffic control, to date most reported research has only focused on the network layer; there are only a few research reports on the application of AI technology to the application layer and the semantic layer that shape the traffic by content recommendation with consideration of user interests (Fu et al, 2018).

The authors of this paper have carried out review of literatures on the development and deployment of artificial intelligence and 5G technologies towards mitigating cyber physical threats, attacks and vulnerabilities on the power system as a cyber-physical system. Important research directions in further development of 5G technology have been highlighted.

REFERENCES

Ahmad, I., Kumar, T., Liyanage, M., Okwuibe, J., Ylianttila, M., & Guvtor, A. (2018). 5G Security: Analysis of Threats and Solutions Artificial Intelligence (AI) for Development. Series Report on AI and IoT in Security Aspects.

Alimi, O. A., Ouahada, K., & Abu-Mahfouz, A. M. (2020). *A Review of Machine Learning Approaches to Power System Security and Stability*. IEEE. doi:10.1109/ACCESS.2020.3003568

Attaran,M. (2021). The Impact of 5G on the Evolution of Intelligent Automation and Industry Digitization. *Journal of Ambient Intelligence and Humanized Computing.*

Ayad, A. (2019). *Cyber-Physical Security of Power Distribution Systems* (Unpublished master thesis). University of Waterloo, Ontario, Canada.

Bisson, D. (2017). Dragonfly 2.0 Attack Campaign Targets Western Energy Sector. *The State of Security*.

Bogale, T.E., Wang, X., & Le, L.B. (2018). Machine Intelligence Techniques for Next-Generation Context-Aware Wireless Networks. *ITU Journal: ICT Discoveries.*

Cardenas, A. (2019). *Cyber-Physical Systems Security*. Academic Press.

Carlsson, A. (2019). *On the Use of 5G for Smart Grid Inter-Substation Control Signalling* (Unpublished master thesis). Karlstad University.

Chatterjee, K., Padmini, V., & Khaparde, S. A. (2017). Review of Cyber Attacks on Power System Operations. *Proceedings of IEEE Region 10 Symposium (TENSYMP)*, 1-6. 10.1109/TENCON-Spring.2017.8070085

Choudhary, G., & Sharma, V. (n.d.). *A Survey on the Security and the Evolution of osmotic and Catalytic Computing for 5G Networks*. Academic Press.

Comiter, M. (2019). Attacking Artificial Intelligence: AI's Security vulnerability and What Policymakers Can Do about It. Harvard Kennedy School, Belfer Center for Science for Science and International Affairs.

Cyber Threat and Vulnerability Analysis of the U.S. Electric Sector. (2016). *Mission Support Center Analysis Report.* Idaho National Laboratory.

Cybersecurity and Infrastructure Security Agency (CISA). (2019). *Overview of Risks Introduced by 5G Adoption in the United States*. CISA.

Diamantoulakis, P. D., Kapinas, V. M., & Karagiannidis, G. K. (2015). *Big Data Analytics for Dynamic Energy Management in Smart Grids.* arXiv: 1504.02424v3 [cs.DB].

Ericsson. (2019). *Employing AI Technologies to Enhance Returns on 5G Network Investments*. Author.

Fonyi, S. (2020). *Overview of 5G Security and Vulnerabilities*. Academic Press.

Fu, Y., Wang, S., Wang, C., Hong, X., & McLaughlin, S. (2018). Artificial Intelligence to Manage Network Traffic of 5G Wireless Networks. *IEEE Network, 32*(November/December). doi:10.1109/MNET.2018.1800115

Gao, B., & Shi, L. (2020, February 10). Modelling ab Attack-Mitigation Dynamic Game-Theoretic Scheme for Security Vulnerability Analysis in a Cyber-Physical Power System. *IEEE Access: Practical Innovations, Open Solutions*, *8*, 30322–30331. Advance online publication. doi:10.1109/ACCESS.2020.2973030

Gawas, A. U. (2015). An Overview on Evolution of Mobile Wireless Communication Networks: 1G–6G. *International Journal on Recent and Innovation Trends in Computing and Communication, 3*(5), 3130-3133.

Global System for Mobile Association (GSMA). (2019). *5G guide a reference for Operators*. GSMA.

Greenberg, A. (2017). Crash Override Malware Took Down Ukraine's Power Grid Last December. *Wired.*

Haider, N., Baig, Z., & Imran, M. (2020). *Artificial Intelligence and Machine Learning in 5G Network Security: Opportunities, Advantages, and Future Research Trends.* arXiv.2007.04490v1[cs.CR]

Harley, R. G., & Liang, J. (2011). Computational Intelligence in Smart Grids. *IEEE Symposium Series on Computational Intelligence (SSCI) 2011-CIASG.*

Hasin, S., Ghafouri, A., Dubey, A., Karsai, G., & Koutsoukos X. (n.d.). *Vulnerability Analysis of Power Systems Based on Cyber-Attacks and Defense Models*. Academic Press.

Hassain, S. (2013). 5G Wireless Communication Systems. *American Journal of Engineering Research, 2*(10), 344–353.

Huawei. (2019). *Position paper on Cyber Security*. Author.

Jiang, H., Wang, Z., & He, H. (2019). An Evolutionary Computation Approach for Smart Grid Cascading Failure vulnerability Analysis. *2019 IEEE Symposium Series on Computational Intelligence (SSCI).*

Joint Radio Company Ltd. (2019). Cutting Through the Hype: 5G and its Potential Impacts on Electric Utilities. doi:10.1109/SSCI44817.2019.9002979

Kafle, V. P., Fukushima, Y., Martinez-Julia, P., & Miyazawa, T. (2018). Consideration on Automation of 5G Network Slicing with Machine Learning. *2018 ITU Kaleidoscope Academic Conference.*

Khan R., Kumar P., Jayakody D. N. K., & Liyanage M. (2019). A Survey on Security and Privacy of 5G Technologies: Potential Solutions, Recent Advancements and Future Directions. *Journal of Latex IEEE Communications Surveys and tutorials.*

Kim, N. Y., Rathore, S., Ryu, J. H., Park, J. H., & Park, J. H. (2018). A Survey on Cyber Physical System Security for IoT: Issues, Challenges, Threats, Solutions. Journal of Information Processing System, 14(6), 1361.

Kim, T. (2010). SCADA Architecture with Mobile Remote Components. *WSEAS Transactions on Systems and Control*, *5*(8).

Krebs, C. (2020). CISA 5G Strategy: Ensuring the Security and Resilence of 5G Infrastructure in our *nation.* US Department of Homeland Security Cybersecurity and Infrastructure Security Agency.

Loghin, D., Cai, S., Chen, G., Dinh, T. T. A., Fan, F., Lin, Q., Ng, J., Ooi, B. C., Sun, X., Ta, Q., Wang, W., Xiao, X., Yang, Y., Zhang, M., & Zhang, Z. (2019). *The Disruptions of 5G on Data-Driven Technologies and Applications.* arXiv: 1909.0806v3 [cs.NI].

Maglaras, L., Ferrag, M. A., Derhab, A., Mukherjee, M., Janicke, H., & Rallis, S. (2019). Threats, Countermeasures and Attribution of Cyber Attacks on Critical infrastructure. *IoT Transactions.* arXiv:1901.03899v1 [cs.CR].

Mavropoulos, O. (2020). D2.1: 5G Security: Current Status and Future Trends. *INSPIRE-5Gplus.*

McClellan, M., Cervello-Pastor, C., & Sallent, S. (2020). Deep Learning at the Mobile Edge: Opportunities for 5G Networks. *Applied Sciences (Basel, Switzerland)*, *2020*(10), 4735. doi:10.3390/app10144735

Mission Support Center. (2016). Cyber Threat and Vulnerability Analysis of the U.S. Electric Sector. Idaho National Laboratory.

Mitra, I., TK, S., Rao, V., Kelkar, V., & Chhaparia, P. (2019). How Cyber-Physical systems are Reshaping the Robotics Landscape. *Cognizant.*

Mockel P., & Makala B. (2019). Artificial Intelligence and 5G Mobile Technology can Drive Investment Opportunities in Emerging Markets. Fresh Ideas about Business in Emerging Markets. *EM Compass.* Note 76.

Mohammed, T., Albeshri, A., Katib, I., & Mehmood, R. (2020). UbiPriSEQ – Deep Reinforcement Learning to Manage Privacy, Security, Energy, and QoS in 5G IoT Hetnets. *Applied Sciences (Basel, Switzerland)*, *2020*(10), 7120. doi:10.3390/app10207120

Morocho-Cayamcela, M. E., & Lim, W. (2018). Artificial Intelligence in 5G Technology: A Survey. *2018 International Conference on Information and Communication Technology Convergence (ICTC)*, 860 – 865. 10.1109/ICTC.2018.8539642

Narayan, A., Kruger, C., Goring, A., Babazadeh, D., Harre, M., Wortelen, B., Luedtke, A., & Lehnhoff, S. (2019). Towards Future SCADA Systems for ICT-Reliant Energy Systems. Internationaler ETG-Kongress 2019.

Nguyen, T. T., & Reddi, V. J. (2019). *Deep Reinforcement Learning for Cyber Security.* arXiv:1906.05799v1 [cs.CR].

Nilsson, N. (1980). *Principles of Artificial Intelligence*. Springer Berlin.

O'Connell E., Moore D., & Newe T. (2020). Challenges Associated with Implementing 5G in Manufacturing. *Telecom MDPI.*

Paganini, P. (2015). Energy Companies Infected by Newly Lazlok Trojan Malware. *Security Affairs.*

Perez, M. G., Celdran, A. H., Giardina, P. G., Bernini, G., Pizzimerti, S., Clemente, F. J. G., Perez, G. M., Festa, G., & Paglianti, F. (n.d.). *Mitigation of Cyber Threats: Protection Mechanics in Federated SDN/NFV Infrastructures for 5G within FIRE+*. Academic Press.

Pollock, J. (n.d.). SCADA: Alive and Well in the Age of IoT. *Microwaves & RF.*

Pour, M. M., Anzalchi, A., & Sarwat, A. (2017). A Review on Cyber Security Issues and Mitigation Methods in Smart Grid Systems. IEEE.

Rappaport, T. S., Muhamed, R., Buehrer, R. M., & Doradla, A. (2000). Mobile and Cellular Radio Communications. In R. C. Dorf (Ed.), *The Engineering Handbook.* CRC Press LLC.

Resilient Energy Platform. (2020). *Cybersecurity and distributed Energy resources.* Author.

Roman, R., Lopez, J., & Mambo, M. (2016). *Mobile Edge Computing, Fog et al: A Survey and Analysis of Security Threats and Challenges.* arXiv: 1602.00484v2 [cs.CR].

Sakhnini, J. (2020). *Security of Smart Cyber-Physical Grids: A Deep learning Approach* (Unpublished master thesis). University of Guelph, Ontario, Canada.

Santofimia-Romero, M., Toro-garcia, X. D., & Lopez-Lopez, J. (2011, October). Artificial Intelligence Techniques for Smart Grid Applications. *CEPIS Upgrade*, *12*(4), 41–48.

Securing 5G's Future: Why Cybersecurity is Key to Realising the Full Promise of 5G Networks. (2020). PWC.

Soldari, D. (n.d.). *5G and the Future of Security in ICT.* Huawei-INTAC-5G-Cyber-Security paper.

Sridhar, S., Hahn, A., & Govindarasu, M. (2012, January). Cyber-Physical System security for the Electric Power Grid. *Proceedings of the IEEE, 100*(1), 210–224. doi:10.1109/JPROC.2011.2165269

Sun, C., Liu, C., & Xie, J. (2016). Cyber- Physical System Security of a Power Grid: State-of-the-Art. *Electronics (Basel)*, *2016*(5), 40. doi:10.3390/electronics5030040

Technology Shapes the Future of Crime and Law Enforcement. (2019). European Union Agency for Law Enforcement Cooperation.

Ten, C., Liu, C., & Manimaran, G. (2008, November). Vulnerability Assessment of Cybersecurity for SCADA Systems. *IEEE Transactions on Power Systems*, *23*(4), 1836–1846. doi:10.1109/TPWRS.2008.2002298

The National Security Challenges of Fifth Generation (5G) Wireless Communications Winning the Race to 5G, Securely. (2019). Intelligence and National Security Alliance (INSA) Cyber Council.

Thirugnanasambandam, V., & Jain, T. (2018). AdaBoost Classifiers for Phasor Measurements-based Security assessment of Power Systems. *The Institution of Engineering and Technology (IET) Journals. IET Generation, Transmission & Distribution, 12*(8), 1747–1753. doi:10.1049/iet-gtd.2017.0013

Thrybom, L., Kaporits, A., & Ekrescom. (Eds.). (2015). *5G PPP White Paper on Energy Vertical Sector*. Version 1.0 30 September 2015.

To-Euro-5G – Supporting the European 5G Initiative. (2019). *The European 5G Annual Journal.*

Trevino, M. (n.d.). Cyber Physical Systems: The Coming Singularity. *Prism – The Journal of Complex Operations, 8*(32019).

Verizon. (2019). *First Principles for Securing 5G: The Design, Deployment, Operation and Innovation of Secure 5G Networks.* White paper by Verizon.

Wang, Y., Chen, C., Wang, J., & Baldick, R. (2015). Research on Resilience of Power Systems under Natural Disasters – A Review. *IEEE Transactions on Power Systems.*

Watts, D. (2003). Security & Vulnerability in Electric Power Systems. In *NAPS 2003, 35th North American Power Symposium.* University of Missouri-Rolla.

Wehenkel, L. (1997). Machine-Learning Approaches to Power-System Security Assessment. *IEEE Expert: AI in Power Systems,* 60-70.

Wurm, J., Jin, Y., Liu, Y., Hu, S., Heffner, K., Rahman, F., & Tehranipoor, M. (2016). *Introduction to Cyber-Physical System Security: A Cross-layer Perspective.* IEEE Transactions on Multi-Scale Computing Systems., doi:10.1109/TMSCS.2016.2569446

Yao, M., Sohul, M., Marojevic, V., & Reed, J. H. (2019). *Artificial Intelligence-Defined 5G Radio Access Networks.* IEEE Communications.

Yohanandhan, R. V., Elavarasan, R. M., Premkumar, M., & Mihet-Popa, L. (2020). Cyber-Physical Power System (CPPS): A Review on Modelling, Simulations, and Analysis with Cyber Security Applications. *IEEE Access : Practical Innovations, Open Solutions.* Advance online publication. doi:10.1109/ACCESS.2020.3016826

You, X., Zhang, C., Tan, X., Jin, S., & Wu, H. (2019). AI for 5G: Research Directions and Paradigms. *Science China. Information Sciences, 62*(2). https://doi.org/10.1007/s11432-018-9596-5

Chapter 18
CLOUDSME:
A Framework for Evaluating SME Adoption of SAAS in the Cloud

Richard Otuka
University of East London, UK

ABSTRACT

Presently, SMEs are finding it difficult to adopt cloud services for their businesses due to various service providers offering similar services. In addition, little work has been carried out in regards to the cloud services adoption process by SMEs. In this chapter, the authors propose CLOUDSME, a novel framework that aids in the adoption process of SaaS cloud services. Accordingly, they implement a decision support system, which includes an ontology of cloud services knowledge within the proposed framework. Analytical hierarchical process (AHP) is used to determine the weight of each cloud service attribute, and a benchmark is set to determine the acceptability of each cloud service based on its ability to meet the acceptable benchmark for each criteria. It can also help in a healthy competition to improve the quality of service among cloud service providers. The CLOUDSME semantic model will guide SME owners in answering user requirements towards decision making in the cloud service adoption process.

1. INTRODUCTION

Cloud computing in recent years is seen as one of the biggest breakthroughs in technology. A large number of cloud service providers exist with each prioritising on various aspects of cloud services data storage, file sharing, CRM, web hosting, programming framework environment, Data centres etc. To remain competitive in cloud technology offerings, these service providers have made an effort towards easy accessibility to their services, which are known to offer tremendous benefits such as reduction in operational cost and eradication of upfront investment for businesses. However, despite this promising technology SMEs are still very slow towards adopting cloud services and this could be due to the growing number of growing number of service providers, Lack of knowledge of cloud services, security issues

DOI: 10.4018/978-1-7998-6992-4.ch018

in the cloud, reliability, interoperability, lack of decision support tools and frameworks(Jagannathan, 2012; Aljabre, 2012 & Ashwini et al, 2012).

The continuous research towards finding a solution to the slow adoption rate of ICT especially cloud computing is the key motivation to this paper. For example in (Verizon, 2009) survey on ICT adoption by small businesses, in its findings reported that 39% of small businesses implemented technology to advertise and promote their business compared to 9% who adopted technology for their business process. Against the above background, the SME owners/ managers who are also willing to adopt this technology are faced with the challenge of selecting service delivery models such as private, public and hybrid clouds (Jagannathan, 2012). At the moment, Potential adopters either go on service provider websites or blogs to manually read about what cloud offerings, price per usage as well as what SLAs are available. This complex approach towards retrieving cloud service knowledge is considered as a barrier to cloud service adoption particularly for SMEs with limited knowledge and experience on services provided.

Generally, cloud services are divided into three layers (Buyya et al, 2009): Software as a Service (SaaS), Platform as a Service (PaaS), Infrastructure as a Service (IaaS).The top layer focus on application services (SaaS) which is usually the cloud interface that allows computer users have access using a web browser and thin computer terminals. The SaaS and PaaS layers, which are, also known as application and software environment layer are built upon the lower layer which is the IaaS layer. The upper layers are mostly developed and provided by a third party service provider while the service suppliers of the IaaS are different as they focus more on the data centre (Fortis et al, 2012). Depending on the needs of an organisation, SaaS services are adopted based on service application requirements, an example is the (CRM) Customer Relationship Management application provided by Salesforce (Cusumano, 2010). PaaS offers a platform for adoption in instances where a business is interested in the development of other applications on top of it, an example is Google App Engine (Ciurana, 2009). While IaaS is adopted as an environment for the deployment, running and management of virtual machines and storages. Finally, it is the IaaS layer that offers on-demand storage in terms of incremental scalability of computer resources (Buyya et al, 2009).

Furthermore, public cloud offerings makes it more difficult in terms of decision making by SME owners who are trying to determine which cloud services fulfil their business requirements. Most cloud service providers offer similar services in different packages, such as pricing, file size restriction and storage, security, operating systems supported, interoperability, fault tolerance, load balancing, programing framework, SLA,. While one service is better on one feature it might not be on another or even more expensive for the same service. Finding the convenient cloud service that satisfies user requirements giving the diversity of cloud service offerings has become a significant challenge especially to SMEs (Wu et al, 2012). It is important that SME managers who want to adopt cloud services are able to evaluate which cloud service is most suitable for their business based on certain criteria and also the ability to choose a cloud service in real time.

Against the above background, we are proposing (CLOUDSME) a framework for evaluating cloud service, which are equipped with an ontology for aiding SME owners/managers in making adequate decisions towards cloud service adoption. Our proposed framework is based on advertised cloud service offerings by service providers and can also be used to compare how various SaaS cloud service providers tend to address user requirements based on individual attributes such as pricing, supported operating systems, file size restriction, storage etc. We go a step further to design an ontology that can act as a decision-making guide towards cloud service adoption by SMEs.

In order to tackle the challenge of outright choosing of one cloud service over another based on only one attribute, which does not show a comprehensive comparison, because they may not be as good in another quality, we adopt the pairwise comparison matrix. We do this by matching SaaS cloud service offerings head to head. The ranking of the cloud service offerings is done using Analytical Hierarchical Process (AHP) (Garg et al, 2013) as a potential decision-making method for ranking cloud services. The application of AHP gives us the ability to prioritise cloud service offerings of each of the cloud service against a descending order list of cloud services in order to select the best cloud service the meets user requirements towards cloud service adoption. A sensitive analysis is performed to check the sensitivity of the final decisions to minor changes in judgement. Finally, protégé an ontology editor which supports description logic reasoning using software like pellet is used to design a cloud service ontology which represents our analytic findings based on a set of rules to give (decision support reasoning)) consistent and accurate feedback. The remaining section of this paper is organised as follows: In Section 2) The CLOUDSME framework architecture Section 3) Using AHP as a decision- making tool Section 4) Case study of advertised cloud services Section 5) cloud service rankings Section 6) System Ontology Section 7) related works 8) conclusion and future works.

Figure 1. CLOUDSME Framework

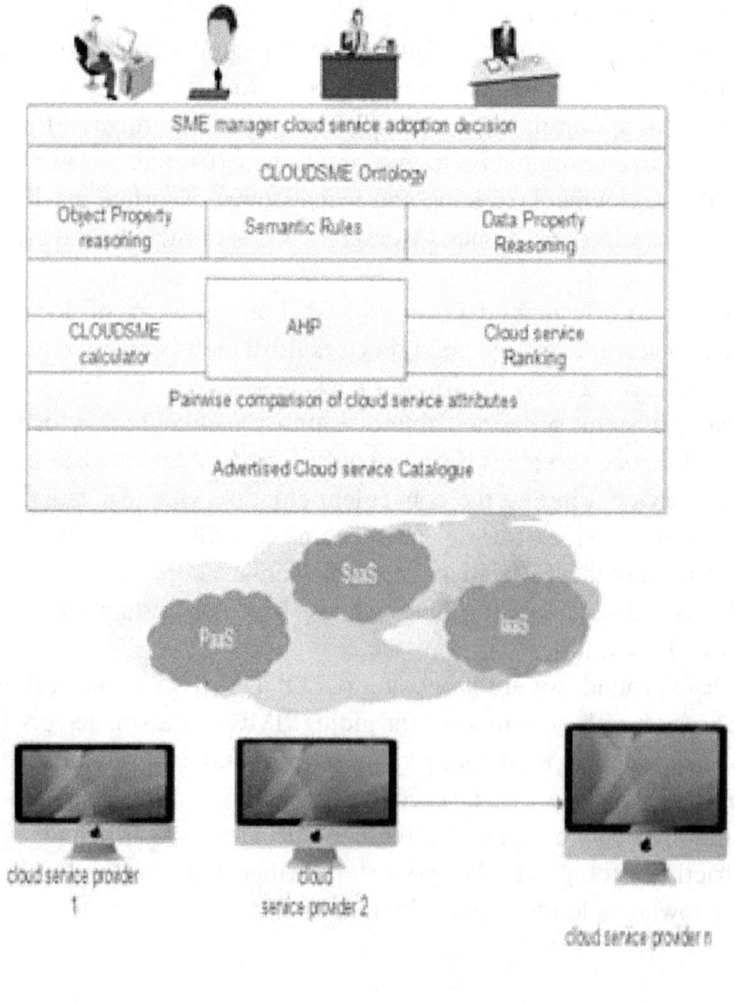

1.2. CLOUDSME Architecture

We propose a decision support framework, which is equipped with an ontology to aid SME owners/ managers in the decision -making the process of SaaS cloud service adoption. The structure is composed of individual features such as advertised cloud service catalogue which shows their characteristics based on users' requirement, the non-functional characteristics of SaaS cloud services are compared depending on their level of user friendliness, price flexibility, compatibility, and storage allowance. A pairwise comparison is used to judge advantages of offerings for one cloud service over another. The weight of the advertised cloud services is calculated using Saaty's steps 3-7 in applying AHP as CLOUDSME calculator. The ranking is done based on the priority of one publicised cloud service characteristics compared to another for each characteristic as shown in table 3. The overall ranking is done based on the combination of all features and finding the SaaS service with the highest priority. While The user decision making can be done based on a SaaS that meet a set benchmark.The framework is equipped with an ontology which is designed based on the object and data properties of each advertised cloud service and the findings from the Analysis. A set of rules see Fig 8 are used to determine the relationships between each of the cloud services towards retrieving user requirement in the cloud service decision-making process.

2. CLOUDSME CALCULATOR

The Analytical Hierarchical Process (AHP) is known as a multi-criteria decision-making method that has the capability to simplify complex and non-structural problems to aid in decision making by presenting the decision factors in the form of a hierarchical structure. It also aims at quantifying relative priorities on a given set of alternatives for a ratio scale. Furthermore, AHP uses the intuitive judgement of a decision- maker and checks the consistency of the comparison of alternatives in the process of decision making. This method was developed by Saaty (Saaty, 2005).In addition, the pairwise comparison aids the decision maker in determining the advantage of one criterion over another. The flexibility, ability to check consistencies and the intuitive appeal to the decision makers gives AHP an edge over other multi-criteria methods (Ramanathan, 2001).

Against the above, we propose a ranking mechanism of advertised SaaS cloud services in a bid to meet user requirements based on AHP (Li et al, 2010). This is based on saaty proposed steps for AHP accordingly, we are using the following steps as the criteria to determine our proposed CLOUDSME calculator.

Step 1: Problem definition and goal determination

Step 2: Determining the decision hierarchy from Top to lowest level. Top (the objective from the decision- makers viewpoint) Intermediate (criteria by which subsequent levels depend on) lowest (level by which subsequent levels depend on).

Step 3: The construction of the pairwise comparison matrices (size n*n) for each of the lower levels with one matrix of each element in the level immediately above by using the relative scale measurement shown in table 1. The pairwise comparison is based on the domination of one element over another.

Step 4: There are n (n-1)/judgements required in the development of the set of matrices in the above step 3. In this step reciprocals are arranged automatically in each pairwise comparison.

Step 5: Here hierarchical synthesis is used to weight the eigenvector by the weight of the criteria and the sum is taken of all weighted vector entries corresponding to those in the next level lower of the hierarchy.

Step 6: After the completion of all the pairwise comparisons, the consistency is determined by using the eigenvalue, λ_{max} to calculate the consistency index, CI as follows:

CI= (λ_{max} - n/n-1), where n is the matrix size. The consistency judgement can be checked taking the consistency ratio (CR) of CI with the appropriate value as shown in table 1. The CR value is acceptable if it is less than or equal to 0.10.if the CR value is more than 0.10 it means the judgement is not consistent and it will need to be improved or reviewed. Table 1: below shows the Average random consistency Metrix which are fixed constant values proportional to the size of metrix

Table 1. Average random consistency (RI)

Size of matrix	1	2	3	4	5	6	7	8	9	10
Random consistency	0	0	0.58	0.9	1.12	1.24	1.32	1.41	1.45	1.49

Step 7: Perform step 3 to step 6 for all the levels of the hierarchy. Table 2 below show pairwise comparison scale for AHP preference when comparing the superiority of one service attribute over another.

Table 2. Pairwise comparison scale for AHP preference

Intensity of importance	Definition	Explanation
1	Equal importance	Two elements contribute equally to the objective
3	Moderate importance	Somewhat more important/better
5	Strong importance	Definitely more important/better
7	Very strong important	Much more important/better
9	Extreme importance	Highest possible order of affirmation of one element over another

3. CASE STUDY OF ADVERTISED CLOUD SERVICES

In the case study, the computation of data is based on advertised cloud service offerings of four major cloud service providers. The four top SaaS cloud services used in this case study are OneDrive, Dropbox, Google Drive and ICloud (Mitroff, 2015).

In the following, we compare each cloud service based on our perceived adoption possibilities by SME owners based on the provider's level of user consideration and flexibility of usage. The relative weight method helps us to determine the relative priority of each attribute for each SaaS cloud service in this case study.

Table 3. Case Study Example

Judgement criteria	One Drive (SaaS)	Dropbox (SaaS	Google Drive (SaaS)	ICloud (SaaS	Value type	Criteria weight
File size Restriction	2GB	-	10GB	15GB	Integer	1.1228
Free Storage	15GB	2GB	15GB	5GB	Integer	0.0578
Payment plan	$2/Month for 100GB, $4/ Month for 200GB	$10/month for 1TB	$2/month 100GB, $10/ Month for 1TB	$20/Month 1TB	integer	0.5619
Operating system supported	Windows, Mac, Android and iOS	Windows, Mac, Linux, Android, iOS, Blackberry Kindle Fire	Windows, Mac, Android and iOS	iOS	String	0.2447

In the following section we will manually apply the AHP method using the following steps below:

- The systhesising of the pairwise comparison matrix
- Calculating the priority vector for each attribute as seen in the case study example Table 4; showing the priority vector for payment through normalization.
- Calculation of Consistency Ratio (CR)
  - Calculation of λ_{max}
- Calculation of CI which is the consistency index
- Selection of appropriate random consistency ratio value from table 2
  - The determination of the consistency of comparison of our pairwise comparison judgement to determine if our judgement was consistent or not.

The calculations for the attributes as shown in the case study based on the steps above is done below:

Table 4. Pair-wise comparison table for payment

Payment	Dropbox	Google Drive	OneDrive	iCloud
Dropbox	1	1/5	1/3	3
Google Drive	5	1	3	7
OneDrive	3	1/3	1	5
iCloud	1/3	1/7	1/5	1

$\lambda_{max} = 4.09105$ CI= 0.03035 CR = 0.0337 < 0.1 OK

The next step which is the synthesizing of the pair-wise comparison matrix is performed by dividing the value of the matrix by its column total, for example the value 0.107 in table 5 is derived by dividing 1 (from table 4) by 9.3, the sum of the column values in Table 3 (1+5+3+ ⅓).

The priority vector in Table 5 can be derived by finding the row averages. For instance, the priority of Dropbox with respect to the criterion ‘payment’ in Table 5 is calculated by dividing the sum total of the rows (0.107 + 0.12 + 0.073 + 0.19) by the number of cloud services (columns) i.e., 4 to obtain the value 0.1225. The priority vector for payment is shown in Table 5 below.

Table 5. Synthesized matrix for payment (Pa)

Payment	Dropbox	Google Drive	OneDrive	iCloud	Priority Vector
Dropbox	0.107	0.12	0.073	0.19	0.1225
Google Drive	0.541	0.60	0.7	0.44	0.570
One Drive	0.322	0.2	0.22	0.31	0.263
iCloud	0.04	0.1	0.04	0.06	0.06

The next step is to estimate the priority consistency ratio and this is done below as follows:

$$0.1225\begin{bmatrix}1\\5\\3\\1/3\end{bmatrix}+0.570\begin{bmatrix}1/5\\1\\1/3\\1/7\end{bmatrix}+0.263\begin{bmatrix}1/3\\3\\1\\1/5\end{bmatrix}+0.06\begin{bmatrix}3\\7\\5\\1\end{bmatrix}=\begin{bmatrix}0.5035\\2.3915\\1.1195\\0.2282\end{bmatrix}\text{(Weighted sum of matrix)}$$

Then we divide the elements of the weighted sum matrices by their corresponding priority vector value as below

$$\frac{0.5035}{0.1225}=4.1102,\frac{2.3915}{0.570}=4.19,\frac{1.1195}{0.263}=4.256,\frac{0.2282}{0.06}=3.803$$

Now we then compute the average of these values to determine λ_{max}.

$$\lambda_{max}=\frac{(4.1102+4.195+4.256+3.803}{4}=4.09105$$

Now we determine the consistency index, CI as below:

$$CI=\frac{\lambda\max-n}{n-1}=\frac{4.09105-4}{4-1}=0.03035.$$

Finally we select the appropriate value of the random consistency ratio RI for a matrix size of four using Table 1, we find RI=0.9. We then determine the consistency ratio, CR, as below:

$$CR = \frac{CI}{RI} = \frac{0.03035}{0.9}$$

CR = 0.0337

The above CR value which is less than 0.1, shows that our judgement is acceptable. Similarly, the pairwise comparison matrices and priority vector for the other SaaS cloud services which is based on the remaining 3 criteria in the case study are shown in Tables 6 – 8, respectively.

Next we determine the pairwise comparison matrices and priority vector for operating system supported following the steps shown above and our result is shown in Table 6 below:

Table 6. Pair –wise comparison matrix for operating system supported (OPS)

OS supported	Dropbox	Google Drive	OneDrive	iCloud	Priority Vector
Dropbox	1	3	3	7	0.521
Google Drive	1/3	1	1	5	0.209
OneDrive	1/3	1	1	5	0.209
iCloud	1/7	1/5	1/5	1	0.051

Next we determine the pairwise comparison matrices and priority vector for File size restriction following the steps shown above and our result is shown in Table 7 below:

Table 7. Pair –wise comparison matrix for File size Restriction (FSR)

File size restriction	Dropbox	Google Drive	OneDrive	iCloud	Priority Vector
Dropbox	1	4	7	3	0.5542
Google Drive	1/4	1	3	1/2	0.1462
OneDrive	1/7	1/3	1	1/4	0.058
iCloud	1/3	2	4	1	0.234

Next we determine the pairwise comparison matrices and priority vector for Free storage following the steps shown above and our result is shown in Table 8 below

Table 8. Pair –wise comparison matrix for Free storage (FS)

Free storage	Dropbox	Google Drive	OneDrive	iCloud	Priority Vector
Dropbox	1	1/5	1/5	1/2	0.0737
Google Drive	5	1	1	3	0.3935
OneDrive	5	1	1	3	0.3935
iCloud	2	1/3	1/3	1	0.1299

Furthermore, we conduct a pairwise comparison procedure for all the 4 criteria based on the importance of one over another in the contribution to the SME owner adoption preference. Table 9 shows the pair-wise comparison matrix, also the priority vector for each of the 4 criteria.

Table 9. Pair –wise comparison matrix and priority vector for the four criteria

-	PA	FS	OPS	FSR	Priority Vector
PA	1	7	3	5	0.5619
FS	1/7	1	1/4	1/3	0.0578
OPS	1/3	4	1	3	0.2447
FSR	1/5	3	1/3	1	0.1228

$\lambda_{max} = 4.0877$ CI= 0.02923 CR = 0.0324 < 0.1 OK

In addition we combine the criterion priorities and the priorities of each SaaS cloud service to determine the overall the overall priority ranking of the cloud services which is represented in Table 10 below:

Table 10. Showing overall Priority Ranking

	PA (0.5619)	FS (0.0578)	OPS (0.2447)	FSR (0.1228)	Overall Priority Vector
Dropbox	0.1225	0.0737	0.5210	0.5542	0.2646
Google Drive	0.5700	0.3935	0.2090	0.1462	0.4116
OneDrive	0.2630	0.3935	0.2090	0.0580	0.2219
iCloud	0.0600	0.1299	0.0510	0.2340	0.0755

Finally, we determine the SaaS cloud services overall priorities as seen in Table 10 as follows: Google drive, Dropbox, OneDrive and ICloud. This indicates that Google Drive has the highest overall priority vector with a weight of 0.4116 based on findings.

The findings in Table 10 is further used to depict a visual representation of our cloud service comparison KPI using kivait graphs in Figure 2.

Figure 2. Showing a visual cloud service comparison based on different CLOUDSME KPIs

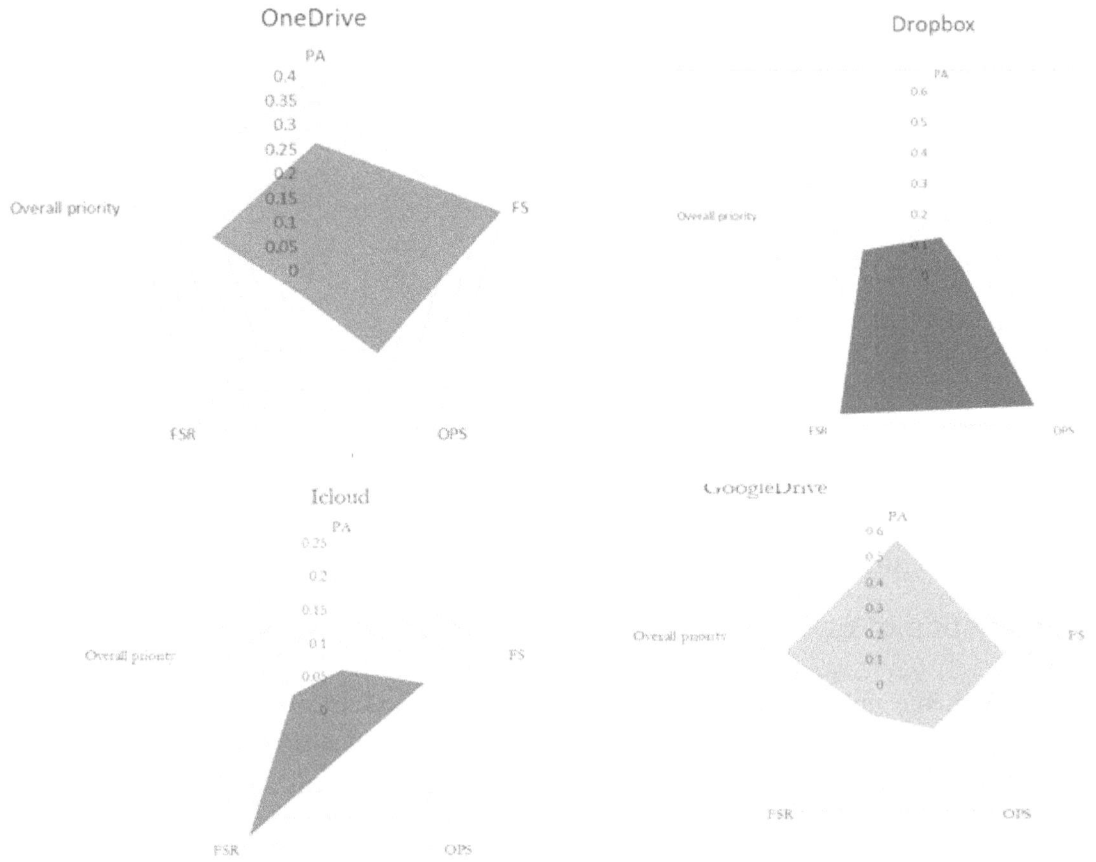

4. DECISION SUPPORT SYSTEM

In the sections above we have shown the sequence of activities that takes place within the Framework in order to get our cloud service rankings. In order to complete our framework, we need to design a decision support system which is equipped with an ontology of cloud services which contains all the information in our case study and also the findings of our cloud service rankings. The ontology will be used to retrieve information based SME owner requirement towards cloud service adoption. The design of the cloud service ontology which has already been tested on protégé software to check consistencies is discussed in the next section.

4.1. Cloudsme DSS Architecture

The decision support system architecture shows the sequence of activities that takes place within the system when a user requirement is queried. In order to complete this project we need to develop an ontology of cloud services. This ontology will hold information about cloud services advertised by service providers and will be used by our system algorithms to retrieve user requirements. The design of the cloud service ontology which has already been tested on protégé software to check consistencies is also explained below.

Figure 3. Showing CLOUDSME DSS Architecture

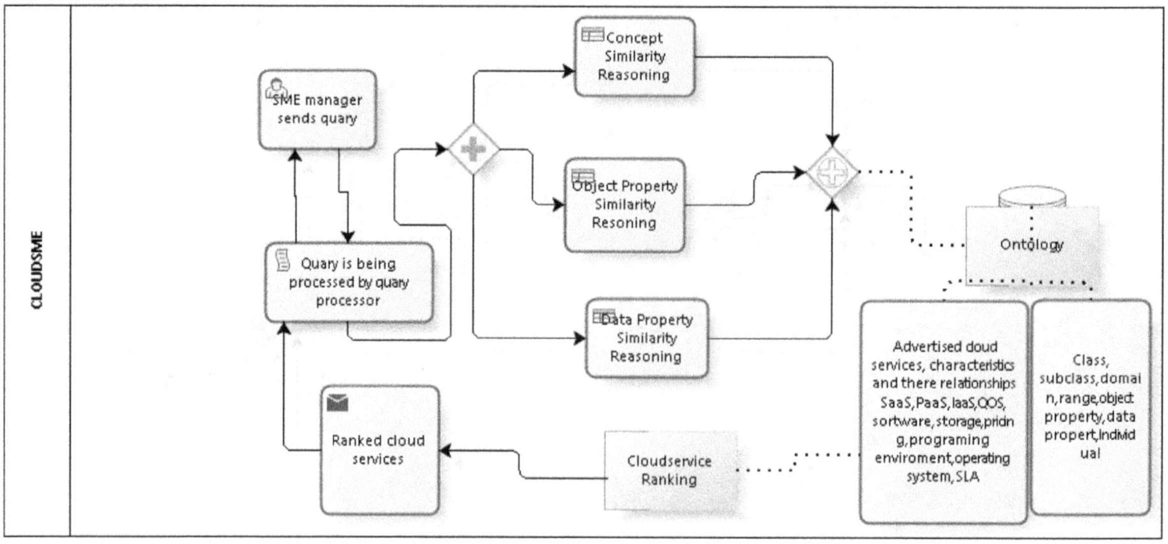

4.2. Characteristics of Each Component of Cloud Service Architecture

The CLOUDSME architecture is shown in fig3 above. Firstly, The SME owner/ manager sends their requirements to CLOUDSME system from the user interface. Then the decision support system carries out the following four functions: 1) Query processing 2) Similarity reasoning 3) Cloud service ranking.

4.2.1. Query Processing: When a user requirement is sent from the SME owner/manager via the user interface, the query processor initiates query processing and converts the query to machine readable form. Then the query processor sends the processed query to the similarity reasoning domain for further processes.

4.2.2. Similarity Reasoning: This is done by consulting the cloud service ontology. Its action is based on the query processed. As mentioned above the main similarity methods as shown in the CLOUDSME system architecture are as follows:

4.2.2.1 Concept similarity reasoning: In a bid to show that our framework which is equipped with a decision support system which includes a semantically designed ontology can be used to answer user requirement. The following example is shown in Fig4 below. In a scenario whereby an SME owner intends to adopt a cloud service for his data storage with a budget 2dollars a month for 100gigabytes.How can CLOUDSME help in his decision making? Based on this requirement, it will trigger the ontology to undergo concept similarity reasoning based on the conceptual modelling of advertised services from service providers as shown in Table 2 above. When the following query is sent (**SaaS and hasPaymentplan1price value 2 and haspaymentPlan1GB value of 100**). Please note that the price value is in USD/Month and the GB (Gigabyte).

Based on the requirement by the SME user and the conceptual modelling of cloud services advertised by service providers as shown in our case study in Table 2 above, the conceptual reasoning result for the above query is One Drive and Google Drive. The SME owner can now narrow his decision on this two SaaS services and can also query another requirement such as Operating System, File size Restriction etc.

Figure 4. Example of conceptual similarity reasoning.

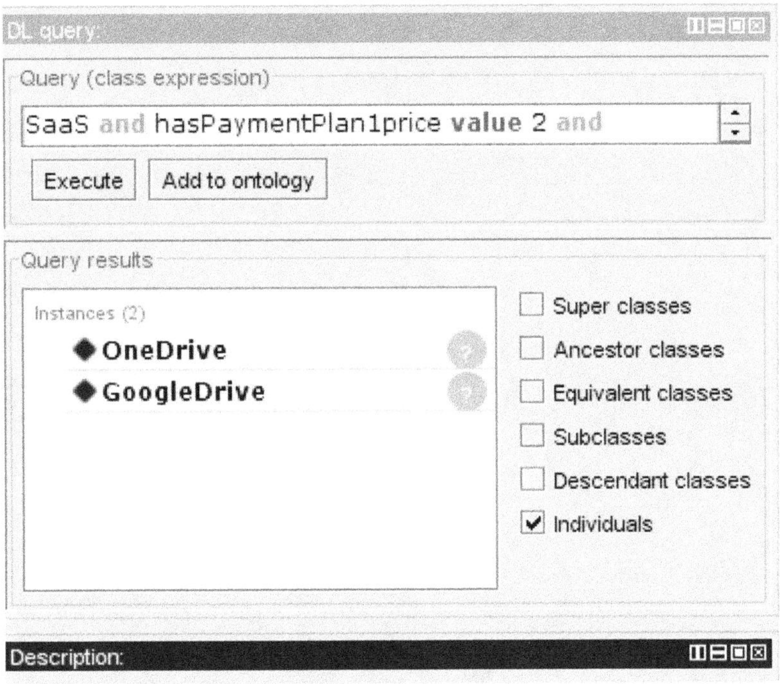

4.2.2.2 Object Property Similarity reasoning: In a bid to show that our decision support system equipped with an ontology within our framework when queried can undergo object property similarity reasoning when an SME user who wants to adopt cloud services but is not sure which service is best for him. In a scenario whereby an SME owner above also needs a cloud service for his company which is supported by android operating system and they consult CLOUDSME. Based on the requirement, the following query is sent (**SaaS and hasOS name3 value "Android"**).Please note that OS stands for operating system.

Based on the modelled ontology of advertised cloud services of service providers as shown in Table 2, the object Property similarity reasoning result for the above query is seen in fig 5 above.

4.2.2.3 Data Property Similarity Reasoning: In a bid to show that our framework which is equipped with a decision support system which includes a semantically designed ontology can be used to answer user requirement in the decision-making process of adopting cloud services by SME. Based on datatype similarity reasoning, if the same SME owner in the two cases as shown above also considers a SaaS cloud service with a file size restriction of data of between 2 and 14 Gigabytes (GB). The following query is sent (**SaaS and hasFileSizeRestrictonGB some int [>1, <15]).**

From the above datatype property, we can see that based on advertised cloud services, OneDrive and Google Drive have file size restriction of data between 2 and 14 Gigabytes.

4.2.3 Individual Similarity Matching

From the above case scenario, we can see that Google Drive and OneDrive keep meeting user requirement and this will still not solve his selection problem. In a bid for CLOUDSME framework to solve his

Figure 5. Example of Object property similarity reasoning

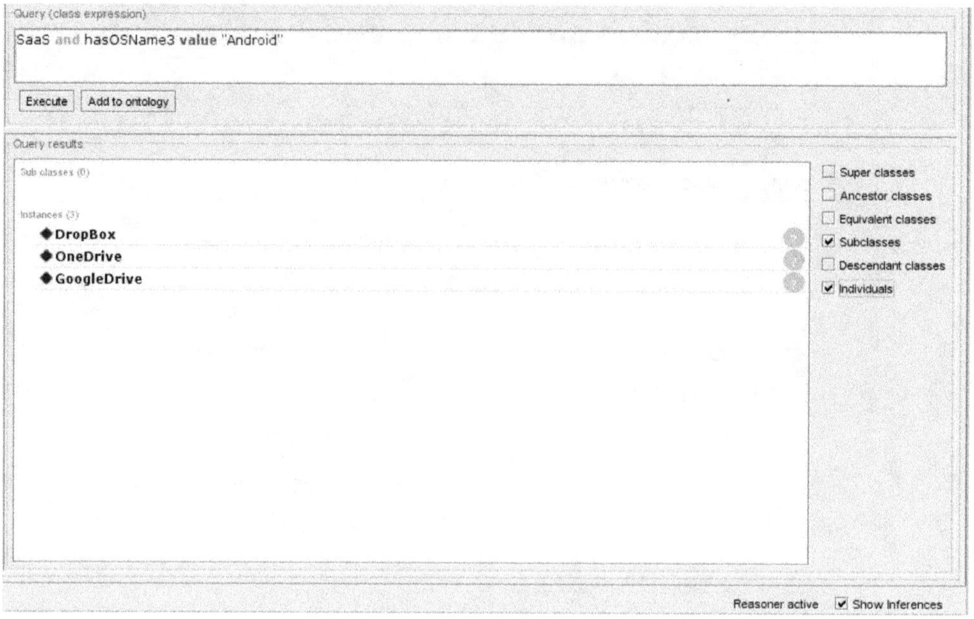

Figure 6. Data Property similarity reasoning

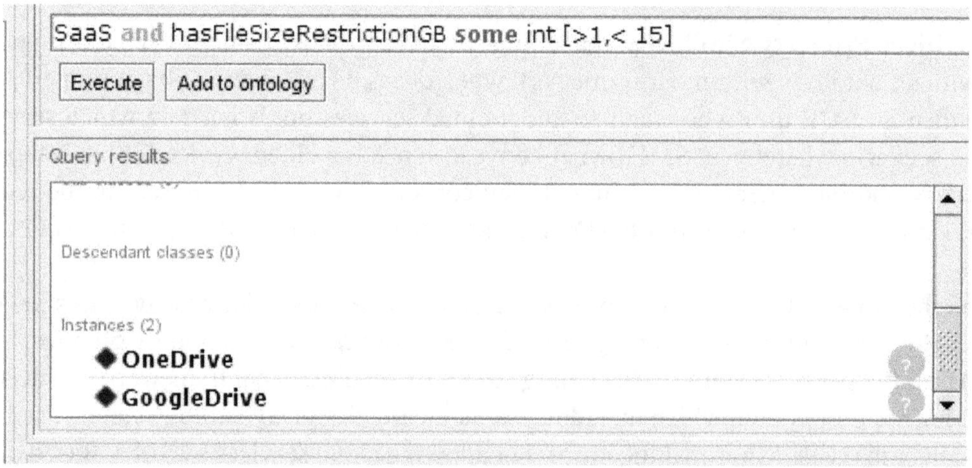

problem, the system will further determine the acceptability of Onedrive and Google Drive to determine if they meet the minimum priority standard by setting the acceptable standard for File size restriction to (0.1228=>1228) as seen in table 9 above. The following rule is being queried. Please see fig 7 for a list of rules.

(SaaS and (hasFSRPriority some int[>=1228]))(?x)) - > FSRacceptible (?x)

Figure 7. Individual Similarity Matching

5 CLOUD SERVICE RANKING

CLOUDSME service ranking is done using the gold, silver, bronze ranking method which is being used in the Olympics and the world cup. This is achieved based on the following the following criteria.

Step 1: Declare the priority of each of the criteria used for our judgement shown in Table 9 above. (E.g. Payment has the highest priority of 0.5619)
Step 2: convert each priority to whole numbers e.g. (0.5619=>5619)
Step 3: The cloud services are ranked as follows:

5.1 Gold: Any cloud service that has a priority vector weight greater than or equal to the priority weight of payment which is the highest from table 9 and has a priority greater than or equal to the priority of any of the other three criteria.

5.2 Silver: Any cloud service that has a priority vector weight higher than or less than the weight of payment and has a priority weight greater than or equal to the priority weight of Operating systems supported and has a priority greater than or equal to the priority weight of either of remaining two criteria.

Figure 8. Showing Rules within the system

5.3Bronze: Any cloud service that has a priority vector weight greater than or equal to the priority vector weight of the 2 remaining judgement criteria.

Step 4: We declare the above step 3 as a set of rules in our system as shown in Fig 8 below.

Step 5: We execute the rule as below

(SaaS and (hasPaymentPriority some int[>=5619) and (hasFSRPriority some int [>=1228]) or (hasOSPriority some int[>= 2447)or (hasFSPriority some int[>= 578)) (? X) –> SaaSGold (? X).

Figure 9. Showing Google Drive as the only SaaS to meet the Gold criteria

Finally, we show the Ranking list of the SaaS cloud services as shown in Table 11 below:

Table 11. Showing ranking of cloud services

Cloud service	Gold	Silver	Bronze
Google Drive	√	-	√
Dropbox	-	√	√
ICloud	-	-	√
One Drive	-	-	-

Google Drive is the highest ranked having attained one gold and one bronze.

6. RELATED WORK

In this section we compare our work with previous research works evaluation and adoption of cloud services.

An increase in awareness of cloud computing has led many researchers to propose various frameworks in a bid to give businesses a better understanding of this new technology. For instance (Laszewski et al, 2012) proposed a framework Geogrid which compares different IaaS and PaaS frameworks such as Nimbus, Eucalyptus, OpenStack and Open Nebula in terms of traditional high-performance computing services and with bare- metal. While the work of (Ebneter et al, 2010) proposes a framework which gives businesses the ability to analyse and determine if cloud computing will have a positive impact on their business operation. Also the work of (Boa et al, 2010) proposes a method for measuring the quality of cloud service measurement based on two significant aspects which are availability and performance of service in terms of customer evaluation of cloud offerings based on ability to meet user quality of service (QOS) requirements. Our work complements these previous works by measuring the QOS offerings advertised by different cloud service providers according to how they meet user needs. Although the evaluation and ranking of various cloud services are in its preliminary stage in cloud computing, in other computing areas such as web services the concept has been widely used. The most related work in the field of cloud services ranking is done by (Tran et al, 2009; Garg et al, 2013). Their research also proposed a similar framework which is based on AHP concept for classifying cloud services. However, they focused more on the Infrastructure as a Service (IaaS) quantitative aspect of cloud services offerings. Our work also adopted the AHP concept but we focused on the Software as a Service (SaaS) qualitative aspect of cloud services and in addition we designed a decision support system equipped with an ontology which is included in our framework to aid SME owners in the decision making towards cloud services adoption. Furthermore, the work of (Zhang et al, 2012) proposed an owl-based ontology, a cloud computing ontology (CoCoon) that defines the functional and non-functional concepts, attributes and relationships of IaaS cloud offerings. While (Ali, et al, 2014) proposed an ontology- based cloud service representation that can be used for cloud service discovery or cloud computing management. Our study is complementary to this works in the design concept of our ontology but we have gone further to rules which aid in solving the issue of complex queries as well as representing our cloud service ranking findings within our ontology.

In summary, according to the knowledge of the authors, our work is the first to apply AHP based ranking of qualitative SaaS cloud services of advertised cloud services as well as representing cloud service ranking with the use of an ontology.

7. CONCLUSION AND FUTURE WORK

Cloud computing is seen as a technology that provides services over the internet just like public utilities. Many cloud service providers present cloud services in the own format as their so standardisation for representing cloud services. Similarly, the presence of dominant players within the IT sector providing this technology shows that cloud computing is important in the world today and this has made it challenging for possible adopters especially in the SME sector to make a decision on which cloud service will best suit their company business process. Therefore, the need for a framework that will aid customer's cloud service decision in terms of how service providers meet their requirement is vital.

Against the above background, this research work presents CLOUDSME, A framework that helps in the SaaS cloud service adoption by systematically measuring the quality of service attributes advertised by SaaS cloud service providers. In the framework, a pairwise comparison was used to judge the importance of one service provider attribute over another based on flexibility, quality of service and pricing as determinants. We also proposed the use of Analytical Hierarchical Processes (AHP) to rank each cloud service attribute ranking over another before finding the overall priority ranking of each cloud service over another. We also proposed the use of a decision support system equipped with a semantic web ontology of advertised SaaS cloud services and cloud service rankings based on our findings.

We believe that our framework does not only present a significant step in the accurate measurement and ranking of SaaS cloud services, but will also aid SME owners in the cloud service adoption decision process especially when the DSS aspect of our framework is consulted as well as give the cloud service providers an avenue to compare their offerings with that of other cloud service providers and identify areas where improvements need to be done.

In future work, we intend to extend our work to the quantifiable quality of cloud services (IaaS &PaaS) in terms of performance from a user perspective as well as developing a system that can be used to aggregate the QOS configuration between cloud service layers in different applications.

REFERENCES

Ali, A., Shamsuddin, M., & Eassa, F. (2014). Ontology-Based Cloud service Representation'. *Research Journal of Applied Sciences, Engineering and Technology*, *8*(1), 83–94.

Aljabre, A. (2012). Cloud Computing for Increased Business Value. *International Journal of Business and Social Science*, *3*(1), 234–239.

Ashwini, R., Sanjay, K., Sanjay, M., & Rahul, T. (2012). Decision Point for Cloud Computing in Small, Medium Enterprises (SMEs). *Proceedings of 7th international Conference for Internet Technology and Secured Transactions (ICITST-2012)*, 688-691.

Boa, D., Xiao, Z., Sun, Y., & Zhao, J. (2010). A Method and Framework for Quality of Cloud Services Measurement. *3rd International Conference on Advanced Computer Theory and Engineering (ICACTE)*, 358-362.

Buyya, R., Yeo, C., Venugopal, J., Broberg, J., & Brandic, I. (2009). Cloud Computing and Emerging IT Platform: Vision, Hype and Reality for Delivering computing as the 5th Utility. *Future Generation Computer Systems*, *25*(6), 599–616. doi:10.1016/j.future.2008.12.001

Ciurana, M. (2009). *Developing with Google App Engine' A press*. Barkley. doi:10.1007/978-1-4302-1832-6

Cusumano, M. (2010). Cloud Computing and SaaS as New Computing Platforms. *Communications of the ACM*, *53*(4), 27–29. doi:10.1145/1721654.1721667

Ebneter, D., Grivas, S., Kumar, T., & Wache, H. (2010). Enterprise Architecture Framework for Enabling Cloud Computing. *IEEE 3rd International Conference on Cloud Computing*, 542-543.

Fortis, T., Munteanu, V., & Negru, V. (2012). Towards an Ontology for Cloud Services. *IEEE 6TH International Conference on Complex, Intelligent and Software Intensive Systems*, 787-792.

Garg, S., Versteeg, S., & Buyya, R. (2013). A Framework for Ranking of Cloud Computing Services. *Future Generation Computer Systems*, *29*(4), 1012–1023.

Jagannathan, S. (2012). *Comparison and Evaluation of Open-source Cloud Management software*. KTH Royal Institute of Technology Stockholm.

Khan, S. (2014). Elements of Cloud Adoption. *IEEE Journal on Cloud Computing*, *1*(1), 71–73. doi:10.1109/MCC.2014.7

Laszewski, G., Diaz, J., Wang, F., & Fox, C. (2012). Quality Comparison of Multiple Cloud Frameworks. *IEEE 5th International Conference on Cloud Computing*, 734-741.

Li, A., Yang, X., Kandula, S., & Zhang, M. (2010). CloudCmp: comparing public cloud providers. *Proceedings of the 10 th Annual Conference on Internet Measurement.*

Mitroff, S. (2015). *OneDrive, Dropbox, Google Drive and Box: Which cloud storage service is right for you?* Available at: www.Cnet.com

Ramanathan, R. (2001). A note on the use of the analytic hierarchy process for environmental impact assessment. *Journal of Environmental Management*, *63*(1), 27–35. doi:10.1006/jema.2001.0455 PMID:11591027

Saaty, T. (2005). *Theory and Applications of Analytic Network Process* (Vol. 4922). RWS Publications.

Tran, V., Tsuji, H., & Masudu, R. (2009). A new QoS ontology and its QoS –based ranking algorithm for web services. *Simulation Modelling Practice and Theory*, *17*(8), 1378–1398.

Wu, J., Chen, L., Xie, Y., & Zhang, Z. (2012). Titan:A system for effective web service discovery. *Proceedings of the 21st International Conference Companion on World Wide Web*, 441-444. 10.1145/2187980.2188069

Zhang, M., Ranjan, A., Georgakopoulos, D., Menzel, M., & Nepal, S. (2012). An ontology –based system for cloud infrastructure services discovery. *Proceedings of 8th International Conference on Collaborative Computing: Networking, Applications and Work sharing (CollaborateCom)*, 524-530.

Chapter 19
Ontology-Based Travel Recommender System

Shi Pu
Nottingham Trent University, China

Isibor Kennedy Ihianle
https://orcid.org/0000-0001-7445-8573
Nottingham Trent University, UK

ABSTRACT

Recommender systems are designed to suggest information to users according to their preferences. The items could be movies, books, or various kinds of products. Most of the existing recommender systems are based on a database with limited advantages. However, in this chapter, the authors propose a knowledge-driven travel recommender system to integrate semantic data built using web ontology language (OWL) ontology to allow users to find suitable destinations that fulfil users' travel preferences. This work aims to develop a travel recommendation tool and to examine the reliability, the usability of the system, and satisfaction rate of users. They are also able to demonstrate that users can obtain desired results through queries on the ontology-based system. The overall evaluation of the system shows that users are happy and satisfied with the recommendation results.

1. INTRODUCTION

Ontology is a formal and shared conceptualization of a certain domain and defines a set of concepts relevant to the domain and the relationships between them in a machine-understandable language - Web Ontology Language (OWL). Classes, instances, properties, and rules are the main components of ontologies. Classes represent the different concepts of a domain abstractly. Web Ontology Language is emerging and popular with knowledge-base and recommender systems (Ruotsalo, Haav, Stoyanov, Rochee, Fanid, Deliaic, Mäkelä, Kauppinen, & Hyvönen, 2013; Codina, 2014; Ihianle, Naeem & Islam, 2017; Ihianle, Naeem, Islam & Tawil 2018). On the other hand, a recommender system is a subclass of information filtering system that seeks to predict the "rating" or "preference" a user would give to an

DOI: 10.4018/978-1-7998-6992-4.ch019

item. Recommender systems have been applied to many different scenarios and they have been designed to predict and deliver the most appropriate or the most desired information to specific users based on analysis of user's preferences (Wei, Jennings, Moreau, & Hall, 2008; Lu, Wu, Mao, Renjith & Anjali, 2014; Wang, Zhang, 2015). With online platforms like Amazon and Netflix, recommender systems are utilized to help customers to find the most appropriate or the most desired products using information filtering system (Gomez-Uribe & Hunt, 2015).

According to Choi, Kim, & Ryu, 2015, providing the most preferred products or services online is one of the solutions travel agencies utilize to increase their competitive advantage. The same authors also noted the challenges and limitations with recommender systems built on databases as redundances and filtering limitations. (Bedi, Agarwal, & Jindal, 2014) claims that in the travel related areas, knowledge-based recommender systems are the most useful. Therefore, ontology-based system with could be a solution to minimize redundancies thereby reducing irrelevant information and delivering high-quality recommender system applications to users (Choi, Kim, & Ryu, 2015). Integrating this to the sematic web which is a mesh of linked data could help represent travel information on the world wide web efficiently and easily processed and accessed by machines.

The paper proposes a content-based recommender system enhanced using the ontological information in the tourism and travel domain. It is hoped that the development of a knowledge-driven recommender system could provide a better solution in delivering relative information to users than the database-based recommender systems could.

This paper starts with an introduction followed by a review of related literature in section 2 and section 3 presents the proposed recommender system, the design and system architecture. Section 5 presents the evaluation and chapter 6 the conclusion and future work.

2 LITERATURE REVIEW

Recent related research work provides many different recommendation processes techniques based on customer's preferences and other information relevant to the user. One of the travel business-related recommender systems was developed by (Artemenko, Pasichnyk, Kunanec, & Tabachyshyn, 2019). This system focused on delivering real-time recommendations of attractions based on users' location information retrieved from mobile devices/sensors and supported by context analysis. The authors discussed ways of improving recommendations by analyzing the contexts. One of the most important views were on the ways of retrieving high quality, desired and contextual information. According to Ricci, Rokach, & Shapira, (2015) contextual information could be provided by:

1. The user directly.

Information and data are provided directly from users' responses or embedded sensors. For instance, user refusing to grant permission for the sharing of location information, the system could get the location information from the questions that the user needs to answer.

2. Implied information.

Contextual data can be collected indirectly. For example, the location information could be obtained from alternative sources like mobile phone when in use even if permission to share location information is not granted.

3. Formulated information.

The contextual information could be formulated from the calculations or using intelligent data analysis (Smirnov, Kashevnik & Shilov, 2014).

However, successful collation of contextual data would lead to generating appropriate recommendation successfully only after detailed analysis. The authors emphasized that it is essential to analyze the data properly to generate high-quality recommendations to particular users.

Figure 1. Architecture of PersonalTour Source: (Lorenzi, Loh, & Abel, 2011)

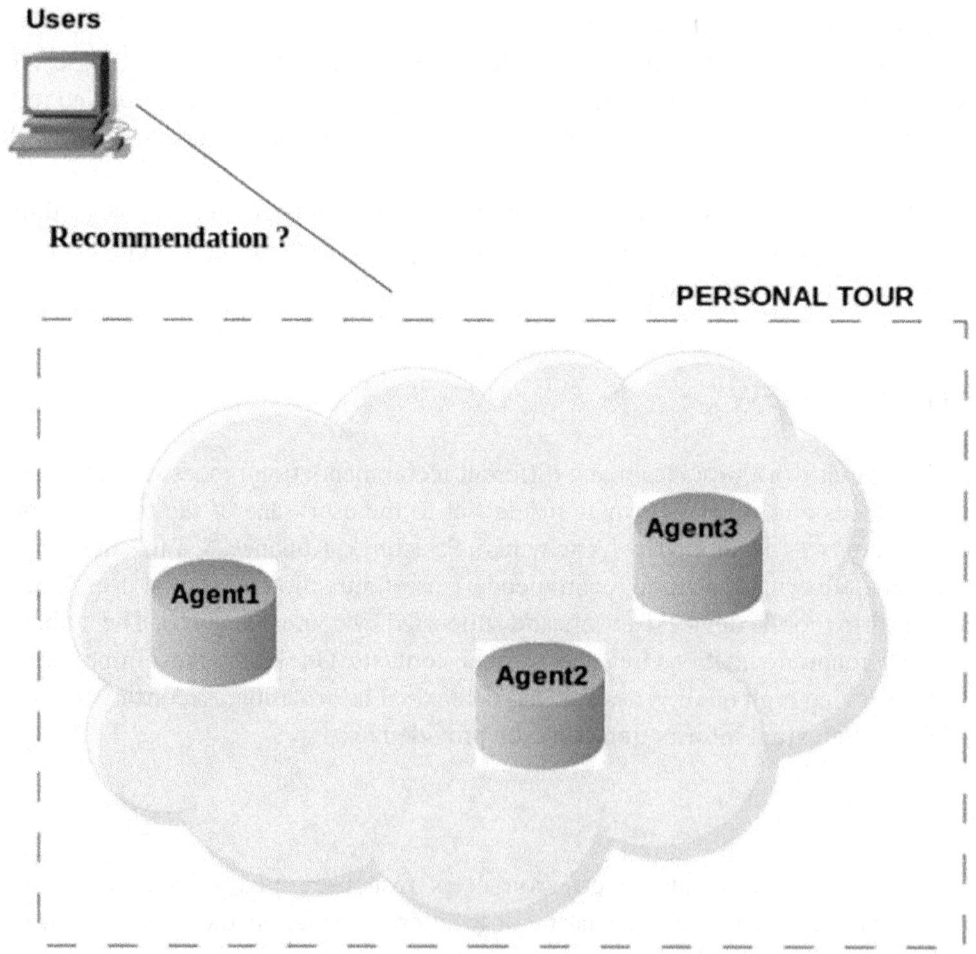

Bedi, Agarwal, & Jindal, 2014., introduced the reputation of items into the recommendation process to create the most preferred recommendations. Unlike the study conducted by Artemenko, Pasichnyk,

Kunanec, & Tabachyshyn, (2019), Bedi, Agarwal, & Jindal, (2014) used multi-agent methods with the recommender system containing Items Reputation Agent (IRA) and uses another agent called Similarity Agent (SA) to calculate the possibilities of users' interests on all items respectively. Artemenko, Pasichnyk, Kunanec, & Tabachyshyn, (2019) suggested using the rating of items to form the recommendations based on the algorithm R: User×Item→Rating, to reduce irrelevant and outdated information. They also stressed that this method might face a problem when it needs to generate recommendations for new users with very limited background data/profile as it is hard to recommend anything without prior data (Bedi, Agarwal, & Jindal, 2014). Under this circumstance, the limitations of Artemenko, Pasichnyk, Kunanec, & Tabachyshyn, (2019) work were obvious.

PersonalTour as illustrated in Figure 1 was proposed by Lorenzi, Loh, & Abel, (2011). It is a multi-agent recommender system designed to help customers to find the best travel packages according to their preferences. Unlike the study by Artemenko, Pasichnyk, Kunanec, & Tabachyshyn, (2019), the key point of this research is sharing the knowledge of each travel agent to recommend some travel package to a customer. Furthermore, it provided a more sophisticated rating system and emphasized the importance of customers in the cycle of recommendation. With the PersonalTour, customers' evaluations and recommendations are dependent on the calculation of confidence degrees of the agent. In the earlier work of Wei, Jennings, Moreau, & Hall, (2008), the user evaluation of a market-based recommender system is also used as a multi-agent approach with a rewarding mechanism. Besides recommendation agents, it has an auctioneer agent. It distributes rewards to certain agents whenever the agent's recommendation is selected by the user, (see Figure 2 below for the process).

Figure 2. The marketplace architecture Sources: (Wei, Jennings, Moreau, & Hall, 2008)

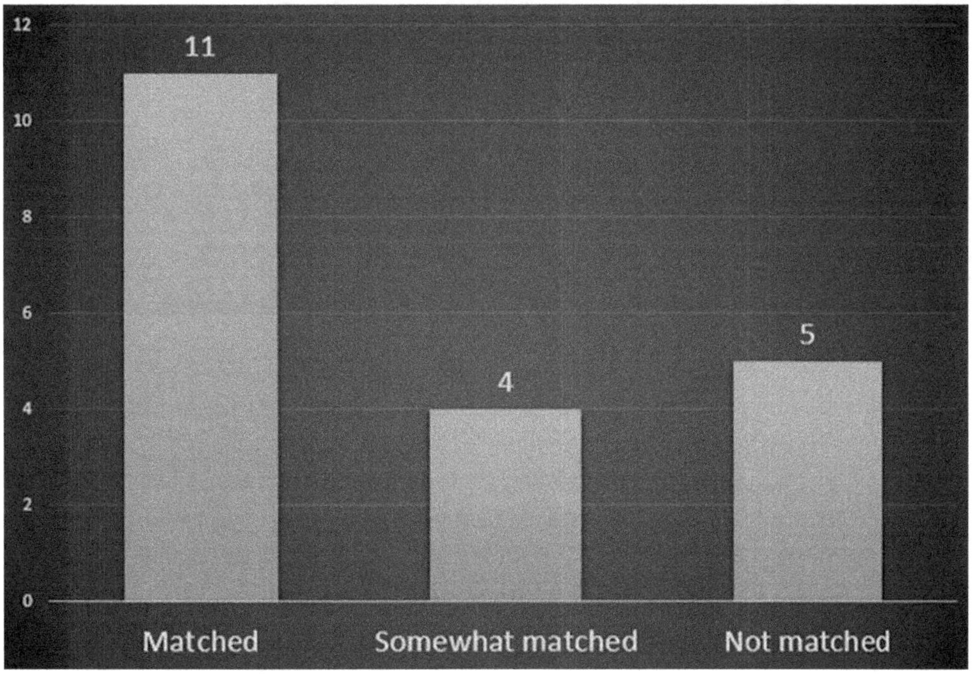

On the contrary, Wei, Jennings, Moreau, & Hall, (2008) study demands users to evaluate each of the received recommendations with the rate of 1 (like this recommendation) or 0 (do not like the recommendation). It is suggested that a more intelligent recommendation process could be the substitute for human tasks (Wei, Jennings, Moreau, & Hall, 2008). In the experiment conducted, 72 out of 94 recommendations provided by the recommender system are rated by users as purchasable (76.59%), and 67 travel packages recommended by human experts are rated as satisfied by users (71.27%). Although it is not a remarkable improvement, it proves that the system provides same high quality of recommendations as human travel agents (Lorenzi, Loh, & Abel, 2011).

Figure 3. Using context analysis in e-tourism mobile location-based recommender system Source: (Artemenko, Pasichnyk, Kunanec, & Tabachyshyn, 2019)

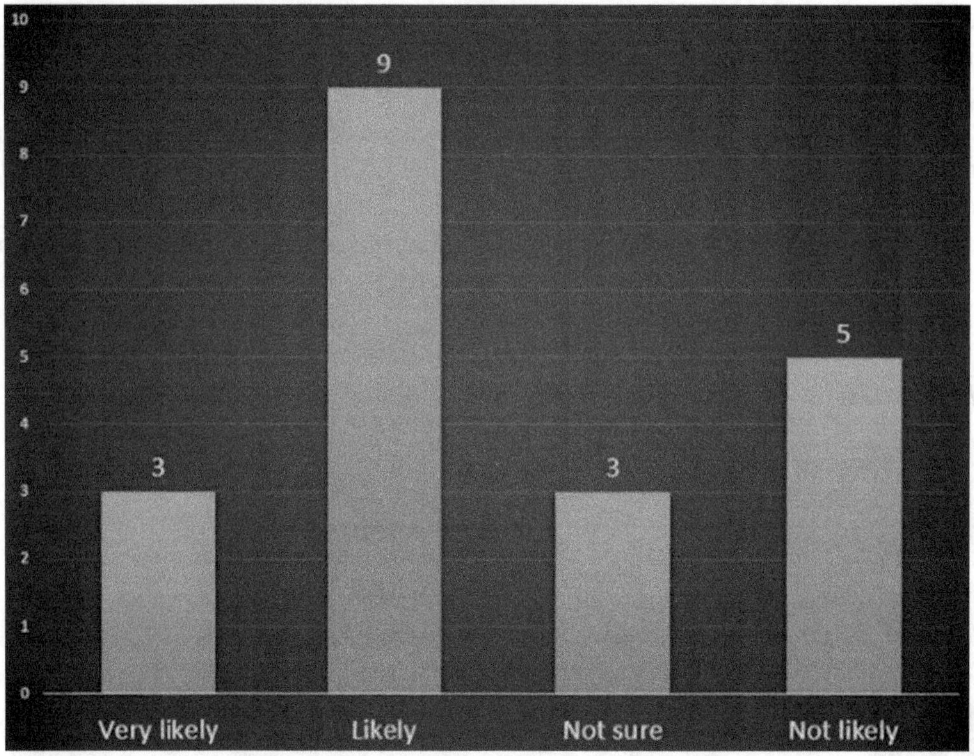

The biggest problem of this development of a recommender system is that the accuracy of recommendations will be updated only based on users' feedback (Maddah & Kavakli, 2013). Both Wei, Jennings, Moreau, & Hall, (2008) and the work by Artemenko, Pasichnyk, Kunanec, & Tabachyshyn, (2019) as shown in Figure 3 discovered the same difficulty which is challenge for recommendations especially for new users. It is found to be difficult to produce information without asking a question and sometimes users could not identify their needs clearly (Lorenzi, Loh, & Abel, 2011).

2.2 Recommendation Techniques/Process Review

The higher quality of recommendations comes from the good process, and there are two process strategies in general - prediction and recommendation (Renjith & Anjali, 2014). Renjith & Anjali, (2014) also stated that the prediction process was used to anticipate users' possible behaviour according to records. The recommendation process produces recommendations based on the prediction process to users. In Artemenko, Pasichnyk, Kunanec, & Tabachyshyn, (2019) paper, the prediction process analyses the context to produce a recommendation. There are three general methods of utilizing context in the recommender system.

1. In the first case, data collected from the user part will be dealt with to remove irrelevant data, then the recommendations will be produced accordingly.
2. In the second case, a number of recommendations are made, but they will be modified based on the received context.
3. In the third case, while producing a recommendation, context will be utilized to calculate the rating of items.

PersonalTour recommendation process starts from users providing preferences like the base information for the recommender system to process. In this system, recommendations consists of travel products such as flights. Agents dealt with parts of the recommendation which start from 'here' and matching recommendations parts by parts till the entire recommendations are complete. The results are sent to users to provide feedback to the recommender system so that the system will update itself to provide even more accurate results (Lorenzi, Loh, & Abel, 2011). Multi-agent recommendation systems use this approach except for the work of Artemenko, Pasichnyk, Kunanec, & Tabachyshyn, (2019), even though, the design of the multi-agent is slightly different. In Wei, Jennings, Moreau, & Hall, (2008) recommender system, agents provide related recommendations to user and rewarded if selected. It is similar to the study of the PersonalTour project (Lorenzi, Loh, & Abel, 2011), but (Bedi, Agarwal, & Jindal, 2014) proposed differently, each agent dealt with a specific domain and function, such as Hotel Recommendation Agent (HRA).

2.3 OWL Ontology

Previous systems such as PersonalTour Lorenzi, Loh, & Abel, (2011) and e-tourism mobile location-based recommender system Artemenko, Pasichnyk, Kunanec, & Tabachyshyn, (2019) are based on database model. The main difference that identified by Sir, Bradac & Fiedler, (2015) between ontologies and databases is that most database application used the Close World Assumption (CWA) system of knowledge representation whereas ontologies use Open World Assumption (OWA). According to McGuinness & Van Harmelen, (2004), the Web Ontology Language (OWL) is used by the OWA application and allows the application to have the ability of content processing rather than plain information presentation to end-users. OWL is used to state the meaning of terms in words and the relationships between those terms exactly and specifically. OWL mainly consists of Resource Description Framework (RDF) Schema features, Property Characteristics and Property Restrictions. A brief description of each feature (McGuinness & Van Harmelen, 2004):

RDF Schema Features

1. ***Class***: It is a group of concepts with properties, for example, Mercedes-Benz AMG and Suzuki Vitara are both cars and belong to ***Car*** class. Generally, in Protégé, ***'Thing'*** is the base class of all OWL classes.
2. ***Rdfs: subClassOf***: In the hierarchy of classes, a class may be a subclass of another, for example, ***Truck*** may be a subclass of class ***Car***. Therefore, the reasoner could deduce that if truck_A is an individual of class ***Truck***, truck_A is also a ***Car***.

Property Characteristics

1. ***Functional***: A Property can be set to have a specific value which means each individual could have maximum 1 value (or 0 value). This is ***FunctionalProperty***. For example, ***hasDriver*** is determined to be a ***FunctionalProperty***, which means a car could have 0 or 1 driver maximum.

Property Restrictions

1. ***someValuesFrom***: A property which is stated to have a someValuesFrom restriction to a class, then some value for the property should have an instance of the class.

Travel related recommender systems provide the customized and preferred information to users according to users' preferences (Borràs, Moreno, & Valls, 2014). As mentioned earlier, e-tourism mobile location-based recommender systems analyses contextual information during the recommendation process and recommendations are generated in real-time. Furthermore, the PersonalTour recommender system proved to be able to deliver good quality recommendations, although, its weaknesses were on providing recommendations with the lack of information. The recommendation process/techniques, of the last two works considered suggested reputation/rating mechanisms. But one of the big differences between them is that the PersonalTour recommender system utilized a multi-agent approach. Although these related works proved to be valuable and useful, their weaknesses should not be ignored. For these reasons, a recommendation system should be developed carefully to overcome the limitations and weaknesses outlined.

3 THE PROPOSED RECOMMENDER SYSTEM

This paper proposes a recommender system designed and built to have an ontology knowledge base. The proposed recommender system as shown in Figure 4. is a content-based recommender system that makes specific recommendations considering user preferences. The proposed recommender system collects data representative of the user preferences and ratings which are then used to define the user profile. It uses ontology language to represent the user profile even with sparse data. It selects the concepts and attributes of the domain ontology to generate an ontology-based user profile. The proposed ontology-based recommendation process is performed and enabled by an ontology-based tree, OWL reasoner and ontology-based query component. Furthermore, the user profile and provides appropriate personalized recommendations over time.

Figure 4. The System Architecture Diagram embedded in protégé

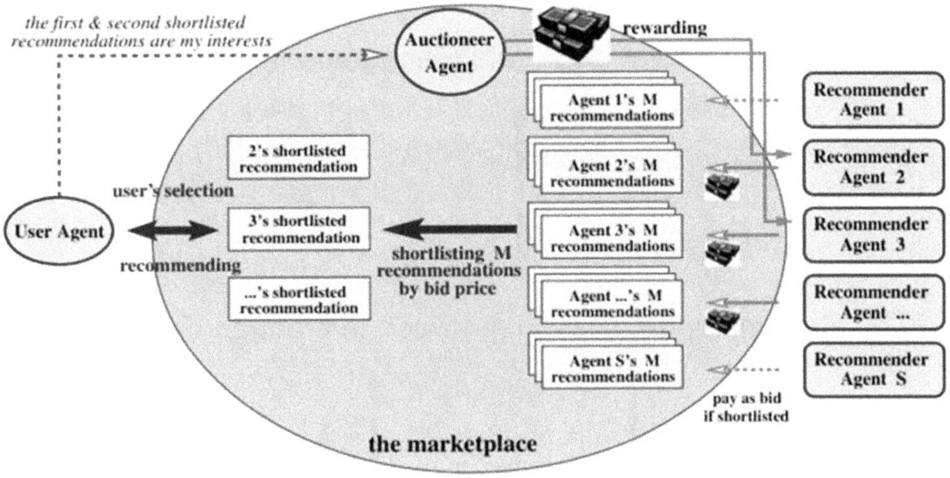

The proposed recommender system is implemented using OWL which differs from the related works in the following ways:

Own Expert System

Unlike previous studies stated above, this recommender system will be like an expert system, mainly focusing on users' preferences which were provided ahead, and then using a reputation mechanism to update the system itself to increase the accuracy of the recommendations.

Users' Preferences and Profiles are Considered as a Significant Role

According to Wei, Jennings, Moreau, & Hall, (2008) and Artemenko, Pasichnyk, Kunanec, & Tabachyshyn, (2019), it is difficult to produce recommendations to a new user when there is no previous data of the user and they tried to solve this by collecting data from alternative sources. The proposed recommender system would focus on users' preferences and profiles provided by the user and with the minimum burden of human tasks make recommendations.

Previous studies utilized context data storage system such as plain text database whereas in the proposed work utilizes ontology tree and a knowledge base repository.

3.1 Ontology Tree Design

The ontological tree design as shown in Figure 5 is represented in a tree-like data structure which shows the specification of the concepts within the domain (Carral, Hitzler, Lapp, & Rudolph, 2017). This allows for recommendations to be made based on user queries. In this particular case, the first objective is to allow users to choose their preferences, and base on the preferences, the system would deduce the most appropriate results back to users. Therefore, the ontology requirements is the need to perform the knowledge representation regarding Accommodation, Attraction and Transportation as class concepts.

Figure 5. Class hierarchy and Property hierarchy

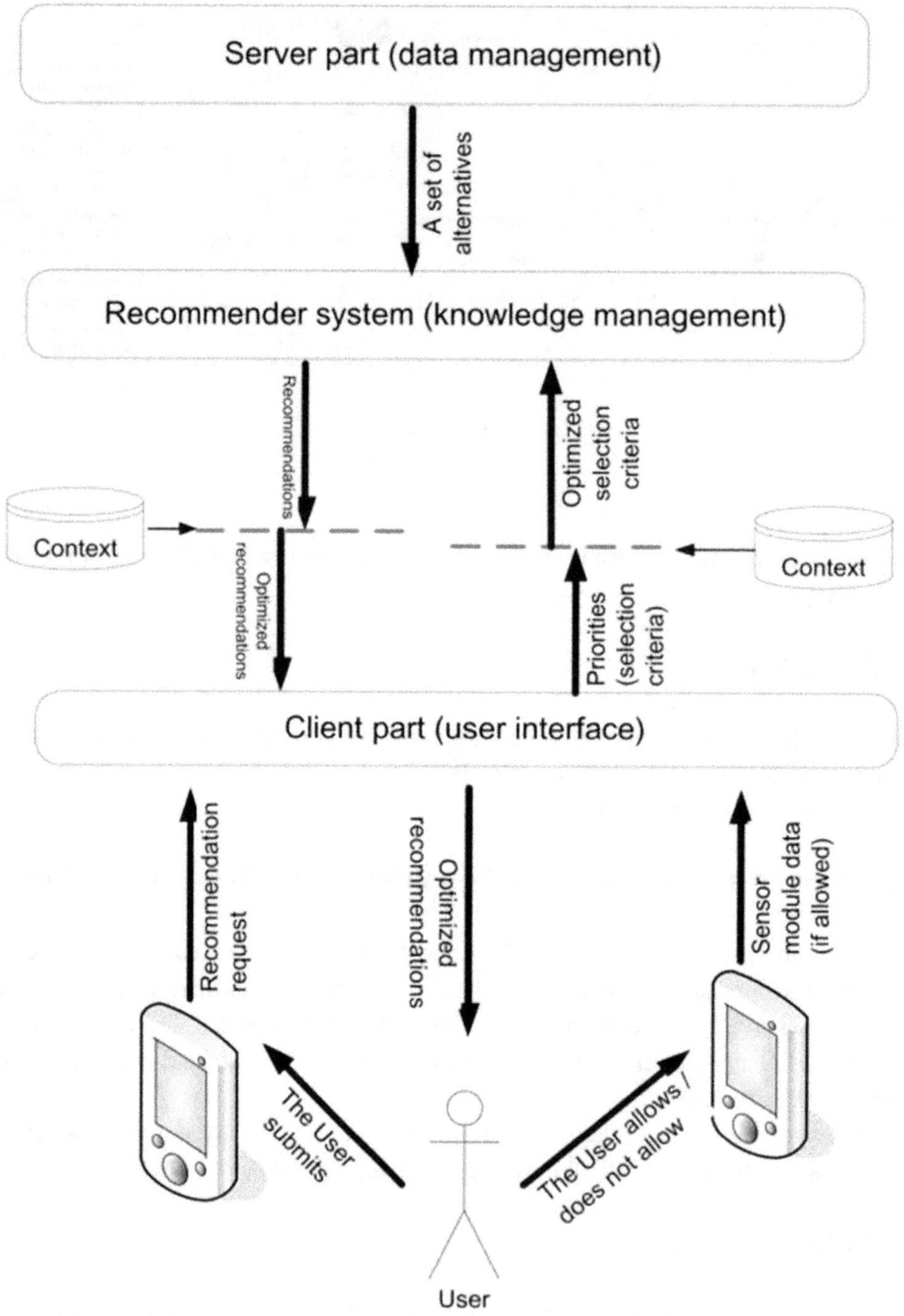

Examples of classes and properties specification for the travel recommender system:

- The ***Destination*** class contains different destination names, such as London, Peak District National Park.
- The ***Accommodation*** class represents different types of staying options located in the destination.
- The ***Attraction*** class represents different individuals that would attract people to pay a visit, such as ancient building, beautiful view.
- The ***Transportation*** class represents the different transportation types that are most popular within that area.
- The ***hasAccommodation*** property has the domain of ***Destination*** class in the range of ***Accomodation*** class.
- The ***hasAttraction*** property has the domain of ***Destination*** class in the range of ***Attraction*** class.
- The ***hasTransportation*** property has the domain of ***Destination*** class in the range of ***Transportation*** class.
- The ***hasHotel*** property has the domain of ***Destination*** class in the range of ***Hotel*** class which is the subclass of ***Accommodation*** class.

To clearly explain the relationships between classes, assumes an end-user who wants to find a place to spend his/her vacation. If this particular user also wants to go to a place that has some museums and has a museum as one of the attractions of a destination. Therefore, the reasoner would deduce the places that contain a museum, and the result will be represented to the end-user.

3.2 OWL API

The proposed ontology recommender system depends on the OWL API (Application Programming Interface) which has gone through a considerable number of developments since its first development (Horridge & Bechhofer, 2011). The OWL API includes key features of axiom-centric abstraction, general-purpose reasoner interface and serialising ontologies in different syntaxes.

As shown in Figure 6, the OWL API provides different implementations to acquire and store the axioms. The OWL Reasoner supports multiple functions such as consistency checking, class hierarchy checking, property hierarchy checking and individual reasoning.

3.3 The System Process and Implementation.

To fulfil the project requirements and objectives, the overall system process needs to be considered before the system is fully constructed. This helps the developer to have a clear thought of the development pattern and to determine the feasibility of the project. Figure 7 which is shown below conceptualizes and describes the overall process of the system with the interaction of end-users. The proposed system was implemented by modifying PizzaFinder application, developed by Matthew Horridge using Java programming language (Horridge & Bechhofer, 2011). The Query Interface (GUI) (see Figure 8 below) read from the configuration file and communicate with the reasoner to produce the list of information regarding ***Travel*** class. The user interface has two main panels, on the left side is the Travel list panel, user can choose one or more accommodation type, attractions and transportations and add them into the right side preference panel after the user made the choices, click the 'Let's Go' button to get the results.

Figure 6. The UML diagram of the management of ontologies in the OWL API (Horridge & Bechhofer, 2011)

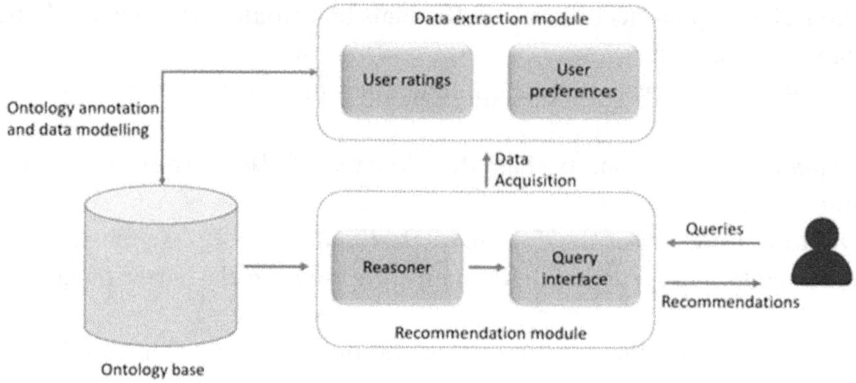

4 EVALUATION

The evaluation is based on the feedback and data collected from the survey. This survey focuses on 20 candidates who have previously used an online travel agency to plan their trip. The survey was anonymously collected, therefore no name, id or anything that might be related to personal identification exists in the collected raw data.

4.1Usability of the Knowledge-driven Travel Recommender System

Based on the responses from users as shown in Figure 9, 11 out of 20 found their expected match in 1 minute, 2 out of 20 could not find the results based on their preference. Furthermore, around 75% of total users found this application is easy to use (see Figure 10). To sum up, most of the candidates found the proposed application is easy to use, and around 80% of them found expected results in 2 minutes. It can be deduced that the level of usability of this system is above adequate.

4.2 The Satisfaction of the Knowledge-driven Travel Recommender System

According to the survey, 5 of the users were not very happy with the results. On the other hand, 15 of them think the results are matched with what they have chosen (see Figure 11). Furthermore, Figure 12 shows the likelihood of the users recommending the system. 10 users were asked about how likely they would recommend this system to others; the results indicated that 12 of them were likely to recommend this recommender system to others.

Figure 7. The System process Diagram embedded in protégé

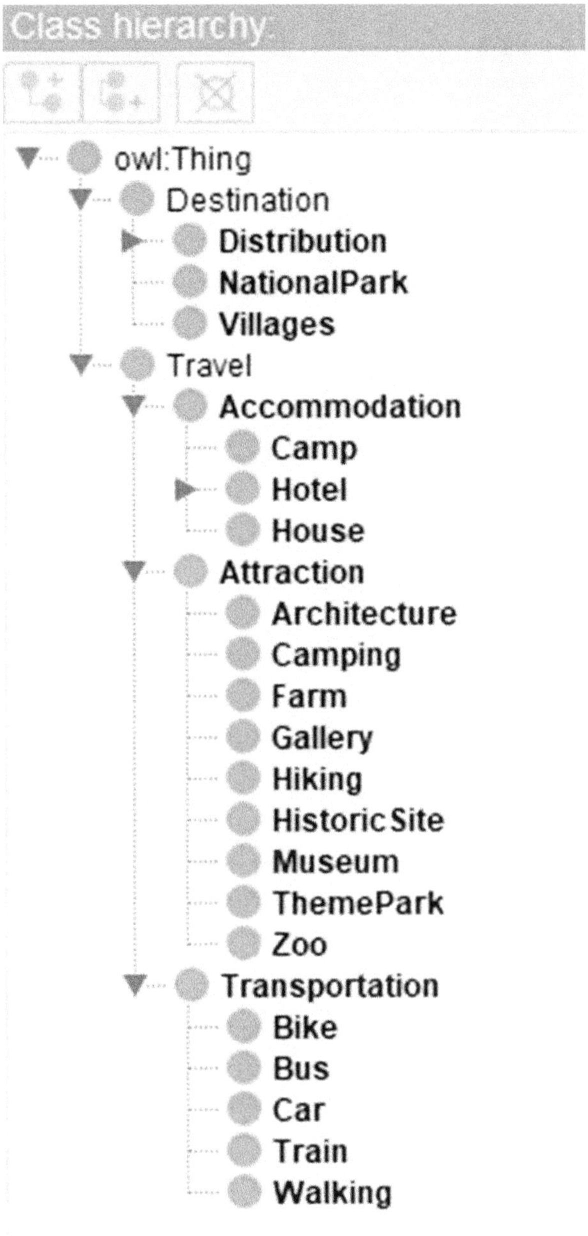

Figure 8. The System Architecture Diagram embedded in protégé

Figure 9. Time that candidates used to find expected match.

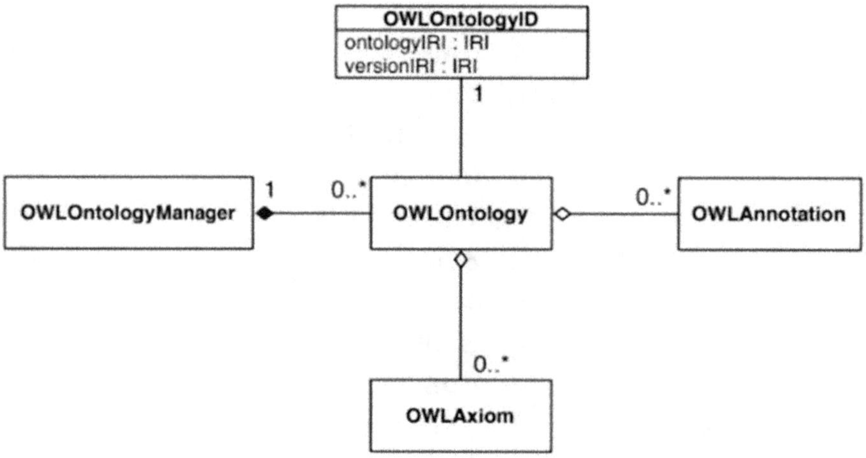

Figure 10. User impression of the system

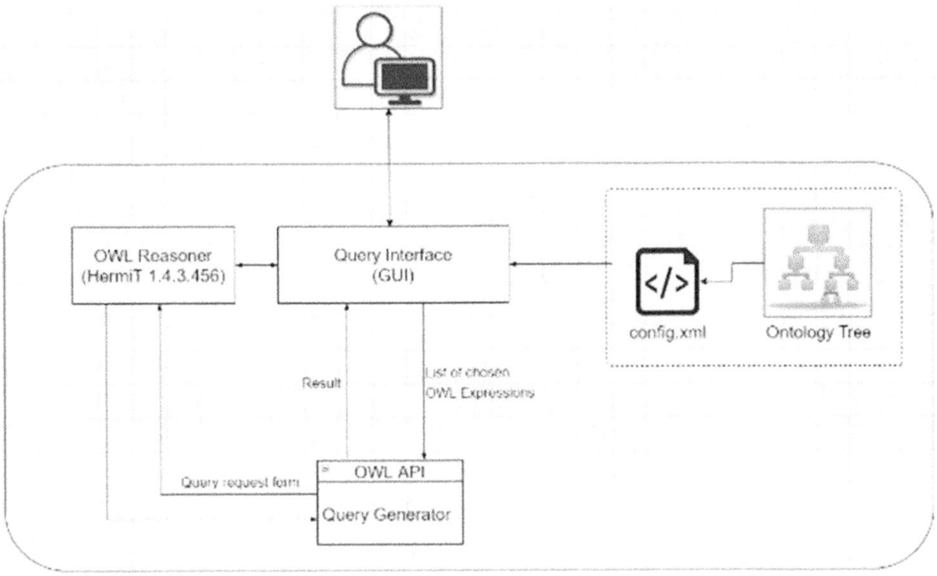

Figure 11.

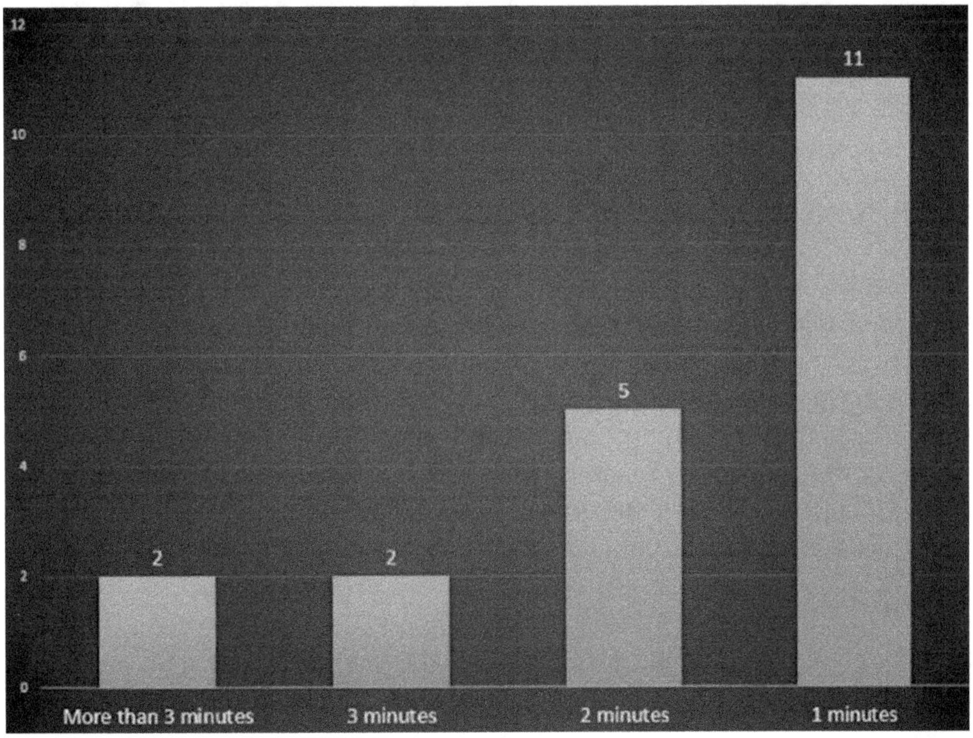

Figure 12. The likelihood of candidates willing to recommend the system to others

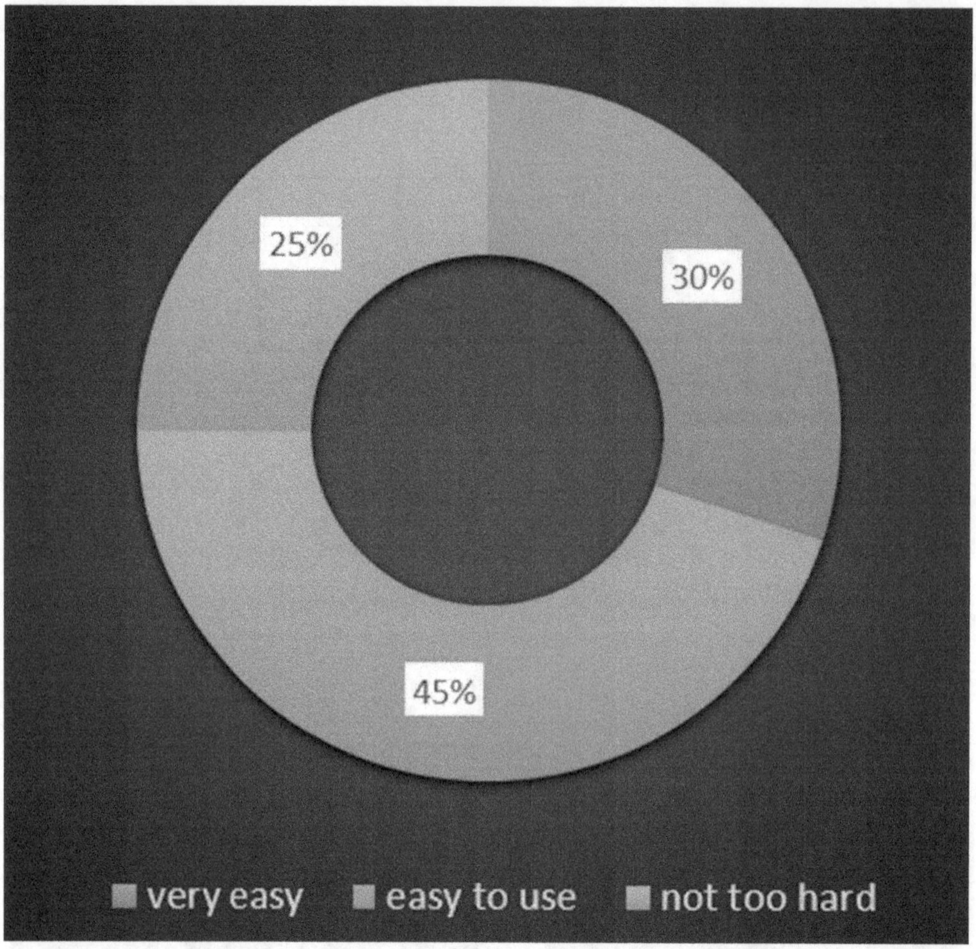

5 CONCLUSION AND FUTURE WORK

This paper proposed an ontology-driven recommender system to allow users to find the expected place to go for a vacation based on their preferences. At the end of development, most of the participant of a survey found the system is adequately useful. It is also noteworthy that some of the candidates suggested that the information the system can provide is very limited, they could not find the exact results even though they found the system is user experience friendly. The reason for this issue is because that the current ontology tree is relatively small, it cannot fit to fix all the real-life problem by now, but it is anticipated that with the expansion of the design of ontologies, the system will be more efficient and satisfying. After the tests and evaluation, few issues are spotted and future work regarding those are expected:

1. To extend the ontology to provide more information for more user preferences.
2. To extend the current system beyond Accommodation, Attraction and Transportation to retrieve the results.

3. The current system is solely based on the Windows Operating System.

In the future work, as were stated above, with the further development of this ontology, much more entities will be added to the ontology.

REFERENCES

Algherairy, A. & Bechhofer, S. (2014). *The Manchester sushi finder*. Academic Press.

Artemenko, O., Pasichnyk, V., Kunanec, N., & Tabachyshyn, D. (2019). Using context analysys for providing real time recommendations in e-tourism mobile location-based recommender systems. In *IEEE 14th International Conference on Computer Sciences and Information Technologies (CSIT)* (Vol. 3, pp. 166-169). IEEE. 10.1109/STC-CSIT.2019.8929822

Bedi, P., Agarwal, S., & Jindal, V. (2014). Marst: Multi-agent recommender system for e-tourism using reputation based collaborative filtering. In *International Workshop on Databases in Networked Information Systems*. Aizu-Wakamatsu: Springer.

Borràs, J., Moreno, A., & Valls, A. (2014). Intelligent tourism recommender systems: A survey. *Expert Systems with Applications*, *41*(16), 7370–7389. doi:10.1016/j.eswa.2014.06.007

Carral, D., Hitzler, P., Lapp, H., & Rudolph, S. (2017). *On the Ontological Modeling of Trees*. arXiv preprint arXiv:1710.05096.

Chen, Nugent, & Wang. (2012). *A Knowledge-Driven Approach to Activity Recognition in Smart Homes*. Academic Press.

Choi, I., Kim, J., & Ryu, Y. (2015). A two-tiered recommender system for tourism product recommendations. In *48th Hawaii International Conference on System Sciences* (pp. 3354-3363). IEEE.

Codina, V. (2014). *Exploiting Distributional Semantics for Content-Based and Context-Aware Recommendation* (Ph.D. Thesis). Universitat Politècnica de Catalunya – BarcelonaTech, Departament de Llenguatges i Sistemes Informàtics.

Gomez-Uribe, C., & Hunt, N. (2015). The netflix recommender system: Algorithms, business value, and innovation. *ACM Transactions on Management Information Systems*, *6*(4), 1–19. doi:10.1145/2843948

Horridge, M., & Bechhofer, S. (2011). The owl api: A java api for owl ontologies. *Semantic Web*, *2*(1), 11–21. doi:10.3233/SW-2011-0025

Ihianle, I. K., Naeem, U., & Islam, S. (2017). Ontology-driven activity recognition from patterns of object use. *Proceedings of the ACM International Joint Conference on Pervasive and Ubiquitous Computing (UbiComp '17) and Proceedings of ACM International Symposium on Wearable Computers.* 10.1145/3123024.3124424

Ihianle, I. K., Naeem, U., Islam, S., & Tawil, A.-R. (2018). A Hybrid Approach to Recognising Activities of Daily Living from Object Use in the Home Environment. *Informatics (MDPI)*, *5*(1), 6. doi:10.3390/informatics5010006

Lorenzi, F., Loh, S., & Abel, M. (2011), PersonalTour: A recommender system for travel packages. In *IEEE/WIC/ACM International Conferences on Web Intelligence and Intelligent Agent Technology* (Vol. 2, pp. 333-336). 10.1109/WI-IAT.2011.69

Lu, J., Wu, D., Mao, M., Wang, W., & Zhang, G. (2015). Recommender system application developments: A survey. *Decision Support Systems*, *74*, 12–32. doi:10.1016/j.dss.2015.03.008

Maddah, A., & Kavakli, M. (2013). System architecture & feature design for engineering a web-based Travel Advisor system. *14th ACIS International Conference on Software Engineering, Artificial Intelligence, Networking and Parallel/Distributed Computing*, 683-688. 10.1109/SNPD.2013.88

McGuinness, D., & Van Harmelen, F. (2004). OWL web ontology language overview. *W3C Recommendation, 10*(10).

Oracle. (2014). *What is Java technology and why do I need it?* Available at: https://java.com/en/download/faq/whatis_java.xml

Renjith, S., & Anjali, C. (2014)., A personalized mobile travel recommender system using hybrid algorithm. *IEEE First International Conference on Computational Systems and Communications (ICCSC)*, 12-17.

Ricci, F., Rokach, L., & Shapira, B. (2015). Recommender systems: introduction and challenges. In *Recommender systems handbook* (pp. 1–34). Springer. doi:10.1007/978-1-4899-7637-6_1

Ruotsalo, T., Haav, K., Stoyanov, A., Rochee, S., Fanid, E., Deliaic, R., Mäkelä, E., Kauppinen, T., & Hyvönen, E. (2013). SMARTMUSEUM: A Mobile Recommender System for the Web of Data. *Journal of Web Semantics*, *20*, 50–67. doi:10.1016/j.websem.2013.03.001

Sir, M., Bradac, Z., & Fiedler, P. (2015). Ontology versus database. *IFAC-PapersOnLine*, *48*(4), 220–225. doi:10.1016/j.ifacol.2015.07.036

Smirnov, A., Kashevnik, A., Shilov, N. & A, P. (2014). Smart Space-Based in-Vehicle Application for e-Tourism. In *Internet of Things, Smart Spaces, and Next Generation Networks and Systems*. Springer.

Wei, Y., Jennings, N., Moreau, L., & Hall, W. (2008). User evaluation of a market-based recommender system. *Autonomous Agents and Multi-Agent Systems*, *17*(2), 251–269. doi:10.100710458-008-9029-x

Compilation of References

GPP TR 23.558 (2020o) "*Architecture for enabling Edge Applications (EA)*" Rel. 17 v2.0.0

GPP TR 23.700-40 (2020f) "*Study on enhancement of network slicing; Phase 2*" Rel. 17 v2.0.0

GPP TR 23.748 (2020n) "*Study on enhancement of support for Edge Computing in 5G Core network (5GC)*" Rel. 17 v17.0.0

GPP TR 23.752 (2020i) "*Study on system enhancement for Proximity based Services (ProSe) in the 5G System (5GS)*" Rel. 17 v2.0.0

GPP TR 23.754 (2020d) "*Study on supporting Unmanned Aerial Systems (UAS) connectivity, Identification and tracking*" Rel. 17 v17.0.0

GPP TR 23.755 (2020e) "*Study on application layer support for Unmanned Aerial Systems (UAS)*" Rel. 16 v2.0.0

GPP TR 23.758 (2019) "*Study on application architecture for enabling Edge Applications*" Rel. 17 v17.0.0

GPP TR 23.948 (2019a) "*Deployment guidelines for typical edge computing use cases in 5G Core network (5GC)*" Rel. 17, no published document, yet

GPP TR 28.811 (2020g) "*Study on network slice management enhancement*" Rel. 17 v0.4.0

GPP TR 36.777 (2018) "*Enhanced LTE support for aerial vehicles*" Rel. 15, v15.0.0

GPP TR 38.836 (2020j) "*Study on NR sidelink relay*" Rel. 17 v2.0.0

GPP TS 22.104 (2020q) "*Service requirements for cyber-physical control applications in vertical domains*" Rel. 17 v17.4.0

GPP TS 22.125 (2020b) "*Unmanned Aerial System (UAS) support in 3GPP*" Rel. 16 v17.2.0

GPP TS 22.261 (2020c) "*Service requirements for the 5G system*" Rel. 15 v18.1.1

GPP TS 22.263 (2020p) "*Service requirements for Video, Imaging and Audio for Professional Applications (VIAPA)*" Rel. 17 v17.3.0

GPP TS 23.273 (2020r) "*5G System (5GS) Location Services (LCS); Stage 2*" Rel. 16 v16.5.0

GPP TS 23.501 (2020) "*System architecture for the 5G System (5GS)*" Rel. 15 v16.7.0

GPP TS 23.501 (2020a) "*System architecture for the 5G System (5GS)*" Rel. 15 v16.7.0

GPP TS 28.313 (2020m) "*Self-Organizing Networks (SON) for 5G networks*" Rel. 16 v17.0.0

GPP TS 37.320 (2020l) "*Universal Terrestrial Radio Access (UTRA) and Evolved Universal Terrestrial Radio Access (E-UTRA); Radio measurement collection for Minimization of Drive Tests (MDT); Overall description; Stage 2*" Rel. 10, v16.3.0

GPP TS 38.101-1, (2020s) "*NR; User Equipment (UE) radio transmission and reception; Part 1: Range 1 Standalone*", Rel. 15 v17.0.0

GPP TS 38.101-2, (2020t) "*NR; User Equipment (UE) radio transmission and reception; Part 2: Range 2 Standalone*", Rel. 15 v17.0.0

GPP TS 38.215 (2020k) "*NR; Physical layer measurements*" Rel. 15 v16.4.0

GPP TS 38.300 (2020h) "*NR; NR and NG-RAN Overall description; Stage-2*" Rel. 15 v16.4.0

Drones Project. (2019). *D2.1 Initial definition of the trial controller architecture mechanisms and APIs*. https://5gdrones.eu/wp-content/uploads/2020/06/D2.1-Initial-definition-of-the-trial-controller-architecture-mechanisms-and-APIs_v1.1.pdf

GPPP. (2020) *5GPPP Empowering Vertical Industries through 5G Networks - Current Status and Future Trends.* https://5g-ppp.eu/wp-content/uploads/2020/09/5GPPP-VerticalsWhitePaper-2020-Final.pdf

Abdullah, K., & Hussain, Z. M. (2007, December). Performance of Fourier-based and wavelet-based OFDM for DVB-T systems. In *2007 Australasian Telecommunication Networks and Applications Conference* (pp. 475-479). IEEE. 10.1109/ATNAC.2007.4665248

AbuHussain, M., & Hasar, U. C. (2020). Design of X-bandpass waveguide Chebyshev filter based on CSRR metamaterial for telecommunication systems. *MDPI Electronics*, *9*(1), 101. doi:10.3390/electronics9010101

Adams, J. (2019). *Mail Online - Britain faces Big Brother surveillance society WORSE than George Orwell's 1984: Lip-reading cameras that will force people to talk 'with hands over mouths' are latest attack on privacy, government watchdog warns.* Retrieved 09 December, 2020, from https://www.dailymail.co.uk/news/article-7399467/Britain-faces-Big-Brother-surveillance-society-WORSE-George-Orwells-1984-watchdog-says.html

Adegoke, V., Chen, D., Banissi, E., & Barsikzai, S. (2020). Enhancing Ensemble Prediction Accuracy of Breast Cancer Survivability and Diabetes Diagnostic Using Optimized EKF-RBFN Trained Prototypes. In A. Madureira, A. Abraham, N. Gandhi, C. Silva, & M. Antunes (Eds.), *Proceedings of the Tenth International Conference on Soft Computing and Pattern Recognition(SoCPaR 2018) Advances in Intelligent Systems and Computing.* Cham: Springer. doi:10.1007/978-3-030-17065-3_6

Agrawal, G. P., & Dutta, N. K. (1993). *Semiconductor Lasers*. Springer.

Ahmad, I., Kumar, T., Liyanage, M., Okwuibe, J., Ylianttila, M., & Guvtor, A. (2018). 5G Security: Analysis of Threats and Solutions Artificial Intelligence (AI) for Development. Series Report on AI and IoT in Security Aspects.

Ahmed, G. H., & Ali, O. (2007). Design and conception of optical links simulator for telecommunication applications under Simulink environment. *WSEAS International Conference on Applied Electromagnetics, Wireless and Optical Communications*, 52-57.

Ahmed, M., Yamada, M., & Saito, M. (2001). Numerical modeling of intensity and phase noise in semiconductor lasers. *IEEE Journal of Quantum Electronics*, *37*(12), 1600–1610. doi:10.1109/3.970907

Ahsan, M. R., Islam, M. T., Habib Ullah, M., Mahadi, W. N. L., & Latef, T. A. (2014). Compact Double-P Slotted Inset-Fed Microstrip Patch Antenna on High Dielectric Substrate. *The Scientific World Journal*, *2014*, 1–6. Advance online publication. doi:10.1155/2014/909854 PMID:25165750

Ahsan, M. R., Islam, M. T., Habib Ullah, M., & Misran, N. (2014). Bandwidth enhancement of a dual band planar monopole antenna using meandered microstrip feeding. *TheScientificWorldJournal*, *2014*, 1–8. Advance online publication. doi:10.1155/2014/856504 PMID:24723832

Akdemir, U., Turaga, P., & Chellappa, R. (2008). An ontology-based approach for activity recog nition from video. *Proceedings of 16th ACM international conference on Multimedia (MM, 08).*

Akyildiz, I. F., Han, C., & Nie, S. (2018). Combating the distance problem in the millimeter wave and Terahertz frequency band. *IEEE Communications Magazine*, *56*, 102–108.

Alan, B. (1938). *Improvements in or relating to high frequency electrical conductors or radiators.* British Patent no. 515684.

Algherairy, A. & Bechhofer, S. (2014). *The Manchester sushi finder*. Academic Press.

Ali, A., Shamsuddin, M., & Eassa, F. (2014). Ontology-Based Cloud service Representation'. *Research Journal of Applied Sciences, Engineering and Technology*, *8*(1), 83–94.

Ali, E., Ismail, M., Nordin, R., & Abdulah, N. F. (2017). Beamforming techniques for massive MIMO systems in 5G: Overview, classification, and trends for future research. *Front Inform Tech EL*, *18*, 753–772.

Alimi, O. A., Ouahada, K., & Abu-Mahfouz, A. M. (2020). *A Review of Machine Learning Approaches to Power System Security and Stability*. IEEE. doi:10.1109/ACCESS.2020.3003568

Aljabre, A. (2012). Cloud Computing for Increased Business Value. *International Journal of Business and Social Science*, *3*(1), 234–239.

Allam, Z., & Dhunny, Z. (2019). *On big data, artificial intelligence and smart cities*. Academic Press.

Allam, Z., & Dhunny, Z. (2019). On big data, artificial intelligence and smart cities. *Cities (London, England)*, *89*, 80–91. doi:10.1016/j.cities.2019.01.032

Alnajjar, H., Alnaser, N. W., & Alnaser, W. E. (2019). Level of Electromagnetic Radiation Emitted from Mobile Phone Base Station during F1 Race in the Kingdom of Bahrain. *International Journal of Physics: Study and Research*, *2*(1), 65–70. doi:10.18689/ijpsr-1000109

Anagnostopoulos, T., Zaslavsy, A., Medvedev, A., & Khoruzhnicov, S. (2015b). Top--k query based dynamic scheduling for IoT-enabled smart city waste collection. In *2015 16th IEEE International Conference on Mobile Data Management* (Vol. 2, pp. 50-55). IEEE.

Anagnostopoulos, T., Kolomvatsos, K., Anagnostopoulos, C., Zaslavsky, A., & Hadjiefthymiades, S. (2015a). Assessing dynamic models for high priority waste collection in smart cities. *Journal of Systems and Software*, *110*, 178–192. doi:10.1016/j.jss.2015.08.049

Anagnostopoulos, T., Zaslavsky, A., Kolomvatsos, K., Medvedev, A., Amirian, P., Morley, J., & Hadjieftymiades, S. (2017). Challenges and opportunities of waste management in IoT-enabled smart cities: A survey. *IEEE Transactions on Sustainable Computing*, *2*(3), 275–289. doi:10.1109/TSUSC.2017.2691049

Ananda, S. K., Oishee, M., Prasanna, K. L., & Subhasis, B. (2017). Omnidirectional Assistive Wheelchair: Design and Control with Isometric Myoelectric Based Intention Classification. *Procedia Computer Science*, *105*, 68–74. doi:10.1016/j.procs.2017.01.200

Anastasi, G., Conti, M., & Di Francesco, M. (2008). Data collection in sensor networks with data mules: An integrated simulation analysis. In *2008 IEEE Symposium on Computers and Communications* (pp. 1096-1102). IEEE. 10.1109/ISCC.2008.4625629

Angelidou, M. (2014). Smart city policies: A spatial approach. *City*, S3–S11. doi:10.1016/j.cities.2014.06.007

Ang, K. L. M., Seng, J. K. P., & Zungeru, A. M. (2017). Optimizing energy consumption for big data collection in large-scale wireless sensor networks with mobile collectors. *IEEE Systems Journal*, *12*(1), 616–626. doi:10.1109/JSYST.2016.2630691

Ang, L. M., Seng, K. P., Ijemaru, G. K., & Zungeru, A. M. (2018). Deployment of IoV for smart cities: Applications, architecture, and challenges. *IEEE Access: Practical Innovations, Open Solutions*, *7*, 6473–6492. doi:10.1109/ACCESS.2018.2887076

Ang, L. M., Seng, K. P., Zungeru, A. M., & Ijemaru, G. K. (2017). Big sensor data systems for smart cities. *IEEE Internet of Things Journal*, *4*(5), 1259–1271. doi:10.1109/JIOT.2017.2695535

Arinze, S. N., Onoh, G. N., & Abonyi, D. O. (2020). Development of A Light Fidelity System for Improving Network Performance in a Wireless Local Area Network. *International Journal of Emerging Trends in Engineering and Development*, *2*(10), 31-41.

Arinze, S. N., Onoh, G. N., & Abonyi, D. O. (2020). Performance of Light Fidelity and Wireless Fidelity Networks in a WLAN. *International Journal of Research in Engineering and Science*, *4*(1), 10-20.

Artemenko, O., Pasichnyk, V., Kunanec, N., & Tabachyshyn, D. (2019). Using context analysys for providing real time recommendations in e-tourism mobile location-based recommender systems. In *IEEE 14th International Conference on Computer Sciences and Information Technologies (CSIT)* (Vol. 3, pp. 166-169). IEEE. 10.1109/STC-CSIT.2019.8929822

Arts, K., Wal, R., & Adams, W. (2015). Digital technology and the conservation of nature. *Ambio*, *44*(S4), 661–673. Advance online publication. doi:10.100713280-015-0705-1 PMID:26508352

Ashwini, R., Sanjay, K., Sanjay, M., & Rahul, T. (2012). Decision Point for Cloud Computing in Small, Medium Enterprises (SMEs). *Proceedings of 7th international Conference for Internet Technology and Secured Transactions (ICITST-2012)*, 688-691.

Atta, L. C. V. (1981). A history of early microwave antenna development. *IEEE Antennas and Propagation Society Newsletter*, *23*(5), 10–14. doi:10.1109/MAP.1981.27564

Attaran,M. (2021). The Impact of 5G on the Evolution of Intelligent Automation and Industry Digitization. *Journal of Ambient Intelligence and Humanized Computing.*

Ayad, A. (2019). *Cyber-Physical Security of Power Distribution Systems* (Unpublished master thesis). University of Waterloo, Ontario, Canada.

Azad, A. R., & Mohan, A. (2018). Substrate integrated waveguide dual-band and wide-stopband bandpass filters. *IEEE Microwave and Wireless Components Letters*, *28*(8), 660–662. doi:10.1109/LMWC.2018.2844103

Baig, S., Asif, H. M., Umer, T., Mumtaz, S., Shafiq, M., & Choi, J. G. (2018). High Data Rate Discrete Wavelet Transform-Based PLC-VLC Design for 5G Communication Systems. *IEEE Access: Practical Innovations, Open Solutions*, *6*, 52490–52499. doi:10.1109/ACCESS.2018.2870138

Balanis, C. (2005). *Antenna theory: Analysis and design.* Wiley-Interscience.

Balanis, C. A. (2015). *Antenna Theory: Analysis and Design* (4th ed.). Wiley.

Balmain, K. G., & Nkeng, J. N. (1976). Asymmetry phenomenon of log-periodic dipole antennas. *IEEE Transactions on Antennas and Propagation*, *24*(4), 402–410. doi:10.1109/TAP.1976.1141380

Bao, L., & Intille, S. (2004). Activity recognition from user annotated acceleration data. *2nd International Conference Pervasive Computing (PERVASIVE, 2004).*

Bao, H., & Shieh, W. (2007). Transmission simulation of coherent optical OFDM signals in WDM systems. *Optics Express*, *15*(8), 4410. doi:10.1364/OE.15.004410 PMID:19532688

Bao, X., Lee, N., Raj, R. B., Rangan, K. P., & Maria, A. (1998). Engineering solder paste performance through controlled stress rheology analysis. *Soldering & Surface Mount Technology*, *10*(2), 26–35. doi:10.1108/09540919810219949

Baratis, E., Petrakis, E., Batsakis, S., Maris, N., & Papadakis, N. (2009). *Toql: Temporal ontology querying language*. Advances in Spatial and Temporal Databases.

Basak, A., Talukder, Z., & Islam, R. (2013). Performance analysis and comparison between Coarse WDM and Dense WDM. *Global Journal of Research in Engineering - Electrical and Electronics Engineering, 13*(6).

Batagejl, B. (2002). Need of knowing non-linear coefficient in optical networks. *WSEAS International Conferences.*

Bath, J. (2007). *Lead-free soldering. XIV.* Hardcover. doi:10.1007/978-0-387-68422-2

BBC. (2017). *BBC -Robot automation will 'take 800 million jobs by 2030' - report.* BBC. Retrieved November 26, 2020, from https://www.bbc.co.uk/news/world-us-canada-42170100

Bedi, P., Agarwal, S., & Jindal, V. (2014). Marst: Multi-agent recommender system for e-tourism using reputation based collaborative filtering. In *International Workshop on Databases in Networked Information Systems.* Aizu-Wakamatsu: Springer.

Berenger, J. P. (1994). A perfectly matched layer for the absorption of electromagnetic waves. *Journal of Computational Physics*, *114*(2), 185–200. doi:10.1006/jcph.1994.1159

Berrilio, P. D. (2014). *Exploiting Vehicular Networks for Data Collection in Smart Cities* (Doctoral dissertation). University of Rennes, France.

Bharadwaj, B., Kumudha, M., & Chaithra, G. (2017, February). Automation of Smart waste management using IoT to support "Swachh Bharat Abhiyan"-a practical approach. In *2017 2nd International Conference on Computing and Communications Technologies (ICCCT)* (pp. 318-320). IEEE.

Bharadwaj, P., Deutsch, B., & Novotny, L. (2009). Optical antennas. *Advances in Optics and Photonics*, *1*, 438–483.

Binh, L.-N. (2010). *Optical fiber communications systems: Theory, practice and Matlab Simulink models.* CRC Press.

BIS. (2014). Smart cities framework – Guide to establishing strategies for smart cities and communities. *PAS*, *181*, 2014.

Bisson, D. (2017). Dragonfly 2.0 Attack Campaign Targets Western Energy Sector. *The State of Security.*

Bjerregaard, T., & Mahadevan, S. (2006). A survey of research and practices of network-on-chip. *ACM Computing Surveys*, *38*, 1–50.

Blei, D., Ng, A., & Jordan, M. (2003). Latent dirichlet allocation. *Machine Learning.*

Boa, D., Xiao, Z., Sun, Y., & Zhao, J. (2010). A Method and Framework for Quality of Cloud Services Measurement. *3rd International Conference on Advanced Computer Theory and Engineering (ICACTE)*, 358-362.

Bogale, T.E., Wang, X., & Le, L.B. (2018). Machine Intelligence Techniques for Next-Generation Context-Aware Wireless Networks. *ITU Journal: ICT Discoveries.*

Bolos, N., Tsoros, A., & Holopainen, A. (2015). Social, innovative and smart cities are happy and resilient': Insights from the WHO EURO 2014 International Healthy Cities Conference. *International Journal of Health Geographics*, *14*(3). PMID:25588543

Borja, J. (2007). Counterpoint: Intelligent cities and innovative cities. *E-Journal on the Knowledge Society*. Retrieved from http://www.uoc.edu/uocpapers/5/dt/eng/mitchell.pdf

Borràs, J., Moreno, A., & Valls, A. (2014). Intelligent tourism recommender systems: A survey. *Expert Systems with Applications*, *41*(16), 7370–7389. doi:10.1016/j.eswa.2014.06.007

Boulos, M., & Al-Shorbaji, N. (2014). On the Internet of Things, smart cities and the WHO Healthy Cities. *International Journal of Health Geographics*, *13*(10). PMID:24669838

Braun, T. F., Fung, B. C. M., Iqbal, F., & Shah, B. (2018). Security and privacy challenges in smart cities. *Sustainable Cities and Society*, *39*, 499–507. doi:10.1016/j.scs.2018.02.039

Brillouin, L. (1922). Origin of radiation resistance. *Radioélectricité*, *3*, 147–152.

Buckus, R., Strukčinskienė, B., Raistenskis, J., Stukas, R., Šidlauskienė, A., Čerkauskienė, R., Isopescu, D., Stabryla, J., & Cretescu, I. (2017). A Technical Approach to the Evaluation of Radiofrequency Radiation Emissions from Mobile Telephony Base Stations. *International Journal of Environmental Research and Public Health*, *14*(3), 244–254. doi:10.3390/ijerph14030244 PMID:28257069

Budiharto, W., & Gunawan, A.A.S. (2018). *Design and algorithm for electric wheelchair using voice recognition and joystick.* . doi:10.24507/icicel.12.11.1171

Buyya, R., Yeo, C., Venugopal, J., Broberg, J., & Brandic, I. (2009). Cloud Computing and Emerging IT Platform: Vision, Hype and Reality for Delivering computing as the 5th Utility. *Future Generation Computer Systems*, *25*(6), 599–616. doi:10.1016/j.future.2008.12.001

CAAC. (2017, May). *Regulations on Real-name Registration of Civil Unmanned Aerial Vehicles.* http://www.caac.gov.cn/en/XWZX/201705/t20170524_44222.html

CAAT. (2018, October 21). *Center for Advanced Automotive Technology*. Retrieved from http://autocaat.org/Technologies/Automated_and_Connected_Vehicles/

Cai, C., Wang, J., Zhu, L., & Wu, W. (2016). A new approach to design a microstrip wideband balun bandpass filter. *IEEE Microwave and Wireless Components Letters*, *26*(2), 116–118. doi:10.1109/LMWC.2016.2516760

Cao, D., Masoud, O., Boley, D., & Papanikolopoulos, N. (2009). Human motion recognition using support vector machines. *Computer Vision and Image Understanding*, *113*(10), 113. doi:10.1016/j.cviu.2009.06.002

Caragliu, A., Del-Bo, C., & Nijkamp, P. (2011). Smart Cities in Europe. *Journal of Urban Technology*, *18*(2), 65–82. doi:10.1080/10630732.2011.601117

Cardenas, A. (2019). *Cyber-Physical Systems Security*. Academic Press.

Carlsson, A. (2019). *On the Use of 5G for Smart Grid Inter-Substation Control Signalling* (Unpublished master thesis). Karlstad University.

Carral, D., Hitzler, P., Lapp, H., & Rudolph, S. (2017). *On the Ontological Modeling of Trees.* arXiv preprint arXiv:1710.05096.

Carter, P. S. (1939). *Wide band, short wave antenna and transmission line system.* U.S. Patent 2,181,870.

Cavin, R. K., Lugli, P., & Zhirnov, V. V. (2012). Science and Engineering Beyond Moore's Law. *Proceedings of the IEEE*, *100*, 1720–1749.

Cebik, L. B. (2007). Log periodic arrays. In *W4RNL, The ARRL Antenna Book* (21st ed., pp. 10-1–10-31). The American Radio Relay League, Inc.

Ceniklioğlu, B., Özen, A., & Develi, I. (2018, May). Zero-tail discrete wavelet transform spread OFDM system for 5G and beyond. In *2018 26th Signal Processing and Communications Applications Conference (SIU)* (pp. 1-4). IEEE.

Čertík, F., & Róka, R. (2012). The nonlinear FWM effect and its influence on optical signals utilized different modulation techniques in the WDM transmission systems. *Optical Communications Conference.*

Čertík, F., & Róka, R. (2014). Nonlinear SPM and XPM effects and their influence on optical signals utilized different modulation techniques in WDM transmission systems. *Optical Communications Conference.*

Chaqfeh, M., Mohamed, N., Jawhar, I., & Wu, J. (2016, December). Vehicular cloud data collection for intelligent transportation systems. In 2016 3rd Smart Cloud Networks & Systems (SCNS) (pp. 1-6). IEEE. doi:10.1109/SCNS.2016.7870555

Chatterjee, K., Padmini, V., & Khaparde, S. A. (2017). Review of Cyber Attacks on Power System Operations. *Proceedings of IEEE Region 10 Symposium (TENSYMP)*, 1-6. 10.1109/TENCONSpring.2017.8070085

Chen, Nugent, & Wang. (2012). *A Knowledge-Driven Approach to Activity Recognition in Smart Homes*. Academic Press.

Chen, L., & Nugent, C. (2009). Ontology based activity recognition in intelligent pervasive environments. *International Journal of Web Information Systems*, *5*(4), 410–430. doi:10.1108/17440080911006199

Chen, L., Nugent, C., Hoey, J., Cook, D., & Yu, Z. (2012). Sensor based activity recognition. *IEEE Transactions on Systems, Man and Cybernetics. Part C, Applications and Reviews*, *42*(6), 790–808. doi:10.1109/TSMCC.2012.2198883

Chen, L., Nugent, C., & Wang, H. (2011). Knowledge driven approach to activity recognition in smart homes. *IEEE Transactions on Knowledge and Data Engineering*.

Chen, S., Zhao, Y., Peng, M., & Wang, Y. (2014). A codesigned compact dual-band filtering antenna with PIN loaded for WLAN applications. *International Journal of Antennas and Propagation*, *2014*, 1–6. Advance online publication. doi:10.1155/2014/826171

Chen, X., Zhao, F., Yan, L., & Zhang, W. (2013). A compact filtering antenna with flat gain response within the passband. *IEEE Antennas and Wireless Propagation Letters*, *12*, 857–860. doi:10.1109/LAWP.2013.2271972

Chiaramello, E., Bonato, M., Fiocchi, S., Tognola, G., Parazzini, M., Ravazzani, P., & Wiart, J. (2019). Radio Frequency Electromagnetic Fields Exposure Assessment in Indoor Environments: A Review. *International Journal of Environmental Research and Public Health*, *16*(6), 955–984. doi:10.3390/ijerph16060955 PMID:30884917

China Copyright and Media. (2014). *Planning outline for the construction of a social credit system (2014-2020).* Retrieved from https://chinacopyrightandmedia.wordpress.com/2014/06/14/planning-outline-forthe-construction-of-a-social-credit-system-2014-2020/

Chi-Sheng, C., Tung-Yung, H., & Tze-Yuan, L. (2014). Design and development of solar power-assisted manual/electric wheelchair. JRRD, 51(9), 1411–1426.

Choi, I., Kim, J., & Ryu, Y. (2015). A two-tiered recommender system for tourism product recommendations. In *48th Hawaii International Conference on System Sciences* (pp. 3354-3363). IEEE.

Choudhary, G., & Sharma, V. (n.d.). *A Survey on the Security and the Evolution of osmotic and Catalytic Computing for 5G Networks*. Academic Press.

Chuang, C. T., & Chung, S. J. (2009). New printed filtering antenna with selectivity enhancement. *European Microwave Week 2009, EuMW 2009: Science, Progress and Quality at Radiofrequencies, Conference Proceedings - 39th European Microwave Conference, EuMC 2009,* 747–750. 10.1109/EUMC.2009.5296120

Chuang, C. T., & Chung, S. J. (2011a). A compact printed filtering antenna using a ground-intruded coupled line resonator. *IEEE Transactions on Antennas and Propagation, 59*(10), 3630–3637. doi:10.1109/TAP.2011.2163777

Chuang, C. T., & Chung, S. J. (2011b). Synthesis and design of a new printed filtering antenna. *IEEE Transactions on Antennas and Propagation, 59*(3), 1036–1042. doi:10.1109/TAP.2010.2103001

Chuang, M., & Wu, M. (2011). Microstrip Diplexer Design Using Common T-Shaped Resonator. *IEEE Microwave and Wireless Components Letters, 21*(11), 583–585. doi:10.1109/LMWC.2011.2168949

Chu, H., Hong, H., Zhu, X., Li, P., & Guo, Y. X. (2018). Implementation of Synthetic Material in Dielectric Resonator-Based Filtering Antennas. *IEEE Transactions on Antennas and Propagation, 66*(7), 3690–3695. doi:10.1109/TAP.2018.2819891

Cisco. (2006). *ONS 15454 - DWDM Engineering and Planning Guide*. Author.

Ciurana, M. (2009). *Developing with Google App Engine' A press*. Barkley. doi:10.1007/978-1-4302-1832-6

Clement, J. (2020). *Worldwide digital population as of October 2020*. Statista. Retrieved November 26, 2020, from https://www.statista.com/statistics/617136/digital-population-worldwide/

Clements, D., Desmulliez, M. P. Y., & Abraham, E. (2007). The evolution of paste pressure during stencil printing. *Soldering & Surface Mount Technology, 19*(3), 9–14. doi:10.1108/09540910710843720

Codina, V. (2014). *Exploiting Distributional Semantics for Content-Based and Context-Aware Recommendation* (Ph.D. Thesis). Universitat Politècnica de Catalunya – BarcelonaTech, Departament de Llenguatges i Sistemes Informàtics.

Colin, C. J., & Joseph, K. (1989). *Solder paste technology: principles and applications. Edition: Illustrated.* TAB Professional and Reference Books.

Collin, R. E. (1985). *Antennas and Radiowave Propagation*. McGraw-Hill.

Comiter, M. (2019). Attacking Artificial Intelligence: AI's Security vulnerability and What Policymakers Can Do about It. Harvard Kennedy School, Belfer Center for Science for Science and International Affairs.

Compañó, R. (2001). Trends in Nanoelectronics. *Nanotechnology, 12*, 85.

Compton, R. T., Jr. (1963). The aperture admittance of a rectangular waveguide radiating into a lossy half-space. Technical Report, 1691-1. Ohio State University.

Contreras-Castillo, J., Zeadally, S., & Guerrero-Ibañez, J. A. (2017). Internet of vehicles: Architecture, protocols, and security. *IEEE Internet of Things Journal, 5*(5), 3701–3709. doi:10.1109/JIOT.2017.2690902

Crassee, I., Levallois, J., Walter, A. L., Ostler, M., Bostwick, A., Rotenberg, E., Seyller, T., Marel, D. d., & Kuzmenko, A. B. (2011). Two-dimensional material nano photonics. *Nature Physics, 1*(7), 48–51.

Čučka, M., Šalík, P., Róka, R., Műnster, P., & Filka, M. (2018). Simulation models of pulse generator for OTDR in Matlab and VPIphotonics. *International Conference on Telecommunications and Signal Processing*. 10.1109/TSP.2018.8441274

Cusumano, M. (2010). Cloud Computing and SaaS as New Computing Platforms. *Communications of the ACM, 53*(4), 27–29. doi:10.1145/1721654.1721667

Cyber Threat and Vulnerability Analysis of the U.S. Electric Sector. (2016). *Mission Support Center Analysis Report*. Idaho National Laboratory.

Cybersecurity and Infrastructure Security Agency (CISA). (2019). *Overview of Risks Introduced by 5G Adoption in the United States*. CISA.

Dahle, R., Laforge, P., & Kuhling, J. (2017). 3-D printed customizable inserts for waveguide filter design at X-band. *IEEE Microwave and Wireless Components Letters*, *27*(12), 1080–1082. doi:10.1109/LMWC.2017.2754345

Dai, L., Wang, Z., & Yang, Z. (2012). Next-generation digital television terrestrial broadcasting systems: Key technologies and research trends. *IEEE Communications Magazine*, *50*(6), 150–158. doi:10.1109/MCOM.2012.6211500

Dainkeh, A., Nwajana, A. O., & Yeo, K. S. K. (2016). Filtered power splitter using microstrip square open loop resonators. *Progress In Electromagnetic Research*, *C*, 64, 133–140. doi:10.2528/PIERC16042005

Dalmia, S., White, G., Sundaram, V., & Swaminathan, M. (2004). Design of quasi-lumped element diplexer for multiple wireless applications using liquid crystalline polymer (LCP) based substrates. *Proceedings of the 34th European Microwave Conference (EuMC)*, 737–740.

Dash, S., & Patnaik, A. (2018). Material selection for THz antennas. *Microwave and Optical Technology Letters*, *60*, 1183–1187.

Dash, S., & Patnaik, A. (2020). Behavior of Graphene Based Planar Antenna at Microwave and Terahertz Frequency, Photonics and Nanostructures - Fundamentals and Applications. *Photonics*, 100800.

Daugherty, D. (2015). *Driving Unconventional Growth through the Industrial Internet of Things*. Accenture.

Del Río Bellisco, J., Hermosilla, J., Monteroy, R., & Espí, P. (2004). Education tool for optical communication engineering teaching. *WSEAS International Conferences*.

Deng, J., & Feng, L. (2019). Dual-band microstrip filtering antennas with symmetrical slots. *Progress in Electromagnetics Research Letters*, *86*(May), 13–19. doi:10.2528/PIERL19050404

Deshmukh, G., & Jagtap, S. (2013). Four wave mixing in DWDM optical system. *International Journal of Computational Engineering Research*, *3*(6).

Devi, J., Datta, P. (n.d.). A Study on Dielectric Properties of Cadmium Sulfide–Zinc Sulfide Core–Shell Nanocomposites for Application as Nanoelectronic Filter Component in the Microwave Domain. *Journal of Elec Materi*, *47*, 3529–3542.

Devi, J., & Datta, P. (2018). Yagi–Udananoantenna For NIR domain. *Journal of Computational Electronics*, *17*, 406–418.

Diamantoulakis, P. D., Kapinas, V. M., & Karagiannidis, G. K. (2015). *Big Data Analytics for Dynamic Energy Management in Smart Grids*. arXiv: 1504.02424v3 [cs.DB].

Dibal, P. Y., Onwuka, E. N., Agajo, J., & Alenoghena, C. O. (2018). Application of wavelet transform in spectrum sensing for cognitive radio: A survey. *Physical Communication*, *28*, 45–57. doi:10.1016/j.phycom.2018.03.004

Dilmaghani, R., & Ghavami, M. (2007, September). Wavelet vs. Fourier based UWB systems. In *2007 IEEE 18th International Symposium on Personal, Indoor and Mobile Radio Communications* (pp. 1-5). IEEE. 10.1109/PIMRC.2007.4394653

Dinh, N. Q., Michishita, N., Yamada, Y., & Nakatani, K. (2011). Deterministic equation for self-resonant structures of very small normal-mode helical antennas. *IEICE Transactions on Communications*, *E94-B*(5), 1276–1279. doi:10.1587/transcom.E94.B.1276

Dirks, S. (2009). *A Vision of Smarter Cities: How Cities Can Lead the Way into a*. IBM Global Business Services.

Dirks, S. G. (2010). *Smarter Cities for Smarter Growth: How Cities Can Optimize*. IBM Global Business Services. Retrieved from ftp://public.dhe.ibm.com/common/ssi/ecm/en/gbe0334 8usen/GBE03348USEN.PDF

Dirks, S. K. (2009). *How Smart is Your City? Helping Cities Measure Progress.* IBM Global Business Services. Retrieved from ftp://public.dhe.ibm.com/common/ssi/ecm/en/gbe0324

Dixon, P. (2017). A failure to do no harm—India's Aadhaar biometric ID program and its inability to protect privacy in relation to measures in Europe and the U.S. *Health and Technology*, *7*(4), 539–567. doi:10.100712553-017-0202-6 PMID:29308348

Djordfević, A. R., Zajić, A. G., Ilic, M. M., & Stuber, G. L. (2006). Optimization of helical antennas. *IEEE Antennas & Propagation Magazine*, *48*(6), 107–115. doi:10.1109/MAP.2006.323359

Dongale, T. D., Magdum, S. K., Goilkar, K. Y., Chougule, N. A., & Ghatage, S. R. (2013). Shifting towards Nanoelectronics: A device level overview. *Journal of Academia and Industrial Research*, *1*, 467–471.

Duan, W., Zhang, X. Y., Pan, Y. M., Xu, J. X., & Xue, Q. (2016). Dual-Polarized Filtering Antenna with High Selectivity and Low Cross Polarization. *IEEE Transactions on Antennas and Propagation*, *64*(10), 4188–4196. doi:10.1109/TAP.2016.2594818

Durairaj, R., Mallik, S., & Ekere, N. N. (2008). Solder paste characterisation: towards. *Soldering & Surface Mount Technology*, *20*(3), 34-40.

Durairaj, R., Mallik, S., Marks, A., Winter, M., Bauer, R., & Ekere, N.N. (2006). *Rheological Characterisation of New Lead-Free Solder Paste Formulations for Flip-Chip Assembly.* Electronics Manufacturing Engineering Research Group, Medway School of Engineering, University of Greenwich Chatham Maritime.

Durairaj, R., Ekere, N. N., & Salam, B. (2004). Thixotropy flow behaviour of solder and conductive adhesive pastes. *Journal of Materials Science Materials in Electronics*, *15*(10), 677–683. doi:10.1023/B:JMSE.0000038923.62923.1e

Durairaj, R., Nguty, T. A., & Ekere, N. N. (2001). Critical Factors Affecting Paste Flow During Stencil Printing of Solder Paste. *Soldering & Surface Mount Technology*, *13*(2), 30–34. doi:10.1108/09540910110385239

Durairaj, R., Ramesh, S., Mallik, S., Seman, A., & Ekere, N. N. (2009). Rheological characterisation and printing performance of Sn/Ag/Cu solder pastes. *Materials & Design*, *30*(3), 3812–3818. doi:10.1016/j.matdes.2009.01.028

Ebneter, D., Grivas, S., Kumar, T., & Wache, H. (2010). Enterprise Architecture Framework for Enabling Cloud Computing. *IEEE 3rd International Conference on Cloud Computing*, 542-543.

Edward, T. C. (2000). *Foundation of Interconnect and Microstrip Design* (3rd ed.). Wiley. doi:10.1002/9781118894514

Eke, J., Okafor, P. U., & Arinze, S. N. (2020). Investigating the Effects of Temperature, Paste Height and Time on the Rolling of Solder Paste. *Innovative Systems Design and Engineering.* www.iiste.org

Ekere, N. N., & Mallik, S. (2008). A methodology for Characterising New Lead-Free Solder Paste Formulations used for Flip-chip Assembly Applications. *Smart Processing Technology*, *2*(1), 59–64.

Electronic Communication Committee. (2018). *ECC Report 268: Technical and Regulatory Aspects and the Needs for Spectrum Regulation for Unmanned Aircraft Systems (UAS).* https://docdb.cept.org/download/efbd2f57-f227/ECCRep268.pdf

Electronic Communication Committee. (2019). *ECC Report 298: Analysis of the suitability and update of the regulatory technical conditions for 5G MFCN and AAS operation in the 1920 -1980 MHz and 2110-2170 MHz band.* https://docdb.cept.org/download/5b21b308-d1e7/ECC%20Report%20298.pdf

Electronic Communication Committee. (2019a). *ECC Report 297: Analysis of the suitability and update of the regulatory technical conditions for 5G MFCN and AAS operation in the 900 MHz and 1800 MHz bands.* https://docdb.cept.org/download/5949a9c0-66bd/ECC%20Report%20297.pdf

Electronic Communication Committee. (2019b). *ECC Decision (05) 05. Harmonised utilization of spectrum for Mobile/Fixed Communications Networks (MFCN) operating within the band 2500-2690 MHz.* https://docdb.cept.org/download/217614c9-9edd/ECCDec0505.pdf

Electronic Communication Committee. (2020). *ECC Report 309: Analysis of the usage of aerial UE for communication in current MFCN harmonised bands.* https://docdb.cept.org/download/9f4305fb-aa77/ECC%20Report%20309.pdf

Electronic Communication Committee. (2020a). *ECC Report 308: Analysis of the suitability and update of the regulatory technical conditions for 5G MFCN and AAS operation in the 2500-2690 MHz band.* https://docdb.cept.org/download/cd436696-88f5/ECC%20Report%20308.pdf

Emerson, D. T. (1995). The gain of the axial-mode helix antenna. *Antenna Compendium, 4*, 64–68.

Engheta, N., Salandrino, A., & Alu, A. (2005). Circuit elements at optical frequencies: Nano-inductors, nanocapacitors and nano-resistors. *Physical Review Letters, 95*, 095504.

Enkhbayar, B., Bang, J. H., Cha, E. J., & Ahn, B. C. (2009). Formula for gain of open-ended rectangular waveguide. *Electronics Letters, 45*(24), 1203–1204. doi:10.1049/el.2009.2237

Er, N. I., Singh, K., & Bonnin, J. M. (2017). On the performance of VDTN routing protocols with V2X communications for data delivery in smart cities. Academic Press.

Ericsson. (2019). *Employing AI Technologies to Enhance Returns on 5G Network Investments.* Author.

Ericsson. (2020). *5G evolution – the latest from 3GPP.* Retrieved October 21, 2020, from Ericsson.com: https://www.ericsson.com/en/future-technologies/5g-evolution

Er, N. I., Singh, K. D., & Bonnin, J. M. (2019). CityView: A Vehicle Based Image and Data Collection for Smart Cities. In *2019 International Conference on Wireless and Mobile Computing, Networking and Communications (WiMob)* (pp. 1-7). IEEE. 10.1109/WiMOB.2019.8923547

Estep, N. A., Sounas, D. L., & Alu, A. (2016). Magnetless Microwave Circulators Based on Spatiotemporally Modulated Rings of Coupled Resonators. *IEEE Transactions on Microwave Theory and Techniques, 1*(64), 502–518.

ETSI. (2004). E. Digital Video Broadcasting (DVB); Framing structure, channel coding and modulation for digital terrestrial television. *ETSI EN, 300*(744), V1.

EUI. (2020). *CitTech: Global Citizenship and Technology.* Retrieved from European University Institute: http://globalcitizenship.eui.eu/research/global-citizenship-technology/

European Commission. (2016). *Communication from the Commission to the European Parliament, the Council, the European Economic and Social Committee and the Committee of the Regions: "5G for Europe: An Action Plan".*

European Commission. (2018). *On the road to automated mobility: An EU strategy for mobility of the future.* https://ec.europa.eu/transport/sites/transport/files/3rd-mobility pack/com20180283_en.pdf

European Commission. (2019). *Commission Implementing Decision (EU) 2019/784 on harmonisation of the 24,25-27,5 GHz frequency band for terrestrial systems capable of providing wireless broadband.* https://eur-lex.europa.eu/legal-content/EN/TXT/?uri=CELEX%3A32019D0784

European Commission. (2019a). *Commission Delegated Regulation (EU) 2019/945 on unmanned aircraft systems and on third-country operators of unmanned aircraft systems.* https://eur-lex.europa.eu/legal-content/EN/TXT/?uri=CELEX:32019R0945

European Commission. (2019b). *Commission Implementing Regulation (EU) 2019/947 on the rules and procedures for the operation of unmanned aircraft electronic communications services in the Union.* https://eur-lex.europa.eu/eli/reg_impl/2019/2072/oj

European Commission. (2020). *Commission Implementing Regulation (EU) 2020/639 amending Implementing Regulation (EU) 2019/947 as regards standard scenarios for operations executed in or beyond the visual line of sight.* https://www.easa.europa.eu/document-library/regulations/commission-implementing-regulation-eu-2020639

European Union Aviation Safety Agency. (2018). *Terms of reference for rulemaking task RMT.0230. Regulatory framework to accommodate unmanned aircraft systems in the European aviation system.* https://www.easa.europa.eu/sites/default/files/dfu/ToR%20RMT.0230%20%E2%80%93%20Issue%202.pdf

European Union Aviation Safety Agency. (2019). *Opinion 05/2019 Standard scenarios for UAS operations in the 'specific' category.* https://www.easa.europa.eu/document-library/opinions/opinion-052019

European Union Aviation Safety Agency. (2020). *Opinion No 01/2020 High-level regulatory framework for the U-space.* https://www.easa.europa.eu/sites/default/files/dfu/Opinio n%20No%2001-2020.pdf

European Union Aviation Safety Agency. (2020a). *Appendix to Opinion No 01/2020. Comment-response document (CRD) to draft Opinion.* https://www.easa.europa.eu/sites/default/files/dfu/Appendix to Opinion No 01-2020.pdf

European Union Aviation Safety Agency. (2020b). *Draft acceptable means of compliance (AMC) and Guidance Material (GM) to Opinion No 01/2020 on a high-level regulatory framework for the U-space.* https://www.easa.europa.eu/sites/default/files/dfu/Draft%20AMC%20%26%20GM%20to%20the%20U-space%20Regulation%20%E2%80%94%20for%20info%20only.pdf

Falconer, G., & Mitchell, S. (2012). *Smart City Framework A Systematic Process for Enabling Smart+Connected Communities.* Cisco. Retrieved October 15, 2020, from https://www.cisco.com/c/dam/en_us/about/ac79/docs/ps/motm/Smart-City-Framework.pdf

Fallavi, K. N., Kumar, V. R., & Chaithra, B. M. (2017, February). Smart waste management using Internet of Things: A survey. In *2017 International Conference on I-SMAC (IoT in Social, Mobile, Analytics and Cloud)(I-SMAC)* (pp. 60-64). IEEE. 10.1109/I-SMAC.2017.8058247

Fatadin, I., Ives, D., & Wicks, M. (2006). Numerical simulation of intensity and phase noise from extracted parameters for CW DFB lasers. *IEEE Journal of Quantum Electronics*, *42*(9), 934–941. doi:10.1109/JQE.2006.880117

Feizi, M., Nayyeri, V., & Ramahi, O. M. (2014). Modeling Magnetized Graphene in the Finite-Difference Time-Domain Method using an Anisotropic Surface Boundary Condition. *IEEE Transactions on Antennas and Propagation*, *1*(66), 233–241.

Ferrari, A. C., & ... . (2015). Science and technology roadmap for graphene, related two –dimensional crystals, and hybrid systems. *Nanoscale*, *7*, 4598–4810.

Fisher, W. (2008). Digital Video and Audio Broadcasting. A Practical Engineering Guide. Springer.

Fonyi, S. (2020). *Overview of 5G Security and Vulnerabilities.* Academic Press.

Forbes. (2020). *Intenet of Things.* Retrieved October 17, 2020, from Forbes: https://www.forbes.com/internet-of-things/#d37e04476e53

Fork, W., & Hines, G. (2015, December 8). *The Internet Of Things: Making Cities -- And The Way They Use Technology -- Smarter.* Retrieved from Forbes: https://www.forbes.com/sites/uhenergy/2015/12/08/the-internet-of-things-making-cities-and-the-way-they-use-technology-smarter/#5a507e8f6f94

Fortis, T., Munteanu, V., & Negru, V. (2012). Towards an Ontology for Cloud Services. *IEEE 6TH International Conference on Complex, Intelligent and Software Intensive Systems*, 787-792.

Fu, Y., Wang, S., Wang, C., Hong, X., & McLaughlin, S. (2018). Artificial Intelligence to Manage Network Traffic of 5G Wireless Networks. *IEEE Network*, *32*(November/December). doi:10.1109/MNET.2018.1800115

Gajda, G., & Stuchly, S. S. (1983). Numerical analysis of open-ended coaxial lines. *IEEE Transactions on Instrumentation and Measurement*, *31*(5), 380–384.

Gallastegi, M., Huss, A., Santa-Marina, L., Aurrekoetxea, J. J., Guxens, M., Birks, L. E., Ibarluzea, J., Guerra, D., Röösli, M., & Jiménez-Zabala, A. (2018). Children's exposure assessment of radiofrequency fields: Comparison between spot and personal measurements. *Environment International*, *118*, 60–69. doi:10.1016/j.envint.2018.05.028 PMID:29803802

Gao, B., & Shi, L. (2020, February 10). Modelling ab Attack-Mitigation Dynamic Game-Theoretic Scheme for Security Vulnerability Analysis in a Cyber-Physical Power System. *IEEE Access: Practical Innovations, Open Solutions*, *8*, 30322–30331. Advance online publication. doi:10.1109/ACCESS.2020.2973030

Gao, J., Li, X., Flucke, J., & Boeck, G. (2004). Direct parameter-extraction method for laser diode rate-equation model. *Journal of Lightwave Technology*, *22*(6), 1604–1609. doi:10.1109/JLT.2004.829211

Garg, S., Versteeg, S., & Buyya, R. (2013). A Framework for Ranking of Cloud Computing Services. *Future Generation Computer Systems*, *29*(4), 1012–1023.

Gawas, A. U. (2015). An Overview on Evolution of Mobile Wireless Communication Networks: 1G– 6G. *International Journal on Recent and Innovation Trends in Computing and Communication*, *3*(5), 3130-3133.

GE. (2020). Retrieved October 2020, 2020, from Everything you need to know about the Industrial Internet of Things: https://www.ge.com/digital/blog/everything-you-need-know-about-industrial-internet-things

Geim, A. K., & Ram Subramaniam, A. (2007). Two-dimensional material nano photonics. *Nature Materials*, *1*(3), 183–191.

Ghosal, A., & Halder, S. (2018). In Z. Mahmood (Ed.), *Building Intelligent Systems for Smart Cities: Issues, Challenges and* (pp. 107–125). Smart Cities. Computer Communications and Networks. doi:10.1007/978-3-319-76669-0_5

Giles, M. (2018). Cybersecurity's insidious new threat: workforce stress. *MIT Technology Review.* Retrieved October 18, 2020, from https://www.technologyreview.com/2018/08/07/141139/cybersecuritys-insidious-new-threat-workforce-stress/

Giudici, G., Milne, A., & Vinogradov, D. (2020). Cryptocurrencies: Market analysis and perspectives. *Economia e Politica Industriale*, *47*(1), 1–18. doi:10.100740812-019-00138-6

Global System for Mobile Association (GSMA). (2019). *5G guide a reference for Operators.* GSMA.

Godage, L. (2019). *Global Unmanned Aerial Vehicle Market (UAV) Industry Analysis and Forecast (2018-2026).* https://www.inkwoodresearch.com/reports/unmanned-aerial-vehicles-market/

Goldman Sachs. (n.d.) *Drones reporting for work.* https://www.goldmansachs.com/insights/technology-driving-innovation/drones/

Gomez-Uribe, C., & Hunt, N. (2015). The netflix recommender system: Algorithms, business value, and innovation. *ACM Transactions on Management Information Systems*, *6*(4), 1–19. doi:10.1145/2843948

Graziosi, P., & Neophytou, N. (2018). Simulation study of ballistic spin-MOSFET devices with ferromagnetic channels based on some Heusler and oxide compounds. *Journal of Applied Physics*, *123*, 084503.

Greenberg, A. (2017). Crash Override Malware Took Down Ukraine's Power Grid Last December. *Wired.*

Greenblum, C. (1956a). Notes on the development of Yagi arrays – Part I. *QST*, 11–17, 114, 116.

Greenblum, C. (1956b). Notes on the development of Yagi arrays – Part II. *QST*, *23–26*, 122.

Greene & Srihari. (2008). *A procedure for determining the high-speed stencil printing performance of solder pastes in an electronic service provider's environment.* Emerald Group Publishing Limited.

Greffet, J. J. (2005). Nanoantennas for light emission. *Science*, *308*, 1561–153.

Gretzel, U., Werthner, H., Koo, C., & Lamsfus, C. (2015). Conceptual foundations for understanding smart tourism. *Computers in Human Behavior*, *50*, 558–563. doi:10.1016/j.chb.2015.03.043

Griffths, T., & Steyvers, M. (2004). Finding scientific topics. *Proceedings of the National Academy of Sciences.*

Gruber, T. (1993). A translation approach to portable ontology specifications. *Knowledge Acquisition*, *5*(2), 5. doi:10.1006/knac.1993.1008

GSM Association. (2020). *Generic Network Slice Template.* https://www.gsma.com/newsroom/wp-content/uploads//NG.116-v3.0.pdf

GSMA. (2020). *Global 5G Landscape, Q3 2020.* https://data.gsmaintelligence.com/research/research/research-2020/global-5g-landscape-q3-2020

Güney, K. (2001). Simple design method for optimum gain pyramidal horns. *International Journal of Electronics and Communications*, *55*(3), 205–208. doi:10.1078/1434-8411-00031

Gurjot S. G., Paramdeep S. Sandeep K. A. (2016). *Electronic Wheelchair for Physically Disabled Persons.* . doi:10.17485/ijst/2016/v9i21/89753

Gutierrez, J. M., Jensen, M., Henius, M., & Riaz, T. (2015). Smart waste collection system based on location intelligence. *Procedia Computer Science*, *61*, 120–127. doi:10.1016/j.procs.2015.09.170

Hacıvelioğlu, F., Alper Uslu, M., & Sevgi, L. (2011). A MATLAB based virtual tool for the electromagnetic wave scattering from a perfectly reflecting wedge. *IEEE Antennas & Propagation Magazine*, *53*(6), 234–243. doi:10.1109/MAP.2011.6157766

Haider, N., Baig, Z., & Imran, M. (2020). *Artificial Intelligence and Machine Learning in 5G Network Security: Opportunities, Advantages, and Future Research Trends.* arXiv.2007.04490v1[cs.CR]

Hammerstad, E. O. (1975). Equations for microstrip circuit design. *Proceedings European Microwave Conference*, 268–272.

Hanson, G. W. (2008). *Fundamentals of Nanoelectronics* (1st ed.). Prentice Hall.

Hao, H., Hui, D., & Lau, D. (2020). Material advancement in technological development for the 5G wireless communications. *Nanotechnology Reviews*, *9*, 683–699.

Hardell, L., Carlberg, M., & Hedendahl, L. K. (2018). Radiofrequency radiation from nearby base stations gives high levels in an apartment in Stockholm, Sweden: A case report. *Oncology Letters*, *15*(5), 7871–7883. doi:10.3892/ol.2018.8285 PMID:29725476

Haribabu, P., Kassa, S. R., Nagaraju, J., Karthik, R., Shirisha, N., & Anila, M. (2017, December). Implementation of a smart waste management system using IoT. In *2017 International Conference on Intelligent Sustainable Systems (ICISS)* (pp. 1155-1156). IEEE. 10.1109/ISS1.2017.8389367

Harley, R. G., & Liang, J. (2011). Computational Intelligence in Smart Grids. *IEEE Symposium Series on Computational Intelligence (SSCI) 2011-CIASG.*

Harper, C. A. (2000). *High performance printed circuit boards. Edition: illustrated.* Published by McGraw-Hill Professional.

Harrington, R. F. (1967). Matrix methods for field problems. *Proceedings of the IEEE*, *55*(2), 136–149. doi:10.1109/PROC.1967.5433

Harrington, R. F. (1968). *Field Computation by Moment Methods*. Macmillan.

Hasan, M. M. (2012). Performance comparison of wavelet and FFT based multiuser MIMO OFDM over wireless Rayleigh fading channel. *International Journal of Energy, Information and Communications*, *3*(4), 1–8.

Hasin, S., Ghafouri, A., Dubey, A., Karsai, G., & Koutsoukos X. (n.d.). *Vulnerability Analysis of Power Systems Based on Cyber-Attacks and Defense Models*. Academic Press.

Hassain, S. (2013). 5G Wireless Communication Systems. *American Journal of Engineering Research*, *2*(10), 344–353.

Hassan, E., Martynenko, D., Wadbro, E., Fischer, G., & Berggren, M. (2019). Compact differential-fed planar filtering antennas. *Electronics (Switzerland)*, *8*(11), 1–15. doi:10.3390/electronics8111241

Hattingh, M. (2020). SMART City and Economy: Bibliographic Coupling and Co-occurrence occurrence. In E-business, E-services, and E-society (pp. 102-113). Skukuza.

Hawkins, A. (2017). Chinese citizens want the government to rank them. *Foreign Policy*. Retrieved from https://foreignpolicy.com/2017/05/24/chinese-citizens-want-the-government-to-rank-them/

Haydey, M., Levy, J., & Thompson, J. (2015). *The SAGE Handbook of Research in International Education* (2nd ed.). SAGE. doi:10.4135/9781473943506

Hemment, D., & Townsend, A. (2013). Future Everything. Academic Press.

Hertz, H. (1888). On the finite velocity of propagation of electromagnetic actions. *Electric Waves*, 110.

Hetzer, D., Muehleisen, M., Kousaridas, A., & Alonso-Zarate, J. (2019, June). *5G Connected and Automated Driving: Use Cases and Technologies in Cross-border Environments*. In *2019 European Conference on Networks and Communications (EuCNC)* (pp. 78-82). IEEE.

He, X., Zhang, Y., Du, M., & Xu, J. (2018). Lightweight and compact high-gain filtering aperture antenna fabricated by three-dimensional printing technology. *IEEE Antennas and Wireless Propagation Letters*, *17*(7), 1141–1144. doi:10.1109/LAWP.2018.2830815

Hillier, W. (2020). *Is Big Data Dangerous?* Retrieved October 25, 2020, from https://careerfoundry.com/en/blog/data-analytics/is-big-data-dangerous/

Hlali, A., Houaneb, Z., & Zairi, H. (2018). Tunable filter based on hybrid metal graphene structures over an ultrawide terahertz band using an improved Wave Concept Iterative Process method. *International Journal for Light and Electron Optics*, *1*(181), 423–431.

Hlali, A., Houaneb, Z., & Zairi, H. (2020). Modeling of Magnetically Biased Graphene Coupler at Terahertz Frequency Through an Improved Anisotropic WCIP Method. *IEEE Transactions on Magnetics*, *1*(56), 1–9.

Hlali, A., Houaneb, Z., & Zairi, H. (2020). Terahertz microstrip circulator based on graphene. *Journal of Electromagnetic Waves and Applications*, *1*(34), 1–6.

Hong, J. S., & Lancaster, M. J. (2001). *Microstrip Filters for RF/microwave applications.* Wiley & Sons. doi:10.1002/0471221619

Hong, J.-S. (2011). *Microstrip filters for RF/microwave applications* (2nd ed.). John Wiley & Sons. doi:10.1002/9780470937297

Horridge, M., & Bechhofer, S. (2011). The owl api: A java api for owl ontologies. *Semantic Web*, *2*(1), 11–21. doi:10.3233/SW-2011-0025

Houben, R. (2018). *Cryptocurrencies and blockchain.* European Parliament - Policy Department for Economic, Scientific and Quality of Life Policies.

Hou, Z., Liu, C., Zhang, B., Song, R., Wu, Z., Zhang, J., & He, D. (2020). Dual-/tri-wideband bandpass filter wit high selectivity and adjustable passband for 5G mid-band mobile communications. *MDPI Electronics*, *9*(2), 205. doi:10.3390/electronics9020205

Howard, H. M. (1995). *Soldering handbook for printed circuits and surface mounting: Design, materials, processes, equipment, trouble-shooting, quality, economy, and line management* (2nd ed.). Published by Springer.

Huang, F., Wang, J., & Zhu, L. (2016). A new approach to design a microstrip dual-mode balun bandpass filter. *IEEE Microwave and Wireless Components Letters*, *26*(4), 252–254. doi:10.1109/LMWC.2016.2538726

Huawei. (2019). *Position paper on Cyber Security.* Author.

Hu, S., Hu, Y., Zheng, H., Zhu, W., Gao, Y., & Zhang, X. (2020). A compact 3.3–3.5 GHz filter based on modified composite right–/left–handed resonator units. *MDPI Electronics*, *9*, 1.

Huynh, D. (2008). *Human activity recognition with wearable sensors* (PhD thesis). Technische Universit Darmstadt.

Huynh, T., & Schiele, B. (2006). Unsupervised discovery of structure in activity data using multiple eigenspaces. *2nd International Workshop on Location and Context Awareness (LoCA, 2006).*

Huynh, T., Fritz, M., & Schiele, B. (2008). Discovery of activity patterns using topic models. *Ubiquitous computing (UbiComp 08).*

Huynh, T., Blanke, U., & Schiele, B. (2007). Scalable recognition of daily activities with wearable sensors. *3rd International Conference on Location and Context awareness (LOCA 07).*

Hwang, J. S. (1989). *Solder Paste in Electronics Packaging.* Van Nostrand Reinhold. doi:10.1007/978-94-011-6050-6

Hwang, J. S. (2004). *Lead-free implementation and production: A manufacturing guide.* McGraw-Hill Professional.

Ian, W. (2018). *First-Ever Temperature Stable Solder Paste Unveiled.* Available online: https://www.henkel-adhesives.com/com/table_images/Technical_Article_--_Temperature_Stable_Solder_Paste_Unveiled_final.pdf

IBM. (2016, October 16). *What Is Big Data?* Retrieved from https://www-01.ibm.com/software/data/bigdata/what-is-big-data.html

ICF. (2015). *What is an Intelligent Community?* Retrieved October 12, 2020, from ICF: https://www.intelligentcommunity.org/what_is_an_intelligent_community

Ihianle, I. K., Naeem, U., & Islam, S. (2017). Ontology-driven activity recognition from patterns of object use. *Proceedings of the ACM International Joint Conference on Pervasive and Ubiquitous Computing (UbiComp '17) and Proceedings of ACM International Symposium on Wearable Computers.* 10.1145/3123024.3124424

Ihianle, I., Naeem, U., & Tawil, A. (2016). Recognition of activities of daily living from topic model. *Procedia Computer Science.*

Ihianle, I., Nwajana, A., Ebenuwa, S. H., Otuka, R. I., Owa, K., & Orisatoki, M. O. (2020). A deep learning approach for human activities recognition from multimodal sensing devices. *IEEE Open Access, 8.*

Ihianle, I. K., Naeem, U., Islam, S., & Tawil, A.-R. (2018). A hybrid approach to recognising activities of daily living from object use in the home environment. *Informatics (MDPI), 5*(1), 5. doi:10.3390/informatics5010006

Ihianle, I., Naeem, U., & Tawil, A. (2015). A dynamic segmentation based activity discovery through topic modelling. *IET International Conference on Technologies for Active and Assisted Living (TechAAL 2015).*

Isbell, D. E. (1960). Log periodic dipole arrays. *IRE Transactions on Antennas and Propagation, 8*(3), 260–267. doi:10.1109/TAP.1960.1144848

Islam, M. S., & Logeeswaran, V. J. (2010). Nanoscale materials and devices for future communication networks. *Communications Magazine, IEEE, 48,* 112–120.

Ismail, N. (2017, December 5). *Smart cities could lead to cost savings of $5 trillion - report suggests.* Retrieved from Information Age: https://www.information-age.com/smart-cities-lead-cost-savings-5-trillion-123469863/

ITU –R. (2015). *IMT Vision – Framework and overall objectives of the future development of IMT for 2020 and beyond.* https://www.itu.int/dms_pubrec/itu-r/rec/m/R-REC-M.2083-0-201509-I!!PDF-E.pdf

ITU-T Telecommunication Standardization Sector. (2003a). *Recommendation G.694.1 - Spectral grids for WDM applications: DWDM wavelength grid.*

ITU-T Telecommunication Standardization Sector. (2003b). *Recommendation G.694.2 - Spectral grids for WDM applications: CWDM wavelength grid.*

ITU-T Telecommunication Standardization Sector. (2009a).: *Recommendation G.652- Characteristics of a single-mode optical fiber and cable.*

ITU-T Telecommunication Standardization Sector. (2009b). *Recommendation G.657 - Characteristics of a bending-loss insensitive single-mode optical fiber and cable for the access network.*

ITU-T Telecommunication Standardization Sector. (2010). *Recommendation G.656 - Characteristics of a fibre and cable with non-zero dispersion for wideband optical transport.*

Jadhav, J. B., & Deore, P. J. (2017). Filtering antenna with radiation and filtering functions for wireless applications. *Journal of Electrical Systems and Information Technology, 4*(1), 125–134. doi:10.1016/j.jesit.2016.10.007

Jagannathan, S. (2012). *Comparison and Evaluation of Open-source Cloud Management software.* KTH Royal Institute of Technology Stockholm.

Jagdish, P., C., Poonam, G. C., Kedar, M. K. (2019). *Design and Fabrication of Lever Operated Wheelchair for Disabled Person with No Legs.* doi:10.17148/IARJSET.2019.611

Jain, A., & Bagherwal, R. (2017, July). Design and implementation of a smart solid waste monitoring and collection system based on Internet of Things. In *2017 8th International Conference on Computing, Communication and Networking Technologies (ICCCNT)* (pp. 1-5). IEEE. 10.1109/ICCCNT.2017.8204165

Jamthe, D. V., & Bhande, S. A. (2017). Nanotechnology in 5G wireless communication network – an approach. *Int Res J Eng Tech., 4*, 58–61.

Jiang, H., Wang, Z., & He, H. (2019). An Evolutionary Computation Approach for Smart Grid Cascading Failure vulnerability Analysis. *2019 IEEE Symposium Series on Computational Intelligence (SSCI).*

Jiang, H., & Wilford, P. A. (2005). A hierarchical modulation for upgrading digital broadcast systems. *IEEE Transactions on Broadcasting, 51*(2), 223–229. doi:10.1109/TBC.2005.847619

Jin, J. Y., Liao, S., & Xue, Q. (2018). Design of Filtering-Radiating Patch Antennas with Tunable Radiation Nulls for High Selectivity. *IEEE Transactions on Antennas and Propagation, 66*(4), 2125–2130. doi:10.1109/TAP.2018.2804661

John, H. L., Wong, C. P., Ning-Cheng, L. S., & Ricky, L. V. (2002). *Electronics manufacturing: with lead-free, halogen-free, and conductive-adhesive materials. Edition: illustrated.* McGraw-Hill Professional.

Joint Radio Company Ltd. (2019). Cutting Through the Hype: 5G and its Potential Impacts on Electric Utilities. doi:10.1109/SSCI44817.2019.9002979

Jull, E. V. (1972). Reflection from the aperture of a long E-plane sectoral horn. *IEEE Transactions on Antennas and Propagation, 20*(1), 62–68. doi:10.1109/TAP.1972.1140137

Jung, E.-Y., & Hwang, H.-Y. (2007). A balun-BPF using a dual-mode ring resonator. *IEEE Microwave and Wireless Components Letters, 17*(9), 652–654. doi:10.1109/LMWC.2007.903442

Kafle, V. P., Fukushima, Y., Martinez-Julia, P., & Miyazawa, T. (2018). Consideration on Automation of 5G Network Slicing with Machine Learning. *2018 ITU Kaleidoscope Academic Conference.*

Karthikeyan, S., Rani, G. S., Sridevi, M., & Bhuvaneswari, P. T. V. (2017, May). IoT enabled waste management system using ZigBee network. In *2017 2nd IEEE International Conference on Recent Trends in Electronics, Information & Communication Technology (RTEICT)* (pp. 2182-2187). IEEE. 10.1109/RTEICT.2017.8256987

Kasteren, T., Englebienne, G., & Krose, B. (2010). *Human activity recognition from wireless sensor network data: Benchmark and software*. Atlantis Press Book.

Katayoun, F., & GaticaPerez, D. (2011). Discovering routines from largescale human locations using probabilistic topic models. *ACM Transactions on Intelligent Systems and Technology, 2*(1).

Kemp, K. (2019). *Digital 2019: Global Intenet use accelerates.* We are social. Retrieved November 26, 2020, from https://wearesocial.com/blog/2019/01/digital-2019-global-internet-use-accelerates

Ketshabetswe, L. K., Zungeru, A. M., Mangwala, M., Chuma, J. M., & Sigweni, B. (2019). Communication protocols for wireless sensor networks: A survey and comparison. *Heliyon, 5*(5), e01591. doi:10.1016/j.heliyon.2019.e01591 PMID:31193432

Khan R., Kumar P., Jayakody D. N. K., & Liyanage M. (2019). A Survey on Security and Privacy of 5G Technologies: Potential Solutions, Recent Advancements and Future Directions. *Journal of Latex IEEE Communications Surveys and tutorials.*

Khan, M., Babar, M., Ahmed, S. H., Shah, S. C., & Han, K. (2017). Smart city designing and planning based on big data analytics. *Sustainable Cities and Society, 35*, 271–279. doi:10.1016/j.scs.2017.07.012

Khan, S. (2014). Elements of Cloud Adoption. *IEEE Journal on Cloud Computing, 1*(1), 71–73. doi:10.1109/MCC.2014.7

Kibria, R., & Austin, M. W. (2015). All optical signal-processing techniques utilizing four wave mixing. *Photonics, 2*(1), 200–213. doi:10.3390/photonics2010200

Kim, N. Y., Rathore, S., Ryu, J. H., Park, J. H., & Park, J. H. (2018). A Survey on Cyber Physical System Security for IoT: Issues, Challenges, Threats, Solutions. Journal of Information Processing System, 14(6), 1361.

Kim, T. (2010). SCADA Architecture with Mobile Remote Components. *WSEAS Transactions on Systems and Control*, *5*(8).

King, H. E., & Wong, J. L. (1980). Characteristics of 1 to 8 wavelength uniform helical antennas. *IEEE Transactions on Antennas and Propagation*, *28*(2), 291–296. doi:10.1109/TAP.1980.1142322

King, R. W. P. (2013). *The Theory of Linear Antennas: With Charts and Tables for Practical Applications*. Harvard University Press. doi:10.4159/harvard.9780674182189

Kitchin, R. (2013). Big data and human geography: Opportunities, challenges and risks. *Dialogues in Human Geography*, *3*(3), 262–267. doi:10.1177/2043820613513388

Kitchin, R. (2015b). The Promise and Perils of Smart Cities. *Society for Computers & Law*, *26*(2), 1–5.

Kleinberg, J. (2017). *Human decisions and machine predictions.* NBER Working Paper.

Kordiboroujeni, Z., & Bornemann, J. (2014). Mode-matching analysis and design of substrate integrated integrated waveguide T-junction diplexer and corner filter. International Journal of Numerical Modelling: Electronic Networks. *Devices and Fields*, *28*(5), 497–507.

Koumaras, H., Makropoulos, G., Batistatos, M., Kolometsos, S., Gogos, A., Xilouris, G., Sarlas, A., & Kourtis, M. A. (2021). 5G-Enabled UAVs with Command and Control Software Component at the Edge for Supporting Energy Efficient Opportunistic Networks. *Energies*, *14*(5), 1480.

Kouyoumjian, R. G., & Pathak, P. H. (1974). A uniform geometrical theory of diffraction for an edge in a perfectly conducting surface. *Proceedings of the IEEE*, *62*(11), 1448–1461. doi:10.1109/PROC.1974.9651

Kraus, J. D. (1947). Helical beam antennas. *Electronics (Basel)*, *20*, 109–111.

Krebs, C. (2020). CISA 5G Strategy: Ensuring the Security and Resilence of 5G Infrastructure in our *nation.* US Department of Homeland Security Cybersecurity and Infrastructure Security Agency.

Ksentini, A., & Frangoudis, P. A. (2020). Towards Slicing-Enabled Multi-access Edge Computing in 5G. *IEEE Network*, *34*(2), 99–105.

Kuklinski, S., Tomaszewski, L., Korzec, P., & Kołakowski, R. (2020). 5G-UASP: 5G-based multi-provider UAV platform architecture. *6th IEEE International Conference on Network Softwarization (NetSoft 2020).*

Kumar, B. S., Varalakshmi, N., Lokeshwari, S. S., Rohit, K., & Sahana, D. N. (2017, December). Eco-friendly IOT based Waste segregation and management. In *2017 International Conference on Electrical, Electronics, Communication, Computer, and Optimization Techniques (ICEECCOT)* (pp. 297-299). IEEE. 10.1109/ICEECCOT.2017.8284686

Kumar, G., & Ray, K. P. (2013). Broadband Microstrip Antennas. In *Microstrip and Printed Antenna Design*. Artech House Inc. doi:10.1049bew048e_ch4

Kumbasar, V., & Kucur, O. (2012). Performance comparison of wavelet based and conventional OFDM systems in multipath Rayleigh fading channels. *Digital Signal Processing*, *22*(5), 841–846. doi:10.1016/j.dsp.2012.02.004

Kushpal, S., Anup, N., Asif, Y., Shravani, G., Sonia, P., & Preeti, V. (2016). Effect of electromagnetic radiations from mobile phone base stations on general health and salivary function. *Journal of International Society of Preventive & Community Dentistry*, *6*(1), 54–59. doi:10.4103/2231-0762.175413 PMID:27011934

Labrador, M., & Yejas, O. (2013). *An ontology based approach for activity recognition from video. In Human Activity Recognition: Using Wearable Sensors and Smartphones*. CRC Press.

Langley, P., Iba, W., & Thompson, K. (1992). An analysis of bayesian classifiers. *10th National Conference on Artificial Intelligence*.

Laszewski, G., Diaz, J., Wang, F., & Fox, C. (2012). Quality Comparison of Multiple Cloud Frameworks. *IEEE 5th International Conference on Cloud Computing*, 734-741.

Lau, F. K. H., & Yeung, V. W. S. (1997). A hierarchical evaluation of the solder paste printing process. *Journal of Materials Processing Technology*, *69*(1-3), 79–89. doi:10.1016/S0924-0136(96)00045-3

Lee, J. G., & Ryu, H. C. (2018, August). Wireless Communication System Based on OFDM System Using Wavelet Transform. In *2018 Progress in Electromagnetics Research Symposium (PIERS-Toyama)* (pp. 2041-2046). IEEE. 10.23919/PIERS.2018.8597769

Lee, N. C. (2002). *Reflow Soldering Processes: SMT, BGA, CSP and Flip-Chip Technologies*. Newnes Publication. Available Online http://books.google.co.uk/books?id=Tv_AuIPrZmEC

Lee, S., & Mase, K. (2002). Activity and location recognition using wearable sensor. *IEEE Pervasive Computing*.

Leo, M., Medioni, G., Trivedi, M., Kanade, T., & Farinella, G. (2017). Computer vision for assistive technologies. *Computer Vision and Image Understanding*, *154*, 115. doi:10.1016/j.cviu.2016.09.001

Leone. (2003). *The War on Our Freedoms: Civil Liberties in an Age of Terrorism*. The Century Foundation.

Li, A., Yang, X., Kandula, S., & Zhang, M. (2010). CloudCmp: comparing public cloud providers. *Proceedings of the 10 th Annual Conference on Internet Measurement*.

Liao, L., Fox, D., & Kautz, H. (2007). Extracting places and activities from gps traces using hierarchical conditional random fields. *The International Journal of Robotics Research*, *26*(1), 119–134. doi:10.1177/0278364907073775

Liebig, T., Rennings, A., Held, S., & Erni, D. (2013). openEMS-a free and open source equivalent-circuit (EC) FDTD simulation platform supporting cylindrical coordinates suitable for the analysis of traveling wave MRI applications. *International Journal of Numerical Modelling: Electronic Networks, Devices and Fields*, *26*(6), 680–696.

Li, J. F., Chen, Z. N., Wu, D. L., Zhang, G., & Wu, Y. J. (2018). Dual-Beam Filtering Patch Antennas for Wireless Communication Application. *IEEE Transactions on Antennas and Propagation*, *66*(7), 3730–3734. doi:10.1109/TAP.2018.2835519

Likotiko, E. D., Nyambo, D., & Mwangoka, J. (2017). *Multi-agent based IoT smart waste monitoring and collection architecture*. arXiv preprint arXiv:1711.03966.

Lin, C. K., & Chung, S. J. (2011). A compact filtering microstrip antenna with quasi-elliptic broadside antenna gain response. *IEEE Antennas and Wireless Propagation Letters*, *10*, 381–384. doi:10.1109/LAWP.2011.2147750

Liu, D. Q., Zhang, M., Luo, H. J., Wen, H. L., & Wang, J. (2018). Dual – band platform - free PIFA for 5G MIMO application of mobile devices. *IEEE Transactions on Antennas and Propagation*, *66*, 6328–6333.

Liu, H., Neal, A. T., Zhu, Z., Luo, Z., Xu, X., Tománek, D., & Ye, P. D. (2014). Phosphorene: An unexplored 2D semiconductor with a high hole mobility. *ACS Nano*, *1*(4), 4033–4041.

Liu, Y. T., Leung, K. W., Ren, J., & Sun, Y. X. (2019). Linearly and Circularly Polarized Filtering Dielectric Resonator Antennas. *IEEE Transactions on Antennas and Propagation*, *67*(6), 3629–3640. doi:10.1109/TAP.2019.2902670

Liu, Y.-T., Leung, K. W., & Yang, N. (2020). Compact absorptive patch antenna. *IEEE Transactions on Antennas and Propagation*, *68*(2), 633–642. doi:10.1109/TAP.2019.2938798

Li, X., Keegan, B., Mtenzi, F., Weise, T., & Tan, M. (2019). Energy-efficient load balancing ant-based routing algorithm for wireless sensor networks. *IEEE Access: Practical Innovations, Open Solutions*, *7*, 113182–113196. doi:10.1109/ACCESS.2019.2934889

Loghin, D., Cai, S., Chen, G., Dinh, T. T. A., Fan, F., Lin, Q., Ng, J., Ooi, B. C., Sun, X., Ta, Q., Wang, W., Xiao, X., Yang, Y., Zhang, M., & Zhang, Z. (2019). *The Disruptions of 5G on Data-Driven Technologies and Applications.* arXiv: 1909.0806v3 [cs.NI].

Lokuliyana, S., Jayakody, A., Dabarera, G. S. B., Ranaweera, R. K. R., Perera, P. G. D. M., & Panangala, P. A. D. V. R. (2018, August). Location based garbage management system with iot for smart city. In *2018 13th International Conference on Computer Science & Education (ICCSE)* (pp. 1-5). IEEE. 10.1109/ICCSE.2018.8468682

Lorenzi, F., Loh, S., & Abel, M. (2011), PersonalTour: A recommender system for travel packages. In *IEEE/WIC/ACM International Conferences on Web Intelligence and Intelligent Agent Technology* (Vol. 2, pp. 333-336). 10.1109/WI-IAT.2011.69

Lotfinezhad, M., & Liang, B. (2005). Energy efficient clustering in sensor networks with mobile agents. In *IEEE Wireless Communications and Networking Conference*, 2005 (Vol. 3, pp. 1872-1877). IEEE. 10.1109/WCNC.2005.1424797

Low, T., Rodin, A. S., Carvalho, A., Jiang, Y., Wang, H., Xia, F., & Neto, A. H. C. (2014). Tunable optical properties of multilayers black phosphorus thin films. *Physical Review*, *1*(7), 075434.

Lu, J., Wu, D., Mao, M., Wang, W., & Zhang, G. (2015). Recommender system application developments: A survey. *Decision Support Systems*, *74*, 12–32. doi:10.1016/j.dss.2015.03.008

Luo, C., Wang, Y., Hong, Y., Chen, W., Ding, X., Zhu, Y., & Li, D. (2019). Minimizing data collection latency with unmanned aerial vehicle in wireless sensor networks. *Journal of Combinatorial Optimization*, *38*(4), 1019–1042. doi:10.100710878-019-00434-w

Luo, G. Q., Hong, W., Tang, H. J., Chen, J. X., Yin, X. X., Kuai, Z. Q., & Wu, K. (2007). Filtenna consisting of horn antenna and substrate integrated waveguide cavity FSS. *IEEE Transactions on Antennas and Propagation*, *55*(1), 92–98. doi:10.1109/TAP.2006.888459

Macchiarella, G. (2011, October). Synthesis of Star-Junction Multiplexers. *IEEE Microwave Magazine*, *12*(6), 101–109. doi:10.1109/MMM.2011.942011

Macchiarella, G., & Tamiazzo, S. (2015). Design of "masthead" combiners. *2015 European Microwave Conference (EuMC)*, 686-689. 10.1109/EuMC.2015.7345856

Maddah, A., & Kavakli, M. (2013). System architecture & feature design for engineering a web-based Travel Advisor system. *14th ACIS International Conference on Software Engineering, Artificial Intelligence, Networking and Parallel/Distributed Computing*, 683-688. 10.1109/SNPD.2013.88

Maglaras, L., Ferrag, M. A., Derhab, A., Mukherjee, M., Janicke, H., & Rallis, S. (2019). Threats, Countermeasures and Attribution of Cyber Attacks on Critical infrastructure. *IoT Transactions*. arXiv:1901.03899v1 [cs.CR].

Maheshwari, D., & Janssen, M. (2014). Reconceptualizing measuring, benchmarking for improving interoperability in smart. *Government Information Quarterly*, *31*, S84–S92. doi:10.1016/j.giq.2014.01.009

Malapela, T. (2017, December 19). *Can artificial intelligence help improve agricultural productivity?* Retrieved October 18, 2020, from e-agriculture: http://www.fao.org/e-agriculture/news/can-artificial-intelligence-help-improve-agricultural-productivity

Malapur, B. S., & Pattanshetti, V. R. (2017, August). IoT based waste management: An application to smart city. In *2017 International Conference on Energy, Communication, Data Analytics and Soft Computing (ICECDS)* (pp. 2476-2486). IEEE. 10.1109/ICECDS.2017.8389897

Mallik S., (2009). *Study of the Time-dependent Rheological Beehaviour of Lead-free Solder Pastes and Flux Mediums used for Flip-Chip Assembly Applications*. Academic Press.

Mallik, S., Ekere, N. N. R., Durairaj, R., & Marks, A. E. (2008). An investigation into the rheological properties of different lead-free solder paste for surface mount application. *Soldering & Surface Mount Technology,* 3-10.

Mansour, G., Lancaster, M. J., Hall, P. S., Gardner, P., & Nugoolcharoenlap, E. (2014). Design of Filtering Microstrip Antenna Using Filter Synthesis Approach. *Progress in Electromagnetics Research, 145*, 59–67. doi:10.2528/PIER14011405

Manyika, J., Lund, S., Chuin, M., Bughin, J., Woetzel, J., Batra, P., . . . Sanhvi, S. (2017). *McKinsy & Company - Jobs lost, jobs gained: What the future of work will mean for jobs, skills, and wages*. Retrieved November 25, 2020, from https://www.mckinsey.com/featured-insights/future-of-work/jobs-lost-jobs-gained-what-the-future-of-work-will-mean-for-jobs-skills-and-wages

Marceau, J. (2008). Introduction: Innovation in the city and innovative cities. *Innovation: Management. Policy*, (10), 136–145.

Marconi, G. (1896). *Transmitting electrical signals*. U.S. Patent 586193.

Marcuvitz, N. (1964). *Waveguide Handbook*. Boston Technical Publishers.

Marks, A., Mallik, S., Durairaj, R., & Ekere, N. N. (2007). Effect of Abandon Time on Print Quality and Rheological Characteristics for Lead-Free Solder Pastes used for Flip-Chip Assembly. *32nd International Electronics Manufacturing Technology Symposium (IEMT)*. 10.1109/IEMT.2007.4417048

Marojevic, V., Guvenc, I., Dutta, R., Sichitiu, M. L., & Floyd, B. A. (2020). Advanced Wireless for Unmanned Aerial Systems: 5G Standardization, Research Challenges, and AERPAW Architecture. *IEEE Vehicular Technology Magazine, 15*(2), 22–30.

Massaguer, D., Fok, C. L., Venkatasubramanian, N., Roman, G. C., & Lu, C. (2006). Exploring sensor networks using mobile agents. In *Proceedings of the fifth international joint conference on Autonomous agents and multiagent systems* (pp. 323-325). 10.1145/1160633.1160688

MATLAB R2015b. (2015). *MATLAB Creating Graphical User Interfaces*. MathWorks.

MATLAB R2020a. (2020). *MATLAB App Building*. MathWorks.

Matthaei, G., Young, L., & Jones, E. M. T. (1980). Microwave Filters, Impedance-matching Networks and Coupling Structure (A. H. Inc, Ed.). Academic Press.

Mattson, A. (2005). Single frequency networks in DTV. *IEEE Transactions on Broadcasting, 51*(4), 413–422. doi:10.1109/TBC.2005.858419

Mavropoulos, O. (2020). D2.1: 5G Security: Current Status and Future Trends. *INSPIRE-5Gplus.*

Maybell, M. J., & Simon, P. S. (1993). Pyramidal horn gain calculation with improved accuracy. *IEEE Transactions on Antennas and Propagation, 41*(7), 884–889. doi:10.1109/8.237618

Mazumder, P., Kulkarni, S., Bhattacharya, M., Sun, J. P., & Haddad, G. I. (1998). Digital Circuit Applications of Resonant Tunneling Devices. *Proceedings of the IEEE*, *86*, 664–686.

McClellan, M., Cervello-Pastor, C., & Sallent, S. (2020). Deep Learning at the Mobile Edge: Opportunities for 5G Networks. *Applied Sciences (Basel, Switzerland)*, *2020*(10), 4735. doi:10.3390/app10144735

McGuinness, D., & Van Harmelen, F. (2004). OWL web ontology language overview. *W3C Recommendation*, *10*(10).

Medvedev, A., Fedchenkov, P., Zaslavsky, A., Anagnostopoulos, T., & Khoruzhnikov, S. (2015). Waste management as an IoT-enabled service in smart cities. In *Internet of Things, Smart Spaces, and Next Generation Networks and Systems* (pp. 104–115). Springer. doi:10.1007/978-3-319-23126-6_10

Meo, S. A., Almahmoud, M., Alsultan, Q., Alotaibi, N., Alnajashi, I., & Hajjar, W. H. (2019). Mobile Phone Base Station Tower Settings Adjacent to School Buildings: Impact on Students' Cognitive Health. *American Journal of Men's Health*, *13*(1), 1–6. doi:10.1177/1557988318816914 PMID:30526242

Mergen, G., Zhao, Q., & Tong, L. (2006). Sensor networks with mobile access: Energy and capacity considerations. *IEEE Transactions on Communications*, *54*(11), 2033–2044. doi:10.1109/TCOMM.2006.884845

Méric, H., Lacan, J., Arnal, F., Lesthievent, G., & Boucheret, M. L. (2012, April). Improving broadcast channel rate using hierarchical modulation. In *European Wireless 2012; 18th European Wireless Conference 2012* (pp. 1-5). VDE.

Méric, H., Lacan, J., Amiot-Bazile, C., Arnal, F., & Boucheret, M. L. (2011, April). Generic approach for hierarchical modulation performance analysis: Application to DVB-SH. In *2011 Wireless Telecommunications Symposium (WTS)* (pp. 1-6). IEEE. 10.1109/WTS.2011.5960857

Microlab. (2014). Nigeria Migrates to the Independent Tower Company Business Model. *Inside Towers.*

Mission Support Center. (2016). Cyber Threat and Vulnerability Analysis of the U.S. Electric Sector. Idaho National Laboratory.

Mitra, I., TK, S., Rao, V., Kelkar, V., & Chhaparia, P. (2019). How Cyber-Physical systems are Reshaping the Robotics Landscape. *Cognizant.*

Mitroff, S. (2015). *OneDrive, Dropbox, Google Drive and Box: Which cloud storage service is right for you?* Available at: www.Cnet.com

Mockel P., & Makala B. (2019). Artificial Intelligence and 5G Mobile Technology can Drive Investment Opportunities in Emerging Markets. Fresh Ideas about Business in Emerging Markets. *EM Compass.* Note 76.

Modayil, J., Bai, T., & Kautz, H. (2008). Improving the recognition of interleaved activities. *10th International Conference on Ubiquitous Computing (UbiComp 2008).*

Mohammed, T., Albeshri, A., Katib, I., & Mehmood, R. (2020). UbiPriSEQ – Deep Reinforcement Learning to Manage Privacy, Security, Energy, and QoS in 5G IoT Hetnets. *Applied Sciences (Basel, Switzerland)*, *2020*(10), 7120. doi:10.3390/app10207120

Mokráň, M., Róka, R., & Čertík, F. (2015). Analysis of possible utilization of phase shift keying modulations for long-haul optical transmission systems. Applied Physics of Condensed Matter.

Mondal, S., Nandi, D., & Bera, R. (2020, February). V2X Communication Test Bed for Smart Electrical Vehicle with 5G IOV Technology. In *2020 URSI Regional Conference on Radio Science (URSI-RCRS)* (pp. 1-4). IEEE.

Moore, G. E. (1965). Cramming more components onto integrated circuits. *Electronics Magazine*, *38*, 4.

Morocho-Cayamcela, M. E., & Lim, W. (2018). Artificial Intelligence in 5G Technology: A Survey. *2018 International Conference on Information and Communication Technology Convergence (ICTC)*, 860–865. 10.1109/ICTC.2018.8539642

Mukherjee, B. (2006). *Optical WDM networks*. Springer Science Business Media Inc.

Mur, G. (1981). Absorbing boundary conditions for the finite-difference approximation of the time-domain electromagnetic field equations. *IEEE Transactions on Electromagnetic Compatibility*, *23*(4), 377–382. doi:10.1109/TEMC.1981.303970

Muzaffar, R., Raffelsberger, C., Fakhreddine, A., Luque, J. L., Emini, D., & Bettstetter, C. (2020, June). First experiments with a 5G-Connected drone. In *Proceedings of the 6th ACM Workshop on Micro Aerial Vehicle Networks, Systems, and Applications* (pp. 1-5). ACM.

Narayan, A., Kruger, C., Goring, A., Babazadeh, D., Harre, M., Wortelen, B., Luedtke, A., & Lehnhoff, S. (2019). Towards Future SCADA Systems for ICT-Reliant Energy Systems. Internationaler ETG-Kongress 2019.

National Safety Council. (2019). *Understanding RF radiation*. https://www.safetyandhealthmagazine.com/articles/18168-understanding-rf-radiation

Nguty, T. A., & Ekere, N. N. (2000). Modelling the effects of temperature on the rheology of solder paste and flux system. *Journal of Materials Science Materials in Electronics*, *11*(1), 39–43. doi:10.1023/A:1008904220118

Nguty, T. A., & Ekere, N. N. (2000). The rheological properties of solder and solar pastes and the effect on stencil printing. *Rheologica Acta*, *39*(6), 607–612. doi:10.1007003970000117

Nguyen, T. T., & Reddi, V. J. (2019). *Deep Reinforcement Learning for Cyber Security*. arXiv:1906.05799v1 [cs.CR].

Nguyen, C. (2001). *Analysis Methods for RF, Microwave, and Millimeter-Wave Planar Transmission Line Structures*. John Wiley & Sons, Inc. doi:10.1088/0957-0233/12/10/705

Nilsson, N. (1980). *Principles of Artificial Intelligence*. Springer Berlin.

Noveck, B. (2015). *Smart citizens, smarter state the technologies of expertise and the future of governing*. Harvard Univ. Press. doi:10.4159/9780674915435

Nwajana, A. (2017). *Analysis and design of a substrate integrated waveguide multi-coupled resonator diplexer* (PhD Thesis). University of East London, London, UK.

Nwajana, A. O., Ijemaru, G. K., Ang, K. L.-M., Seng, J. K. P., & Yeo, K. S. K. (2021). Unbalanced two-way filtering power splitter for wireless communication systems. *Electronics MDPI*, *10*(5), 617.

Nwajana, A. O., Yeo, K. S. K., & Dainkeh, A. (2016). Low cost SIW Chebyshev bandpass filter with new input/output connection. *Proceedings of the 16th Mediterranean Microwave Symposium (MMS)*, 1–4.

Nwajana, A. O. (2020). Circuit modelling of bandpass/channel filter with microstrip implementation. *Indonesian Journal of Electrical Engineering, and Informatics*, *8*(4), 696–705.

Nwajana, A. O. (2020). *Dual-band microwave filter for WiMax application*. Lambert Academic Publishing.

Nwajana, A. O., Dainkeh, A., & Yeo, K. S. K. (2017). Substrate integrated waveguide (SIW) bandpass filter with novel microstrip-CPW-SIW input coupling. *Journal of Microwaves, Optoelectronics and Electromagnetic Applications*, *16*(2), 393–402. doi:10.1590/2179-10742017v16i2793

Nwajana, A. O., Okafor, P. U., & Arinze, S. N. (2018). T-junction diplexer for RF front end of a cellular base station. *International Journal of Microwave and Optical Technology*, *15*(6), 539–545.

Nwajana, A. O., & Yeo, K. S. K. (2016). Microwave diplexer purely based on direct synchronous and asynchronous coupling. *Wuxiandian Gongcheng*, *25*(2), 247–252. doi:10.13164/re.2016.0247

Nwajana, A. O., & Yeo, K. S. K. (2016). Multi-coupled resonator microwave diplexer with high isolation. *Proceedings of the 46th European Microwave Conference (EuMC)*, 1167–1170. 10.1109/EuMC.2016.7824556

O'Connell E., Moore D., & Newe T. (2020). Challenges Associated with Implementing 5G in Manufacturing. *Telecom MDPI.*

O'Leary, S. (1997). Hierarchical transmission and COFDM systems. *IEEE Transactions on Broadcasting*, *43*(2), 166–174. doi:10.1109/11.598365

OECD. (2015). *Data-Driven Innovation: Big Data for Growth and Well-Being.* Paris: OECD. Retrieved from https://www.oecd-ilibrary.org/science-and-technology/data-driven-innovation_9789264229358-en

Ogata, S. (1991, August). Rheology and Printability of Solder Paste. *IEICE Transactions*, *E74*(8), 2378–2383.

Ogbodo, E. A., Ngobigha, F., Ali, N. O., & Atuba, S. (2021). A Diplexer with Direct Coupled Resonant Junction. Microwave and Optical Technology Letters. *Microstrip diplexers with dual-mode patch resonant junctions. 46th European Microwave Conference (EuMC)*, 1155-1158.

Ogbodo, E. A., Wu, Y., & Wang, Y. (2017). *Dual-Path Dual-Band Filters Based On Patch Resonators.* International Workshop on Electromagnetism (iWEM), London, UK. 10.1109/iWEM.2017.7968779

Ogbodo, E. A. (2017). *Novel microwave devices based on multi-port filtering networks.* University of Greenwich.

Ogbodo, E. A., Aly, M. G., Ali, A. E. T., Wu, Y., & Wang, Y. (2018). Dual-band filtering antenna using dual-mode patch resonators. *Microwave and Optical Technology Letters*, *60*(10), 2564–2569. doi:10.1002/mop.31396

Ogbodo, E. A., Ngobigha, F., Obadiah Ali, N., & Atuba, S. A. (2021). diplexer with direct coupled resonant junction. *Microwave and Optical Technology Letters*, 1–5.

Ogbodo, E. A., Wang, Y., & Yeo, K. S. K. (2016). Microstrip dual-band bandpass filter using u-shaped resonators. *Progress in Electromagnetics Research Letters*, *59*, 1–6. doi:10.2528/PIERL15072303

Ogbodo, E. A., Wu, Y., Callaghan, P., & Wang, Y. (2017). Asynchronous Coupling Resonant Junctions for Compact Diplexers and Multi-Port Filtering Networks. *Microwave and Optical Technology Letters*, *59*(12), 3046–3051. doi:10.1002/mop.30877

Ogbodo, E. A., Wu, Y., Callaghan, P., & Wang, Y. (2018). Masthead Combiner Employing Asynchronously Coupled Resonant Junctions. *48th European Microwave Conference (EuMC).* 10.23919/EuMC.2018.8541754

Ohno, H., Stiles, M. D., & Dieny, B. (2016). Spintronics. *Proc IEEE Inst Electr Electron Eng.*, *104*, 1782–1786.

Ohnsman, F. (2018, May 31). *Forbes*. Retrieved October 18, 2020, from Waymo dramatically expanding autonomous taxi fleet, eyes sales to individuals: https://www.forbes.com/sites/alanohnsman/2018/05/31/waymo-adding-up-to-62000-minivans-to-robot-fleet-may-supply-tech-for-fca-models/

Okafor, P.U., Arinze, S.N., & Onah, O. I. (2020). Artificial Neural Network Inverse Model (AIM) Based Position Control Of DC Motor. *International Journal of Engineering, Science and Mathematics*, *9*(2).

Okafor, P. U., & Eneh, I. I. (2013, September-October). Investigation of Printing Performance of Solder Paste at Different Temperatures. *International Journal of Engineering Research and Applications*, *3*(5), 818–829. www.ijera.com

Okeyo, G., Chen, L., Wang, H., & Sterritt, R. (2012b). A hybrid ontological and temporal approach for composite activity modelling. *IEEE 11th International Conference on Trust, Security and Privacy in Computing and Communications.*

Okeyo, G., Chen, L., Wang, H., & Sterritt, R. (2012a). Dynamic sensor data segmentation for realtime activity recognition. *Pervasive and Mobile Computing.*

Opiah, A. (2020). *The 5G and IoT Revolution.* Retrieved October 28, 2020, from https://www.capacitymedia.com/articles/3825285/the-5g-and-iot-revolution

Oracle. (2014). *What is Java technology and why do I need it?* Available at: https://java.com/en/download/faq/whatis_java.xml

Ordonez, J., Toledo, P., & Sanchis, A. (2013). *Activity recognition using hybrid genera tive/discriminative models on home environments using binary sensors.* MDPI Sensor Open Access.

Oskooi, A. F., Roundy, D., Ibanescu, M., Bermel, P., Joannopoulos, J. D., & Johnson, S. G. (2010). MEEP: A flexible free-software package for electromagnetic simulations by the FDTD method. *Computer Physics Communications, 181*(3), 687–702. doi:10.1016/j.cpc.2009.11.008

Oxfam. (2020). *What is global citizenship?* Retrieved October 17, 2020, from Oxfam: https://www.oxfam.org.uk/education/who-we-are/what-is-global-citizenship/

Pacha, S., Srinivasan, S., & Gnana, S. (2020). Mobile controlled automated wheel chair for disabilities. *International Journal of Innovative Technology and Exploring Engineering, 9*(1S).

Padilla, D. D. (2020). The Internet of Things (IoT) Making Waste Collection Part of a Smarter Future. *Looking Ahead – IoT, Smart Cities and Refuse Collection.* https://txswana.org/images/meeting/040917/Presentations/don_diego_padilla_the_internet_of_things__lot_.pdf

Paganini, P. (2015). Energy Companies Infected by Newly Lazlok Trojan Malware. *Security Affairs.*

Page, M. (2020, October 16). *Covid-19 news: Remdesivir has little effect on survival, finds WHO.* Retrieved October 17, 2020, from NewScientist: https://www.newscientist.com/article/2237475-covid-19-news-remdesivir-has-little-effect-on-survival-finds-who/

Panwar, N., Sharma, S., & Singh, A. K. (2016). A survey on 5G: The next generation of mobile communication. *Physical Communication, 18*, 64–84.

Pan, Y. M., Hu, P. F., Leung, K. W., & Zhang, X. Y. (2018). Compact Single-/Dual-Polarized Filtering Dielectric Resonator Antennas. *IEEE Transactions on Antennas and Propagation, 66*(9), 4474–4484. doi:10.1109/TAP.2018.2845457

Papas, C. H., & King, R. (1949). Input impedance of wide-angle conical antennas fed by a coaxial line. *Proceedings of the IRE, 37*(11), 1269–1271. doi:10.1109/JRPROC.1949.234607

Papas, C. H., & King, R. (1951). Radiation from wide-angle conical antennas fed by a coaxial line. *Proceedings of the IRE, 39*(1), 49–51. doi:10.1109/JRPROC.1951.230420

Patil, G. R., & Wankhade, P. S. (2014). 5G wireless technology. *Int J Comput Sci Mobile Comput, 3*, 203–207.

Patterson, D., Liao, L., Fox, D., & Kautz, H. (2003). Inferring high level behavior from low level sensors. *5th Conference on Ubiquitous Computing (UbiComp, 03).*

Patterson, D., Kautz, D., & Philipose, M. (2005). Fine-grained activity recognition by aggregating abstract object usage. *9th IEEE International Symposium on Wearable Computers.* 10.1109/ISWC.2005.22

Peisa, J. (2020). *5G evolution - 3GPP reases a6 & 17 overview.* Stockholm: Ericsson. Retrieved October 18, 2020, from https://www.ericsson.com/49bdd9/assets/local/reports-papers/ericsson-technology-review/docs/2020/5g-nr-evolution.pdf

Perez, M. G., Celdran, A. H., Giardina, P. G., Bernini, G., Pizzimerti, S., Clemente, F. J. G., Perez, G. M., Festa, G., & Paglianti, F. (n.d.). *Mitigation of Cyber Threats: Protection Mechanics in Federated SDN/NFV Infrastructures for 5G within FIRE+*. Academic Press.

Perkins, C. E., & Royer, E. M. (1999). Ad-hoc on-demand distance vector routing. In *Proceedings WMCSA'99. Second IEEE Workshop on Mobile Computing Systems and Applications* (pp. 90-100). IEEE.

Physio-pedia.com. (2020). *Role of the Wheelchair*. Available Online: https://www.physio-pedia.com/Role_of_the_Wheelchair

Pittman, J. (2017). Exploring Global Citizenship Theories to Advance Educational, Social,Economic and Environmental Justice. *Journal of Tourism & Hospitality (Los Angeles, Calif.)*, *6*(6). Advance online publication. doi:10.4172/2167-0269.1000326

Pollock, J. (n.d.). SCADA: Alive and Well in the Age of IoT. *Microwaves & RF.*

Pour, M. M., Anzalchi, A., & Sarwat, A. (2017). A Review on Cyber Security Issues and Mitigation Methods in Smart Grid Systems. IEEE.

Powell, J. R. (2008). The quantum limit to Moore's law. *Proceedings of the IEEE*, *96*, 1247–1248.

Pozar, D. M. (2012). *Microwave Engineering* (4th ed.). Wiley.

PwC. (2016). *Global report on the commercial applications of drone technology.* https://www.pwc.pl/pl/pdf/clarity-from-above-pwc.pdf

PWC. (2020, October 25). *Making sense of bitcoin, cryptocurrency and blockchain.* Retrieved from PWC - United States: https://www.pwc.com/us/en/industries/financial-services/fintech/bitcoin-blockchain-cryptocurrency.html

Qian, H., Mao, Y., Xiang, W., & Wang, Z. (2010). Recognition of human activities using svm multiclass classification. *Pattern Recognition Letters*, *31*(2), 100–111. doi:10.1016/j.patrec.2009.09.019

Qian, J. F., Chen, F. C., & Chu, Q. X. (2018). A Novel Tri-Band Patch Antenna With Broadside Radiation and Its Application to Filtering Antenna. *IEEE Transactions on Antennas and Propagation*, *66*(10), 5080–5585. doi:10.1109/TAP.2018.2853160

Qui, L.-F., Wu, L.-S., Xie, B., Yin, W.-Y., & Mao, J.-F. (2018). Substrate integrated waveguide filter with flat passband based on complex couplings. *IEEE Microwave and Wireless Components Letters*, *28*(6), 494–496. doi:10.1109/LMWC.2018.2829349

Ramanathan, R. (2001). A note on the use of the analytic hierarchy process for environmental impact assessment. *Journal of Environmental Management*, *63*(1), 27–35. doi:10.1006/jema.2001.0455 PMID:11591027

Rana, N., Luthra, S., Mangla, S. K., Islam, R., Roderick, S., & Dwivedi, Y. K. (2019). Barriers to the Development of Smart Cities in Indian Context. *Information Systems Frontiers*, *21*(3), 503–525. doi:10.100710796-018-9873-4

Rao, P. H., Selvan, K. T., & Das, S. K. (1997). Finite-range gain prediction of pyramidal horn with improved accuracy. *IEEE Transactions on Electromagnetic Compatibility*, *39*(4), 374–376. doi:10.1109/15.649840

Rappaport, T. S., Muhamed, R., Buehrer, R. M., & Doradla, A. (2000). Mobile and Cellular Radio Communications. In R. C. Dorf (Ed.), *The Engineering Handbook*. CRC Press LLC.

Rathor, S., & Saxena, D. (2020). Energy management system for smart grid: An overview and key issues. *International Journal of Energy Research, 44*(6), 4067–4109. doi:10.1002/er.4883

Renjen, P. (2019). *Harvard Business Review - How Leaders Are Navigating the Fourth Industrial Revolution.* Retrieved March 26, 2020, from https://hbr.org/sponsored/2019/03/how-leaders-are-navigating-the-fourth-industrial-revolution

Renjith, S., & Anjali, C. (2014)., A personalized mobile travel recommender system using hybrid algorithm. *IEEE First International Conference on Computational Systems and Communications (ICCSC)*, 12-17.

Resilient Energy Platform. (2020). *Cybersecurity and distributed Energy resources.* Author.

Riboni, D., Pareschi, L., Radaelli, L., & Bettini, C. (2011). Is ontology based activity recognition really effective? *IEEE International Conference on Pervasive Computing and Communications Workshops (PERCOM Workshops).* 10.1109/PERCOMW.2011.5766927

Ricci, F., Rokach, L., & Shapira, B. (2015). Recommender systems: introduction and challenges. In *Recommender systems handbook* (pp. 1–34). Springer. doi:10.1007/978-1-4899-7637-6_1

Robertson, I. D., Karacaoglu, U., & Sánchez-Hernández, D. (1996). CAD techniques for microwave circuits. *IEEE Electronics & Communication Engineering Journal, 8*(6), 245–256. doi:10.1049/ecej:19960601

Rojas, J. P., Singh, D., Inayat, S. B., Sevilla, G. T., Fahad, H. M., & Hussain, M. M. (2017). Review – micro and nano –engineering enabled new generation of thermoelectric generator devices and applications. *ECS Journal of Solid State Science and Technology: JSS, 6*, 3036–3044.

Róka, R., & Čertík, F. (2015a). Simulation tools for broadband passive optical networks. In Simulation technologies in networking and communications: Selecting the best tool for the test. CRC Press, Taylor and Francis Group.

Róka, R., & Čertík, F. (2015b). Modeling of negative influences at the signal transmission in the optical medium. Applied Physics of Condensed Matter.

Róka, R., & Mokráň, M. (2019). Simulation platform for analyzing environmental effects in multi-wavelength transmission systems in Matlab Simulink. *Przeglad elektrotechniczny, 95*(6), 9-14.

Róka, R., & Šalík, P. (2020). An effective integration of the PM 16-QAM modulation in enhanced metropolitan networks with the EDFA amplification. *Journal of Electrical Engineering - EČ, 71*(5), 317-325.

Róka, R. (2012). Fixed transmission media. In *Technology and engineering applications of Simulink.* InTech. doi:10.5772/37442

Róka, R. (2015a). The simulation of negative influences in the environment of fixed transmission media. *International Conference on Applied Physics, Simulation and Computers*, 58-68.

Róka, R. (2015b). The environment of fixed transmission media and their negative influences in the simulation. *International Journal of Mathematics and Computers in Simulation, 9*, 190–205.

Róka, R., & Čertík, F. (2012). Modeling of environmental influences at the signal transmission in the optical transmission medium. *International Journal of Communication Networks and Information Security, 4*(3), 146–162.

Róka, R., & Čertík, F. (2015c). Simulation and analysis of the signal transmission in the optical transmission medium. *International Conference on Simulation and Modeling Methodologies, Technologies and Application*, 219-226. 10.5220/0005569602190226

Róka, R., & Mokráň, M. (2015). Modeling of the PSK utilization at the signal transmission in the optical transmission medium. *International Journal of Communication Networks and Information Security, 7*(3), 187–196.

Róka, R., & Mokráň, M. (2018). Effect of the FWM influence on the CWDM signal transmission in the optical transmission media. *International Journal of Circuits. Systems and Signal Processing*, *12*, 42–47.

Róka, R., Mokráň, M., & Šalík, P. (2017). Simulation of negative influences on the CWDM signal transmission in the optical transmission media. *International Journal of Circuits. Systems and Signal Processing*, *11*, 75–80.

Róka, R., & Šalík, P. (2017). Impact of environmental influences on multilevel modulation formats at the signal transmission in the optical transmission medium. *International Journal of Communication Networks and Information Security*, *9*(1), 76–87.

Roman, R., Lopez, J., & Mambo, M. (2016). *Mobile Edge Computing, Fog et al: A Survey and Analysis of Security Threats and Challenges.* arXiv: 1602.00484v2 [cs.CR].

Ruotsalo, T., Haav, K., Stoyanov, A., Rochee, S., Fanid, E., Deliaic, R., Mäkelä, E., Kauppinen, T., & Hyvönen, E. (2013). SMARTMUSEUM: A Mobile Recommender System for the Web of Data. *Journal of Web Semantics*, *20*, 50–67. doi:10.1016/j.websem.2013.03.001

Saaty, T. (2005). *Theory and Applications of Analytic Network Process* (Vol. 4922). RWS Publications.

Sakhnini, J. (2020). *Security of Smart Cyber-Physical Grids: A Deep learning Approach* (Unpublished master thesis). University of Guelph, Ontario, Canada.

Šalík, P., & Róka, R. (2018). Analysis and simulation of dynamic properties for the DFB laser. *Przeglad elektrotechniczny, 94*(7), 17-20.

Šalík, P., & Róka, R. (2017). Analysis of possibilities for numerical simulations of continues wave DFB laser. *International Congress on Ultra-Modern Telecommunications and Control Systems.*

Šalík, P., Róka, R., & Tomáš, G. (2018). Simulation platform of optical transmission system in Matlab Simulink. *Procedia Computer Science*, *134*, 196–203. doi:10.1016/j.procs.2018.07.162

Samsuzzaman, M., Islam, M. T., Mandeep, J. S., & Misran, N. (2014). Printed wide-slot antenna design with bandwidth and gain enhancement on low-cost substrate. *TheScientificWorldJournal*, *2014*, 1–10. Advance online publication. doi:10.1155/2014/804068 PMID:24696661

Sánchez-Hernández, D., Karacaoglu, U., Gokdemir, T., Khalid, H. A., Rezazadeh, A. A., Ferrando, M., Reyes, E., & Robertson, I. D. (1997). CAD techniques for the electromagnetic design of monolithic millimetre-wave integrated circuits. In G. Pujolle (Ed.), *Annals of Telecommunications* (pp. 164–173). Springer Nature.

Sandler, S. S., & King, R. W. P. (1994). Compact conical antennas for wide-band coverage. *IEEE Transactions on Antennas and Propagation*, *42*(3), 436–439. doi:10.1109/8.280735

Santofimia-Romero, M., Toro-garcia, X. D., & Lopez-Lopez, J. (2011, October). Artificial Intelligence Techniques for Smart Grid Applications. *CEPIS Upgrade*, *12*(4), 41–48.

Sarkar, S. (2014). The unique identity (UID) project, biometrics and re-imagining governance in India. *Oxford Development Studies*, *42*(4), 516–533. doi:10.1080/13600818.2014.924493

Schaller, R. R. (1997). Moore's Law: Past, Present, and Future. *IEEE Spectrum*, *34*(6), 52–59. doi:10.1109/6.591665

Schoelkopf, R. J., Wahlgren, P., Kozhevnikov, A. A., Delsing, P. D., & Prober, E. (1998). The Radio-Frequency Single-Electron Transistor (RF-SET): A Fast and Ultrasensitive Electrometer. *Science*, *280*, 1238–1242.

Securing 5G's Future: Why Cybersecurity is Key to Realising the Full Promise of 5G Networks. (2020). PWC.

Selvan, K. T. (1998). Simple formulas for the gain and far-field of open-ended rectangular waveguides. *IEE Proceedings. Microwaves, Antennas and Propagation, 145*(1), 80–84. doi:10.1049/ip-map:19981552

Selvan, K. T. (1999). Accurate design method for optimum gain pyramidal horns. *Electronics Letters, 35*(4), 249–250. doi:10.1049/el:19990240

Selvan, K. T., Rao, P. H., & Geetha, S. (1999). High-frequency reflection from the aperture of a long E-plane sectoral horn. *IEEE Antennas & Propagation Magazine, 41*(3), 58–59. doi:10.1109/74.775247

Serrano, D. C., Diaz, J. S. G., Alù, A., & Melcón, A. Á. (2015). Electrically and magnetically biased graphene-based cylindrical waveguides: Analysis and applications as reconfigurable antennas. *IEEE Transactions on Terahertz Science and Technology, 1*(5), 951–960.

SESAR Joint Undertaking. (2017). *U-space: Blueprint.* https://op.europa.eu/en/publication-detail/-/publication/f8613e25-cf38-11e7-a7df-01aa75ed71a1/language-en

SESAR Joint Undertaking. (2019). *Concept of Operations for U-space.* https://www.sesarju.eu/node/3411

SESAR Joint Undertaking. (2020). *U-space.* https://www.sesarju.eu/U-space

Seuwou, P., Banissi, E., & Ubakanma, G. (2019). The Future of Mobility with Connected and Autonomous Vehicles in Smart Cities. *Digital Twin Technologies and Smart Cities, Internet of Things (IoT).*

Seuwou, P., Banissi, E., & Ubakanma, G. (2020). The Future of Mobility with Connected and Autonomous Vehicles in Smart Cities. In Digital Twin Technologies and Smart Cities. Internet of Things (Technology, Communications and Computing (pp. 37-52). Springer. doi:10.1007/978-3-030-18732-3_3

Seuwou, P., Chrysoulas, C., Banissi, E., & Ubakanma, G. (2020). Measuring Consumer Behavioural Intention to Accept Technology: Towards Autonomous Vehicles Technology Acceptance Model (AVTAM). *Advances in Intelligent Systems and Computing*, 507-516.

Seuwou, P., Patel, D., & Ubakanma, G. (2014). Vehicular Ad hoc Network Applications and Security: A Study into the Economic and the Legal implications. *International Journal of Electronic Security and Digital Forensics.*

Seuwou, P., Patel, D., & Ubakanma, G. (2014). Vehicular ad hoc network applications and security: A study into the economic and the legal implications. *International Journal of Electronic Security and Digital Forensics, 6*(2), 115–129. doi:10.1504/IJESDF.2014.063107

Shahin, S. (2018). Big data and the illusion of choice: Comparing the evolution of India's Aadhaar and China's Social Credit. *Social Science Computer Review.* Advance online publication. Retrieved October 10, 2020, from. doi:10.1177/0894439318789343

Shah, R. C., Roy, S., Jain, S., & Brunette, W. (2003). Data mules: Modeling and analysis of a three-tier architecture for sparse sensor networks. *Ad Hoc Networks, 1*(2-3), 215–233. doi:10.1016/S1570-8705(03)00003-9

Sharma, J., Anbarasu, M., Chandan, C., & Shanmugasundaram, M. (2017). *Iris movement based wheel chair control using raspberry pi.* . doi:10.1088/1757-899X/263/5/052049

Sharma, V., & Kaur, D. (2015). Review on multiplexing techniques in optical communication systems. *European Scientific Journal*, 2.

Sharma, V., & Singh, G. (2014). On BER assessment of conventional-and wavelet-OFDM over AWGN channel. *Optik (Stuttgart), 125*(20), 6071–6073. doi:10.1016/j.ijleo.2014.07.073

Shea, C., Arora, S., & Brown, S. (2007). *Wave Soldering Flux Selection for Pb-Free Assembly*. Circuit Assembly. Available Online: http://circuitsassembly.com/cms/cms/content/view/4511/95/

Shi, Y., Zhang, J., Zhou, M., Feng, W., Cao, B., & Che, W. (2019). Miniaturized W-band gap waveguide bandpass filter using the MEMS technique for both waveguide and surface mounted packaging. *IEEE Transactions on Circuits and Wystems. II, Express Briefs*, *66*(6), 938–942. doi:10.1109/TCSII.2018.2873236

Shyam, G. K., Manvi, S. S., & Bharti, P. (2017, February). Smart waste management using Internet-of-Things (IoT). In *2017 2nd international conference on computing and communications technologies (ICCCT)* (pp. 199-203). IEEE.

Silva, R., Bonnin, J. M., & Ernst, T. (2017). *An ITS-based Architecture for Opportunistic Networking*. Academic Press.

Silver, S. (1949). *Microwave Antenna Theory and Design*. McGraw-Hill.

Silvester, P. P. (1969). Finite element solution of homogeneous waveguide problems. *Alta Frequenza*, *38*, 313–317.

Si-Mohammed, S., Bouaziz, M., Hellaoui, H., Bekkouche, O., Ksentini, A., Taleb, T., Tomaszewski, L., Lutz, T., Srinivasan, G., Jarvet, T., & Montowtt, P. (2020). Supporting UAV Services in 5G Networks: New High-Level Architecture integrating 5G with U-space. *IEEE Vehicular Technology Magazine*.

Sir, M., Bradac, Z., & Fiedler, P. (2015). Ontology versus database. *IFAC-PapersOnLine*, *48*(4), 220–225. doi:10.1016/j.ifacol.2015.07.036

Sivani, S., & Sudarsanam, D. (2013). Impacts of radio-frequency electromagnetic field (RF-EMF) from cell phone towers and wireless devices on biosystem and ecosystem. *Biology and Medicine (Aligarh)*, *4*(4), 202–216.

Smirnov, A., Kashevnik, A., Shilov, N. & A, P. (2014). Smart Space-Based in-Vehicle Application for e-Tourism. In *Internet of Things, Smart Spaces, and Next Generation Networks and Systems*. Springer.

SNG. (2017, July 13). *Report: Smart city investment will double by 2021*. Retrieved October 13, 2020, from StateScoop: https://statescoop.com/report-smart-city-investment-will-double-by-2021/

Soh, Z. H. C., Husa, M. A. A. H., Abdullah, S. A. C., & Shafie, M. A. (2019, April). Smart Waste Collection Monitoring and Alert System via IoT. In *2019 IEEE 9th Symposium on Computer Applications & Industrial Electronics (ISCAIE)* (pp. 50-54). IEEE.

Soldari, D. (n.d.). *5G and the Future of Security in ICT.* Huawei-INTAC-5G-Cyber-Security paper.

Song, X., Liu, Z., Xiang, Y., & Aydin, K. (2018). Two-dimensional material nano photonics. *Optics Express*, *1*(26), 5469–5477.

Son, H., & Pyo, C. (2005). Design of RFID tag antennas using an inductively coupled feed. *Electronics Letters*, *41*(18), 994. doi:10.1049/el:20051536

Sridhar, S., Hahn, A., & Govindarasu, M. (2012, January). Cyber-Physical System security for the Electric Power Grid. *Proceedings of the IEEE*, *100*(1), 210–224. doi:10.1109/JPROC.2011.2165269

Steer, M. B., Bandler, J. W., & Snowden, C. M. (2002). Computer-aided design of RF and microwave circuits and systems. *IEEE Transactions on Microwave Theory and Techniques*, *50*(3), 996–1005. doi:10.1109/22.989983

Stoner, L. (2013). The Importance of Global Citizenship to Higher Education: The Role of Short-Term Study Abroad. *British Journal of Education Society & Behavioural Science*, *3*(2), 184–194. doi:10.9734/BJESBS/2013/2910

Stratton, J. A. (1941). *Electromagnetic Theory*. McGraw-Hill.

Sun, C., Liu, C., & Xie, J. (2016). Cyber- Physical System Security of a Power Grid: State-of-the-Art. *Electronics (Basel), 2016*(5), 40. doi:10.3390/electronics5030040

TachoverD. (2020). Retrieved from https://childrenshealthdefense.org/news/what-you-should-know-about-5g-satellites-how-musks-sci-fi-dreams-are-becoming-our-living-nightmare

Tai, C. T., & Long, S. A. (2007). Dipoles and monopoles. In J. L. Volakis (Ed.), *Antenna Engineering Handbook* (pp. 4-3–4-32). McGraw-Hill Inc.

Tamiazzo, S., & Macchiarella, G. (2013). Design and realization of filtering units in a triple-band masthead amplifier system for BTS of mobile communications. *IEEE MTT-S Int. Microwave Symp. Dig.*

Tapia, E., Intille, S., & Larson, K. (2004). Activity recognition in the home using simple and ubiquitous sensor. *2nd International Conference in Pervasive Computing (PERVASIVE 2004).* 10.1007/978-3-540-24646-6_10

Tapia, E., Intille, S., Larson, K., Haskell, W., Wright, J. A. K., & Friedman, R. (2007). Realtime recognition of physical activities and their intensities using wireless accelerometers and a heart rate monitor. *11th IEEE International Symposium on Wearable Computers.*

Technology Shapes the Future of Crime and Law Enforcement. (2019). European Union Agency for Law Enforcement Cooperation.

Ten, C., Liu, C., & Manimaran, G. (2008, November). Vulnerability Assessment of Cybersecurity for SCADA Systems. *IEEE Transactions on Power Systems, 23*(4), 1836–1846. doi:10.1109/TPWRS.2008.2002298

Teng, H., Liu, Y., Liu, A., Xiong, N. N., Cai, Z., Wang, T., & Liu, X. (2019). A novel code data dissemination scheme for Internet of Things through mobile vehicle of smart cities. *Future Generation Computer Systems, 94*, 351–367. doi:10.1016/j.future.2018.11.039

The National Security Challenges of Fifth Generation (5G) Wireless Communications Winning the Race to 5G, Securely. (2019). Intelligence and National Security Alliance (INSA) Cyber Council.

Thiele, G. A. (1969). Analysis of Yagi-Uda-type antennas. *IEEE Transactions on Antennas and Propagation, 17*(1), 24–31. doi:10.1109/TAP.1969.1139356

Thirugnanasambandam, V., & Jain, T. (2018). AdaBoost Classifiers for Phasor Measurements-based Security assessment of Power Systems. *The Institution of Engineering and Technology (IET) Journals. IET Generation, Transmission & Distribution, 12*(8), 1747–1753. doi:10.1049/iet-gtd.2017.0013

Thrybom, L., Kaporits, A., & Ekrescom. (Eds.). (2015). *5G PPP White Paper on Energy Vertical Sector.* Version 1.0 30 September 2015.

To-Euro-5G – Supporting the European 5G Initiative. (2019). *The European 5G Annual Journal.*

Tomaszewski, L., Kołakowski, R., & Korzec, P. (2020b, June). On 5G support of cross-border UAV operations. In *2020 IEEE International Conference on Communications Workshops (ICC Workshops)* (pp. 1-6). IEEE.

Tomaszewski, L., Kołakowski, R., & Kukliński, S. (2020a, June). Integration of U-space and 5GS for UAV services. In *2020 IFIP Networking Conference (Networking)* (pp. 767-772). IEEE.

Tomaszewski, L., Kukliński, S., & Kołakowski, R. (2020, June). A new approach to 5G and MEC integration. In *IFIP International Conference on Artificial Intelligence Applications and Innovations* (pp. 15-24). Springer.

Tomita, A., & Suzuki, A. (1991). A new density matrix theory for semiconductor lasers, including non- Markovian intraband relaxation and its application to nonlinear gain. *IEEE Journal of Quantum Electronics*, *27*(6), 1630–1641. doi:10.1109/3.89987

Toness, R. V. (2014). *India building database to unite records for 1.2 billion.* Retrieved from https://www.bloomberg.com/news/articles/2014-12-10/india-building-database-to-unite-records-for-12-billion

Tong, L., Zhao, Q., & Adireddy, S. (2003). Sensor networks with mobile agents. In *IEEE Military Communications Conference, 2003. MILCOM 2003.* (Vol. 1, pp. 688-693). IEEE. 10.1109/MILCOM.2003.1290187

Toppeta, D. (2010). The Smart City Vision: How Innovation and ICT Can Build Smart, "Livable". *The Innovation Knowledge Foundation.* Retrieved from http://www.thinkinnovation.org/file/research/23/en/Top peta_Report_005_2010.pdf

Tran, V., Tsuji, H., & Masudu, R. (2009). A new QoS ontology and its QoS –based ranking algorithm for web services. *Simulation Modelling Practice and Theory*, *17*(8), 1378–1398.

Tredinnick, L. (2019). Cryptocurrencies and the blockchain. *Journal Indexing and Metrics*, *36*(1).

Trevino, M. (n.d.). Cyber Physical Systems: The Coming Singularity. *Prism – The Journal of Complex Operations, 8*(32019).

U.N. (2020a). *The Impact of Digital Technologies.* Retrieved November 26, 2020, from United Nations: https://www.un.org/en/un75/impact-digital-technologies

U.N. (2020b). *The Age of Digital Interdependence.* United Nations.

UN. (2020a). *United Nations - The Impact of Digital Technologies.* Retrieved November 26, 2020, from https://www.un.org/en/un75/impact-digital-technologies

UNICEF. (2013). *Global citizenship: a high school educator's guide (grades 9-12).* UNICEF. Retrieved October 13, 2020, from https://www.gcedclearinghouse.org/resources/global-citizenship-high-school-educators-guide-grades-9-12

Unique Identification Authority of India. (n.d.). *About Aadhaar.* Retrieved from https://uidai.gov.in/youraadhaar/about-aadhaar.html

United, N. (2018). *Revision of World Urbanization Prospects.* United Nations. Retrieved October 17, 2020, from https://www.un.org/development/desa/publications/2018-revision-of-world-urbanization-prospects.html

Venkatesulu, S., Vijai Bhaskar, P., & Vijayasree, R. (2018). Mathematical Analysis Of Electromagnetic Radiation Emit From A Cellular Base Station. *International Journal of Advance Engineering and Research Development,* 5(2).

Venkatesulu, V. S., Giri, P. M. N., & Venkata, R. P. (2014). Monitoring Of Electromagnetic Radiation for Cellular Base Stations Using Arm Processor. *International Journal of Innovative Research in Computer and Communication Engineering*, *2*(6), 4603–4609.

Verizon. (2019). *First Principles for Securing 5G: The Design, Deployment, Operation and Innovation of Secure 5G Networks.* White paper by Verizon.

Wadell, B. C. (1991). *Transmission Line Design Handbook.* Artech House.

Wafa, B., & Yusra, A. (2020). Decentralized Motion Control for Omnidirectional Wheelchair Tracking Error Elimination Using PD-Fuzzy-P and GA-PID Controllers. *Sensors (Basel)*, *3525*(12), 3525. Advance online publication. doi:10.339020123525

Wallace, R. J., Robbins, J. M., Loffi, J. M., Holliman, J. K., Metscher, D. S., & Rogers, T. R. (2020). Evaluating LAANC Utilization & Compliance for Small Unmanned Aircraft Systems in Controlled Airspace. *International Journal of Aviation, Aeronautics, and Aerospace, 7*(2), 4.

WallSteet. (2018, October 26). *Bit coin the next world currency.* Retrieved from WallStreet: https://wall-street.com/bit-coin-the-next-world-currency/

Wang, Y., Chen, C., Wang, J., & Baldick, R. (2015). Research on Resilience of Power Systems under Natural Disasters – A Review. *IEEE Transactions on Power Systems.*

Warren, C., Giannopoulos, A., & Giannakis, I. (2016). gprMax: Open source software to simulate electromagnetic wave propagation for Ground Penetrating Radar. *Computer Physics Communications, 209*(12), 163–170. doi:10.1016/j.cpc.2016.08.020

Warrier, A. (2018). *Smart Transportation & Smart Waste Management.* https://medium.com/bolt-iot/smart-transportation-smart-waste-management-d35f1e141629

Washburn, D. S. (2010). *Helping CIOs Understand "Smart City" Initiatives: Defining the Smart City, Its Drivers, and the Role of the CIO.* Forrester Research, Inc. Retrieved from http://public.dhe.ibm.com/partnerworld/pub/smb/smart

Watts, D. (2003). Security & Vulnerability in Electric Power Systems. In *NAPS 2003, 35th North American Power Symposium.* University of Missouri-Rolla.

Wehenkel, L. (1997). Machine-Learning Approaches to Power-System Security Assessment. *IEEE Expert: AI in Power Systems,* 60-70.

Wei, F., Zhao, X.-B., & Shi, X. W. (2019). A balanced filtering quasi-Yagi antenna with low cross-polarization levels and high common-mode suppression. *IEEE Access: Practical Innovations, Open Solutions, 7,* 100113–100119. doi:10.1109/ACCESS.2019.2931141

Weiland, T. (1977). A discretization method for the solution of Maxwell's equations for six-component fields. *Electronics and Communications, AEU., 31*(3), 116–120.

Weiner, M. M. (2003). *Monopole Antennas.* Marcel Dekker. doi:10.1201/9780203912676

Wei, Y., Jennings, N., Moreau, L., & Hall, W. (2008). User evaluation of a market-based recommender system. *Autonomous Agents and Multi-Agent Systems, 17*(2), 251–269. doi:10.100710458-008-9029-x

Wen, Z., Hu, S., De Clercq, D., Beck, M. B., Zhang, H., Zhang, H., Fei, F., & Liu, J. (2018). Design, implementation, and evaluation of an Internet of Things (IoT) network system for restaurant food waste management. *Waste Management (New York, N.Y.), 73,* 26–38. doi:10.1016/j.wasman.2017.11.054 PMID:29242117

West, A. (2014). *In US, government hacking is '40 years ahead' of the language in privacy laws.* The World. Available at https://www.pri.org/stories/2014-03-28/us-government-hacking-40-years-ahead-language-privacy-laws

Wheeler, H. A. (1965). Transmission-line properties of parallel strips separated by a dielectric sheet. *IEEE Transactions on Microwave Theory and Techniques, 13*(2), 172–185. doi:10.1109/TMTT.1965.1125962

Williamson, B. (2015). Educating the smart city: Schooling smart citizens through computational urbanism. *Big Data & Society, 2*(2), 1–13. doi:10.1177/2053951715617783

Winkowka, J., Szpilko, D., & Pejic, S. (2019). Smart city concept in the light of the literature review. *Internationa Society for Manufaturin, service and Management Engineering,* 70-86.

Wired. (2017). *Big data meets big brother as China moves to rate its citizens*. Retrieved from https://www.wired.co.uk/article/chinese-government-social-credit-score-privacy-invasion

Wong, S.-W., Wang, K., Chen, Z.-N., & Chu, X.-Q. (2014). Electric coupling structure of substrate integrated waveguide (SIW) for the application of 140-GHz bandpass filter on LTCC. *IEEE Transactions on Components, Packaging, and Manufacturing Technology*, *4*(2), 316–322. doi:10.1109/TCPMT.2013.2285388

Woods, B., & Watson, N. (2013). The social and technological history of wheelchairs. *International Journal of Therapy and Rehabilitation*, *11*(9), 407–410. doi:10.12968/ijtr.2004.11.9.19587

Worldbank.org. (2020). *Disability Inclusion*. Available online: https://www.worldbank.org/en/topic/disability

Wu, Y., Wang, Y., & Ogbodo, E. A. (n.d.). Microstrip wideband diplexer with narrow guard band based on all-resonator structures. *46th European Microwave Conference (EuMC),* 1163-1166. doi:10.1049/el:20051536

Wu, J., Chen, L., Xie, Y., & Zhang, Z. (2012). Titan:A system for effective web service discovery. *Proceedings of the 21st International Conference Companion on World Wide Web*, 441-444. 10.1145/2187980.2188069

Wu, J., Shen, Y., Reinhardt, K., Szu, H., & Dong, B. (2013). A Nanotechnology Enhancement to Moore's Law. *Applied Computational Intelligence and Soft Computing*, 1–14.

Wurm, J., Jin, Y., Liu, Y., Hu, S., Heffner, K., Rahman, F., & Tehranipoor, M. (2016). *Introduction to Cyber-Physical System Security: A Cross-layer Perspective*. IEEE Transactions on Multi-Scale Computing Systems., doi:10.1109/TMSCS.2016.2569446

Wyllie, D. (2013, August 20). *How 'Big Data' is helping law enforcement*. Retrieved October 19, 2020, from Police1.com: https://www.police1.com/police-products/software/data-information-sharing-software/articles/how-big-data-is-helping-law-enforcement-LLLcFsngeq28DKVz/

Xia, F., Wang, H., Xiao, D., Dubey, M., & Ram Subramaniam, A. (2014). Two-dimensional material nano photonics. *Nature Photonics*, *1*(8), 89–907.

Xu, W., Zhou, H., Cheng, N., Lyu, F., Shi, W., Chen, J., & Shen, X. (2017). Internet of vehicles in big data era. *IEEE/CAA Journal of Automatica Sinica, 5*(1), 19-35.

Yaghjian, A. D. (1984). Approximate formulas for the far fields and gain of open ended rectangular waveguide. *IEEE Transactions on Antennas and Propagation*, *32*(4), 378–384. doi:10.1109/TAP.1984.1143332

Yagi, H., & Uda, S. (1926). Projector of the sharpest beam of electric waves. *Proceedings of the Imperial Academy (Japan)*, *2*(2), 49–52. doi:10.2183/pjab1912.2.49

Yamada, N., Sakamoto, K., Kunito, G., Yamazaki, K., Isoda, Y., & Tanaka, S. (2007). Applying ontology and probabilistic model to human activity recognition from surrounding things. *IPSJ Digital Courier, 3.*

Yamada, Y. (2011). Design of a very small antenna for metal-proximity applications. In S. Preradovic (Ed.), *Advanced Radio Frequency Identification Design and Applications* (pp. 77–114). InTech Open Access. doi:10.5772/14734

Yao, M., Sohul, M., Marojevic, V., & Reed, J. H. (2019). *Artificial Intelligence-Defined 5G Radio Access Networks*. IEEE Communications.

Yee, K. S. (1966). Numerical solution of initial boundary value problems involving Maxwell's equations in isotropic media. *IEEE Transactions on Antennas and Propagation*, *14*(3), 302–307. doi:10.1109/TAP.1966.1138693

Yeo, K. S. K., & Nwajana, A. O. (2013). A novel microstrip dual-band bandpass filter using dual-mode square patch resonators. *Progress In Electromagnetic Research*, *C*, 36, 233–247. doi:10.2528/PIERC12120312

Yohanandhan, R. V., Elavarasan, R. M., Premkumar, M., & Mihet-Popa, L. (2020). Cyber-Physical Power System (CPPS): A Review on Modelling, Simulations, and Analysis with Cyber Security Applications. *IEEE Access : Practical Innovations, Open Solutions*. Advance online publication. doi:10.1109/ACCESS.2020.3016826

You, K. Y. (2016). *RF Coaxial Slot Radiators: Modeling, Measurements and Applications*. Artech House.

You, K. Y. (2017). Materials characterization using microwave waveguide system. In S. Goudos (Ed.), *Microwave Systems and Applications* (pp. 341–358). InTech Open Access. doi:10.5772/66230

You, K. Y. (2019). Analytical and calibration models for slot antennas. In J. Zbitou & A. Errkik (Eds.), *Emerging Innovations in Microwave and Antenna Engineering* (pp. 57–100). IGI Global. doi:10.4018/978-1-5225-7539-9.ch003

You, K. Y., Abbas, Z., Khalid, K., & Kong, N. F. (2009). Improved formulation for admittance of thin and short monopole driving from coaxial line into dissipative media. *IEEE Antennas and Wireless Propagation Letters*, *8*, 1246–1249. doi:10.1109/LAWP.2009.2035645

You, K. Y., Abbas, Z., Malek, M. F. A., & Cheng, E. M. (2014). Non-destructive dielectric measurements and calibration for thin materials using waveguide-coaxial adaptors. *Measurement Science Review*, *14*(1), 16–24. doi:10.2478/msr-2014-0003

You, X., Zhang, C., Tan, X., Jin, S., & Wu, H. (2019). AI for 5G: Research Directions and Paradigms. *Science China. Information Sciences*, *62*(2). https://doi.org/10.1007/s11432-018-9596-5

Yu, C., Hong, W., Kuai, Z., & Wang, H. (2012). Ku-band linearly polarized omnidirectional planar filtenna. *IEEE Antennas and Wireless Propagation Letters*, *11*, 310–313. doi:10.1109/LAWP.2012.2191259

Yusuf, Y., & Gong, X. (2011). Compact low-loss integration of high-Q 3-D filters with highly efficient antennas. *IEEE Transactions on Microwave Theory and Techniques*, *59*(4 Part 1), 857–865. doi:10.1109/TMTT.2010.2100407

Zacharia, K. P. (1998). Radiation resistance of resonant normal mode helices. *IETE Journal of Education*, *39*(1), 5–7. doi:10.1080/09747338.1998.11689746

Zanella, A., Bui, N., Castellani, A., Vangelista, L., & Zorzi, M. (2014). Internet of things for smart cities. *IEEE Internet of Things Journal*, *1*(1), 22–32. doi:10.1109/JIOT.2014.2306328

Zhang, B., & Xue, Q. (2018). Filtering antenna with high selectivity using multiple coupling paths from source/load to resonators. *IEEE Transactions on Antennas and Propagation*, *66*(8), 4320–4325. doi:10.1109/TAP.2018.2839968

Zhang, H., & James, G. (1998, January). A broadband T-junction diplexer with integrated iris filters. *Microwave and Optical Technology Letters*, *17*(1), 69–72. doi:10.1002/(SICI)1098-2760(199801)17:1<69::AID-MOP19>3.0.CO;2-6

Zhang, M., Ranjan, A., Georgakopoulos, D., Menzel, M., & Nepal, S. (2012). An ontology –based system for cloud infrastructure services discovery. *Proceedings of 8th International Conference on Collaborative Computing: Networking, Applications and Work sharing (CollaborateCom)*, 524-530.

Zhang, P., Liu, L., Chen, D., Weng, M.-H., & Yang, R.-Y. (2020). Application of a stub-loaded square ring resonator for wideband bandpass filter design. *MDPI Electronics*, *9*(1), 176. doi:10.3390/electronics9010176

Zhang, Y., & Fromherz, M. (2004). Message-initiated constraint-based routing for wireless ad-hoc sensor networks. In *First IEEE Consumer Communications and Networking Conference, 2004. CCNC 2004* pp. 648-650 IEEE. 10.1109/CCNC.2004.1286943

Zhang, Y., Kuhn, L. D., & Fromherz, M. P. (2004). Improvements on ant routing for sensor networks. In *International Workshop on Ant Colony Optimization and Swarm Intelligence* (pp. 154-165). Springer. 10.1007/978-3-540-28646-2_14

Zhao, P., & Wu, K. (2014, December). An Iterative and Analytical Approach to Optical Synthesis of a Multiplexer with aStar-Junction. *IEEE Transactions on Microwave Theory and Techniques*, *62*(12), 3362–3369. doi:10.1109/TMTT.2014.2364222

Zhao, Q., & Tong, L. (2004). Distributed opportunistic transmission for wireless sensor networks. In *2004 IEEE International Conference on Acoustics, Speech, and Signal Processing* (vol. 3, pp. iii-833). IEEE.

Zhao, Q., & Tong, L. (2005). Energy efficiency of large-scale wireless networks: Proactive versus reactive networking. *IEEE Journal on Selected Areas in Communications*, *23*(5), 1100–1112. doi:10.1109/JSAC.2005.845411

Zheludev, N. I., & Kivshar, Y. S. (2012). From metamaterials to metadevices. *Nature Materials*, *11*, 917–924.

Zhirnov, V. V., Cavin, R. K., Hutchby, J. A., & Bourianoff, G. I. (2003). Limits to Binary Logic Switch Scaling—A Gedanken Model. *Proceedings of the IEEE*, *91*, 1934–1939.

Zhou, J. (2010). Microwave Filters. In *Microwave and Millimeter Wave Technologies from Photonic Bandgap Devices to Antenna and Applications*. IntechOpen. doi:10.5772/9052

Zuo, J., Chen, X., Han, G., Li, L., & Zhang, W. (2009). An integrated approach to RF antenna-filter co-design. *IEEE Antennas and Wireless Propagation Letters*, *8*, 141–144. doi:10.1109/LAWP.2009.2012732

About the Contributors

Augustine O. Nwajana holds a PhD in Electrical and Electronic Engineering from the University of East London UK, and a research CPD certificate in Practical Antenna Design: from Theory to Practice, from the University of Oxford UK. He was with Siemens AG as a Telecoms Engineer for four years (2005 to 2009), where his experience spanned many countries including the USA, UK, UAE, Germany, South Africa, Ghana, and Nigeria. He was also with Coventry University (as an associate lecturer) and with the University of East London (as a part-time lecturer). Dr. Nwajana is currently a senior lecturer with the University of Greenwich where his current research interest includes RF energy harvesting; battery-less RF/microwave systems; analysis and design of RF/microwave devices (e.g., SIW and microstrip filters, diplexers/multiplexers, power dividers/combiners, couplers, antennas, etc.) for modern communication systems. He has authored/co-authored five books, several journal articles, book chapters, and conference papers. Augustine is currently serving as editor for IGI Global USA, and Electronics MDPI Switzerland. He has served as keynote/Invited speaker and TPC member for several conferences. He is a Senior Member of the IEEE, and a Fellow of the UK Higher Education Academy. He was a recipient of the Federal Government of Nigeria Scholarship Award.

Isibor Kennedy Ihianle received his B.Sc. and Ph.D. degrees in computer science from the University of East London specialising in the recognition of activities of daily living. He has over 10 years of teaching experience spanning across the areas of software engineering, Object-Oriented Programming, Database Design and Development, machine learning techniques, Knowledge-based systems, Ontology models, Internet Science and Data Analytics. He is currently is a senior lecturer and course leader Level 7 Digital Technology and Solutions at the Nottingham Trent University, United Kingdom. He is also a Fellow Higher Education Academy, Member of the British Computing Society and a Member of the IEEE. He has published several conference and journal papers. His research interests include human activity recognition from multi-modal sensing devices, health systems optimization, context awareness, knowledge-based systems with ontology, topic model and its applications and data analytic.

* * *

Abiodun Ernest Amoran obtained Bachelor of Engineering degree in Electrical and Electronics Engineering from University of Ado now Ekiti State University in 2010 and proceeded to Ladoke Akintola University of Technology Ogbomoso where he obtained his Masters of Technology (M. Tech) in Electronic and Electrical Engineering in 2015. Engr Amoran is a Registered Engineer with the Council for the Regulation of Engineering in Nigeria (COREN). His research interest ranges from Pattern Recognition, Artificial Intelligence, Machine Learning, Digital Signal Processing to Biomedical Imaging.

Destiny Chidi Anthony is a masters student at Torrens University, Brisbane campus, Australia. He previously obtained a masters in Exploration Geophysics from the university of Benin, Nigeria and a BSc in Physics Electronics from the University of Portharcourt. He presently lives in Brisbane Australia.

Ndidi Arinze received her Doctor of Philosophy in Engineering at Enugu State University of Science and Technology. She is a lecturer at Enugu State University of Science and Technology. Her research areas include Wireless Communication, Light Fidelity Technology, Internet of Things, Electronics, Control Engineering, Quality of Service and Antenna.

Jutika Devi is working as a Scientist under WOS-A Scheme, Government of India at Gauhati University, India. She has completed her PhD at 2018 in the domain of Nanoelectronics for Next-Generation Communication Devices. She was associated with Indian Institute of Technology Kanpur,India as a researcher. She has five years of teaching experience in postgraduate and phd level at Gauhati University. She has publishes many SCI Index research articles and delivered invited talks on nanophotonics at international level. Currently she is working on core/shell nanocompiste for biomedical imaging application. She is recognized as an innovator by Worldwide Academia Industry Network on the focus area of communication and medical technology.

Piotr Dybiec is an experienced business advisor and technology expert with more than 25 years of professional experience in telecommunications, IT and business consulting gathered on various expert and management positions in companies like Orange, KPMG, Oracle as well as professional associations and small startups (e.g. co-funding of PINmaster.net brand). He holds M.Sc. in telecommunication and MBA in Entrepreneurship. Recently also involved in U-Space related European projects acting as senior consultant for DroneRadar, one of the most advanced Polish U-space related solution provider (PansaUTM, DroneRadar flight planning mobile application, etc.).

Solomon H. Ebenuwa received the B.Ed in physics, the M.Sc. degree in advanced computing, and the Ph.D. degree in computer science from the School of Architecture, Computing, and Engineering, University of East London (UEL). He is currently a Lecturer at the University of Greenwich, London. His research interest are in the area of algorithm development used for performance optimisation, data mining, machine learning, decisions-based systems, and programming data structure in relation to automation and control. He is a Fellow of the U.K. Higher Education Academy. He has received many academic awards for innovative teaching and student support.

Gerald K. Ijemaru is currently a researcher at the School of Science, Technology & Engineering at University of the Sunshine Coast (USC) Australia. He was an Assistant Professor for more than 8 years at the Department of Electrical & Electronics Engineering at the Federal University Oye-Ekiti, Nigeria. He is currently researching on large-scale Multimedia Internet of Things (IoT) for Smart Cities.His major research interests are in Multimedia Internet of Things (MIoT), Internet of Vehicles (IoV), Multimedia processing and communications, Smart devices, Intelligent systems, Big data analytics, Embedded systems, Wireless multimedia sensor systems, Wireless sensor networks, Mobile applications and development, Wireless communications, Energy Harvesting/Wireless Power Transfer technologies, and the development of innovative technologies for real-world systems including smart cities, engineering, etc. Professionally, Gerald is an Engineer, and a member of several professional bodies including the Institute of Electrical and Electronics Engineers (IEEE), Council for the Regulation of Engineering in Nigeria (COREN), International Society for Development & Sustainability and the Nigeria Society of Engineers (NSE). He has a cutting-edge research in modern technological advances for humanity and the profession which are evinced in his research output. He has authored two published academic books & over 20 International research articles in several reputable journals, including IEEE Internet of Things Journal, IEEE Access, International Journal Ad Hoc and Ubiquitous Computing (IJAHUC), with over 177 citations and Hi-Index of 5. He has an excellent peer review records and is an academic reviewer for highly reputable journals including IEEE Internet of Things Journal, IEEE Transactions on Sustainable Computing (T-SUC) and International Journal of Distributed Sensor Networks (IJDSN). Owing to his research excellence, Gerald has bagged several awards including the highly competitive and prestigious Griffith Sciences International Postgraduate Research Scholarship (GSC IPRS), Coventry University International Merit Scholarship, University of the Sunshine Coast Research Scholarship, USC Research Integrity & Ethics award, and so on.

Chandra Kala K. is Assistant professor in Electronics and Communication Engineering Department of Noorul Islam Centre for Higher Education, India. She received BE degree in Electronics and Communication Engineering and ME degree in Communication Systems from Anna University. Her research interest is in wireless communication, image processing and neural networks.

Harilaos Koumaras is a research assistant professor at the Institute of Informatics and Telecommunications of NCSR "Demokritos". He received his PhD in 2007 from the Department of Informatics and Telecommunications of National and Kapodistrian University of Athens (NKUA), having granted the four-year scholarship of NCSR "Demokritos". Dr. Koumaras has been also scholar of the non-profit organization "Alexander S. Onassis". Till today, he has published more than 100 articles in peer-reviewed journals, international conferences and book chapters, while he has served as a reviewer for high-quality journals and magazines. His current research interests are focused in 5G/B5G systems, network virtualization, SDN/NFV, MANO, and quality of service/experience in media delivery. Currently, Dr. Koumaras is the project coordinator of H2020 5G-PPP 5GENESIS project, which develops a 5G experimentation infrastructure across Europe, including Athens, Principal Investigator of H2020 5G-PPP 5G!Drones project, focusing on UAV and drones experimentation and trial on 5G networks and Technical Manager of EVOLVED-5G project, which develops the framework for the design and development of 5G-enabled applications and services in the area of smart manufacturing.

Jaya Kumari is Professor in Electronics and Communication Engineering Department of Mar Baselios College of Engineering and Technology, Kerala, India. She received BE degree in Electronics and Communication Engineering from MS University; Master degree in Applied Electronics and Instrumentation from Kerala University and PhD from Kerala University in 2009. She has teaching experience of 22 years, research experience of 14 years and administrative experience of 16 years. She had published several papers in international journals. Her research interest includes wireless communication and networking, signal and image processing, detection & estimation theory, spread spectrum systems and error correcting codes. She is a Fellow of Institution of Electronics and Telecommunication Engineers

(IETE), Institution of Engineers (India) and council of Engineering and Technology also life member of Indian Society for Technical Education (ISTE), and Senior Member of Institution of Electrical and Electronics Engineers (IEEE).

Thomas Lutz currently serves as Senior Architect and is active in the area of safe unmanned traffic management. He has 20+ years experience on large scale distributed systems in different domains like water management, finance, maritime, air traffic management and public safety, always with a strong focus on information management.

George Makropoulos is a graduate in Electronics engineering from Technological Educational Institute of Athens and received his MSc degree in Advanced Informatics and Computing Systems from University of Piraeus in Greece. He is currently a PhD candidate (SDN networks and 5G technology) in Informatics and Telecommunications Department of National and Kapodistrian University of Athens. He has working experience as a Research Engineer and was involved in several EU-Funded projects in the fields of IoT and Telecommunications.

Pawel Montowtt is a telecom engineer and certified PM. He graduated from Gdansk University of Technology in 1995. Over the last 20 years he worked for planning and optimization of 2G/3G/4G for T-Mobile, Telia and Orange and suppliers: Alcatel-Lucent, Nokia. Since 2019 working for INVOLI and 5G!Drones project.

Ericmoore T. Ngharamike obtained a bachelor degree in Computer Science from Federal University of Technology Owerri, Imo state, Nigeria in 2012, and a Master degree in Network Computing from Coventry University, United Kingdom in 2015. He is currently a lecturer in the department of Computer Science, Federal University Oye-Ekiti. His research interest is in the area of wireless sensor networks, network security, internet of things (IoT), machine learning, artificial intelligence and data analytics.

Ebenezer Nnajiofo Ogbodo is a researcher of microelectronics and electronics hardware embedded programmer, electrical and electronics circuit designer, etc.

Eugene A. Ogbodo graduated with B.Eng from Enugu State University of Science and Technology (ESUT) Enugu, Nigeria in 2004. He achieved his MSc in 2011 at the University of East London, UK, before moving to the University of Greenwich, UK, where he achieved his PhD in 2017 after researching on Novel microwave devices based on multi-port filtering networks. He concentrated on the components used in satellite and base station communication technologies. His career has spanned across different academic institutions where he held different post as lecturer, teaching assistant, course director and researcher. His research interest is in the RF & Microwave Engineering and not limited to Filters, amplifiers, splitters, antennas, and propagations as well as the integration of different components.

Emmanuel Oleka holds a Ph.D. in Electrical Engineering from the North Carolina A&T State University, and is a Senior Member of the IEEE.

Paul Olulope holds BEng degree from Ekiti State University Ado Ekiti, Nigeria in 2000 and M.Eng. from University of Benin, Edo-State, Nigeria in 2015. He completed his Ph.D. from University of Cape

Town South Africa in 2014. He has been lecturing at Ekiti State University, Ado Ekiti, Nigeria since 2002. He is a registered member of the Institute of Electrical and Electronic Engineers (IEEE), a registered member of the Council for the Regulation of Engineering in Nigeria (COREN) and the Nigeria Society of Engineers (NSE). His Research interest is on the application of Computational intelligence to power system stability, control, hybrid renewable energy, integration of renewable energy and distributed generation into power grid.

Kehinde Oluwafemi Olusuyi obtained his Bachelor of Engineering and Master of Engineering degrees in Electrical and Electronics Engineering from the Federal University of Technology, Akure (FUTA) in 1998 and 2014 respectively. He is currently pursuing a PhD programme in Electrical and Electronics Engineering at Ekiti State University, Ado Ekiti. Kehinde is a Corporate Member of the Nigerian Society of Engineers (NSE) and a Registered Engineer with the Council for the Regulation of Engineering in Nigeria (COREN). Some of his research interests include Power System Engineering, Power System Security and Stability Studies, Artificial Intelligence, Machine Learning, Electric Vehicles and Computational Intelligence. He is happily married with three children.

Osondu Onah, MNSE, MSPE, MIEEE, MIET, and COREN registered, is a researcher, with rich field background in industrial controls and automation with interest in the impact of rapid telecommunication breakthrough for Multinational companies having large Industrial Control Systems.

Richard I. Otuka (Member, IEEE) received the B.Sc. and M.Sc. degrees in information systems and the Ph.D. degree in computer science from the School of Architecture, Computing, and Engineering, University of East London (UEL). He is currently an IT Lecturer with the Oaklands College of Further Education. He is also a Researcher in cloud computing and semantic technology. He has received many academic awards for innovative teaching and student support.

Eno Edet Peter is a Graduate Assistant in the department of Electrical/Electronics Engineering, Federal University Oye-Ekiti (FUOYE). She obtained her B.Eng. from Niger Delta University and is currently running her Masters in University of Ibadan, Nigeria. Her research interest includes Internet-of-Things (IoT), Cognitive Radio Network, Big Data, Telecommunication, Digital Communications, Artificial Intelligence.

Rastislav Róka (Assoc. Prof.) was born in Šaľa, Slovakia on January 27, 1972. He received his MSc. and PhD. degrees in Telecommunications from the Slovak University of Technology, Bratislava, in 1995 and 2002. Since 1997, he has been working as a senior lecturer at the Institute of MICT, FEI STU, Bratislava. Since 2009, he is working as an associated professor at this institute. At present, his research activity is focused on the signal transmission through metallic access networks by means of xDSL and PLC technologies and through optical transport, metropolitan and access networks by means of new WDM and TDM technologies using various techniques of the optical and digital signal processing. A main effort is dedicated to effective utilization of the optical fiber's transmission capacity by means of DBA and DWA algorithms. Assoc. Prof. Róka is the IEEE Senior Member since 2016.

Patrice Seuwou received his PhD in Computer Science & Informatics from London South Bank University, specialising on Autonomous vehicles adoption. Patrice master's degree was in Innovative

Technology from the University of East London in the United Kingdom, specialising in Forensics science. He is a senior lecturer in Digital Business, programme leader for the BSc (Hons) Digital Business and MBA deputy programme leader. Prior to that, he was a programme leader for a portfolio of undergraduate courses in the private sector in partnership with Plymouth University, a lecturer at London South Bank University and also Academic Registrar & Quality Manager working for a large private institution in partnership with Gloucestershire University. He has published in the area of Vehicular Ad hoc Networks, Actor-Network Theory, Self-driving car and his research interests include Autonomous vehicles, Management Information Systems, VANET security, Authentication and privacy, computer networks including wired and wireless networks, mobile and ubiquitous computing, Linux systems, RFID technology, protocols simulation and legislation in computer science. Patrice is currently exploring the experiences of Black Asian and Minority Ethnic (BAME) students in Higher education through the lens of Critical Race Theory and the Theory of Practice by Pierre Bourdieu. Patrice is a Senior Fellow of the Higher Education Academy (HEA), Reviewer for the International Journal of Electronic Security and Digital Forensics (IJESDF), External Examiner for Pearson. Patrice is a Microsoft certified professional, Member of the Association of Computer Professional (ACP), a member of the Institution of Engineering and Technology (IET), a member of the Chartered Management Institute (CMI) and also Member of the Chartered Institute for IT(BCS). He also Qualified as Enterprise Solution Architect with the British Computer Society.

Adrian K. Stavrakis received his B.Sc. in Electronics Engineering from International Hellenic University (ex. ATEITH) in 2016. Following that he acquired a postgraduate qualification in the field of Smart Systems Integration, and between 2018 - 2020 was an Erasmus+ scholar receiving dual diplomas, an M.Sc. in Smart Telecom and Sensing Networks from Aston University, UK and an M.Eng. in Engineering, Information Science and Technology, option : Electrical Engineering by Telecom SudParis. He is currently engaged as a research associate with Horizon 2020 projects in the field of flexible and textile electronics.

Sreeja T. K. is Assistant professor in Electronics and Communication Engineering Department of Noorul Islam Centre for Higher Education, India. She received BE degree in Electronics and Communication Engineering and ME degree in Communication Systems from Anna University. Her research interest is in antenna design, wireless communication.

Kok Yeow You was born in 1977. He obtained his B.Sc. Physics (Honours) degree in Universiti Kebangsaan Malaysia (UKM) in 2001. He pursued his M.Sc. in Microwave at the Faculty of Science in 2003 and his Ph.D. in Wave Propagation at the Institute for Mathematical Research in 2006 in Universiti Putra Malaysia (UPM), Serdang, Selangor, Malaysia. Recently, he is a senior lecturer at the School of Electrical Engineering, Faculty of Engineering, Universiti Teknologi Malaysia (UTM), Skudai, Johor, Malaysia. His main personnel research interest is in the theory, simulation, and instrumentation of electromagnetic wave propagation at microwave frequencies focusing on the development of microwave passive devices and sensors for medical and agricultural applications.

Index

R

S

T

U

V

W

Premier Reference Source
Emerging GIS Applications for Emergency and Disaster Management

Premier Reference Source
Managerial Strategies and Green Solutions for Project Sustainability

Premier Reference Source
Comparative Approaches to Using R and Python for Statistical Data Analysis

Premier Reference Source
Solutions for High-Touc
Communications in a
High-Tech World

www.ingramcontent.com/pod-product-compliance
Lightning Source LLC
LaVergne TN
LVHW080620300926
849162LV00005B/1220
* 9 7 8 1 7 9 9 8 6 9 9 2 4 *